W9-BGZ-036

CONTEMPORARY

Engineering Economics, 3/E

CONTEMPORARY
Engineering
Economics

THIRD EDITION

CHAN S. PARK
Auburn University

PRENTICE HALL, UPPER SADDLE RIVER, NEW JERSEY 07458

Library of Congress Cataloging-in-Publication Data

Park, Chan S.

    Contemporary engineering economics / Chan S. Park.—3rd ed.

    p. cm.

    Includes bibliographical references and index.

    ISBN 0-13-089310-2

    1. Engineering economy.   I. Title.

TA177.4.P369 2001

658.15—dc21   00-049197

Vice President and Editorial Director, ECS: *Marcia J. Horton*

Acquisitions Editor: *Laura Curless*

Editorial Assistant: *Erin Katchmar*

Vice President and Director of Production and Manufacturing, ESM: *David W. Riccardi*

Executive Managing Editor: *Vince O'Brien*

Managing Editor: *David A. George*

Production Editor: *Patty Donovan*

Director of Creative Services: *Paul Belfanti*

Creative Director: *Carole Anson*

Art Director: *Jonathan Boylan*

Art Editor: *Xiaohong Zhu*

Cover Designer: *Marjory Dressler*

Manufacturing Manager: *Trudy Pisciotti*

Manufacturing Buyer: *Lynda Castillo*

Marketing Manager: *Holly Stark*

 © 2002, 1997 Prentice-Hall, Inc.
Upper Saddle River, New Jersey 07458

All rights reserved. No part of this book may be reproduced, in any form or by any means without permission in writing from the publisher.

Printed in the United States of America

10 9 8 7 6 5 4 3 2 1

ISBN 0-13-089310-2

Prentice-Hall International (UK) Limited, *London*

Prentice-Hall of Australia Pty. Limited, *Sydney*

Prentice-Hall Canada Inc., *Toronto*

Prentice-Hall Hispanoamericana, S.A., *Mexico City*

Prentice-Hall of India Private Limited, *New Delhi*

Prentice-Hall of Japan, Inc., *Tokyo*

Pearson Education Asia Pte. Ltd., *Singapore*

Editora Prentice-Hall Do Brasil, Ltda., *Rio De Janeiro*

For my wife, Inkyung (Kim) and my children, Michael and Edward.

Preface

What is "Contemporary" About Engineering Economics?

Decisions made during the engineering design phase of product development determine the majority of the costs of manufacturing that product (some say 85%). As design and manufacturing processes become more complex, the engineer is making decisions that involve money more than ever before. Thus, the competent and successful engineer in the twenty-first century must have an improved understanding of the principles of science, engineering, and economics, coupled with relevant design experience. Increasingly, in the new world economy, successful businesses will rely on engineers with such expertise.

Economic and design issues are inextricably linked in the product/service life cycle. Therefore, one of my strongest motivations for writing this text was to bring the realities of economics and engineering design into the classroom and to help students integrate these issues when contemplating many engineering decisions.

With the advent of information technology, the Internet becomes an indispensable tool in exchanging information. Accordingly, we have developed a comprehensive companion website to the book to provide numerous teaching and learning aids. I believe that our website is a critical resource for transitioning the teaching of engineering economy into the twenty-first century.

Of course my underlying motivation for writing this book was not simply to address contemporary needs, but to address as well the ageless goal of all educators: to help students to learn. Thus, thoroughness, clarity, and accuracy of presentation of essential engineering economics was my aim at every stage in the development of the text.

Changes in the Third Edition

In the complex and changing world of a global economy in an age of information, the practice of engineering economics is dynamic and as new developments occur, they should be incorporated into a textbook such as this one. In addition, the author and publisher are constantly seeking ways of improving the book in terms of clarity and understanding. As a result, we have made several important changes in this edition, including the following:

- The overall book design style has been changed. The new design format allows us to use a second color more effectively to highlight the most important information and separate examples from the main text reading, providing an effective study and review tool for students.
- There are 17 chapters, including three new chapters. Each chapter is classified into one of five parts:
  - Part I: Financial and Cost Information;
  - Part II: Money and Investing;
  - Part III: Evaluating Business and Engineering Assets;

- Part IV: Development of Project Cash Flows; and
- Part V: Special Topics in Engineering Economics.

- All sections were updated to reflect the latest tax laws, interest rates, and other financial developments.

- A discussion on end-of-chapter "Computer Notes" is consolidated into the website for Contemporary Engineering Economics:

    http://www.prenhall.com/park or http://www.eng.auburn.edu/~Park/cee.html

    This allows us to remove all Microsoft Excel spreadsheet discussions from the main text. Now students can download various spreadsheet templates from the website and open directly in Excel for Windows. The obvious benefit is that it is no longer necessary to enter the spreadsheets by hand. Users can then modify the basic templates for the specific problem at hand.

- About one third of examples and self-test questions in the main chapters are either new or revised ones to reflect the contemporary nature of economic decision problems.

- In Chapters 9 and 14, some of the advanced topics (or optional materials) such as the multiple-rates of return problems and risk simulation have been removed from the main text and placed in the Appendix. This separation will allow the instructors to budget their lecture hours more effectively according to their audience and curriculum.

- Chapter 1 (Engineering Economic Decisions) is completely revised to reflect the ever-expanding role of engineers in the new economy. The main purpose of the opening chapter is to provide students with the general scope of their respective roles in making a variety of engineering as well as business decisions.

- Chapter 2 (Understanding Financial Statements) is a new chapter for the third edition. The main purpose of this chapter is to introduce the basics of business language, known as financial accounting, so that engineers can understand and speak in a common language when it comes to making a variety of business decisions. Engineers should understand some of the basics of accounting, as they are constantly involved in a variety of business decisions.

- Chapter 3 (Cost Concepts and Behaviors), also a new chapter, covers the various cost definitions as well as their behaviors in decision making. In particular, it discusses the marginal concept, which is the basis for any economic decision. The section on short-term operational economic decisions (commonly known as Present Economic Studies) is detailed in this chapter.

- Chapter 4 (Time is Money) retains most of examples as well as the presentations in Chapter 2 of the second edition. Several new examples are introduced and the topics involving equivalence concepts are streamlined.

- Chapter 5 (Understanding Money and Its Management) retains much of the materials in Chapter 3 of the second edition, but contains new sections on personal finance such as credit cards, commercial loans, and home mortgages.

- Chapter 6 (Principles of Investing) is a new chapter geared toward unraveling the mysteries of the financial markets—the language, the players, the strategies, and above all, the risks and rewards of investments, as well as their ups and downs. Even though all the examples are drawn from the financial markets related to personal investments, the same principles should govern general corporate investment decisions.

- Chapter 7 (Present Worth Analysis) is much the same as Chapter 4 of the second edition, except that the principles of comparing mutually exclusive projects have been covered in greater detail.
- Chapter 12 (Developing of Project Cash Flows) is equivalent to Chapter 9 of the second edition, except that the subjects of working capital investment and the generalized cash flow approach have been streamlined in terms of presentation.
- Chapter 14 (Project Risk and Uncertainty) has been expanded to include decision tree analysis.
- Chapter 15 (Replacement Decisions) has been divided into two parts: The first part introduces the basic replacement decision problems without considering the effects of income taxes, whereas the second part revisits the same decision problems with income tax consideration. This treatment allows students to learn this important concept without the additional complications of income taxes.

Overview of the Text

Although it contains little advanced math and few truly difficult concepts, the introductory engineering economics course is often a curiously challenging one for the sophomores, juniors, and seniors who take it. There are several likely explanations for this difficulty.

1. The course is the student's first analytical consideration of money (a resource with which he or she may have had little direct contact beyond paying for tuition, housing, food, and textbooks).
2. An emphasis on theory—while critically important to forming the foundation of a student's understanding—may obscure for the student the fact that the course aims, among other things, to develop a very practical set of analytical tools for measuring project worth. This is unfortunate since, at one time or another, virtually every engineer—not to mention every individual—is responsible for the wise allocation of limited financial resources.
3. The mixture of industrial, civil, mechanical, electrical, and manufacturing engineering, and other undergraduates who take the course often fail to "see themselves" in the skills the course and text are intended to foster. This is perhaps less true for industrial engineering students, whom many texts take as their primary audience, but other disciplines are often motivationally shortchanged by a text's lack of applications that appeal directly to them.

Goal of the Text

This text aims not only to provide sound and comprehensive coverage of the concepts of engineering economics but also to address the difficulties of students outlined above, all of which have their basis in an inattentiveness to the practical concerns of engineering economics. More specifically, this text has the following chief goals:

1. To build a thorough understanding of the theoretical and conceptual basis upon which the practice of financial project analysis is built.
2. To satisfy the very practical needs of the engineer toward making informed financial decisions when acting as a team member or project manager for an engineering project.
3. To incorporate all critical decision-making tools—including the most contemporary, computer-oriented onesæthat engineers bring to the task of making informed financial decisions.
4. To appeal to the full range of engineering disciplines for which this course is often required: industrial, civil, mechanical, electrical, computer, aerospace, chemical, and manufacturing engineering, as well as engineering technology.

Prerequisites

The text is intended for undergraduate engineering students at the sophomore level or above. The only mathematical background required is elementary calculus. For Chapter 14, a first course in probability or statistics is helpful but not necessary, since the treatment of basic topics there is essentially self-contained.

Content and Approach

Educators generally agree upon what comprises the proper contents and organization of an engineering economics text. A glance at the table of contents will demonstrate that this text matches the standard embraced by most instructors and that reflected in competing texts. However, one of my driving motivations was to supersede the standard in terms of the depth of coverage and care with which difficult concepts are presented. Accordingly, the content and approach of *Contemporary Engineering Economics* reflect the following goals.

Understanding the Role of Engineers in Business

Explaining early in the book how business operates, and how engineering project decisions are made within business, helps students see how engineering decisions can affect the bottom line (profit) of the firm (Chapters 1–3). Most students—even those who do not plan to run their own businesses—are generally interested in personal finance and investment in general. Since people's ability to learn a subject is a function of their interest and motivation, this text begins by showing how a typical project idea evolves, how we measure the success of a typical engineering decision, and how to communicate the results in common business language (accounting).

Thorough Development of the Concept of the Time Value of Money

The notion of the time value of money and the interest formula that model it form the foundation upon which all other topics in engineering economics are built. Because of their great importance, and because many students are being exposed to an analytical approach to money for the first time, interest topics are carefully and thoroughly developed in Chapters 4–6.

Thorough, Reasonably Paced Coverage of the Major Analysis Methods

The equivalence methods—present worth, annual worth, and future worth—and rate of return analysis are the bedrock methods of project evaluation and comparison. This text carefully develops these topics in Chapters 7–9, pacing them for maximum student comprehension of the subtleties, strengths, and weaknesses of each method.

1. A separate, dedicated chapter (Chapter 8) on annual worth is presented to emphasize the circumstances in which that method of project analysis is preferred over other methods. In particular, topics related to design economics are explored in detail.
2. The difficulties and exceptions associated with rate of return analysis are thoroughly covered in Chapter 9. Coverage of internal rate of return for non-simple projects is placed in Appendix A as optional for those who wish to avoid this complication in an introductory course.

Increased Emphasis on Developing After-Tax Cash Flows

Estimating and developing project cash flows is the first critical step in conducting an engineering economic analysis for most practicing engineers—further analysis, comparison of projects, and decision making all depend on intelligently developed project cash flows. A particularly important goal of this text is to instill confidence in developing after-tax cash flows.

1. Chapter 12 is a unique synthesis of previously developed topics (analysis methods, depreciation, and income taxes) and is dedicated to building skill and confidence in developing after-tax cash flows for a series of fairly complex projects.
2. To account for the ever-changing nature of tax systems, the website for *Contemporary Engineering Economics* is created and maintained by the author so that any new changes in depreciation and tax rates are posted in this increasingly popular Internet tool.

Complete Coverage of the Special Topics that Round Out a Comprehensive Introduction to Engineering Economics

A number of special topics are important to a comprehensive understanding of introductory engineering economics. Chapters 14–17 cover the topics such as (1) project risk and uncertainty, (2) replacement analysis, (3) capital budgeting, and (4) public sector analysis.

Recognizing that time availability and priorities vary from course to course and instructor to instructor, each one of these chapters is sufficiently self-contained that it may be skipped or covered out of sequence, as needed.

Addressing Educational Challenges

The features of *Contemporary Engineering Economics* were selected and shaped to address key educational challenges. It is the observation of both the author and the publisher—based on many conversations with engineering educators—that certain challenges consistently frustrate both instructors and students across the engineering curriculum. Low student motivation and enthusiasm, difficulties on te part of students in developing problem-solving skills and intuition, challenges in integrating technology without shortchanging fundamental concepts and traditional methods, and difficulties students experience in prioritizing and retaining enormous amounts of information were among the key educational challenges that drove the care with which the features in *Contemporary Engineering Economics* were designed.

Building Problem-Solving Skills and Confidence

The examples in the text are formatted to maximize their usefulness as guides to problem solving. Further, they are intended to stimulate student curiosity and inspire students to look beyond the mechanics of problem solving to "what if" issues, alternative solution methods, and interpretation of solutions. Each example in the text is formatted as follows.

- **Example** titles promote ease of student reference and review.
- **Discussion** sections at the beginning of complex examples help students begin organizing a problem-solving approach.
- **Given** and **Find** heads in the Solutions sections help students identify critical data. This convention is employed in Chapters 4–10, and then omitted in Chapters 11–14 after student confidence in setting up solution procedures has been established.
- **Comments** sections at the ends of examples add additional insights: an alternative solution method, a short cut, an interpretation of the numerical solution, and extend the educational value of the example.

Capturing the Student's Imagination

Students want to know how the conceptual and theoretical knowledge they are acquiring will be put to use. *Contemporary Engineering Economics* incorporates real-world applications and contexts in a number of ways to stimulate student enthusiasm and imagination.

1. *A real-world, conceptual overview of engineering economics is established in Chapter 1*: This is geared toward providing an engaging introduction to engineering economics via examples of its practical use.
2. *Chapter-opening scenarios*: These are included with the hope of establishing interest and "need-to-know" chapter concepts within the context of a practical application.
3. *Self-test questions*: These end-of-chapter problem sets should help students affirm their understanding of important concepts presented in the chapter.

These self-test questions can be used as a comprehensive FE exam review, as they are already in multiple-choice format.

4. *An abundance of homework problems involving real engineering projects*: This feature is aimed toward stimulating student interest and motivation with actual engineering investment projects, many taken from today's headlines.

5. *A full range of engineering disciplines represented in problems, examples, chapter openers, and case studies*: This diverse approach illustrates the many disciplines that require engineering economics. Industrial, chemical, civil, electrical, mechanical, and manufacturing engineering and other areas are all represented.

Taking Advantage of the Internet

The integration of computer use is another important feature of *Contemporary Engineering Economics*. Students have greater access to and familiarity with the various spreadsheet tools, and instructors have greater inclination either to treat these topics explicitly in the course or to encourage students to experiment independently.

A remaining concern is that the use of computers will undermine true understanding of course concepts. This text does not promote the trivial or mindless use of computers as a replacement for genuine understanding of and skill in applying traditional solution methods. Rather, it focuses on the computer's productivity-enhancing benefits for complex project cash flow development and analysis. Specifically, *Contemporary Engineering Economics* includes a robust introduction to computer automation in the form of Computer Notes, which appear on the book's website:

http://www.prenhall.com/park or http://www.eng.auburn.edu/~park/cee.html

Spreadsheets are introduced via Microsoft Excel examples. For spreadsheet coverage, the emphasis is on demonstrating a chapter concept that embodies some complexity that can be much more efficiently resolved on a computer than by traditional long-hand solutions.

Internet Tool: The *Contemporary Engineering Economics* Website (http://www.prenhall.com/park)

The companion website has been created and maintained by the author. This text takes advantage of the Internet as a tool that becomes an increasingly important resource medium to access a variety of information on the web. This website contains a variety of resources for both instructors and students, including sample test questions, supplemental problems, EzCash software, a Case Study Library, and lecture notes. As you type the address and click the open button, you will see the *Contemporary Engineering Economics* Home Page, as shown in Fig. P.1. As you will note from the figure, several menus are available. Each menu item is explained as follows:

- Study Guide: Click this menu to find out what typical resource materials are available on the website. This site includes (1) sample text questions, (2) solutions to chapter problems, (3) interest tables, and (4) computer notes with Excel files of selected example problems in the text.

- Analysis Tools: This site includes (1) EzCash software and (2) a collection of various financial calculators available on the Internet. EzCash is an integrated computer software package that was developed under the auspices of a grant from the National Science Foundation. This software can be downloaded from the website free of charge. The software includes the most frequently used methods of economic analysis. It is menu driven for convenience and flexibility, and it provides (1) a flexible and easy-to-use cash flow editor for data input and modifications, (2) an electronic spreadsheet-like data entry facility for after-tax cash flow analysis, and (3) an extensive array of computational modules and user selected graphic outputs.

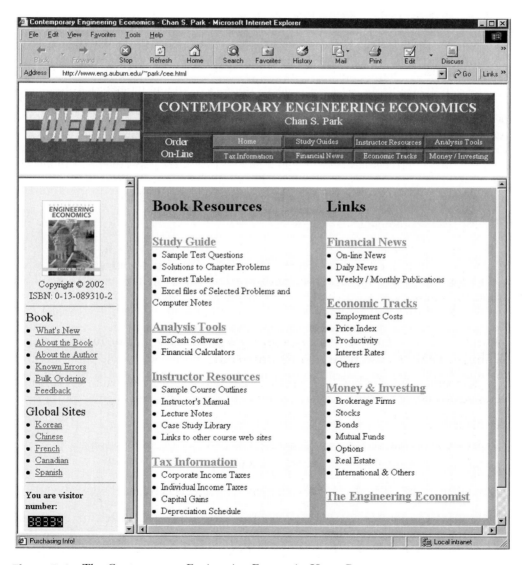

Figure P-1 The *Contemporary Engineering Economics* Home Page

- Instructor Resources: Any information useful to instructors who teach an engineering economic course with the *Contemporary Engineering Economics* text is found in this location. Typically, course outlines based on a quarterly as well as a semester system are provided as an aid to instructors who adopt this text for the first time. A collection of well-designed engineering economic case studies is listed at this location. Initially, only a few case problems will be provided, but as new ones are developed or found, the case library will expanded in volume as well as in variety. You will also find lecture notes developed by the author.

- Tax Information: This section will serve as a clearinghouse in terms of disseminating ever-changing tax information, including personal as well as corporate income taxes. Links are provided to various tax sites on the web, so you will find the most up-to-date information on depreciation schedules as well as capital gains taxes.

- Financial News: You can access various financial news outlets on the web by visiting this location. The site divided news outlets into on-line news, and daily, weekly, and monthly publications.

- Economic Tracks: Any cost and price information as well as most recent interest rate trends are found here. In particular, the consumer price indices, productivity figures, and employment cost indices are some of representative economic data provided.

- Money & Investing: Click this menu to find a gateway to a variety of information useful to conducting engineering economic analysis. For example, a direct link is provided to the most up-to-date stock prices, options, and mutual funds performances.

Flexibility of Coverage

For a typical three-credit-hour, one-semester course, the majority of topics in the text can be covered in the depth and breadth in which they are presented. For other arrangements—quarter terms or fewer credit hours—Chapters 1–12 present the essential topics, with subsequent chapters presenting optional coverage. By varying the depth of coverage and supplementing the reading with *Contemporary Engineering Economics* Case Studies, there are enough materials for a continuing, two-term engineering economics course. Sample syllabi for a three-credit and a two-credit semester (or three-credit quarter system) course in engineering economy are provided in Table P.1.

Supplements

Much of the ancillary materials are supplied free of charge via the companion website. Other items are available free of charge to adopting instructors.

1. Instructor's Manual: A comprehensive manual is available to instructors who adopt the book. The manual contains answers to all end-of-chapter problems.
2. *Contemporary Engineering Economics* Case Studies: A collection of actual cases, two personal-finance and six industry-based, is now available on the companion website. The investment projects detailed in the cases relate to a

Chapter	Topic(s)	3-Credit Hour Semester Course	2-Credit Hour Semester/ Quarter Course
1	Engineering Economic Decisions	x	x
2	Understanding Financial Statements	Optional	
3	Cost Concepts and Behaviors	x	Optional
4	Time Is Money	x	x
5	Understanding Money and Its Management	x	x
6	Principles of Investing	x	Optional
7	Present Worth Analysis	x	x
8	Annual Equivalent Worth Analysis	x	x
8	Rate of Return Analysis	x	x
10	Depreciation	x	x
11	Corporate Income Taxes	x	x
12	Development of Project Cash Flows	x	x
13	Inflation and Its Impact on Project Cash Flows	x	Optional
14	Project Risk and Uncertainty	x	x (only up to Sec. 14.3)
15	Replacement Analysis	x	x (only up to Sec. 15.3)
16	Capital Budgeting Decisions	x	
17	Economic Analysis in the Public Sector	Optional	

Table P.1
Sample Syllabi for Courses in Engineering Economy

variety of engineering disciplines. Each case is based on multiple text concepts, thus encouraging students to synthesize their understanding in the context of complex, real-world investments. Each case begins with a list of engineering economic concepts utilized in the case and concludes with discussion questions to test students' conceptual understanding.

3. Microsoft PowerPoint Slides on CD-ROM: A comprehensive set of lecture notes in PowerPoint format is available on CD-ROM to instructors who adopt the text.

Acknowledgments

This book reflects the efforts of a great many individuals over a number of years. In particular, I would like to recognize the following individuals, whose reviews and comments on prior editions have contributed to this edition. Once again, I would like to thank each of them:

Kamran Abedini, *California Polytechnic-Pomona*
James Alloway, *Syracuse University*
Mehar Arora, *U. Wisconsin-Stout*
Joel Arthur, *California State University-Chico*
Robert Baker, *University of Arizona*
Robert Barrett, *Cooper Union and Pratt Institute*
Tom Barta, *Iowa State University*
Charles Bartholomew, *Widener University*
Richard Bernhard, *North Carolina State University*
Bopaya Bidanda, *University of Pittsburgh*
James Buck, *University of Iowa*
Philip Cady, *The Pennsylvania State University*
Tom Carmichal, *Southern College of Technology*
Jeya Chandra, *The Pennsylvania State University*
Max C. Deibert, *Montana State University*
Stuart E. Dreyfus, *University of California-Berkeley*
W. J. Foley, *RPI*
Jane Fraser, *University of Southern Colorado*
Anil K. Goyal, *RPI*
Bruce Hartsough, *University of California-Davis*
Carl Hass, *University of Texas-Austin*
John Held, *Kansas State University*
T. Allen Henry, *University of Alabama*
R.C. Hodgson, *University of Notre Dame*
Scott Iverson, *University of Washingto*
Peter Jackson, *Cornell University*
Philip Johnson, *University of Minnesota*
Harold Josephs, *Lawrence Tech*
Henry Kallsen, *University of Alabama*
W. J. Kennedy, *Clemson University*
Oh Keytack, *University of Toledo*
Wayne Knabach, *South Dakota State University*
Stephen Kreta, *California Maritime Academy*
John Krogman, *University of Wisconsin-Platteville*
Dennis Kroll, *Bradley University*
Michael Kyte, *University of Idaho*
William Lesso, *University of Texas-Austin*
Martin Lipinski, *Memphis State University*
Robert Lundquist, *Ohio State University*
Richard Lyles, *Michigan State University*

Abu S. Masud, *The Wichita State University*
Sue McNeil, *Carnegie-Mellon University*
James Milligan, *University of Idaho*
Richard Minesinger, *University of Massachusett*s, Lowell
Gary Moynihan, *The University of Alabama*
James S. Noble, *University of Missouri, Columbia*
Wayne Parker, *Mississippi State University*
Elizabeth Pate'-Cornell, *Stanford University*
Cecil Peterson, *GMI*
George Prueitt, *U.S. Naval Postgraduate School*
J.K. Rao, *California State University-Long Beach*
Susan Richards, *GMI*
Bruce A. Reichert, *Kansas State University*
Mark Roberts, *Michigan Tech*
John Roth, *Vanderbilt University*
Paul L. Schillings, *Montana State University*
Bill Shaner, *Colorado State University*
Fred Sheets, *California Polytechnic-Pomona*
Dean Shup, *University of Cincinnati*
Milton Smith, *Texas Tech*
David C. Slaughter, *University of Californi*a, Davis
Charles Stavridge, *FAMU/FSU*
Junius Storry, *South Dakota State University*
Frank E. Stratton, *San Diego State University*
George Stukhart, *Texas A&M University*
Donna Summers, *University of Dayton*
Joe Tanchoco, *Purdue University*
Deborah Thurston, *University of Illinois at Urbana-Champaign*
Lt. Col. James Treharne, *U.S. Army*
L. Jackson Turaville, *Tennessee Technological University*
Thomas Ward, *University of Louisville*
Theo De Winter, *Boston University*

Special Acknowledgement

Personally, I wish to thank the following individuals for their additional inputs to the third edition: Professor Bruce Hartsough, *University of California, Davis*, who reviewed the entire manuscript and provided detailed comments and suggestions for improving the technical contents as well as the organization. Both Professor Yoo Yang,

Cal Poly State University and Professor Philip A. Farrington, *University of Alabama at Huntsville*, also reviewed an earlier draft and provided valuable suggestions for improving the manuscript; Michael D. Park (my son), who read the first three new chapters from a student's point of view and offered valuable insights; two of my students, Eric Cross who did an excellent job in developing a user-friendly book website and Liz Griffin, who assisted me in developing PowerPoint slides for lecture notes; Laura Curless, my editor at Prentice Hall, who assumed responsibility for the overall project; Liz Vequist at Prentice Hall, who obtained all required permissions for me; Patty Donovan, production editor at Pine Tree Composition, who oversaw the entire book production. Finally, I would like to thank Dr. Alice E. Smith, Chair of Industrial & Systems Engineering at *Auburn University*, who provided me with the student resources.

CHAN S. PARK
AUBURN, ALABAMA

Table of Contents

PART THREE
Evaluating Business and Engineering Assets

CHAPTER SEVEN
Present Worth Analysis

CHAPTER EIGHT
Annual Equivalent Worth Analysis

CHAPTER NINE
Rate of Return Analysis

CHAPTER FIFTEEN
Replacement Decisions   692

CHAPTER SIXTEEN
Capital Budgeting Decisions   756

CHAPTER SEVENTEEN
Economic Analysis in the Public Sector

CONTEMPORARY
Engineering Economics, 3/E

Financial and Cost Information

CHAPTER 1

Engineering Economic Decisions

Bose Packs Concert Acoustics into Home-Speaker Systems[1] Dr. Amar Bose, an MIT professor and chairman of speaker-maker Bose Corporation, defied the conventional wisdom of consumer electronics. Dr. Bose grew up poor in Philadelphia, where his father immigrated from India and worked as an importer until he lost the business during World War II. While his mother worked as a teacher, he set up a radio-repair business at the age of 14 in the basement that soon became the family's main support. He entered MIT and never left, earning a doctorate degree in 1956. As a reward for finishing his research, he decided to buy himself a stereo system. Although he had done his homework on the hi-fi's engineering specifications, he was profoundly disappointed with his purchase. Mulling why something that looked good on paper sounded bad in the open air, Dr. Bose concluded the answer was directional. In a concert hall, sound waves radiate outward from the instruments and bounce back at the audience from the walls. However, home stereo speakers aimed sound only forward. Therefore, Dr. Bose began tinkering to develop a home speaker that could reproduce the concert experience.

In 1964, he formed Bose Corporation, and four years later introduced his first successful speaker, the 901. Based on the principle of reflected sound, the speaker bounces sounds off walls and ceilings to surround the listener. In 1968, he pioneered the use of "reflected sound" in an effort to bring concert-hall quality to home-speaker systems. A decade later, he convinced General Motors Corp. to let his company design a high-end speaker system for the Cadillac Seville, helping to push car stereos beyond the mediocre. Recently, he introduced a compact radio system that can produce rich bass sound. In the process, closely held Bose has become the world's No.1 speaker maker, with annual sales of more than $700 million, and one of the few U.S. firms that beats the Japanese in consumer electronics. The success has vaulted Dr. Bose into Forbes magazine's list of the 400 wealthiest Americans, with a net worth estimated at $500 million.

---

[1] Bose Packs Concert Acoustics into Home-Speaker Systems, *The Wall Street Journal,* December 31, 1996 (by William M. Bulkeley) and Bose Corporation—History of Company on its web site (*http://www.bose.com*).

It is not a typical story of how Dr. Bose got motivated in inventing a directional home speaker and eventually transformed his invention to a multi-million dollar business. Companies such as Dell, Microsoft, and Yahoo all produce computer-related products and have market values of several billion dollars. These companies were all started by highly motivated young college students just like Dr. Bose. One thing that is also common to all these successful businesses is that they have capable and imaginative engineers who constantly generate good ideas for capital investment, execute them well, and obtain good results. You might wonder about what kind of role these engineers play in making such business decisions. In other words, what specific tasks are assigned to these engineers, and what tools and techniques are available to them for making such capital investment decisions? We explore these throughout the book.

1.1 The Rational Decision-Making Process

We, as individuals or businesspersons, constantly make decisions in our daily lives. Most of them may take place by default, without consciously recognizing that we are actually following some sort of a logical decision flowchart. Rational decision making can be a complex process that contains a number of essential elements. Instead of presenting some rigid rational decision making processes, we will provide examples of how two engineering students approached their financial as well as engineering design problems. By reviewing these examples, we will be able to identify some essential elements common to any rational decision making. The first example, by Monica, illustrates how she narrowed down her choice between two competing alternatives in buying an automobile. The second example, by Sonya, illustrates how a typical class project idea evolves and how she approached the design problem by following a logical method of analysis.

1.1.1 How Do We Make a Typical Personal Decision?

For Monica Russell, a senior at the University of Washington, the future holds a new car. Her 1990 Civic Honda has clocked almost 110,000 miles, and she wants to replace it soon. But how to do it—buy or lease? In either case, "car payments would be difficult," said the engineering major who works as a part-time cashier at a local supermarket. "I have never leased before, but I am leaning toward it this time to save on the down payment. I also don't want to worry about major repairs," she said. For Monica, leasing would provide the warranty protection she wants, along with a new car every 3 years. On the other hand, she would be limited to a specified amount of miles, usually 12,000 per year, after which she would have to pay 20 cents or more per mile. Monica knew too well that choosing the right vehicle is an important decision to make, and so is choosing the best possible financing. At this point, Monica was not sure what was ahead of her if she decided to lease the car instead of traditional buying.

Establishing the Goal or Objective

Monica decided to survey the local papers and the Internet for the latest lease programs, including factory-subsidized "sweetheart" deals, and special incentive packages. Within her budget, either the 2001 Saturn SC1 or the 2001 Honda Civic DX Coupe appeared to be equally attractive in terms of style, price, and options. Monica finally decided to visit the dealers' lots to see how both models looked and to take them for a test drive. Both gave her very satisfactory driving experiences. Monica thought that it would be important to examine carefully many technical as well as safety features of the automobiles. It seemed both models were equally identical in terms of reliability, safety features, and quality.

Evaluation of Feasible Alternatives

Monica figured that her 1990 Honda could be traded in at around $2,000. This would be just enough to make any down payment required in leasing the automobile. In the process, Monica also learned that, if she decided to lease the vehicle, there were two types of leases—open-end and closed-end. The most popular by far was closed-end because open-end leases exposed the consumer to possible higher payments at the end of the lease if the car had depreciated faster than expected. Closed-end means Monica could return the vehicle at the end of the lease and "walk away" to lease or buy another vehicle. However, she would have to pay for extra mileage or excess wear or damage. She thought that since she would not be a pedal-to-the-metal driver, lease-end charges would not be problem for her.

To get the best financial deal she could with a lease, she obtained some financial facts from both dealers for their best offers. With each offer, she added all the costs of leasing, from the down payment to the disposition fee. This would determine the total cost, not counting routine items such as oil changes and other maintenance. It appeared that, with the Saturn SC1, Monica could save about $401 in lease payments ($35 \times \$11 + \$16$), plus a $250 disposition fee, or total savings[2] of $651. However, if she drove any additional miles over the limit, her savings would reduce by 5 cents for each additional mile over the limit. This came out about 13,020 extra miles over the limit to lose all the savings. Without knowing her driving needs after graduation, her conclusion was to lease the Honda Civic DX. Certainly, any monetary savings would be important, but she preferred having some flexibility in her future driving needs.

Knowing Other Opportunities

Even though Monica was not interested in buying the car, it could be even more challenging to determine precisely whether she would be better off buying. To make a comparison of leasing versus buying, Monica could consider what she likely would pay for the same vehicle under both scenarios. If she would own the car for as long as she would lease it, she could sell the car and use the proceeds to pay off any outstanding loan. If finances are her only consideration, her choice will depend on the specifics of the deal. But beyond finances, she needs to consider the positives and negatives of her personal preferences. By leasing, she would never experience the "joy" of the last payment—but she would have a new car every 3 years.

Review of Monica's Decision Process

Now we may revisit the decision making process in a more structured way. While difficult to isolate into a finite number of logical steps, the analysis can be thought of as including the following steps.

- Step 1: Recognize a decision problem.
   "Monica needs a new car."

- Step 2: Define the goals or objectives.
   "Mechanical security and lower monthly payments."

- Step 3: Collect all the relevant information.
   "Gathering technical as well as financial data."

- Step 4: Identify a set of feasible decision alternatives.
   Option 1: Lease the 2001 Saturn SC1.
   Option 2: Lease the 2001 Honda Civic Coupe DX.

---

[2] If Monica considered the time value of money concept in her comparison, the amount of actual savings will be less than $651, which we will demonstrate in Chapter 4.

Table 1.1
Financial
Data for
Auto Leas-
ing: Saturn
versus
Honda

	Auto Leasing		
	Saturn	Honda	Difference
1. Manufacturer's suggested retail price (MSRP)	$14,205	$13,775	$430
2. Lease length	36 months	36 months	
3. Allowed mileage	36,000 miles	36,000 miles	
4. Monthly lease payment	$189	$200	–$11
5. Mileage surcharge over 36,000 miles	$0.20 per mile	$0.15 per mile	+$0.05
6. Disposition fee at lease end	$0	$250	–$250
7. Total due at signing:			
• First month lease	$189	$200	
• Down payment	$1,100	$1,600	
• Administrative fee	$495	$0	
• Refundable security deposit	$200	$200	
Total	$1,984	$2,000	–$16

Models compared: The 2001 Saturn SC1with automatic transmission and A/C, and the 2001 Honda Civic DX Coupe with automatic transmission and A/C.

Disposition fee: This is another paperwork charge for getting the vehicle ready for resale.

- Step 5: Select the decision criterion to use.
  "Satisfying the current and future driving needs subject to specified constraints."
- Step 6: Select the best alternative.
  "Option 2."

These six steps are known as the "rational decision making process." Certainly, we do not always follow these six steps in every decision problem. Some decision problems may not require much of our time and effort. Quite often, we even make our decisions solely based on emotional reasons. However, for any complex economic decision problems, a structured decision framework such as outlined proves to be worthwhile.

1.1.2 How Do We Approach an Engineering Design Problem?

The idea of design and development is what most distinguishes engineering from science, which concerns itself principally with understanding the world as it is. Decisions made during the engineering design phase of a product's development determine the majority of the costs of manufacturing that product. As design and manufacturing processes become more complex, the engineer, increasingly, will be called upon to make decisions that involve money. In this section, we will provide an example of how engineers would get from "thought" to "thing." The following story of how an electrical engineering student approached her design problem and exercised her judgment has much to teach us about some of the fundamental characteristics of the human endeavor known as engineering decision.[3]

---

[3] Background material from 1991 Annual Report, GWC Whiting School of Engineering, Johns Hopkins University (with permission).

Getting an Idea: Necessity is the Mother of Invention

Most consumers abhor lukewarm beverages, especially during the hot days of summer, but throughout history, necessity has been the mother of invention. Several years ago, Sonya Talton, an electrical engineering student at Johns Hopkins University, had a revolutionary idea—self-chilling soda can!

Picture this. It's one of those sweltering, hazy August afternoons. Your friends have finally gotten their acts together for a picnic at the lake. Together you go over the list of stuff you need—blankets, radio, sunscreen, sandwiches, chips, and soda. You wipe the sweat from your neck, reach for a soda, and realize that it's about the same temperature as the 90°F afternoon. Great start. Everyone's just dying to make another trip back to the store for ice. Why can't they come up with a soda container that can chill itself, anyway?

Setting Design Goals and Objectives

Sonya decided to take on the topic of a soda container that can chill itself as a term project in her engineering graphics and design course. The professor stressed innovative thinking and urged students to consider unusual or novel concepts. The first thing Sonya needed to do was to establish some goals for the project:

- Get the soda as cold as possible in the shortest possible time.
- Keep the container design simple.
- Keep the size and weight of the newly designed container similar to that of the traditional soda can. (This would allow beverage companies to use existing vending machines and storage equipment.)
- Keep the production cost low.
- Make the product environmentally safe.

Evaluating Design Alternatives

With these goals in mind, Sonya had to think of a practical yet innovative way of chilling the can. Ice was the obvious choice—practical, but not innovative. Sonya had a great idea: What about a chemical ice pack? Sonya's next question was, what's inside? The answer was ammonium nitrate (NH_4NO_3) and a water pouch. When the needed pressure is applied to the chemical ice pack, the water pouch breaks and mixes with the NH_4NO_3, creating an endothermic reaction (the absorption of heat). The NH_4NO_3 draws the heat out of the soda, causing it to chill. How much water is in the water pouch? Sonya measured it: 135 ml. She wondered what would happen if she reduced the amount of water. After several trials involving different amounts of water, Sonya found that she could chill the soda can from 80°F to 48°F in a 3-minute period. At this point, she needed to determine how cold a refrigerated soda gets. She put a can in the fridge for 2 days and found out it chilled to 41°F. Sonya's idea was definitely feasible. But was it economically marketable?

Gauging Product Costing and Pricing

In her engineering graphics and design course, the topic of how economic feasibility plays a major role in the engineering design process was discussed. The professor emphasized the importance of marketing surveys and cost/benefit analyses as ways to gauge a product's potential. Sonya surveyed approximately 80 people. She asked them only two questions: their age and how much would they be willing to pay for a self-chilling can of soda. The under-21 group was willing to pay the most, 84 cents. The 40-plus bunches only wanted to pay 68 cents. Overall, the surveyed group would be willing to shell out 75 cents for a self-chilling soda. (It was hardly a scientific market survey, but it did give Sonya a feel for what would be a reasonable price for her product.)

The next hurdle was to determine the existing production cost of one can of soda. How much more would it cost to produce the self-chiller? Would it be profitable? She went to the library, and there she found the bulk cost of chemicals and materials she would need. Then she calculated how much she would require for one unit. She couldn't believe it! It only costs 12 cents to manufacture one can of soda, including transportation. Her can would cost 2 or 3 cents more. That wasn't bad, considering the average consumer was willing to pay up to 25 cents more for the self-chilling can than for the traditional one.

Considering Green Engineering

The only two constraints left to consider were possible chemical contamination and recyclability. Theoretically it should be possible to build a machine that would drain the solution from the can and recrystallize it. The ammonium nitrate could then be reused in future soda cans, including the plastic outer can. Chemical contamination, the only remaining restriction, was a big concern. Unfortunately, there was absolutely no way to ensure that the chemical and the soda would never come in contact with one another inside the cans. To ease consumer fears, Sonya decided a color or odor indicator could be added to alert the consumer to contamination if it occurred.

Sonya's conclusion? The self-chilling beverage can would be an incredible technological advancement. The product would be convenient for the beach, picnics, sporting events, and barbecues. Its design would incorporate consumer convenience while addressing environmental concerns. It would be innovative, yet inexpensive, and it would have an economic, as well as a social impact, on society.

1.2 Economic Decisions

The economic decisions that engineers make in business differ very little from those made by Sonya, except for the scale of the concern. Suppose, for example, that a firm is using a lathe that was purchased 12 years ago to produce pump shafts. As the production engineer in charge of this product, you expect demand to continue into the foreseeable future. However, the lathe has begun to show its age: It has broken frequently during the last 2 years and has finally stopped operating altogether. Now you have to decide whether to replace or repair it. If you expect a more efficient lathe to be available in the next 1 or 2 years, you might repair it instead of replacing it. The major issue is whether you should make the considerable investment in a new lathe now or later. As an added complication, if demand for your product begins to decline, you may have to conduct an economic analysis to determine whether declining profits from the project offset the cost of a new lathe.

Let us consider a real world engineering decision problem of a much larger scale. Public concern about poor air quality is increasing, particularly that caused by gasoline-powered automobiles. With requirements looming in a number of jurisdictions for automakers to produce electric vehicles, General Motors Corporation has decided to build an advanced electric car to be known as Impact. The biggest question remaining about the feasibility of the vehicle concerns its battery. With its current experimental battery design, Impact's monthly operating cost would be roughly twice that of a conventional automobile. The primary advantage of the design, however, is that Impact does not emit any pollutants, a feature that could be very appealing at a time when government air-quality standards are becoming more stringent, and consumer interest in the environment is ever growing.

Engineers at General Motors have stated that the total annual demand for Impact would need to be 100,000 cars to justify production. Although General Motors management has already decided to build the battery-powered electric car, the engineers involved in making the engineering economic decision were still debating whether the demand for such a car would be sufficient to justify its production.

Obviously, this level of engineering decision is more complex and more significant to the company than a decision about when to purchase a new lathe. Projects of this nature involve large sums of money over long periods of time, and it is difficult to predict market demand accurately. An erroneous forecast of product demand can have serious consequences: With any overexpansion, unnecessary expenses will have to be paid for unused raw materials and finished products. In the case of Impact, if the improved battery design never materializes, demand may remain insufficient to justify the project.

In this book, we will consider many investment situations, personal investments as well as business investments. The focus, however, will be on evaluating engineering projects on the basis of economic desirability and on investment situations that face a typical firm.

1.3   Predicting the Future

Economic decisions differ in a fundamental way from the types of decisions typically encountered in engineering design. In a design situation, the engineer utilizes known physical properties, the principles of chemistry and physics, engineering design correlations, and engineering judgment to arrive at a workable and optimal design. If the judgment is sound, the calculations are done correctly, and we ignore technological advances, the design is time invariant. In other words, if the engineering design to meet a particular need is done today, next year, or in five years time, the final design would not change significantly.

In considering economic decisions, the measurement of investment attractiveness, which is the subject of this book, is relatively straightforward. However, information required in such evaluations always involves predicting or forecasting product sales, product selling price, and various costs over some future time frame—5 years, 10 years, 25 years, etc.

All such forecasts have two things in common. First, they are never completely accurate when compared with the actual values realized at future times. Second, a prediction or forecast made today is likely to be different than one made at some point in the future. It is this ever-changing view of the future which can make it necessary to revisit and even change previous economic decisions. Thus, unlike engineering design, the conclusions reached through economic evaluation are not necessarily time invariant. Economic decisions have to be based on the best information available at the time of the decision and a thorough understanding of the uncertainties in the forecasted data.

1.4   Role of Engineers in Business

Apple Computer, Microsoft Corporation, and Sun Microsystems produce computer products and have a market value of several billion dollars. These companies were all started in the late 1970s or early 1980s by young college students with technical backgrounds. When they went into the computer business, these students initially organized their companies as proprietorships. As the businesses grew, they became partnerships and were eventually converted to corporations. This chapter will introduce the three primary forms of business organization and briefly discuss the role of engineers in business.

1.4.1   Types of Business Organization

As an engineer, it is important to understand the nature of the business organization with which you are associated. This section will present some basic information about the type of organization you should choose should you decide to go into business for yourself.

The three legal forms of business, each having certain advantages and disadvantages are proprietorships, partnerships, and corporations.

Proprietorships

A **proprietorship** is a business owned by one individual. This person is responsible for a firm's policies, owns all its assets, and is personally liable for its debts. A proprietorship has two major advantages. First, it can be formed easily and inexpensively. No legal and organizational requirements are associated with setting up a proprietorship, and organizational costs are therefore virtually nil. Second, the earnings of a proprietorship are taxed at the owner's personal tax rate, which may be lower than the rate at which corporate income is taxed. Apart from personal liability considerations, the major disadvantage of a proprietorship is that it cannot issue stocks and bonds, making it difficult to raise capital for any business expansion.

Partnerships

A **partnership** is similar to a proprietorship except that it has more than one owner. Most partnerships are established by a written contract between the partners, which normally specifies salaries, contributions to capital, and the distribution of profits and losses. A partnership has many advantages, among which are its low cost and ease of formation. Because more than one person makes contributions, a partnership typically has a larger amount of capital available for business use. Since the personal assets of all the partners stand behind the business, a partnership can borrow money more easily from a bank. Each partner pays only personal income tax on his or her share of a partnership's taxable income.

On the negative side, under partnership law, each partner is liable for a business's debts. This means that the partners must risk all their personal assets, even those not invested in the business. And while each partner is responsible for his or her portion of the debts in the event of bankruptcy, if any partner cannot meet his or her pro rata claim, the remaining partners must take over the unresolved claims. Finally, a partnership has a limited life insofar as it must be dissolved and reorganized if one of the partners quits.

Corporations

A **corporation** is a legal entity created under provincial or federal law. It is separate from its owners and managers. This separation gives the corporation four major advantages: (1) It can raise capital from a large number of investors by issuing stocks and bonds; (2) it permits easy transfer of ownership interest by trading shares of stock; (3) it allows limited liability—personal liability is limited to the amount of the individual's investment in the business; and (4) it is taxed differently than proprietorships and partnerships, and under certain conditions, the tax laws favor corporations. On the negative side, it is expensive to establish a corporation. Furthermore, a corporation is subject to numerous governmental requirements and regulations.

As a firm grows, it may need to change its legal form because the form of a business affects the extent to which it has control of its own operations and its ability to acquire funds. The legal form of an organization also affects the risk borne by its owners in case of bankruptcy and the manner in which a firm is taxed. Apple Computer, for example, started out as a two-man garage operation. As the business grew, the owners felt constricted by this form of organization: It was difficult to raise capital for business expansion; they felt that the risk of bankruptcy was too high to bear; and as their business income grew, their tax burden grew as well. Eventually, they found it necessary to convert the partnership into a corporation.

In the United States, the overwhelming majority of business firms are proprietorships followed by corporations and partnerships. However, in terms of total business volume (dollars of sales), the quantity of business done by proprietorships and partnerships is several times less than that of corporations. Since most business is conducted by companies of the corporation form, this text will generally address economic decisions encountered in corporations.

1.4.2   Engineering Economic Decisions

What role do engineers play within a firm? What specific tasks are assigned to the engineering staff, and what tools and techniques are available to it to improve a firm's profits? Engineers are called upon to participate in a variety of decision making processes, ranging from manufacturing, through marketing, to financing decisions. We will restrict our focus, however, to various economic decisions related to engineering projects. We refer to these decisions as **engineering economic decisions.**

In manufacturing, engineering is involved in every detail of a product's production, from the conceptual design to the shipping. In fact, engineering decisions account for the majority (some say 85%) of product costs. Engineers must consider the effective use of capital assets such as buildings and machinery. One of the engineer's primary tasks is to plan for the acquisition of equipment (**capital expenditure**) that will enable the firm to design and produce products economically.

With the purchase of any fixed asset, equipment for example, we need to estimate the profits (more precisely, cash flows) that the asset will generate during its service period. In other words, we have to make capital expenditure decisions based on predictions about the future. Suppose, for example, you are considering the purchase of a deburring machine to meet the anticipated demand for hubs and sleeves used in the production of gear couplings. You expect the machine to last 10 years. This purchase decision thus involves an implicit 10-year sales forecast for the gear couplings, which means that a long waiting period will be required before you will know whether the purchase was justified.

An inaccurate estimate of asset needs can have serious consequences. If you invest too much in assets, you incur unnecessarily heavy expenses. Spending too little on fixed assets is also harmful, for then the firm's equipment may be too obsolete to produce products competitively and, without an adequate capacity, you may lose a portion of your market share to rival firms. Regaining lost customers involves heavy marketing expenses and may even require price reductions and/or product improvements, all of which are costly.

1.4.3 Personal Economic Decisions

In the same way that an engineer can play a role in the effective utilization of corporate financial assets, each of us is responsible for managing our personal financial affairs. After paying for nondiscretionary or essential needs such as housing, food, clothing, and transportation, any remaining money is available for discretionary expenditures on items such as entertainment, travel, investment, etc. For money which we choose to invest, we want to maximize the economic benefit at some acceptable risk. The investment choices are unlimited including savings accounts, guaranteed investment certificates, stocks, bonds, mutual funds, registered retirement savings plans, rental properties, land, business ownership, etc.

How do you choose? Analysis of personal investment opportunities utilizes the same techniques used for engineering economic decisions. Again, the challenge is predicting the performance of an investment into the future. Choosing wisely can be very rewarding while choosing poorly can be disastrous. Some investors in the gold-mining stock BreX who sold prior to the fraud investigation became millionaires. Others, who did not sell, lost everything.

A wise investment strategy is one which manages risk by diversifying investments. You have a number of different investments, which range from very low to very high risk, and are in a number of business sectors. Since you do not have all your money in one place, the risk of losing everything is significantly reduced. (We discuss some of these important issues in Chapter 6.)

1.5   Large-Scale Engineering Projects

In the development of any product, a company's engineers are called upon to translate an idea into reality. A firm's growth and development largely depends upon a constant flow of ideas for new products, and for the firm to remain competitive, it has to make existing products better or produce them at a lower cost. Traditionally, a marketing department would propose a product and pass the recommendation to the engineering department. The engineering department would work up a design and pass it on to manufacturing, which would make the product. With this type of product development cycle, a new product normally takes several months (or even years) to reach the market. A typical toolmaker, for example, would take 3 years to develop and market a new machine tool. However, the Ingersoll-Rand Company, a leading toolmaker, was able to cut down the normal development time of its products by one-third. How did the company do this? A group of engineers examined the current product development cycle to find out why things dragged on. They learned how to compress the crippling amount of time it took to bring products to life. In the next section, we present an example of how a design engineer's idea eventually turned into an innovative automotive product.

1.5.1   How a Typical Project Idea Evolves

The Toyota Motor Corporation introduced the world's first mass-produced car powered by a combination of gasoline and electricity, known as Prius. The Prius would be the first of a new generation of Toyota cars whose engines would cut air pollution dramatically and boost fuel efficiency to significant levels. Toyota, in short, wants to launch and dominate a new "green" era for automobiles—and it will spend $1.5 billion to do it. Developed for the Japanese market initially, the Prius uses both a gasoline engine and an electric motor as motive power sources. The Prius emits less pollution than ordinary cars, and it gets more mileage, which means less output of carbon dioxide. Moreover, the Prius provides highly responsive performance and smooth acceleration. The following information from *Business Week*,[4] illustrates how a typical strategic business decision is made by a group of engineering staff in a larger company. Additional information has been provided by Toyota Motor Corporation.

Why Going for a Greener Car?

Toyota first started to develop the Prius in 1995. Four engineers were assigned to figure out what types of cars people would drive in the 21st century. After a year of research, a chief engineer concluded that Toyota would have to sell cars better suited to a world with scarce natural resources. He considered electric motors. But an electric car can only travel 215 km before it must be recharged. Another option was fuel-cell cars that run on hydrogen. But he suspected mass production might not be possible for another 15 years. So the engineer finally settled on a hybrid system powered by an

---

[4] By Emily Thornton in Tokyo, with Keith Naughton in Detroit and David Woodruff, "Japan's Hybrid Cars—Toyota and rivals are betting on pollution fighters—Will they succeed?" *Business Week,* December 4, 1997.

electric motor, a nickel-metal hydride battery, and a gasoline engine. From Toyota's perspective, it is a big bet as oil and gasoline prices are bumping along at record lows. Many green cars remain expensive and require trade-offs in terms of performance. No carmaker has ever succeeded in selling consumers en masse something they have never wanted to buy—*cleaner air*. Even in Japan, where a liter of regular gas can cost as much as $1, carmakers have trouble pushing higher fuel economy and lower carbon emissions. Toyota has several reasons for going green. In the next century, as millions of new car owners in China, India, and elsewhere take to the road, Toyota predicts that gasoline prices will rise worldwide. At the same time, Japan's carmakers expect pollution and global warming to become such threats that governments will enact tough measures to clean the air.

What is So Unique in Design?

It took Toyota engineers 2 years to develop the current power system in the Prius. The car comes with a dual engine powered by both an electric motor and a newly developed 1.5-liter gasoline engine. When the engine is in use, a special "power split device" sends some of the power to the drive shaft to move the car's wheels. The device also sends some of the power to a generator, which in turn creates electricity, to either drive the motor or recharge the battery. Thanks to this variable transmission, the Prius can switch back and forth between motor and engine, or employ both, without creating any jerking motion. The battery and electric motor propel the car at slow speeds. At normal speeds, the electric motor and gasoline engine work together to power the wheels. At higher speeds, the battery kicks in extra power if the driver must pass another car or zoom up a hill.

When the car decelerates and brakes, the motor converts the vehicle's kinetic energy into electricity and sends it through an inverter to be stored in the battery (see Figure 1.1). So the car's own movement as well as the power from the gasoline engine provide the electricity needed. The energy created and stored by deceleration boosts the car's efficiency. So does the fact that the engine shuts down automatically when the car stops at a light. At higher speeds and during acceleration, the companion electric motor works harder, allowing the gas engine to run at peak efficiency. Toyota claims that in tests, its hybrid car has boosted fuel economy by 100% and engine efficiency by 80%. The Prius emits about half as much carbon dioxide, and about one-tenth as much carbon monoxide, hydrocarbons, and nitrous oxide as conventional cars.

Is It Safe to Drive on a Rainy Day?

Yet, major hurdles remain to creating a mass market for green vehicles. Car buyers are not anxious enough about global warming to justify a "sea level change" in automakers' marketing. Many of Japan's innovations run the risk of becoming impressive technologies meant for the masses but only bought by the elite. The unfamiliarity of green technology can also frighten consumers. The Japanese government sponsors festivals where people can test drive alternative-fuel vehicles. But some turned down that chance on a rainy weekend in May because they feared riding in an electric car might electrocute them. An engineer points out, "It took 20 years for automatic transmission to become popular in Japan." It certainly would take that long for many of these technologies to catch on.

- **Starting from Standstill Slow Speeds**
  The engine operates only in its most efficient range. So when the Prius is starting from rest to moving slowly, the engine shuts down and the electric motor drives the wheel (AI).

- **Normal Driving**
  In normal driving, a power-split device drives the engine output between the wheels and the generator, which supplies electricity for the motor (C).

- **Full Throttle Acceleration**
  During full throttle acceleration or under heavy load, the motor gets a power boost from the battery (AI).

Energy flow

Starting from rest/low speeds • Full-throttle acceleration
Normal driving • Deceleration/braking

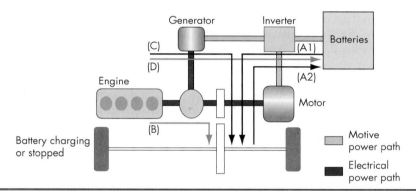

- **Deceleration or Braking**
  During deceleration or braking, the motor functions as a generator turning kinetic energy from the wheels into electricity to charge the battery (A2).

- **Battery Charging**
  The computer-controlled charging system maintains a constant charge in the battery (D).

- **Stopped**
  When the vehicle is at a standstill, the engine automatically stops (B).

Figure 1.1   Design concept adopted by Prius (Courtesy of Toyota Corporation)

How Much Would It Cost?

The biggest question remaining about the mass production of the vehicle concerns its production cost. Costs will need to come down for Toyota's hybrid to be competitive around the world, where gasoline prices are lower than in Japan. Economies of scale will help as production volumes increase. But further advances in engineering also will be essential. With its current design, Prius' monthly operating cost would be roughly twice that of a conventional automobile. Toyota, in contrast, believes it will sell more than 12,000 Prius models to Japanese drivers during the first year. To do that, it is charging just $17,000 a car. The company insists it will not lose any money on the Prius. But rivals and analysts estimate that at current production levels, the cost of building a Prius could be as much as $32,000. If so, Toyota's low-price strategy will generate annual losses of more than $100 million on the new compact.

Will There Be Enough Demand?

Why buy a $17,000 Prius when a $13,000 Corolla is more affordable? It does not help Toyota that it is launching the Prius in the middle of a sagging domestic car market. Nobody really knows how big the final market for hybrids will be. Toyota forecasts that hybrids will account for a third of the world's auto market as early as 2005. But Japan's Ministry of International Trade & Industry expects 2.4 million alternative fuel vehicles, including hybrids, to roam Japan's back streets by 2010. Nonetheless, the Prius has set a new standard for Toyota's competitors. There seems to be no turning back for the Japanese. The government may soon tell carmakers to slash carbon dioxide emissions by 20% by 2010. And it wants them to cut nitrous oxide, hydrocarbon, and carbon monoxide emissions by 80%. The government may also soon give tax breaks to consumers who buy green cars.

The prospect for Prius started looking good—the total sales for Prius reached over 7,700 units as of June 1998. With this encouraging sales trend, Toyota finally announced that the Prius would be introduced in the North American and European markets by the year 2000. The total sales volume will be approximately 20,000 units per year in the North American and European market combined. As for the 2000 North American and European introduction, Toyota is planning to use the next two years to develop a hybrid vehicle optimized to the usage conditions of each market.

What is the Business Risk in Marketing the Prius?

Engineers at Toyota Motors have stated that California would be the primary market outside Japan, but they added that an annual demand of 50,000 cars would be necessary to justify production. Despite Toyota management's decision to introduce the hybrid electric car to the U.S. market, the Toyota engineers were still uncertain whether there would be such a demand in the U.S. market. Furthermore, competitors including the U.S. automakers just do not see how it can achieve the economies of scale needed to produce green cars at a profit. The primary advantage of the design, however, is that Prius can cut any pollutants in half. This is a feature that could be very appealing at a time when government air-quality standards are becoming more rigorous, and consumer interest in the environment is strong. However, in the case of Prius, if a significant reduction in production cost never materializes, demand may remain insufficient to justify the investment in the green car.

1.5.2 Impact of Engineering Projects on Financial Statements

Engineers must understand the business environment in which a company's major business decisions are made. It is important for an engineering project to generate profits, but it also must strengthen the firm's overall financial position. How do we measure Toyota's success in the Prius project? Will enough Prius models be sold, for example, to keep the green-engineering business as Toyota's major source of profits? While the Prius project will provide comfortable, reliable, low-cost driving for the customers, the bottom line is its financial performance over the long run.

Regardless of a business's form, each company has to produce basic financial statements at the end of each operating cycle (typically a year). These financial statements provide the basis for future investment analysis. In practice, we seldom make investment decisions based solely on an estimate of a project's profitability because we must also consider its overall impact on the financial strength and position of the company.

Suppose that you were the president of the Toyota Corporation. Let us further suppose that you even hold some shares in the company, which also makes you one of the company's many owners. What objectives would you set for the company? While all firms are in business in hopes of making a profit, what determines the market value of a company is not profits per se, but cash flow. It is, after all, available cash that determines the future investments and growth of the firm. Therefore, one of your objectives should be to increase a company's value to its owners (including yourself) as much as possible. The market price of your company's stock to some extent represents the value of your company. Many factors affect your company's market value: present and expected future earnings, the timing and duration of these earnings, as well as risks associated with them. Certainly, any successful investment decision will increase a company's market value. Stock price can be a good indicator of your company's financial health and may also reflect the market's attitude about how well your company is managed for the benefit of its owners.

Investors liked the new hybrid car, which resulted in an increased demand for the company's stock. This, in turn, caused stock prices, and hence shareholder wealth, to increase. In fact, this new, heavily promoted, green-car turned out to be a market leader in its class and contributed to sending Toyota's stock to an all-time high in early 2000. This caused Toyota's market value to continue to increase well into 2000. Any successful investment decision on Prius's scale will tend to increase a firm's stock prices in the marketplace and promote long-term success. Thus, in making a large-scale engineering project decision, we must consider its possible effect on the firm's market value. (In Chapter 2, we discuss the financial statements in detail and show how to use them in our investment decision making.)

1.6   Types of Strategic Engineering Economic Decisions

The story of how the Toyota Corporation successfully introduced a new product and regained the hybrid electric car market share is typical: Someone had a good idea, executed it well, and obtained good results. Project ideas such as the Prius can originate from many different levels in an organization. Since some ideas will be good, while others will not, we need to establish procedures for screening projects. Many large companies have a specialized project analysis division that actively searches for new ideas, projects, and ventures. Once project ideas are identified, they are typically classified as (1) equipment and process selection, (2) equipment replacement, (3) new product and product expansion, (4) cost reduction, and (5) service or quality improvement. This classification scheme allows management to address key questions: Can the existing plant, for example, be used to achieve the new production levels? Does the firm have the knowledge and skill to undertake this new investment? Does the new proposal warrant the recruitment of new technical personnel? The answers to these questions help firms screen out proposals that are not feasible given a company's resources.

Prius project represents a fairly complex engineering decision that required the approval of top executives and the board of directors. Virtually all big businesses at some time face investment decisions of this magnitude. In general, the larger the investment, the more detailed is the analysis required to support the expenditure. For example, expenditures to increase output of existing products, or to manufacture a new product, would invariably require a very detailed economic justification. Final decisions on new products and marketing decisions are generally made at a high level within the company. On the other hand, a decision to repair damaged equipment can be made at a lower level within a company. In this section, we will provide many real examples to illustrate each class of engineering economic decisions. At this point, our intention is not to provide the solution to each example, but rather to describe the nature of decision problems that a typical engineer would face in the real world.

1.6.1   Equipment and Process Selection

What we have just described is a class of engineering decision problems that involve selecting the best course of action when several ways to meet a project's requirements exist. Which of several proposed items of equipment shall we purchase for a given purpose? The choice often turns on which item is expected to generate the largest savings (or return on the investment). Example 1.1 illustrates a process selection problem.

Example 1.1 Material Selection from an Automotive Exterior Body

Engineers at General Motors want to investigate alternative materials and processes for the production of automotive exterior body panels. The engineers have identified two types of material: sheet metal and glass fiber reinforced polymer, known as plastic sheet molding compound (SMC), as shown in Figure 1.2. Exterior body panels are traditionally made of sheet steel. With a low material cost, this sheet metal also lends itself to the stamping process, a very high-volume, proven manufacturing process. On the other hand, reinforced polymer easily meets the functional requirements of body panels (such as strength and resistance to corrosion). There is considerable debate among engineers as to the relative economic merits of steel as opposed to plastic panels. Much of the debate stems from the dramatically different cost structures of the two materials.

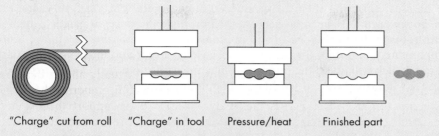

"Charge" cut from roll   "Charge" in tool   Pressure/heat   Finished part

Figure 1.2   Sheet molding compound process (Courtesy of Dow Plastics, a business group of the Dow Chemical Company)

Description	Plastic SMC	Steel Sheet Stock
Material cost ($/kg)	$1.65	$0.77
Machinery investment	$2.1 million	$24.2 million
Tooling investment	$0.683 million	$4 million
Cycle time (minute/part)	2.0	0.1

Since plastic is petroleum based, it is inherently more expensive than steel, and because the plastic-forming process involves a chemical reaction, it has a slower cycle time. However, both machinery and tool costs for plastic are lower than for steel because of relatively low-forming pressures, lack of tool abrasion, and single-stage pressing involved in handling. Thus, the plastic would require a lower initial investment, but would incur higher material costs. Neither material is obviously superior economically. What, then, is the required annual production volume that would make the plastic material more economical?

Comments: The choice of material will dictate the manufacturing process for the body panels. Many factors will affect the ultimate choice of the material, and engineers should consider all major cost elements, such as machinery and equipment, tooling, labor, and material. Other factors may include press and assembly, production and engineered scrap, the number of dies and tools, and the cycle times for various processes.

1.6.2   Equipment Replacement

This category of investment decisions involves considering the expenditure necessary to replace worn-out or obsolete equipment. For example, a company may purchase 10 large presses expecting them to produce stamped metal parts for 10 years. After 5 years, however, it may become necessary to produce the parts in plastic, which would require retiring the presses early and purchasing plastic molding machines. Similarly, a company may find that, for competitive reasons, larger and more accurate parts are required, which will make the purchased machines obsolete earlier than expected. Example 1.2 will provide a real-world example in this category.

Example 1.2 Replacing PCB Transformers

A large power generation and transmission company in the eastern United States has a nuclear power plant. This facility was constructed with 44 electrical transformers that utilize a fluid containing polychlorinated biphenyl (PCB) as the liquid portion of the transformer insulation systems. The transformers use the fluid containing PCB as a coolant. As illustrated in Figure 1.3, the liquid, driven by convection currents, flows around the transformer coils, and the "hot" liquid travels outside of the transformer,

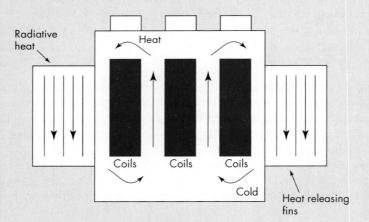

Figure 1.3   A liquid-cooled transformer (the arrows indicate the flow of the liquid and the direction of convection currents)

where it is cooled; it then returns to the coils and the process is repeated. In the event of a fire or a transformer explosion, PCBs may form carcinogenic compounds which are extremely dangerous to human health. Government regulations now ban the production of PCBs. PCB transformers are being replaced or converted to alternate coolants. An internal risk assessment study by the company indicated a slight probability of an explosion with existing PCB transformers. The potential clean-up costs of a PCB transformer fire would range from $15 to $100 million. The study also identified the transformers offering the greatest risk to be those in the turbine building because of a large open area, which could be contaminated in a violent failure.

The company was faced with making a decision as to whether it should continue to live with the well known PCB environmental risks or to reduce the risks by implementing one of the following options:

- Option 1: Replace all 44 transformers with nonhazardous liquid filled transformers (LFTs). The high dielectric strength of LFTs provides greater design flexibility in optimizing specific load requirements so that lower operating costs can be achieved.

- Option 2: Replace 24 transformers in the open area with LFTs and retain the remaining transformers. By only replacing transformers in open areas, 20 transformers in other areas still remain at risk.

- Option 3: Reclassify all the transformers by replacing the PCB fluid with a silicone fluid, which is nontoxic and environmentally safe. The challenge associated with reclassification is the possibility of derating the transformers.

The LFTs would cost $30,000 each, and the reclassified transformers $14,000 each. Annual maintenance costs including parts and labor are estimated at $500 per reclassified transformer and $300 per replaced transformer. A conservative estimate for clean-up in the event of an accident was made at $80 million over a period of 20 years. Knowing that an individual transformer failure rate is estimated at 0.6% over its lifetime, engineers predict that it is highly likely that one incident would be encountered over a 20-year operation.

Comments: This example involves weighing an expenditure to replace serviceable, but potentially hazardous, equipment in case of an accident. In this problem, we need to consider the economic consequences in the case of an accident, even though its probability is relatively small. With ever-increasing awareness of transformer safety, we need to consider environmental as well as safety elements in an engineering decision. No adverse environmental effects associated with silicon liquid use, production, or disposal have been identified. In fact, silicon liquid is used extensively in hairspray, skincare products, antacids, and even in the food processing industry. In addition, silicon liquid can be reconditioned and reused.

1.6.3 New Product and Product Expansion

Investments in this category are those that increase the revenues of a company if output is increased. A description of two common types of expansion decision problems follows: The first category includes decisions about expenditures to increase the output of existing production or distribution facilities. In these situations, we are basically asking, "Shall we build or otherwise acquire a new facility?" The expected future cash inflows in this investment category are the profits from the goods and services produced in the new facility. The second type of decision problem includes considering expenditures necessary to produce a new product or to expand into a new geographic area. These projects normally require large sums of money over long periods. Example 1.3 illustrates an investment decision problem associated with introducing a new product.

Example 1.3 Marketing a Rapid Prototype System

DuPont, a leading chemical company, has set up a separate venture, Somos, to develop and market a rapid prototype system. This computer-driven technology allows design engineers to design and create plastic prototypes of complicated parts in just a few hours. An example of a rapid prototype system is shown in Figure 1.4.

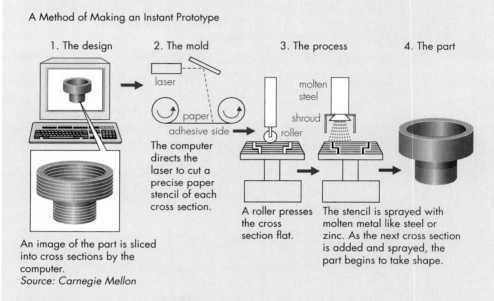

A Method of Making an Instant Prototype

1. The design   2. The mold   3. The process   4. The part

The computer directs the laser to cut a precise paper stencil of each cross section.

An image of the part is sliced into cross sections by the computer.
Source: Carnegie Mellon

A roller presses the cross section flat.

The stencil is sprayed with molten metal like steel or zinc. As the next cross section is added and sprayed, the part begins to take shape.

Figure 1.4   A typical rapid prototype system

Rapid prototyping has allowed designers to complete projects that formerly took 6 months, or more, in less than 3 weeks. Assignments that took several weeks may now be done in a day or two. Rapid prototyping also saves tens of thousands of dollars per part in modeling costs, compared to traditional methods.

The technology might also give birth to true just-in-time manufacturing for some businesses. An auto parts replacement shop, for example, might simply stock metal and plastic powders along with a library of computer programs that would allow it to build any part a customer needs on the spot. The venture will require an investment of $40 million on the part of DuPont, and the prototype system once developed would be sold for $385,000.

Comments: In this example, we are asking the basic questions involved in introducing a new product: Is it worth spending $40 million to market the rapid prototyping? How many sales are required to recover the investment? The future expected cash inflows in this example are the profits from the goods and services produced with the new product. Are these profits large enough to warrant the investment in equipment and the costs required to make and introduce the product?

1.6.4 Cost Reduction

A cost-reduction project is one that attempts to lower a firm's operating costs. Typically we need to consider whether a company should buy equipment to perform an operation now done manually or spend money now in order to save more money later. The expected future cash inflows on this investment are savings resulting from lower operating costs. Example 1.4 illustrates a cost-reduction project.

Example 1.4 Lone Star Trucking Company

Lone Star Trucking, a Dallas-based company with 50 trucks, wishes to convert this fleet to use a fuel other than gasoline. If possible, it would like to have 30% or more of its fleet vehicles operating on an alternative fuel by September 1, 2001, 50% or more of the fleet operating on alternative fuel by September 1, 2003, and 90% or more of the fleet using alternative fuels by September 1, 2005.

Two different types of fuels are under consideration: Compressed Natural Gas (CNG) and Liquid Petroleum Gas (LPG). Both CNG and LPG engines are considered to be economically more efficient than conventional gasoline engines. However, the initial conversion cost for the CNG option will be $305,520. The use of CNG will save the company $16,520 in fuel costs per year for the first 2 years, $27,535 per year for the next 2 years, and $49,560 each year for the remaining 26 years of the project.

To use the LPG option will require an initial investment of $92,520. Every 3 years, each LPG-converted engine must be overhauled at a rate of $350. The savings in fuel costs for LPG are $11,440 a year for the first 2 years, $19,100 a year for the next 2 years, and $34,368 a year for the remaining 26 years.

The major drawback in the CNG project is the high cost of the required compressor and the relatively high cost of the conversion kits. Lone Star needs to decide whether the large initial cost of the CNG project more than offsets the increased fuel savings.

Comments: The choice of whether or not to change over to the alternative fuels and which fuel to switch to is affected by many factors. Among these factors are the number of trucks used by the company and the future outlook of the alternative fuel price. As the size of the fleet increases, so do the savings with the alternative fuel. Therefore, a small company with only a few trucks is less likely to change its fleet of trucks over to an alternative fuel. In other words, a small company does not travel enough cumulative miles to save the money required to recover the initial investment.

1.6.5   Service Improvement

Most of the examples in the previous sections were related to economic decisions in the manufacturing sector. The decision techniques we develop in this book are also applicable to various economic decisions involved in improving services. Example 1.5 which follow, will illustrate this.

Example 1.5 Levi's Personal Pair of Jeans

Levi Strauss & Company, the world's largest jeans maker, found that women complained more than men about difficulties in finding off-the-rack jeans that fit properly. This led to a consideration of whether to sell made-to-order Levi's for women. As shown in Figure 1.5, the idea is that sales clerks at Original Levi's Stores can use a personal computer and the customer's measurements to create

A sales clerk measures the customer using instructions from a computer as an aid.

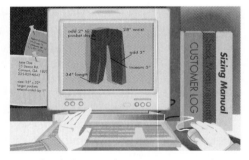

The clerk enters the measurements and adjusts the data based on the customer's reaction to samples.

The final measurements are relayed to a computerized fabric cutting machine at the factory.

Bar codes are attached to the clothing to track it as it is assembled, washed, and prepared for shipment.

Figure 1.5   Making customized blue jeans for women from Data to Denim—A new computerized system being installed at some Original Levi's Stores allows women to order customized blue jeans. Levi Strauss declined to have its factory photographed, so here is an artist's conception of how the process works.

what amounts to a digital blue jeans blueprint. When transmitted electronically to a Levi's factory, the computer file would instruct a robotic tailor to cut a bolt of denim precisely to the woman's measurements. The finished Levi's, about $10 more than a mass-produced pair, are shipped back to the store within 3 weeks— or directly to the customer by Federal Express for an additional $5 fee. The service being market-tested by Levi is the first of its type in the clothing industry. It has the potential to change the way people buy clothes, and it will allow stores to cut down on inventory. It is part of an emerging industrial trend toward so-called mass customization, in which computerized instructions enable factories to modify mass-market products one at a time to suit the needs of individual customers.

Levi Strauss currently offers this service at 15 locations and plans to offer the service at more than 30 Original Levi's Stores throughout North America. This marketing thrust would require a considerable investment in robotic tailors and realignments of the current sewing operation and the assembly process.

Comments: Levi said each computer-designed pair of jeans was basically a customized product. Eventually, this could mean no inventory and no markdowns. A company following this route is not mass producing a product and hoping it sells. This is no small consideration in the nation's apparel industry, in which an estimated $25 billion worth of manufactured clothing each year either goes unsold or sells only after severe markdowns. Levi's main problem is to determine how much demand the women's line would generate. How many more jeans would Levi need to sell to justify the cost of additional robotic tailors? This analysis should involve a comparison of the cost of operating the additional robotic tailors with the additional revenue generated by selling more jeans.

1.7 Summary

- This chapter has provided an overview of a variety of engineering economic problems that commonly are found in business world. We examined the place of engineers in the firm, and we saw that the engineers have been playing an increasingly important role in the organization as evidenced in the development of a hybrid vehicle known as Prius at Toyota. Commonly, engineers are called upon to participate in a variety of strategic business decisions ranging from product design to marketing.

- The term **engineering economic decision** refers to all investment decisions relating to engineering projects. The facet of an economic decision of most interest from an engineer's point of view is the evaluation of costs and benefits associated with making a capital investment

- The five main types of engineering economic decisions are (1) equipment and process selection, (2) equipment replacement, (3) new product and product expansion, (4) cost reduction, and (5) service improvement.

- The factors of **time** and **uncertainty** are the defining aspects of any investment project.

CHAPTER 2

Understanding Financial Statements

B	C	D	E	F
			Interest (%)	Equivalen Option
Option 1	Option 2		0	$700
			1	$721
			2	$743
$100,000			3	$766
$100,000			4	$789
$100,000			5	$814
$100,000			6	$839
$100,000			7	$865
$100,000			8	$892
$100,000			9	$920
	$100,000		10	$948
	$100,000		11	$979
	$100,000			
	$100,000			

Suppose you have $50,000 to invest. What would you do? You could at least deposit the money in a savings account that earns some interest. Your investment would be safe because deposits in U.S. banks are insured, but it will not grow very fast in a bank account. As an alternative, you might be interested in buying land near the Interstate on the outskirts of town. There have been some rumors that a European automobile company may locate a factory in the vicinity. If that is the case, your investment may double or triple in value. However, it is also equally risky, as the automobile company may not locate near your property. Besides, others also know of the carmaker's plan, and you will have to pay a premium price for the land. A third possible investment is for you to buy stock in a fast-growing company such as Dell Computer Corporation. This investment is more risky than depositing money in a bank account but less risky than buying land in hopes of selling to the carmaker at a premium price. If you want to explore investing in Dell stock, what information would you go by?

You would certainly prefer that Dell have a record of accomplishment of profitable operations—earning a profit (net income) year after year. The company would need a steady stream of cash coming in and a manageable level of debt. How would you determine whether the company meets these criteria? Investors commonly use the financial statements contained in the annual report as a staring point in forming expectations about the future levels of earnings and about the firm's riskiness.

Before making any financial decision, it is good to understand an elementary fact of your financial situation—one that you'll also need for retirement planning, estate planning, and, more generally, to get an answer to that question, "How am I doing?" It is called your **net worth.** If you decided to invest that $50,000 in buying Dell stocks, how will it affect your net worth? You may need this information for your own financial planning, but it is routinely required whenever you have to borrow a large sum of money from any financial institutions. For example, when you are buying a home, you need to apply for a mortgage. Invariably, the bank will ask you to submit your net worth statement as a part of loan processing. Your net worth statement is a snapshot of where you stand financially at a given point in time. The bank will determine how credit-worthy you are by examining your net worth. In the similar way, a corporation prepares the same kind of information for their financial planning or reporting their financial health to stockholders or investors. We will first review the basics of figuring out the personal net worth, and then illustrate how any investment decision will affect this net worth statement. Understanding the relationship between the net worth and investing decisions will enhance the overall understanding of how a company manages their assets in business operations.

2.1   Accounting: The Basis of Decision Making

We need financial information when we are making business decisions. Virtually all business and most individuals keep accounting records to aid in making decisions. As illustrated in Figure 2.1, accounting is the information system that measures business activities, processes that information into reports, and communicates the results to decision makers. For this reason, we call accounting "the language of business." The better you understand this language, the better you can manage your financial well-being, and the better your financial decisions will be.

Personal financial planning, education expenses, loans, car payments, income taxes, and investments are all based on the information system we call accounting. The use of accounting information is diverse and varied:

- **Individual People** use accounting information in their day-to-day affairs to manage bank accounts, to evaluate job prospects, to make investments, and to decide whether to rent an apartment or buy a house.

- **Business Mangers** use accounting information to set goals for their organizations, to evaluate progress toward those goals, and to take corrective actions if necessary. Decisions based on accounting information may include which building or equipment to purchase, how much merchandise inventory to keep on hand, and how much cash to borrow.

- **Investors and Creditors** provide the money a business needs to begin operations. To decide whether to help start a new venture, potential investors evaluate what income they can expect on their investment. This means analyzing the financial

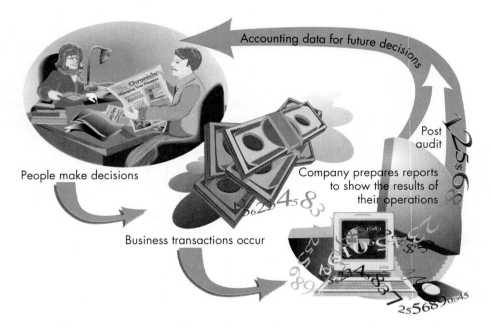

Figure 2.1   The accounting system, which illustrates the flow of information

statements of the business. Before making a loan, banks determine the borrower's ability to meet scheduled payments. This evaluation includes a projection of future operations and revenue, which is based on accounting information.

An essential product of an accounting information system is a series of financial statements that allow people to make informed decisions. For personal use, the net worth statement is a snapshot of where you stand financially at a given point time. For business use, financial statements are the documents that report financial information about a business entity to decision makers. They tell us how a business is performing and where it stands financially. Our purpose is not to present bookkeeping aspects of accounting, but to acquaint you with financial statements and to give you the basic information you need to make sound engineering economic decisions through the reminder of the book.

2.2   Net Worth: How Well Am I Doing?

Most of the materials in this chapter describe business situations, but the principles of accounting apply to the financial affairs of individuals as well. The two most important statements in monitoring your financial progress are a cash flow statement—how much is it costing you to live—and net worth statement. A cash flow statement tells you how much you have earned and spent during a specified period, while net worth tells you what assets you have to help meet that need and where you stand with respect to your financial goals. In this respect, many financial institutions require the net worth statement more commonly when they evaluate an individual's loan application. Let us first examine how we might develop a net worth statement for an individual.

2.2.1   The Starting Point

As part of deciding how to pursue your financial goals, you should look at where you stand right now. You do that by adding your assets—such as cash, investments, and pension plans—in one column and your liabilities—or debts—in the other. Then subtract your liabilities from your assets to find your net worth. In other words, your net worth is what you would be left over with if you sold everything and paid off all you owe. (While you are in college, your net worth would be relatively small or even negative, depending upon your financial status. However, as you leave campus, find a full-time job, and begin to save money, your net worth will start to accumulate.) For illustrative purposes, we will use the net worth statement prepared for a typical individual. (See Table 2.1.)

Assets: What You Own

On the asset side, cash, stocks, bonds and the like are easy to value. When you are calculating what your personal assets are worth, the number to use is their *fair market value*. That is the price a buyer would pay for things you are willing to sell.

- **Cash reserve assets** are cash, or the equivalent of cash, that you can use on short notice to cover an emergency or make an investment. These include the money

Table 2.1
A Typical Net Worth Statement for an Individual (as of December 31, 2000)

Assets (current estimated value):	Amount	Liabilities	Amount
Cash reserve assets:		Short-term debts:	
Cash in banks	$24,000	Car loans	
Money market accounts (CD)		Charge accounts	$850
Life insurance		Margin loans	
(cash surrender value)	45,000	(stocks/bonds)	140,000
Investment assets:		Bank loans	
Stocks/bonds	390,000	Alimony	
Mutual funds		Taxes owed:	
IRA & Keogh accounts	55,000	income	—
Pension & 401(k) accounts	275,000	real estate	1,000
Personal assets:		other	
Cars	20,000	Insurance	1,150
Computers, electronic			
equipment	4,000	Long-term debts:	
Other personal property	26,000	Home mortgages	155,000
Real estate:		College loan	
home	400,000	Business loan	
land & other		Estimates taxes:	
Business intrest		**Total liabilities**	**$298,000**
Total assets	**$ 1,239,000**	**Net worth**	**$941,000**

in your checking, savings, and money market accounts, certificate of deposits (CDs), Treasury bills, and the cash value of your life insurance policy.

- **Investment assets,** like stocks, bonds, and mutual funds, produce income and growth. We consider retirement plans and annuities to be long-term investments.

- **Personal assets** are your possessions. Some—like antiques, stamp collections, and art—may appreciate, or increase in value, making them investments as well. Others—like cars, boats, computers, and electronic equipment—depreciate or decrease in value over time.

- **Real estate** is a special asset because you can use it yourself, rent it, or perhaps sell it for a profit.

- **Business interests** include such assets as partnership interests, shares in closely held corporations, and the like. They can be especially tough to value and might require an expert to value.

Liabilities: What You Owe Currently and Long-Term

On the debt side, divide your liabilities into unsecured and secured. Unsecured loans include credit cards, overdraft checking, or signature loans. Secured loans like mortgages and car loans are ones where the lender can take the items, or collateral, if you do not pay.

- **Unsecured loans or short-term debts** are your current bills—credit card purchases, installment and personal loans, income and real estate taxes, and insurance premiums. Your present credit card balances are generally included, even if you regularly pay your entire bill each month.

- **Secured loans or long-term debts** are mortgages and other loans that you repay in installments over several years such as home mortgage.

- **Estimated taxes** are the amount that you have to pay quarterly if you run a business.

   Once you have it all together, add up both sides, subtract the liabilities from the assets, and you have your net worth. In our example in Table 2.1, the individual's net worth is $941,000 as of December 31, 2000.

2.2.2   Using Net Worth Statements

Figuring your net worth is not only a critical first step in financial planning, but it will also come in handy in many financial situations. Once you have the numbers, you can do other things, such as separating out those assets that produce income and those that do not. While everything is included for estate taxes, things like your house and cars do not help much in planning retirement finances. You could sell them, of course, but you have to live somewhere, and most people need a car.

- Mortgage lenders require a statement of your assets and liabilities as part of the application. When potential lenders assess your loan application—a net worth statement—to decide whether you qualify for a loan, they look at what you already owe.

- College financial aid is based on your net worth, so you will have to show your assets and liabilities.

- Loan and line-of-credit applications usually require net worth statements.

- Certain high-risk investments may require that you have a minimum net worth—say $1 million or more.

   With an analysis of your net worth,[1] you can begin to match your likely retirement income needs with possible sources of investment income. Planners strongly suggest you hang on to the figures and repeat the exercise in succeeding years. It will tell you whether you are gaining or losing, and ultimately, whether your goals are within reach or whether you have to make adjustments.

---

[1] Most people think real wealth comes from inheritance, or from striking it rich on Wall Street or in Silicon Valley. According to the book entitled, *The Millionaire Next Door* by Thomas Stanley and William Danko (Longstreet: 1997), most millionaires do not fit that stereotype at all. Instead, they are hard-working, hard-saving folks who live next door. Their secret is not so much earning a lot of money as it is saving and not spending. According to the book, there is a big difference between income and wealth, and most people do not recognize it. Wealth is not income. You cannot retire on income! Wealth is your assets less your liabilities, which is the definition of "net worth" we discussed in this section.

2.3 Financial Status for Businesses

Just like figuring out your personal wealth, all businesses must prepare their financial status. Of the various reports corporations issue to their stockholders, the annual report is by far the most important. This annual report contains basic financial statements as well as management's opinion of the past year's operations and the firm's future prospects. What would managers and investors want to know about a company at the end of fiscal year? Four basic questions that managers or investors are likely to ask are:

- What is the company's financial position at the end of fiscal period?
- How well did the company operate during the fiscal period?
- Where did the company decide to use their profits for?
- How much cash did the company generate and spend during the period?

As illustrated in Figure 2.2, the answer to each question is provided by one of financial statements: the balance sheet statement, the income statement, the statement of retained earnings, and the cash flow statement. The fiscal year (or operating cycle) can be any 12-month term, but is usually January 1 through December 31 of a calendar year.

As mentioned in Section 1.4.2, one of the primary responsibilities of engineers in business is to plan for the acquisition of equipment (capital expenditure) that will enable the firm to design and produce products economically. This will require estimation of savings and costs associated with the equipment acquisition and the degree of risk associated with project execution. This will affect the business' bottom line (profitability), which will eventually affect the firm's stock price in the marketplace. Therefore, engineers should understand meanings of various financial statements to

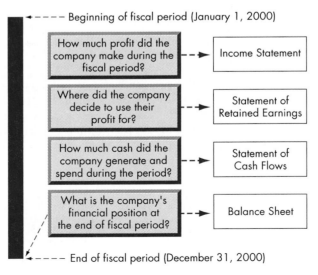

Figure 2.2 Information reported on the financial statements

communicate with upper management regarding the nature of project's profitability. This is summarized in Figure 2.3.

For illustration purposes, we use data taken from Dell Computer Corporation, a manufacturer of a wide range of computer systems including desktops, notebooks, and workstations, to discuss the basic financial statements. Formed in 1984, Michael Dell began his computer business at The University of Texas-Austin, often hiding his IBM PC in his roommate's bathtub when his family came to visit. His dorm-room business officially became Dell Computer Corporation on May 3, 1984. In 1998, Dell became the No. 2 and fastest-growing among all major computer-systems companies world-wide, with 20,800 employees around the globe. Dell's pioneering "direct model" is a simple concept of selling personal computer systems directly to customers. It offers in-person relationships with corporate and institutional customers; telephone and Internet purchasing (the latter now averaging $6 million a day); build-to-order computer systems; phone and on-line technical support; and next day, on-site product service.

The company's revenue for the last four quarters totaled $25.3 billion. The increase in net revenues for both fiscal 2000 and fiscal 1999 were principally because of increased units sold. Unit shipments grew 50% and 64% for fiscal years 2000 and 1999, respectively. In its 2000 annual report, management painted a more optimistic picture for the future, stating that Dell will continue to invest in information systems, research, development, and engineering activities, to support its growth and to provide for new competitive products. Of course, there is no assurance that Dell's revenue will continue to grow at the annual rate of 50% in the future.

What can individual investors make of all this? Actually, we can make quite a bit. As you will see, investors use the information contained in an annual report to form expectations about future earnings and dividends. Therefore, the annual report is certainly of great interest to investors.

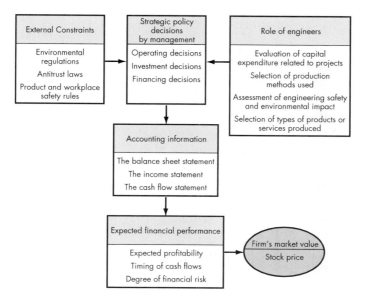

Figure 2.3   Summary of major factors affecting stock price

2.3.1 The Balance Sheet

What is the company's financial position at the end of reporting period? We find the answer to this question in the company's **balance sheet statement.** A company's balance sheet, sometimes called its **statement of financial position,** reports three main categories of items: assets, liabilities, and stockholders' equity. As shown in Table 2.2, the first-half portion of Dell's year-end 2000 and 1999 balance sheets lists the firm's assets, while the remaining portion shows the liabilities and equity, or claims against

Table 2.2
Consolidated Statements of Financial Positions

Balance Sheet Statement—Dell Computer Corporation				
Assets (in millions)	28-Jan-00	29-Jan-99	Change	Percent
Current assets:				
Cash	$ 3,809	$ 1,726	$ 2,803	121%
Short-term investments	323	923	(600)	−65%
Account receivables, net	2,608	2,094	514	25%
Inventories	391	273	118	43%
Other	550	791	(241)	−30%
Total current assets	7,681	5,807	1,874	32%
Property, plant and equipment, net	765	523	242	46%
Long-term investments	1,048	532	516	97%
Equity securities and other investments	1,673	—	1,673	
Goodwill and others	304	15	289	1927%
Total assets	$ 11,471	$ 6,877	$ 4,594	67%
LIABILITIES AND STOCKHOLDERS' EQUITY				
Current liabilities:				
Accounts payable	$ 3,538	$ 2,397	$ 1,141	48%
Accrued and other	1,654	1,298	356	27%
Total current liabilities	5,192	3,695	1,497	41%
Long-term debt	508	512	(4)	−1%
Other	463	349	114	33%
Total liabilities	6,163	4,556	1,607	35%
Stockholders' equity:				
Preferred stock	—	—		
Common stock and capital in excess of $0.01 par value	3,583	1,781	1,802	101%
Retained earnings	1,260	606	654	108%
Other	465	(66)	531	
Total stockholders' equity	5,308	2,321	2,987	129%
Total liabilities and stockholders' equity	$ 11,471	$ 6,877	$ 4,594	67%

these assets. The financial statements are based on the most basic tool of accounting, the accounting equation. The **accounting equation** shows the relationship among assets, liabilities, and owners' equity.

$$\text{Assets} = \text{Liabilities} + \text{Owners' Equity}$$

Every business transaction, no matter how simple or how complex, can be expressed in terms of its effect on the accounting equation. Regardless of whether a business grows or contracts, this equality between the assets and the claims against the assets is always maintained. In other words, any change in the amount of total assets is necessarily accompanied by an equal change on the other side of the equation, that is, by an increase or decrease in either the liabilities or the owners' equity.

Assets

The dollar amount shown under the Assets column represents how much the company owns at the time of reporting. We list the asset items in the order of their "liquidity," or the length of time it takes to convert them to cash.

- **Current assets** can be converted to cash or its equivalent in less than one year. These generally include three major accounts.

  1. The first is *cash*. A firm typically has a cash account at a bank to provide for the funds needed to conduct day-to-day business. Although we all state the assets in terms of dollars, only cash represents actual money. Cash equivalent items are also listed. These include "marketable securities" and "short term investments."

  2. The second account is *accounts receivable*, which is money that is owed to the firm but has not yet been received. For example, when Dell receives an order from a retail store, the company will send an invoice along with the shipment to the retailer. Then the unpaid bill immediately falls into a category called "accounts receivable." As this bill is paid, it will be deducted from the accounts receivable account and placed into the cash category. Normally, a typical firm will have a 30- to 45-day accounts receivable, depending on the frequency of its bills and the payment terms for customers.

  3. The third account is *inventories*, which show dollars Dell has invested in raw materials, work-in-process, and finished goods available for sale.

- **Fixed assets** are relatively permanent and take time to convert them into cash. Fixed assets reflect the amount of money Dell paid for its plant and equipment when it acquired those assets at some time in the past. The most common fixed assets include the physical investment in the business, such as land, buildings,[2] factory machinery, office equipment, and automobiles. With the exception of land, most fixed assets have a limited useful life. For example, buildings and equipment are used up over a period of years. Each year, a portion of the usefulness of these assets expires, and a portion of their total cost should be recognized as depreciation expense. The term *depreciation* means the accounting

---

[2] Land and buildings are commonly called **real assets** to distinguish from equipment and machinery.

process for this gradual conversion of fixed assets into expenses. The "*property, plant and equipment, net*" thus represents the current book value of these assets after deducting such depreciation expenses.

- **Other assets** are listed finally. Typical assets in this category include investments made in other companies and intangible assets such as goodwill, copyrights, franchises, and so forth. Goodwill appears on the balance sheet only when an operating business is purchased in its entirety. This indicates any additional amount paid for the business above the fair market value of the business. (Here, the fair market value is defined as the price that a buyer is willing to pay when the business is offered for sale.)

Liabilities and Stockholders' Equity (Owners' Net Worth)

The claims against assets are of two types—liabilities and the stockholders' equity. The liabilities of a company indicate where the company obtained the funds to acquire the assets and to operate the business. Here, the liability means money the company owes. Stockholders' equity indicates the portion of the assets of a company, which are provided by the investors (owners). Therefore, stockholders' equity is also the liability of the company to its owners.

- **Current liabilities** are those debts that must be paid in the near future (normally payable within one year). The major current liabilities include accounts and notes payable within a year. They also include accrued expenses (wages, salaries, interest, rent, taxes, etc., owed but not yet due for payment), and advance payments and deposits from customers.
- **Other liabilities** include *long-term liabilities* such as bonds, mortgages, and long-term notes, which are due and payable more than one year in the future.
- **Stockholders' equity** represents the amount that is available to the owners after all other debts have been paid. It generally consists of preferred and common stock, treasury stock, capital surplus, and retained earnings. Preferred stock is a hybrid between common stock and debt. In case of bankruptcy, preferred stock ranks below debt but above common stock. The preferred dividend is fixed, so preferred stockholders do not benefit if the company's earnings grow. In fact, many firms do not use any preferred stock. The common stockholders' equity, or **net worth**, is a residual:

  Assets − Liabilities − Preferred stock = Common stockholders' equity
  $11,471 − $6,163   − $0   = $5,308

- **Common stock** is the aggregate par value of the company's stock issued. Companies rarely issue stocks at a discount (i.e., at an amount below the stated par). Corporations normally set the par value low enough so that in practice stock is usually sold at a premium.
- **Paid-in capital** (capital surplus) is the amount of money received from the sale of stock more than the par value of the stock. Outstanding stock is the number of shares issued that actually is held by the public. If the corporation buys back part of its own issued stock, it is listed as *Treasury Stock* on the balance sheet.

- **Retained earnings** represent the cumulative net income of the firm since its beginning, less the total dividends that have been paid to stockholders. In other words, retained earnings indicate the amount of assets that have been financed by plowing profits back into the business. Therefore, these retained earnings belong to the stockholders.

2.3.2 The Income Statement

The second financial report is the **income statement** that indicates whether the company is making or losing money during a stated *period*. Most businesses prepare quarterly and monthly income statements as well. The company's accounting period refers to the period covered by an income statement. For Dell, the accounting period begins on February 1 and ends at January 31 of the following year. Table 2.3 gives the 2000 and 1999 income statements for Dell.

Reporting Format

Typical items that are itemized in the income statement are as follows:

- **Revenue** is the price of goods sold and services rendered during a given accounting period.
- The **net sales** figure represents the gross sales less any sales return and allowances.
- Shown on the next several lines, we list the expenses and costs of doing business as deductions from the revenues. The largest expense for a typical manufacturing firm is its production expense for making a product (such as labor, materials, and overhead), called **cost of goods sold** (or cost of revenue).
- Net sales less the cost of goods sold indicates the **gross margin**.
- Next, we subtract any other operating expenses from the operating income. These other operating expenses are items such as interest, lease, selling, and administration expenses. This results in the **operating income**.
- Finally, we determine the **net income** (or net profit) by subtracting the income taxes from the taxable income. This net income is also commonly known as the *accounting income*.

Earnings Per Share

Another important financial information provided in the income statement is the **earnings per share** (EPS).[3] In simple situations, we compute this by dividing the "available earnings to common stockholders" by the number of shares of common

---

[3] In reporting EPS, the firm is required to distinguish between "basic EPS" and "diluted EPS." The basic EPS means that the net income of the firm is divided by the number of shares of common stock outstanding. On the other hand, the diluted EPS includes all common stock equivalents (convertible bonds, preferred stock, warrants, and rights) along with common stock. Therefore, the diluted EPS will be usually less than the basic EPS.

Table 2.3
Income
Statement
for Dell
Computer
Corporation

Income Statement—Dell Computer Corporation		
(in millions, except per share amount)	Fiscal Year Ended	
	28-Jan-00	29-Jan-99
Net revenue	$25,265	$18,243
Cost of revenue	20,047	14,137
Gross margin	5,218	4,106
Operating expenses:		
Selling, general and administrative	2,387	1,788
Research, development and engineering	568	272
Total operating expenses	2,955	2,060
Operating income	2,263	2,046
Other income	188	38
Income before income taxes	2,451	2,084
Provision for income taxes	785	624
Net income	$1,666	$1,460
Earnings per common share:		
Basic	$0.66	$0.58
Diluted	$0.61	$0.53
Weighted average shares outstanding:		
Basic	2,536	2,531
Diluted	2,728	2,772
Retained Earnings:		
Balances at beginning of period	606	607
Net income	1,666	1,460
Repurchase of common stocks	(1,012)	(1,461)
Balances at end of period	$1,260	$606

stock outstanding. Stockholders and potential investors want to know what their share of profits is, not just the total dollar amount. Presentation of profits on a per share basis allows the stockholders to relate earnings to what he or she paid for a share of stock. Naturally, companies want to report a higher EPS to their investors as a means of summarizing how well they managed their businesses for the benefits of owners. Dell earned $0.61 per share in 2000, up from $0.53 in 1999, but it paid no dividend.

Retained Earnings

As a supplement to the income statement, many corporations also report their retained earnings during the accounting period. When a corporation makes some profits, it has to decide what to do with these profits. The corporation may decide to pay out some of the profits as dividends to its stockholders. Alternatively, it may retain the remaining profits in business to finance expansion or support other business activities.

When the corporation declares dividends, preferred stock has priority over common as the receipt of dividends. Preferred stock pays a stated dividend, much like the interest payment on bonds. The dividend is not a legal liability until the board of directors has declared it. However, many corporations view the dividend payments to preferred stockholders as liability. Therefore, "available for common stockholders" reflects the net earnings of the corporation less the preferred stock dividends. When preferred and common stock dividends are subtracted from net income, the remainder is retained earnings (profits) for the year. As mentioned previously, these retained earnings are reinvested into the business.

Example 2.1 Understanding The Dell's Balance Sheet and Income Statements

With revenue of $25.3 billion for fiscal year 2000, the Company is the world's leading direct computer systems company. With Tables 2.2 and 2.3, discuss how Dell generated the net income during the fiscal year.

Solution

The Balance Sheet: The $11,471 million of total assets shown in Table 2.2 were necessary to support these sales of $25,265 million.

- Dell obtained the bulk of the funds used to buy assets.
  1. By buying on credit from their suppliers (accounts payable)
  2. By borrowing from financial institutions (notes payable and long-term bonds)
  3. By issuing common stock to investors
  4. By plowing earnings into business as reflected in the retained earnings account
- The net increase in fixed assets is $242 million ($765–$523); however, this net amount is after a deduction for the year's depreciation expenses. We should add depreciation expense back to show the increase in gross fixed assets. From the company's cash flow statement later in Table 2.4, we see that 2000 depreciation expense is $156 million; thus, the acquisition of fixed assets equals $398 million.
- Dell had a total of long-term debt of $508 million that consists of the several bonds issued previous years. The interest payments associated with these long-term debts were about $34 million.
- Dell had 2,728 million shares of common stock outstanding. Investors actually have provided the company with a total capital of $3,583 million. However, Dell has retained the current as well as previous earnings of $1,260 million since it was incorporated. These earnings belong to Dell's common stockholders. The combined net stockholder's equity was $5,308.
- Stockholders on the average have a total investment of $1.95 per share ($5,308 million/2,728 million shares) in the company. This figure of $1.95 is known as

the stock's book value. In the summer of 2000, the stock traded in the general range of $35 to $52 per share. Note that this market price is quite different from the stock's book value. Many factors affect the market price—most importantly how investors expect the company to do in the future. Certainly, the company's direct made-to-order business practices have had a major influence on the market value of its stock.

Income Statement: Net sales were $25,265 million in 2000, compared with $18,243 million in 1999, a gain of 38%. Profits from operations (operating income) rose 11% to $2,263 million, and net income was up 14% to $1,666 million.

- Dell issued no preferred stock, so there is no required cash dividend. Therefore, the entire net income of $1,666 million belongs to the common stockholders.

- Earnings per common share climbed at a faster pace to $0.61, an increase of 15%. Dell could retain this income fully for reinvestment in the firm, or it could pay it as dividends to the common stockholders. Instead, Dell repurchased and retired 56 million common stocks for $1,012 million. We can see that Dell had earnings available to common stockholders of $1,666 million. The beginning balance of the retained earnings was $606 million. Therefore, the total retained earnings grew to $2,272 million, and of this amount, Dell paid out $1,012 million to repurchase 56 million shares of their own shares. The resulting ending balance was $1,260 million for the retained earning account.

2.3.3 The Cash Flow Statement

The income statement explained in the previous section only indicates whether the company was making or losing money during the reporting period. Therefore, the emphasis was on determining the net income (profits) of the firm, namely for the operating activities. However, the income statement ignores two other important business activities for the period—financing and investing activities. Therefore, we need another financial statement, the cash flow statement that details how the company generated the cash and how the company used the cash during the reporting period.

Sources and Uses of Cash

The difference between the sources (inflows) and uses (outflows) of cash represent the net cash flow during the reporting period. This is a very important piece of information because investors determine the value of an asset (or a whole firm) by the cash flows it generates. Certainly, a firm's net income is important, but cash flows are even more important. This is particularly true because we need cash to pay dividends and to purchase the assets required for continuing operations. As we mentioned in the previous section, the goal of the firm should be to maximize the price of its stock. Since the value of any asset depends on the cash flows produced by the asset, managers want to maximize cash flows available to investors over the long run. Therefore, we should make investment decisions based on cash flows rather than profits. For such investment decisions, it is necessary to convert profits (as determined in the income statement) to cash flows. Table 2.4 is Dell's statement of cash flows, as it would appear in the company's annual report.

Reporting Format

In preparing the cash flow statement such as Table 2.4, many companies identify the sources and uses of cash according to the types of business activities. There are three types of activities:

- **Operating activities:** We start with the net change in operating cash flows from the income statement. Here operating cash flows represent those cash flows related to production and sales of goods or services. All non-cash expenses are

Table 2.4
Cash Flow Statement for Dell Computer Corporation

Cash Flow Statement—Dell Computer Corporation		
	Fiscal Year Ended	
(in millions)	28-Jan-00	29-Jan-99
Cash flows from operating activities:		
Net income	$1,666	$1,460
Depreciation and amortization	156	103
Changes in working capital	2,104	873
Net cash provided by operating activities	3,926	2,436
Cash flows from investing activities:		
Marketable securities:		
Purchase	(3,101)	(1,938)
Sales	2,319	1,304
Capital expenditures	397	(296)
Net cash used in investing activities	(1,183)	(930)
Cash flows from financing activities:		
Purchase of common stock	(1,061)	(1,518)
Issuance of common stock under employee plans	289	212
Proceeds from issuance of long-term debt	20	494
Cash received from sale of equity options and other	63	
Repayment of borrowings	(6)	
Net cash used in financing activities	(695)	(812)
Effect of exchange rate changes on cash	35	(10)
Net increase in cash	$2,083	$684
Cash at beginning of period	1,726	1,042
Cash at end of period	$3,809	$1,726

simply added back to net income (or after-tax profits). For example, an expense such as depreciation is only an accounting expense (bookkeeping entry). While we may charge them against current income as an expense, they do not involve an actual cash outflow. The actual cash flow may have occurred when the asset was purchased. Any adjustments in **working capital**[4] terms will also be listed here.

- **Investing activities:** Once we determine the operating cash flows, we next consider any cash flow transactions related to the investment activities. Investment activities include such as purchasing new fixed assets (cash outflow), reselling old equipment (cash inflow), or buying and selling financial assets.

- **Financing activities:** Finally, we detail cash transactions related to financing any capital used in business. For example, the company could borrow or sell more stock, resulting in cash inflows. Paying off existing debt will result in cash outflows.

By summarizing cash inflows and outflows from three activities for a given accounting period, we obtain the net changes in cash flow position of the company.

Example 2.2 Understanding the Dell's Cash Flow Statement

As shown in Table 2.4, cash flow from operations amounted to $2,083 million. Note that this is significantly more than $1,666 million earned during the reporting period. Where did all the rest of monies come?

Solution

- The main reason for the difference is because of the accrual-basis accounting principle used by the Dell Corporation. In **accrual-basis accounting**, an accountant recognizes the impact of a business event as it occurs. When the business performs a service, makes a sale, or incurs an expense, the accountant enters the transaction into the books, whether or not cash has been received or paid. For example, an increase in accounts receivable ($2,608 – $2,094 million = $514 million) during 2000 represents the amount of total sales on credit. Since this figure was included in the total sales in determining the net income, we need to subtract this figure to determine the true cash position. After adjustments, the net cash provided from operating activities is $3,926 million.

- For the investment activities, there was an investment outflow for $397 million in new plant and equipment. Dell sold $2,319 million worth of stocks and bonds during the period, and reinvested $3,101 million in the various financial securities. The net cash flow provided from these investing activities amounted to – $1,183 million, which means an outflow.

---

[4] The difference between the increase in current assets and the spontaneous increase in current liabilities is the **net change in net working capital.** If this change is positive, this indicates that additional financing, over and above the cost of the fixed assets, is needed to fund the increase in current assets. This will further reduce the cash flow from the operating activities.

- On the other hand, the financing decision produced a net outflow of $695 million including repurchase of own stocks. (This repurchase of their own stock is equivalent to investing their idle cash from operation in the stock market. Dell could buy other company's stock such as IBM or Microsoft with the money. However, they liked their own stock better than any other stocks on the market, so they ended up buying their own.) Finally, there was the effect of exchange rate changes on cash for the foreign sales. This amounts to a net increase of $35 million. Together, the three types of activities generated a total cash flow of $2,083 million. With the initial cash balance of $1,726 million, the ending cash balance now increases to $3,809 million. This same amount denotes the change in Dell's cash position as shown in the cash accounts in the balance sheet.

2.4   Using Ratios to Make Business Decisions[5]

As we have seen in Dell's financial statements, the purpose of accounting is to provide information for decision making. They tell what has happened during a particular point in time. In that sense, the financial statements are essentially historical documents. However, most users of financial statements are concerned about what will happen in the future. For example,

- Stockholders are concerned with future earnings and dividends.
- Creditors are concerned with the company's ability to repay its debts.
- Managers are concerned with the company's ability to finance future expansion.
- Engineers are concerned with planning actions that will influence the future course of business events.

Despite the fact that financial statements are historical documents, they can still provide valuable information bearing on all of these concerns. An important part of financial analysis is the calculation and interpretation of various financial ratios. In this section, we consider some of the widely used ratios that analysts use in attempting to predict the future course of events in business organizations. We may group these ratios in five categories (debt management, liquidity, asset management, profitability, and market trend) as outlined in Figure 2.4. In all financial ratio calculations, we will use the 2000 financial statements for Dell Computer Corporation, as summarized in Table 2.5.

2.4.1   Debt Management Analysis

All businesses need assets to operate. To acquire assets, the firm must raise capital. When the firm finances its long-term needs externally, it may obtain funds from the capital markets. Capital comes in two forms, **debt** and **equity**. Debt capital refers to

---

[5] This optional topic may be omitted at the instructor's discretion.

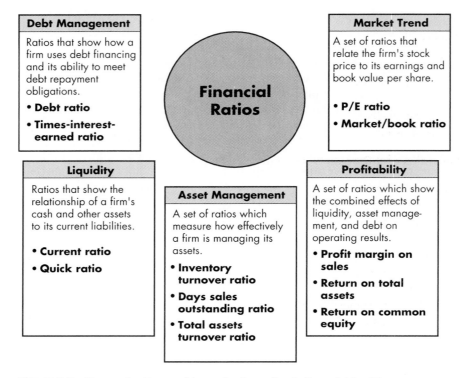

Figure 2.4   Types of ratios used in evaluating a firm's financial health

the borrowed capital from financial institutions. Equity capital refers to the capital obtained from the owners of the company.

The basic methods of debt financing include bank loan and bond sales. For example, a firm needs $10,000 to purchase a computer. In this situation, the firm would borrow money from a bank and repay the loan with the specified interests in a few years, known as a *short-term debt financing*. Suppose that the firm needs $100 million for a construction project. It would normally be very expensive (or require a substantial amount of mortgage) to borrow the money directly from a bank. In this situation, it would go public to borrow money on a long-term basis. When investors lend capital to a company and the company agrees to repay the loan at an agreed interest rate, the investor is creditor of the corporation. The document that records the nature of the arrangement between the issuing company and the investor is called a **bond**. The way of raising capital through issuing a bond is called a long-*term debt financing*.

Similarly, there are different types of equity capital. For example, the equity of a proprietorship represents the money provided by the owner. For a corporation, equity capital comes in two forms: *preferred* and *common stock*. Investors provide capital to a corporation and the company agrees to provide the investor with fractional ownership in the corporation. Preferred stock pays a stated *dividend*, much like the interest payment on bonds. However, the dividend is not a legal liability until the company declares it. Preferred stock has preference over common as to the receipt of dividends,

Balance Sheet—January 28, 2000	
Assets	
Cash	$3,809
Other current assets	3,872
Total current assets	7,681
All other assets	3,790
Total assets	**11,471**
Liabilities and Stockholders' Equity	
Current liabilities	5,192
Other liabilities	971
Total liabilities	6,163
Stockholders' equity:	
Common stock	3,583
Retained earnings	1,260
Other equity	(465)
Total stockholders' equity	5,308
Total liabilities and stockholders' equity	**$11,471**

Income Statement—Fiscal Year 2000	
Net revenue	25,265
Expenses (including income taxes)	23,599
Net income	**$1,666**

Statement of Retained Earnings—Fiscal Year 2000	
Beginning retained earnings	606
Net income	1,666
Purchase and retirement of 56 million shares	(1,012)
Ending retained earnings	**$1,260**

Statement of Cash Flows—Fiscal Year 2000	
Net cash flow from operating activities	$3,926
Net cash flow from investing activities	(1,183)
Net cash flow from financing activities	(695)
Effect of exchange rate change on cash	(35)
Beginning cash	1,726
Ending cash	**$3,804**

Table 2.5
Key Summary of Dell's Financial Statements

as to assets in case of liquidation. We can examine the extent to which a company uses debt financing (or financial leverage) in business operation.

- Check the balance sheet to determine the extent to which borrowed funds have been used to finance assets, and
- Review the income statement to see the extent to which fixed charges (interests) are covered by operating profits.

Two essential indicators of a business's ability to pay long-term liabilities are the *debt ratio* and the *times-interest-earned ratio*.

Debt Ratio

The relationship between total liabilities and total assets, generally called the **debt ratio**, tells us the proportion of the company's assets that it has financed with debt:

$$\text{Debt ratio} = \frac{\text{Total debt}}{\text{Total assets}}$$

$$= \frac{\$6,163}{\$11,471} = 53.73\%$$

Total debt includes both current liabilities and long-term debt. If the debt ratio is one, then the company has used debt to finance all the assets. As of January 28, 2000, Dell's debt ratio is 53.73%; this means that its creditors have supplied close to 54% of the firm's total financing. Certainly, most creditors prefer low debt ratios because the lower the ratio, the greater the cushion against creditors' losses in case of liquidation. If a company seeking financing already has large liabilities, then additional debt payments may be too much for the business to handle. For this highly leveraged company, creditors generally charge higher interest rates on new borrowing to help protect themselves.

Times-Interest-Earned Ratio

The most common measure of the ability of a company's operations to provide protection to the long-term creditor is the times-interest-earned ratio. We find this ratio by dividing earnings before interest and income taxes (EBIT) by the yearly interest charges that must be met. Dell issued $500 million worth of senior notes and long-term bonds with a combined interest rate of 6.8%. This results in $34 million interest expenses in year 2000:

$$\text{Times}-\text{interest}-\text{earned ratio} = \frac{\text{EBIT}}{\text{Interest expense}}$$

$$= \frac{(\$2,451 + \$34)}{\$34} = 73 \text{ times}$$

The ratio measures the extent to which operating income can decline before the firm is unable to meet its annual interest costs. Failure to meet this obligation can bring legal action by the firm's creditors, possibly resulting in bankruptcy. Note that we use the earnings before interest and income taxes, rather than net income, in the numerator. Because Dell must pay interest with pre-tax dollars, Dell's ability to pay current interest is not affected by income taxes. Only those earnings remaining after all interest charge are subject to income taxes. For Dell, the times-interest-earned ratio for 2000 would be 73 times. This ratio is exceptionally higher when compared with its competitor, Compaq of 7.31 times, during the same operating period.

2.4.2 Liquidity Analysis

If you were one of the many suppliers to Dell, your primary concern would be whether Dell is able to pay off its debts as they come due over the next year or so. Short-term creditors want to be repaid on time. Therefore, they focus on Dell's cash flows and on its working capital, as these are the company's primary sources of cash in the near future. The excess of current assets over current liabilities is known as **working capital**. This figure indicates the extent to which current assets can be converted to cash to meet current obligations. Therefore, we view a firm's net working capital as a measure of its *liquidity* position. The amount of working capital available to a firm is of considerable interest to short-term creditors. In general, the larger the working capital, the better able is the business to pay its debt.

Current Ratio

We calculate the current ratio by dividing current assets by current liabilities:

$$\text{Current ratio} = \frac{\text{Current assets}}{\text{Current liabilities}}$$

$$= \frac{\$7,681}{\$5,192} = 1.48 \text{ times}$$

If a company is getting into financial difficulty, it begins paying its bills (accounts payable) more slowly, borrowing from its bank, and so on. If current liabilities are rising faster than current assets, the current ratio will fall, and this could spell trouble. What is an acceptable current ratio? The answer depends on the nature of the industry. The general rule of thumb calls for a current ratio of 2 to 1. This rule, of course, is subject to many exceptions, depending heavily on the composition of the assets involved.

Quick (Acid Test) Ratio

The quick ratio tells us whether the company could pay all its current liabilities if they came due immediately. We calculate the quick ratio by deducting inventories from current assets and then dividing the remainder by current liabilities:

$$\text{Quick ratio} = \frac{\text{Current assets} - \text{Inventories}}{\text{Current liabilities}}$$

$$= \frac{\$7,681 - \$391}{\$5,192} = 1.40 \text{ times}$$

The quick ratio measures how well a company can meet its obligations without having to liquidate or depend too heavily on its inventory. Inventories are typically the least liquid of a firm's current assets; hence, they are the assets on which losses are most likely to occur in case of liquidation. Although Dell's current ratio may look below the average for its industry, 2.5, its liquidity position is relatively strong as it carried a very little amount of inventory in its current assets (only $391 million out of $7,681 million of current assets). We often compare against industry average figures, and should note at this point that an industry average is not an absolute number that all firms should strive to maintain. In fact, some very well-managed firms will be above the average, while other good firms will be below it. However, if we consider a firm's ratios to be quite different from the average for its industry, we should examine the reason why this variance occurs.

2.4.3 Asset Management Analysis

The ability to sell inventory and collect accounts receivables is fundamental to business success. Therefore, the third group of ratios measures how effectively the firm is managing its assets. We will review three ratios related to a firm's asset management. They are (1) inventory turnover ratio, (2) days sale outstanding ratio, and (3) total asset turnover ratio. The purpose of these ratios is to answer this question: Does the total amount of each type of asset as reported on the balance sheet seem reasonable in view of current and projected sales levels? Any asset acquisition requires the use of funds. If a firm has too many assets, its cost of capital will be too high; hence, its profits will be depressed. On the other hand, if assets are too low, the firm is likely lose profitable sales.

Inventory Turnover

The inventory turnover ratio measures how many times the company sold and replaced its inventory during the year. We compute the ratio by dividing sales by the average level of inventories on hand. We compute the average inventory figure by taking the average of the beginning and ending inventory figures. Since Dell has a beginning inventory figure of $273 million and an ending inventory figure of $391 million, its average inventory for the year would be $332 million. Then, we compute the Dell's inventory turnover for 2000 as follows:

$$\text{Inventory turnover ratio} = \frac{\text{Sales}}{\text{Average inventory balance}}$$

$$= \frac{\$25,265}{\$332} = 76.10 \text{ times}$$

As a rough approximation, Dell was able to sell and restock its inventory 76.10 times per year. Dell's turnover of 76.10 times is much faster than its competitor Compaq, 19.2 times. This suggests that Compaq is holding excessive stocks of inventory; excess stocks are, of course, unproductive, and they represent an investment with a low or zero rate of return.

Days Sales Outstanding (Accounts Receivable Turnover)

The days sales outstanding (DSO) is a rough measure of how many times a company's accounts receivable have been turned into cash during the year. We determine the ratio, also called the "**average collection period**," by dividing accounts receivable by average sales per day. In other words, this ratio indicates the average length of time the firm must wait after making a sale before receiving cash.

$$\text{DSO} = \frac{\text{Receivables}}{\text{Average sales per day}} = \frac{\text{Receivables}}{\text{Annual sales}/360}$$

$$= \frac{\$2,608}{\$25,265/360} = \frac{\$2,608}{\$70.18}$$

$$= 37.16 \text{ days}$$

For Dell, on average, it takes 37 days to collect on a credit sale. During the same period, Compaq's average collection periods were 64 days. Whether the average of 38 days taken to collect an account is good or bad depends on the credit terms Dell is offering its customers. If credit terms are 30 days, we can say that Dell's customers, on the average, are not paying their bills on time. In order to improve their working capital position, most customers tend to withhold payment for as long as the credit terms will allow and may even go over a few days. The long collection period may signal that customers are in financial trouble or the company has a poor credit management.

Total Assets Turnover

The total assets turnover ratio measures how effectively the firm uses its total assets in generating its revenues. It is the ratio of sales to all the firm's assets:

$$\text{Total assets turnover ratio} = \frac{\text{Sales}}{\text{Total assets}}$$

$$= \frac{\$25,265}{\$11,471} = 2.20 \text{ times}$$

Dell's ratio of 2.20 times, when compared with Compaq's 1.41, is almost 56% faster, indicating that Dell is using its total assets about as 56% more intensively as is Compaq. In fact, Dell's total investment in plant and equipment is about one-fifth of Compaq. If we view Dell's ratio as the industry average, we can say that Compaq has too much investment in inventory, plant and equipment compared to the size of sale.

2.4.4 Profitability Analysis

One of the most important goals for any business is to earn a profit. The ratios examined thus far provide useful clues as to the effectiveness of a firm's operations, but the profitability ratios show the combined effects of liquidity, asset management, and debt on operating results. Therefore, ratios that measure profitability play a large role in decision-making.

Profit Margin on Sales

We calculate the profit margin on sales by dividing net income by sales. This ratio indicates the profit per dollar of sales:

$$\text{Profit margin on sales} = \frac{\text{Net income available to common stockholders}}{\text{Sales}}$$

$$= \frac{\$1,666}{\$25,265} = 6.59\%$$

For Dell, its profit margin is equivalent to 6.59 cents for each sales dollar generated. Dell's profit margin is greater than the Compaq's profit margin of 1.48%. This indicates that, although Compaq's sales are about 50% more than Dell's revenue during the same operating period, Compaq's operation is less efficient than Dell's operation. Compaq's low profit margin is also a result of its heavy use of debt and carrying a very high volume of inventory. Recall that net income is income after taxes. Therefore, if two firms have identical operations in the sense that their sales, operating costs, and earnings before income tax are the same, but if one company uses more debt than the other, it will have higher interest charges. Those interest charges will pull net income down, and since sales are constant, the result will be a relatively low profit margin.

Return on Total Assets

The return on total assets (ROA), or simply return on assets, measures a company's success in using its assets to earn a profit. The ratio of net income to total assets measures the return on total assets after interest and taxes:

$$\text{Return on total assets} = \frac{\text{Net income} + \text{Interest expense}(1 - \text{tax rate})}{\text{Average total assets}}$$

$$= \frac{(\$1,666 + \$34(1 - 0.3251)}{(\$11,471 + \$6,877)/2} = 18.42\%$$

Adding interest expenses back to net income results in an adjusted earnings figure that shows what earnings would have been if the assets had been acquired solely by selling shares of stock. (Note that Dell's effective tax rate is 32.51% in year 2000.) With this adjustment, we may be able to compare the return on total assets for companies with differing amounts of debt. Again, Dell's 18.42% return on assets is well above the 2.62% for Compaq. This high return results from (1) the company's high basic earning power plus (2) its low use of debt, both of which cause its net income to be relatively high.

Return on Common Equity

Another popular measure of profitability is rate of return on common equity. This ratio shows the relationship between net income and common stockholders' investment in the company. That is, how much income is earned for every $1 invested by the common shareholders. To compute this ratio, we first subtract preferred dividends from net income, which is known as "net income available to common stockholders." We then divide this net income available to common stockholders by the average common stockholders' equity during the year. We compute average common equity by using the beginning and ending balances. At the beginning of fiscal 2000, Dell's common equity balance is $2,321 million, and its ending balance is $5,308 million. The average balance is then simply $3,814.50 million.

$$\text{Return on common equity} = \frac{\text{Net income available to common stockholders}}{\text{Average common equity}}$$

$$= \frac{\$1,666}{(5,308 + 2,321)/2}$$

$$= \frac{1,666}{\$3,814.50} = 43.68\%$$

The rate of return on common equity for Dell is 43.68% during 2000. During the same period, Compaq's return on common equity amounts to 8.85%, which is considered a poor performance compared for the computer industry (15.9% in year 2000) in general.

2.4.5 Market Value Analysis

When you purchase a company's stock, what would be your primary factors in valuing the stock? In general, investors purchase stock to earn a return on their investment. This return consists of two parts: (1) gains (or losses) from selling the stock at a price that differs from the investors' purchase price, and (2) dividends, the periodic distributions of profits to stockholders. The market value ratios, such as price/earnings ratio and market/book ratio, relate the firm's stock price to its earnings and book value per share. These ratios give management an indication of what investors think of the company's past performance and future prospects. If the firm's asset as well as debt management is sound and its profit is rising, then its market value ratios will be high, and its stock price will probably as high as can be expected.

Price/Earnings Ratio

The price/earnings (P/E) ratio shows how much investors are willing to pay per dollar of reported profits. Dell's stock sells for $38.5 in early February of 2000, so with an EPS of $0.61 its P/E ratio is 63.11:

$$\text{P/E ratio} = \frac{\text{Price per share}}{\text{Earnings per share}}$$

$$= \frac{\$38.5}{\$0.61} = 63.11$$

That is, the stock was selling for about 63.11 times its current earnings per share. In general, P/E ratios are higher for firms with high growth prospects, other things held constant, but they are lower for firms with lower expected earnings. Dell's expected annual increase in operating earnings is 30% over the next 3 to 5 years. Since Dell's ratio is greater than 27%, the average for other computer industry, this suggests that investors value Dell's stock more highly than most, as having excellent growth prospects, or both. However, all stocks with high P/E ratios will also carry high risk whenever the expected growths never materialize. Any slight earning disappointment tends to punish the market price significantly.

Book Value per Share

Another ratio frequently used in assessing the well-being of the common stockholders is book value per share. The book value per share measures the amount that would be distributed to holders of each share of common stock if all assets were sold at their balance sheet carrying amounts and if all creditors were paid off. We compute the book value per share for Dell's common stock as follows:

$$\text{Book value per share} = \frac{\text{Total stockholder's equity-preferred stock}}{\text{Shares outstanding}}$$

$$= \frac{\$5,308}{2,728} = \$1.95$$

If we compare this book value with the current market price of $38.50, then we may say that the stock appears to be overpriced. Once again, market prices reflect expectations about future earnings and dividends, whereas book value largely reflects the results of events that occurred in the past. Therefore, the market value of a stock tends to exceed its book value.

2.4.6 Limitations of Financial Ratios in Business Decisions

Business decisions are made in a world of uncertainty. As useful as ratios are, they have limitations. We can draw an analogy between their use in decision making and a physician's use of a thermometer. A reading of 102°F indicates that something is wrong with the patient, but the temperature alone does not indicate what the problem is or how to cure it. In other words, ratio analysis is useful, but analysts should be aware of ever-changing market conditions and make adjustments as necessary. It is also difficult to generalize about whether a particular ratio is "good" or "bad." For example, a high current ratio may indicate a strong liquidity position, which is good, but holding too much cash in a bank account (which will increase the current ratio) may not be the best utilization of funds. Ratio analysis based on any one year may not represent the true business condition. It is important to analyze trends in various financial ratios as well as their absolute levels, for trends give clues as to whether the financial situation is likely to improve or to deteriorate. To do a **trend analysis**, one simply plots a ratio over time. As a typical engineering student, your judgment in interpreting a set of financial ratios is understandably weak at this point, but it will improve as you encounter many facets of business decisions in the real world. Again, accounting is a language of business, and as you speak more often, it can provide useful insights into a firm's operations.

2.5   Summary

The primary purposes of this chapter were (1) to describe the basic financial statements, (2) to present some background information on cash flows and corporate profitability, and (3) to discuss techniques used by investors and managers to analyze financial statements. We list the concepts covered below:

- Before making any major financial decisions, it would be important to understand the impact on your net worth. Your net worth statement is a snapshot of where you stand financially at a given point in time.

- The three basic financial statements contained in the annual report are the balance sheet, the income statement, and the statement of cash flows. Investors use the information provided in these statements to form expectations about the future levels of earnings and dividends, and about the firm's riskiness.

- A firm's balance sheet shows a snapshot of the firm's financial position at a particular point in time.

- A firm's income statement reports the results of operations over a period of time, and it shows earnings per share as its "bottom line."

- A firm's statement of cash flows reports the impact of operating, investing, and financing activities on cash flows over an accounting period.

- The purpose of calculating a set of financial ratios is two/fold: (1) to examine the relative strengths and weakness of a company as compared with other companies in the same industry, and (2) to show whether its position has been improving or deteriorating over time.

- Liquidity ratios show the relationship of a firm's current assets to its current liabilities, and thus its ability to meet maturing debts. Two commonly used liquidity ratios are the current ratio and the quick (acid-test) ratio.

- Asset management ratios measure how effectively a firm is managing its assets. Some of the major ratios include inventory turnover, fixed assets turnover, and total assets turnover.

- Debt management ratios reveal (1) the extent to which the firm is financed with debt and (2) its likelihood of defaulting on its debt obligations. In this category, we may include the debt ratio and times-interest-earned.

- Profitability ratios show the combined effects of liquidity, asset management, and debt management policies on operating results. Profitability ratios include the profit margin on sales, the return on total assets, and the return on common equity.

- Market value ratios relate the firm's stock price to its earnings and book value per share, and they give management an indication of what investors think of the company's past performance and future prospects. Market value ratios include the price/earnings ratio and book value per share.

- Trend analysis, where one plots a ratio over time, is important, because it reveals whether the firm's ratios are improving or deteriorating over time.

Self-Test Questions

2s.1 True-False questions:

[ ] 1. The balance sheet statement summarizes how much the firm owns as well as owes for a typical operating period.

[ ] 2. The income statement summarizes the net income produced by the corporation at a specified reporting date.

[ ] 3. The cash flow statement summarizes how the corporation generated cash during the operating period.

[ ] 4. Working capital measures the company's ability to repay current liabilities using only current assets.

2s.2 Definitional problems: Listed below are several terms that relate to financial statements.

1. The balance sheet
2. Income statement
3. Cash flow statement
4. Operating activities
5. Investment activities
6. Financing activities
7. Treasury account
8. Capital account

• You as outsider investor would view a firm's _____ report to be the most important in learning the quality of earnings.

• Retained earnings as reported on the _____ statement represent income earned by the firm in past years that has not been paid out as dividends.

• The _____ statement is designed to show how the firm's operations have affected its cash position with actual net cash flows into or out of the firm during some specified period.

• Typically, a firm's cash flow statement is categorized into three activities: cash flow associated with _____ activities, _____ activities, and _____ activities.

• When you issue stock, the money raised beyond the par value is shown in the _____ account in the balance sheet statement.

2s.3 Definitional problems: Listed below are several terms that relate to ratio analysis.

1. Book value per share
2. Inventory turnover
3. Debt-to-equity ratio
4. Average collection period
5. Average sales period
6. Return on common on stockholders' equity
7. Earnings per share
8. Price-earnings ratio
9. Return on total assets
10. Current ratio
11. Account receivable turnover

Choose the financial ratio or term above that most appropriately describes the following situations. In doing so, just identify the item number in the blank space.

• The _____ tends to have an effect on the market price per share, as reflected in the price-earnings ratio.

• The _____ indicates whether a stock is relatively cheap or relatively expensive in relation to current earnings.

- The _____ measures the amount that would be distributed to holders of common stock if all assets were sold at their balance sheet carrying amount and if all creditors were paid off.

- The _____ is a rough measure of how many times a company's accounts receivable have been turned into cash during the year.

- The _____ is a measure of the amount of assets being provided by creditors for each dollar of assets being provided by the stockholders.

- The _____ measures how well management has employed assets.

2s.4 The following data apply to Fisher & Company (millions of dollars except ratio figures):

- Cash and marketable securities $100
- Fixed assets $280
- Sales $1,200
- Net income $358
- Inventory $180
- Current ratio   3.2
- Average collection period   45 days
- Average common equity $500

Find Fisher's

Accounts receivable _____	Current Liabilities _____
Current assets _____	Total assets _____
Long-term debt _____	Profit margin _____

Problems

Financial Statements

2.1 Consider the following balance sheet entries for War Eagle Corporation:

Balance Sheet Statement as of December 31, 2000		
Assets:		
Cash		$150,000
Marketable Securities		200,000
Accounts Receivables		150,000
Inventories		50,000
Prepaid taxes and Insurance		30,000
Manufacturing Plant		
at Cost	$600,000	
Less: Accumulated		
Depreciation	100,000	
Net Fixed Assets		500,000
Goodwill		20,000
Liabilities and Shareholders' Equity:		
Notes Payable		50,000
Accounts payable		100,000
Income Taxes Payable		80,000
Long-Term Mortgage Bonds		400,000
Preferred Stock, 6%, $100 par		
value (1,000 shares)		100,000
Common Stocks, $15 par value		
(10,000 shares)		150,000
Capital Surplus		150,000
Retained Earnings		70,000

(a) Compute the firm's
  Current Assets: $_____
  Current Liabilities: $_____
  Working Capital: $_____
  Shareholders' Equity $_____

(b) If the firm had the net income after taxes of $500,000, what is the earnings per share?

(c) When the firm issued its common stock, what was the market price of the stock per share?

2.2 A chemical processing firm is planning on adding a duplicate polyethylene plant at another location. The financial information for the first project year is provided as follows:

Sales		$1,500,000
Manufacturing Costs		
Direct Materials	$150,000	
Direct Labor	200,000	
Overhead	100,000	
Depreciation	200,000	
Operating Expenses		150,000
Equipment Purchase		400,000
Borrowing to finance the		
equipment		200,000
Increase in inventories		100,000
Decrease in accounts		
receivable		20,000
Increase in wages payable		30,000
Decrease in notes payable		40,000
Income taxes		272,000
Interest payment on financing		20,000

(a) Compute the working capital requirement during this project period.

(b) What is the taxable income during this project period?

(c) What is the net income during this project period?

(d) Compute the net cash flow from this project during the first year.

Financial Ratio Analysis

2.3   Consider the following financial statements for Juniper Networks, Inc. The closing stock price for Juniper Network was $56.67 (split adjusted) on December 31, 1999. Based on the financial data presented, compute the various financial ratios and make your informed analysis on Juniper's financial health.

	Dec. 1999 US $ (000) (YEAR)	Dec. 1998 US $ (000) (YEAR)
Balance Sheet Summary		
Cash	158,043	20,098
Securities	285,116	0
Receivables	24,582	8,056
Allowances	632	0
Inventory	0	0
Current Assets	377,833	28,834
Property and Equipment, Net	20,588	10,569
Depreciation	8,172	2,867
Total Assets	513,378	36,671
Current Liabilities	55,663	14,402
Bonds	0	0
Preferred Mandatory	0	0
Preferred Stock	0	0
Common Stock	2	1
Other Stockholders' Equity	457,713	17,064
Total Liabilities and Equity	513,378	36,671
Income Statement Summary		
Total Revenues	102,606	3,807
Cost of Sales	45,272	4,416
Other Expenses	31,954	31,661
Loss Provision	0	0
Interest Income	8,011	1,301
Income Pre Tax	−6,609	−69

Income Statement Summary (*continued*)

Income Tax	2,425	2
Income Continuing	−9,034	−30,971
Discontinued	0	0
Extraordinary	0	0
Changes	0	0
Net Income	**−9,034**	**−30,971**
EPS Primary	**$−0.1**	**$−0.80**
EPS Diluted	$−0.10	$−0.80
	$−0.05	$−0.40
	(split adjusted)	(split adjusted)

(a) Debt ratio

(b) Times-Interest-Earned Ratio

(c) Current ratio

(d) Quick (Acid Test) ratio

(e) Inventory turnover ratio

(f) Days sales outstanding

(g) Total assets turnover

(h) Profit margin on sales

(i) Return on total assets

(j) Return on common equity

(k) Price/Earning ratio

(l) Book value per share

2.4   The following financial statements summarize the financial conditions for Solectron, Inc., an electronic outsourcing contractor, for the fiscal year of 1999. Unlike Juniper Network Corporation in Problem 2.3, the company made profits for several years. Compute the various financial ratios and interpret the firm's financial health during the fiscal year of 1999.

	Aug. 1999 US $ (000) (12-MOS)	Aug. 1998 US$ (000) (YEAR)
Balance Sheet Summary		
Cash	1,325,637	225,228
Securities	362,769	83,576
Receivables	1,123,901	674,193
Allowances	5,580	−3,999
Inventory	1,080,083	788,519
Current Assets	3,994,084	1,887,558
Property and Equipment, Net	1,186,885	859,831
Depreciation	533,311	−411,792
Total Assets	4,834,696	2,410,568
Current Liabilities	1,113,186	840,834
Bonds	922,653	385,519
Preferred Mandatory	0	0
Preferred Stock	0	0
Common Stock	271	117
Other Stockholders' Equity	2,792,820	1,181,209
Total Liabilities and Equity	4,834,696	2,410,568
Income Statement Summary		
Total Revenues	8,391,409	5,288,294
Cost of Sales	7,614,589	4,749,988
Other Expenses	335,808	237,063
Loss Provision	2,143	2,254
Interest Expense	36,479	24,759
Income Pre Tax	432,342	298,983
Income Tax	138,407	100,159
Income Continuing	293,935	198,159
Discontinued	0	0
Extraordinary	0	0
Changes	0	0
Net Income	**293,935**	**198,159**
EPS Primary	**$1.19**	**$1.72**
EPS Diluted	$1.13	$1.65

(a) Debt ratio

(b) Times-Interest-Earned Ratio

(c) Current ratio

(d) Quick (Acid Test) ratio

(e) Inventory turnover ratio

(f) Days sales outstanding

(g) Total assets turnover

(h) Profit margin on sales

(i) Return on total assets

(j) Return on common equity

(k) Price/Earning ratio

(l) Book value per share

2.5 J. C. Olson & Co. had earnings per share of $8 in year 2001, and it paid a $4 dividend. Book value per share at year end was $80. During the same period, the total retained earnings increased by $24 million. Olson has no preferred stock, and no new common stock was issued during the year. If Olson's year-end debt (which equals its total liabilities) was $240 million, what was the company's year-end debt/asset ratio?

Short Case Studies

2.6 Consider the two companies: Cisco Systems and Lucent Technologies. Both companies compete each other in the network equipment sector. Lucent enjoys strong relationships among Bay Bells in the telephone equipment area and Cisco has a dominant role in the network routers and switching equipment area. You may first get their annual report from their web sites, and answer the following questions. [Note: You can visit the firms' web sites (look for "Investors' Relations") to download their annual reports— http://www.cisco.com and http://www.lucent .com.]

(a) Based on the most recent financial statements, comment on each company's financial performance in the following areas:

- Asset management
- Liquidity
- Debt management
- Profitability
- Market trend

(b) Check the current stock prices for both companies. The stock ticker symbol is CSCO for Cisco and LU for Lucent. Based on your analysis in 2.6(a), which company would you bet on your money and why?

CHAPTER 3

Cost Concepts and Behaviors

Each American consumes a yearly average of 5.8[1] gallons of ice cream or equivalent ice cream products, and consumption is the highest in July and August. About 98% of all U.S. households purchase ice cream. Supermarkets account for almost a third of ice cream sales—selling 2 gallons of ice cream per capita a year. So where else do Americans buy their ice cream? Convenience stores and scoop shops such as The 31 Baskin Robbins, of course, where a single dip ranges from $1.40 to as much as $3. Costs and profits vary widely. Here is a look at why it costs $2.50 for a single-dip ice cream cone at a typical store in Washington, D.C. The annual sales volume (the number of ice cream cones sold) averages around 185,000 cones, or $462,500. This is equivalent to selling more than 500 cones a day assuming a 7-day operation.

Unit Price of an Ice Cream Cone*

Items	Total Cost	Unit Price	% of Price
Ice cream (cream, sugar, milk and milk solids)	$120,250	$0.65	26%
Cone	9,250	0.05	2%
Rent	112,850	0.61	24%
Wages	46,250	0.25	10%
Payroll taxes	9,250	0.05	2%
Sales taxes	42,550	0.23	9%
Business taxes	14,800	0.08	3%
Debt service	42,550	0.23	9%
Supplies	16,650	0.09	4%
Utilities	14,800	0.08	3%
Other expenses (insurance, advertising, fees, heating & lighting for shop)	9,250	0.05	2%
Profit	24,050	0.13	5%
Total	**$462,500**	**$2.50**	**100%**

*Based on annual volume of 185,000 cones.

[1] *Source:* Ice Cream Consumption in America, MakeIceCream.com.

E ven though the annual sales volume is very close to $0.5 million, the net profit margin would be only 5% of the sales price. If the shop were able to increase the unit price from $2.50 to $2.75, there would be additional revenue of $46,250. On the other hand, with the unit price fixed, any increase in cost items will directly eat into the profit margin. Knowing the cost components and their behaviors, the shop owner would be able to control the business expenses and target the areas of profit improvement more effectively.

Before we study the various engineering economic decision problems, we need to understand the concept of various costs. At the level of plant operations, engineers must make decisions involving materials, plant facilities, and the in-house capabilities of company personnel. Let us consider as an example the manufacture of food processors. In terms of material selection, several of the parts could be made of plastic, whereas others must be made of metal. Once materials have been chosen, engineers must consider the production methods, the shipping weight, and the method of packaging necessary to protect the different types of material. In terms of actual production, parts may be made in-house or purchased from an outside vendor depending on the availability of machinery and labor.

All these operational decisions (commonly known as **present economic studies** in traditional engineering economic texts) require estimating the costs associated with various productions or manufacturing activities. As these costs provide the basis for developing successful business strategies and planning future operations, it is important to understand how various costs respond to changes in levels of business activity. In this chapter, we discuss many of the possible uses of cost data. We also discuss how to define and classify them for each use. Our first task in doing so is to explain how costs are classified for making various engineering economic decisions.

3.1   General Cost Terms

In engineering economics, the term **cost** is used in many different ways. Because there are many types of costs, each is classified differently according to the immediate needs of management. For example, engineers may want cost data to prepare external reports, to prepare planning budgets, or to make decisions. Also, each different use of cost data demands a different classification and definition of cost. For example, the preparation of external financial reports requires the use of historical cost data, whereas decision making may require current cost or estimated future cost data.

Our initial focus in this chapter is on manufacturing companies, because their basic activities (such as acquiring raw materials, producing finished goods, marketing, etc.) are commonly found in most other businesses. Therefore, the understanding of costs in a manufacturing company can be very helpful in understanding costs in other types of business organizations.

3.1.1   Manufacturing Costs

The several types of manufacturing costs incurred by a typical manufacturer are illustrated in Figure 3.1. In converting raw materials into finished goods, a manufacturer incurs various costs of operating a factory. Most manufacturing companies divide manufacturing costs into three broad categories: direct materials, direct labor, and manufacturing overhead.

Direct Materials

Direct raw materials refer to any materials that are used in the final product and that can be easily traced into it. Some examples are wood in furniture, steel in bridge construction, paper in printing firms, and fabric for clothing manufacturers. It is also im-

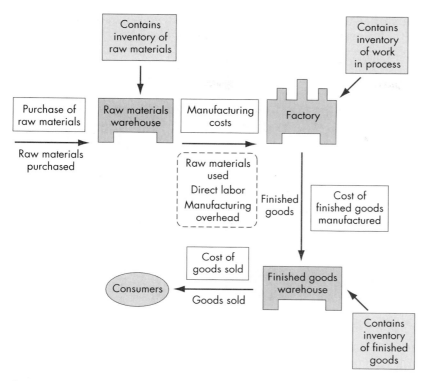

Figure 3.1   Various types of manufacturing costs incurred by a manufacturer

portant to note that the finished product of one company can become the raw materials of another company. For example, the computer chips produced by Intel are a raw material used by Dell Computer in its personal computers.

Direct Labor

Like the term direct material, direct labor is for those labor costs that go into the production of a product. The labor costs of assembly-line workers, for example, would be direct labor costs, as would the labor costs of welders in metal fabricating industries, carpenters or bricklayers in home building, and machine operators in various manufacturing operations.

Manufacturing Overhead

Manufacturing overhead, the third element of manufacturing cost, include all costs of manufacturing except direct materials and direct labor. In particular, they include such items as indirect materials; indirect labor; maintenance and repairs on production equipment; heat and light, property taxes, depreciation, and insurance on manufacturing facilities; and overtime premium. The most important thing to note about manufacturing overhead is the fact that unlike direct materials and direct labor, it is not easily traceable to specific units of output. In addition, many manufacturing overhead costs do not change as output changes, as long as the production volume stays within the capacity. For example, depreciation of factory buildings is unaffected by the amount of production during any particular period. If a new building is required to meet any increased production, however, the manufacturing overhead will certainly increase.

- Sometimes it may not be worth the effort to trace the costs of relatively insignificant materials to the finished products. Such minor items would include the solder used to make electrical connections in a computer circuit board or the glue used to bind this textbook. Materials such as solder and glue are called *indirect materials* and are included as part of manufacturing overhead.

- Sometimes, we may not be able to trace some of the labor costs to the creation of product. We treat this type of labor cost as a part of manufacturing overhead, along with indirect materials. *Indirect labor* includes the wages of janitors, supervisors, material handlers, and night security guards. Although the efforts of these workers are essential to production, it would be either impractical or impossible to trace their costs to specific units of product. Therefore, we treat such labor costs as indirect labor.

3.1.2 Nonmanufacturing Costs

There are two additional costs in supporting any manufacturing operations. They are (1) marketing or selling costs and (2) administrative costs. Marketing or selling costs include all costs necessary to secure customer orders and get the finished product or service into the hands of the customer. Cost breakdowns of these types provide data for control over selling and administrative functions in the same way that manufacturing cost breakdowns provide data for control over manufacturing functions. For example, a company incurs costs for:

- Overhead: Heat and light, property taxes, depreciation or similar items, associated with its selling and administrative functions.

- Marketing: Advertising, shipping, sales travel, sales commissions, and sales salaries. Administrative costs include all executive, organizational, and clerical costs associated with the general management of an organization.

- Administrative functions: Executive compensation, general accounting, public relations, and secretarial support.

3.2 Classifying Costs for Financial Statements

For purposes of preparing financial statements, we often classify costs as either period costs or product costs. To understand the difference between period costs and product costs, we must introduce the matching concept essential to any accounting studies. In financial accounting, the **matching principle** states that *the costs incurred to generate particular revenue should be recognized as expenses in the same period that the revenue is recognized.* This matching principle is the key to distinguishing between period costs and product costs. Some costs are matched against periods and become expenses immediately. Other costs, however, are matched against products and do not become expenses until the products are sold, which may be in the following accounting period.

3.2.1   Period Costs

Period costs are those costs that are charged to expenses in the period in which they are incurred. The underlying assumption is that the associated benefits are received in the same period as the cost is incurred. Some specific examples are all general and administrative expenses, selling expenses, insurance and income taxes expense. Therefore, advertising, executive salaries, sales commissions, public relations, and other nonmanufacturing costs discussed earlier would all be period costs. Such costs are not related to the production and flow of manufactured goods, but are deducted from revenue in the income statement. In other words, period costs will appear on the income statement as expenses in the time in which they occur.

3.2.2   Product Costs

Some costs are better matched against products than they are against periods. Costs of this type—called **product costs**—consist of the costs involved in the purchase or manufacturing of goods. In the case of manufactured goods, these costs consist of direct materials, direct labor, and manufacturing overhead. Product costs are not viewed as expenses; rather, they are the cost of creating inventory. Thus, product costs are considered an asset until the related goods are sold. At this point of sale, the costs are released from inventory as expenses (typically called cost of goods sold) and matched against sales revenue. Since product costs are assigned to inventories, they are also known as *inventory costs*. In theory, product costs include all manufacturing costs—that is, all costs relating to the manufacturing process. Product costs appear on financial statements when the inventory, or final goods, is sold, not when the product is manufactured as shown in Figure 3.2.

To explain the classification scheme for financial statements, consider Example 3.1.

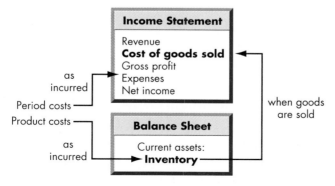

Figure 3.2   How the period costs and product costs flow through financial statements from manufacturing floor to sales

Example 3.1 Classifying Costs for Uptown Ice Cream Shop

Consider the operating cost breakdown given in the chapter opening for the ice cream shop in Washington, D.C. If you were to classify the operating costs into either product cost or period cost, how would you do it?

Unit Price of an Ice Cream	
Ice cream (cream, sugar, milk, and milk solids)	$ 0.65
Cone	0.05
Rent	0.61
Wages	0.25
Payroll taxes	0.25
Sales Taxes	0.23
Business taxes	0.08
Debt service	0.23
Supplies	0.09
Utilities	0.08
Other (insurance, advertising, fees, and heating/lighting for shop)	0.05
Profit	0.13
	$ 2.50

Solution:

Product costs: Costs incurred in preparing 185,000 ice cream cones per year

Raw materials:

Ice cream @ $0.65	$120,250
Cone @ $0.05	9,250

Labor:

Wages @ $0.25	46,250

Overhead:

Supplies @ $0.09	16,650
Utilities @ $0.08	14,800
Total product cost	$207,200

Period costs: Costs incurred in running the shop regardless of sales volume

Business taxes:

Payroll taxes @ $0.05	$9,250

Sales Taxes @ $0.23	42,550
Business taxes @ $0.08	14,800
Operating expenses:	
Rent @ $0.61	112,850
Debt service @ $0.23	42,550
Other @ $0.05	9,250
Total period cost	$231,250

3.2.3   Cost Flows and Classifications in a Manufacturing Company

To understand product costs more fully, we may look briefly at the flow of costs in a manufacturing company. By doing so, we will be able to see how product costs move through the various accounts and affect the balance sheet and the income statement in the course of manufacture and sale of goods. The flows of period costs and product costs through the financial statements are illustrated in Figure 3.3. All product costs filter through the balance sheet statement in the name of "inventory cost." If the

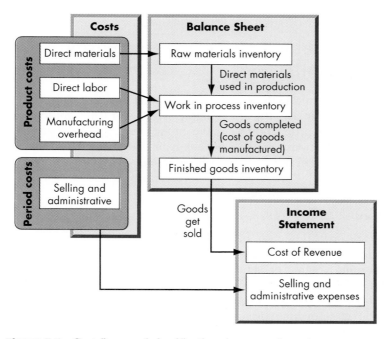

Figure 3.3   Cost flows and classifications in a manufacturing company

product gets sold, the inventory costs in the balance sheet statement are transferred to the income statement in the name of "cost of goods sold."

- **Raw materials inventory**—This account in the balance sheet statement represents the unused portion of the raw materials on hand at the end of the fiscal year.

- **Work-in-process inventory**—This balance sheet entry consists of the partially completed goods on hand in the factory at year-end. When raw materials are used in production, their costs are transferred to the Work-in-Process inventory account as direct materials. Note that direct labor cost and manufacturing overhead cost are also added directly to Work-in-Process. Work-in-Process can be viewed as the assembly line in a manufacturing plant, where workers are stationed and where products slowly take shape as they move from one end of the assembly line to the other.

- **Finished goods inventory**—This account shows the cost of finished goods on hand and awaiting sale to customers at year-end. As goods are completed, accountants transfer the cost in the Work-in-Process account into Finished Goods. Here the goods await sale to a customer. As goods are sold, their cost is then transferred from Finished Goods into Cost of Goods Sold (or, cost of revenue). At this point, we finally treat the various material, labor, and overhead costs that were involved in the manufacture of the units being sold as *expenses* in the income statement.

3.3   Cost Classification for Predicting Cost Behavior

In engineering economic analysis, we need to predict how a certain cost will behave in response to a change in activity. For example, a manager will want to estimate the impact a 5% increase in production will have on the company's total wages, before a decision to alter production is made. **Cost behavior** describes how a cost item will react or respond to changes in the level of business activity.

3.3.1   Volume Index

In general, the operating costs of any company are likely to respond in some way to changes in its operating volume. In studying cost behavior, we need to determine some measurable volume or activity that has a strong influence on the amount of cost incurred. The unit of measure used to define "volume" is called a **volume index**. A volume index may be based on production inputs, such as tons of coal processed, direct labor hours used, or machine-hours worked; or it may be based on production outputs, such as number of kilowatt-hours generated. For a vehicle, the number of miles driven per year may be used as a volume index. Once we identify a volume index, we try to find out how costs change in response to changes in this volume index.

3.3.2 Cost Behaviors

Accounting systems typically record the cost of resources acquired and track their subsequent usage. Fixed costs and variable costs are the two most common cost behavior patterns. An additional category known as "mixed (semivariable) costs" contain two parts—the first part of cost is fixed, and the other part is variable as the volume of output varies.

Fixed Costs

The costs of providing a company's basic operating capacity are known as its **fixed cost** or **capacity cost**. For a cost item to be classified as fixed, it must have a relatively wide span of output where costs are expected to remain constant. This span is called the **relevant range**. In other words, fixed costs do not change within a given time period although volume may change. For our automobile example, the annual insurance premium, property tax and license fee are fixed costs since they are independent of the number of miles driven per year. Some typical examples would be building rents, depreciation of buildings, machinery and equipment, and salaries of administrative and production personnel. In our Uptown Scoop Ice Cream Store example, we may classify expenses such as rent, business taxes, debt service, and other (insurance, advertising, professional fees) as fixed costs.

Variable Costs

In contrast to fixed operating costs, **variable operating costs** have a close relationship to the level of volume. If, for example, volume increases 10%, a total variable cost will also increase by approximately 10%. Gasoline is a good example of a variable automobile cost, as fuel consumption is directly related to miles driven. Similarly, the tire replacement cost will also increase as a vehicle is driven more. In a typical manufacturing environment, direct labor and material costs are major variable costs. In our Uptown Scoop example, the variable costs would include ice cream and cone (direct materials), wages, payroll taxes, sales taxes, and supplies. Both payroll and sales taxes are related to sales volume. In other words, if the store becomes busy, more servers are needed, which will increase the payroll as well as taxes.

Mixed Costs

Some costs do not fall precisely into either the fixed or the variable category, but contain elements of both. We refer to these as mixed costs (or **semi-variable costs**). In our automobile example, **depreciation** (loss of value) is a mixed cost. Some depreciation occurs simply from passage of time, regardless of how many miles a car is driven, and this represents the fixed portion of depreciation. On the other hand, the more miles an automobile is driven a year, the faster it loses its market value, and this represents the variable portion of depreciation. A typical example of a mixed cost in manufacturing is the cost of electric power. Some components of power consumption, such as lighting are independent of operating volume, while other components may likely vary directly with volume (e.g., number of machine-hours operated). In our Uptown Scoop example, the utility cost can be a mixed cost item—some lighting and heating requirement might stay the same, but the use of power mixers will be in proportion to sales volume.

3.3.3 Average Unit Cost

The foregoing description of fixed, variable, and semi-variable costs was expressed in terms of total volume for a period. We often use the term **average cost** to express activity cost on a per unit basis. In terms of unit costs, the description of cost is quite different:

- The variable cost per unit of volume is a constant.
- Fixed cost per unit varies with changes in volume: As volume increases, the fixed cost per unit decreases.
- The mixed cost per unit also changes as volume changes, but the amount of change is smaller than that for fixed costs.

To explain the behavior of the fixed and variable costs in relation to volume, we consider Example 3.2.

Example 3.2 Calculating Average Cost per Mile

Consider the somewhat simplified data to describe the cost of owning and operating a typical passenger car in Dallas, Texas. Runzheimer & Co.,[2] a consulting firm, compiled the following operating data for a 1998 midsize sedan. Using the data, develop a cost-volume chart and calculate the average cost per mile as a function of annual mileage.

---

[2] The cost estimates are from the annual study conducted for the American Automobile Association (AAA) by Runzheimer & Co. (*http://www.Runzheimer.com*). In trying to figure out what it costs to own and operate a vehicle, keep in mind that the Runzheimer study does not take into account such expenses as tolls and /or parking fees. These could easily add hundreds more to the annual cost. In addition, insurance costs were based on personal use of the vehicles driven less than 10 miles to or from work each day with no youthful drivers on the policy. In our example, we will assume that the insurance cost would remain unchanged for an annual driving up to 20,000 miles.

Cost Classification	Reference	Cost
Variable Costs:		
Standard miles per gallon	20 miles/gallon	
Average fuel price per gallon	$1.34/gallon	
Fuel and oil per mile		$0.0689
Maintenance per mile		$0.0360
Tires per mile		$0.0141
Annual Fixed Costs:		
Insurance:		
Comprehensive	$250 Deductible	$90
Collision	$500 Deductible	$147
Body injury & Property damage		$960
License & Registration		$95
Property tax		$372
Mixed Costs: Depreciation		
Fixed portion per year		$3,703
Variable portion per mile		$0.04

Table 3.1
Cost Classification of Owning and Operating a Passenger Car

Solution

In Table 3.1, we itemize the cost entries by fixed, variable, and mixed classes. In Table 3.2, we also summarize the costs of owning and operating the automobile at various annual operating volumes from 5,000 to 20,000 miles. Once the total cost figures are available at specific volumes, we can calculate the effect of volume on unit (per-mile) costs by converting the total cost figures to average unit costs.

Volume Index (miles)	5,000	10,000	15,000	20,000
Variable costs (11.90 cents per mile)	$595	$1,190	$1,785	$2,380
Mixed costs:				
Variable portion	200	400	600	800
Fixed portion	3,703	3,703	3,703	3,703
Fixed costs:	467	467	467	467
Total variable cost	795	1,590	2,385	3,180
Total fixed cost	4,170	4,170	4,170	4,170
Total costs	$4,965	$5,760	$6,555	$7,350
Cost per mile	$0.9930	$0.5760	$0.4370	$0.3675

Table 3.2
Operating Costs as Function of Mileage Driven

To determine the estimated annual costs for any assumed mileage, we construct a **cost-volume diagram** as shown in Figure 3.4(a). Further we can show the relation between volume (miles driven per year) and the three types of cost class separately as in Figure 3.4 (b) through (d). We can use these cost-volume graphs to estimate both the separate and combined costs of operating at other possible volumes. For example, an owner who expects to drive 12,000 miles in a given year may estimate the total cost at $6.078, or 50.65 cents per mile. In Figure 3.4, all costs more than those necessary to operate at the zero level are known as variablecosts. Since the fixed cost is $4,170 a year, the remaining $1,908 is variable cost. By combining all the fixed and variable elements of cost, we can state simply

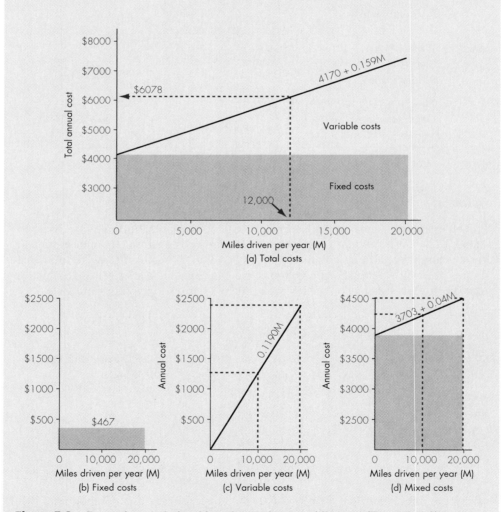

Figure 3.4   Cost-volume relationships of annual automobile costs (Example 3.2)

that the cost of owning and operating an automobile is $4,170 per year, plus 15.90 cents per mile driven during the year.

Figure 3.5 illustrates graphically the average unit cost of operating the automobile. The average fixed cost, represented by the height of the curved line in Figure 3.5 will decline steadily as the volume increases. The average unit cost is high when volume is low because the total fixed cost is spread over a relatively few units of volume. In other words, the total fixed costs remain the same regardless of the number of miles driven, but the average fixed cost decreases on a per-mile basis as the miles driven increase.

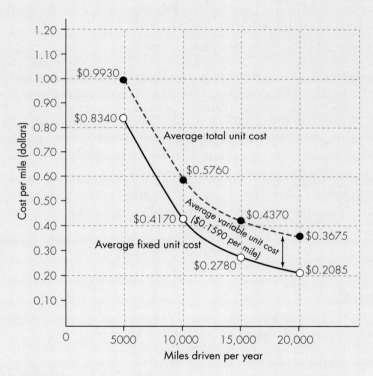

Figure 3.5   Average cost per mile for owning and operating a car (Example 3.2)

3.4   Cost Concepts Relevant to Decision Making

Costs are an important feature of many business decisions. In making decisions, it is essential to have a firm grasp of the concepts differential cost, opportunity cost, and sunk cost.

3.4.1   Differential Cost and Revenue

As we will see throughout the text, decisions involve choosing among alternatives. In business decisions, each alternative will have certain costs and benefits that must be compared to the costs and benefits of the other available alternatives. A difference in costs between any two alternatives is known as a **differential cost**. Similarly, a difference in revenues between any two alternatives is known as **differential revenue**. A differential cost is also known as an incremental cost, although technically an incremental cost should refer only to an increase in cost from one alternative to another.

Cost-volume relationships based on differential costs find many engineering applications. They are useful in making a variety of short-term operational decisions. Many short-run problems have the following characteristics:

- The base case is the status quo (current operation or existing method), and we propose an alternative to the base case. If we find the alternative to have lower costs than the base case, we accept the alternative, assuming that non-quantitative factors do not offset the cost advantage. The **differential (incremental) cost** is the difference in total cost that results from selecting one alternative instead of another. If several alternatives are possible, we select the one with the maximum savings from the base. Problems of this type are often called trade-off problems because one type of cost is traded off for another.
- New investments in physical assets are not required.
- The planning horizon is relatively short (a week, a month, but less than a year).
- Relatively few cost items are subject to change by management decision.

Some common examples include method changes, operations planning, and make or buy decisions.

Method Change

We may derive the best information about the future costs from an analysis of historical costs. The proposed alternative is to consider some new method of performing an activity. If the differential costs of the proposed method are significantly lower than the current method, we adopt the new method. In a typical manufacturing environment, when demand is high, managers are interested in whether to use a one-shift plus overtime operation or to add a second shift. When demand is low, it is equally possible to explore whether to operate temporarily at very low volume or to shut down until operations at normal volume become economical. We may analyze other production decisions in terms of differential costs. In a chemical plant, several routes exist for scheduling products through the plant. The problem is which route provides the largest differential cost. To illustrate how engineers may use cost-volume analysis in a typical operational analysis, let us consider Example 3.3.

Example 3.3 Differential Cost Associated with Adopting a New Production Method

The engineering department at an auto-parts manufacturer recommends that the current dies (base) be replaced with higher quality dies (alternative), which would result in substantial savings in manufacturing one of the company's products. The higher cost of materials would be more than offset by the savings in machining time and electricity. If estimated monthly costs under the two alternatives are as follows, what is the differential cost for going with better dies?

	Current Dies	Better Dies	Differential Cost
Variable costs:			
Materials	$150,000	$170,000	+$20,000
Machining labor	85,000	64,000	−21,000
Electricity	73,000	66,000	−7,000
Fixed costs:			
Supervision	25,000	25,000	0
Taxes	16,000	16,000	0
Depreciation	40,000	43,000	+3,000
Total	$392,000	$387,000	−$5,000

Solution

In this problem, the differential cost is thus −$5,000 a month. With the differential cost being negative, it indicates a saving rather than an addition to total cost. An important point to remember is that differential costs will usually include variable costs but fixed costs are affected only if the decision involves going outside of the relevant range. Although the production volume remains unchanged in our example, the slight increase in depreciation expense is due to the acquisition of new machine tools (dies). All other items of fixed costs remained within their relevant ranges.

Operations Planning

In a typical manufacturing environment, when demand is high, managers are interested in whether to use a one-shift plus overtime operations or to add a second shift. When demand is low, it is equally possible to explore whether to operate temporarily at very low volume or to shut down until operations at normal volume become economical. In a chemical plant, several routes exist for scheduling products through the plant. The problem is which route provides the lowest cost.

To illustrate how engineers may use cost-volume analysis in a typical operational analysis, consider Example 3.4.

Example 3.4 Break-Even Volume Analysis

Sandstone Corporation has one of its manufacturing plants operating on a single-shift 5-day week. The plant is operating at its full capacity (24,000 units of output per week) without the use of overtime or extra-shift operation. Fixed costs for single-shift operation amount to $90,000 per week. The average variable cost is a constant $30 per unit, at all output rates, up to 24,000 units per week. The company has received an order to produce extra 4,000 units per week beyond the current single-shift maximum capacity. Two options are being considered to fill the new order.

- Option 1: Increase the plant's output to 36,000 units a week by adding overtime or by adding Saturday operations or both. No increase in fixed costs is entailed, but the variable cost is $36 per unit for any output in excess of 24,000 units per week, up to a 36,000-unit capacity.
- Option 2: Operate a second shift. The maximum capacity of the second shift is 21,000 units per week. The variable cost on the second shift is $31.50 per unit, and operation of a second shift entails additional fixed costs of $13,500 per week.

Determine the range of operating volume that will make Option 2 profitable.

Solution

In this example, the operating costs related to the first shift operation will remain unchanged if one alternative is chosen instead of another. Therefore, these costs are irrelevant to the current decision and can safely be left out of the analysis. Therefore we need to examine only the increased total cost due to the additional operating volume under each option (known as incremental analysis). Let Q denote the additional operating volume:

- Option 1: Overtime and Saturday operation: $36 Q
- Option 2: Second-shift operation: $13,500 + $31.50 Q

We can find the break-even volume (Q_b) by equating the incremental cost functions and solving for Q:

$$36Q = 13,500 + 31.5Q$$

$$4.5Q = 13,500$$

$$Q_b = 3,000 \text{ units}$$

If the additional volume exceeds 3,000 units, the second-shift operation becomes more efficient than overtime or Saturday operation. A break-even (or cost-volume) graph based on the above data is shown in Figure 3.6. The horizontal scale represents additional volume per week. The upper limit of the relevant volume range for Option 1 is 12,000 units, whereas the upper limit for Option 2 is 21,000 units. The vertical scale is in dollars of costs. The operating savings expected at any volume may be read from the cost-volume graph. For example, the break-even

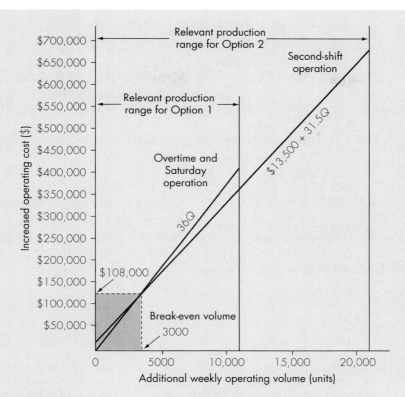

Figure 3.6 Cost-volume relationships of operating overtime and a Saturday operation versus second-shift operation beyond 24,000 units (Example 3.4)

point (zero savings) is 3,000 units per week. Option 2 is a better choice, since the additional weekly volume from the new order exceeds 3,000 units.

Make or Buy Decision

In business, the make-or-buy decision arises on a fairly frequent basis. Many firms perform certain activities using their own recourses, and they pay outside firms to perform certain other activities. It is a good policy to constantly seek to improve the balance between these two types of activities by asking whether we should outsource some function that we are now performing ourselves to other firms or vice versa. Example 3.5 illustrates how we may obtain this type of trade-off decision based on the differential cost concept.

Example 3.5 Make-Or-Buy Decision

Consider Benson Company, a farm equipment manufacturer that currently produces 20,000 units of gas filters annually for use in its lawnmower production. The expected annual production cost of the gas filters is summarized below:

Variable costs:	
Direct materials	$100,000
Direct labor	190,000
Power and water	35,000
Fixed costs:	
Heating and light	20,000
Depreciation	100,000
Total cost	**$445,000**

Tompkins Company has offered to sell Benson 20,000 units of gas filters for $17.00 per unit. If Benson accepts the offer, some of the manufacturing facilities currently used to manufacture the gas filters could be rented to a third party at an annual rent of $35,000. Should Benson accept Tompkins's offer, and why?

Solution

This problem is unusual in the sense that the buy option would generate a rental fee of $35,000. In other words, Benson could rent out the current manufacturing facilities if it were to purchase the gas filters from Tompkins. To compare the two options, we need to examine the cost for each option.

	Make Option	Buy Option	Differential Cost
Variable cost			
Direct materials	$100,000		–$100,000
Direct labor	190,000		–190,000
Power and water	35,000		–35,000
Gas filters		340,000	340,000
Fixed costs			
Heating light	20,000	20,000	0
Depreciation	100,000	100,000	0
Rental income		–35,000	–35,000
Total cost	$445,000	$425,000	–$20,000
Unit cost	$22.25	$21.25	–$1.00

The buy option has a lower unit cost and saves $1 for each gas-filter use. If the rental income were not considered, however, the decision would favor the make option.

3.4.2 Opportunity Costs

Opportunity cost may be defined as the potential benefit that is given up as you seek an alternative course of action. In fact, virtually every alternative has some opportunity cost associated with it. For example, suppose you have a part-time job that pays you $200 per week while attending college. You would like to spend a week at the beach during spring break, and your employer has agreed to give you the week off. What would be the opportunity cost of taking the time off to be at the beach? The $200 in lost wages would be an opportunity cost.

In economic sense, opportunity cost could mean the contribution to income that is foregone by not using a limited resource in its best use. Or, we may view opportunity costs as cash flows that could be generated from an asset the firm already owns, provided they are not used for the alternative in question. In general, accountants do not post opportunity cost in the accounting records of an organization. However, this opportunity cost must be explicitly considered in every decision.

> **Current Practice**: *An Opportunity Lost—What Is Real Cost?* PepsiCo, the world's No. 2 soft-drink company, recently got a lesson in the economic concept known as "opportunity costs." PepsiCo's Venezuelan bottler and distributor, Oswaldo Cisneros, wanted to realign his business interests and asked PepsiCo to buy part of his company. Pepsi executives decided Cisneros was demanding too much, and they stalled the negotiations. A month ago, Cisneros announced he had cut a deal with Coca-Cola instead. Overnight, Pepsi went from owning half of the soft-drink market in Venezuela to zero. It suffered enormous public and investor embarrassment, and created doubts among its other bottlers and distributors around the world as to how committed the company is to supporting them. Given those costs, you have to wonder whether Pepsi properly evaluated the price Cisneros was asking. It may have seemed like an outrageous sum initially. But, what will it cost Pepsi to rebuild in Venezuela? And, how much time and effort will have to be invested in regaining distributor and investor confidence? That is the essence of **opportunity costs**.
>
> Source: "An Opportunity Lost—What Is Real Cost?" by John Koenig, *Orlando Sentinel,* September 22, 1996.

3.4.3 Sunk Costs

A sunk cost is one that has already been incurred by past actions. Sunk costs are not relevant to decisions because they cannot be changed regardless of what decision is made now or in the future. The only costs relevant to a decision are those costs that vary among the alternative courses of action being considered. To illustrate a sunk cost, suppose that you have a very old car that requires frequent repairs. You want to sell the car, and figure that the current market value would be about $1,200 at best. While you are in the process of advertising the car, you find that the car's water-pump is leaking. You decided to have the water-pump repaired, which cost you $200. A friend of yours is interested in buying your car, and has offered $1,300. Would you take the offer? Or, decline the offer simply because you cannot recoup the repair cost with that offer. In this example, the $200 repair cost is a sunk cost. You cannot change this repair cost whether or not you keep the car. Since your friend's offer is $100 more than the best market value, it would be better to accept the offer.

Current Practice: *Good manager knows to stop pouring good money after bad.* To illustrate, after spending millions of dollars halfway through a faulty project, a manager often feels obliged to see it through to the end, even if the project has little chance of turning profit. . . . John Hickenlooper, owner of Wynkoop Brewing Co. in Denver, knows how hard it is to ignore those **sunk costs**. Hoping to open a brewpub in Colorado Springs, he "carried" a building for three years, spending $100,000 and falling five months behind on the mortgage payments. "Another restaurant chain came along and offered me enough to get my money out of it," Hickenlooper said. "It was a tough decision. I came so close to walking away." He almost fell victim to the first classification of sunk cost victims, those who see sunk costs as a signal that it is time to "stop throwing good money after bad." Those victims shy from a reasonable business opportunity. The second kind of victim sees sunk costs as a sort of commitment. That victim remains tied to an unreasonable business venture. The secret, says Kieran Brady, director at Cap Gemini America in Denver, is to determine the value of a business opportunity from today, not from some point in the past. Let us say you have spent $3 million on a project that requires another $2 million for completion, leaving you $5 million in the hole. You recently determined the project would bring in $3 million. Do you proceed? Absolutely, says Brady. That first $3 million is gone no matter what you do. The next $2 million will, in effect, generate a 50% return.

Source: "Good manager knows to stop pouring good money after bad," by Steve Caulk, *Rocky Mountain News,* April 28, 1996.

3.4.4   Marginal Costs

Another cost term useful in cost-volume analysis is marginal cost. We define **marginal cost** as the added cost that would result from increasing the rate of output by a single unit. The accountant's differential cost concept can be compared to the economist's marginal cost concept. In speaking of changes in cost and revenue, the economist employs the terms marginal cost and marginal revenue. The revenue that can be obtained from selling one more unit of product is called **marginal revenue**. The cost involved in producing one more unit of product is called **marginal cost**.

Example 3.6 Marginal Costs Versus Average Costs

Consider a company that has an available electric load of 37 horsepower and that purchases its electricity at the following rates:

kWh/Month	@$/kWh	Average Cost ($/kWh)
First 1,500	$0.050	$0.050
Next 1,250	0.035	$\dfrac{\$75 + 0.0350(X - 1,500)}{X}$
Next 3,000	0.020	$\dfrac{\$118.75 + 0.020(X - 2,750)}{X}$
All over 5,750	0.010	$\dfrac{\$178.25 + 0.010(X - 5,750)}{X}$

In this rate schedule, the unit variable cost in each rate class represents the marginal cost per kilowatt-hours (kWh). On the other hand, we may determine the average costs in the third column by finding the cumulative total cost and dividing it by the total number of kWh (X). Suppose that the current monthly consumption of electric power averages 3,200 kWh. Based on this rate schedule, determine the marginal cost of adding one more kWh, and for a given operating volume (3,200 kWh), the average cost per kWh.

Solution

The marginal cost of adding one more kWh is $0.020. The average variable cost per kWh is calculated as follows:

kWh	Rate ($/kWh)	Cost
First 1,500	0.050	$75.00
Next 1,250	0.035	43.75
Remaining 450	0.020	9.00
Total		$127.75

The average variable cost per kWh is $127.75/3,200kWh = $0.0399kWh. Or we can find the value by using the formulas in the third column in the rate schedule.

$$\frac{\$118.75 + 0.020(3,200 - 2,750)}{3,200} = \frac{\$127.75}{3,200\text{kWh}} = \$0.0399\text{kWh}$$

Changes in the average variable cost per unit are the result of changes in the marginal cost. As shown in Figure 3.7, the average variable cost continues to fall because the marginal cost is lower than the average variable cost over the entire volume.

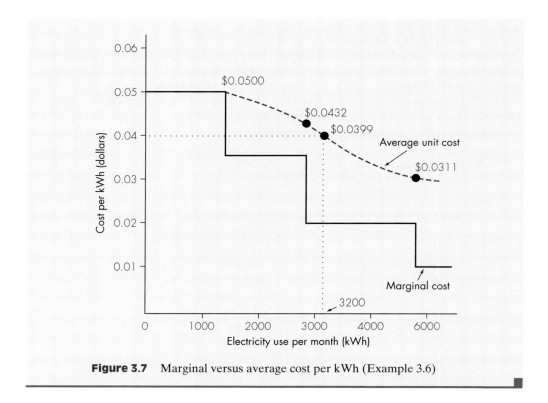

Figure 3.7   Marginal versus average cost per kWh (Example 3.6)

3.5   Thinking on the Margin: Fundamental Economic Decision Making

We make decisions every day, as entrepreneurs, professionals, executives, investors, and consumers, with little thought as to where our motivations come from or how our assumptions of logic fit a particular academic regimen. As you will see, the engineering economic decisions that we will describe throughout the text owe much to English economist Alfred Marshall (1842–1924) and his concepts of **marginalism**. In our daily quest for material gain, rarely do we recall the precepts of microeconomics or marginal utility, yet the ideas articulated by Marshall remain some of the most useful core principles guiding economic decision making.

3.5.1   Marginal Analysis

The fundamental concept of marginal thinking is that you are where you are, and what is past is irrelevant. (This means that we should ignore the sunk cost.) The point is whether you move forward, and you will do so if the benefits outweigh the costs, even if they do so by a smaller margin than before. You will continue to produce a product or service until the cost of doing so equals the revenue derived. Microeconomics suggests that businesspersons and consumers measure progress not in great leaps, but in small incremental steps. Rational persons will reevaluate strategies and take new actions if benefits exceed costs on a marginal basis. If there is more than one alternative to choose, do so always with the alternative with greatest marginal benefit.

As you may learn from an introductory microeconomics class, the economist's marginal cost concept is the same as the accountant's differential concept. For a business problem on profit-maximization, economists place a great deal of emphasis on the importance of marginal analysis—rational individuals and institutions should do the most cost effective activities first. In specific,

- When you examine marginal costs, you want to do the least expensive methods first and the most expensive last (if at all).

- Similarly, when you examine marginal revenue (or demand) the most revenue-enhancing methods should be done first, the least revenue enhancing last.

- If we need to consider the revenue along with the cost, the volume at which the total revenue and total cost are the same is known as the break-even point. This break-even point may also be calculated if the cost and revenue function are assumed to be linear, by using the following formula:

$$\text{Break-even volume} = \frac{\text{Fixed costs}}{\text{Sales price per unit} - \text{Variable cost per unit}}$$

$$= \frac{\text{Fixed costs}}{\text{Marginal contribution per unit}}$$

The difference between the unit sales price and the unit variable cost is the producer's **marginal contribution,** also known as **marginal income.** This means each unit sold contributes toward absorbing the company's fixed costs.

Current Practice: *Thinking on the Margin*. In college, my roommate and I, on a road trip, were caught in a snowstorm in western Pennsylvania. We pulled into a motel and asked about room rates. When the clerk quoted the price, we pointed out the late hour, and the fact that the motel was not full. Furthermore, if we drove to a competing facility in search of a cheaper price, the clerk was unlikely to rent the room. We offered to pay half of what the clerk quoted, and went on to explain marginal utility—a room rented for half the standard price was a contribution toward profit, whereas an empty room with zero revenue has no utility at all. On the margin, our offer was preferable to the alternative. (As long as the marginal revenue exceeds the marginal cost, it will be profitable.) We were delighted that the clerk agreed to our bargain, but, more than that, we were amazed that we had finally found a practical use for the arcane material being thrust upon us as hapless students in ECON 101.

Source: "Making Sense of Marginalism: Alfred Marshall's Brilliance Is with Us Today," by Lewis J. Walker, *On Wall Street,* copyright 1997 Securities Data Publishing.

Example 3.7 Profit-Maximization Problem: Marginal Analysis[3]

Suppose you are a chief executive officer (CEO) of a small pharmaceutical company that manufactures generic aspirin. You want the company to profit-maximize. You can sell as many aspirin as you make at the prevailing market price. You have only one manufacturing plant, which is the constraint. You have the plant working at full capacity Monday through Saturday, but you close the plant on Sunday because on Sundays, you have to pay workers overtime rates, and it is not worth it. The marginal costs of production are constant Monday through Saturday. Marginal costs are higher on Sundays, only because labor costs are higher.

Now you obtain a long-term contract to manufacture a brand-name aspirin. The cost of making the generic aspirin or the branded aspirin is identical. In fact, there is no cost or time involved in switching from the manufacture of one to another. You will make much larger profits from the branded aspirin, but the demand is limited. One day of manufacturing each week will permit you to fulfill the contract. You can manufacture both the brand name and the generic aspirin. Compared with the situation before you obtained the contract, your profits will be much higher if you now begin to manufacture on Sundays—even if you have to pay overtime wages.

- Each day you can make 1,000 cases of generic aspirin. You can sell as many as you make, for the market price of $10 per case. Every week you have fixed costs of $5,000 (land tax and insurance). No matter how many cases you manufacture, the cost to you for materials and supplies is $2 per case; the cost labor is $5 per case, except on Sundays, when it is $10 per case.

- Your order for the branded aspirin requires that you manufacture 1,000 cases per week, which you sell for $30 per case. The costs for the branded aspirin are identical to the cost of generic aspirin.

What do you do?

Solution

The marginal costs are constant Monday through Saturday; they rise substantially on Sunday and are above the marginal revenue from manufacturing generic aspirin. In other words, your company should manufacture the branded aspirin first. Your marginal revenue is the highest the "first" day, when you manufacture the branded aspirin. It then falls and remains constant for the rest of the week. On the seventh day (Sunday), the marginal revenue from manufacturing the generic aspirin is still below the marginal cost. You should manufacture for six days—one day for the branded aspirin and five days for the generic, and on Sundays, the plant should close.

---

[3] Source: "Profit Maximization Problem," by David Hemenway and Elon Kohlberg, *Economic Inquiry*, October 1, 1997, Page 862, Copyright 1997 Western Economic Association International.

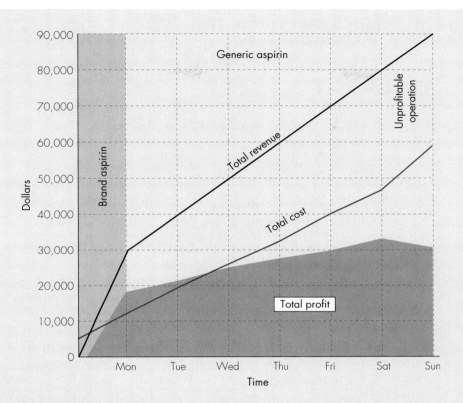

Figure 3.8   Weekly profits as a function of time. Sunday operation becomes un-profitable as the marginal revenue stays at $10 per case whereas the marginal cost increases to $12 per case (Example 3.7)

- The marginal revenue from manufacturing on Sunday is $10,000 (1,000 cases times $10 per case).
- The marginal cost from manufacturing on Sunday is $12,000 (1,000 cases times $12 per case—$10 labor plus $2 materials).
- Profits will be $2,000 lower if the plant operates on Sunday, as shown in Figure 3.8.

3.5.2   Looking Ahead

The operational decision problem described in Example 3.7 tends to have a relatively short-term horizon (weekly or monthly). That is, decisions do not commit a firm to a certain course of action over a considerable period. If operational decision problems significantly affect the amount of funds that must be invested in a firm, fixed costs will have to increase. For example, if the daily aspirin production increases to 10,000 cases per day, exceeding the current production capacity, the firm must make new invest-ment in plant and equipment. Since the benefits of the expansion will continue to occur over an extended period, we need to consider the economic effects of the fixed costs over the life of the assets. This requires understanding the concept of **time value of money**, which we will discuss in the next chapter.

3.6 Summary

In this chapter, we examined several ways in which managers classify costs. How the costs will be used will dictate how the costs will be classified.

- Most manufacturing companies divide **manufacturing costs** into three broad categories: *direct materials, direct labor*, and *manufacturing overhead*. **Nonmanufacturing costs** are classified into two categories: *marketing* or *selling costs*, and *administrative costs*.

- For purpose of valuing inventories and determining expenses for the balance sheet and income statement, costs are classified as either **product costs** or **period costs**.

- For purposes of predicting **cost behavior**—how costs will react to changes in activity—managers commonly classify costs into two categories: variable and fixed costs.

- An understanding of the various **cost-volume relationships** is essential to developing successful business strategies and to planning future operations.
    Fixed operating costs: Costs that do not vary with production volume.
    Variable operating costs: Costs that vary depending on the level of production or sales.
    Average costs: Costs expressed in terms of units obtained by dividing total costs by the total volumes.
    Differential (incremental) costs: Costs that represent differences in total costs, which results from selecting one alternative instead of another.
    Opportunity costs are the benefits obtained by following other courses of action.
    Sunk costs are past costs not relevant to decisions because they cannot be changed no matter what actions are taken.
    Marginal costs: Added costs that result from increasing rates of outputs, usually by single units.
    Marginal analysis: In economic analysis, we need to answer the apparently trivial question, "Is it worthwhile?" Whether the action in question will add sufficiently to the benefits enjoyed by the decision maker to make it worth the cost. This is the heart of marginal–decision making—the statement that an action merits performance if and only if, as a result, we can expect to be better off than we were before.

- At the level of plant operations, engineers must make decisions involving materials, production processes, and the in-house capabilities of company personnel. Most of these operating decisions do not require any investments in physical asset and, therefore, they solely depend on the cost and volume of business activity, without a consideration of the time value of money.

Self-Test Questions

3s.1 If each of the following statements is true, write in "T"; otherwise write in "F."

[ ] 1. The marginal cost is defined as the incremental cost to produce one additional unit.

[ ] 2. The sunk cost should not be considered in making any investment decisions.

[ ] 3. Since depreciation is not an actual cash expense, we should not consider this expense in pricing a manufactured product.

[ ] 4. If you have more than one product to produce, your manufacturing strategy should be producing the product with the highest profit margin first.

[ ] 5. Opportunity costs are not generally recognized in cost accounting, so these costs should not be considered in pricing a product.

3s.2 Definitional problem: Listed below are several terms related to cost terms and accounting information. (This problem is adapted from *Managerial Accounting*, R. H. Garrison and E. W. Noreen, 8th edition, Irwin, 1997, copyright © Richard D. Irwin, p. 68.)

- Period cost
- Variable cost
- Opportunity cost
- Fixed cost
- Marginal cost
- Product cost
- Prime cost
- Sunk cost
- Indirect cost
- Inventory cost
- Cost of goods sold

Choose the cost term or terms above that most appropriately describe the costs identified in each of the following situations. The same cost term can be used more than once.

1. Von Hoffman Press (a printer for Addison Wesley) prints a college textbook entitled *Contemporary Engineering Economics*. The paper going into the manufacture of the book would be called direct materials and classified as a _____ cost. In terms of cost behavior, the paper could also be described as a _____ cost.

2. We may classify the paper and other materials used in the manufacture of the book, combined with the direct labor cost involved, as _____ cost.

3. There are two types of depreciation accounts. The first account may list all depreciation expenses related to equipment directly used in book production. The other account would list all other equipment to support general business operation. For example, Von Hoffman would classify depreciation on the equipment used to print the book as a _____ cost. However, depreciation on any equipment used by the company in selling and administrative activities would be classified as a _____ cost.

4. In terms of cost behavior, depreciation would probably be classified as a _____ cost.

5. An inventorial cost is also known as a _____ cost, since such costs go into the Work-in-Process (WIP) inventory account. Note that the WIP inventory will eventually go into the Finished Goods inventory account before appearing on the income statement as part of _____.

6. We also classify the combined cost such as the direct labor cost, direct material cost and the manufacturing overhead cost involved in the manufacture of the book as _____ cost.

7. Addison Wesley sells the book through college sales reps who are paid a commission on each book sold. The company

would classify these commissions as a
_____ cost. In terms of
cost behavior, commissions as a
_____ cost.

8. In any publishing business, the cost of unsold copies must be recognized. Several hundred copies of the book were left over from the previous edition and are stored in a warehouse. The amount invested in these books would be called a _____ cost.

9. Costs can be classified in several ways. For example, Von Hoffman pays $4000 rent each month for the building that houses its printing press. The rent would be part of manufacturing overhead. In terms of cost behavior, it would be classified as a _____ cost. The rent can also be classified as a _____ cost in terms of cost object.

10. The salary of Von Hoffman Press's president would be classified as a _____ cost, since the salary will appear on the income statement as an expense in the time period in which it is incurred.

3s.3 What effect would the following changes have on the break-even point and expected profit?

Cost or Revenue Items	Break-even Point	Profit
(a) increase in fixed cost	(increase, decrease, no change)	(increase, decrease, no change)
(b) increase in labor cost	(increase, decrease, no change)	(increase, decrease, no change)
(c) increase in unit price	(increase, decrease, no change)	(increase, decrease, no change)
(d) increase in financing cost	(increase, decrease, no change)	(increase, decrease, no change)
(e) increase in sales expense	(increase, decrease, no change)	(increase, decrease, no change)

Problems

Classifying Costs

3.1 Identify which of the following transactions and events are product costs and which are period costs:

- Storage and material handling costs for raw materials
- Gains or losses on disposal of factory equipment
- Lubricants for machinery and equipment used in production
- Depreciation of a factory building
- Depreciation of manufacturing equipment
- Depreciation of the company president's automobile
- Leasehold costs for land on which factory buildings stand
- Inspection costs of finished goods
- Direct labor cost
- Raw materials cost
- Advertising expenses

Cost Behavior

3.2 Identify which of the following costs are fixed and which are variable:

- Wages paid to temporary workers
- Property taxes on factory building
- Property taxes on administrative building
- Sales commission
- Electricity for machinery and equipment in the plant
- Heat and air-conditioning for the plant
- Salaries paid to design engineers
- Regular maintenance on machinery and equipment
- Basic raw materials used in production
- Factory fire insurance

3.3 The following figures, on top of page 86, are a number of cost behavior patterns that might be found in a company's cost structure. The vertical axis on each graph represents total cost, and the horizontal axis on each graph represents level of activity (volume). For each of the following situations, identify the graph that illustrates the cost pattern involved. Any graph may be used more than once. (Adapted originally from CPA exam, and the same materials are also found in *Managerial Accounting*, R. H. Garrison and E. W. Noreen, 8th edition, Irwin, 1997, copyright © Richard D. Irwin, p. 271).

[ ] a. Electricity bill—a flat rate fixed charge, plus a variable cost after a certain number of kilowatt-hours are used.

[ ] b. City water bill, which is computed as follows:

First 1,000,000 gallons ------ $1,000 flat
or less   rate

Next 10,000 gallons----------$0.003 per
                             gallon used

Next 10,000 gallons----------$0.006 per
                             gallon used

Next 10,000 gallons----------$0.009 per
                             gallon used

Etc.------------------------------etc.

[ ] c. Depreciation of equipment, where the amount is computed by the straight-line method. When the depreciation rate was established, it was anticipated that the obsolescence factor would be greater than the wear-and-tear factor.

[ ] d. Rent on factory building donated by the city, where the agreement calls for a fixed fee payment unless 200,000 labor-hours or more are

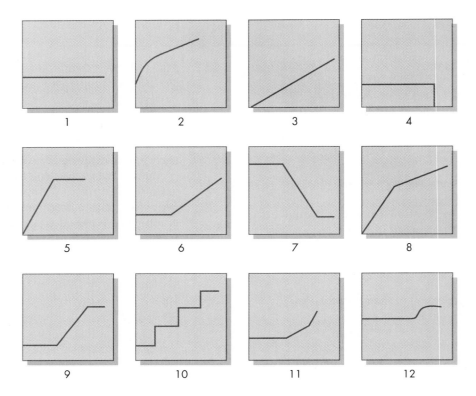

worked, in which case no rent need be paid.

[ ] e. Cost of raw materials, where the cost decreases by 5 cents per unit for each of the first 100 units purchased, after which it remains constant at $2.50 per unit.

[ ] f. Salaries of maintenance workers, where one maintenance worker is needed for every 1,000 hours of machine-hours or less (that is, 0 to 1,000 hours requires one maintenance worker, 1,001 to 2,000 hours requires two maintenance workers, etc.).

[ ] g. Cost of raw material used.

[ ] h. Rent on factory building donated by the county, where the agreement calls for rent of $100,000 less $1 for each direct labor-hour worked in excess of 200,000 hours, but a minimum rental payment of $20,000 must be paid.

[ ] i. Use of a machine under a lease, where a minimum charge of $1,000 be paid for up to 400 hours of machine time. After 400 hours of machine time, an additional charge of $2 per hour is paid up to a maximum charge of $2,000 per period.

3.4 Harris Company manufactures a single product. Costs for the year 2001 for output levels of 1,000 and 2,000 units are as follows:

	1,000	2,000
Units produced	1,000	2,000
Direct labor	$30,000	$30,000
Direct materials	20,000	40,000
Overhead:		
Variable portion	12,000	24,000
Fixed portion	36,000	36,000
Selling & Administrative costs:		
Variable portion	5,000	10,000
Fixed portion	22,000	22,000

At each level of output, compute the following:

(a) Total manufacturing costs

(b) Manufacturing costs per unit

(c) Total variable costs

(d) Total variable costs per unit

(e) Total costs that have to be recovered if the firm is to make a profit

Cost-Volume-Profit Relationships

3.5 Bragg & Stratton Company manufactures a specialized motor for chain saws. The company expects to manufacture and sell 30,000 motors in year 2001. It can manufacture an additional 10,000 motors without adding new machinery and equipment. Its projected total costs for the 30,000 units are as follows:

Direct materials	$150,000
Direct labor	300,000
Manufacturing overhead:	
Variable portion	100,000
Fixed portion	80,000
Selling and Administrative costs:	
Variable portion	180,000
Fixed portion	70,000

The selling price for the motor is $80.

(a) What is the total manufacturing cost per unit if 30,000 motors are produced?

(b) What is the total manufacturing cost per unit if 40,000 motors are produced?

(c) What is the break-even price on the motors?

3.6 The following chart shows the expected monthly profit or loss of Cypress Manufacturing Company within the range of its monthly practical operating capacity. Using the information provided in the chart, answer the flowing questions:

(a) What is the company's break-even sales volume?

(b) What is the company's marginal contribution rate?

(c) What effect would a 5% decrease in selling price have on the break-even point in (a)?

(d) What effect would a 10% increase in fixed costs have on the marginal contribution rate in (b)?

(e) What effect would a 6% increase in variable costs have on the break-even point in (a)?

(f) If the chart also reflects $20,000 monthly depreciation expenses, compute the sale at break-even point for cash costs.

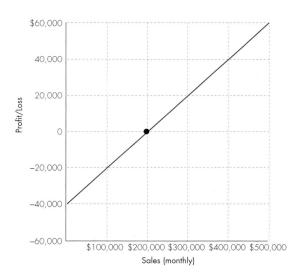

3.7 The next graph represents a cost-volume-profit graph.

In the graph, identify the following line segments or points:

(a) Line EF represents

_____.

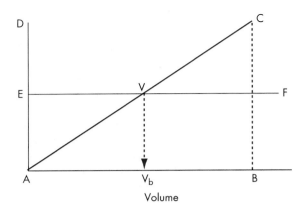

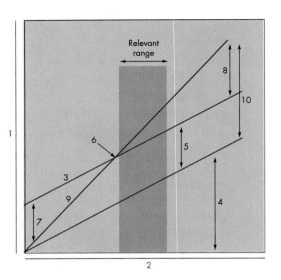

(b) The horizontal axis AB represents _____, and the vertical axis AD represents _____.

(c) Point V represents _____.

(d) The distance CB divided by the distance AB is _____.

(e) The point V_b is a break-even _____.

3.8 A cost-volume-profit graph, as illustrated below, is a useful technique for showing relationships between costs, volume, and profits in an organization.

(a) Identify the numbered components in the CVP graph.

No.	Description	No.	Description
1		6	
2		7	
3		8	
4		9	
5		10	

(b) Using a typical CVP relationship above, fill in the missing amounts in each of the four situations below.

Each case is independent of the others.

Case	Units Sold	Sales	Variable Expenses	Contribution Margin per Unit	Fixed Expenses	Net Income (Loss)
A	9,000	$270,000	$162,000		$90,000	
B		$350,000		$15	$170,000	$40,000
C	20,000		$280,000	$6		$35,000
D	5,000	$100,000			$82,000	($12,000)

Cost Concepts Relevant to Decision Making

3.9 An executive from a large merchandising firm has called your VP for Production to get a price quote for an additional 100 units of a given product. The VP has asked you to prepare a cost estimate. The number of hours required to produce a unit is 5. The average labor rate is $12 per hour. The materials cost $14 per unit. Overhead for an additional 100 units is estimated at 50% of direct labor cost. If the company wants to have a 30% profit margin, what should be the unit price to quote?

3.10 The Morton Company produces and sells two products, A and B, respectively. Financial data related to producing these two products are summarized as follows:

	Product A	Product B
Selling price	$10.00	$12.00
Variable costs	5.00	10.00
Fixed costs	$600	$2000

(a) If these products are sold in the ratio of 4A for 3B, what is the break-even point?

(b) If the product mix has changed to 5A for 5B, what would happen to the break-even point?

(c) In order to maximize the profit, which product mix should be pushed?

(d) If both products must go through the same manufacturing machine and there are only 30,000 machine hours available per period, which product should be pushed? Assume product A requires 0.5 hour per unit and B requires 0.25 hour per unit.

3.11 Pearson Company manufactures a variety of electronic PCBs (printed circuit board) that go into cellular phones. The company has just received an offer from an outside supplier to provide the electrical soldering for the company's Motorola product line (Z-7 PCB, slim-line). The quoted price is $4.80 per unit. Person is interested in this offer, since its own soldering operation of the PCB is at its peak capacity.

- Outsourcing Option: The company estimates that if the supplier's offer were accepted, the direct labor and variable

overhead costs of the Z-7 slim-line would be reduced by 15% and the direct material cost would be reduced by 20%.

- In-house Production Option: Under the present operations, Pearson Company manufactures all of its own PCBs from start to finish. The Z-7 slim-lines are sold through Motorola at $20 per unit. Fixed overhead charges to the Z-7 slim-line total $20,000 each year. The further breakdown of producing one unit is given below:

Direct materials	$7.50
Direct labor	5.00
Manufacturing Overhead	4.00
Total cost	$16.00

The manufacturing overhead of $4.00 per unit includes both variable and fixed manufacturing overhead, based on production of 100,000 units each year.

(a) Should Pearson Company accept the outside supplier's offer?

(b) What is the maximum unit price that Pearson Company would be willing to pay the outside supplier?

Short Case Studies

3.12 The Hamilton Flour Company is currently operating its mill 6 days a week, 24 hours a day, on three shifts. The company could easily obtain a sufficient volume of sales at current prices to take the entire output of a seventh day of operation each week. The mill's practical capacity is 6,000 hundredweight of flour per day.

- Flour sells for $12.40 a hundredweight (cwt.) and the price of wheat is $4.34 a bushel. About 2.35 bushels of wheat are required per cwt. of flour. Fixed costs now average $4,200 a day, or $0.70 per cwt. The average variable cost of mill operation, almost entirely wages, is $0.34 per cwt.

- With Sunday operation, wages would be doubled for Sunday work, which would bring the variable cost of Sunday Operation to $0.66 per cwt. Total fixed costs per week would increase by $420 (or $29,820) if the mill were to operate on Sunday.

(a) Using the information provided above, compute the break-even volumes for 6-day and 7-day operation.

(b) What are the marginal contribution rates for 6-day and 7-day operations?

(c) Compute the average total cost per cwt. for 6-day operation and the net profit margin before taxes, per cwt.

(d) Would it be economical for the mill to operate on Sundays? (Justify your answer numerically.)

PART TWO

Money And Investing

CHAPTER 4

Time Is Money

Take or Not To Take the Offer[1] Rosalind Setchfield, the Yuma, Arizona, resident won $1.3 million in a 1987 Arizona lottery drawing, to be paid in 20 annual installments of $65,277. However, in 1989, her husband, a construction worker, suffered serious injuries when a crane dropped a heavy beam on him. The couple's medical expenses and debt mounted. Six years later, in early 1995, a prize broker from Singer Asset Finance Co. telephoned Mrs. Setchfield with a promising offer. Singer would immediately give her $140,000 for one-half of each of her next 9 prize checks, an amount equal to 48% of the $293,746 she had coming over that period. A big lump sum had obvious appeal to Mrs. Setchfield at that time, and she ended up with signing a contract with Singer. Did she make a right decision?

Year	Installment	Year	Installment	Reduced Payment
1988	$65,277	1995	$65,277	$32,639
1989	65,277	1996	65,277	32,639
1990	65,277	1997	65,277	32,639
1991	65,277	1998	65,277	32,639
1992	65,277	1999	65,277	32,639
1993	65,277	2000	65,277	32,639
1994	65,277	2001	65,277	32,639
		2002	65,277	32,639
		2003	65,277	32,639
		2004	65,277	
		2005	65,277	
		2006	65,277	
		2007	65,277	

---

[1] Source: Fidelity, "Others Are Relishing Big Returns From the Lottery," by Vanessa O'Connell, *The Wall Street Journal*, September 23, 1997

A lthough Mrs. Setchfield's decision to accept the deal was understandable under extreme financial stress, she may be surprised by how her decision stands up under normal economic analysis. First, most people familiar with investments would tell her that giving up $32,639 a year for 9 years is likely to prove a far poor deal than receiving $140,000 today. We can also use the techniques presented in this chapter to show that her deal only makes sense for those individuals who can invest their moneys at an interest rate of 18.10%. This is rather a high rate of return for normal investors. Enhance Financial Services Group made the windfall for Singer possible, a large municipal-bond reinsurer in New York. Enhance had instructed Singer to buy this and other prizes at a substantial discount and then arrange to have lotteries send Enhance the prize money, or partial payments, that would otherwise be mailed to winners. Enhance had determined that it would pay $196,000 for the stake Singer acquired from Mrs. Setchfield for $140,000. To arrive at its price, the Enhance Group's treasurer calculated the return the company wanted to earn—then about 9.5% interest, compounded annually—and applied that rate in reverse to the $293,745.60 sum Enhance stood to collect over 9 years. For bringing Mrs. Setchfield and Enhance together, Singer earned $56,000, keeping 29% of what Enhance offered. Singer says the deals it strikes with winners apply a basic tenet of all financial transactions, the **time value of money**: A dollar in hand today is worth more than one that will be paid to you in the future.

In engineering economic analysis, the principles discussed in this chapter are regarded as the underpinning for nearly all project investment analysis. This is because we always need to account for the effect of interest operating on sums of cash over time. Interest formulas allow us to place different cash flows received at different times in the same time frame and to compare them. As will become apparent, almost our entire study of engineering economic analysis is built on the principles introduced in this chapter.

4.1 Interest: The Cost of Money

Most of us are familiar in a general way with the concept of interest. We know that money left in a savings account earns interest so that the balance over time is greater than the sum of the deposits. We know that borrowing to buy a car means repaying an amount over time, that it includes interest, and that it is therefore greater than the amount borrowed. What may be unfamiliar to us is the idea that, in the financial world, money itself is a commodity, and like other goods that are bought and sold, money costs money.

The cost of money is established and measured by an **interest rate**, a percentage that is periodically applied and added to an amount (or varying amounts) of money over a specified length of time. When money is borrowed, the interest paid is the charge to the borrower for the use of the lender's property; when money is loaned or invested, the interest earned is the lender's gain from providing a good to another. **Interest**, then, may be defined as the cost of having money available for use. In this section, we examine how interest operates in a free-market economy and establish a basis for understanding the more complex interest relationships that follow later on in the chapter.

4.1.1 The Time Value of Money

The "time value of money" seems like a sophisticated concept, yet it is one that you grapple with every day. Should you buy something today or save your money and buy it later? Here is a simple example of how your buying behavior can have varying results. Pretend you have $100, and you want to buy a $100 refrigerator for your dorm room. If you buy it now, you are broke. However, if you invest money at 6% interest, in a year you can still buy the refrigerator, and you will have $6 left over. Well, if the price of the refrigerator increases at an annual rate of 8% due to inflation, you will not have enough money ($2 short) to buy the refrigerator a year from now. In that case, you probably are better off buying the refrigerator now. Clearly your earning interest rate should be higher than the inflation rate to make any economic sense of the delayed purchase. In other words, under inflationary economy, your purchasing power will continue to decrease as you further delay the purchase of the refrigerator. In order to make up this future loss in purchasing power, your earning interest rate should be sufficiently larger than the anticipated inflation rate. After all, time, like money, is a finite resource. There are only 24 hours in a day, so time has to be budgeted, too. What this example illustrates is that we must connect the "earning power" and the "purchasing power" to the concept of time.

When we deal with large amounts of money, long periods of time, or high interest rates, the change in the value of a sum of money over time becomes extremely significant. For example, at a current annual interest rate of 10%, $1 million will earn $100,000 in interest in a year; thus, to wait a year to receive $1 million clearly involves a significant sacrifice. When deciding among alternative proposals, we must take into account the operation of interest and the time value of money to make valid comparisons of different amounts at various times.

The way interest operates reflects the fact that money has a time value. This is why amounts of interest depend on lengths of time; interest rates, for example, are typically given in terms of a percentage per year. We may define the principle of the time value of money as follows: The economic value of a sum depends on when it is received. Because money has both **earning** as well as **purchasing power** over time (it can be put to work, earning more money for its owner), a dollar received today has a greater value than a dollar received at some future time.

When lending or borrowing interest rates are quoted by financial institutions on the marketplace, those interest rates reflect the desired earning, as well as any protection from loss in the future purchasing power of money because of inflation. (If we want to know the true earning in isolation of inflation, we can determine the real interest rate. We consider this issue in Chapter 13. The earning power of money and its loss of value because of inflation represent different analytical techniques.) In the meantime, we will assume that the interest rate used in this book (unless otherwise mentioned) reflects the market interest rate, which considers the earning power, as well as the effect of inflation perceived in the marketplace. We will also assume that all cash flow transactions are given in terms of actual dollars where the effect of inflation, if any, is reflected in the amount.

4.1.2   Elements of Transactions Involving Interest

Many types of transactions involve interest—e.g., borrowing or investing money, or purchasing machinery on credit—but certain elements are common to all of these types of transactions.

1. An initial amount of money in transactions involving debt or investments is called the **principal**.
2. The **interest rate** measures the cost or price of money and is expressed as a percentage per period of time.
3. A period of time, called the **interest period**, determines how frequently interest is calculated. (Note that even though the length of time of an interest period can vary, interest rates are frequently quoted in terms of an annual percentage rate. We will discuss this potentially confusing aspect of interest in Chapter 5.)
4. A specified length of time marks the duration of the transaction and thereby establishes a certain **number of interest periods**.
5. A **plan for receipts or disbursements** that yields a particular cash flow pattern over a specified length of time. (For example, we might have a series of equal monthly payments that repay a loan.)
6. A **future amount of money** results from the cumulative effects of the interest rate over a number of interest periods.

For the purposes of calculation, these elements are represented by the following variables:

A_n = A discrete payment or receipt occurring at the end of some interest period.

i = The interest rate per interest period.

N = The total number of interest periods.

P = A sum of money at a time chosen for purposes of analysis as time zero, sometimes referred to as the **present value** or **present worth**.

F = A future sum of money at the end of the analysis period. This sum may be specified as F_N.

A = An end-of-period payment or receipt in a uniform series that continues for N periods. This is a special situation where $A_1 = A_2 = \ldots = A_N$.

V_n = An equivalent sum of money at the end of a specified period n that considers the effect of time value of money. Note that $V_0 = P$ and $V_N = F$.

Because frequent use of these symbols will be made in this text, it is important that you become familiar with them. Note, for example, the distinction between A, A_n, and A_N. The symbol A_n refers to a specific payment or receipt, at the end of period n, in any series of payments. A_N is the final payment in such a series because N refers to the total number of interest periods. A refers to any series of cash flows where all payments or receipts are equal.

Example of an Interest Transaction

As an example of how the elements we have just defined are used in a particular situation, let us suppose that an electronics manufacturing company buys a machine for $25,000 and borrows $20,000 from a bank at a 9% annual interest rate. In addition, the company pays a $200 loan origination fee when the loan commences. The bank offers two repayment plans, one with equal payments made at the end of every year for the next 5 years, and the other with a single payment made after the loan period of 5 years. These payment plans are summarized in Table 4.1.

In Plan 1 the principal amount, P, is $20,000, and the interest rate, i, is 9%. The interest period is 1 year, and the duration of the transaction is 5 years, which means

Table 4.1 Repayment Plans for Example Given in Text (for $N = 5$ years and $i = 9\%$)

End of Year	Receipts	Payments Plan 1	Payments Plan 2
Year 0	$20,000.00	$ 200.00	$ 200.00
Year 1		5,141.85	0
Year 2		5,141.85	0
Year 3		5,141.85	0
Year 4		5,141.85	0
Year 5		5,141.85	30,772.48

$P = \$20,000$, $A = \$5,141.85$, $F = \$30,772.48$

Note: You actually borrow $19,800 with the origination fee of $200, but you pay back based on $20,000.

there are five interest periods ($N = 5$). It bears repeating that whereas 1 year is a common interest period, interest is frequently calculated at other intervals: monthly, quarterly, or semi-annually, for instance. For this reason, we used the term **period** rather than **year** when we defined the preceding list of variables. The receipts and disbursements planned over the duration of this transaction yield a cash flow pattern of five equal payments, A, of $5,141.85 each, paid at year-end during years 1 through 5. (You'll have to accept these amounts on faith for now—the following section presents the formula used to arrive at the amount of these equal payments, given the other elements of the problem.)

Plan 2 has most of the elements of Plan 1, except that instead of five equal repayments we have a grace period followed by a single future repayment, F, of $30,772.78.

Cash Flow Diagrams

Problems involving the time value of money can be conveniently represented in graphic form with a cash flow diagram (Figure 4.1). **Cash flow diagrams** represent time by a horizontal line marked off with the number of interest periods specified. The cash flows over time are represented by arrows at relevant periods: Upward arrows represent positive flows (receipts) and downward arrows represent negative flows (disbursements). Note, too, that the arrows actually represent **net cash flows**: Two or more receipts or disbursements made at the same time are summed and shown as a single arrow. For example, $20,000 received during the same period as a $200 payment would be recorded as an upward arrow of $19,800. The lengths of the arrows can also suggest the relative values of particular cash flows.

Cash flow diagrams function in a manner similar to free body diagrams or circuit diagrams, which most engineers frequently use: Cash flow diagrams give a convenient summary of all the important elements of a problem as well as a reference point to determine whether a problem statement has been converted into its appropriate para-

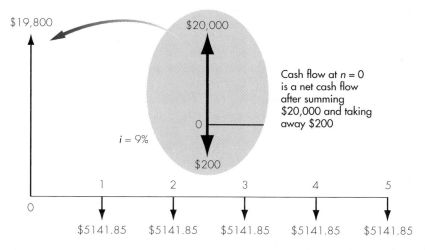

Figure 4.1   A cash flow diagram for plan 1 of the loan repayment example summarized in Table 4.1

meters. The text frequently uses this graphic tool, and you are strongly encouraged to develop the habit of using well-labeled cash flow diagrams as a means to identify and summarize pertinent information in a cash flow problem. Similarly, a table such as the one shown in Table 4.1 can help you organize information in another summary format.

End-of-Period Convention

In practice, cash flows can occur at the beginning or in the middle of an interest period, or at practically any point in time. One of the simplifying assumptions we make in engineering economic analysis is the **end-of-period convention**, which is the practice of placing all cash flow transactions at the end of an interest period. This assumption relieves us of the responsibility of dealing with the effects of interest within an interest period, which would greatly complicate our calculations.

It is important to be aware of the fact that, like many of the simplifying assumptions and estimates we make in modeling engineering economic problems, the end-of-period convention inevitably leads to some discrepancies between our model and real-world results.

Suppose, for example, that $100,000 is deposited during the first month of the year in an account with an interest period of 1 year and an interest rate of 10% per year. In such a case, the difference of 1 month would cause an interest income loss of $10,000. This is because, under the end-of-period convention, the $100,000 deposit made during the interest period is viewed as if the deposit were made at the end of the year as opposed to 11 months earlier. This example gives you a sense of why financial institutions choose interest periods that are less than 1 year, even though they usually quote their rate in terms of annual percentage.

Armed with an understanding of the basic elements involved in interest problems, we can now begin to look at the details of calculating interest.

4.1.3 Methods of Calculating Interest

Money can be loaned and repaid in many ways, and, equally, money can earn interest in many different ways. Usually, however, at the end of each interest period, the interest earned on the principal amount is calculated according to a specified interest rate. The two computational schemes for calculating this earned interest are said to yield either **simple interest** or **compound interest**. Engineering economic analysis uses the compound interest scheme almost exclusively.

Simple Interest

The first scheme considers interest earned on only the principal amount during each interest period. In other words, under simple interest, the interest earned during each interest period does not earn additional interest in the remaining periods, *even though you do not withdraw it.*

In general, for a deposit of P dollars at a simple interest rate of i for N periods, the total earned interest I would be

$$I = (iP)N. \tag{4.1}$$

The total amount available at the end of N periods, F, thus would be

$$F = P + I = P(1 + iN). \tag{4.2}$$

Simple interest is commonly used with add-on loans or bonds; these are reviewed in Chapter 5.

Compound Interest

Under a compound interest scheme, the interest earned in each period is calculated based on the total amount at the end of the previous period. This total amount includes the original principal plus the accumulated interest that has been left in the account. In this case, you are in effect increasing the deposit amount by the amount of interest earned. In general, if you deposited (invested) P dollars at interest rate, i, you would have $P + iP = P(1 + i)$ dollars at the end of one period. If the entire amount (principal and interest) is reinvested at the same rate, i, for another period, you would have, at the end of the second period,

$$P(1 + i) + i[P(1 + i)] = P(1 + i)(1 + i)$$
$$= P(1 + i)^2.$$

Continuing, we see that the balance after period three is

$$P(1 + i)^2 + i[P(1 + i)^2] = P(1 + i)^3.$$

This interest-earning process repeats, and after N periods, the total accumulated value (balance) F will grow to

$$F = P(1 + i)^N. \tag{4.3}$$

Example 4.1 Compound Interest

Suppose you deposit $1,000 in a bank savings account that pays interest at a rate of 8%, compounded annually. Assume that you don't withdraw the interest earned at the end of each period (year), but let it accumulate. How much would you have at the end of year 3?

Solution

Given: $P = \$1,000$, $N = 3$ years, and $i = 8\%$ per year

Find: F

Applying Eq. (4.3) to our 3-year, 8% case, we obtain

$$F = \$1000(1 + 0.08)^3 = \$1259.71.$$

The total interest earned is $259.71, which is $19.71 more than was accumulated under the simple interest method (Figure 4.2). We can keep track of the interest accruing process more precisely as follows:

Period	Amount at Beginning of Interest Period	Interest Earned for Period	Amount at End of Interest Period
1	$1000.00	$1000(0.08)	$1080.00
2	1080.00	1080(0.08)	1166.40
3	1166.40	1166.40(0.08)	1259.71

Comments: At the end of the first year, you would have $1,000 plus $80 in interest, or a total of $1,080. In effect, at the beginning of the second year, you would be depositing $1,080, rather than $1,000. Thus, at the end of the second year, the interest earned would be 0.08($1,080) = $86.40, and the balance would be $1,080 + $86.40 = $1,166.40. This is the amount you would be depositing at the beginning of the third year, and the interest earned for that period would be 0.08($1,166.40) = $93.31. With a beginning principal amount of $1,166.40 plus the $93.31 interest, the total balance would be $1,259.71 at the end of year 3. If the total balance were then withdrawn, the net cash flow in this case would appear as

Year	Cash Flow
0	–$1,000
1	0
2	0
3	1,259.71

which is the same value worked out previously.

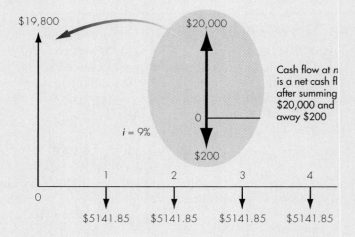

Figure 4.2 Two methods of calculating the balance when $1000 at 8% interest is deposited for 3 years (Example 4.1)

4.1.4   Simple Interest versus Compound Interest

From Eq. (4.3), the total interest earned over N periods is

$$I = F - P = P[(1 + i)^N - 1]. \tag{4.4}$$

When compared with the simple interest scheme, the additional interest earned with compound interest is

$$\Delta I = P[(1 + i)^N - 1] - (iP)N \tag{4.5}$$

$$= P[(1 + i)^N - (1 + iN)]. \tag{4.6}$$

As either i or N becomes large, the difference in interest earnings also becomes large, so the effect of compounding is further pronounced. Note that, when $N = 1$, compound interest is the same as simple interest. Example 4.2 will illustrate the comparison.

Example 4.2 Comparing Simple to Compound Interest

In 1626, Peter Minuit of the Dutch West India Company paid $24 to purchase Manhattan Island in New York from the Indians. In retrospect, if Mr. Minuit had invested the $24 in a savings account that earns 8% interest, how much would it be worth in 2000?

Solution

Given: $P = \$24$, $i = 8\%$ per year, $N = 374$ years

Find: F based on (a) 8% simple interest and (b) 8% compound interest

(a) With 8% simple interest:

$$F = \$24[1 + (0.08)(374)] = \$742.08$$

(b) With 8% compound interest:

$$F = \$24(1 + 0.08)^{374} = \$75,979,388,483,896$$

Comments: The significance of compound interest is obvious in this example. Many of us can hardly comprehend the magnitude of $75 trillion. In 2000, the total population in the United States was estimated to be around 274 million. If the money were distributed equally among the population, each individual would receive $277,297. Certainly, there is no way of knowing exactly how much Manhattan Island is worth today, but most real estate experts would agree that the value of the island is not anything near 1 trillion dollars. (Note that the U.S. national debt as of May 15, 2001 was 5.64 trillion dollars.)

4.2   Economic Equivalence

The observation that money has a time value leads us to an important question: If receiving $100 today is not the same as receiving $100 at any future point, how do we measure and compare various cash flows? How do we know, for example, whether we should prefer to have $20,000 today, and $50,000 ten years from now, or $8000 each year for the next 10 years (Figure 4.3)? In this section, we will describe the basic analytical techniques for making these comparisons. Then, in Section 4.3, we will use these techniques to develop a series of formulas that can greatly simplify our calculations.

4.2.1   Definition and Simple Calculations

The central question in deciding among alternative cash flows involves comparing their economic worth. This would be a simple matter if, in the comparison, we did not need to consider the time value of money: We could simply add the individual payments within a cash flow, treating receipts as positive cash flows and payments (disbursements) as negative cash flows. The fact that money has a time value makes our calculations more complicated. We need to know more than just the size of a payment in order to determine its economic effect completely. In fact, as we will discuss in this section, we need to know several things:

- Its magnitude.
- Its direction—is it a receipt or a disbursement?

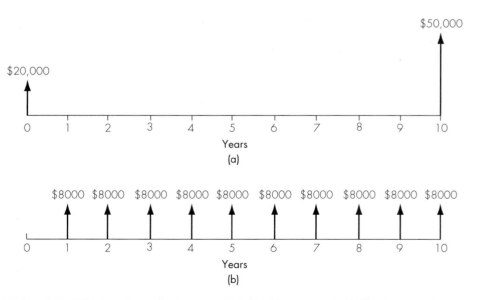

Figure 4.3   Which option would you prefer? (a) Two payments ($20,000 now and $50,000 at the end of 10 years) or (b) ten equal annual receipts in the amount of $8000

- Its timing—when is the payment made?
- The interest rate in operation during the time period under consideration.

It follows that to assess the economic impact of a series of payments, we must consider the impact of each payment.

Calculations for determining the economic effects of one or more cash flows are based on the concept of economic equivalence. **Economic equivalence** exists between cash flows that have the same economic effect and could therefore be traded for one another in the financial marketplace, which we assume to exist.

Economic equivalence refers to the fact that a cash flow—whether a single payment or a series of payments—can be converted to an *equivalent* cash flow at any point in time. For example, we could find the equivalent future value, F, of a present amount, P, at interest rate, i, at period, n; or we could determine the equivalent present value, P, of N equal payments, A.

The strict conception of equivalence, which limits us to converting a cash flow into another equivalent cash flow, may be extended to include the comparison of alternatives. For example, we could compare the value of two proposals by finding the equivalent value of each at any common point in time. If financial proposals that appear to be quite different turn out to have the same monetary value, then we can be *economically indifferent* to choosing between them: In terms of economic effect, one would be an even exchange for the other so no reason exists to prefer one over the other in terms of their economic value.

A way to see the concepts of equivalence and economic indifference at work in the real world is to note the variety of payment plans offered by lending institutions for consumer loans. Table 4.2 extends the example we developed earlier to include three different repayment plans for a loan of $20,000 for 5 years at 9% interest. You will notice, perhaps to your surprise, that the three plans require significantly different repayment patterns and different total amounts of repayment. However, because money has time value, these plans are equivalent, and economically, the bank is indifferent to a consumer's choice of plan. We will now discuss how such equivalence relationships are established.

	Repayments			
	Plan 1	Plan 2	Plan 3	
Year 1	$5,141.85	0	$1,800.00	
Year 2	5,141.85	0	1,800.00	
Year 3	5,141.85	0	1,800.00	
Year 4	5,141.85	0	1,800.00	
Year 5	5,141.85	$30,772.48	21,800.00	
Total of payments	$25,709.25	$30,772.48	$29,000.00	
Total interest paid	$5,709.25	$10,772.48	$9,000.00	

Table 4.2 Typical Repayment Plans for a Bank Loan of $20,000 (for $N = 5$ years and $i = 9\%$)

Plan 1: Equal annual installments; Plan 2: End-of-loan-period repayment of principal and interest; Plan 3: Annual repayment of interest and end-of-loan repayment of principal

Equivalence Calculations: A Simple Example

Equivalence calculations can be viewed as an application of the compound interest relationships we developed in Section 4.1. Suppose, for example, that we invest $1000 at 12% annual interest for 5 years. The formula developed for calculating compound interest, $F = P(1 + i)^N$ (Eq. 4.3), expresses the equivalence between some present amount, P, and a future amount, F, for a given interest rate, i, and a number of interest periods, N. Therefore, at the end of the investment period, our sums grow to

$$\$1000(1 + 0.12)^5 = \$1762.34.$$

Thus we can say that at 12% interest, $1,000 received now is equivalent to $1,762.34 received in 5 years, and that we could trade $1,000 now for the promise of receiving $1,762.34 in 5 years. Example 4.3 further demonstrates the application of this basic technique.

Example 4.3 Equivalence

Suppose you are offered the alternative of receiving either $3,000 at the end of 5 years or P dollars today. There is no question that the $3,000 will be paid in full (no risk). Because you have no current need for the money, you would deposit the P dollars in an account that pays 8% interest. What value of P would make you indifferent to your choice between P dollars today and the promise of $3,000 at the end of 5 years?

Discussion: Our job is to determine the present amount that is economically equivalent to $3,000 in 5 years given the investment potential of 8% per year. Note that the problem statement assumes that you would exercise the option of using the earning power of your money by depositing it. The "indifference" ascribed to you refers to economic indifference; i.e., within a marketplace where 8% is the applicable interest rate, you could trade one cash flow for the other.

Solution

Given: $F = \$3,000$, $N = 5$ years, $i = 8\%$ per year
Find: P

Equation: Eq. (4.3), $F = P(1+i)^N$
Rearranging to solve for P,

$$P = F/(1 + i)^N.$$

Substituting,

$$P = \$3,000/(1 + 0.08)^5 = \$2,042.$$

We can summarize the problem graphically as in Figure 4.4.

Comments: In this example, it is clear that, if P is anything less than $2,042, you would prefer the promise of $3,000 in 5 years to P dollars today; if P were greater than $2,042, you would prefer P. As you may have already guessed, at a lower interest rate, P must be higher to be equivalent to the future amount. For example, at $i = 4\%$, $P = \$2,466$.

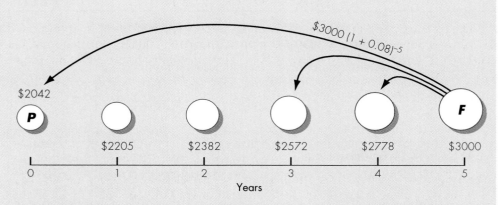

Figure 4.4 Various dollar amounts that will be economically equivalent to $3000 in 5 years, given an interest rate of 8% (Example 4.3)

4.2.2 Equivalence Calculations: General Principles

In spite of their numerical simplicity, the examples we have developed reflect several important general principles, which we will now explore.

Principle 1: Equivalence Calculations Made to Compare Alternatives Require a Common Time Basis

Just as we must convert fractions to common denominators to add them together, we must also convert cash flows to a common basis to compare their value. One aspect of this basis is the choice of a single point in time at which to make our calculations. In Example 4.3, if we had been given the magnitude of each cash flow, and had been asked to determine whether they were equivalent, we could have chosen any reference point and used the compound interest formula to find the value of each cash flow at that point. As you can readily see, the choice of $n = 0$ or $n = 5$ would make our problem simpler because we only need to make one set of calculations: At 8% interest, either convert $2,042 at time 0 to its equivalent value at time 5, or convert $3,000 at time 5 to its equivalent value at time 0. (To see how to choose a different reference point, take a look at Example 4.4.)

When selecting a point in time at which to compare the value of alternative cash flows, we commonly use either the present time, which yields what is called the **present worth** of the cash flows or some point in the future, which yields their **future worth**. The choice of the point in time often depends on the circumstances surrounding a particular decision, or it may be chosen for convenience. For instance, if the present worth is known for the first two of three alternatives, all three may be compared by simply calculating the present worth of the third.

Example 4.4 Equivalent Cash Flows are Equivalent at Any Common Point in Time

In Example 4.3, we determined that, given an interest rate of 8% per year, receiving $2,042 today is equivalent to receiving $3,000 in 5 years. Are these cash flows also equivalent at the end of year 3?

Discussion: This problem is summarized in Figure 4.5. The solution consists of solving two equivalence problems: (1) What is the future value of $2,042 after 3 years at 8% interest (part (a) of the solution)? (2) Given the sum of $3,000 after 5 years and an interest rate of 8%, what is the equivalent sum after 3 years (part (b) of the solution)?

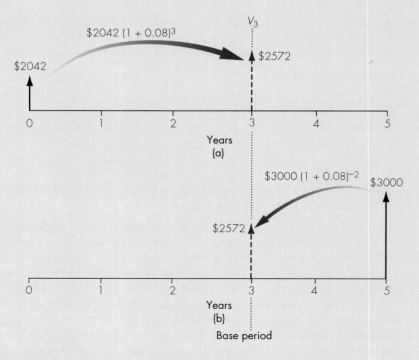

Figure 4.5   Selection of a base period for an equivalence calculation (Example 4.4)

Solution

Given:

(a) $P = \$2{,}042$; $i = 8\%$ per year; $N = 3$ years

(b) $F = \$3{,}000$; $i = 8\%$ per year; $N = 5 - 3 = 2$ years

Find: (1) V_3 for part (a); (2) V_3 for part (b); (3) Are these two values equivalent?
Equation:

$$(a)\ F = P(1 + i)^N$$

$$(b)\ P = F(1 + i)^{-N}$$

Notation: The usual terminology of F and P is confusing in this example, since the cash flow at $n = 3$ is considered a future sum in Part (a) of the solutions and a past cash flow in Part (b) of the solution. To simplify matters, we are free to arbitrarily designate a reference point, $n = 3$, and understand that it need not to be now or the present. Therefore we assign a single variable, V_3, for the equivalent cash flow at $n = 3$.

1. The equivalent worth of $2,042 after 3 years is

$$V_3 = 2{,}042(1 + 0.08)^3$$

$$= \$2{,}572.$$

2. The equivalent worth of the sum $3000, 2 years earlier is:

$$V_3 = F(1 + i)^{-N}$$

$$= \$3{,}000(1 + 0.08)^{-2}$$

$$= \$2{,}572.$$

(Note that $N = 2$ because that is the number of periods during which discounting is calculated in order to arrive back at year 3.)

3. While our solution doesn't strictly prove that the two cash flows are equivalent at any time, they will be equivalent at any time as long as we use an interest rate of 8%.

Principle 2: Equivalence Depends on Interest Rate

The equivalence between two cash flows is a function of the magnitude and timing of individual cash flows and the interest rate or rates that operate on those cash flows. This is easy to grasp in relation to our simple example: $1,000 received now is equivalent to $1,762.34 received 5 years from now only at a 12% interest rate. Any change in the interest rate will destroy the equivalence between these two sums as we will demonstrate in Example 4.5.

Example 4.5 Changing the Interest Rate Destroys Equivalence

In Example 4.3, we determined that, given an interest rate of 8% per year, receiving $2,042 today is equivalent to receiving $3,000 in 5 years. Are these cash flows equivalent at an interest rate of 10%?

Solution

Given: $P = \$2,042$; $i = 10\%$ per year, $N = 5$ years
Find: F: Is it equal to $3,000?

We first determine the base period where an equivalence value is computed. Since we can select any period as the base period, let's select $N = 5$. Then, we need to calculate the equivalent value of $2,042 today 5 years from now.

$$F = \$2,042(1 + 0.10)^5 = \$3,289.$$

Since this amount is greater than $3,000, the change in interest rate destroys the equivalence between the two cash flows.

Principle 3: Equivalence Calculations May Require the Conversion of Multiple Payment Cash Flows to a Single Cash Flow

In all the examples presented thus far, we have limited ourselves to the simplest case of converting a single payment at one time to an equivalent single payment at another time. Part of the task of comparing alternative cash flow series involves moving each individual cash flow in the series to the same single point in time and summing these values to yield a single equivalent cash flow. We perform such a calculation in Example 4.6.

Example 4.6 Equivalence Calculations with Multiple Payments

Suppose that you borrow $1,000 from a bank for 3 years at 10% annual interest. The bank offers two options: (1) Repaying the interest charges for each year at the end of that year and repaying the principal at the end of year 3 or (2) repaying the loan all at once (including both interest and principal) at the end of year 3. The repayment schedules for the two options are as follows:

Options	Year 1	Year 2	Year 3
Option 1: End-of-year repayment of interest and principal repayment at end of loan	$100	$100	$1,100
Option 2: One end-of-loan repayment of both principal and interest	0	0	1,331

Determine whether these options are equivalent, assuming that the appropriate interest rate for our comparison is 10%.

Discussion: Since we pay the principal after 3 years in either plan, the repayment of principal can be removed from our analysis. This is an important point: *We can ignore the common elements of alternatives being compared so that we can focus entirely on comparing the interest payments.* Notice that under Option 1, we will pay a total of $300 interest, whereas under Option 2, we will pay a total of $331. Before concluding that we prefer Option 1, remember that a comparison of the two cash flows is based on a *combination of payment amounts and timing of those payments.* To make our comparison, we must compare the equivalent value of each option at a single point in time. Since Option 2 is already a single payment at $n = 3$ years, it is simplest to convert the Option 1 cash flow pattern to a single value at $n = 3$. To do this, we must convert the three disbursements of Option 1 to their respective equivalent values at $n = 3$. At that point, since they share a time in common, we can simply sum them in order to compare them to the $331 sum in Option 2.

Solution

Given: The cash flow diagrams in Figure 4.6; $i = 10\%$ per year

Find: A single future value, F, of the flows in Option 1

Equation: $F = P(1+i)^N$, applied to each disbursement in the cash flow diagram

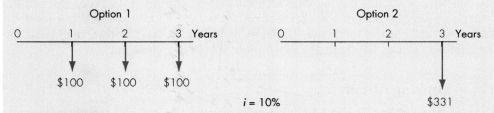

Figure 4.6   Equivalent cash flow diagram for Option 1 and Option 2 (excluding the common principal payment $1,000 at the end of year 3) (Example 4.5)

N in Eq. (4.3) is the number of interest periods upon which interest is in effect; n is the period number (i.e., for year 1, $n = 1$). We determine its value by finding the interest period for each payment. Thus, for each payment in the series, N, can be calculated by subtracting n from the total number of years of the loan (3). That is, $N = 3 - n$. Once the value of each payment has been found, we sum them:

$$F_3 \text{ for } \$100 \text{ at } n = 1 : \$100(1 + .10)^{3-1} = \$121$$

$$F_3 \text{ for } \$100 \text{ at } n = 2 : \$100(1 + .10)^{3-2} = \$110$$

$$F_3 \text{ for } \$100 \text{ at } n = 3 : \$100(1 + .10)^{3-3} = \underline{\$100}$$

$$\text{Total} = \$331.$$

> By converting the cash flow in Option 1 to a single future payment at year 3, we can compare it to Option 2. We see that the two interest payments are equivalent. Thus, the bank would be economically indifferent to a choice between the two plans. Note that the final interest payment in Option 1 does not accrue any compound interest.

Principle 4: Equivalence is Maintained Regardless of Point-of-View

As long as we use the same interest rate in equivalence calculations, equivalence can be maintained regardless of point of view. In Example 4.6, the two options were equivalent at an interest rate of 10% from the banker's point of view. What about from a borrower's point of view? Suppose you borrow $1,000 from a bank and deposit it in another bank that pays 10% interest annually. Then, you make future loan repayments out of this savings account. Under Option 1, your savings account at the end of year 1 will show a balance of $1,100 after the interest earned during the first period has been credited. Now you withdraw $100 from this savings account (the exact amount required to pay the loan interest during the first year), and you make the first year interest payment to the bank. This leaves only $1,000 in your savings account. At the end of year 2, your savings account will earn another interest payment in the amount of $1,000(0.10) = $100, making an end-of-year balance of $1,100. Now you withdraw another $100 to make the required loan interest payment. After this payment, your remaining balance will be $1,000. This balance will grow again at 10%, so you will have $1,100 at the end of year 3. After making the last loan payment ($1,100), you will have no money left in either account. For Option 2, you can keep track of the yearly account balances in a similar fashion. You will find that you reach a zero balance after making the lump sum payment of $1,331. If the borrower had used the same interest rate as the bank, the two options would be equivalent.

4.2.3 Looking Ahead

The preceding examples should have given you some insight into the basic concepts and calculations involved in the concept of economic equivalence. Obviously, the variety of financial arrangements possible for borrowing and investing money is extensive, as is the variety of time-related factors (e.g., maintenance costs over time, increased productivity over time, etc.) in alternative proposals for various engineering projects. It is important to recognize that even the most complex relationships incorporate the basic principles we have introduced in this section.

In the remainder of this chapter, we will represent all cash flow diagrams in the context of an initial deposit, with a subsequent pattern of withdrawals, or an initial borrowed amount with a subsequent pattern of repayments. A cash flow diagram representation of a more complicated equivalence calculation appears in Figure 4.7, which summarizes the payment options offered to the 2000 winner of the Publishers Clearing House Ten Million Dollar Super Prize. If we were limited to the methods developed in this section, a comparison between the two payment options would involve a large number of calculations. Fortunately, in the analysis of many transactions, certain cash flow patterns emerge that may be categorized. For many of these cash flow

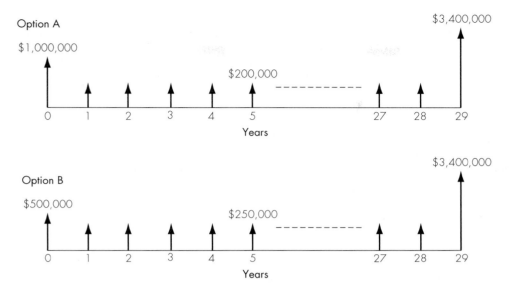

Figure 4.7 Two payment options provided by the 2000 Publishers' Clearing House, $10 Million SuperPrize: Option A, $1,000,000 cash now, $200,000 yearly plus a $3,400,000 final payment at the end of 29 years. Option B, $500,000 cash now, $250,000 a year thereafter, plus $2,500,000 at the end of 29 years

patterns we can derive formulas that can be used to simplify our work. In Section 4.3, we develop these formulas.

4.3 Development of Interest Formulas

Now that we have established some working assumptions and notations and have a preliminary understanding of the concept of equivalence, we will develop a series of interest formulas for use in more complex comparisons of cash flows.

As we begin to compare series of cash flows instead of single payments, the required analysis becomes more complicated. However, when patterns in cash flow transactions can be identified, we can take advantage of these patterns by developing concise expressions for computing either the present or future worth of the series. We will classify five major categories of cash flow transactions, develop interest formulas for them, and present several working examples of each type. Before we present the details, however, these five types of cash flows will be described briefly in the following section.

4.3.1 The Five Types of Cash Flows

Whenever we identify patterns in cash flow transactions, we may use these patterns to develop concise expressions for computing either the present or future worth of the series. For this purpose, we will classify cash flow transactions into five categories: (1)

Single cash flow, (2) uniform series, (3) linear gradient series, (4) geometric gradient series, and (5) irregular series. To simplify the description of various interest formulas, we will use the following notation:

1. **Single Cash Flow:** The simplest case involves the equivalence of a single present amount and its future worth. Thus, the single-cash flow formulas deal with only two amounts: A single present amount, P, and its future worth, F, (Figure 4.8a). You have already seen the derivation of one formula for this situation in Section 4.1.3, which gave us Eq. (4.3):

$$F = P(1 + i)^N.$$

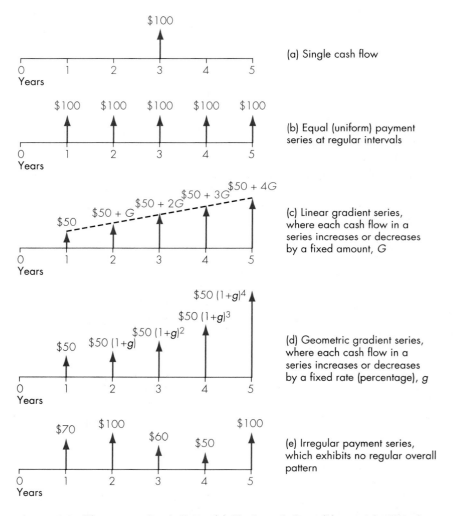

Figure 4.8 Five types of cash flows: (a) Single cash flow, (b) equal (uniform) payment series, (c) linear gradient series, (d) geometric gradient series, and (e) irregular payment series

2. **Equal (Uniform) Series:** Probably the most familiar category includes transactions arranged as a series of equal cash flows at regular intervals, known as an **equal-payment series** (or **uniform series**) (Fig. 4.8b). For example, this describes the cash flows of the common installment loan contract, which arranges repayment of a loan in equal periodic installments. The equal-cash flow formulas deal with the equivalence relations of P, F, and A (the constant amount of the cash flows in the series).

3. **Linear Gradient Series:** While many transactions involve series of cash flows, the amounts are not always uniform; they may, however, vary in some regular way. One common pattern of variation occurs when each cash flow in a series increases (or decreases) by a fixed amount (Fig. 4.8c). A 5-year loan repayment plan might specify, for example, a series of annual payments that increase by $500 each year. We call this type of cash flow pattern, a **linear gradient series** because its cash flow diagram produces an ascending (or descending) straight line, as you will see in Section 4.3.5. In addition to P, F, and A, the formulas used in such problems involve a *constant amount*, G, of the change in each cash flow.

4. **Geometric Gradient Series:** Another kind of gradient series is formed when the series in cash flow is determined, not by some fixed amount like $500, but by some fixed *rate*, expressed as a percentage. For example in a 5-year financial plan for a project, the cost of a particular raw material might be budgeted to increase at a rate of 4% per year. The curving gradient in the diagram of such a series suggests its name which is a **geometric gradient series** (Fig. 4.8d). In the formulas dealing with such series, the rate of change is represented by a lowercase g.

5. **Irregular Series:** Finally, a series of cash flows may be irregular, in that it does not exhibit a regular overall pattern. Even in such a series, however, one or more of the patterns already identified may appear over segments of time in the total length of the series. The cash flows may be equal, for example, for five consecutive periods in a ten-period series. When such patterns appear, the formulas for dealing with them may be applied and their results included in calculating an equivalent value for the entire series.

4.3.2   Single Cash Flow Formulas

We begin our coverage of interest formulas by considering the simplest of cash flows: Single cash flows.

Compound Amount Factor

Given a present sum, P, invested for N interest periods at interest rate, i, what sum will have accumulated at the end of the N periods? You probably noticed right away that this description matches the case we first encountered in describing compound interest. To solve for F (the future sum) we use Eq. (4.3):

$$F = P(1 + i)^N$$

Because of its origin in the compound interest calculation, the factor $(1 + i)^N$ is known as the **compound amount factor**. Like the concept of equivalence, this factor is one of the foundations of engineering economic analysis. Given this factor, all the other important interest formulas can be derived.

This process of finding F is often called the **compounding process**. The cash flow transaction is illustrated in Figure 4.9. (Note the time scale convention. The first period begins at $n = 0$ and ends at $n = 1$.) If a calculator is handy, it is easy enough to calculate $(1 + i)^N$ directly.

Interest Tables

Interest formulas such as the one developed in Eq. (4.3), $F = P(1 + i)^N$, allow us to substitute known values from a particular situation into the equation and to solve for the unknown. Before the hand calculator was developed, solving these equations was very tedious. With a large value of N, for example, one might need to solve an equation such as $F = \$20,000(1 + 0.12)^{15}$. More complex formulas required even more involved calculations. To simplify the process, tables of compound interest factors were developed, and these allow us to find the appropriate factor for a given interest rate and the number of interest periods. Even with hand calculators, it is still often convenient to use these tables, and they are included in this text in Appendix D. Take some time now to become familiar with their arrangement and, if you can, locate the compound interest factor for the example just presented, in which we know P, and to find

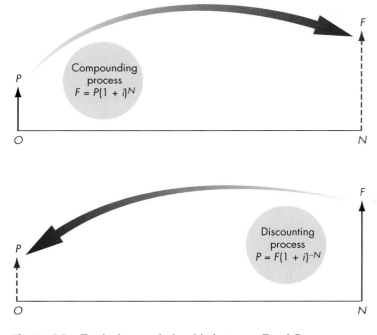

Figure 4.9   Equivalence relationship between F and P

F we need to know the factor by which to multiply \$20,000 when the interest rate, i, is 12%, and the number of periods is 15:

$$F = \$20,000 \underbrace{(1 + 0.12)^{15}}_{5.4736} = \$109,472.$$

Factor Notation

As we continue to develop interest formulas in the rest of this chapter, we will express the resulting compound interest factors in a conventional notation that can be substituted in a formula to indicate precisely which table factor to use in solving an equation. In the preceding example, for instance, the formula derived as Eq. (4.3) is $F = P(1 + i)^N$. In ordinary language, this tells us that to determine what future amount, F, is equivalent to a present amount, P, we need to multiply P by a factor expressed as 1 plus the interest rate, raised to the power given by the number of interest periods. To specify how the interest tables are to be used, we may also express that factor in a functional notation as follows: $(F/P, i, N)$, which is read as "Find F, Given P, i, and N." This is known as the **single-payment compound amount factor**. When we incorporate the table factor in the formula, it is expressed as follows:

$$F = P(1+i)^N = P(F/P, i, N).$$

Thus, in the preceding example, where we had $F = \$20,000(1.12)^{15}$, we can write $F = \$20,000(F/P, 12\%, 15)$. The table factor tells us to use the 12% interest table and find the factor in the F/P column for $N = 15$. Because using the interest tables is often the easiest way to solve an equation, this factor notation is included for each of the formulas derived in the following sections.

Example 4.7 Single Amounts: Find *F*, Given *i*, *N*, *P*

If you had \$2,000 now and invested it at 10%, how much would it be worth in 8 years (Figure 4.10)?

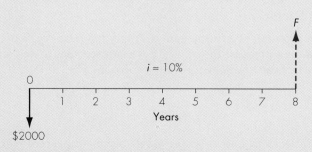

Figure 4.10   Cash flow diagram from the investor's point of view (Example 4.7)

Solution

Given: $P = \$2,000$, $i = 10\%$ per year, and $N = 8$ years

Find: F

We can solve this problem in any of three ways:

1. Using a calculator: You can simply use a calculator to evaluate the $(1 + i)^N$ term. (Financial calculators are pre-programmed to solve most future-value problems.)

$$F = \$2,000(1 + 0.10)^8$$

$$= \$4,287.18.$$

2. Using compound interest tables: The interest tables can be used to locate the compound amount factor for $i = 10\%$ and $N = 8$. The number you get can be substituted into the equation. Compound interest tables are included as Appendix D of this book.

$$F = \$2,000\,(F/P, 10\%, 8) = \$2,000(2.1436) = \$4,287.20.$$

This is essentially identical to the value obtained by the direct evaluation of the single cash flow compound amount factor. This slight deviation is due to rounding errors.

3. Using a computer: Many financial software programs for solving compound interest problems are available for use with personal computers. As summarized in Appendix C, many spreadsheet programs such as Excel also provide financial functions to evaluate various interest formulas. With Excel, the future worth calculation looks like the following:

$$= FV\,(10\%, 8, 2,000, .0)$$

Present Worth Factor

Finding the present worth of a future sum is simply the reverse of compounding and is known as **discounting process**. In Eq. (4.3), we can see that if we were to find a present sum, P, given a future sum, F, we simply solve for P:

$$P = F\left[\frac{1}{(1 + i)^N}\right] = F(P/F, i, N). \qquad (4.7)$$

The factor $1/(1 + i)^N$ is known as the **single payment present worth factor**, and is designated $(P/F, i, N)$. Tables have been constructed for P/F factors and for various values of i and N. The interest rate i and the P/F factor are also referred to as the **discount rate** and **discounting factor**, respectively.

Example 4.8 Single Amounts: Find *P*, Given *F*, *i*, *N*

Suppose that $1,000 is to be received in 5 years. At an annual interest rate of 12%, what is the present worth of this amount (Figure 4.11)?

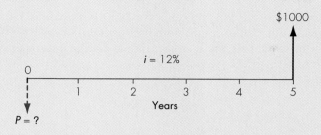

Figure 4.11   Cash flow diagram (Example 4.8)

Solution:

Given: F = $1,000, i = 12% per year, and N = 5 years
Find: P

$$P = \$1,000(1 + 0.12)^{-5} = \$1,000(0.5674) = \$567.40.$$

Using a calculator may be the best way to make this simple calculation. To have $1,000 in your savings account at the end of 5 years, you must deposit $567.40 now.

We can also use the interest tables to find that:

$$P = \$1,000 \overbrace{(P/F, 12\%, 5)}^{(0.5674)} = \$567.40$$

Again, you could use a financial calculator or computer to find the present worth. With Excel, the present value calculation looks like the following:

$$= \text{PV}(12\%, 5, 1,000,,0).$$

Solving for Time and Interest Rates

At this point, you should realize that the compounding and discounting processes are reciprocals of one another, and that we have been dealing with one equation in two forms.

$$\text{Future value form: } F = P(1 + i)^{N}$$

$$\text{Present value form: } P = F(1 + i)^{-N}$$

There are four variables in these equations—*P, F, N*— and *i*. If you know the values of any three, you can find the value of the fourth. Thus far, we have always given you the interest rate *(i)* and the number of years *(N)* plus either the *P* or the *F*. In many situations, though, you will need to solve for *i* or *N,* as we discuss below.

Example 4.9 Solving for *i*

Suppose you buy a share for $10 and sell it for $20, your profit is $10. If that happens within a year, your rate of return is an impressive 100% ($10/$10 = 1). If that takes 5 years, what would be the average annual rate of return on your investment? (See Figure 4.12)

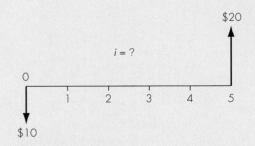

Figure 4.12   Cash flow diagram (Example 4.9)

Solution:

Given: $P = \$10$, $F = \$20$, $N = 5$

Find: *i*

Here we know *P*, *F,* and *N,* but we do not know *i,* the interest rate you will earn on your investment. This type of rate of return is a lot easier to calculate as you make only a one-time lump-sum investment. Problems such as this are solved as follows:

$$F = P(1 + i)^N$$

$$\$20 = \$10(1 + i)^5. \text{ Solve for } i.$$

- **Method 1**   Go through a trial-and-error process in which you insert different values of *i* into the equation until you find a value that "works" in the sense that the right-hand-side of the equation equals $20. The solution value is $i = 14.87\%$. The trial-and-error procedure is extremely tedious and inefficient for most problems, so it is not widely practiced in the real world.

- **Method 2**   You can solve the problem by using the interest tables in Appendix E. Now look across the $N = 5$ row under $(F/P, i, 5)$ column until you can locate the value of 2.

$$\$20 = \$10(1 + i)^5$$

$$2 = (1 + i)^5 = (F/P,i,5)$$

This value is closely located at 15% interest table with $(F/P, 15\%, 5) = 2.0114$, so the interest rate at which \$10 grows to \$20 over 5 years is very close to 15%. This procedure will be very tedious for fractional interest rates or where N is not a whole number, as you may have to approximate the solution by linear interpolation.

- **Method 3** Most practical approach is to use either a financial calculator or an electronic spreadsheet such as Excel. A financial function such as RATE(N,0,P,F) allows us to calculate an unknown interest rate. The precise command statement would be as follows:

$$= \textbf{RATE}(5, 0, -10, 20) = 14.87\%$$

Note that we enter the present value (P) as a negative number indicating a cash outflow in Excel format.

Example 4.10 Single Amounts: Find *N*, Given *P*, *F*, *i*

You have just purchased 100 shares of General Electric stock at \$60 per share. You will sell the stock when its market price has doubled. If you expect the stock price to increase 20% per year, how long do you expect to wait before selling the stock (Figure 4.13)?

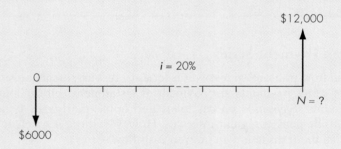

Figure 4.13 Cash flow diagram (Example 4.10)

Solution

Given: $P = \$6,000$, $F = \$12,000$, $i = 20\%$ per year

Find: N (years)

Using the single-payment compound amount factor, we write

$$F = P(1 + i)^N = P(F/P, i, N)$$

$$\$12,000 = \$6,000(1 + 0.20)^N = \$6,000(F/P, 20\%, N)$$

$$2 = (1.20)^N = (F/P, 20\%, N).$$

Again, we could use a calculator or a computer spreadsheet program to find N.

1. Using a calculator: Solving for N gives

$$\log 2 = N \log 1.20$$

$$N = \frac{\log 2}{\log 1.20}$$

$$= 3.80 \approx 4 \text{ years.}$$

2. Using a spreadsheet program: Within Excel, the financial function $NPER(i,0,P,F)$ computes the number of compounding periods it will take an investment (P) to grow to a future value (F), earning a fixed interest rate (i) per compounding period. In our example, the Excel command would look like this:

$$= NPER(20\%, 0, -6{,}000, 12{,}000)$$

$$= 3.801784.$$

Comments: A very handy rule of thumb, called the Rule of 72, can determine approximately how long it will take for a sum of money to "double." The rule states that, to find the time it takes for the present sum of money to grow by a factor of 2, we divide 72 by the interest rate. For our example, the interest rate is 20%. Therefore, the Rule of 72 indicates $72/20 = 3.60$ or roughly 4 years for a sum to double. This is, in fact, relatively close to our exact solution.

4.3.3 Uneven Payment Series

A common cash flow transaction involves a series of disbursements or receipts. Familiar examples of series payments are payment of installments on car loans and home mortgage payments. Payments on car loans and home mortgages typically involve identical sums to be paid at regular intervals. However, there is no clear pattern over the series, we call the transaction an uneven cash-flow series.

We can find the present worth of any uneven stream of payments by calculating the present value of each individual payment and summing the results. Once the present worth is found, we can make other equivalence calculations, e.g., future worth can be calculated by using the interest factors developed in the previous section.

Example 4.11 Present Value of an Uneven Series by Decomposition into Single Payments

Wilson Technology, a growing machine shop, wishes to set aside money now to invest over the next 4 years in automating its customer service department. The company can earn 10% on a lump sum deposited now, and it wishes to withdraw the money in the following increments:

Year 1: $25,000 to purchase a computer and database software designed for customer service use;

Year 2: $3,000 to purchase additional hardware to accommodate anticipated growth in use of the system;

Year 3:   No expenses; and

Year 4: $5,000 to purchase software upgrades.

How much money must be deposited now to cover the anticipated payments over the next 4 years?

Discussion: This problem is equivalent to asking what value of P would make you indifferent in your choice between P dollars today and the future expense stream of ($25,000, $3,000, $0, $5,000). One way to deal with an uneven series of cash flows is to calculate the equivalent present value of each single cash flow and to sum the present values to find P. In other words, the cash flow is broken into three parts as shown in Figure 4.14.

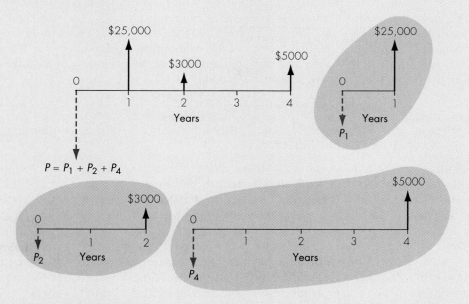

Figure 4.14   Decomposition of uneven cash flow series (Example 4.11)

Solution

Given: Uneven cash flow in Fig. 4.14, $i = 10\%$ per year

Find: P

$$P = \$25,000(P/F, 10\%, 1) + \$3,000(P/F, 10\%, 2) + \$5,000(P/F, 10\%, 4)$$

$$= \$28,622.$$

Comments: To see if indeed $28,622 is sufficient amount, let's calculate the balance at the end of each year. If you deposit $28,622 now, it will grow to (1.10)($28,622) or $31,484 at the end of year 1. From this balance, you pay out $25,000. The remaining balance, $6,484, will again grow to (1.10)($6,484) or $7,132 at the end of year 2. Now you make the second payment ($3,000) out of this balance, which will leave you with only $4,132 at the end of year 2. Since no payment occurs in year 3, the balance will grow to $(1.10)^2($4,132) or $5,000 at the end of year 4. The final withdrawal in the amount of $5000 will deplete the balance completely.

Example 4.12 Calculating the Actual Worth of a Long-Term Contract

On December 11, 2000, The Texas Rangers and superstar baseball shortstop Alex Rodriguez agreed to a 10-year deal worth $252 million, including a signing bonus of $10 million, making Rodriguez, the highest paid player in team sports history. The agreement calls for annual salaries of $21 million from 2001 to 2004, $25 million from 2005 to 2006, and $27 million from 2007 to 2010. The $10 million signing bonus is prorated over the first 5 years of the contract, $2 million each year from 2001 to 2005. With the salary and signing bonus paid at the beginning of each season, the net annual payment schedule looks like the following:

Beginning of Season	Contract Salary	Prorated Signing Bonus	Total Annual Payment
2001	$21,000,000	$2,000,000	$23,000,000
2002	21,000,000	2,000,000	23,000,000
2003	21,000,000	2,000,000	23,000,000
2004	21,000,000	2,000,000	23,000,000
2005	25,000,000	2,000,000	27,000,000
2006	25,000,000		25,000,000
2007	27,000,000		27,000,000
2008	27,000,000		27,000,000
2009	27,000,000		27,000,000
2010	27,000,000		27,000,000

(a) How much is Alex's contract actually worth at the time of signing? Assume that Alex's interest rate is 6% per year.

(b) For the signing bonus portion, suppose that the Texas Rangers allow Alex to take either the prorated payment option as described above or a lump sum payment option in the amount of $9 million at the time of contract. Should Alex take the lump sum option instead of the prorated one?

Solution

Given: Payment series given in Figure 4.15, $i = 6\%$ per year
Find: P

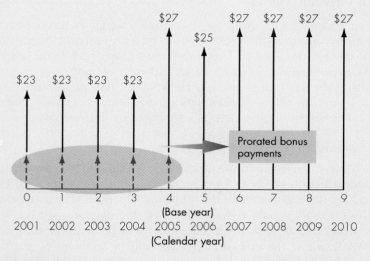

Figure 4.15 Alex Rodriguez's $252 million Texas Rangers contract (Example 4.12)

(a) Actual worth of the contract at the time of signing:

$$P_{\text{contract}} = \$23,000,000 + \$23,000,000(P/F, 6\%, 1) + \$23,000,000(P/F, 6\%, 2)$$
$$+ \$23,000,000(P/F, 6\%, 3) + \ldots + \$27,000,000(P/F, 6\%, 9)$$
$$= \$215,755,986$$

(b) Choice between the prorated payment option and the lump sum payment: The equivalent present worth of the prorated payment option is

$$P_{\text{bonus}} = \$2,000,000 + \$2,000,000(P/F, 6\%, 1) + \$2,000,000(P/F, 6\%, 2)$$
$$+ \$2,000,000(P/F, 6\%, 3) + \$2,000,000(P/F, 6\%, 4)$$
$$= \$8,930,211$$

which is smaller than $9,000,000. Therefore, Alex would be better off taking the lump sum option if, and only if, his money could be invested at 6% or higher.

Comments: Note that the actual contract is worth less than $252 million as published. This "brute force" approach of breaking cash flows into single amounts will always work, but it is slow and subject to error because of the many factors that must be included in the calculation. We develop more efficient methods in the text sections for cash flows with certain patterns.

4.3.4   Equal Payment Series

As we learned in Example 4.12, the present worth of a stream of future cash flows can always be found by summing the present worth of each individual cash flow. However, if cash flow regularities are present within the stream (such as we just saw in the prorated bonus payment series in Example 4.12) the use of short-cuts, such as finding the present worth of a uniform series may be possible. We often encounter transactions in which a uniform series of payments exists. Rental payments, bond interest payments, and commercial installment plans are based on uniform payment series.

Compound Amount Factor—Find *F*, given *A*, *i*, *N*

Suppose we are interested in the future amount, *F*, of a fund to which we contribute *A* dollars each period and on which we earn interest at a rate of *i* per period. The contributions are made at the end of each of the following *N* periods. These transactions are graphically illustrated in Figure 4.16. Looking at this diagram, we see that, if an amount, *A*, is invested at the end of each period, for *N* periods, the total amount, *F*, that can be withdrawn at the end of *N* periods will be the sum of the compound amounts of the individual deposits.

As shown in Figure 4.17, the *A* dollars we put into the fund at the end of the first period will be worth $A(1 + i)^{N-1}$ at the end of *N* periods. The *A* dollars we put into the fund at the end of the second period will be worth $A(1 + i)^{N-2}$, and so forth. Finally, the last *A* dollars that we contribute at the end of the *N*th period will be worth exactly *A* dollars at that time. This means there exists a series in the form:

$$F = A(1 + i)^{N-1} + A(1 + i)^{N-2} + \ldots + A(1 + i) + A,$$

or expressed alternatively,

$$F = A + A(1 + i) + A(1 + i)^2 + \ldots + A(1 + i)^{N-1}. \tag{4.8}$$

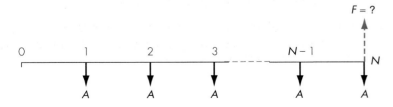

Figure 4.16   Cash flow diagram of the relationship between *A* and *F*

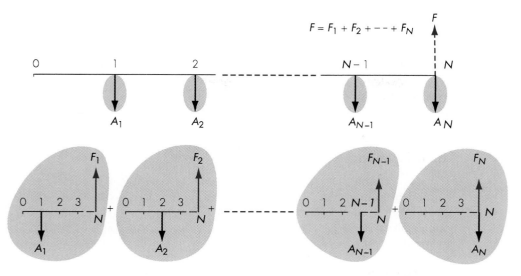

Figure 4.17   The future worth of a cash flow series obtained by summing the future worth figures of each individual flow

Multiplying Eq. (4.8) by $(1 + i)$ results in

$$(1 + i)F = A(1 + i) + A(1 + i)^2 + \ldots + A(1 + i)^N. \qquad (4.9)$$

Subtracting Eq. (4.8) from Eq. (4.9) to eliminate common terms gives us:

$$F(1 + i) - F = -A + A(1 + i)^N.$$

Solving for F yields

$$F = A\left[\frac{(1 + i)^N - 1}{i}\right] = A(F/A, i, N). \qquad (4.10)$$

The bracketed term in Eq. (4.10) is called the **equal payment series compound amount factor**, or the **uniform series compound amount factor**; its factor notation is $(F/A, i, N)$. This interest factor has been calculated for various combinations of i and N in the interest tables.

Example 4.13 Uniform Series: Find *F*, Given *i*, *A*, *N*

Suppose you make an annual contribution of $3,000 to your savings account at the end of each year for 10 years. If your savings account earns 7% interest annually, how much can be withdrawn at the end of 10 years (Figure 4.18)?

Solution

Given: $A = \$3,000$, $N = 10$ years, and $i = 7\%$ per year

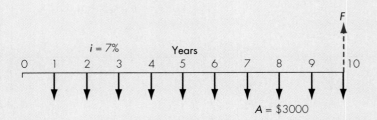

Figure 4.18   Cash flow diagram (Example 4.13)

Find: F

$$F = \$3000(F/A, 7\%, 10)$$
$$= \$3000(13.8164)$$
$$= \$41,449.20.$$

To obtain the future value of the annuity on Excel, we may use the following financial command:

$$= \text{FV}(7\%,10,3000,,0)$$

Example 4.14 Handling Time Shifts in a Uniform Series

In Example 4.13, the first deposit of the 10-deposit series was made at the end of period 1 and the remaining nine deposits were made at the end of each following period. Suppose that all deposits were made at the *beginning* of each period instead. How would you compute the balance at the end of period 10?

Solution

Given: Cash flow as shown in Figure 4.19, $i = 7\%$ per year

Find: F_{10}

Compare Fig. 4.19 to Fig. 4.18: Each payment has been shifted to 1 year earlier; thus each payment would be compounded for 1 extra year. Note that with the end-of-year deposit, the ending balance (F) was $41,449.20. With the beginning-of-year deposit, the same balance accumulates by the end of period 9. This balance can earn interest for one additional year. Therefore, we can easily calculate the resulting balance by

$$F_{10} = \$41,449.20(1.07) = \$44,350.64.$$

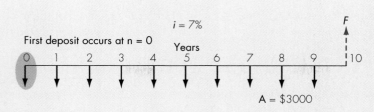

Figure 4.19   Cash flow diagram (Example 4.14)

Annuity due can be easily evaluated using the following financial command available on Excel:

$$= \mathbf{FV}(7\%,10,3000,,1)$$

Comments: Another way to determine the ending balance is to compare the two cash flow patterns. By adding the $3,000 deposit at period 0 to the original cash flow and subtracting the $3,000 deposit at the end of period 10, we obtain the second cash flow. Therefore, the ending balance can be found by making adjustment to the $41,449.20:

$$F_{10} = \$41,449.20 + 3,000(F/P, 7\%, 10) - 3,000 = \$44,350.64.$$

Sinking-Fund Factor—Find A, given F, i, N

If we solve Eq. (4.10) for A, we obtain

$$A = F\left[\frac{i}{(1 + i)^N - 1}\right] = F(A/F, i, N). \tag{4.11}$$

The term within the brackets is called the **equal payment series sinking-fund factor**, or **sinking-fund factor**, and is referred to by the notation (A/F, i, N). A sinking fund is an interest-bearing account into which a fixed sum is deposited each interest period; it is commonly established for the purpose of replacing fixed assets.

Example 4.15 Combination of a Uniform Series and a Single Present and Future Amount

To help you reach a $5,000 goal 5 years from now, your father offers to give you $500 now. You plan to get a part-time job and make five additional deposits at the end of each year. (The first deposit is made at the end of the first year.) If all your money is deposited in a bank that pays 7% interest, how large must your annual deposit be?

Discussion: If your father reneges on his offer, the calculation of the required annual deposit is easy because your five deposits fit the standard end-of-period pattern for a uniform series. All you need to evaluate is

$$A = \$5,000(A/F, 7\%, 5) = \$5,000(0.1739) = \$869.50.$$

If you do receive the $500 contribution from your father at $n = 0$, you may divide the deposit series into two parts: One contributed by your father at $n = 0$ and five equal annual deposit series contributed by yourself. Then you can use the F/P factor to find how much your father's contribution will be worth at the end of year 5 at a 7% interest rate. Let's call this amount F_c. The future value of your five annual deposits must then make up the difference, $\$5,000 - F_c$.

Solution

Given: cash flow as shown in Figure 4.20, $i = 7\%$ per year, and $N = 5$ years

Find: A

$$\begin{aligned}
A &= (\$5,000 - F_C)(A/F, 7\%, 5) \\
&= [\$5,000 - \$500(F/P, 7\%, 5)](A/F, 7\%, 5) \\
&= [\$5,000 - \$500(1.4026)](0.1739) \\
&= \$747.55.
\end{aligned}$$

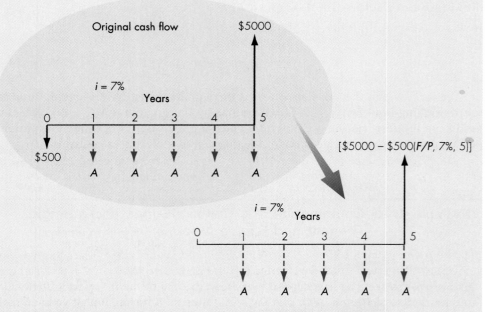

Figure 4.20   Equivalent cash flow diagram (Example 4.15)

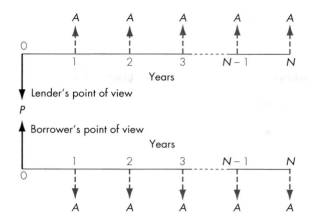

Figure 4.21   Cash flow diagram of the relationship between P and A, where P is the amount borrowed and A is a series of payments of a fixed amount for N periods

Capital Recovery Factor (Annuity Factor)—Find A, given P, i, N

We can determine the amount of a periodic payment, A, if we know P, i, and N. Figure 4.21 illustrates this situation. To relate P to A, recall the relationship between P and F in Eq. (4.3), $F = P(1 + i)^N$. By replacing F in Eq. (4.11) by $P(1 + i)^N$ we get

$$A = P\,(1 + i)^N \left[\frac{i}{(1 + i)^N - 1} \right],$$

or

$$A = P \left[\frac{i(1 + i)^N}{(1 + i)^N - 1} \right] = P\,(A/P, i, N). \tag{4.12}$$

Now we have an equation for determining the value of the series of end-of-period payments, A, when the present sum, P, is known. The portion within the brackets is called the **equal-payment series capital recovery factor**, or simply **capital recovery factor**, which is designated $(A/P, i, N)$. In finance, this A/P factor is referred to as the **annuity factor**. The annuity factor indicates a series of payments of a fixed, or constant, amount for a specified number of periods.

Example 4.16 Uniform Series: Find *A*, Given *P, i, N*

BioGen Company, a small biotechnology firm, has borrowed $250,000 to purchase laboratory equipment for gene splicing. The loan carries an interest rate of 8% per year and is to be repaid in equal installments over the next 6 years. Compute the amount of this annual installment (Figure 4.22).

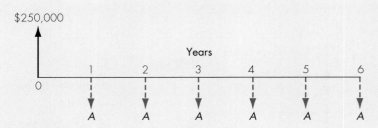

Figure 4.22   A loan cash flow diagram from BioGen's point of view (Example 4.16)

Solution

Given: $P = \$250,000$, $i = 8\%$ per year, $N = 6$ years
Find: A

$$A = \$250,000(A/P, 8\%, 6)$$
$$= \$250,000(0.2163)$$
$$= \$54,075.$$

Excel solution using annuity function commands:

$$= \mathbf{PMT}(i,N,P)$$
$$= \mathbf{PMT}(8\%,6,250000)$$
$$= \$54,075$$

Example 4.17 Deferred loan repayment

In Example 4.16, suppose that BioGen wants to negotiate with the bank to defer the first loan repayment until the end of year 2 (but still desires to make six equal installments at 8% interest). If the bank wishes to earn the same profit as in Example 4.16, what should be the annual installment (Figure 4.23)?

Solution

Given: $P = \$250,000$, $i = 8\%$ per year, $N = 6$ years, but the first payment occurs at the end of year 2
Find: A

By deferring 1 year, the bank will add the interest during the first year to the principal. In other words, we need to find the equivalent worth of $250,000 at the end of year 1, P':

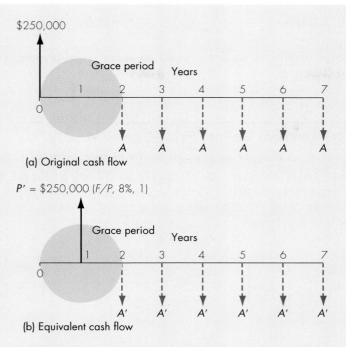

Figure 4.23   A deferred loan cash flow diagram from Bio-Gen's point of view (Example 4.17)

$$P' = \$250,000(F/P, 8\%, 1)$$
$$= \$270,000.$$

In fact, BioGen is borrowing $270,000 for 6 years. To retire the loan with six equal installments, the deferred equal annual payment, P', will be

$$A' = \$270,000(A/P, 8\%, 6)$$
$$= \$58,401.$$

By deferring the first payment for one year, BioGen needs to make additional $4,326 payments in each year.

Present Worth Factor—Find P, given A, i, N

What would you have to invest now in order to withdraw A dollars at the end of each of the next N periods? We face just the opposite of the equal payment capital recovery factor situation—A is known, but P has to be determined. With the capital recovery factor given in Eq. (4.12), solving for P gives us

$$P = A\left[\frac{(1 + i)^N - 1}{i(1 + i)^N}\right] = A\,(P/A, i, N). \tag{4.13}$$

The bracketed term is referred to as the **equal payment series present worth factor** and is designated $(P/A, i, N)$.

Example 4.18 Uniform Series: Find *P*, given *A*, *i*, *N*

Let us revisit the lottery problem (Ms. Rosalind Setchfield) introduced in the chapter opening. Recall that Mrs. Setchfield gave up $32,639 a year for 9 years to receive a cash lump sum of $140,000. If she could invest her money at 8% interest, did she make a right decision? What should have been the fair amount to give up her nine future lottery receipts?

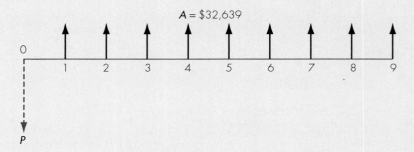

Figure 4.24   Cash flow diagram (Example 4.18)

Solution

Given: $i = 8\%$ per year, $A = \$32,639$, and $N = 9$ years
Find: P

- **Tabular Solution:**

$$P = \$32,639(P/A, 8\%, 9) = \$32,639(6.4269)$$
$$= \$203,893$$

- **Excel solution:**

$$= \text{PV}(8\%,9,32639,,0) = \$203,893$$

Comments: Clearly we can tell her that giving up $32,639 a year for 9 years to receive $140,000 today is a losing proposition if she can earn only 8% return on her investment. At this point, we may be interested in knowing at what rate of return her deal (receiving $140,000) would make sense. Since we know $P = \$140,000$, $N = 9$, $A = \$32,639$, we solve for i.

If you know the cash flows and the *PV* (or *FV*) of a cash flow stream, you can determine the interest rate. In our case, you are looking for the interest rate that caused the *P/A* factor to equal $(P/A,i,9) = (\$140,000/\$32,639) = 4.2893$. Since we are dealing with an annuity, we could proceed as follows:

- With a financial calculator, enter $N = 9$, $PV = 140,000$, $PMT = -32,639$, and then press the *i* key to find $i = 18.0955\%$
- To use the tables, first recognize that $\$140,000 = \$32,639(P/A,i,9)$ or $(P/A,i,9) = 4.2893$. Look up 4.2893 or a close value in Appendix E. In the *P/A* column with $N = 9$ in Table E. 23, you will find that $(P/A,18\%,9) = 4.3030$, indicating that the interest rate should be very close to 18%. If the factor did not appear in the table, this would indicate that the interest rate was not a whole number. In that case, you could not use this procedure to find the exact rate. In practice, this is not a problem, because in business people use financial calculators to find interest rates.
- To use the Excel's financial command, you simply evaluate the following command to solve the unknown interest rate problem for an annuity.

$$= \textbf{RATE}(N,A,P,F,type,guess)$$

$$= \textbf{RATE}(9,32639,140000,0,0,10\%)$$

$$= 18.0955\%$$

For Mrs. Setchfield, it is not likely she will find a financial investment that provides this high rate of return. Therefore, she should not accept the deal at the outset.

Example 4.19 Composite Series that Requires Both (*P/F, i, N*) and (*P/A, i, N*) Factors

Reconsider the 2000 Publishers' Clearing House, Ten Million Dollar Super Prize payment options in Figure 4.7. Which option would you prefer with an interest rate of 8%?

- Option A: $1,000.000 now,$200,000 yearly, plus a $3,400,000 final payment at the end of year 29.
- Option B: $500,000 now, $250,000 a year thereafter, plus $2,500,000 final payment at the end of year 29.

Note that 30 payments are required in either option.

Solution

Given: Cash flows given as in Fig. 4.7, $i = 8\%$ per year

Find: *P* for each option

Both payment options consist of a uniform series and two lump-sum payments during the first and the last year. Therefore, we can compute the equivalent present worth in two steps: (1) Find the equivalent present worth of the two lump-sum payment using $(P/F, i, n)$ factor, and (2) find the equivalent present worth of the uniform series by using the $(P/A, i, n)$ factor. Note also that the yearly payment begins at the beginning of each year. Then, the equivalent worth calculation would look like the following:

$$P_{\text{Option A}} = \$1,000,000 + \$200,000(P/A, 8\%, 28)$$
$$+ \$3,400,000(P/F, 8\%, 29)$$
$$= \$3,575,129.$$
$$P_{\text{Option B}} = \$500,000 + \$250,000(P/A, 8\%, 28)$$
$$+ \$2,500,000(P/F, 8\%, 29)$$
$$= \$3,531,089.$$

Comments: The difference between the two options is \$44,040, favoring option A. The result indicates that having an additional \$500,000 during the first year and \$900,000 during the last year will allow more wealth accumulation than receiving an additional \$50,000 each year for 28 years. As you can verify, the result will be the reverse whenever your interest rate is lower than 6.18%

4.3.5 Linear Gradient Series

Engineers frequently encounter situations involving periodic payments that increase or decrease by a constant amount (G) from period to period. This situation occurs often enough to warrant the use of special equivalence factors that relate the arithmetic gradient to other cash flows. Figure 4.25 illustrates a **strict gradient series**, $A_n = (n-1)G$. Note that the origin of the strict gradient series is at the end of the first period with a zero value. The gradient G can be either positive or negative. If $G > 0$, the

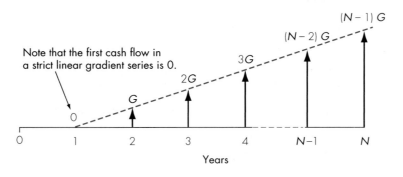

Figure 4.25 Cash flow diagram of a strict gradient series

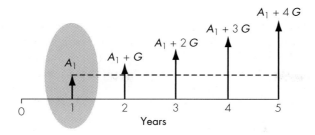

Figure 4.26   Cash flow diagram for a typical problem involving linear gradient series. Note that a nonzero cash flow occurs at the end of first period

series is referred to as an *increasing* gradient series. If $G < 0$, it is a *decreasing* gradient series.

Unfortunately, the strict form of the increasing or decreasing gradient series does not correspond with the form that most engineering economic problems take. A typical problem involving a linear gradient includes an initial payment during period 1 that increases by G during some number of interest periods, a situation illustrated in Figure 4.26. This contrasts with the strict form illustrated in Figure 4.26 in which no payment is made during period 1, and the gradient is added to the previous payment beginning in period 2.

Gradient Series as Composite Series

In order to utilize the strict gradient series to solve typical problems we must view cash flows as shown in Figure 4.27 as a **composite series**, or a set of two cash flows, each corresponding to a form that we can recognize and easily solve. Figure 4.28 illustrates that the form in which we find a typical cash flow can be separated into two components: A uniform series of N payments of amount, A_1, and a gradient series of increments of constant amount, G. The need to view cash flows, which involve linear gradient series as composites of two series, is very important for the solution of problems, as we shall now see.

Present Worth Factor—Linear Gradient: Find P, given G, N, i

How much would you have to deposit now to withdraw the gradient amounts specified in Figure 4.25. To find an expression for the present amount P, we apply the single-payment present worth factor to each term of the series and obtain

$$P = 0 + G/(1 + i)^2 + 2G/(1 + i)^3 + \ldots + (N - 1)G/(1 + i)^N,$$

or

$$P = \sum_{n=1}^{N} (n - 1)G(1 + i)^{-n}. \qquad (4.14)$$

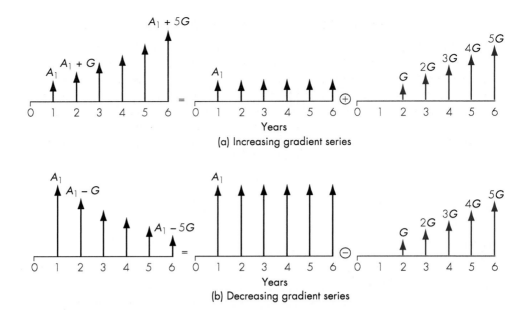

Figure 4.27   Two types of linear gradient series as composites of a uniform series of N payments of A_1 and the gradient series of increments of constant amount G

Letting $G = a$ and $1/(1 + i) = x$ yields

$$P = 0 + ax^2 + 2ax^3 + \ldots + (N - 1)ax^N$$

$$= ax \left[0 + x + 2x^2 + \ldots + (N - 1)x^{N-1}\right]. \tag{4.15}$$

Since an arithmetic-geometric series $\{0, x, 2x^2, \ldots, (N-1)x^{N-1}\}$ has the finite sum of

$$0 + x + 2x^2 + \ldots + (N - 1)x^{N-1} = x \left[\frac{1 - Nx^{N-1} + (N - 1)x^N}{(1 - x)^2}\right].$$

we can rewrite Eq. (4.15) as

$$P = ax^2 \left[\frac{1 - Nx^{N-1} + (N - 1)x^N}{(1 - x)^2}\right]. \tag{4.16}$$

Replacing the original values for A and x, we obtain

$$P = G \left[\frac{(1 + i)^N - iN - 1}{i^2(1 + i)^N}\right] = G(P/G, i, N). \tag{4.17}$$

The resulting factor in brackets above is called the **gradient series present worth factor** and for which we use the notation $(P/G, i, N)$.

Example 4.20 Linear Gradient: Find P, Given A_1, G, i, N

A textile mill has just purchased a lift truck that has a useful life of 5 years. The engineer estimates that the maintenance costs for the truck during the first year will be $1,000. Maintenance costs are expected to increase as the truck ages at a rate of $250 per year over the remaining life. Assume that the maintenance costs occur at the end of each year. The firm wants to set up a maintenance account that earns 12% annual interest. All future maintenance expenses will be paid out of this account. How much does the firm have to deposit in the account now?

Solution

Given: $A_1 = \$1,000$, $G = \$250$, $i = 12\%$ per year, and $N = 5$ years

Find: P

This is equivalent to asking what is the equivalent present worth for this maintenance expenditure if 12% interest is used. The cash flow may be broken into its two components as shown in Figure 4.28.

The first component is an equal-payment series (A_1), and the second is linear gradient series (G).

$$P = P_1 + P_2$$
$$P = A_1(P/A, 12\%, 5) + G(P/G, 12\%, 5)$$
$$= \$1,000(3.6048) + \$250(6.397)$$
$$= \$5,204.$$

Note that the value of N in the gradient factor is 5, not 4. This occurs because, by definition of the series, the first gradient value begins at period 2.

Comments: As a check, we can compute the present worth of the cash flow by using the $(P/F, 12\%, n)$ factors:

The slight difference is caused by a rounding error.

Period (n)	Cash Flow	(P/F, 12%, n)	Present Worth
1	$1,000	0.8929	$892.90
2	1,250	0.7972	996.50
3	1,500	0.7118	1,067.70
4	1,750	0.6355	1,112.13
5	2,000	0.5674	1,134.80
			Total $5204.03

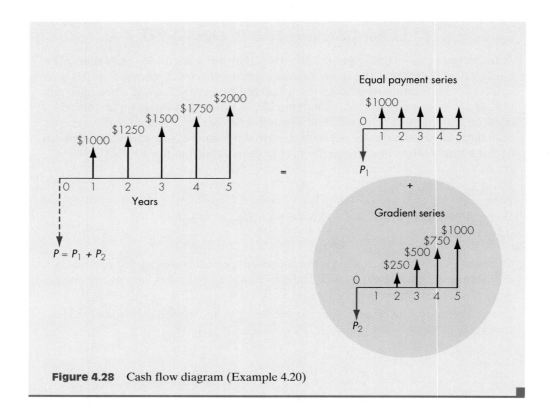

Figure 4.28   Cash flow diagram (Example 4.20)

Gradient-to-Equal-Payment Series Conversion Factor—Find A, given G, i, N

We can obtain an equal payment series equivalent to the gradient series, as depicted in Figure 4.29 by substituting Eq. (4.17) into Eq. (4.12) for P to obtain

$$A = G\left[\frac{(1+i)^N - iN - 1}{i[(1+i)^N - 1]}\right] = G(A/G, i, N), \tag{4.18}$$

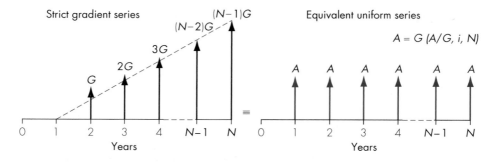

Figure 4.29   Converting a gradient series into an equivalent uniform series

where the resulting factor in brackets is referred to as the **gradient-to-equal payment series conversion factor** and is designated $(A/G, i, N)$.

Example 4.21 Linear gradient: Find *A*, Given *A*₁, *G*, *i*, *N*

John and Barbara have just opened two savings accounts at their credit union. The accounts earn 10% annual interest. John wants to deposit $1,000 in his account at the end of the first year and increase this amount by $300 for each of the following 5 years. Barbara wants to deposit an equal amount each year for next 6 years. What should be the size of the Barbara's annual deposit so that the two accounts would have equal balances at the end of 6 years (Figure 4.30)?

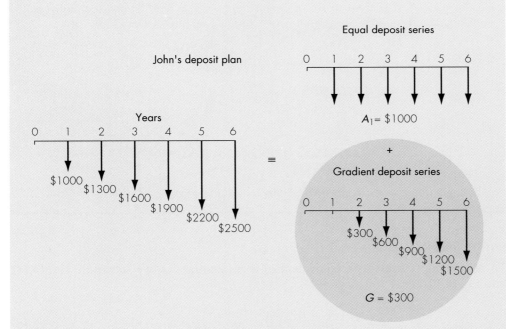

Figure 4.30   John's deposit series viewed as a combination of uniform and gradient series (Example 4.21)

Solution

Given: $A_1 = \$1,000$, $G = \$300$, $i = 10\%$, and $N = 6$

Find: A

Since we use the end-of-period convention unless otherwise stated, this series begins at the end of the first year, and the last contribution occurs at the end of the sixth year.

We can separate the constant portion of $1,000 from the series, leaving the gradient series of 0, 0, 300, 600, . . . , 1,500.

To find the equal payment series beginning at the end of year 1 and ending at year 6 that would have the same present worth as that of the gradient series, we may proceed as follows:

$$A = \$1,000 + \$300(A/G, 10\%, 6)$$

$$= \$1,000 + \$300(2.22236)$$

$$= \$1,667.08.$$

Barbara's annual contribution should be $1,667.08.

Comments: Alternatively, we can compute Barbara's annual deposit by first computing the equivalent present worth of John's deposits and then finding the equivalent uniform annual amount. The present worth of this combined series is

$$P = \$1,000(P/A, 10\%, 6) + \$300(P/G, 10\%, 6)$$

$$= \$1,000(4.3553) + \$300(9.6842)$$

$$= \$7,260.56.$$

The equivalent uniform deposit is

$$A = \$7,260.56(A/P, 10\%, 6) = \$1,667.02.$$

(The slight difference in cents is caused by a rounding error).

Future Worth Factor—Find F, given G, i, N

To obtain the future worth equivalent of a gradient series, we substitute Eq. (4.18) into Eq. (4.10) for A:

$$F = \frac{G}{i}\left[\frac{(1 + i)^N - 1}{i} - N\right] = G(F/G, i, N). \qquad (4.19)$$

Example 4.22 Declining Linear Gradient: Find F, Given A_1, G, i, N

Suppose that you make a series of annual deposits into a bank account that pays 10% interest. The initial deposit at the end of the first year is $1,200. The deposit amounts decline by $200 in each of the next 4 years. How much would you have immediately after the 5th deposit?

Solution

Given: Cash flow shown in Figure 4.31, $i = 10\%$ per year, $N = 5$ years
Find: F

The cash flow includes a decreasing gradient series. Recall that we derived the linear gradient factors for an increasing gradient series. For a decreasing gradient series, the solution is most easily obtained by separating the flow into two components: A uniform series and an increasing gradient which is *subtracted* from the uniform series (Figure 4.31). The future value is

$$F = F_1 - F_2$$
$$= A_1(F/A, 10\%, 5) - \$200(P/G, 10\%, 5)(F/P, 10\%, 5)$$
$$= \$1,200(6.105) - \$200(6.862)(1.611)$$
$$= \$5,115.$$

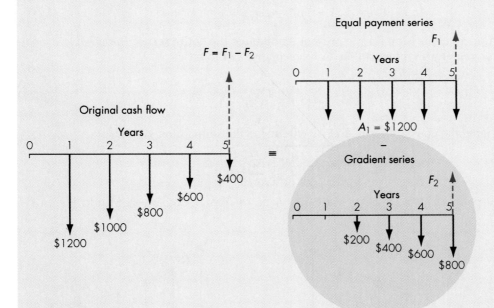

Figure 4.31 A series of decreasing gradient deposits viewed as a combination of uniform and gradient series (Example 4.22)

4.3.6 Geometric Gradient Series

Many engineering economic problems, particularly those relating to construction costs, involve cash flows that increase or decrease over time, not by a constant amount (linear gradient), but rather by a constant percentage (**geometric**), which is called **compound growth**. Price changes caused by inflation are a good example of such a geometric series. If we use g to designate the percentage change in a payment from one period to the next, the magnitude of the nth payment, A_n, is related to the first payment A_1 as expressed by

$$A_n = A_1(1 + g)^{n-1}, n = 1, 2, \ldots, N. \qquad (4.20)$$

The g can take either a positive or a negative sign depending on the type of cash flow. If $g > 0$, the series will increase, and if $g < 0$, the series will decrease. Figure 4.32 illustrates the cash flow diagram for this situation.

Present Worth Factor—Find P, Given A_1, g, i, N

Notice that the present worth, P_n, of any cash flow A_n at interest rate i is

$$P_n = A_n(1 + i)^{-n} = A_1(1 + g)^{n-1}(1 + i)^{-n}.$$

To find an expression for the present amount for the entire series, P, we apply the **single payment present worth factor** to each term of the series:

$$P = \sum_{n=1}^{N} A_1(1 + g)^{n-1}(1 + i)^{-n}. \qquad (4.21)$$

Bringing the constant term $A_1(1 + g)^{-1}$ outside the summation yields

$$P = \frac{A_1}{(1 + g)} \sum_{n=1}^{N} \left[\frac{1 + g}{1 + i} \right]^n. \qquad (4.22)$$

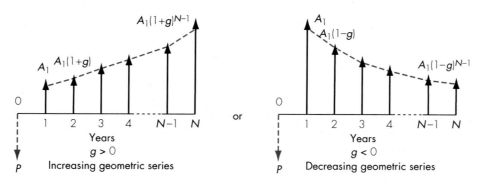

Figure 4.32 A geometrically increasing or decreasing gradient series at a constant rate g

Let $a = \dfrac{A_1}{1 + g}$ and $x = \dfrac{1 + g}{1 + i}$. Then, rewrite Eq. (4.22) as

$$P = a(x + x^2 + x^3 + \ldots + x^N). \tag{4.23}$$

Since the summation in Eq. (4.23) represents the first N terms of a geometric series, we may obtain the closed-form expression as follows: First, multiply Eq. (4.23) by x:

$$xP = a(x^2 + x^3 + x^4 + \ldots + x^{N+1}). \tag{4.24}$$

Then, subtract Eq. (4.24) from Eq. (4.23):

$$P - xP = a(x - x^{N+1})$$
$$P(1 - x) = a(x - x^{N+1})$$
$$P = \frac{a(x - x^{N+1})}{1 - x}, \text{ where } x \neq 1. \tag{4.25}$$

If we replace the original values for a and x, we obtain

$$P = \begin{cases} A_1 \left[\dfrac{1 - (1 + g)^N(1 + i)^{-N}}{i - g} \right], & \text{if } i \neq g \\ NA_1/(1+i) & \text{if } i = g \end{cases} \tag{4.26}$$

or $$P = A_1(P/A_1, g, i, N).$$

The factor within brackets is called the **geometric-gradient-series present worth factor** and designated $(P/A_1, g, i, N)$. In the special case where $i = g$, Eq. (4.22) becomes $P = [A_1/(1 + i)]N$.

Example 4.23 Geometric Gradient: Find P, Given A_1, g, i, N

Ansell Inc., a medical device manufacturer, uses compressed air in solenoids and pressure switches in its machines to control various mechanical movements. Over the years, the manufacturing floor has changed layouts numerous times. With each new layout, more piping was added to the compressed air delivery system to accommodate new locations of manufacturing machines. None of the extra, unused old pipe was capped or removed; thus the current compressed air delivery system is inefficient and fraught with leaks. Because of the leaks in the current system, the compressor is expected to run 70% of the time that the plant will be in operation during the upcoming year. This will require 260 kWh of electricity at a rate of $0.05/kWh. (The plant runs 250 days a year, 24 hours per day.) If Ansell continues to operate the current air delivery system, the compressor run time will increase by 7% per year for the next 5 years because of ever-worsening leaks.

(After 5 years, the current system will not be able to meet the plant's compressed air requirement, so it will have to be replaced.) If Ansell decides to replace all of the old piping now, it will cost $28,570. The compressor will still run the same number of days; however, it will run 23% less (or $70\%(1 - 0.23) = 53.9\%$ usage during the day) because of the reduced air pressure loss. If Ansell's interest rate is 12%, is it worth fixing now?

Solution

Given: Current power consumption, $g = 7\%$, $i = 12\%$, $N = 5$ years

Find: A_1 and P

Step 1: We need to calculate the cost of power consumption of the current piping system during the first year. The power consumption is equal to the following:

$$\begin{aligned} \text{Power cost} =\ & \%\ \text{of day operating} \\ & \times\ \text{days operating per year} \\ & \times\ \text{hours per day} \\ & \times\ \text{kWh} \times \$/\text{kWh} \\ =\ & (70\%) \times (250\ \text{days/year}) \times (24\ \text{hours/day}) \\ & \times (260\ \text{kWh}) \times (\$0.05/\text{kWh}) \\ =\ & \$54,440. \end{aligned}$$

Step 2: Each year the annual power cost will increase at the rate of 7% over the previous year's cost. The anticipated power cost over the 5-year period is summarized in Figure 4.33. The equivalent present lump sum cost at 12% for this geometric gradient series is

$$\begin{aligned} P_{\text{Old}} &= \$54,440(P/A_1, 7\%, 12\%, 5) \\ &= \$54,440 \left[\frac{1 - (1 + 0.07)^5(1 + 0.12)^{-5}}{0.12 - 0.07} \right] \\ &= \$222,283. \end{aligned}$$

Step 3: If Ansell replaces the current compressed air system with the new one, the annual power cost will be 23% less during the first year and will remain at that level over the next 5 years. The equivalent present lump sum cost at 12% is

$$\begin{aligned} P_{\text{New}} &= \$54,440(1 - 0.23)(P/A, 12\%, 5) \\ &= \$41,918.80(3.6048) \\ &= \$151,109. \end{aligned}$$

Step 4: The net cost for not replacing the old system now is $71,174 (= $222,283 − $151,109). Since the new system costs only $28,570, the replacement should be made now.

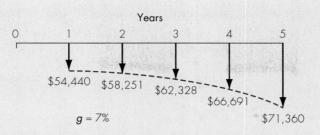

Figure 4.33 If repair is not performed, expected power expenditure over next 5 years due to deteriorating leaks will increase at a rate of 7% per year (Example 4.23)

Comments: In this example, we assumed that the cost of removing the old system was included in the cost of installing the new system. If the removed system has some salvage value, replacing it will result in even greater savings. We will consider many types of replacement issues in Chapter 15.

Future Worth Factor—Find F, Given A_1, g, N, i

The future worth equivalent of the geometric series can be obtained by multiplying Eq. (4.26) by the F/P factor, $(1 + i)^N$.

$$F = \begin{cases} A_1\left[\dfrac{(1 + i)^N - (1 + g)^N}{i - g}\right], & \text{if } i \neq g \\ NA_1(1 + i)^{N-1}, & \text{if } i = g \end{cases} \tag{4.27}$$

or

$$F = A_1(F/A_1, g, i, N).$$

Example 4.24 Geometric gradient: Find A_1, given F, g, i, N

A self-employed individual, Jimmy Carpenter, is opening a retirement account at a bank. His goal is to accumulate $1,000,000 in the account by the time he retires from work in 20 years time. A local bank is willing to open a retirement account that pays 8% interest, compounded annually, throughout the 20 years. Jimmy expects his annual income will increase at a 6% annual rate during his working career. He wishes to start with a deposit at the end of year 1 (A_1) and increase the deposit at a rate of 6% each year thereafter. What should be the size of his first deposit(A_1)? The first deposit will occur at the end of year 1, and subsequent deposits will be made at the end of each year. The last deposit will be made at the end of year 20.

Solution

Given: $F = \$1,000,000$, $g = 6\%$ per year, $i = 8\%$ per year, and $N = 20$ years
Find: A_1 as in Figure 4.34.

$$F = A_1(F/A_1, g, i, N)$$
$$= A_1(F/A_1, 6\%, 8\%, 20)$$
$$= A_1(72.6911).$$

Solving for A_1 yields

$$A_1 = \$1,000,000/72.6911 = \$13,757$$

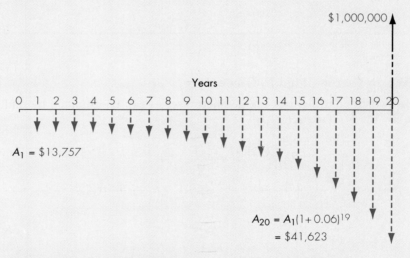

Figure 4.34 Jimmy Carpenter's retirement plan (Example 4.24)

Table 4.3 summarizes the interest formulas developed in this section and the cash flow situations in which they should be used. Recall that all the interest formulas developed in this section are only applicable to situations where the interest (compounding) period is the same as the payment period (e.g., annual compounding with annual payment). Also, in this table we present some useful interest factor relationships.

4.4 Unconventional Equivalence Calculations

Throughout the preceding section, we occasionally presented two or more methods of attacking example problems even though we had standard interest factor equations by which to solve them. It is important that you become adept at examining problems

Flow Type	Factor Notation	Formula	Excel Command	Cash Flow Diagram
S I N G L E	Compound amount $(F/P, i, N)$	$F = P(1 + i)^N$	=FV($i,N,P,,0$)	
	Present worth $(P/F, i, N)$	$P = F(1 + i)^{-N}$	=PV($i,N,F,,0$)	
E Q U A L	Compound amount $(F/A,i,N)$	$F = A\left[\dfrac{(1 + i)^N - 1}{i}\right]$	=FV($i,N,A,,0$)	
P A Y M E N T	Sinking fund $(A/F,i,N)$	$A = F\left[\dfrac{i}{(1 + i)^N - 1}\right]$	=PMT $(i,N,P,F,0)$	
S E R I E S	Present worth $(P/A,i,N)$	$P = A\left[\dfrac{(1 + i)^N - 1}{i(1 + i)^N}\right]$	=PV($i,N,A,,0$)	
	Capital recovery $(A/P,i,N)$	$A = P\left[\dfrac{i(1 + i)^N}{(1 + i)^N - 1}\right]$	=PMT(i,N,P)	
G R A D I E N T	Linear gradient Present worth $(P/G,i,N)$	$P = G\left[\dfrac{(1 + i)^N - iN - 1}{i^2(1 + i)^N}\right]$		
S E R I E S	Geometric gradient Present worth $(P/A_1,g,i,N)$	$P = \left[\begin{array}{l} A_1\left[\dfrac{1 - (1 + g)^N(1 + i)^{-N}}{i - g}\right] \\ \dfrac{NA_1}{1 + i}(if\ i = g) \end{array}\right.$		

Table 4.3
Summary of Discrete Compounding Formulas with Discrete Payments

from unusual angles and that you seek out unconventional solution methods, because not all cash flow problems conform to the neat patterns for which we have discovered and developed equations. Two categories of problems that demand unconventional treatment are composite (mixed) cash flows and problems in which we must determine the interest rate implicit in a financial contract. We will begin this section by examining instances of composite cash flows.

4.4.1 Composite Cash Flows

Although many financial decisions do involve constant or systematic changes in cash flows, many investment projects contain several components of cash flows that do not exhibit an overall pattern. We can take advantage of these in problem-solving. Consequently, it is necessary to expand our analysis to deal with these mixed types of cash flows.

To illustrate, consider the cash flow stream shown in Figure 4.35. We want to compute the equivalent present worth for this mixed payment series at an interest rate of 15%. Three different methods are presented.

Method 1: A "brute force" approach is to multiply each payment by the appropriate $(P/F, 10\%, n)$ factors and then to sum these products to obtain the present worth of the cash flows, $543.72. Recall that this is exactly the same proce-

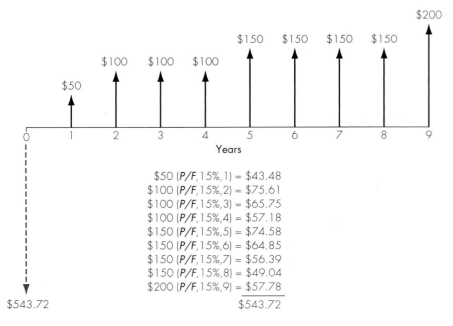

$$50 (P/F, 15\%, 1) = \$43.48$$
$$100 (P/F, 15\%, 2) = \$75.61$$
$$100 (P/F, 15\%, 3) = \$65.75$$
$$100 (P/F, 15\%, 4) = \$57.18$$
$$150 (P/F, 15\%, 5) = \$74.58$$
$$150 (P/F, 15\%, 6) = \$64.85$$
$$150 (P/F, 15\%, 7) = \$56.39$$
$$150 (P/F, 15\%, 8) = \$49.04$$
$$200 (P/F, 15\%, 9) = \$57.78$$
$$\overline{\$543.72}$$

Figure 4.35 Equivalent present worth calculation using P/F factors (Method 1— "Brute Force Approach")

dure we used to solve the category of problems called the uneven payment series, which were described in Section 4.3.3. Fig. 4.35 illustrates this computational method.

Method 2: We may group the cash flow components according to the type of cash flow pattern that they fit, such as the single payment, equal payment series and so forth, as shown in Figure 4.36. Then, the solution procedure involves the following steps:

- Group 1: Find the present worth of $50 due in year 1:

$$\$50(P/F, 15\%, 1) = \$43.48.$$

- Group 2: Find the equivalent worth of a $100 equal-payment series at year 1 (V_1) and then bring this equivalent worth at year 0 again:

$$\underbrace{\$100(P/A, 15\%, 3)}_{V_1}(P/F, 15\%, 1) = \$198.54.$$

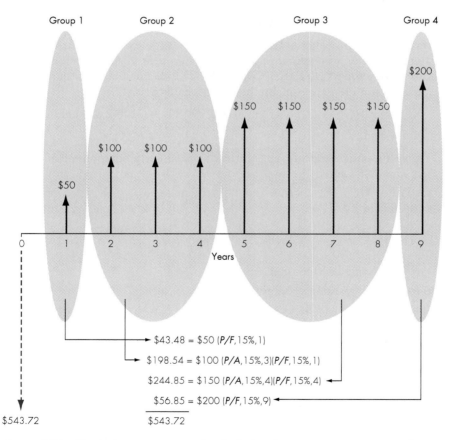

Figure 4.36 Equivalent present worth calculation for an uneven payment series using *P/F* and *P/A* factors (Method 2—"Grouping Approach")

- Group 3: Find the equivalent worth of a $150 equal-payment series at year 4 (V_4) and then bring this equivalent worth at year 0.

$$\underbrace{\$150(P/A, 15\%, 4)}_{V_4}(P/F, 15\%, 4) = \$244.85.$$

- Group 4: Find the equivalent present worth of the $200 due in year 9:

$$\$200(P/F, 15\%, 9) = \$56.85$$

- Group total—sum the components:

$$P = \$43.48 + \$198.54 + \$244.85 + \$56.85 = \$543.72.$$

A pictorial view of this computational process is given in Fig. 4.36.

Method 3: In computing the present worth of the equal-payment series components, we may use an alternative method.

- Group 1: Same as in Method 2
- Group 2: Recognize that a $100 equal payment series will be received during years 2 through 4. Thus, we could determine the value of a 4-year annuity, subtract from it the value of a 1-year annuity, and have remaining the value of a 4-year annuity, whose first payment is due in year 2. This result is achieved by subtracting the (P/A, 15%, 1) for a 1-year, 15% annuity from that for a 4-year annuity, and then multiplying the difference by $100:

$$\$100[(P/A, 15\%, 4) - (P/A, 15\%, 1)] = \$100(2.8550 - 0.8696)$$
$$= \$198.54.$$

Thus, the equivalent present worth of the annuity component of the uneven stream is $198.54.

- Group 3: We have another equal-payment series which starts in year 5 and ends in year 8.

$$\$150[(P/A, 15\%, 8) - (P/A, 15\%, 4)] = \$150(4.4873 - 2.8550)$$
$$= \$244.85.$$

- Group 4: Same as Method 2
- Group total—Sum the components:

$$P = \$43.48 + \$198.54 + \$244.85 + \$56.85 = \$543.72.$$

Either the "brute force" method in Fig. 4.35 or the method utilizing both (P/A, i, n) and (P/F, i, n) factors can be used to solve problems of this type. Either Method 2 or Method 3 is much easier if the annuity component runs for many years, however. For example, the alternative solution would be clearly superior for finding the equivalent present worth of a stream consisting of $50 in year 1, $200 in years 2 through 19, and $500 in year 20.

Also, note that in some instances we may want to find the equivalent value of a stream of payments at some point other than the present (year 0). In this situation, we proceed as before, but compound and discount to some other points in time, say year 2, rather than year 0. Example 4.25 illustrates the situation.

Example 4.25 Cash Flows with Subpatterns

The two cash flows in Figure 4.37 are equivalent at an interest rate of 12%, compounded annually. Determine the unknown value, C.

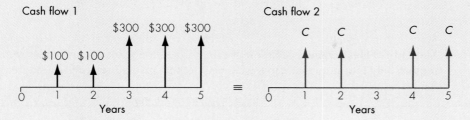

Figure 4.37   Equivalence calculation (Example 4.25)

Solution

Given: Cash flows as in Fig. 4.37, $i = 12\%$ per year
Find: C

- *Method 1:* Compute the present worth of each cash flow at time 0.

$$P_1 = \$100(P/A, 12\%, 2) + \$300(P/A, 12\%, 3)(P/F, 12\%, 2)$$

$$= \$743.42$$

$$P_2 = C(P/A, 12\%, 5) - C(P/F, 12\%, 3)$$

$$= 2.8930C.$$

Since the two flows are equivalent, $P_1 = P_2$.

$$743.42 = 2.8930C.$$

Solving for C, we obtain $C = \$256.97$.

- *Method 2:* We may select a time point other than 0 for comparison. The best choice of a base period is largely determined by the cash flow patterns. Obviously, we want to select a base period that requires the minimum number of interest factors for the equivalence calculation. Cash flow 1 represents a combined series of two equal payment cash flows, whereas cash flow 2 can be viewed as an equal payment series with the third payment missing. For cash flow 1, computing the equivalent worth at period 5 will require only two interest factors:

$$V_{5,1} = \$100(F/A, 12\%, 5) + \$200(F/A, 12\%, 3)$$

$$= \$1310.16.$$

For cash flow 2, computing the equivalent worth of the equal payment series at period 5 will also require two interest factors:

$$V_{5,2} = C(F/A, 12\%, 5) - C(F/P, 12\%, 2)$$
$$= 5.0984C.$$

Therefore, the equivalence would be obtained by letting $V_{5,1} = V_{5,2}$:

$$\$1310.16 = 5.0984C.$$

Solving for C yields $C = \$256.97$, which is the same result obtained from Method 1. The alternative solution of shifting the time point of comparison will require only four interest factors whereas Method 1 would require five interest factors.

Example 4.26 Establishing a College Fund

A couple with a newborn daughter want to save for their child's college expenses in advance. The couple can establish a college fund that pays 7% annual interest. Assuming that the child enters college at age 18, the parents estimate that an amount of $40,000 per year will be required to support the child's college expenses for 4 years. Determine the equal annual amounts the couple must save until they send their child to college. (Assume that the first deposit will be made on the child's first birthday and the last deposit on the child's 18th birthday. The first withdrawal will also be made at the beginning of the freshmen year, which is the child's 18th birthday.)

Solution

Given: Deposit and withdrawal series shown in Figure 4.38, $i = 7\%$ per year

Find: Unknown annual deposit amount (X)

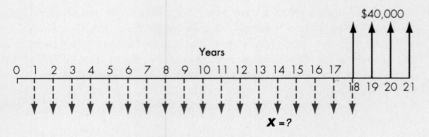

Figure 4.38   Establishing a college fund (Example 4.26)

- Method 1: Establish economic equivalence at period 0:

Step 1: Find the equivalent single lump sum deposit now:

$$P_{\text{Deposit}} = X(P/A, 7\%, 18)$$
$$= 10.0591X.$$

Step 2: Find the equivalent single lump sum withdrawal now:

$$P_{\text{Withdrawal}} = \$40{,}000(P/A, 7\%, 4)(P/F, 7\%, 17)$$
$$= \$42{,}892.$$

Step 3: Since the two amounts are equivalent, by equating $P_{\text{Deposit}} = P_{\text{Withdrawal}}$, we obtain X:

$$10.0591X = \$42{,}892$$
$$X = \$4{,}264.$$

- Method 2: Establish the economic equivalence at the child's 18th birthday.

  Step 1: Find the accumulated deposit balance at the child's 18th birthday:

  $$V_{18} = X(F/A, 7\%, 18)$$
  $$= 33.9990X.$$

  Step 2: Find the equivalent lump sum withdrawal at the child's 18th birthday:

  $$V_{18} = \$40{,}000 + \$40{,}000(P/A, 7\%, 3)$$
  $$= \$144{,}972.$$

  Step 3: Since the two amount must be the same, we obtain

  $$33.9990X = \$144{,}972$$
  $$X = \$4{,}264.$$

The computational steps are also summarized in Figure 4.39. In general, the second method is the more efficient way to obtain an equivalence solution to this type of decision problem.

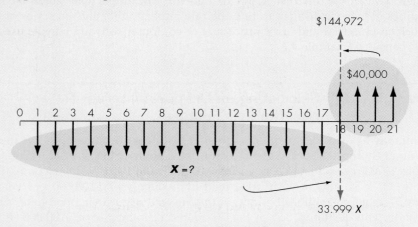

Figure 4.39   An alternative equivalence calculation (Example 4.26)

Comments: To verify if the annual deposits of $4,264 over 18 years would be sufficient to meet the child's college expenses, we can calculate the actual year-by-year balances: With the 18 annual deposits of $4,264, the balance at the child's 18th birthday is

$$\$4,264(F/A, 7\%, 18) = \$144,972.$$

From this balance, the couple will make four annual tuition payments:

Year n	Beginning Balance	Interest Earned	Tuition Payment	Ending Balance
Freshman	$144,972	$0	$40,000	$104,972
Sophomore	104,972	7,348	40,000	72,320
Junior	72,320	5,062	40,000	37,382
Senior	37,382	2,618	40,000	0

4.4.2 Determining an Interest Rate to Establish Economic Equivalence

Thus far, we have assumed that, in equivalence calculations, a typical interest rate is given. Now we can use the same interest formulas that we developed earlier to determine interest rates explicit in equivalence problems. For most commercial loans, interest rates are already specified in the contract. However, when making some investments in financial assets, such as stocks, you may want to know what rate of growth (or rate of return) your asset is appreciating over the years. (This kind of calculation is the basis of rate-of-return analysis, which is covered in Chapter 9.) Although we can use interest tables to find the rate implicit in single payments and annuities, it is more difficult to find the rate implicit in an uneven series of payments. In such cases, a trial-and-error procedure or computer software may be used. To illustrate, consider Example 4.27.

Example 4.27 Calculating an Unknown Interest Rate with Multiple Factors

You may have already won $2 million!!! Just peel the game piece off the Instant Winner Sweepstakes ticket, and mail it to us along with your orders for subscriptions to your two favorite magazines. As a Grand Prize Winner you may choose between a $1 million cash prize paid immediately or $100,000 per year for 20 years—that's $2 million!!! Suppose, that instead of receiving one lump sum of $1 million, you decide to accept the 20 annual installments of $100,000. If you are like most jackpot winners, you will be tempted to spend your winnings to improve your lifestyle during the first several years. Only after you get this type of spending "out of your system," will you save later sums for investment purposes. Suppose that you are considering the following two options:

Option 1: You save your winnings for the first 7 years and then spend every cent of the winnings in the remaining 13 years.

Option 2: You do the reverse and spend for 7 years and then save for 13 years.

If you can save winnings at 7% interest, how much would you have at the end of 20 years and what interest rate on your savings will make these two options equivalent? (Cash flows into savings for the two options are shown in Figure 4.40.)

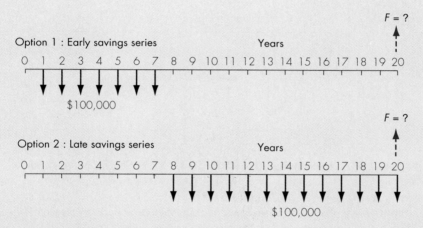

Figure 4.40   Equivalence calculation (Example 4.27)

Solution

Given: Cash flows in Fig. 4.40

Find: (a) F and (b) i at which the two flows are equivalent

(a) In Option 1, the net balance at the end of year 20 can be calculated in two steps: Find the accumulated balance at the end of year 7 (V_7) first, and second, find the equivalent worth of V_7 at the end of year 20. For Option 2, find the equivalent worth of the 13 equal annual deposits at the end of year 20:

$$F_{\text{Option 1}} = \$100,000(F/A, 7\%, 7)(F/P, 7\%, 13)$$

$$= \$2,085,485$$

$$F_{\text{Option 2}} = \$100,000(F/A, 7\%, 13)$$

$$= \$2,014,064.$$

Option 1 accumulates $71,421 more than the Option 2.

(b) To compare the alternatives, we may compute the present worth for each option at period 0. By selecting time period 7, however, we can establish the same economic equivalence with fewer interest factors. As shown in Figure 4.41, we calculate the equivalent value for each option at the end of period 7, (V_7), remembering that the end of period 7 is also the beginning of period 8. (Recall from Example 2.4 that the choice of the point in time at which to compare two cash flows for equivalence is arbitrary.)

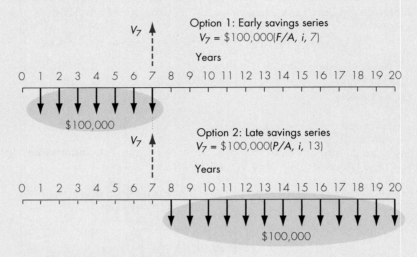

Figure 4.41   Establishing an economic equivalence at period 7 (Example 4.27)

For Option 1:

$$V_7 = \$100{,}000(F/A, i, 7)$$

For Option 2:

$$V_7 = \$100{,}000(P/A, i, 13).$$

To be equivalent, these values must be the same.

$$\$100{,}000(F/A, i, 7) = \$100{,}000(P/A, i, 13)$$

$$\frac{(F/A, i, 7)}{(P/A, i, 13)} = 1.$$

Here, we are looking for an interest rate that makes the ratio 1. When using the interest tables we need to resort to a trial-and-error method. Suppose that we guess the interest rate to be 6%. Then

$$\frac{(F/A, 6\%, 7)}{(P/A, 6\%, 13)} = \frac{8.3938}{8.8527} = 0.9482.$$

This is less than 1. To increase the ratio, we need to use an i value such th creases the $(F/A, i, 7)$ factor value, but decreases the $(P/A, i, 13)$ value. This pen if we use a larger interest rate. Let's try $i = 7\%$.

$$\frac{(F/A, 7\%, 7)}{(P/A, 7\%, 13)} = \frac{8.6540}{8.3577} = 1.0355.$$

Now the ratio is greater than 1.

Interest Rate	(F/A, i, 7)/(P/A, i, 13)
6%	0.9482
?	1.0000
7%	1.0355

As a result, we find that the interest rate is between 6% and 7% and may be approximated by linear interpolation as shown in Figure 4.42:

$$i = 6\% + (7\% - 6\%) \left[\frac{1 - 0.9482}{1.0355 - 0.9482} \right]$$

$$= 6\% + 1\% \left[\frac{0.0518}{0.0873} \right]$$

$$= 6.5934\%.$$

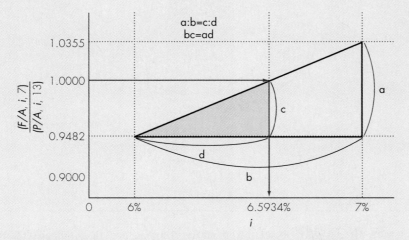

Figure 4.42 Linear interpolation to find unknown interest rate (Example 4.27)

At 6.5934% interest, the two options are equivalent, and you may decide to indulge your desire to spend like crazy for the first 7 years. However, if you could obtain a higher interest rate, you would be wiser to save for 7 years and spend for the next 13.

Comments: This example demonstrates that finding an interest rate is an iterative process, which is more complicated and generally less precise than the problem of finding an equivalent worth at a known interest rate. Since computers and financial calculators can speed the process of finding unknown interest rates, these tools are highly recommended for these types of problem solving. With a computer, a more precise break-even value of 6.60219% is found.

Figure 4.43 shows a radar chart that compares the two savings plans at various interest rates. Note that, as the interest rate increases, the smaller the economic difference between the two options up to the point at $i = 6.60\%$. It then increases again.

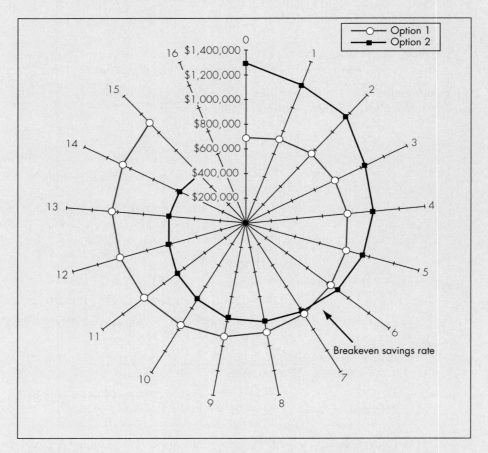

Figure 4.43 Radar chart obtained from an Excel spreadsheet that compares the two savings plans at various interest rates

4.5 Summary

- Money has a time value because it can earn more money over time. A number of terms involving the time value of money were introduced in this chapter as follows:

  Interest is the cost of money. More specifically, it is a cost to the borrower and an earning to the lender, above and beyond the initial sum borrowed or loaned.

  Interest rate is a percentage periodically applied to a sum of money to determine the amount of interest to be added to that sum.

  Simple interest is the practice of charging an interest rate only to an initial sum.

  Compound interest is the practice of charging an interest rate to an initial sum *and* to any previously accumulated interest that has not been withdrawn from the initial sum. Compound interest is by far the most commonly used system in the real world.

  Economic equivalence exists between individual cash flows and/or patterns of cash flows that have the same value. Even though the amounts and timing of the cash flows may differ, the appropriate interest rate makes them equal.

- The compound interest formula is perhaps the single most important equation in this text:

$$F = P(1 + i)^N,$$

  where P is a present sum, i is the interest rate, N is the number of periods for which interest is compounded, and F is the resulting future sum. All other important interest formulas are derived from this one.

- **Cash flow diagrams** are visual representations of cash inflows and outflows along a timeline. They are particularly useful for helping us detect which of the five patterns of cash flow is represented by a particular problem.

- The five patterns of cash flow are as follows:

  1. Single payment: A single present or future cash flow.

  2. Uniform series: A series of flows of equal amounts at regular intervals.

  3. Linear gradient series: A series of flows increasing or decreasing by a fixed amount at regular intervals.

  4. Geometric gradient series: A series of flows increasing or decreasing by a fixed percentage at regular intervals.

  5. Irregular series: A series of flows exhibiting no overall pattern. However, patterns might be detected for portions of the series.

- **Cash flow patterns** are significant because they allow us to develop **interest formulas**, which streamline the solution of equivalence problems. Table 4.3 summarizes the important interest formulas that form the foundation for all other analyses you will conduct in engineering economic analysis.

Self-Test Questions

4s.1 If you make the following series of deposits at an interest rate of 10%, compounded annually, what would be the total balance at the end of 10 years?

End of Period	Amount of Deposit
0	$800
1–9	$1500
10	0

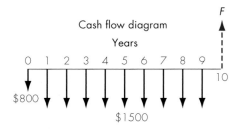

Cash flow diagram

(a) $F = \$22,256$ (b) $F = \$24,481$
(c) $F = \$24,881$ (d) $F = \$25,981.$

4s.2 In computing the equivalent present worth of the following cash flow series at period 0, which of the following expressions is incorrect?

End of Period	Payment
0	
1	
2	
3	
4–7	$100

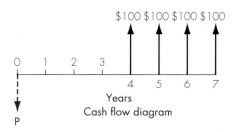

Cash flow diagram

(a) $P = \$100(P/A, i, 4)(P/F, i, 4)$
(b) $P = \$100(F/A, i, 4)(P/F, i, 7)$
(c) $P = \$100(P/A, i, 7) - \$100(P/A, i, 3)$
(d) $P = \$100[(P/F, i, 4) + (P/F, i, 5) + (P/F, i, 6) + (P/F, i, 7)].$

4s.3 To withdraw the following $1,000 payment series, determine the minimum deposit (P) you should make now if your deposits earn an interest rate of 10%, compounded annually. Note that you are making another deposit at the end of year 7 in the amount of $500. With the minimum deposit P, your balance at the end of year 10 should be zero.

End of Period	Deposit	Withdrawal
0	P	
1–6		$1000
7	$500	
8–10		1000

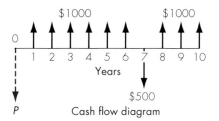

Cash flow diagram

(a) $P = \$4,912$ (b) $P = \$4,465$
(c) $P = \$5,374$ (d) $P = \$5,912.$

4s.4 Consider the following cash flow series.

What value of C makes the deposit series equivalent to the withdrawal series at an interest rate of 12%, compounded annually?

(a) $C = \$200$ (b) $C = \$282.70$
(c) $C = \$394.65$ (d) $C = \$458.90.$

End of Period	Deposit	Withdrawal
0	$1,000	
1	800	
2	600	
3	400	
4	200	
5		
6		C
7		2C
8		3C
9		4C
10		5C

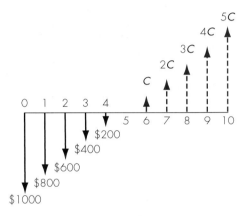

4s.5 A couple is planning to finance their 5-year-old daughter's college education. They established a college fund that earns 10%, compounded annually. What annual deposit must be made from the daughter's 5th birthday (now) to her 16th birthday to meet the future college expenses shown in the following table. Assume that today is her 5th birthday.

Birthday	Deposit	Withdrawal
5–16	A	
17		
18		$25,000
19		27,000
20		29,000
21		31,000

(a) A = $3,742   (b) A = $1,978
(c) A = $4,115   (d) A = $3,048.

4s.6 John and Susan just opened savings accounts at two different banks. They deposited $1,000, respectively. John's bank pays a simple interest at an annual rate of 10%, whereas Susan's bank pays compound interest at an annual rate of 9.5%. No interest would be taken out for a period of 3 years. At the end of 3 years, whose balance would be greater and by how much (in nearest dollars)?

(a) Susan's balance, by $30
(b) John's balance, by $1
(c) Susan's balance, by $13
(d) John's balance, by $18.

4s.7 If you invest $2,000 today in a savings account at an interest rate of 12%, compounded annually, how much principal and interest would you accumulate in 7 years?

(a) $4,242   (b) $4,422
(c) $2,300   (d) $1,400.

4s.8 How much do you need to invest in equal annual amounts for the next 10 years if you want to withdraw $5,000 at the end of the eleventh year and increase the annual withdrawal by $1,000 each year thereafter until year 25? The interest rate is 6%, compounded annually.

(a) $20,000   (b) $106,117
(c) $8,054   (d) $5,000.

4s.9 What is the present worth of the following income strings at an interest rate of 10%? (All cash flows occur at year end.)

n	Net Cash Flow
1	$11,000
2	12,100
3	0
4	14,641

(a) $37,741   (b) $30,000

(c) $43,923   (d) $32,450.

4s.10 You want to find the equivalent present worth for the following cash flow series at an interest rate of 15%. Which of the following statements is *incorrect*?

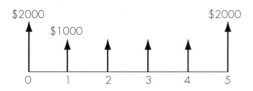

(a) $1,000(P/A, 15%, 4) + $2,000 + $2,000(P/F, 15%, 5)

(b) $1,000(P/F, 15%, 5) + $1,000(P/A, 15%, 5) + $2,000

(c) [$1,000(F/A, 15%, 5) + $1,000] × (P/F, 15%, 5) + $2,000

(d) [$1,000(F/A, 15%, 4) + $2,000] × (P/F, 15%, 4) + $2,000.

4s.11 If you make the following series of deposits at an interest rate of 10%, compounded annually, what would be the total balance at the end of 8 years?

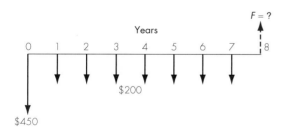

(a) $2,550   (c) $3,052

(b) $2,862   (d) $3,252

4s.12 What value of F3 would be equivalent to the payments shown in the cash flow diagram below? Assume the interest rate is 10%, compounded annually.

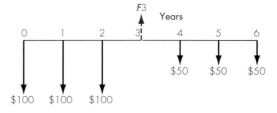

(a) $450   (c) $457

(b) $462   (d) $488

4s.13 Consider the cash flow series shown below. Determine the required annual deposits (end of year) that will generate the cash flows from years 4 to 7. Assume the interest rate is 10%, compounded annually.

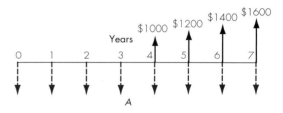

(a) $555   (c) $518

(b) $568   (d) $698

4s.14 What is the amount of interest earned on $2,000 for 5 years at 10% simple interest per year?

(a) $3,000   (c) $1,000

(b) $3,221   (d) $1,221

4s.15 What annual interest rate would make these two cash flows equivalent?

(a) 25%   (c) 15%

(b) 20%   (d) 10%

4s.16 If you borrow $10,000 at an interest rate of 15%, compounded annually, with the repayment schedule as follows, what is the amount A?

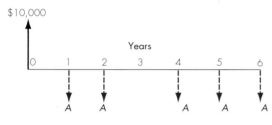

(a) $2,643   (c) $4,130
(b) $3,591   (d) $3,198

4s.17 Compute the value of V in the following cash flow diagram. Assume $i = 12\%$.

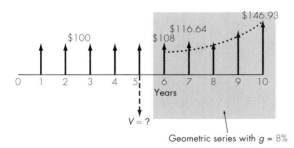

(a) $1,083   (c) $1,260
(b) $1,131   (d) $1,360

4s.18 What value of C makes the two annual cash flows equivalent at an annual rate of 10%?

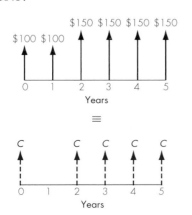

(a) $99   (c) $160
(b) $150   (d) $220

4s.19 Find the value of X so that the two cash flows below are equivalent for an interest rate of 10%.

$$\equiv$$

(a) $464   (c) $524
(b) $494   (d) $554

4s.20 Compute the value of F in the following cash flow diagram. Assume $i = 10\%$, compounded annually.

(a) $1,220   (c) $1,517
(b) $1,320   (d) $1,488

Problems

Types of Interest

4.1   You deposit $3,000 in a savings account that earns 9% simple interest per year. To double your balance, you wait at least (?) years. But if you deposit the $3,000 in another savings account that earns 8% interest, compounded yearly, it will take (?) years to double your balance.

4.2   Compare the interest earned by $1,000 for 10 years at 7% simple interest with that earned by the same amount for 10 years at 7%, compounded annually.

4.3   You are considering investing $1,000 at interest rate of 6%, compounded annually, for 5 years or investing the $1,000 at 7% per year simple interest rate for 5 years. Which option is better?

4.4   You are about to borrow $3,000 from a bank at an interest rate of 9%, compounded annually. You are required to make three equal annual repayments in the amount of $1,185.16 per year, with the first repayment occurring at the end of year 1. In each year, show the interest payment and principal payment.

Equivalence Concept

4.5   Suppose you have the alternative of receiving either $10,000 at the end of 5 years or P dollars today. Currently you have no need for money, so you would deposit the P dollars in a bank that pays 6% interest. What value of P would make you indifferent in your choice between P dollars today and the promise of $10,000 at the end of 5 years?

4.6   Suppose that you are obtaining a personal loan from your uncle in the amount of $10,000 (now) to be repaid in 2 years to cover some of your college expenses. If your uncle always earns 10% interest (annually) on his money invested in various sources, what minimum lump sum payment 2 years from now would make your uncle happy?

Single Payments (Use of F/P or P/F Factors)

4.7   What will be the amount accumulated by each of these present investments?
   (a) $7,000 in 8 years at 9%, compounded annually
   (b) $1,250 in 12 years at 4%, compounded annually
   (c) $5,000 in 31 years at 7%, compounded annually
   (d) $20,000 in 7 years at 6%, compounded annually.

4.8   What is the present worth of these future payments?
   (a) $4,500 — 6 years from now at 7%, compounded annually
   (b) $6,000 — 15 years from now at 8%, compounded annually
   (c) $20,000 — 5 years from now at 9%, compounded annually
   (d) $12,000 — 8 years from now at 10%, compounded annually.

4.9   For an interest rate of 8%, compounded annually, find
   (a)   How much can be loaned now if $6,000 will be repaid at the end of 5 years?
   (b)   How much will be required in 4 years to repay a $15,000 loan now?

4.10 How many years will it take an investment to triple itself if the interest rate is 9%, compounded annually?

4.11 You bought 200 shares of Motorola stock at $3,800 on December 31, 2000. Your intention is to keep the stock until it doubles in value. If you expect 15% annual growth for Motorola stock, how many years do you expect to hold on the stock? Compare the solution obtained by the Rule of 72 (discussed in Example 4.8).

4.12 From the interest tables in the text, determine the following value of the factors by interpolation. Then compare these answers with those obtained by evaluating the *F/P* factor or the *P/F* factor.

(a) The single-payment compound amount factor for 38 periods at 6.5% interest.

(b) The single-payment present worth factor for 57 periods at 8% interest.

Uneven Payment Series

4.13 If you desire to withdraw the following amounts over the next 5 years from the savings account which earns a 7% interest compounded annually, how much do you need to deposit now?

n	Amount
2	$32,000
3	33,000
4	36,000
5	38,000

4.14 If $1,000 is invested now, $1,500 two years from now, and $2,000 four years from now at an interest rate of 6% compounded annually, what will be the total amount in 10 years?

4.15 A local newspaper headline blared: "Bo Smith signed for $30 Million." A reading of the article revealed that on April 1, 2000, Bo Smith, the former record-breaking running back from Football University, signed a $30 million package with the Dallas Rangers. The terms of the contract were $3 million immediately: $2.4 million per year for the first 5 years (first payment after 1 year) and $3 million per year for the next 5 years (first payment at year 6). If Bo's interest rate is 8% per year, what would be his contract worth at the time of contract signing?

4.16 How much invested now at 6% would be just sufficient to provide three payments with the first payment in the amount of $3,000 occurring 2 years hence, $4,000 five years hence, and $5,000 seven years hence?

Equal Payment Series

4.17 What is the future worth of a series of equal year-end deposits of $2,000 for 10 years in a savings account that earns 9%, annual interest, if

(a) All deposits were made at the *end* of each year?

(b) All deposits were made at the *beginning* of each year?

4.18 What is the future worth of the following series of payments?

(a) $1,000 at the end of each year for 5 years at 7%, compounded annually

(b) $2,000 at the end of each year for 10 years at 8.25%, compounded annually

(c) $3,000 at the end of each year for 30 years at 9%, compounded annually

(d) $5,000 at the end of each year for 22 years at 10.75%, compounded annually.

4.19 What equal annual series of payments must be paid into a sinking fund to accumulate the following amount?

(a) $12,000 in 13 years at 5%, compounded annually

(b) $25,000 in 8 years at 9%, compounded annually

(c) $15,000 in 25 years at 7%, compounded annually

(d) $8,000 in 8 years at 11%, compounded annually.

4.20 Part of the income that a machine generates is put into a sinking fund to replace the machine when it wears out. If $2,000 is deposited annually at 7% interest, how many years must the machine be kept before a new machine costing $30,000 can be purchased?

4.21 A no-load (commission free) mutual fund has grown at a rate of 13%, compounded annually, since its beginning. If it is anticipated that it will continue to grow at this rate, how much must be invested every year so that $10,000 will be accumulated at the end of 5 years?

4.22 What equal annual payment series is required to repay the following present amounts?

(a) $15,000 in 5 years at 8% interest, compounded annually

(b) $5,500 in 4 years at 9.5% interest, compounded annually

(c) $8,000 in 3 years at 15% interest, compounded annually

(d) $60,000 in 20 years at 9% interest, compounded annually.

4.23 You have borrowed $20,000 at an interest rate of 12%. Equal payments will be made over a 3-year period (first payment at the end of the first year). The annual payment will be (?) and the interest payment for the second year will be (?).

4.24 What is the present worth of the following series of payments?

(a) $900 at the end of each year for 12 years at 6%, compounded annually

(b) $1,500 at the end of each year for 10 years at 9%, compounded annually

(c) $500 at the end of each year for 5 years at 7.25%, compounded annually

(d) $5,000 at the end of each year for 8 years at 8.75%, compounded annually.

4.25 From the interest tables in Appendix D determine the following value of the factors by interpolation. Then compare the results with those obtained from evaluating the A/P and P/A interest formulas.

(a) The capital recovery factor for 36 periods at 6.25% interest.

(b) The equal-payment series present worth factor for 125 periods at 9.25% interest.

Linear Gradient Series

4.26 An individual deposits an annual bonus into a savings account that pays 6% interest, compounded annually. The size of the bonus increases by $1,000 each year, and the initial bonus amount was $3000. Determine how much will be in the account immediately after the 5th deposit.

4.27 Five annual deposits in the amounts of ($1,200, $1,000, $800, $600, and $400) are made into a fund that pays interest at a rate of 9%, compounded annually. Determine the amount in the fund immediately after the 5th deposit.

4.28 Compute the value of P in the cash flow diagram below. Assume $i = 8\%$.

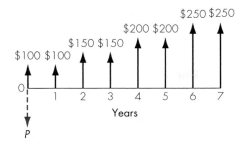

4.29 What is the equal payment series for 10 years that is equivalent to a payment series of $12,000 at the end of the first year, decreasing by $1000 each year over 10 years. Interest is 8%, compounded annually.

Geometric Gradient Series

4.30 Suppose that an oil well is expected to produce 100,000 barrels of oil during its first production year. However, its subsequent production (yield) is expected to decrease by 10% over the previous year's production. The oil well has a proven reserve of 1,000,000 barrels.

(a) Suppose that the price of oil is expected to be $30 per barrel for next several years. What would be the present worth of the anticipated revenue stream at an interest rate of 12%, compounded annually, over next 7 years?

(b) Suppose that the price of oil is expected to start at $30 per barrel during the first year, but to increase at the rate of 5% over the previous year's price, what would be the present worth of the anticipated revenue stream at an interest rate of 12%, compounded annually, over next 7 years?

(c) Reconsider (b) above. After 3-years' production, you decide to sell the oil well. What would be the fair price for the oil well?

4.31 A city engineer has estimated the annual toll revenues from a newly proposed highway construction over 20 years as follows:

$$A_n = (\$2,000,000)(n)(1.06)^{n-1},$$

$$n = 1, 2, \ldots, 20.$$

For the bond validation, the engineer was asked to present the estimated total present value of toll revenue at an interest rate of 6%. Assuming annual compounding, find the present value of the estimated toll revenue.

4.32 What is the amount of 10 equal annual deposits that can provide five annual withdrawals, when a first withdrawal of $3,000 is made at the end of year 11, and subsequent withdrawals increase at the rate of 6% per year over the previous year's, if

(a) The interest rate is 8%, compounded annually?

(b) The interest rate is 6%, compounded annually?

Various Interest Factor Relationships

4.33 By using only those factors given in interest table, find the values of the following factors, which are not given in your tables. Show the relationship between the factors by using the factor notation and calculate the value of the factor. Then compare the solution obtained by using the factor formulas to directly calculate the factor values.

Example: $(F/P, 8\%, 10) =$
$(F/P, 8\%, 4)(F/P, 8\%, 6) = 2.159$

(a) $(P/F, 8\%, 67)$

(b) $(A/P, 8\%, 42)$

(c) $(P/A, 8\%, 135)$.

4.34 Prove the following relationships among interest factors:

(a) $(F/P, i, N) = i(F/A, i, N) + 1$

(b) $(P/F, i, N) = 1 - (P/A, i, N)i$

(c) $(A/F, i, N) = (A/P, i, N) - i$

(d) $(A/P, i, N) = i/[1 - (P/F, i, N)]$.

(e) $(P/A, i, N \to \infty) = 1/i$

(f) $(A/P, i, N \to \infty) = i$

(g) $(P/G, i, N \to \infty) = 1/i^2$

Equivalence Calculations

4.35 Find the present worth of the cash receipts where $i = 10\%$, compounded annually, with only four interest factors.

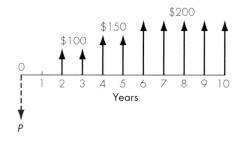

4.36 Find the equivalent present worth of the cash receipts where $i = 10\%$. In other words, how much do you have to deposit now (with the second deposit in the amount of $200 at the end of the first year) so that you will be able to withdraw $200 at the end of second year, $120 at the end of third year, and so forth, where the bank pays you a 10% annual interest on your balance?

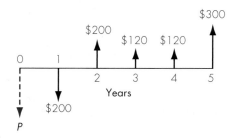

4.37 What value of A makes the two annual cash flows equivalent at 10% interest, compounded annually?

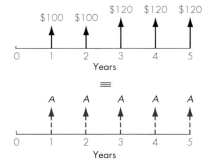

4.38 The two cash flow transactions shown in the cash flow diagram are said to be equivalent at 10% interest, compounded annually. Find the unknown X value which satisfies the equivalence.

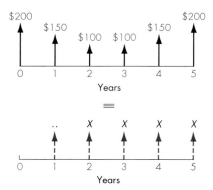

4.39 Solve for the present worth of this cash flow using at most three interest factors at 10% interest, compounded annually.

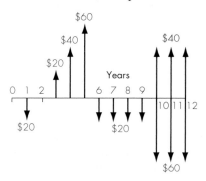

4.40 From the cash flow diagram, find the value of C that will establish the economic equivalence between the deposit series and the withdrawal series at an interest rate of 8%, compounded annually.

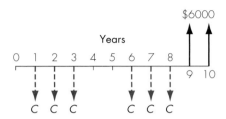

(a) $1,335   (b) $862

(c) $1,283   (d) $828.

2.41 The following equation describes the conversion of a cash flow into an equivalent equal payment series with $N = 10$. Given the equation, reconstruct the original cash flow diagram:

$$A = [800 + 20(A/G, 6\%, 7)] \times (P/A, 6\%, 7)(A/P, 6\%, 10) + [300(F/A, 6\%, 3) - 500](A/F, 6\%, 10).$$

4.42 Consider the following cash flow. What value of C makes the inflow series equivalent to the outflow series at an interest rate of 12%?

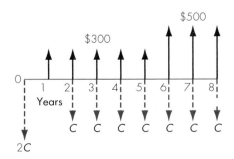

4.43 Find the value of X so that the two cash flows in the figure are equivalent for an interest rate of 10%.

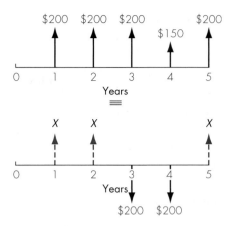

4.44 What single amount at the end of the fifth year is equivalent to a uniform annual series of $3,000 per year for 10 years, if the interest rate is 6%, compounded annually?

4.45 In computing either the equivalent present worth (P) or future worth (F) for the cash flow, at $i = 10\%$, identify all the correct equations from the list below to compute them.

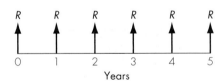

(1) $P = R(P/A, 10\%, 6)$

(2) $P = R + R(P/A, 10\%, 5)$

(3) $P = R(P/F, 10\%, 5) + R(P/A, 10\%, 5)$

(4) $F = R(F/A, 10\%, 5) + R(F/P, 10\%, 5)$

(5) $F = R + R(F/A, 10\%, 5)$

(6) $F = R(F/A, 10\%, 6)$

(7) $F = R(F/A, 10\%, 6) - R$.

4.46 On the day his baby was born, a father decided to establish a savings account for the child's college education. Any money that is put into the account will earn an interest rate of 8%, compounded annually. He will make a series of annual deposits in equal amounts on each of his child's birthdays from the first through the child's 18th, so that his child can make four annual withdrawals from the account in the amount of $20,000 on each of his birthdays. Assuming that the first withdrawal will be made on the child's 18th birthday, which of the following statements are correct to calculate the required annual deposit?

(1) $A = (\$20,000 \times 4)/18$

(2) $A = \$20,000(F/A, 8\%, 4) \times (P/F, 8\%, 21)(A/P, 8\%, 18)$

(3) $A = \$20,000(P/A, 8\%, 18) \times (F/P, 8\%, 21)(A/F, 8\%, 4)$

(4) $A = [\$20,000(P/A, 8\%, 3) + \$20,000](A/F, 8\%, 18)$

(5) $A = \$20,000[(P/F, 8\%, 18) + (P/F, 8\%, 19) + (P/F, 8\%, 20) + (P/F, 8\%, 21)](A/P, 8\%, 18)$.

4.47 Find the equivalent equal-payment series (A) using an A/G factor, such that the two cash flows are equivalent at 10%, compounded annually.

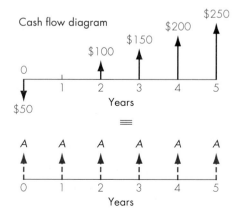

4.48 Consider the following cash flow:

Year End	Payment
0	$500
1–5	1000

In computing F at the end of year 5 at an interest rate of 12%, which of the following statements is incorrect?

(a) $F = \$1,000(F/A, 12\%, 5) - \$500(F/P, 12\%, 5)$

(b) $F = \$500(F/A, 12\%, 6) + \$500(F/A, 12\%, 5)$

(c) $F = [\$500 + \$1,000(P/A, 12\%, 5)] \times (F/P, 12\%, 5)$

(d) $F = [\$500(A/P, 12\%, 5) + \$1,000] \times (F/A, 12\%, 5)$.

4.49 Consider the cash flow series given. In computing the equivalent worth at $n = 4$, which of the following statements is incorrect?

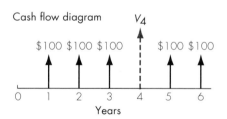

(a) $V_4 = [\$100(P/A, i, 6) - \$100(P/F, i, 4)](F/P, i, 4)$

(b) $V_4 = \$100(F/A, i, 3) + \$100(P/A, i, 2)$

(c) $V_4 = \$100(F/A, i, 4) - \$100 + \$100(P/A, i, 2)$

(d) $V_4 = [\$100(F/A, i, 6) - \$100(F/P, i, 2)](P/F, i, 2)$.

4.50 Henry Cisco is planning to make two deposits, $25,000 now and $30,000 at the end of year 6. He wants to withdraw C each year for the first 6 years and $(C + \$1000)$ each year for the next 6 years. Determine the value of C if the deposits earn 10% interest, compounded annually.

(a) $7,711
(b) $5,794
(c) $6,934
(d) $6,522.

Solving for Unknown Interest Rate

4.51 At what rate of interest, compounded annually, will an investment double itself in 5 years?

4.52 Determine the interest rate (i) that makes the pairs of cash flows economically equivalent.

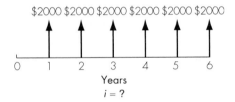

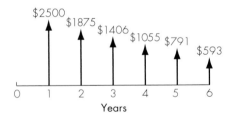

4.53 You have $10,000 available for investment in stock. You are looking for a growth stock whose value can grow to $35,000 over 5 years. What kind of growth rate are you looking for?

Short Case Studies

4.54 Read the following letter from a magazine publisher:

Dear Parent:

Currently your Growing Child/Growing Parent subscription will expire with your 24-month issue. To renew on an annual basis until your child reaches 72 months would cost you a total of $63.84 ($15.96 per year). We feel it is so important for you to continue receiving this material until the 72nd month that we offer you an opportunity to renew now for $57.12. Not only is this a savings of 10% over the regular rate, it is an excellent inflation hedge for you against increasing rates in the future. Please act now by sending $57.12.

(a) If your money is worth 6% per year, determine whether this offer can be of any value.

(b) What rate of interest would you be indifferent between the two renewal options?

4.55 The State of Florida sold a total of 36.1 million lottery tickets at $1 each during the first week of January 2000. As prize money, a total of $41 million will be distributed ($1,952,381 at the *beginning* of each year) over the next 21 years. The distribution of the first-year prize money occurs now, and the remaining lottery proceeds will be put into the state's educational reserve funds, which earns an interest at the rate of 6%, compounded annually. After making the last prize distribution (at the beginning of Year 21), how much would be left over in the reserve account?

4.56 A local newspaper carried the following story.

Texas Cowboys quarterback James Olson, the number-one pick in the National Football League draft, will earn either $11,406,000 over 12 years or $8,600,000 over 6 years. Olson must declare which plan he prefers. The $11 million package is deferred through the year 2011, while the nondeferred arrangement ends after the 2005 season. Regardless of which plan is chosen, Olson will be playing through the 2005 season.

Deferred Plan		Nondeferred Plan	
2000	$2,000,000	2000	$2,000,000
2001	566,000	2001	900,000
2002	920,000	2002	1,000,000
2003	930,000	2003	1,225,000
2004	740,000	2004	1,500,000
2005	740,000	2005	1,975,000
2006	740,000		
2007	790,000		
2008	540,000		
2009	1,040,000		
2010	1,140,000		
2011,	1,260,000		
Total	$11,406,000	Total	$8,600,000

(a) As it happens, Olson ended up with the non-deferred plan. In retrospect, if Olson's interest rate were 6%, did he make a wise decision in 2000?

(b) At what interest rate would the two plans be economically equivalent?

4.57 Fairmont Textile has a plant in which employees have been having trouble with Carpal Tunnel Syndrome (CTS) (inflammation of the nerves that pass through the carpal tunnel, a tight space at the base of the palm), resulting from long-term repetitive activities, such as years of sewing operation. It seems as if 15 of the employees working in this facility developed signs of CTS over the last 5 years. DeepSouth, the company's insurance firm, has been increasing Fairmont's liability insurance steadily because of this CTS problem. Deep-South is willing to lower the insurance premiums to $16,000 a year (from the current $30,000 a year) for the next 5 years if Fairmont implements an acceptable CTS-prevention program that includes making the employees aware of CTS and how to reduce chances of it developing. What would be the maximum amount that Fairmont should invest in the CTS prevention program to make this program worthwhile? The firm's interest rate is 12%, compounded annually.

4.58 Kersey Manufacturing Co., a small plastic fabricator, needs to purchase an extrusion molding machine for $120,000. Kersey will borrow money from a bank at an interest rate of 9% over 5 years. Kersey expects its product sales to be slow during the first year, but to increase subsequently at an annual rate of 10%. Kersey therefore arranges with the bank to pay off the loan on a "balloon scale," which results in the lowest payment at the end of first year, each subsequent payment to be just 10% over the previous one. Determine these five annual payments.

4.59 The Research and Development (R&D) department of Boswell Electronics Company (BEC) has developed a voice-recognition system that could lead to greater acceptance of personal computers among the Japanese. Currently, the complicated and voluminous set of Japanese characters required for a keyboard make it unwieldy and bulky in comparison to the trim, easily portable keyboards, to which Westerners are accustomed. Boswell has used voice-recognition systems in more traditional settings, such as medical reporting in the English language. However, adapting this technology to the Japanese language could result in a breakthrough new level of acceptance of personal computers in Japan. The investment required to develop a full-scale commercial version would cost BEC $10 million, which will be financed at an interest rate of 12%. The system will sell for about $4,000 (or net cash profit of $2,000) and run on high-powered PCs. The product will have a 5-year market life. Assuming that the annual demand for the product remains constant over the market life, how many units does BEC have to sell each year to pay off the initial investment and interest?

4.60 *Millionaire Babies—How to Save our Social Security System.* It sounds a little wild, but that is probably the point. Senator Bob Kerrey, D-Nebraska, has proposed giving every newborn baby a $1,000 government savings account at birth, followed by five annual contributions of $500 each. If left untouched in an investment account, Kerrey says, by the time the baby reaches age 65, which $3,500 contribution per child would grow to $600,000 over the years, even at medium returns for a thrift-savings plan. At about 9.4%, the balance would grow to be $1,005,132. (How would you calculate this number?) With about 4 million babies born each year, the proposal would cost the federal government $4 billion annually. Kerrey offered this idea in a speech devoted to tackling Social Security reform. About 90% of the total annual Social Security tax collections of more than $300 billion are used to pay current beneficiaries, which is the largest federal program. The remaining 10% is invested in interest-bearing government bonds that finance the day-to-day expenses of the federal government. Discuss the economics of Senator Bob Kerrey's Social Security savings plan.

4.61 Troy Aikman, the Dallas Cowboys' quarterback, agreed to an 8-year, $50 million contract that at the time made him the highest-paid player in professional football history.[3] The contract includes a signing bonus of $11 million.

The agreement called for annual salaries of $2.5 million in 1993, $1.75 million in 1994, $4.15 million in 1995, $4.90 million in 1996, $5.25 million in 1997, $6.2 million in 1998, $6.75 million in 1999, an $7.5 million in the year 2000. The $11-million sighing bonus was prorated over the course of the contract, so that an addition $1.375 million was paid each year over the 8-year contract period. With the salary paid at the beginning of each season, the net annual payment schedule looked like the following:

Beginning of Season	Prorated Contact Salary	Actual Signing Bonus	Annual Payment
1993	$2,500,00	$1,375,000	$3,875,000
1994	1,750,000	1,375,000	3,125,000
1995	4,150,000	1,375,000	5,525,000
1996	4,900,000	1,375,000	6,275,000
1997	5,250,000	1,375,000	6,625,000
1998	6,200,000	1,375,000	7,575,000
1999	6,750,000	1,375,000	8,125,000
2000	7,500,000	1,375,000	8,875,000

(a) How much was Troy's actual contract actually worth at the time of signing?

(b) For the signing bonus portion, suppose that the Dallas Cowboys allowed Troy to take either the prorated payment option as described above or a lump sum payment option in the amount of $8 million at the time of contract. Should Troy have taken the lump sum option instead of the prorated one?

Assume that Troy could invest his money at 6% interest

---

[3] *The Sporting News.* (Reprinted with permission.)

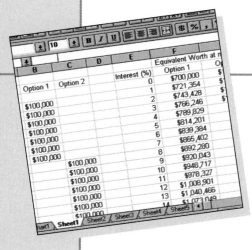

Understanding Money and Its Management

Financing Your Automobile—Installment Loan Versus Up-Front Cash Payment: I got an e-mail from Eric Cressler, one of my students in class on Sunday, May 10, 1998. The e-mail reads as follows:

Dear Professor

This afternoon my parents decided to purchase a 1998 Chevy Blazer. They were presented with the following pair of options for purchase.

Option A: Purchase the vehicle at the normal price of $26,200 and pay for the vehicle over 3 years with equal monthly payments at 1.9% APR (annual percentage rate) financing.

Option B: Purchase the vehicle paying a discount price of $24,048 to be paid immediately. The funds, which would be used to purchase the vehicle, are presently earning 5% annual interest compounded monthly.

It strikes me that the problem is one of the textbook problems, but the interest rate throws an interesting twist. I haven't finished solving it myself yet (the decision was already made before I even found out in any case). Anyway, I thought I would share this example with you, as it seems particularly applicable to what we are presently studying.

Eric Cressler

Now put yourself in the position of Eric's parents. First, what is the meaning of the annual percentage rate (APR) of 1.9% quoted by the dealer? Under what circumstances would you prefer to go with the dealer's financing? Which interest rate (the dealer's interest rate or the savings rate) would you use in comparing the two options?

In this chapter, we will consider several concepts crucial to managing money. In Chapter 4, we examined how time affects the value of money, and we developed various interest formulas for this purpose. Using these basic formulas, we will now extend the concept of equivalence to determine interest rates implicit in many financial contracts. To this end, we will introduce several examples in the area of loan transactions. For example, many commercial loans require that interest compound more frequently than once a year—for instance monthly or quarterly. To consider the effect of more frequent compounding, we must begin with an understanding of the concepts of nominal and effective interest.

5.1 Nominal and Effective Interest Rates

In all the examples in Chapter 4, the implicit assumption was that payments are received *once a year,* or *annually.* However, some of the most familiar financial transactions in both personal financial matters and engineering economic analysis involve payments not based on one annual payment; for example, monthly mortgage payments and quarterly earnings on savings accounts. Thus, if we are to compare different cash flows with different compounding periods, we need to evaluate them on a common basis. The need has led to the development of the concepts of **nominal interest rate** and **effective interest rate.**

5.1.1 Nominal Interest Rates

Take a closer look at the billing statement from any credit card. Or if you financed a new car recently, examine the loan contract. You will typically find the interest that the bank charges on your unpaid balance. Even if a financial institution uses a unit of time other than a year—a month or a quarter (e.g., when calculating interest payments), the institution usually quotes the interest rate on an *annual basis.* Many banks, for example, state the interest arrangement for credit cards in this way:

"18% compounded monthly."

The credit card statement above simply means that each month the bank will charge 1.5% interest on an unpaid balance. We say 18% is the **nominal interest rate** or **annual percentage rate** (APR), and the compounding frequency is monthly (12). As shown in Figure 5.1, to obtain the interest rate per compounding period, we divide 18% by 12, for example, 1.5% per month.

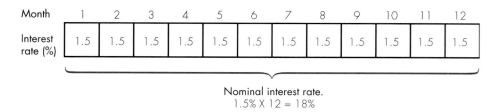

Figure 5.1 The nominal interest rate is determined by summing the individual interest rates per period

Although the annual percentage rate, or APR, is commonly used by financial institutions, and is familiar to many customers, the APR does not explain precisely the amount of interest that will accumulate in a year. To explain the true effect of more frequent compounding on annual interest amounts, we will introduce the term effective interest rate.

5.1.2   Effective Annual Interest Rates

The **effective annual interest rate** is the one rate that truly represents the interest earned in a year. For instance, in our credit card example, the bank will charge 1.5% interest on unpaid balance at the end of each month. Therefore, the 1.5% rate represents the effective interest rate per month. On a yearly basis, you are looking for a cumulative rate—1.5% each month for 12 times. This cumulative rate predicts the actual interest payment on your outstanding credit card balance.

Suppose you deposit $10,000 in a savings account that pays you at an interest rate of 9%, compounded quarterly. Here, 9% represents the nominal interest rate, and the interest rate per quarter is 2.25% (9%/4). The following is an example of how interest is compounded when it is paid quarterly:

First quarter	Base amount	$10,000.00
	+ Interest (2.25%)	+ $225.00
Second quarter	= New base amount	$10,225.00
	+ Interest (2.25%)	+ $230.06
Third quarter	=New base amount	$10,455.06
	+ Interest (2.25%)	+ $235.24
Fourth quarter	=New base amount	$10,690.30
	+ Interest (2.25%)	+ $240.53
	= Value after one year	$10,930.83

Clearly, you are earning more than 9% on your original deposit. In fact, you are earning 9.3083% ($930.83/$10,000). We could calculate the total annual interest payment for a principal amount of $10,000 using the formula given in Eq. (4.2). If $P = \$10,000$, $i = 2.25\%$, and $N = 4$, we obtain

$$F = P(1 + i)^N$$
$$= \$10,000(1 + 0.0225)^4$$
$$= \$10,930.83.$$

The implication is that, for each dollar deposited, you are earning an equivalent annual interest of 9.38 cents. In terms of an effective annual interest rate (i_a), the interest payment can be rewritten as a percentage of the principal amount:

$$i_a = \$930.83/\$10,000 = 0.093083, \text{ or } 9.3083\%.$$

In other words, earning a 2.25% interest per quarter for 4 quarters is equivalent to earning 9.3083% interest just one time each year.

Table 5.1
Nominal and
Effective
Interest
Rates with
Different
Compound-
ing Periods

Nominal Rate	Effective Rates				
	Compounding Annually	Compounding Semi-annually	Compounding Quarterly	Compounding Monthly	Compounding Daily
4%	4.00%	4.04%	4.06%	4.07%	4.08%
5	5.00	5.06	5.09	5.12	5.13
6	6.00	6.09	6.14	6.17	6.18
7	7.00	7.12	7.19	7.23	7.25
8	8.00	8.16	8.24	8.30	8.33
9	9.00	9.20	9.31	9.38	9.42
10	10.00	10.25	10.38	10.47	10.52
11	11.00	11.30	11.46	11.57	11.62
12	12.00	12.36	12.55	12.68	12.74

Table 5.1 shows effective interest rates at various compounding intervals for 4%–12% APR. As you can see, depending on the frequency of compounding, the effective interest earned or paid by the borrower can differ significantly from the APR. Therefore, truth-in-lending laws require that financial institutions quote both nominal interest rate and compounding frequency, i.e., effective interest, when you deposit or borrow money.

Certainly, more frequent compounding increases the amount of interest paid over a year at the same nominal interest rate. Assuming that the nominal interest rate is r, and M compounding periods occur during the year, then, the effective annual interest rate (i_a) can be calculated as follows:

$$i_a = (1 + r/M)^M - 1. \tag{5.1}$$

When $M = 1$, we have the special case of annual compounding. Substituting $M = 1$ in Eq. (5.1), reduces it to $i_a = r$. That is, when compounding takes place once annually, effective interest is equal to nominal interest. Thus, in the examples in Chapter 4, where only annual interest was considered, we were, by definition, using effective interest rates.

Example 5.1 Determining a Compounding Period

Consider the following bank advertisement that appeared in a local newspaper: "Open a California Federal Certificate of Deposit (CD) and get a guaranteed rate of return on as little as $1,000. It's a smart way to manage your money for months."

CD Type	6-month	1-year	2-year	5-year
Rate (nominal)	8.00%	8.50%	9.20%	9.60%
Yield (effective)	8.16%	8.87%	9.64%	10.07%
Minimum deposit	$1,000	$1,000	$10,000	$20,000

In this advertisement, no mention is made of specific interest compounding frequencies. Find the compounding period for each CD.

Solution

Given: $r = 8\%$ per year, i_a

Find: M

First we will consider the 6-month CD. The nominal interest rate is 8% per year, and the effective annual interest rate (yield) is 8.16%. Using Eq. (5.1), we obtain the expression

$$0.0816 = (1 + 0.08/M)^M - 1$$

or

$$1.0816 = (1 + 0.08/M)^M.$$

By trial and error we find $M = 2$, which indicates semi-annual compounding. Thus, the 6-month CD earns 8% interest, compounded semi-annually. If the CD is not cashed at maturity, normally it will be renewed automatically at the original interest rate. For example, if you leave the CD in the bank for another 6 months, your CD will earn 4% interest on $1040:

$$\text{CD value after 1-year deposit} = \$1040(1 + 0.04)$$

$$= \$1081.60.$$

Note: This is equivalent to earning 8.16% interest on $1000 for 1 year.

Similarly, we can find the interest periods for the other CDs: For the 1-year CD, we note that the difference between nominal and effective interest is greater than in the case of the 6-month CD, where the compounding period was semi-annual. A greater difference between nominal and effective rates suggests a greater number of compounding periods, so let us guess at the other end of the spectrum, i.e., daily compounding. This guess proves to be correct, as $i_a = (1 + 0.085/365)^{365} - 1 = 8.87\%$. The same reasonable guess will prove true for the 2-year and 5-year CD—both are compounded daily.

Example 5.2 Future Value of a Certificate of Deposit

Suppose you purchase the 5-year CD described in Example 5.1. How much would the CD be worth when it matures at the end of 5 years?

Solution

Given: $P = \$20,000$, $r = 9.6\%$ per year, $M = 365$ periods per year
Find: F

If you purchase the 5-year CD in Example 5.1, it will earn 9.60% interest, compounded daily. This means that your CD earns an effective annual interest of 10.07%:

$$
\begin{aligned}
F &= P(1 + i_a)^N \\
&= \$20,000(1 + 0.1007)^5 \\
&= \$20,000(F/P, 10.07\%, 5) \\
&= \$32,313.
\end{aligned}
$$

5.1.3 Effective Interest Rates per Payment Period

We can generalize the result of Eq. (5.1) to compute the effective interest rate for *any time duration.* As you will see later, the effective interest rate is usually computed based on the payment (transaction) period. For example, if cash flow transactions occur quarterly, but interest is compounded monthly, we may wish to calculate the effective interest rate on a quarterly basis. To consider this, we may redefine Eq. (5.1) as

$$
\begin{aligned}
i &= (1 + r/M)^C - 1 \\
&= (1 + r/CK)^C - 1,
\end{aligned}
\tag{5.2}
$$

where

M = The number of interest periods per year,

C = The number of interest periods per payment period,

K = The number of payment periods per year.

Note that $M = CK$ in Eq. (5.2).

Example 5.3 Effective Rate per Payment Period

Suppose that you make quarterly deposits in a savings account which earns 9% interest, compounded monthly. Compute the effective interest rate per quarter.

Solution

Given: $r = 9\%$, C = three interest periods per quarter, and K = four quarterly payments per year, and $M = 12$ interest periods per year

Find: i

Using Eq. (5.2), we compute the effective interest rate per quarter as

$$i = (1 + 0.09/12)^3 - 1$$
$$= 2.27\%.$$

Comments: For the special case of annual payments with annual compounding, we obtain $i = i_a$ with $C = M$ and $K = 1$. Figure 5.2 illustrates the relationship between the nominal and effective interest rates per payment period.

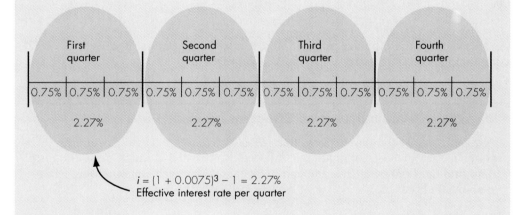

Figure 5.2   Functional relationships among r, i, and i_a where interest is calculated based on 9%, compounded monthly, and payments occur quarterly (Example 5.3)

5.1.4 Continuous Compounding

To be competitive on the financial market, or to entice potential depositors, some financial institutions offer frequent compounding. As the number of compounding periods (M) becomes very large, the interest rate per compounding period (r/M) becomes very small. As M approaches infinity and r/M approaches zero, we approximate the situation of **continuous compounding.**

By taking limits on both sides of Eq. (5.2), we obtain the effective interest rate per payment period as

$$
\begin{aligned}
i &= \lim_{CK \to \infty} [(1 + r/CK)^C - 1] \\
&= \lim_{CK \to \infty} (1 + r/CK)^C - 1 \\
&= \lim_{CK \to \infty} [(1 + r/CK)^{CK}]^{1/K} - 1 \\
&= (e^r)^{1/K} - 1.
\end{aligned}
$$

Therefore,

$$
i = e^{r/K} - 1. \tag{5.3}
$$

To calculate the effective annual interest rate for continuous compounding, we set K equal to 1, resulting in

$$
i_a = e^r - 1. \tag{5.4}
$$

As an example, the effective annual interest rate for a nominal interest rate of 12% compounded continuously is $i_a = e^{0.12} - 1 = 12.7497\%$.

Example 5.4 Calculating an Effective Interest Rate

Assuming a $1,000 initial deposit, find the effective interest rate per quarter, at a nominal rate of 8%, compounded (a) weekly, (b) daily, and (c) continuously. Also find the final balance at the end of 3 years under each compounding scheme as shown in Figure 5.3.

Solution

Given: $r = 8\%$, M, C, $K = 4$ quarterly payments per year, $P = \$1000$, $N = 12$ quarters
Find: i, F

(a) Weekly compounding:

$r = 8\%$, $M = 52$, $C = 13$ interest periods per quarter, $K = 4$ payments per year;

$$
i = (1 + 0.08/52)^{13} - 1 = 2.0186\%.
$$

With $P = \$1,000$, $i = 2.0187\%$, and $N = 12$ (quarters) in Eq. (5.3),

$$
F = \$1,000(F/P, 2.0186\%, 12) = \$1271.03.
$$

(b) Daily compounding:

$r = 8\%$, $M = 365$, $C = 91.25$ days per quarter, $K = 4$;

$$i = (1 + 0.08/365)^{91.25} - 1 = 2.0199,$$

$$F = \$1,000(F/P, 2.0199\%, 12) = \$1271.21.$$

(c) Continuous compounding:

$r = 8\%$, $M \to \infty$, $C \to \infty$, $K = 4$; using Eq. (5.3)

$$i = e^{0.08/4} - 1 = 2.0201\%,$$

$$F = \$1,000(F/P, 2.0201\%, 12) = \$1271.23.$$

Comments: Note that the difference between daily compounding and continuous compounding is often negligible. Many banks offer continuous compounding to entice deposit customers, but the extra benefits are small.

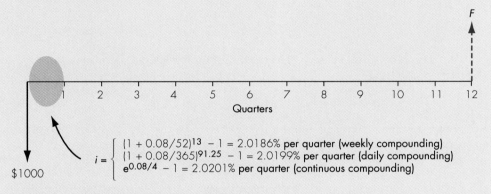

Figure 5.3   Calculation of effective interest rate per quarter (Example 5.4)

5.2 Equivalence Calculations when Payment Periods and Compounding Periods Coincide

All the examples in Chapter 4 assumed annual payments and annual compounding. Whenever a situation occurs where the compounding and payment periods are equal ($M = K$), whether compounded annually or at some other interval, this solution method can be used:

1. Identify the number of compounding periods (M) per year.
2. Compute the effective interest rate per payment period, i.e., using Eq. (5.2), and with $C = 1$ and $K = M$, we have

$$i = r/M.$$

3. Determine the number of compounding periods:

$$N = M \times (\text{number of years}).$$

Example 5.5 Calculating Auto Loan Payments

Suppose you want to buy a car. You have surveyed the dealers' newspaper advertisements, and the following one has caught your attention:

> 8.5% Annual Percentage Rate! 48-month financing on all Mustangs in stock. 60 to choose from.
> ALL READY FOR DELIVERY! Prices Starting As Low As $21,599.
> You just add sales tax and 1% for dealer's freight. We will pay the tag, title, and license.
>
> Add 4% sales tax = $863.96
> Add 1% dealer's freight = $215.99
> Total purchase price = $22,678.95.

You can afford to make a down payment of $2,678.95, so the net amount to be financed is $20,000. What would the monthly payment be? (Figure 5.4).

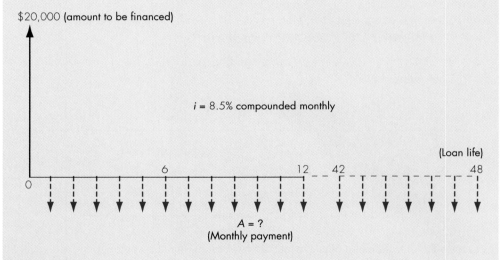

Figure 5.4   A car loan cash transaction (Example 5.5)

Solution

The advertisement does not specify a compounding period, but in automobile financing, the interest and the payment periods are almost always monthly. Thus, the 8.5% APR means 8.5%, compounded monthly.

Given: $P = \$20,000$, $r = 8.5\%$ per year, $K = 12$ payments per year, $N = 48$ months, $M = 12$ interest periods per year

Find: A

In this situation, we can easily compute the monthly payment using Eq. (4.12):

$$i = 8.5\%/12 = 0.7083\% \text{ per month,}$$

$$N = (12)(4) = 48 \text{ months,}$$

$$A = \$20,000(A/P, 0708\%, 48) = \$492.97.$$

5.3 Equivalence Calculations when Payment Periods and Compounding Periods Differ

A number of situations involve cash flows that occur at intervals not the same as the compounding intervals. Whenever payment and compounding periods differ from each other, *one or the other must be transformed so that both conform to the same unit of time.* For example, if payments occur quarterly and compounding occurs monthly, the most logical procedure is to calculate the effective interest rate per quarter. On the other hand, if payments occur monthly and compounding occurs quarterly, we may be able to find the equivalent monthly interest rate. The bottom line is that to proceed with equivalency analysis, the compounding and payment periods must be in the same order.

5.3.1 Compounding More Frequent than Payments

The computational procedure for compounding periods and payment periods that cannot be compared is as follows:

1. Identify the number of compounding periods per year (M), the number of payment periods per year (K), and the number of interest periods per payment period (C);
2. Compute the effective interest rate per payment period:
    - For discrete compounding, compute
    $$i = (1 + r/M)^c - 1$$
    - For continuous compounding, compute
    $$i = e^{r/K} - 1.$$

3. Find the total number of payment periods:
    $$N = K \times (\text{number of years}).$$

4. Use i and N in the appropriate formulas in Table 4.3.[1]

Example 5.6 Compounding More Frequent than Payments (Discrete Compounding Case)

Suppose you make equal quarterly deposits of $1,000 into a fund that pays interest at a rate of 12% compounded monthly. Find the balance at the end of year two. (Figure 5.5).

---

[1] Park CS, Sharp-Bette GP. *Advanced Engineering Economics.* New York: John Wiley & Sons, 1990. (Reprinted by permission of John Wiley & Sons Inc.)

Solution

Given: $A = \$1,000$ per quarter, $r = 12\%$ per year, $M = 12$ compounding periods per year, $N = 8$ quarters

Find: F

We follow the procedure for noncomparable compounding and payment periods as described above.

1. Identify the parameter values for M, K, and C, where

   $$M = 12 \text{ compounding periods per year,}$$

   $$K = \text{four payment periods per year,}$$

   $$C = \text{three interest periods per payment period.}$$

2. Use Eq. (5.2) to compute effective interest:

   $$i = (1 + 0.12/12)^3 - 1$$
   $$= 3.030\% \text{ per quarter.}$$

3. Find the total number of payment periods, N, where

   $$N = K(\text{number of years}) = 4(2) = 8 \text{ quarters.}$$

4. Use i and N in the appropriate equivalence formulas:

   $$F = \$1,000(F/A, 3.030\%, 8) = \$8901.81.$$

Comment: No 3.030% interest table appears in Appendix D, but the interest factor can still be evaluated by $F = \$1,000(A/F, 1\%, 3)(F/A, 1\%, 24)$, where the first interest factor finds its equivalent monthly payment and the second interest factor converts the monthly payment series to an equivalent lump-sum future payment.

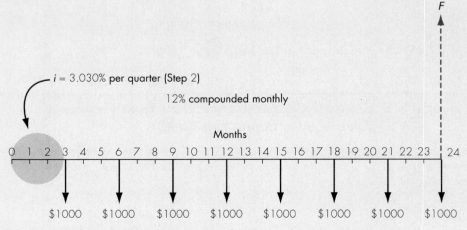

Figure 5.5   Quarterly deposits with monthly compounding (Example 5.6)

Example 5.7 Compounding More Frequent than Payments (Continuous Compounding Case)

A series of equal quarterly receipts of $500 extends over a period of 5 years. What is the present worth of this quarterly payment series at 8% interest compounded continuously? (Figure 5.6).

Discussion: The equivalent problem statement is "How much do you need to deposit now in a savings account that earns 8% interest, compounded continuously, so that you can withdraw $500 at the end of each quarter for 5 years?" Since the payments are quarterly, we need to compute i per quarter for the equivalence calculations.

$$i = e^{r/K} - 1$$

$$= e^{0.08/4} - 1$$

$$= 0.0202 = 2.02\% \text{ per quarter,}$$

$$N = \text{(four payment periods per year)(5 years)}$$

$$= 20 \text{ quarter periods.}$$

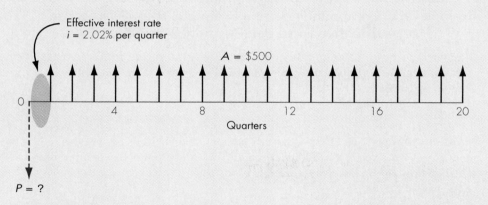

Figure 5.6   A present worth calculation for an equal-payment series with an interest rate of 8%, compounded continuously (Example 5.7)

Solution

Given: $i = 2.02\%$ per quarter, $N = 20$ quarters, and $A = \$500$ per quarter
Find: P

Using the $(P/A,i,N)$ factor with $i = 2.02\%$ and $N = 20$, we find that

$$P = A(P/A,2.02\%,20)$$

$$= \$500(16.3199)$$

$$= \$8,159.96.$$

5.3.2 Compounding Less Frequent than Payments

The two following examples contain identical parameters for savings situations in which compounding occurs less frequently than payments. However, two different underlying assumptions govern how interest is calculated. In Example 5.8, the assumption is that, whenever a deposit is made, it starts to earn interest. In Example 5.9, the assumption is that the deposits made within a quarter do not earn interest until the end of that quarter. As a result, in Example 5.8, we transform the compounding period to conform to the payment period. In Example 5.9, we lump several payments together to match the compounding period. In the real world, which assumption is applicable depends on the transactions and the financial institutions involved. The accounting methods used by many firms record cash transactions that occur within a compounding period as if they had occurred at the end of that period. For example, when cash flows occur daily, but the compounding period is monthly, the cash flows within each month are summed (ignoring interest) and treated as a single payment on which interest is calculated.

Note: *In this textbook, we assume that whenever the timing of a cash flow is specified, one cannot move it to another time point without considering the time value of money, i.e., the practice demonstrated in Example 5.8 should be followed.*

Example 5.8 Compounding Less Frequent than Payments: Effective Rate per Payment Period

Suppose you make $500 monthly deposits to a tax-deferred retirement plan that pays interest at a rate of 10%, compounded quarterly. Compute the balance at the end of 10 years.

Solution

Given: r = 10% per year, M = 4 quarterly compounding periods per year, K = 12 payment periods per year, A = $500 per month, and N = 120 months, and interest is accrued on flow during the compounding period

Find: i, F

As in the case of Example 5.6, the procedure for noncomparable compounding and payment periods is followed:

1. The parameter values for M, K, and C are

   M = 4 compounding periods per year,

   K = 12 payment periods per year,

   C = 1/3 interest periods per payment period.

2. As shown in Figure 5.7, the effective interest rate per payment period is calculated using Eq. (5.2):

$$i = (1 + 0.10/4)^{1/3} - 1$$

$$= 0.826\% \text{ per month.}$$

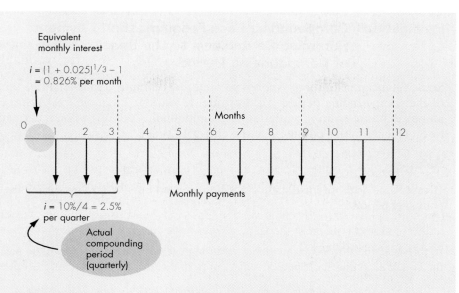

Figure 5.7 Calculation of equivalent monthly interest when the quarterly interest rate is specified (Example 5.8)

3. Find N

$$N = (12)(10) = 120 \text{ payment periods.}$$

4. Use i and N in the appropriate equivalence formulas (Figure 5.8):

$$F = \$500(F/A, 0.826\%, 120)$$
$$= \$101{,}907.89.$$

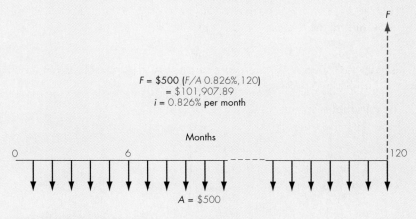

Figure 5.8 Cash flow diagram (Example 5.8)

Example 5.9 Compounding Less Frequent than Payment: Summing Cash Flows to the End of Compounding Period

Some financial institutions will not pay interest on funds deposited after the start of the compounding period. To illustrate, reconsider Example 5.8. Assume that money deposited during a quarter (the compounding period) will not earn any interest (Figure 5.9). Compute F at the end of 10 years.

Solution

Given: Same as those for Example 5.8; however, no interest on flow during the compounding period

Find: F

In this case, the three monthly deposits during each quarterly period will be placed at the end of each quarter. Then the payment period coincides with the interest period:

$$i = 10\%/4 = 2.5\% \text{ per quarter,}$$

$$A = 3(\$500) = \$1,500 \text{ per quarter,}$$

$$N = 4(10) = 40 \text{ payment periods,}$$

$$F = \$1,500(F/A, 2.5\%, 40) = \$101,103.83.$$

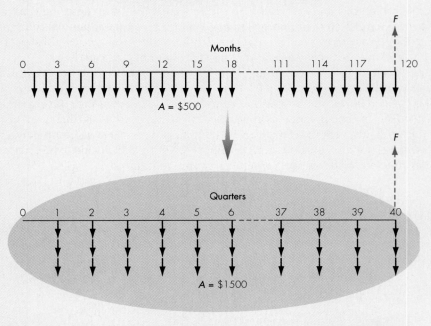

Figure 5.9   Transformed cash flow diagram created by summing monthly cash flows to the end of the quarterly compounding period (Example 5.9)

Comments: In Example 5.9, the balance will be $804.06 less than with the compounding situation in Example 5.8, a fact that is consistent with our understanding that increasing the frequency of compounding increases future value. Some financial institutions follow the practice illustrated in Example 5.8. As an investor, you should reasonably ask yourself whether it makes sense to make deposits in an interest-bearing account more frequently than interest is paid. In the interim between interest compounding, you may be tying up your funds prematurely and foregoing other opportunities to earn interest.

Figure 5.10 illustrates a decision chart that allows you to sum up how you can proceed to find the effective interest rate per payment period given the various possible compounding/interest arrangements.

5.4 Equivalence Calculations with Continuous Payments

As we have seen so far, interest can be compounded annually, semi-annually, monthly, or even continuously. Discrete compounding is appropriate for many financial transactions; mortgages, bonds, and installment loans, which require payments or receipts at discrete times, are good examples. In most businesses, however, transactions occur continuously throughout the year. In these circumstances, we may describe the financial transactions as having a continuous flow of money, for which continuous compounding and discounting are more realistic. This section illustrates how one establishes the economic equivalence between cash flows under continuous compounding.

Continuous cash flows represent situations where money flows continuously and at a known rate throughout a given time period. In business, many daily cash flow transactions can be viewed as continuous cash flows. An advantage of the continuous flow approach is that it more closely models the realities of business transactions. Costs for labor, carrying inventory, and operating and maintaining equipment are typical examples. Others include capital improvement projects that conserve energy, water, or process steam, whose savings can occur continuously.

5.4.1 Single-Payment Transactions

First we will illustrate how single-payment formulas for continuous compounding and discounting are derived. Suppose that you invested P dollars at a nominal rate of r % interest for N years. If interest is compounded continuously, the effective annual interest is $i = e^r - 1$. The future value of the investment at the end of N years is obtained with the F/P factor by substituting $e^r - 1$ for i:

$$F = P(1 + i)^N$$
$$= P(1 + e^r - 1)^N$$
$$= Pe^{rN}.$$

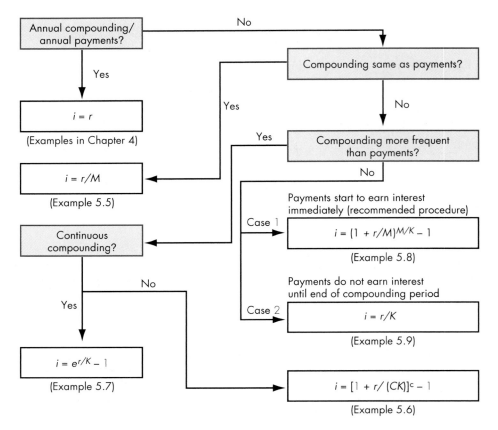

Figure 5.10   A decision flowchart demonstrating how to compute effective interest rate per payment period (i)

This implies that $1 invested now at an interest rate of $r\%$, compounded continuously, accumulates to e^{rN} dollars at the end of N years.

Correspondingly, the present value of F due N years from now and discounted continuously at an interest rate of $r\%$ is equal to

$$P = Fe^{-rN}.$$

We can say that the present value of $1 due N years from now, discounted continuously at an annual interest rate of $r\%$, is equal to e^{-rN} dollars.

5.4.2   Continuous Funds-Flow

Suppose that an investment's future cash flow per unit of time (e.g., per year) can be expressed by a continuous function ($f(t)$), which can take any shape. Assume also that the investment promises to generate cash of $f(t)\Delta t$ dollars between t and $t + \Delta t$, where

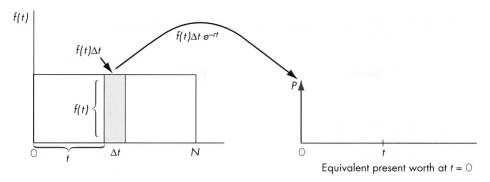

Figure 5.11 Finding an equivalent present worth of a continuous flow payment function $f(t)$ at a nominal rate of $r\%$

t is a point in the time interval $0 \le t \le N$ (Figure 5.11). If the nominal interest rate is constant r during this time interval, the present value of this cash stream is given approximately by the expression

$$\Sigma(f(t)\Delta t)e^{-rt},$$

where e^{-rt} is the discounting factor that converts future dollars into present dollars. With the project's life extending from 0 to N, we take the summation over all subperiods (compounding periods) in the interval from 0 to N. As the division of the interval becomes smaller and smaller, that is, as Δt approaches zero, we obtain the present value expression by the integral

$$P = \int_0^N f(t)e^{-rt}dt. \tag{5.5}$$

Similarly, the future value expression of the cash flow stream is given by the expression

$$F = Pe^{rN} = \int_0^N f(t)e^{r(N-t)}dt, \tag{5.6}$$

where $e^{r(N-t)}$ is the compounding factor that converts present dollars into future dollars. It is important to observe that the time unit is *year,* because the effective interest rate is expressed in terms of a year. Therefore, all time units in equivalence calculations must be converted into years. Table 5.2[2] summarizes some typical continuous cash functions that can facilitate equivalence calculations.

---

[2] Park CS, Sharp-Bette GP. *Advanced Engineering Economics.* New York: John Wiley & Sons, 1990. (Reprinted by permission of John Wiley & Sons Inc.)

Table 5.2
Summary of
Interest
Factors for
Typical
Continuous
Cash Flows
with
Continuous
Compound-
ing

Type of Cash Flow	Cash Flow Function	Parameters Find	Given	Algebraic Notation	Factor Notation
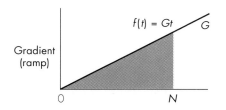Uniform (step)	$f(t) = \bar{A}$	P	$\bar{A}$	$\bar{A}\left[\dfrac{e^{rN}-1}{re^{rN}}\right]$	$(P/\bar{A}, r, N)$
		$\bar{A}$	P	$P\left[\dfrac{re^{rN}}{e^{rN}-1}\right]$	$(\bar{A}/P, r, N)$
		F	$\bar{A}$	$\bar{A}\left[\dfrac{e^{rN}-1}{r}\right]$	$(F/\bar{A}, r, N)$
		$\bar{A}$	F	$F\left[\dfrac{r}{e^{rN}-1}\right]$	$(\bar{A}/P, r, N)$
Gradient (ramp)	$f(t) = Gt$	P	G	$\dfrac{G}{r^2}(1 - e^{-rN}) - \dfrac{G}{r}(Ne^{-rN})$	
Decay	$f(t) = ce^{-jt}$ j^t = decay rate with time	P	c, j	$\dfrac{c}{r+j}(1 - e^{-(r+j)N})$	

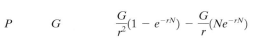

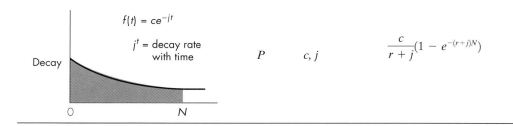

Example 5.10 Comparison of Daily Flows and Daily Compounding with Continuous Flows and Continuous Compounding

Consider a situation where money flows daily. Suppose you own a retail shop and generate $200 cash each day. You establish a special business account and deposit these daily cash flows in an account for 15 months. The account earns an interest rate of 6%. Compare the accumulated cash values at the end of 15 months, assuming

(a) Daily compounding, and

(b) Continuous compounding, respectively.

Solution

(a) With daily compounding:

Given: A = $200 per day, r = 6% per year, M = 365 compounding periods per year, N = 455 days

Find: F

Assuming there are 455 days in the 15-month period, we find

$$
\begin{aligned}
i &= 6\%/365 \\
&= 0.01644\% \text{ per day,} \\
N &= 455 \text{ days.}
\end{aligned}
$$

The balance at the end of 15 months will be

$$
\begin{aligned}
F &= \$200(F/A, 0.01644\%, 455) \\
&= \$200(472.4095) \\
&= \$94,482.
\end{aligned}
$$

(b) With continuous compounding:

Now we approximate this discrete cash flow series by a uniform continuous cash flow function as shown in Figure 5.12. Here we have the situation where an amount flows at the rate of $\bar{A}$ per year for N years.

Note: *Our time unit is a year.* Thus, a 15-month period is 1.25 years. Then, the cash flow function is expressed as

$$
\begin{aligned}
f(t) &= \bar{A}, 0 \leq t \leq 1.25 \\
&= \$200(365) \\
&= \$73,000 \text{ per year.}
\end{aligned}
$$

Given: $\bar{A}$ = $73,000 per year, r = 6% per year, compounded continuously,

N = 1.25 years

Find: F

Substituting these values back into Eq. (5.6) yields

$$
F = \int_{0}^{1.25} 73,000\, e^{0.06(1.25 - t)}dt
$$

$$
= \$73,000 \left[\frac{e^{0.075} - 1}{0.06} \right]
$$

$$
= \$94,759.
$$

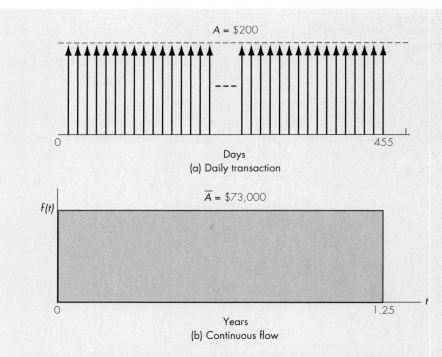

Figure 5.12   Comparison between daily transaction and continuous fund flow transaction (Example 5.10)

The factor in the bracket is known as the **funds flow compound amount factor** and is designated $(F/\bar{A},\ r,\ N)$ as shown in Table 5.2. Also notice that the difference between the two methods is only \$277 (less than 0.3%).

Comments: As shown in this example, the differences between discrete (daily) and continuous compounding have no practical significance in most cases. Consequently, as a mathematical convenience, instead of assuming that money flows in discrete increments at the end of each day, we could assume that money flows continuously during the time period at a uniform rate. This type of cash flow assumption is a common practice in the chemical industry.

5.5 Changing Interest Rates

Up to this point, we have assumed a constant interest rate in our equivalence calculations. When an equivalence calculation extends over several years, more than one interest rate may be applicable to properly account for the time value of money. This is to say, over time, interest rates available in the financial marketplace fluctuate, and a financial institution, committed to a long-term loan, may find itself in the position of losing the opportunity to earn higher interest because some of its holdings are tied up in a lower interest loan. A financial institution may attempt to protect itself from such lost earning opportunities by building gradually increasing interest rates into a long-term loan at the outset. Adjustable-rate mortgage (ARM) loans are perhaps the most common examples of variable interest rates. In this section, we will consider variable interest rates in both single-payment and a series of cash flows.

5.5.1 Single Sums of Money

To illustrate the mathematical operations involved in computing equivalence under changing interest rates, first consider the investment of a single sum of money, P, in a savings account for N interest periods. If i_n denotes the interest rate appropriate during time period N, the future worth equivalent for a single sum of money can be expressed as

$$F = P\,(1 + i_1)(1 + i_2)\ldots(1 + i_{N-1})(1 + i_N), \tag{5.7}$$

and solving for P yields the inverse relation

$$P = F\,[(1 + i_1)(1 + i_2)\ldots(1 + i_{N-1})(1 + i_N)]^{-1}. \tag{5.8}$$

Example 5.11 Changing Interest Rates with Lump-Sum Amount

You deposit $2,000 in an Individual Retirement Account (IRA) that pays interest at 6%, compounded monthly, for the first 2 years and 9%, compounded monthly, for the next 3 years. Determine the balance at the end of 5 years (Figure 5.13).

Solution

Given: P = $2,000, r = 6% per year for first 2 years, 9% per year for last 3 years, M = 4 compounding periods per year, N = 20 quarters

Find: F

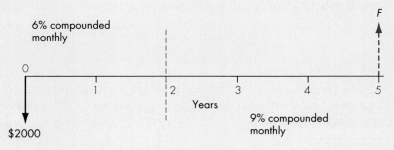

Figure 5.13   Changing interest rates (Example 5.11)

We will compute the value of F in two steps. First we will compute the balance at the end of 2 years, B_2. With 6%, compounded quarterly

$$i = 6\%/12 = 0.5\%$$

$$N = 12(2) = 24 \text{ months}$$

$$B_2 = \$2,000(F/P, 0.5\%, 24)$$

$$= \$2,000\ (1.12716)$$

$$= \$2,254.$$

Since the fund is not withdrawn, but reinvested at 9%, compounded monthly, as a second step, we compute the final balance as follows:

$$i = 9\%/12 = 0.75\%$$

$$N = 12(3) = 36 \text{ months}$$

$$F = B_2(F/P, 0.75\%, 36)$$

$$= \$2,254(1.3086)$$

$$= \$2,950$$

5.5.2 Series of Cash Flows

This consideration of changing interest rates can be easily extended to a series of cash flows. In this case, the present worth of a series of cash flows can be represented as

$$P = A_1(1 + i_1)^{-1} + A_2[(1 + i_1)^{-1}(1 + i_2)^{-1}] + \dots$$
$$+ A_N[(1 + i_1)^{-1}(1 + i_2)^{-1}\dots(1 + i_N)^{-1}].$$

(5.9)

The future worth of a series of cash flows is given by the inverse of Eq. (5.9):

$$F = A_1[(1 + i_2)(1 + i_3)\dots(1 + i_N)]$$
$$+ A_2[(1 + i_3)(1 + i_4)\dots(1 + i_N)] + \dots + A_N.$$

(5.10)

The uniform series equivalent is obtained in two steps. First, the present worth equivalent of the series is found using Eq. (5.9). Then, A, is solved for after establishing the following equivalence equation:

$$P = A(1 + i_1)^{-1} + A[(1 + i_1)^{-1}(1 + i_2)^{-1}] + \dots$$
$$+ A\,[(1 + i_1)^{-1}(1 + i_2)^{-1}\dots(1 + i_N)^{-1}].$$

(5.11)

Example 5.12 Changing Interest Rates with Uneven Cash Flow Series

Consider the cash flow in Figure 5.14 with the interest rates indicated and determine the uniform series equivalent for the cash flow series.

Discussion: In this problem and many others, the easiest approach involves collapsing the original flow into a single equivalent amount, for example, at time zero and then converting the single amount into the final desired form.

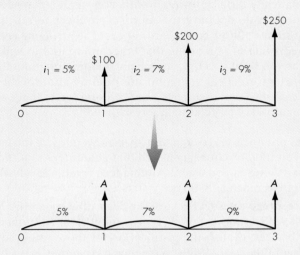

Figure 5.14 Equivalence calculation with changing interest rates (Example 5.12)

Solution

Given: Cash flows and interest rates as shown in Figure 5.14, $N = 3$

Find: A

Using Eq. (5.9), we find the present worth:

$$P = \$100(P/F, 5\%, 1) + \$200(P/F, 5\%, 1)(P/F, 7\%, 1)$$
$$+ \$250(P/F, 5\%, 1)(P/F, 7\%, 1)(P/F, 9\%, 1)$$
$$= \$477.41.$$

Then, we obtain the uniform-series equivalent as follows:

$$\$477.41 = A(P/F, 5\%, 1) + A(P/F, 5\%, 1)(P/F, 7\%, 1)$$
$$+ A(P/F, 5\%, 1)(P/F, 7\%, 1)(P/F, 9\%, 1)$$
$$= 2.6591A$$
$$A = \$179.54.$$

5.6 Debt Management

Credit card debt and commercial loans are among the most significant financial transactions involving interest. Many types of loans are available, but here we will focus on those most frequently used by individuals and in business.

5.6.1 Borrowing with Credit Cards

When credit cards were introduced in 1959, people were able to handle their personal finances in a dramatically different way. From a consumer's perspective, using credit cards means cardholders do not have to wait for a paycheck to reach the bank before they can make a purchase. Most credit cards operate as *revolving credit.* With revolving credit, you have a line of borrowing that you can tap into at will and pay back as quickly or slowly as you want—as long as you pay the minimum required each month. Your monthly bill is an excellent source of information about what your card really costs. Four things affect your card-based credit costs: annual fees, finance charge, grace period, and method of calculating interest.

- **Annual fees** are once-yearly charges for using a particular card. You can often avoid them entirely by choosing a card that guarantees that there will be no annual fee for as long as you use it. Annual fees can range anywhere from $0 (no annual fee) to as much as $300 per year.

- **Annual percentage rates (APR)** are the costs of using credit. Many issuers still charge 18% APR or more a year (1.5% a month) on your outstanding balances and cash advances, though the justification for those rates dates to a period of high inflation in the early 1980s. A finance charge may be as high as 21% or as low as 5.9%, but when the rate is stated as an APR, you can compare one loan to another.

- **Grace period** is the time between when you are billed and when you have to pay. This is the amount of time a lender allows before charging you interest on the balance due. Standard is 25 days from the date of the bill before they assess finance charges. If there is no grace period, finance charges will accrue the moment you make a purchase with your credit card—even though you haven't received a bill.

- **Minimum payment** is set by some issuers which will set a high minimum if they are uncertain of your ability to pay, however, most issuers require you to pay a minimum payment of 2% of the outstanding balance.

- **Finance charges (interest)** will depend on the method used to calculate your finance charge—each one figures its charges and fees according to the agreement you made with the bank when you signed and used its card. For example, suppose you pay an 18% annual finance charge (1.5% per month) on amounts you owe. Your previous balance is $3,000, and you pay $1,000 on the 15th day of a 30-day period. Table 5.3 compares how banks calculate interests under three different methods.

Current Practice: Paying Off Cards Saves a Bundle. Investors are always looking for a sure thing, a risk-free way to get a big return for their money. Yet they often ignore one of the best and safest ways to make the most of their money: paying off credit card bills. To make

Method	Description	Interest You owe
Adjusted Balance	The bank subtracts the amount of your payment from the beginning balance and charges you interest on the remainder. This method costs you the least.	Your beginning balance is $3,000. With the $1,000 payment, your new balance will be $2,000. You pay 1.5% on this new balance, which will be $30.
Average Daily Balance	The bank charges you interest on the average of the amount you owe each day during the period. So the larger the payment you make, the lower the interest you pay.	Your beginning balance is $3,000. With your $1,000 payment at the 15th day, your balance will be reduced to $2,000. Therefore, your average balance will be ($3,000 + $2,000)/2 = $2,500. The bank will charge 1.5% on this amount: (1.5%)($2,500) = $37.50
Previous Balance	The bank does not subtract any payments you make from your previous balance. You pay interest on the total amount you owe at the beginning of the period. This method costs you the most.	Regardless of your payment size, the bank will charge 1.5% on your beginning balance $3,000: (1.5%)($3,000) = $45

Table 5.3
Methods of Calculating Interests on Your Credit Card

matters worse for consumers, minimum payments required by card issuers have been going down from 3% to 2%. Say, for example, that you owe $2,000 on a credit card that charges 18% APR, and you make the minimum 2% payment every month. It would take more than 30 years to pay off the debt.

Pay the minimum, pay for years

Making minimum payments on your credit cards can cost you a bundle over a lot of years. Here's what would happen if you paid the minimum—or more—every month on a $2,705 card balance, with a 18.38% interest rate.

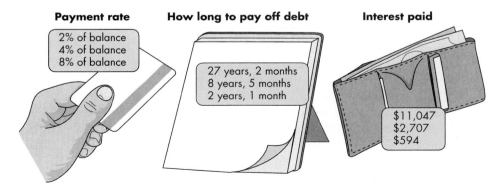

Payment rate	How long to pay off debt	Interest paid
2% of balance 4% of balance 8% of balance	27 years, 2 months 8 years, 5 months 2 years, 1 month	$11,047 $2,707 $594

(Source: *USA Today,* April 21, 1998, © *USA Today,* used with permission)

5.6.2 Commercial Loans

One of the most important applications of compound interest involves loans that are paid off in **installments** over time. If the loan is to be repaid in equal periodic amounts (weekly, monthly, quarterly, or annually), it is said to be an **amortized loan.** Examples include automobile loans, loans for appliances, home mortgage loans, and most business debts other than very short-term loans. Most commercial loans have interest that is compounded monthly. With a loan, a local bank or a dealer advances you the money to pay for the car, and you repay the principal plus interest in monthly installments, usually over a period of 3 to 5 years. The car is your collateral. If you don't keep up with your payments, the lender can repossess, or take back the car and keep all the payments you have made.

Two things determine what borrowing will cost you: the finance charge and the length of the loan. The cheapest loan is not one with the lowest payments, or even the one with the lowest interest rate. Instead, you have to look at the total cost of borrowing, which depends on the interest rate plus fees, and the term, or length of time it takes you to repay. While you probably cannot influence the rate and fees, you may be able to arrange for a shorter term.

- **Annual percentage rate** (APR) is set by lenders which are required to tell you what a loan will actually cost per year, expressed as an annual percentage rate. Some lenders charge lower interest but add high fees; others do the reverse. The APR allows you to compare them on equal terms. It combines the fees with a year of interest charges to give you the true annual interest rate.
- **The fees** are the expenses the lender will charge in lending the money. Application fee covers processing expenses. Attorney fees pay the lender's attorney. Credit search fee covers researching your credit history. Origination fee covers administrative costs, and sometimes appraisal fees. They add up very quickly and can substantially increase the cost of your loan.
- **Finance charges** are the cost of borrowing. For most loans, it includes all the interest, fees, service charges, points, credit-related insurance premiums, and any other charges.
- **Periodic interest rate** is the interest the lender will charge on the amount you borrow. If lender also charges fees, this will not be the true interest rate.
- **The term of your loan** is crucial when determining cost. Shorter terms mean squeezing larger amounts into fewer payments. However, they also mean paying interest for few years, which saves a lot of money.

Amortized Loans

So far, we have considered many instances of amortized loans in which we calculated present or future values of the loans or the amounts of the installment payments. An additional aspect of amortized loans, which will be of great interest to us, is calculating the amount of interest versus the portion of the principal that is paid off in each installment. As we shall explore more fully in Chapter 12, the interest paid on a loan is an important element in calculating taxable income and has repercussions for both personal and business loan transactions. For now, we will focus on several methods of calculating interest and principal paid at any point in the life of a loan.

In calculating the size of monthly installment, lending institutions may use two types of schemes. The first one is the conventional amortized loan based on compound interest method, and the other one is the add-on loan based on the simple interest concept. We explain each method in the following text, but it should be understood that the amortized loan is the most common in various commercial lending. The add-on loan is common in financing appliances, as well as furniture.

In a typical amortized loan, the amount of interest owed for a specified period is calculated based on the remaining balance of the loan at the beginning of the period. It is possible to develop a set of formulas to compute the remaining loan balance, interest payment, and principal payment for a specified period. Suppose we borrow an amount, P, at an interest rate, i, and agree to repay this principal sum, P, including interest, in equal payments, A over N periods. The payment size (A) is $A = P(A/P, i, N)$. Each payment A is divided into an amount that is interest and a remaining amount for the principal payment. Let

B_n = Remaining balance at the end of period n, with $B_0 = P$,

I_n = Interest payment in period n, where $I_n = B_{n-1}i$,

P_n = Principal payment in period n.

Then, each payment can be defined as

$$A = P_n + I_n. \tag{5.12}$$

The interest and principal payments for an amortized loan can be determined in several ways; two methods are presented here. No clear-cut reason is available to prefer one method over the other. Method 1, however, may be easier to adopt when the computational process is automated through a spreadsheet application, whereas method 2 may be more suitable for obtaining a quick solution when a time period is specified. You should become comfortable with at least one of these methods for use in future problem solving. Pick the one that comes most naturally to you.

Method 1—Tabular Method The first method is tabular. The interest charge for a given period is computed progressively based on the remaining balance at the beginning of that period. The interest due at the end of the first period will be

$$I_1 = B_0 i = Pi.$$

Thus, the principal payment at that time will be

$$P_1 = A - Pi,$$

and the balance remaining after the first payment will be

$$B_1 = B_0 - P_1 = P - P_1.$$

At the end of the second period, we will have

$$I_2 = B_1 i = (P - P_1)i,$$
$$P_2 = A - (P - P_1)i = (A - Pi) + P_1 i = P_1(1 + i),$$
$$B_2 = B_1 - P_2 = P - (P_1 + P_2).$$

If we continue, we can show that, at the nth payment,

$$
\begin{aligned}
B_n &= P - (P_1 + P_2 + \ldots + P_n) \\
&= P - [P_1 + P_1(1 + i) + \ldots + P_1(1 + i)^{n-1}] \\
&= P - P_1(F/A, i, n) \\
&= P - (A - Pi)(F/A, i, n).
\end{aligned}
$$

Then, the interest payment during the nth payment period is expressed as

$$I_n = (B_{n-1})i. \tag{5.13}$$

The portion of payment A at period n that can be used to reduce the remaining balance is

$$P_n = A - I_n, \tag{5.14}$$

Example 5.13 Loan Balance, Principal, and Interest— Tabular Method

Suppose you secure a home improvement loan in the amount of $5000 from a local bank. The loan officer computes your monthly payment as follows:

Contract amount	= $5,000,
Contract period	= 24 months,
Annual percentage rate	= 12%,
Monthly installments	= $235.37.

Figure 5.15 is the cash flow diagram. Construct the loan payment schedule by showing the remaining balance, interest payment, and principal payment at the end of each period over the life of the loan.

Solution

Given: $P = \$5,000$, $A = \$235.37$ per month, $r = 12\%$ per year, $M = 12$ compounding periods per year, $N = 24$ months

Find: B_n, I_n for $n = 1$ to 24

We can easily see how the bank calculated the monthly payment of $235.37. Since the effective interest rate per payment period on this loan transaction is 1% per month, we establish the following equivalence relationship:

$$\$235.37(P/A, 1\%, 24) = \$235.37(21.2431) = \$5,000,$$

Figure 5.15 Cash flow diagram of home improvement loan with an APR of 12% (Example 5.13)

The loan payment schedule can be constructed as in Table 5.4. The interest due at $n = 1$ is $50.00, 1% of the $5000 outstanding during the first month. The $185.37 left over is applied to the principal, reducing the amount outstanding in the second month to $4,814.63. The interest due in the second month is 1% of $4,814.63, or $48.15, leaving $187.22 for repayment of the principal. At $n = 24$, the last $235.37 payment is just sufficient to pay the interest on the unpaid loan principal and to repay the remaining principal. Figure 5.16 illustrates the ratios between the interest and principal payments over the life of the loan.

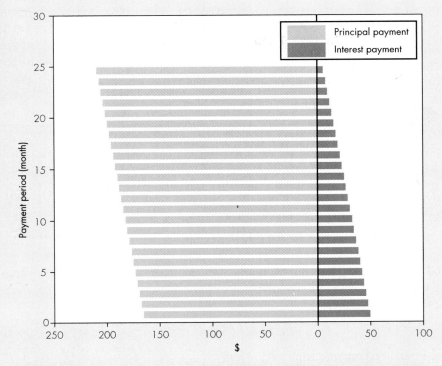

Figure 5.16 The proportions of principal and interest payments over the life of the loan (monthly payment = $235.37) (Example 5.13)

Table 5.4
Creating a
Loan Repayment Schedule Using Excel (Example 5.13)

Payment No.	Payment Size	Principal Payment	Interest Payment	Loan Balance
1	$235.37	$185.37	$50.00	$4,814.63
2	235.37	187.22	48.15	4,627.41
3	235.37	189.09	46.27	4,438.32
4	235.37	190.98	44.38	4,247.33
5	235.37	192.89	42.47	4,054.44
6	235.37	194.82	40.54	3,859.62
7	235.37	196.77	38.60	3,662.85
8	235.37	198.74	36.63	3,464.11
9	235.37	200.73	34.64	3,263.38
10	235.37	202.73	32.63	3,060.65
11	235.37	204.76	30.61	2,855.89
12	235.37	206.81	28.56	2,649.08
13	235.37	208.88	26.49	2,440.20
14	235.37	210.97	24.40	2,229.24
15	235.37	213.08	22.29	2,016.16
16	235.37	215.21	20.16	1,800.96
17	235.37	217.36	18.01	1,583.60
18	235.37	219.53	15.84	1,364.07
19	235.37	221.73	13.64	1,142.34
20	235.37	223.94	11.42	9,18.40
21	235.37	226.18	9.18	6,92.21
22	235.37	228.45	6.92	4,63.77
23	235.37	230.73	4.64	233.04
24	235.37	233.04	2.33	0.00

Comments: Certainly, generation of a loan repayment schedule such as that in Table 5.4 can be tedious and time-consuming process unless a computer is used. As you can see in our book web site, you can download an Excel file that creates the loan repayment schedule, where you can make any adjustment to solve a typical loan problem of your choice.

Method 2—Remaining Balance Method   Alternatively, we can derive B_n by computing the equivalent payments remaining after the nth payment. Thus, the balance with $N - n$ payments remaining is

$$B_n = A(P/A, i, N - n), \qquad (5.15)$$

and the interest payment during period n is

$$I_n = (B_{n-1})i = A(P/A, i, N - n + 1)i, \qquad (5.16)$$

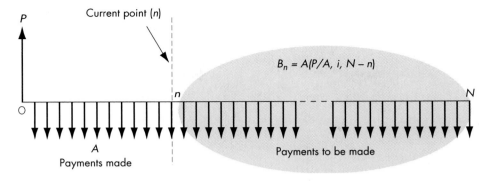

Figure 5.17 Calculating the remaining loan balance based on method 2

where $A(P/A, i, N - n + 1)$ is the balance remaining at the end of period $n - 1$, and

$$P_n = A - I_n = A - A(P/A, i, N - n + 1)i$$
$$= A[1 - (P/A, i, N - n + 1)i].$$

Knowing the interest factor relationship from Table 4.3, $(P/F, i, n) = 1 - (P/A, i, n)i$, we obtain

$$P_n = A(P/F, i, N - n + 1). \tag{5.17}$$

As we can see in Figure 5.17, this latter method provides more concise expressions for computing the loan balance.

Example 5.14 Loan Balances, Principal, and Interest— Remaining Balance Method

Consider the home improvement loan in Example 5.13 and

  (a) For the sixth payment, compute both the interest and principal payments.
  (b) Immediately after making the sixth monthly payment, you would like to pay off the remainder of the loan in a lump sum. What is the required amount of this lump sum?

Solution

(a) Interest and principal payments for the sixth payment.

Given: (as for Example 5.13)

Find: I_6 and P_6

Using Eqs. (5.16) and (5.17), we compute I_6 and P_6 as follows:

$$I_6 = \$235.37(P/A, 1\%, 19)(0.01)$$
$$= (\$4,054.44)(0.01)$$
$$= \$40.54.$$
$$P_6 = \$235.37(P/F, 1\%, 19) = \$194.82,$$

or simply subtract the interest payment from the monthly payment as follows:

$$P_6 = \$235.37 - \$40.54 = \$194.83.$$

(b) Remaining balance after sixth payment: The lower half of Figure 5.18 shows the cash flow diagram that applies to this part of the problem. We can compute the amount you owe after you make the sixth payment by calculating the equivalent worth of the remaining 18 payments at the end of sixth month, with the time scale shifted by 6:

Given: $A = \$235.37$, $i = 1\%$ per month, $N = 18$ months

Find: Remaining balance after 6 months (B_6)

$$B_6 = \$235.37(P/A, 1\%, 18) = \$3,859.62.$$

If you desire to pay off the remainder of the loan at the end of sixth payment, you must come up with $3,859.62. To verify our results, compare the answer to the value given in Table 5.3.

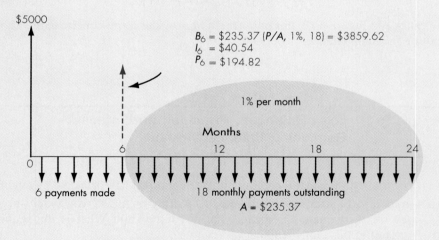

Figure 5.18 Computing the outstanding loan balance after making the sixth payment for the home improvement loan (Example 5.14)

Add-On Loans

The add-on loan is totally different from the popular amortized loan. In this type of loan, the total interest to be paid is precalculated and added to the principal. The principal plus this precalculated interest amount is then paid in equal installments. In such a case, the interest rate quoted is not the effective interest rate, but what is known as **add-on interest**. If you borrow P for N years at an add-on rate of i, with equal payments due at the end of each month, a typical financial institution might compute the monthly installment payments as follows:

$$\text{Total add-on interest} = P(i)(N),$$
$$\text{Principal plus add-on interest} = P + P(i)(N) = P(1 + iN) \qquad (5.18)$$
$$\text{Monthly installments} = P(1 + iN)/(12 \times N).$$

Notice that the add-on interest is *simple interest*. Once the monthly payment is determined, the financial institution computes the APR based on this payment, and you will be told what this value will be. Even though the add-on interest is specified along with the APR value, many ill-informed borrowers think that they are actually paying the add-on rate quoted for this installment loan. To see how much you actually pay interest under a typical add-on loan arrangement, consider Example 5.15.

Example 5.15 Effective Interest Rate for an Add-On Loan

Reconsider the home improvement loan problem in Example 5.13. Suppose that you borrow $5,000 with an add-on rate of 12% for 2 years. You will make 24 equal monthly payments.

(a) Determine the amount of the monthly installment.

(b) Compute the nominal and the effective annual interest rate on the loan.

Solution

Given: Add-on rate = 12% per year, loan amount $(P) = \$5,000$, $N = 2$ years

Find: (a) A and (b) i_a and i

(a) First we determine the amount of add-on interest:

$$iPN = (0.12)(\$5,000)(2) = \$1,200.$$

Then we add this simple interest amount to the principal and divide the total amount by 24 months to obtain A:

$$A = (\$5,000 + \$1,200)/24 = \$258.33$$

(b) Putting yourself in the lender's position, compute the APR value of the loan just described. Since you are making monthly payments, with monthly compounding, you need to find the effective interest rate that makes the present $5,000 sum equivalent to 24 future monthly payments of $258.33. In this situation, we are solving for i in the equation

$$\$258.33 = \$5,000(A/P, i, 24)$$

or

$$(A/P, i, 24) = 0.0517.$$

You know the value of the A/P factor, but you do not know the interest rate, i. As a result, you need to look through several interest tables and determine i by interpolation as shown in Figure 5.19. The interest rate falls between 1.75% and 2.0% and may be computed by a **linear interpolation**:

(A/P, i, 24)	i
0.0514	1.75%
0.0517	?
0.0529	2.00%

$$i = 1.75\% + (0.25\%)\left[\frac{0.0517 - 0.0514}{0.0529 - 0.0514}\right] = 1.8\% \text{ per month.}$$

The nominal interest rate for this add-on loan is $1.8 \times 12 = 21.60\%$, and the effective annual interest rate is $(1 + 0.018)^{12} - 1 = 23.87\%$, rather than the 12% quoted add-on interest. When you take out a loan, you should not confuse the add-on interest rate stated by the lender with the actual interest cost of the loan.

As an alternative to linear interpolation based on interest tables, you can also calculate the effective interest rate for the add-on loan by the many financial calculators or spreadsheet programs currently available. If you solved the equation on either a fi-

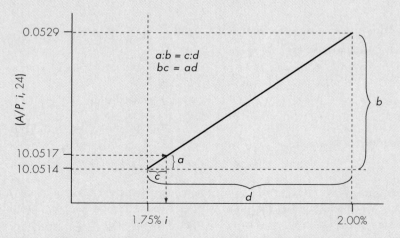

Figure 5.19   Linear interpolation to find unknown interest rate (Example 5.15)

nancial calculator or a computer, the exact value would be 1.79749%. In practice, the small error introduced by this type of interpolation is rarely of significance.

Comments: In the real world, truth-in-lending laws require that APR information always be provided in mortgage and other loan situations—you would not have to calculate nominal interest as a prospective borrower (although you might be interested in calculating actual or effective interest). However, in later engineering economic analyses, you will discover that solving for implicit interest rates, or rates of return on investment, is regularly part of the solution procedure. Our purpose is to periodically give you some practice with this type of problem, even though the example/problem scenario does not exactly model the real-world information you would be given.

5.6.3 Loan Versus Lease

When you choose a car, for example, you also choose how to pay for it. If you do not have the cash on hand to buy a new car outright—and most of us don't—you can consider taking a loan or leasing the car to spread the payments over time. Deciding whether to pay cash, take a loan, or sign a lease depends on a number of personal as well as economic factors. Leasing is an option that lets you pay for the portion of a vehicle you expect to use over a specified term, plus rent charge, taxes, and fees. For example: You might want a $20,000 vehicle. Assume that vehicle might be worth about $9,000 at the end of your 3-year lease (residual value).

- If you have enough money to buy the car, you could purchase the car in cash. If you pay cash, however, you will lose the opportunity to earn interest on the amount you spend. That could be substantial if your investment is paying a good return.

- If you purchase the vehicle using debt financing, your monthly payments will be based on the entire $20,000 value of the vehicle. You will continue to own the vehicle at the end of your financing term, but the interest you will pay on the loan will drive up the real cost of the car.

- If you lease the same vehicle, your monthly payments will be based on the amount of the vehicle you expect to "use up" over the lease term. This value ($11,000, in our example) is the difference between original cost ($20,000) and the estimated value at lease end ($9,000). With leasing, the length of your lease agreement, the monthly payments, and the yearly mileage allowance can be tailored to your driving needs. Greatest financial appeal for leasing is low initial outlay costs: Usually you pay a leasing administrative fee, one month's lease payment, and a refundable security deposit. The terms of your lease will include a specific mileage allowance; if you put additional miles on your car, you will have to pay more for each extra mile.

Example 5.16 compares three financing options—traditional financing, paying cash, and lease financing.

Example 5.16 **Paying Cash Versus Taking a Loan**

Reconsider the car payment option described in the chapter opening.

Option A: Purchase the vehicle at the normal price of $26,200 and pay for the vehicle over 36 months with equal monthly payments at 1.9% APR (annual percentage rate) financing.

Option B: Purchase the vehicle paying a discount price of $24,048 to be paid immediately. The funds, which would be used to purchase the vehicle, are presently earning 5% annual interest compounded monthly.

Which option is a more economically sound one?

Discussion: In calculating the net cost of financing the car, we need to decide which interest rate to use in discounting the loan repayment series. Note that the 1.9% APR represents the dealer's interest rate to calculate the loan payments. With the 1.9% interest, your monthly payments will be A = $26,200(A/P, 1.9%/12, 36) = $749.29. On the other hand, the 5% APR represents your earning opportunity rate. In other words, if you do not buy the car, your money continues to earn 5% APR. Therefore, this 5% rate represents your opportunity cost of purchasing the car. Which interest rate to use in your analysis? Clearly, the 5% rate is the appropriate interest to use in your comparison.

Solution

Given: Loan payment series shown in Figure 5.20, $r = 5\%$, payment period = monthly, compounding period = monthly

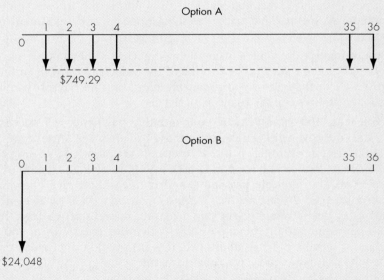

Figure 5.20   Paying cash (option B) versus taking a loan (Option A) (Example 5.16)

Find: The most economical financing option

For each option, we will calculate the net equivalent cost at $n = 0$. Since the loan payments occur monthly, we need to determine the effective interest rate per month, which is 5%/12.

- Option A: Conventional financing:

  Equivalent present cost of the total loan repayments:

  $$P_A = \$749.29(P/A, 5\%/12, 36)$$
  $$= \$25,000$$

- Option B: Cash payment:

  $$P_B = \$24,048.$$

There would be $952 savings in present value with the cash payment option.

Example 5.17 **Buy versus Lease Decision**

Saturn Corporation offers the following two types of financing options for their 2001 Saturn SW2 lines. Calculations are based on special financing programs available at participating Saturn retailers for a limited time.

	Option 1: Traditional Financing	Option 2: Lease Financing
Price of the vehicle	$14,695	$14,695
Down payment	$2,000	0
APR (%)	3.6%	
Monthly payment	$372.55	$236.45
Length	36 months	36 months
Fees	$0	$495
Cash due at lease end		$300
Purchase option at lease end		$8,673.10
Cash due at signing	$2,000	$731.45

For each option, license, title, registration fees, taxes, and insurance are extra. For lease option, the lessee must come up with $731.45 at signing. This cash due at signing includes the first month's lease payment of $236.45 and $495 administrative fee. No security deposit is required. However, $300 disposition fee is due at lease end. The lessee has the option to purchase at lease end for $8,673.10. Lessee is also responsible for excessive wear and use. If your earning interest rate is 6% compounded monthly, which financing option is a better choice?

Discussion: With a lease payment, you pay for the portion of the vehicle you expect to use. At the end of the lease, you simply return the vehicle to the dealer and pay the agreed-upon disposal fee. With traditional financing, your monthly payment is based on the entire $14,695 value of the vehicle, and you will own the vehicle at the end of your financing terms. Since you are comparing the options over 3 years, you must explicitly consider the unused portion (resale value) of the vehicle at the end of the term. In other words, you must consider the resale value of the vehicle to figure out the net cost of owning the vehicle. *As the resale value, you could use $8,673.10 quoted by the dealer* in the lease option. Then you have to ask yourself if you can get that kind of resale value after 3-year ownership.

Solution

Given: Lease payment series shown in Figure 5.21, $r = 6\%$, payment period = monthly, compounding period = monthly

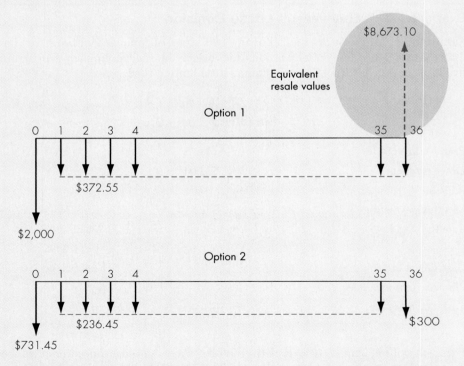

Figure 5.21   Comparing financing options: Tradition financing (Option 1) versus lease financing (Option 2) (Example 5.17)

Find: The most economical financing option with the assumption that you will be able to sell the vehicle for $8,673.10 at the end of 3 years

For each option, we will calculate the net equivalent total cost at $n = 0$. Since the loan payments occur monthly, we need to determine the effective interest rate per month, which is 0.5%.

- **Lease financing**:

    Equivalent present cost of the total lease payments:

    $$P_1 = \$731.45 + \$236.45(P/A, 0.5\%, 35)$$
    $$= \$731.45 + \$7,574.76$$
    $$= \$8,306.21.$$

    Equivalent present cost of disposal fee:

    $$P_2 = \$300(P/F, 0.5\%, 36) = \$250.69.$$

    Equivalent net lease cost:

    $$P = \$8,306.21 + \$250.69$$
    $$= \$8,556.90.$$

- **Conventional financing**

    Equivalent present cost of the total loan payments:

    $$P_1 = \$2,000 + \$372.55(P/A, 0.5\%, 36)$$
    $$= 14,246.10$$

    Equivalent present worth of the resale value:

    $$P_2 = \$8,673.10(P/F, 0.5\%, 36) = \$7,247.63.$$

    Equivalent net financing cost:

    $$P = \$14,246.10 - \$7,247.63$$
    $$= \$6,998.47.$$

It appears that the traditional financing program is more economical at 6% interest compounded monthly.

Comments: By varying the resale value (S), we can find the break-even resale value that makes the traditional financing equivalent to the lease financing.

$$\$8,556.90 = \$14,246.10 - S(P/F, 0.5\%, 36)$$
$$S = (\$14,246.10 - \$8,556.90)/0.8356$$
$$= \$6,808.15$$

At resale value greater than $6,808.15, the conventional financing plan would be the more economical choice.

5.6.4   Home Mortgage

The term **mortgage** refers to a special type of loan for buying a piece of property, such as a house. The most common mortgages are the fixed-rate amortized loans, adjustable-rate loans, and graduated payment loans. Like many of the amortized loans we have considered so far, most home mortgages involve monthly payments and monthly compounding. However, as the names of these types of mortgages suggest, a variety of ways determine how monthly payments and interest rates can be structured over the life of the mortgage.

The Cost of Mortgage

The cost of a mortgage depends on the amount you borrow, the interest you pay, and how long you take to repay. Since monthly payments spread the cost of a mortgage over a long period, it is easy to forget the total expense. For example, if you borrow $100,000 for 30 years at 8.5% interest, your total repayment will be around $277,000, more than two and a half times the original loan. Minor differences in the interest rate—8.5% vs. 8%—can add up to a lot of money over 30 years. At 8% the total repaid would be $264,240, almost $13,000 less than at the 8.5% rate. Other than the interest rate, any of the following factors will increase the overall cost, but a higher interest rate and longer term will have the greatest impact.

- **Loan amount**: The amount you actually borrow after fees and points are deducted. It is the basis for figuring the real interest, or APR, on the money you are borrowing.

- **Loan term**: With a shorter term, you will pay less interest overall, and your monthly payments will be somewhat larger. A 15-year mortgage, as opposed to 30-year mortgage for the same amount, can cut your costs by more than 55%.

- **Payment frequency**: You can pay your mortgage biweekly instead of monthly, or you can make an additional payment each month. With biweekly payments you make 26 regular payments instead of 12 every year. The mortgage is paid off in a little more than half the time, and you pay a little more than half the interest. With an additional payment each month, you can reduce your principal. With a fixed-rate mortgage, you pay off the loan quicker, but regular monthly payments remain the same.

- **Points** (prepaid interest): Interest that you prepay at the closing. Each point is 1% of the loan amount. For example, on a $100,000 loan with 3 points, you would prepay $3,000. This is equivalent to financing $97,000 but your payments are based on $100,000 loan.

- **Fees**: Fees include application fees, loan origination fees, and other initial costs imposed by the lender.

Lenders might be willing to raise the rate by a fraction (say 1/8% or 1/4%) and lower the points—or the reverse—as long as they make the same profit. The advantage of fewer points are lower closing costs and laying out less money when you are apt to need it most. However, if you plan to keep the house longer than 5 to 7 years, paying more points to get a lower interest rate will reduce your long-term cost. Example 5.18 shows the effect of points on the cost of borrowing.

Example 5.18 Points or No Points?

When you are shopping a home mortgage loan, you frequently encounter various types of borrowing options: Pay points up front, or pay no points and accept a slightly higher interest rate. Suppose you want to finance a home mortgage of $100,000 at a 15-year fixed interest rate. Countrywide, a leading independent home lender offers the following two options with no origination fees:

- Option 1: Pay one point with 6.375% interest
- Option 2: Pay no point with 6.75% interest

A point is equivalent to 1% of the face value of the mortgage. In other words, for Option 1, you are borrowing only $99,000, but your lender will calculate your monthly payments based on $100,000. Compute the APR for option 1?

Discussion: Discount fees or points are a fact of life with mortgages. Point is a fee charged by a lender to increase the lender's effective yield on the money borrowed. Points are charged in lieu of interest, the more points paid, the lower the rate of interest required on the loan to provide the same yield or return to the lender. One point equals 1% of the loan amount. Origination fees are the fees charged by a lender to prepare loan document, make credit checks, inspect, and sometimes appraise a property. The fees are usually computed as a percentage of the face value of the mortgage.

Solution

Given: $P = \$100,000$, $r = 6.375\%$, $N = 180$ months, and discount point = 1 point
Find: APR

We first need to find out how much the lender will calculate your monthly payments. Since the mortgage payments will be based on the face value of the loan, your monthly payment will be as follows:

$$A = \$100,000(A/P, 6.375\%/12, 180) = \$863.98$$

Since you have to pay $1,000 to borrow $100,000, you are actually borrowing $99,000. Therefore, you need to find out what kind of interest you are actually paying. In other words, you borrow $99,000 and you make $863.98 monthly payments over 15 years. To find the interest rate, you can set up the following equivalence equation and solve for the unknown interest rate.

$$\$99,000 = \$863.98(P/A, i, 180)$$

$$i = 0.545\% \text{ per month}$$

$$\text{APR} = 0.545\% \times 12 = 6.54\%$$

Note that with one point fee, the lender was able to raise their effective yield from 6.375% to 6.54%. However, it is still earning less than Option 2.

Variable-Rate Mortgages

Mortgage can have either fixed or adjustable rates. As we mentioned earlier, interest rates in the financial marketplace rise and fall over time. If there is a possibility that market rates will rise above the fixed rate, some lenders may be reluctant to lock a loan into a fixed interest rate over a long-term. They thus have to forego the opportunity to earn better rates because their assets are tied up in a loan that earns a lower rate of interest. The variable rate mortgage is meant to encourage lending institutions to commit funds over long periods. Typically, the rate rises gradually each year for the first several years of the mortgage and then settles into a single rate for the remainder of the loan. Example 5.19 illustrates how a lender would calculate the monthly payments with varying interest rates.

Example 5.19 Variable-Rate Mortgage

Consider the following advertisement for selling a beachfront condominium at SunDestin, Florida:

> 95% Financing 8 1/8% interest!!
>
> 5% Down Payment. Own a Gulf-Front Condominium for only $100,000 with a 30-year variable-rate mortgage. We're providing incredible terms: $95,000 mortgage (30 years), year 1 at 8.125%, year 2 at 10.125%, year 3 at 12.125%, and years 4 through 30 at 13.125%.

Compute the monthly payments for each year.

Solution

Given: $P = \$95,000$, r = variable as shown, $M = 12$ compounding periods per year, $N = 360$ months

Find: Monthly payments

(a) Monthly Payment Calculation: During the first year, you borrow $95,000 for 30 years at 8 1/8%. To compute the monthly payment, use $i = (8.125\%)/12 = 0.6771\%$ per month and $N = 360$ months:

$$A_1 = \$95,000(A/P, 0.6771\%, 360) = \$705.37.$$

The remaining balance of the loan after making the 12th payment will be

$$B_{12} = \$705.37(P/A, 0.6771\%, 348) = \$94,223.82.$$

During the second year, the interest rate changes to 10.125%, or 0.8438% per month, but the remaining term is only 348 months. Therefore, the new monthly payment would be

$$A_2 = \$94,223.82(A/P, 0.8438\%, 348) = \$840.19.$$

After making the 24th payment, the remaining mortgage balance is

$$B_{24} = \$840.19(P/A, 0.8438\%, 336) = \$93,656.42.$$

During the third year, the interest rate changes to 12.125%, or 1.0104% per month. The new monthly payment and the remaining balance, after making 36th payment are then

$$A_3 = \$93,656.42(A/P, 1.0104\%, 336) = \$979.75$$
$$B_{36} = \$979.75(P/A, 1.0104\%, 324) = \$93,232.25.$$

For years 4 through 30 years, the interest rate is 13.125%, or 1.0938% per month. The new monthly payment is

$$A_4 = \$93,232.25(A/P, 1.0938\%, 324) = \$1,050.69.$$

As summarized in Figure 5.22, your monthly payments over the life of the loan are

Year	Month	Monthly Payment
1	1–12	$705.37
2	13–24	840.19
3	25–36	979.75
4–30	37–360	1,050.69

(b) Effective Annual Interest Rate Calculation: Once you determine the monthly payments over the life of the loan, you may be interested in knowing the equivalent effective annual fixed interest rate for this payment plan. To do this, you can first find the interest rate per month by solving the following equivalence relationship:

$$\$95,000 = \$705.37(P/A, i, 12) + \$840.19(P/A, i, 12)(P/F, i, 12)$$
$$+ \$979.75(P/A, i, 12)(P/F, i, 24)$$
$$+ \$1,050.69(P/A, i, 324)(P/F, i, 36)$$
$$= (P/A, i, 12)[\$705.37 + 840.19(P/F, i, 12)$$
$$+ \$979.75(P/F, i, 24)]$$
$$+ \$1,050.69(P/A, i, 324)(P/F, i, 36).$$

You now need to solve for i. While interest tables can be used, the many factors involved make it difficult to find the interest rate implicit in an uneven series of payments. You can use a trial-and-error procedure or spreadsheet software. If using a trial-and-error approach, try $i = 1.00\%$ per month:

$$P = (P/A, 1\%, 12)[\$705.37 + \$840.19(P/F, 1\%, 12)$$
$$+ \$979.75(P/F, 1\%, 24)]$$
$$+ \$1,050.69(P/A, 1\%, 324)(P/F, 1\%, 36)$$
$$= \$95,524.29.$$

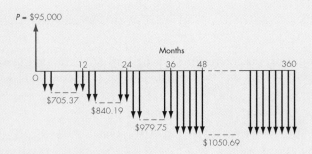

Figure 5.22 A graduated payment schedule for a $95,000 mortgage over 30 years (Example 5.19)

The equivalent present worth amount is too high. Try a higher interest rate, say, $i = 1.1\%$:

$$P = (P/A, 1.1\%, 12) \, [\$705.37 + \$840.19(P/F, 1.1\%, 12)$$
$$+ \, \$979.75(P/F, 1.1\%, 24)]$$
$$+ \, \$1{,}050.69(P/A, 1.1\%, 324)(P/F, 1.1\%, 36)$$
$$= \, \$87{.}119.59.$$

The value of i must be between 1% and 1.1%. By interpolation

P	i
$95,524.29	1%
95,000.00	?
87,119.59	1.1%

$$i = 1\% + (0.1\%) \left[\frac{\$95{,}524.29 - \$95{,}000.00}{\$95{,}524.29 - \$87{,}119.59} \right]$$
$$= 1.0062\% \text{ per month.}$$

(A more precise value obtained by use of the Excel spreadsheet is 1.005839%.) The nominal interest rate (or APR) is $12 \times 1.0062\% = 12.0749\%$. The effective annual interest rate is

$$i_a = (1 + 0.010062)^{12} - 1 = 12.77\%.$$

When considering an amortized loan, with a fixed interest rate of 1.0062%, the equivalent monthly payments would be

$$A = \$95{,}000(A/P, 1.0062\%, 360) = \$982.63,$$

which is about 39% higher than the first-year's monthly payment ($705.37) obtained under the terms of a variable payment plan.

5.7 Summary

- Interest is most frequently quoted by financial institutions as an **APR** or **annual percentage rate**. However, compounding often occurs more often than once annually, and the APR does not account for the effect of this more frequent compounding. The situation leads to the distinction between nominal and effective interest:

- **Nominal interest** is a stated rate of interest for a given period (usually a year).

- **Effective interest** is the actual rate of interest, which accounts for the interest amount accumulated over a given period. The **effective rate** is related to the APR by the following equation:

$$i = (1 + r/M)^M - 1,$$

where r = APR, M = number of compounding periods, and i = effective interest rate.

  In any equivalence problem, the interest rate to use is the effective interest rate per payment period:

$$i = [1 + r/(CK)]^C - 1,$$

where C = number of interest periods per payment period, K = number of payment periods per year, and r/K = nominal interest rate per payment period. Figure 5.10 outlines the possible relationships between compounding and payment periods and indicates which version of the effective interest formula to use.

- The equation for determining the effective interest of continuous compounding is as follows:

$$i = e^{r/K} - 1.$$

The difference in accumulated interest between continuous compounding and very frequent compounding ($M > 50$) is minimal.

- Cash flows, as well as compounding, can be continuous. Table 5.2 shows the interest factors to use for continuous cash flows with continuous compounding.

- Nominal (and hence effective) interest rates may fluctuate over the life of a cash flow series. Some forms of home mortgages and bond yields are typical examples.

- **Amortized loans** are paid off in equal installments over time, and most of these loans have interest that is compounded monthly.

- Under a typical **add-on loan,** the lender precalculates the total simple interest amount and adds it to the principal. The principal plus this precalculated interest amount is then paid in equal installments.

- The term **mortgage** refers to a special type of loan for buying a piece of property, such as a house and commercial buildings. The cost of the mortgage will depend on many factors such as loan amount, loan term, payment frequency, points, and fees.

- Two types of mortgages are common: (1) fixed-rate mortgages and (2) variable-rate mortgages. Fixed-rate mortgages refer to those loans where the borrowing interest rates are fixed over the contract periods, whereas the variable-rate mortgages refer to those whose borrowing interest rates fluctuate depending on the market conditions. In general, the initial interest rate is lower for those variable rate mortgages as the lenders have the flexibility to adjust the cost of loans over the contract period.

Self-Test Questions

5s.1 You have just received credit card applications from two banks, A and B. The interest terms on your unpaid balance are stated as follows:

(1) Bank A: 15%, compounded quarterly

(2) Bank B: 14.8%, compounded daily.

Which of the following statements is incorrect?

(a) The effective annual interest rate for Bank A is 15.865%.

(b) The nominal interest rate for Bank B is 14.8%.

(c) Bank B's term is a better deal because you will pay less interest on your unpaid balance.

(d) Bank A's term is a better deal because you will pay less interest on your unpaid balance.

5s.2 What is the future worth of an equal quarterly payment series of $2,500 for 10 years, if the interest rate is 9%, compounded monthly?

(a) $F = \$158,653$ (b) $F = \$151,930$
(c) $F = \$154,718$ (d) $F = \$160,058$.

5s.3 John secured a home improvement loan from a local bank in the amount of $10,000 at an interest rate of 9%, compounded monthly. He agreed to pay back the loan in 60 equal monthly installments. Immediately after the 24th payment, John decides to pay off the remainder of the loan in a lump sum. What will be the size of this payment?

(a) $P = \$7,473$ (b) $P = \$6,000$
(c) $P = \$6,528$ (d) $P = \$7,710$.

5s.4 You want to open a savings plan for your future retirement. You are considering two options as follows:

- Option 1: You deposit $1,000 at the end of each quarter for the first 10 years. At the end of 10 years, you make no further deposits, but you leave the amount accumulated at the end of 10 years for the next 15 years.

- Option 2: You do nothing for the first 10 years. Then you deposit $6,000 at the end of each year, for the next 15 years.

Compare the amounts accumulated in Option 1 and Option 2. If your deposits or investments earn an interest rate of 6%, compounded quarterly, and you choose Option 2 over Option 1, which of the following statements is correct? At the end of 25 years from now, I will accumulate

(a) $7,067 more   (b) $8,523 more
(c) $14,757 less   (d) $13,302 less.

5s.5 Vi Wilson is interested in buying an automobile priced at $18,000. From her personal savings, she can come up with a down payment in the amount of $3,000. The remaining balance will be financed by the dealer over a period of 36 months at an interest rate of 6.25%, compounded monthly. Which of the following statements is correct?

(a) The dealer's annual percentage rate (APR) is 6.432%.

(b) The monthly payment can be calculated by using $A = \$15,000$

$(A/P, 6.25\%, 3)/12$.

(c) The monthly payment can be calculated by using $A = \$15,000$

$(A/P, 0.5208\%, 36)$.

(d) The monthly payment can be calculated by using $A = \$15,000$

$(A/P, 6.432\%/12, 3)$.

5s.6 A series of equal semi-annual payments of $1,000 for 3 years is equivalent to what present amount at an interest rate of 12%, compounded annually? (All answers are rounded to nearest dollars.)

(a) $4,944

(b) $4,804

(c) $4,500

(d) $5,401.

5s.7 At what rate of interest, compounded quarterly, will an investment double itself in 5 years?

(a) 14.87%

(b) 3.72%

(c) 3.53%

(d) 14.11%.

5s.8 A series of equal quarterly deposits of $1000 extends over a period of 3 years. What is the future worth of this quarterly deposit series at 9% interest, compounded monthly?

(a) $13,160

(b) $12,590

(c) $13,615

(d) $13,112.

5s.9 You have been offered a credit card by an oil company that charges interest at 1.8% per month, compounded monthly. What is

the effective annual interest rate that this oil company charges?

(a) 21.60%

(b) 22.34%

(c) 23.87%

(d) 18.00%.

5s.10 A series of equal quarterly receipts of $1000 extends over a period of 5 years. What is the present worth of this quarterly payment series at 8% interest, compounded *continuously*?

(a) $16,351

(b) $16,320

(c) $15,971

(d) $18,345.

5s.11 You are considering purchasing a piece of industrial equipment that costs $30,000. You decide to make a down payment in the amount of $5,000 and to borrow the remainder from a local bank at an interest rate of 9%, compounded monthly. The loan is to be paid off in 36 monthly installments. What is the amount of the monthly payment?

(a) $954

(b) $833

(c) $795

(d) $694.

5s.12 A building is priced at $100,000. If a down payment of $30,000 is made, and a payment of $1,000 every month thereafter is required, how many months will it take to pay for the building? Interest is charged at a rate of 12%, compounded monthly.

(a) 70

(b) 100

(c) 121

(d) 132.

5s.13 You obtained a loan of $20,000 to finance the purchase of an automobile. Based on monthly compounding for 24 months, the end-of-the-month equal payment was figured out to be $922.90. Immediately after making the 12th payment, you decide you want to pay off the loan in lump sum. The size of this lump sum amount is nearest to which of the following?

(a) $10,498

(b) $11,075

(c) $12,044

(d) $12,546.

5s.14 To raise money for your business, you need to borrow $20,000 from a local bank. If the bank asks you repay the loan in five equal annual installments of $5548.19, determine the bank's annual interest rate on this loan transaction.

(a) 11%

(b) 11.5%

(c) 12%

(d) 27.74%.

5s.15 How many years will it take for an investment to double itself if the interest rate is 9%, compounded quarterly?

(a) $7 < N \leq 8$ years

(b) $8 < N \leq 9$ years

(c) $9 < N \leq 10$ years

(d) $10 < N \leq 11$ years.

5s.16 Consider the following bank advertisement appearing in a local newspaper: "Open a Decatur National Bank Certificate of Deposit (CD), and you get a guaranteed rate of return (effective annual yield) of 8.87%." If there are 365 compounding periods per year, what is the nominal interest rate (annual percentage rate) for this CD?

(a) 8.00%

(b) 8.23%

(c) 8.50%

(d) 8.87%.

5s.17 You borrowed $1,000 at 8%, compounded annually. The loan was repaid according to the following schedule.

n	Repayment Amount
1	$100
2	$300
3	$500
4	X

Find X, the amount that is required to pay off the loan at the end of year 4.

(a) $108

(b) $298

(c) $345

(d) $460.

5s.18 Suppose you deposit $C at the end of each month for 10 years at an interest rate of 12%, compounded continuously. What equal end-of-year deposit over 10 years would accumulate the same amount at the end of 10 years under the same interest compounding?

(a) $A = [\$12C(F/A, 12.75\%, 10)] \times (A/F, 12.75\%, 10)$

(b) $A = \$C(F/A, 1.005\%, 12)$

(c) $A = [\$C(F/A, 1\%, 120)] \times (A/F, 12\%, 10)$

(d) $A = \$C(F/A, 1.005\%, 120) \times (A/F, 12.68\%, 10)$

(e) None of the above.

5s.19 Which of the following banks offers you a better interest deal for your deposit?

Bank A: 8.5%, compounded *quarterly*
Bank B: 8.4%, compounded *continuously*

(a) Bank A

(b) Bank B

(c) Indifferent

(d) Not sufficient information to decide.

5s.20 Compute the value of F, if the interest rate is 8%, compounded quarterly.

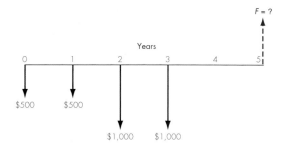

(a) $3,840

(c) $3,900

(b) $3,870

(d) $3,930

5s.21 Compute the lump sum amount required at the end of year 4 to repay an amount of

$20,000 borrowed today at an interest rate of 12%, compounded monthly.

(a) $20,812   (c) $31,470
(b) $27,812   (d) $32,244

5s.22 Suppose you borrowed $10,000 at an interest rate of 12%, compounded monthly over 36 months. At the end of the first year (after 12 payments), you want to negotiate with the bank to pay off the remainder of the loan in 8 equal quarterly payments. What is the amount of this quarterly payment, if the interest rate and compounding frequency remain the same?

(a) $875   (c) $996
(b) $925   (d) $1,006

5s.23 Compute the lump sum amount required at the end of year 4 to repay an amount of $20,000 borrowed today at an interest rate of 12%, compounded quarterly.

(a) $29,600   (c) $31,470
(b) $32,094   (d) $32,244

5s.24 Henry Jones is planning to retire in 15 years. He wishes to deposit an equal amount (A) every 6 months until he retires so that, beginning one year following his retirement, he will receive annual payments of $30,000 for the next 15 years. Determine the value of A which he should deposit bi-annually if the interest rate is 8%, compounded semi-annually.

(a) $2,138   (c) $4,534
(b) $3,249   (d) $5,891

5s.25 Suppose you deposit $1,000 at the end of each quarter for 5 years at an interest of 8% compounded continuously. What

equal end-of-year deposit over 5 years would accumulate the same amount at the end of 5 years under the same interest compounding (8%, compounded continuously)? To answer the question, which of the following is *correct*?

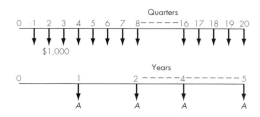

(a) $A = \$1,000 \times (F/A, 2\%, 20) \times (A/F, 8\%, 5)$

(b) $A = \$1,000 \times (F/A, e^{0.02} - 1, 4)$

(c) $A = \$1,000 \times (F/A, e^{0.02} - 1, 20) \times (A/F, 8\%, 5)$

(d) None of the above

5s.26 Assume you deposited $10,000 in a savings account that pays 6%, compounded monthly interest. You wish to withdraw $200 at the end of each month. How many months will it take to deplete the balance?

(a) less than 48 months
(b) between 49 and 52 months
(c) between 53 and 56 months
(d) between 57 and 60 months.

5s.27 Find the value of P.

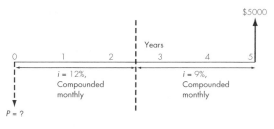

(a) $2,965   (c) $3,110
(b) $3,010   (d) $3,210

Problems

Nominal and Effective Interest Rates

5.1 A loan company offers money at 1.5% per month, compounded monthly.

(a) What is the nominal interest rate?

(b) What is the effective annual interest rate?

(c) How many years will it take an investment to triple itself if interest is compounded monthly?

(d) How many years will it take an investment to triple itself if the nominal rate is compounded continuously?

5.2 A department store has offered you a credit card that charges interest at 0.95% per month, compounded monthly. What is the nominal interest (annual percentage) rate for this credit card? What is the effective annual interest rate?

5.3 A California bank, Berkeley Savings and Loan, advertised the following information: Interest 7.55%—effective annual yield 7.842%. No mention is made of the interest period in the advertisement. Can you figure out the compounding scheme used by the bank?

5.4 War Eagle Financial Sources, which makes small loans to college students, offers to lend $400. The borrower is required to pay $26.61 at the end of each week for 16 weeks. Find the interest rate per week. What is the nominal interest rate per year? What is the effective interest rate per year?

5.5 A financial institution is willing to lend you $40. However, $45 is repaid at the end of one week.

(a) What is the nominal interest rate?

(b) What is the effective annual interest rate?

5.6 The Cadillac Motor Car Company is advertising a 24-month lease of a Cadillac Deville for $520 payable at the beginning of each month. The lease requires a $2,500 down payment, plus a $500 refundable security deposit. As an alternative, the company offers a 24-month lease with a single up-front payment of $12,780, plus a $500 refundable security deposit. The security deposit will be refunded at the end of the 24-month lease. Assuming an interest rate of 6%, compounded monthly, which lease is the preferred one?

5.7 As a typical middle-class consumer, you are making monthly payments on your home mortgage (9% annual interest rate), car loan (12%), home improvement loan (14%), and past-due charge accounts (18%). Immediately after getting a $100 monthly raise, your friendly mutual fund broker tries to sell you some investment funds, with a guaranteed return of 10% per year. Assuming that your only other investment alternative is a savings account, should you buy?

Compounding More Frequent than Annually

5.8 What will be the amount accumulated by each of these present investments?

(a) $4,638 in 10 years at 6%, compounded semi-annually

(b) $6,500 in 15 years at 8%, compounded quarterly

(c) $28,300 in 7 years at 9%, compounded monthly.

5.9 How many years will it take an investment to triple itself if the interest rate is 8% compounded

(a) Quarterly

(b) Monthly

(c) Continuously?

5.10 A series of equal quarterly payments of $5,000 for 12 years is equivalent to what present amount at an interest rate of 9%, compounded as follows:

(a) Quarterly

(b) Monthly

(c) Continuously?

5.11 What is the future worth of an equal payment series of $5000 each for 5 years if the interest rate is 8%, compounded continuously?

5.12 Suppose that $1,000 is placed in a bank account at the end of each quarter over the next 20 years. What is the future worth at the end of 20 years when the interest rate is 8%, compounded as follows:

(a) Quarterly

(b) Monthly

(c) Continuously?

5.13 A series of equal quarterly deposits of $1000 extends over a period of 3 years. It is desired to compute the future worth of this quarterly deposit series at 12%, compounded monthly. Which of the following equations is correct?

(a) $F = 4(\$1,000)(F/A, 12\%, 3)$,

(b) $F = \$1,000(F/A, 3\%, 12)$,

(c) $F = \$1,000(F/A, 1\%, 12)$,

(d) $F = \$1,000(F/A, 3.03\%, 12)$.

5.14 If the interest rate is 7.5%, compounded continuously, what is the required quarterly payment to repay a loan of $10,000 in 4 years?

5.15 What is the future worth of a series of equal monthly payments of $3,000, if the series extends over a period of 6 years at 12% interest, compounded as follows:

(a) Quarterly

(b) Monthly

(c) Continuously?

5.16 What will be the required quarterly payment to repay a loan of $20,000 in 5 years, if the interest rate is 8%, compounded continuously?

5.17 A series of equal quarterly payments of $1,000 extends over a period of 5 years. What is the present worth of this quarterly payment series at 9.75% interest, compounded continuously?

5.18 Suppose you deposit $500 at the end of each quarter for 5 years at an interest rate of 8%, compounded monthly. What equal end-of-year deposit over 5 years would accumulate the same amount at the end of 5 years under the same interest compounding? To answer the question, which of the following is correct?

(a) $A = [\$500(F/A, 2\%, 20)] \times (A/F, 8\%, 5)$,

(b) $A = \$500(F/A, 2.013\%, 4)$,

(c) $A = \$500(F/A, \dfrac{8\%}{12}, 20)$ $\times (A/F, 8\%, 5)$

(d) None of the above.

5.19 A series of equal quarterly payments of $2,000 for 15 years is equivalent to what future lump-sum amount at the end of 10 years at an interest rate of 8%, compounded continuously?

5.20 What is the future worth of the following series of payments?

(a) $3,000 at the end of each 6-month period for 10 years at 6%, compounded semi-annually

(b) $4,000 at the end of each quarter for 6 years at 8%, compounded quarterly

(c) $7,000 at the end of each month for 14 years at 9%, compounded monthly.

5.21 What equal series of payments must be paid into a sinking fund to accumulate the following amount?

(a) $12,000 in 10 years at 6%, compounded semi-annually, when payments are semi-annual

(b) $7,000 in 15 years at 9%, compounded quarterly, when payments are quarterly

(c) $34,000 in 5 years at 7.55%, compounded monthly, when payments are monthly?

5.22 James Hogan is purchasing a $24,000 automobile, which is to be paid for in 48 monthly installments of $583.66. What effective annual interest is being paid for this financing arrangement?

5.23 A loan of $12,000 is to be financed to assist in buying an automobile. Based upon monthly compounding for 30 months, the end-of-the-month equal payment is quoted as $435. What nominal interest rate in percentage is being charged?

5.24 You are purchasing a $9,000 used automobile, which is to be paid for in 36 monthly installments of $288.72. What nominal interest rate is being paid on this financing arrangement?

5.25 Suppose a young newlywed couple are planning to buy a home 2 years from now. To save the down payment required at the time of purchasing a home worth $220,000 (let's assume this is 10% of the sales price, $22,000), they have decided to set aside some money from their salaries at the end of each month. If they can earn 6% interest (compounded monthly) on their savings, determine the equal amount this couple must deposit each month until they reach the point when they can buy the home.

5.26 What is the present worth of the following series of payments?

(a) $500 at the end of each 6-month period for 10 years at 8%, compounded semi-annually

(b) $2,000 at the end of each quarter for 5 years at 8%, compounded quarterly

(c) $3,000 at the end of each month for 8 years at 9%, compounded monthly?

5.27 What is the amount of the quarterly deposits, A, such that you will be able to withdraw the amounts shown in the cash flow diagram, if the interest rate is 8%, compounded quarterly?

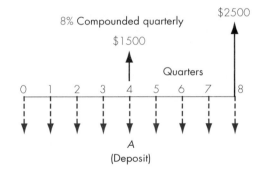

5.28 Georgi Rostov deposits $5,000 in a savings account that pays 6% interest, compounded monthly. Three years later he deposits $4,000. Two years after the $4,000 deposit, he makes another deposit in the amount of $2,500. Four years after the $2,500 deposit, half of the accumulated funds is transferred to a fund that pays 8% interest, compounded quarterly. How much money will be in each account 6 years after the transfer?

5.29 A man is planning to retire in 25 years. He wishes to deposit a regular amount every 3 months until he retires so that, beginning 1 year following his retirement, he will receive annual payments of $32,000 for the next 10 years. How much must he deposit if the interest rate is 8%, compounded quarterly?

5.30 A building is priced at $125,000. If a down payment of $25,000 is made, and a payment of $1,000 every month thereafter is required, how many months will it take to pay for the building? Interest is charged at a rate of 9%, compounded monthly.

5.31 You obtained a loan of $20,000 to finance an automobile. Based on monthly compounding over 24 months, the end-of-the-month equal payment was figured to be $922.90. What is the APR used for this loan?

5.32 *The Engineering Economist* (a professional journal) offers three types of subscriptions payable in advance: 1 year at $27, 2 years at $49, and 3 years at $65. If money can earn 6% interest, compounded monthly, which subscription should you take? (Assume that you plan to subscribe the journal over the next 3 years.)

5.33 A couple is planning to finance their 3-year-old son's college education. Money can be deposited at 6%, compounded quarterly. What quarterly deposit must be made from the son's 3rd birthday to his 18th birthday to provide $50,000 on each birthday from the 18th to the 21st? (Note that the last deposit is made on the date of the first withdrawal.)

5.34 Sam Salvetti is planning to retire in 15 years. Money can be deposited by 8%, compounded quarterly. What quarterly deposit must be made at the end of each quarter until he retires so that he can make a withdrawal of $25,000 semiannually over the 5 years after his retirement? Assume that his first withdrawal occurs at the end of 6 months after his retirement.

5.35 Emily Lacy received $500,000 from an insurance company after her husband's death. Emily wants to deposit this amount in a savings account which earns interest at a rate of 6%, compounded monthly. Then, she would like to make 60 equal monthly withdrawals over the 5-year deposit period, such that, when she makes the last withdrawal, the savings accounts will have a balance of zero. How much can she withdraw each month?

5.36 Anita Tahani, who owns a travel agency, bought an old house to use as her business office. She found that the ceiling was poorly insulated and that the heat loss could be cut significantly if 6 inches of foam insulation were installed. She estimated that with the insulation she could cut the heating bill by $40 per month and the air conditioning cost by $25 per month. Assuming that the summer season is 3 months (June, July, August) of the year and that the winter season is another 3 months (December, January, and February) of the year, how much can she spend on insulation if she expects to keep the property for 5 years? Assume that neither heating nor air conditioning would be required during the fall and spring seasons. If she decides to install the insulation, it will be done at the beginning of May. Anita's interest rate is 9%, compounded monthly.

Continuous Payments with Continuous Compounding

5.37 A new chemical production facility, which is under construction, is expected to be in full commercial operation 1 year from now. Once in full operation, the facility will generate $63,000 cash profit daily over the plant service life of 12 years. Determine the equivalent present worth of the future cash flows generated by the facility at the beginning of commercial operation, assuming

(a) 12% interest, compounded daily, with the daily flows

(b) 12% interest, compounded continuously, with the daily flow series approximated by a uniform continuous cash flow function.

Also compare the difference between (a) discrete (daily) and (b) continuous compounding above.

5.38 Income from a project is projected to decline at a constant rate from an initial value of $500,000 at time 0 to a final value of $40,000 at the end of year 3. If interest is compounded continuously at a nominal annual rate of 11%, determine the present value of this continuous cash flow.

5.39 A sum of $50,000 will be received uniformly over a 5-year period beginning 2 years from today. What is the present value of this deferred funds flow if interest is compounded continuously at a nominal rate of 9%?

5.40 A small chemical company, a producer of an epoxy resin, expects its production volume to decay exponentially according to the relationship

$$y, = 5e^{-0.25t},$$

where y_t is the production rate at time t. Simultaneously, the unit price is expected to increase linearly over time at the rate of

$$u_t = \$55(1 + 0.09t).$$

What is the expression for the present worth of sales revenues from $t = 0$ to $t = 20$ at 12% interest, compounded continuously?

Changing Interest Rates

5.41 Consider the cash flow diagram, which represents three different interest rates applicable over the 5-year time span.

(a) Calculate the equivalent amount P at the present.

(b) Calculate the single-payment equivalent to F at $n = 5$.

(c) Calculate the equal-payment-series cash flow A that runs from $n = 1$ to $n = 5$.

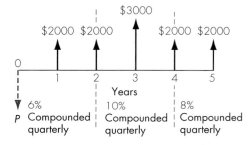

5.42 Consider the cash flow transactions depicted in the cash flow diagram, with the changing interest rates specified.

(a) What is the equivalent present worth? (In other words, how much do you have to deposit now, so that you can withdraw $300 at the end of year 1, $300 at the end of year 2, $500 at the end of year 3, and $500 at the end of year 4)?

(b) What is the single effective annual interest rate over 4 years?

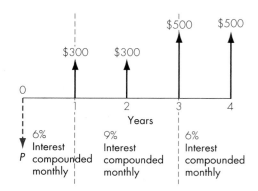

5.43 Compute the future worth for the cash flows with the different interest rates specified. The cash flows occur at the end of each year over 4 years.

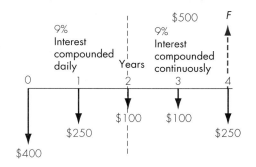

Amortized Loans

5.44 An automobile loan of $15,000 at a nominal rate of 9%, compounded monthly, for 48 months requires equal end-of-month payments of $373.28. Complete the following table for the first six payments, as you would expect a bank to calculate the values:

End of Month (n)	Interest Payment	Repayment of Principal	Remaining Loan Balance
1			$14,739.22
2			
3		$264.70	
4	$106.59		
5	104.59		
6			13,405.71

5.45 Mr. Smith wants to buy a new car which will cost $18,000. He will make a down payment in the amount of $4,000. He would like to borrow the remainder from a bank at an interest rate of 9%, compounded monthly. He agrees to pay the loan payment monthly for a period of 2 years. Select the correct answer for the following questions.

(a) What is the amount of the monthly payment (A)?
   (i) $A = \$10,000(A/P, 0.75\%, 24)$,
   (ii) $A = \$10,000(A/P, 9\%, 2)/12$,
   (iii) $A = \$10,000(A/F, 0.75\%, 24)$,
   (iv) $A = \$12,500(A/F, 9\%, 2)/12$?

(b) Mr. Smith has made 12 payments and wants to figure out the remaining balance immediately after 12th payment. What is the remaining balance?
   (i) $B_{12} = 12A$,
   (ii) $B_{12} = A\,(P/A, 9\%, 1)/12$,
   (iii) $B_{12} = A\,(P/A, 0.75\%, 12)$,
   (iv) $B_{12} = 10,000 - 12A$?

5.46 Talhi Hafid is considering the purchase of a used automobile. The price including the title and taxes is $8,260. Talhi is able to make a $2,260 down payment. The balance, $6,000, will be borrowed from his credit union at an interest rate of 9.25%, compounded daily. The loan should be paid in 48 equal monthly payments. Compute the monthly payment. What is the total amount of interest he has to pay over the life of the loan?

5.47 Suppose you are in the market for a new car worth $18,000. You are offered a deal to make a $1,800 down payment now and to pay the balance in equal end-of-month payments, $421.85, over a 48-month period. Consider the following situations:

(a) Instead of going through the dealer's financing, you want to make a down payment of $1,800 and take out an auto loan from a bank at 11.75% compounded monthly. What would be your monthly payment to pay off the loan in 4 years?

(b) If you were to accept the dealer's offer, what would be the effective rate of interest per month the dealer charges on your financing?

5.48 Bob Pearson borrowed $20,000 from a bank at an interest rate of 12%, compounded monthly. This loan will be repaid in 36 equal monthly installments over 3 years. Immediately after his 20th payment, Bob desires to pay the remainder of the loan in a single payment. Compute the total amount he must pay.

5.49 David Kapamagian borrowed money from a bank to finance a small fishing boat. The bank's loan terms allowed him to defer payments (including interest) for 6 months and to make 36 equal end-of-month payments thereafter. The original bank note was for $4,800 with an interest rate of 12%, compounded monthly. After 16 monthly payments, David found himself in a financial bind and went to a loan company for assistance in lowering his monthly payments. Fortunately, the loan company has offered to pay his debts in one lump sum, if he will pay the company $104 per month for the next 36 months. What monthly rate of interest is the loan company charging on this transaction?

5.50 You are buying a home for $190,000. If you make a down payment of $40,000 and take out a mortgage on the rest at 8.5%, compounded monthly, what will be your monthly payment to retire the mortgage in 15 years?

5.51 With a $250,000 home mortgage loan with a 20-year term at 9% APR, compounded monthly, compute the total payments on principal and interest over first 5 years of ownership.

5.52 A lender requires that monthly mortgage payments be no more than 25% of gross monthly income with a maximum term of 30 years. If you can make only a 15% down payment, what is the minimum monthly income needed to purchase a $200,000 house when the interest rate is 9%, compounded monthly?

5.53 If to buy a $150,000 house, you take out a 9% (APR) mortgage for $120,000, and 5 years later you sell the house for $185,000 (after all other selling expenses), what equity (the amount that you can keep before any tax) would you realize with a 30-year repayment term?

5.54 Just before their 15th payment

- Family *A* had a balance of $80,000 on a 9%, 30-year mortgage;
- Family *B* had a balance of $80,000 on a 9%, 15-year mortgage; and
- Family *C* had a balance of $80,000 on a 9%, 20-year mortgage.

How much interest did each family pay on their 15th payment?

5.55 Home mortgage lenders usually charge points on a loan to avoid exceeding a legal limit on interest rates or to be competitive with other lenders. As an example for a 2-point loan, the lender would loan only $98 for each $100 borrowed. The borrower would receive only $98, but would have to make payments just as if he or she had received $100. Suppose that you receive a loan of $130,000 payable at the end each month for 30 years with an interest rate of 9%, compounded monthly, but you have been charged 3 points. What is the effective interest rate on this home mortgage loan?

5.56 A restaurant is considering purchasing a lot adjacent to its business to provide adequate parking space for its customers. The restaurant needs to borrow $35,000 to secure the lot. A deal has been made between a local bank and the restaurant, so that the restaurant would pay the loan back over a 5-year period with the following payment terms: 15%, 20%, 25%, 30%, and 35% of the initial loan at the end of first, second, third, fourth, and fifth years, respectively.

(a) What rate of interest is the bank earning from this loan transaction?

(b) What would be the total interest paid by the restaurant over the 5-year period?

5.57 Don Harrison's current salary is $60,000 per year, and he is planning retirement 25 years from now. He anticipates that his annual salary will increase by $3,000 each year (first year, $60,000, second year, $63,000, third year, $66,000, and so forth), and he plans to deposit 5% of his yearly salary into a retirement fund that earns 7% interest, compounded daily. What will be the amount accumulated at the time of his retirement?

Add-On Loans

5.58 Katerina Unger wants to purchase a set of furniture worth $3,000. She plans to finance the furniture for 2 years. The furniture store tells Katerina that the interest rate is only 1% per month, and her monthly payment is computed as follows:

- Installment period = 24 months,

- Interest = $24(0.01)(\$3,000) = \$720$,

- Loan processing fee = $25,

- Total amount owed = $3,000 + $720 + $25 = $3,745,

- Monthly payment = $3,745/24 = $156.04 per month.

(a) What is the annual effective interest rate, i_a, Katerina is paying for her loan transaction? What is the nominal interest (annual percentage rate) for this loan?

(b) Katerina bought the furniture and made 12 monthly payments. Now she wants to pay off the remaining installments in one lump sum payment (at the end of 12 months). How much does she owe the furniture store?

5.59 You purchase a piece of furniture worth $5000 on credit through a local furniture store. You are told that your monthly payment will be $146.35 at a 10% add-on interest rate over 48 months, which includes an acquisition fee of $25. After making 15 payments, you have decided to pay off the balance. Compute the remaining balance based on the conventional amortized loan.

Loans with Variable Payments

5.60 Paula Hunt bought a new car for $18,400. A dealer's financing was available through a local bank at an interest rate of 13.5%, compounded monthly. Dealer financing required a 10% down payment and 48 equal monthly payments. Because the interest rate was rather high, she checked her credit union for possible financing. The loan officer at the credit union quoted on 10.5% interest for a new car loan and 12.25% for a used car. But to be eligible for the loan, Paula has to be a member of the union at least for 6 months. Since she joined the union 2 months ago, she had to wait 4 more months to apply for the loan. Now she decided to go ahead with the dealer's financing, and 4 months later refinanced the balance through the credit union at an interest rate of 12.25%.

(a) Compute the monthly payment to the dealer.

(b) Compute the monthly payment to the union.

(c) What is the total interest payment for each loan transaction?

5.61 A house can be purchased for $85,000, and you have $17,000 cash for a down payment. You are considering the following two financing options:

- Option 1: Getting a new standard mortgage with a 10% (APR) interest, and a 30-year term.

- Option 2: Assuming the seller's old mortgage, which has an interest rate of 8.5% (APR), a remaining term of 25 years (original term of 30 years), a remaining balance of $35,394, and payments of $285 per month. You can obtain a second mortgage for the remaining balance ($32,606) from your credit union at 12% (APR), with a 10-year repayment period.

(a) What is the effective interest rate of the combined mortgage?

(b) Compute the monthly payments for each option over the mortgage life.

(c) Compute the total interest payment for each option.

(d) What homeowner's interest rate makes the two financing options equivalent?

Loans with Variable Interest Rates

5.62 A loan of $10,000 is to be financed over a period of 24 months. The agency quotes a nominal rate of 8% for the first 12 months and a nominal rate of 9% for any remaining unpaid balance after 12 months, compounded monthly. Based on these rates, what equal end-of-the-month payment for 24 months would be required to repay the loan with interest?

5.63 Robert Carré financed his office furniture from a furniture dealer. The dealer's terms allowed him to defer payments (including interest) for 6 months and to make 36 equal end-of-month payments thereafter. The original note was for $12,000 with interest at 12%, compounded monthly. After 26 monthly payments, Robert found himself in a financial bind, and he goes to a loan company for assistance. The loan company offered to pay his debts in one lump sum if he will pay them $204 per month for the next 30 months.

(a) Determine the original monthly payment made to the furniture store.

(b) Determine the lump sum payoff amount the loan company will make.

(c) What monthly rate of interest is the loan company charging on this loan?

5.64 If you borrow $120,000 with a 30-year term, 9% (APR) variable rate, and the interest rate can be changed every 5 years,

(a) What is the initial monthly payment?

(b) If at the end of 5 years, the lender's interest rate is 9.75% (APR), what will the new monthly payments be?

Short Case Studies

5.65 Jim Norton, an engineering junior, has received in the mail two guaranteed line of credit applications from two different banks. Each bank offers a different annual fee and finance charge.

Jim expects his average monthly balance after payment to be $300 and plans to keep the card he chooses for only 24 months (after graduation he will apply for a new card). Jim's interest rate (on his savings account) is 6%, compounded daily.

Terms	Bank A	Bank B
Annual fee	$20	$30
Finance charge	1.55% monthly interest rate	16.5% annual percentage rate

(a) Compute the effective annual interest rate for each card.

(b) Which bank's credit card should Jim choose?

5.66 You are considering buying a new car worth $15,000. You can finance the car by either withdrawing cash from your savings account, which earns 8% interest or by borrowing $15,000 from your dealer for 4 years at 11%. You could earn $5635 in interest from your savings account for 4 years if you left the money in the account. If you borrow $15,000 from your dealer, you only pay $3609 in interest over 4 years, so it makes sense to borrow for your new car and keep your cash in your savings account. Do you agree or disagree with the statement above? Justify your reasoning with a numerical calculation.

5.67 The following is the actual promotional pamphlet prepared by Trust Company Bank in Atlanta, Georgia. The pamphlet reads as follows:

"Lower your monthly car payments as much as 48%." Now you can buy the car you want and keep the monthly payments as much as 48% lower than they would be if you financed with a conventional auto loan. Trust Company's *Alternative Auto Loan* (AAL)^{SM} makes the difference. It combines the lower monthly payment advantages of leasing with tax and ownership of a conventional loan. And if you have your monthly payment deducted automatically from your Trust Company checking account, you will save 1/2% on your loan interest rate. Your monthly payments can be spread over 24, 36 or 48 months.

Amount Financed	Financing Period (months)	Monthly Alternative Auto Loan	Payment Conventional Auto Loan
	24	$249	$477
$10,000	36	211	339
	48	191	270
	24	498	955
$20,000	36	422	678
	48	382	541

The amount of the final payment will be based on the residual value of the car at the end of the loan. Your monthly payments are kept low because you make principal payments on only a portion of the loan and not on the residual value of the car. Interest is computed on the full amount of the loan. At the end of the loan period you may:

1. Make the final payment and keep the car.
2. Sell the car yourself, repay the note (remaining balance), and keep any profit you make.
3. Refinance the car.
4. Return the car to Trust Company in good working condition and pay only a return fee.

So, if you've been wanting a special car, but not the high monthly payments that could go with it, consider the *Alternative Auto Loan*. For details, ask at any Trust Company branch.

Note 1: The chart above is based on the following assumptions. Conventional auto loan 13.4% annual percentage rate. *Alternative Auto Loan* 13.4% annual percentage rate.

Note 2: The residual value is assumed to be 50% of sticker price for 24 months; 45% for 36 months. The amount financed is 80% of sticker price.

Note 3: Monthly payments are based on principal payments equal to the depreciation amount on the car and interest in the amount of the loan.

Note 4: The residual value of the automobile is determined by a published residual value guide in effect at the time your Trust Company's *Alternative Auto Loan* is originated.

Note 5: The minimum loan amount is $10,000 (Trust Company will lend up to 80% of the sticker price). Annual household income requirement is $50,000.

Note 6: Trust Company reserves the right of final approval based on customer's credit history. Offer may vary at other Trust Company banks in Georgia.

(a) Show how the monthly payments were computed for the *Alternative Auto Loan* by the bank.

(b) Suppose that you have decided to finance a new car for 36 months from Trust Company. Assume also that you are interested in owning the car (not leasing it). If you decided to go with the *Alternative Auto Loan*, you will make the final payment and keep the car at the end of 36 months. Assume that your opportunity cost rate (personal interest rate) is an interest rate of 8%, compounded monthly. (You may view this opportunity cost rate as an interest rate at which you can invest your money in some financial instruments such as a savings account.) Compare this alternative option with the conventional option and determine your choice.

5.68 In Year 1988, the Michigan Legislature enacted the nation's first state-run program, the *Pay-Now, Learn-Later Plan*, to guarantee college tuition for students whose families invested in a special tax-free trust fund. The minimum deposit is now $1,689 for each year of tuition that sponsors of a newborn want to prepay. The yearly amount to buy into the plan increases with the age of the child: The parents of infants pay the least, and parents of high school seniors paying the most—$8,800 this year. This is because high school seniors will go to college sooner. The State Treasurer, Mr. Robert A. Bowman, contends that the educational trust is a better deal than putting money into a certificate of deposit (CD) or a tuition prepayment plan at a bank because the state promises to stand behind the investment. "Regardless of how high tuition goes, you know it's paid for," he said, "The disadvantage of a CD or a savings account is you have to hope and cross your fingers that tuition won't outpace the amount you save." At the newborns'

rate, $6,756 will prepay 4 years of college, which is 25% less than the state-wide average public-college cost of $9,000 for 4 years (in 1988). In 2006, when a child born this year will be old enough for college, 4 years of college would cost $94,360 at a private institution and $36,560 at a state school, if costs continued to rise the expected average of at least 7% a year. The Internal Revenue Service issued its opinion, ruling that the person who sets aside the money would not be taxed on the amount paid into the fund. The agency said that the student would be subject to federal tax on the difference between the amount paid in and paid out. Assuming that you are interested in the program for a newborn, would you join this program?

5.69 Suppose you are going to buy a home worth $110,000 and you make a down payment in the amount of $50,000. The balance will be borrowed from the Capital Savings and Loan Bank. The loan officer offers the following two financing plans for the property:

- Option 1: A conventional fixed loan at an interest rate of 13% over 30 years with 360 equal monthly payments.

- Option 2: A graduated payment schedule (FHA 235 plan) at 11.5% interest with the following monthly payment schedule.

Year (n)	Monthly Payment	Monthly Mortgage Insurance
1	$497.76	$25.19
2	522.65	25.56
3	548.78	25.84
4	576.22	26.01
5	605.03	26.06
6–30	635.28	25.96

For the FHA 235 plan, mortgage insurance is a must.

(a) Compute the monthly payment for option 1.

(b) What is the effective annual interest rate you are paying for option 2?

(c) Compute the outstanding balance for each option at the end of 5 years.

(d) Compute the total interest payment for each option.

(e) Assuming that your only investment alternative is a savings account that earns an interest rate of 6%, compounded monthly, which option is a better deal?

5.70 Ms. Kennedy borrowed $4,909 from a bank to finance a car at an add-on interest rate of 6.105%. The bank calculated the monthly payments as follows:

- Contract amount = $4,909,

- Contract period = 42 months,

- Add-on interest at 6.105% = $4,909(0.06105)(3.5) = $1,048.90,

- Acquisition fee = $25,

- Loan charge = $1,048.90 + $25 = $1,073.90,

- Total of payments = $4,909 + 1,073.90 = $5,982.90,

- Monthly installment = $5,982.90/42 = $142.45.

After making the 7th payment, Ms. Kennedy wants to pay off the remaining balance. The following is the letter from the bank explaining the net balance Ms Kennedy owes:

(a) Compute the effective annual interest rate for this loan.

(b) Compute the annual percentage rate (APR) for this loan.

(c) Show how would you derive the rebate factor (0.6589).

(d) Verify the payoff amount using the Rule of 78ths formula.

Dear Ms. Kennedy
The following is an explanation of how we arrived at the payoff amount on your loan account.

Original note amount	$5,982.90
Less 7 payments @ $142.45 each	997.15
	4,985.75
Loan charge (interest)	1,073.90
Less acquisition fee	25.00
	$1,048.90

Rebate factor from Rule of 78ths chart is 0.6589 (loan ran 8 months on a 42 month term).

$1,048.90 multiplied by 0.6589 = $691.12.

$691.12 represents the unearned interest rebate.

Therefore:

Balance	$4,985.75
Less unearned interest rebate	691.12
Payoff amount	$4,294.63

If you have any further questions concerning these matters, please contact us.
Sincerely,

S. Govia
Vice President

(e) Compute the payoff amount using the interest factor $(P/A, i, N)$.

Hint: The Rule of 78ths is used by some financial institutions to determine the outstanding loan balance. According to the Rule of 78ths, the interest charged during a given month is figured out by applying a changing fraction to the total interest over the loan period. For example, in the case of a 1-year loan, the fraction used in figuring the interest charge for the first month would be 12/78, 12 being the number of remaining months of the loan, and 78 being the sum of $1 + 2 + ... + 11 + 12$. For the second month, the fraction would be 11/78, and so on. In the case of 2-year loan, the fraction during the first month is 24/300, because there are 24 remaining payment periods, and the sum of the loan periods is $300 = 1 + 2 + ... + 24$.

CHAPTER 6

Principles of Investing

The Berkshire Bunch[1] In 1952 a 21-year-old aspiring money manager, named Warren Buffett, placed a small ad in an Omaha newspaper inviting people to attend a class on ***investing.*** He figured it would be a way to accustom him to appearing before audiences. To prepare, he even spent $100 for a Dale Carnegie course on public speaking. Five years later Dr. Carol Angle, a young pediatrician, signed up for the class. She had heard somewhere that the instructor was a bright kid, and she wanted to hear what he had to say. Only some 20 others showed up that day in 1957. "Warren had us calculate how money would grow, using a slide rule," Dr. Angle, now 71, recalls. "He brainwashed us to truly believe in our hearts in the miracle of ***compound interest.***" Persuaded, she and her husband, William, also a doctor, invited 11 other doctors to a dinner to meet young Warren. Buffett remembers Bill Angle getting up at the end of the dinner and announcing: "I'm putting $10,000 in. The rest of you should, too." They did. Dr. Angle still practices medicine, as director of clinical toxicology at the University of Nebraska Medical Center. However, she does not work for the money. Her family's holdings in Buffett's Berkshire Hathaway have multiplied into a fortune of $300 million.

---

[1] "The Berkshire Bunch: Chance meetings with an obscure young investment counselor made many people widely rich. Without knowing it, they were buying into the greatest compound interest machine ever built." By Dolly Setton, *Forbes,* October 12, 1998 © Copyright 1998 Forbes Inc.

When you invest, you are trying to increase your income and build the value of your assets. Simply stated, investing means putting your money to work earning more money. Done wisely, it can help you meet your financial goals: buying a home, paying for a college education, or enjoying a comfortable retirement. You do not have to be wealthy to be an investor. As you learned in Chapter 4, it is never too soon to start thinking about investing. Investing even a small amount can produce considerable rewards over the long term, especially if you do it regularly. However, investing means you have to make decisions about how much you want to invest and where to invest it. To choose wisely, you need to know what options you have and what **risks** you take when you invest in the different ones. In this chapter, we will look into investing in financial assets such as stocks and bonds. However, the same principles and techniques are used in evaluating business assets in later chapters.

6.1 Investing in Financial Assets

Most individual investors have three basic investment opportunities in financial assets: stocks, bonds, and cash. Cash investments include money in bank accounts, certificates of deposit (CD), and U.S. Treasury bills. You can invest directly in any or all of the three, or indirectly, by buying mutual funds that pools your money with money from other people and then invest it. If you want to invest in financial assets, you have plenty of opportunities. In the United States alone, there are more than 9,000 stocks, 7,500 mutual funds, and thousands of corporate and government bonds to choose from. Even though we will discuss the investment basics in the framework of financial assets in this chapter, the same technical concepts are applicable to any business assets that will follow in later chapters.

6.1.1 Investment Basics

Selecting the best investment for you depends on your personal circumstances as well as general market conditions. For example, a good investment strategy for a long-term retirement plan may not be a good strategy for a short-term college savings plan. In each case, the right investment is a balance of three things: liquidity, safety, and return.

- **Liquidity—How accessible is your money?** If your investment money must be available to cover financial emergencies, you will be concerned about liquidity—how easily you can convert it to cash. Money-market funds and savings accounts are very liquid; so are investments with short maturity dates like CDs. However, if you are investing for longer-term goals, liquidity is not a critical issue. What you are after in that case is growth, or building your assets. We normally consider certain stocks and stock mutual funds as growth investments.

- **Risk—What is the safety involved?** Risk is the chance you take of making or losing money on your investment. For most investors, the biggest risk is losing money, so they look for investments they consider safe. Usually that means putting money into bank accounts and U.S. Treasury bills, as these investments are either insured or default-free. The opposite, but equally important risk, is that your investments will not provide enough growth or income to offset the impact of inflation, the gradual increase in the cost of living. There are additional risks as well, including how the economy is doing. However, *the biggest risk is not investing at all.*

- **Return—How much profit will you be able to expect from your investment?** Safe investments often promise a specific, though limited, return. Those that involve more risk offer the opportunity to make—or lose—a lot of money. Both risk and reward are time dependent. As time progresses, low-yielding investments become more risky because of inflation. On the other hand, the returns associated with higher-risk investments could become more stable and predictable over time, thereby reducing the perceived level of risk.

6.1.2 How to Determine Your Expected Return

Return is what you get back in relation to the amount you invested. Return is one way to evaluate how your investments in financial assets are doing in relation to each other and to the performance of investments in general. Let us look first at how we may derive rates of return.

Basic Concepts

Conceptually, the rate of return that we realistically expect to earn on any investment is a function of three components:

- Risk-free real return

- + Inflation factor

- + Risk premiums

Suppose you want to invest in stock. First, you should expect to be rewarded in some way for not being able to use your money while you are holding the stock. Then, you would be compensated for decreases in purchasing power between the time you invest it and the time it is returned to you. Finally, you would demand additional rewards for any chance that you would not get your money back or that it will have declined in value while invested.

For example, if you were to invest $1,000 in risk-free U.S. Treasury bills for a year, you would expect a real rate of return of about 2%. Your risk premium would be also zero. You probably think that it does not sound like much. However, to that you have to add an allowance for inflation. If you expect inflation to be about 4% during that investment period, you should realistically expect to earn 6% during that interval (2% real return + 4% inflation factor + 0% for risk premium). Here is what it looks like in tabular form:

Real Return	2%
Inflation (loss of purchasing power)	4%
Risk premium (U.S. Treasury Bills)	0%
Total expected return	6%

How would it work out for a riskier investment, say an Internet stock such as Amazon.com? As you consider the stock to be a very volatile one, you would increase the risk premium to something like this:

Real Return	2%
Inflation (loss of purchasing power)	4%
Risk premium (Amazon.com)	20%
Total expected return	26%

So you will not invest your money in Amazon.com unless you are reasonably confident of having it grow at an annual rate of 26%. Again, the risk premium of 20% is a perceived value that can vary from one investor to another.

Return on Investment over Time

If you start out with $1,000 and end up with $2,000, your return is $1,000 on that investment, or 100%. If a similar investment grows to $1,500, your return is $500, or 50%. However, unless you held those investments for the same period, you cannot determine which has a better performance. What you need to compare your return on one investment with the return on another investment is the **annual return**, the average percentage that you have gained on each investment over a series of one-year periods. For example, if you buy a share for $15 and sell it for $20, your profit is $5. If that happens within a year, your rate of return is an impressive 33.33%. If it takes 5 years, your return (compounded) will be closer to 5.92%, since you can spread the profit over a 5-year period. Mathematically, you are solving the following equivalence equation for i.

$$\$20 = \$15(1 + i)^5, \text{ and } i = 5.92\%.$$

Figuring out the actual return on your portfolio investment is not always that simple. There are several reasons:

1. The amount of your investment changes. Most investment portfolios are active, with money moving in and out.
2. The method of computing return can vary. For example, performance can be averaged or compounded, which changes the rate of return significantly, as we will demonstrate in Example 6.1.
3. The time you hold specific investments varies. When you buy or sell can have a dramatic effect on overall return.

Example 6.1 Figuring Average Versus Compound Return

Consider the following six different cases of investment performance of $1,000 investment over a 3-year holding period. Compute the average versus compound return for each case.

	Annual Investment Yield					
Investment	Case 1	Case 2	Case 3	Case 4	Case 5	Case 6
Year 1	9%	5%	0%	0%	–1%	–5%
Year 2	9%	10%	7%	0%	–1%	–8%
Year 3	9%	12%	20%	27%	29%	40%

Solution

Given: 3-year's worth annual investment yield data

Find: Compound versus average rate of return

As an illustration, consider Case 6 for an investment of $1,000. At the end of the first year, the value of investment decreases to $950; at the end of second year, it decreases again to $950(1 − 0.08) = $874; at the end of third year, it increases to $874(1 + 0.40) = $1,223.60. Therefore, one way you can look at the investment is what annual interest rate the initial $1,000 investment would grow to $1,223.60 over 3 years. This is equivalent to solving the following equivalence problem:

$$\$1,223.60 = \$1,000 (1 + i)^3$$
$$i = 6.96\%$$

If someone evaluates the investment based on the average annual rate of return, the investor might proceed the following:

$$i = (-5\% - 8\% + 40\%)/3 = 9\%$$

If you calculate the remaining cases, you will observe that all six cases have the same average annual rate of return, although their compound rates of return vary from 6.96% to 9%.

	Compound Versus Average Rate of Return					
Investment	**Case 1**	**Case 2**	**Case 3**	**Case 4**	**Case 5**	**Case 6**
Average return	9.00%	9.00%	9.00%	9.00%	9.00%	9.00%
Balance at the end of year 3	$1,295	$1,294	$1,284	$1,270	$1,264	$1,224
Compound return (or rate of return)	9.00%	8.96%	8.69%	8.29%	8.13%	6.96%

Your immediate question is "Are they the same indeed?" Certainly not—you will have the most money with Case 1, which also has the highest compound rate of return. The average rate of return figure is easy to calculate, but it ignores the basic principle of time value of money. In other words, according to the average rate of return concept, we may view all six cases as indifferent. However, the amount of money available at the end of year 3 would be different for each case. Although the average rate of return is popular for comparing investments with their yearly performance, it is not a correct measure in comparing the performance for investments over multiyear period.

Comments: You can evaluate the performance of your portfolio by comparing it to standard indexes and averages that are widely reported in the financial press. If you own stocks, you can compare with the performance of the Dow Jones Industrial Average (DJIA), perhaps one of the best-known measures of stock market performance in the world. If you own bonds, you can identify an index that tracks the type you own: corporate, government, or agency. If your investments are in cash, you can follow the movement of interest rates on Treasury bills, CDs, and similar investments. In addition, total return figures for mutual funds performance are reported regularly. You can compare how well your investments are doing against those numbers. Another factor to take into account when evaluating return is the current inflation rate. Certainly, your return needs to be higher than the inflation rate if your investments are going to have real growth.

6.1.3   How to Determine Expected Financial Risk

In this section, we take a brief look at how investment risk is measured and how it affects returns on investments. We will first define more precisely what the term *risk* means as it relates to investments. **Risk** refers to the chance that some unfavorable event will occur. If you are a race car driver, you take a chance with your life—high-speed car racing is certainly risky. If you bet on a horse, you are risking your money. If you invest in a speculative stock, you take a risk in the hope of making a significant return. In terms of financial risk, you need to take two kinds of risk into account: volatility and the effect of changing market conditions, like a recession, an oil embargo, or high inflation.

Volatility

Volatility means sudden swings in value—from high to low, or the reverse. The more volatile an investment is, the more profit you can expect. This is because there can be a big spread between what you paid and what you sell it for. However, you have to be prepared for the price to drop by the same amount. All investments fluctuate in value or relative rate of return. It does not matter if we are talking about stocks, bonds, money market instruments, artwork, or real estates. There are times when each one appears to be either a better or a worse investment than the others are. Volatility of value is common to them all. So the first major component of risk is volatility: the "fluctuation factor."

Certainly, knowing a range of returns is useful. However, how does that help us determine what we can realistically expect from our investments now and in the future? Do we simply expect our U.S. stocks to fluctuate between −26% and +38%? Just knowing the fluctuation range would not be very useful. Even in our quest to be only approximately right, we do want some measure of precision. Perhaps an analogy might help make this clearer. Think of the weather temperature. For any particular day of the year, there is a "normal high" and "normal low" temperature for your locale. You hear it referred to every morning on the weather channel. What the weather forecasters are saying is that for that time of the year they "expect" temperatures to fall within a certain range. We also know that they are not always right. That bright, sunny day on which they convince us to leave home without an umbrella is often the one that suddenly turns rainy and leaves us shivering all the way home.

	Average Annual Return 1970–1997	**Best year**	**Worst year**
U.S. Stocks	13.0%	37.6% (1995)	–26.5% (1974)
International Stocks	12.7%	39.4% (1993)	–26.2% (1974)
Cash Equivalent	6.8%	14.1% (1981)	3.0% (1991)
Real Estate	8.8%	20.5% (1979)	–5.6% (1991)
U.S. Bonds	9.3%	33.5% (1982)	–5.6% (1994)

Table 6.1
Returns from Various Investment Classes, 1970–1997

Measure of Market Risk

Looking to the past can provide some clues as shown in Table 6.1. Over several decades, for example, investors who put up with the stock market's gyrations earned returns for in excess of bonds and cash investments like Treasury bills. Although we have always known, for example, that stock markets move up and down, they did so in a somewhat predictable manner and within a moderate range.

In Table 6.1, what does the annual average of 13% return on U.S. stocks really mean? If you are a golfer, you might prefer an example using golf scores and handicaps. For a 10-handicap golfer, you would normally expect to shoot around 82 (average figure)—the normal 72-par for a championship course. However, your experience tells you that your score varies from round to round, say, from 78 (normal low) to 86 (normal high)—but still with an average of 82. Occasionally, of course, we all have some exceptional games—with scores, say, shooting 74. For some 10-handicap golfers, their scores may vary from 80 to 84; we call them "consistent players," meaning that we can predict their scores with reasonable confidence. In a typical golf tournament, you may prefer to have these consistent players as your playing partners as you generally know what to expect from them.

	Investment (rate of return)	**Golf scores (a golfer with a 10-handicap)**
Expected value	13%	82
Normal "low"	–6%	78
Normal "high"	32%	89

With investment returns, we "expect" them to fall within a certain range, just like our own golf scores. However, occasionally, unpredictably, and for reasons that we can only explain afterwards, they are higher or lower. That's what the mathematician means when he/she refers to **standard deviation.** (In Chapter 14, we look at some mathematical aspects of risk measure. However, at this point, we will limit our risk discussion at the descriptive level.) In our golf example, a player with a scoring range of 80 to 84 will have a smaller standard deviation (less fluctuation) compared with a player in the 78 to 86 range.

Current Practice: Here are definitions of three common statistical risk ratings used in the financial market.[2]

Volatility is known as the "fluctuation or range factor." It measures the deviation from the expected value. If a stock price fluctuates between $100 and $150 with the average price at $125 over a certain period, we can say the stock volatility is ± $25.

Standard deviation is an open-ended measure of price volatility when you have the probabilistic information about the uncertain event. The higher the standard deviation, the more violent the fund's price swings—up and down—the past three years.

Beta measures how closely a fund's performance correlates with broader stock market movements. That would be the Standard & Poor's 500 stock index for most funds. A fund whose performance matches its index would have a beta of 1. A fund with a beta of 1.15 could be expected to do 15% better than its index in a rising market, and 15% worse than the index in a falling market.

Alpha shows whether a fund is producing better or worse returns than expected, given the risk it takes. The higher the alpha, the better. A negative alpha means that the fund has performed worse than expected, given the risk it takes.

6.2 Investment Strategies

Once you understand the implications of market risk, you need to come up with an investment strategy that tells you what to do as far as putting together an appropriate investment portfolio. The next question is how do you go about actually implementing the decisions you have made. Because investing is an inexact science, it is better to be approximately right than precisely wrong. This is the approach taken in this chapter. Two techniques are commonly practiced in financial investment. They are "dollar-cost-averaging" concept and diversification.

6.2.1 Trade-Off Between Risk and Reward

When it comes to investing, trying to weigh risk and reward can be a challenging task. Investors do not know the actual returns that financial assets will deliver, or the difficulties that will occur along the way. Risk and reward are the two key words that will form the foundation for much of this chapter. For this is what investing is all about—the trade-off between the opportunity to earn higher returns and the consequences of trying to do so. The greater the risk, the more you stand to gain or lose. There is no such thing as a risk-free investment. So the real task is not to try to find "risk-free" investments. The challenge is to decide what level of risk you are willing to assume and then, having decided on your risk tolerance, to understand the implications of that choice. Your range of investment choices—their relative risk factors—may be classified into three types of investment groups: cash, debt, and equities.

---

[2] You can find these statistics for all funds in the Morningstar database at www.morningstar.net.

- **Cash—the least risky with the lowest returns:** Cash investment includes bank accounts, CDs, money-market mutual funds, Treasury bills, and other "short-term" instruments. These investments guarantee that you will get your money back, plus interest. Cash-type investments offer *liquidity*. You can access your money relatively easily. They offer a known rate of return, albeit only for a short time. If rates fall, you have to reinvest at the lower rates. On the other hand, if rates are rising, you can ride up with them because you don't have your money "locked in" for too long a period of time. The greatest risk with cash holdings is that, over the long term, they do not generate enough profit to offset inflation and allow the asset to grow in real purchasing power. From an investment perspective, cash asset's most useful features are that it is readily available for opportunities or emergencies.

- **Debt—moderately risky with moderate returns:** Debt investment comprises all contracts between a borrower and a lender, including bonds, mortgages, and personal loans. Debt assets have value in that, assuming they are kept to maturity, you will know exactly how much they will be worth at that time. On the down side, these investments can fluctuate in value during the term. Furthermore, they have not done a good job of keeping pace with inflation over the long term.

- **Equities—the most risky but offering the greatest payoff:** Equities are assets that increase or decrease in value depending on what someone is willing to pay for them at any particular moment. Common examples of equities are stocks, real estate, and growth mutual funds. Although generally looked upon as investments to grow in value over time, they may also provide income. For example, stocks pay dividends and real estate generates rent.

We may explain the relationship between risk and return such that no investment will be made unless the expected rate of return is high enough to compensate the investor for the increased risk of the investment. In this example, it is clear that few, if any, investors would be willing to buy an Internet stock if its expected return were the same as that of the T-bill. If you take no risk, you run into the chance of coming out short. The more you have in the safest investments like CDs, bank accounts, and Treasury bills, the smaller your chance of substantial reward. There is also the risk of outliving your assets because they will not keep up with inflation. Similarly, the more you speculate, the greater the risk of losing your principal entirely. Often the more investors understand about the principles of investing, the more apt they are to take the moderate approach, balancing their risks to get the kind of results they want. That way, they can also afford to invest a small amount in speculative opportunities and keep some assets safe—and liquid—to meet immediate cash needs.

6.2.2 Dollar-Cost Averaging Concept

Dollar-cost-averaging is the process of purchasing securities over a period by periodically investing a predetermined amount at regular time intervals. The goal of dollar cost averaging is to reduce the effect of price fluctuations on the performance of your investment. When the market is rising, additional shares will benefit from the price in-

Table 6.2
An Example of How the Dollar-Cost-Averaging Concept Works

Timing	Amount Invested	Fund Unit Price	No. of Units Purchased	Ending Fund Balance
Month 1	$1,000	$5.00	200	$1,000
Month 2	$1,000	$4.00	250	$1,800
Month 3	$1,000	$2.50	400	$2,125
Month 4	$1,000	$3.75	267	$4,189
Month 5	$1,000	$5.00	200	$6,585
Totals	$5,000		1,317	Profit $1,585

creases. When the market is declining, you can purchase the additional shares at lower prices. This will yield more shares per dollar invested.

Table 6.2 illustrates an example of dollar-cost-averaging at work, assuming you wanted to contribute $1,000 each month into a stock mutual fund. Here is a possible outcome of a five-month purchase plan where the fund unit price fluctuates between $2.50 and $5.00:

In Month 1, for example, your $1,000 bought 200 units at $5.00 per unit. As the price fell through Months 2 and 3, the same dollar amount purchased more units. On the other hand, as the price began to rise, your $1,000 bought fewer units. By Month 5, you owned 1,317 units (or shares), for which you paid a total of $5,000. That averages out to be $3.80 per unit, which demonstrates the major advantage of dollar-cost-averaging. In other words, by making regular purchases of equal amounts, you purchase more units when prices are low and fewer when they go up. This works best for assets with fluctuating prices, such as common stocks. Although the unit value declined immediately after you started your scheduled buys and never recovered to a level any higher than the one at which you started, you still managed to make a profit of $1,585.

Here, we have purposely chosen simple numbers to illustrate the dollar-cost-averaging concept. However, the theory is valid as long as the unit value does not decline indefinitely. If that happens, it is simply a bad investment choice and no matter how many units you buy at any price, you will continue to lose money.

6.2.3 Broader Diversification Reduces Risk

Even if you find risk exciting sometimes, you will probably sleep better if you have your money spread among different assets; i.e., Don't put all your eggs in one basket. Your best protection against risk is diversification—spreading your investments around instead of investing in only one thing. For example, you can balance cash investments like CDs and money market funds with stocks, bonds, and mutual funds. Even within equity investments, you can buy stocks of small-growth companies while also investing in large and well-established companies. What do we gain from this diversification practice? Well, you hope to reduce market volatility. Usually when the return is down in one area, you would like to see the return up in another area. We may explain the concept of diversification graphically as shown in Figure 6.1.

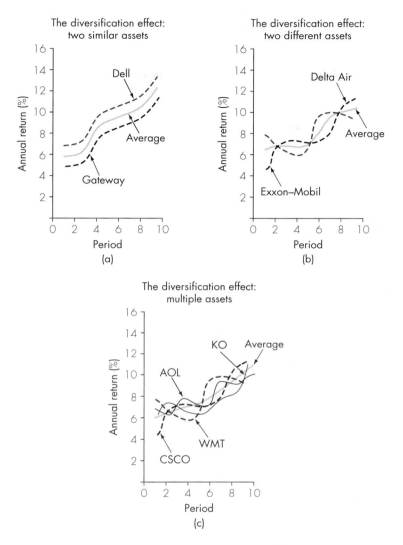

Figure 6.1 Reducing investment risks by asset. (a) Dell and Gateway are two well-known mail-order personal computer (PC) makers. Depending on the anticipated future PC demand, both stocks tend to move in the same direction. (b) Typically both airline and oil industry's profit are greatly influenced by the fuel price. If the fuel price goes up, the airline's profit suffers but the oil company's profit soars. Therefore, stock prices for these two industries tend to move in opposite direction. (c) When your portfolio includes industries such as cyclical (KO and WMT) as well as growth companies (CSCO and AOL), your overall return tends to be stabilized since sector rotation in the stock market place (money moves from one industrial sector to another) would not greatly affect the total asset value in any particular period.

- **Case 1—Invest in two assets with similar return characteristics:** Suppose you have two types of investments in Figure 6.1(a). These two investments are similar in pattern of return, that is, fluctuate to the same degree and as one goes up or down so does the other. In other words, you will experience a great deal of volatility while you are in the market. If you keep both investments, your expected rate of return is simply the weighted average of their returns.

- **Case 2—Invest in two assets with dissimilar return characteristics**: Suppose you find another set of investments in Figure 6.1(b). They have the same potential return but whose returns come at opposite times. That is, as one goes down, the other goes up. Again, the return would be the weighted average of the returns of the two investments, but you may control the risk (fluctuation) considerably. All the negative returns of one would be offset by the positive returns of the other.

- **Case 3—Invest in multiple assets with dissimilar return characteristics**: Of course, in the real world, assets are not likely to observe either case scenario. The more likely situation is that larger portfolios will have a number of assets in them with differing, but not necessarily opposite, patterns of return. Then, the results could be like the one in Figure 6.1(c). The overall yield of the portfolio is the weighted average of the individual assets, but the fluctuation—the risk—is dampened. It is, therefore, possible to achieve a higher rate of return without increasing considerable risk by building a multiple-asset portfolio. This is exactly what we hope to achieve in asset investing through diversification.

6.2.4 Broader Diversification Increases Expected Return

As we observed in Figure 6.1, diversification reduces risk. However, there is far more to the power of diversification than simply spreading your assets over a number of investments. Well-diversified portfolios contain various mixes of stocks, bonds, mutual funds, and cash equivalents like Treasury bills. With it, over lengthy periods, you do not have to sacrifice much in the way of returns to get that reduced volatility. Finding the right mix depends on your assets, your age, and your risk tolerance. Diversification also means regularly evaluating your assets and realigning the investment mix. For example, if your stocks increase in value, they will make up a larger percentage of your portfolio. To maintain a certain level of risk tolerance, you may want to decrease your stock holdings and increase your cash holdings, or bond holdings. Example 6.2 illustrates what a difference an asset allocation makes for a long-term investor.

Example 6.2 Broader Diversification Increases Return

Suppose you have $10,000 in cash and are considering the following two options for investing the money.

- Option 1: You put the entire $10,000 into a secure investment mutual fund consisting of a long-term U.S. Treasury bond with a yield of 7%.

- Option 2: You split the $10,000 into equal amounts of $2,000 and diversify among five investment opportunities with varying degrees of risk from, say, extremely risky to very conservative and with potential returns ranging from −100% to 15%:

Amount	Investment	Expected Return
$2,000	Buying lottery tickets	−100% (?)
$2,000	Under the mattress	0%
$2,000	Term deposit (CD)	5%
$2,000	Corporate bond	10%
$2,000	Mutual fund (stocks)	15%

Given these two opportunities, which would you choose?

Discussion: The first $10,000 bond investment earning 7% poses virtually no risk. With the diversified approach, you are going to lose the first $2,000 with virtual certainty, make nothing on the next $2,000, and only 5% on the third, 10% on the fourth, and 15% on the last $2,000. At first glance, you might think Option 1 is a more rational strategy with the very limited risk potential. Indeed, that might well be the best alternative for many short-term investors. However, we will add one more element to our equation—our time horizon is 25 years. Would that make a difference?

Solution

Given: Rates of return for each investment options

Find: Value at the end of 25 years

As you will see, the longer the time horizon the better choice certain investments, such as stocks (represented by the mutual fund in this example) become. First, we can find the value of the government bond in 25 years as follows:

$$F = \$10,000(F/P, 7\%, 25) = \$54,274$$

Similarly, you can find the value for each investment class in Option 2 (see table top of page 252). At the outset, Option 2 appears to be a losing proposition, but you would end up with about 77% more money despite the fact that the first two choices you made were, at best, unproductive. Of course, you can come up a counter example where Option 1 would be a better choice. However, the message is clear: Diversification among properly chosen assets can increase return without taking unnecessary risks as long as you keep the assets invested in the market over a long period of time.

Option	Amount	Investment	Expected Return	Value in 25 years
1	$10,000	Bond	7%	$54,274
	$2,000	Lottery tickets	−100%	$0
	$2,000	Mattress	0%	$2,000
2	$2,000	Term deposit (CD)	5%	$6,773
	$2,000	Corporate bond	10%	$21,669
	$2,000	Mutual fund (stocks)	15%	$65,838
				$96,280

Comments: On an extreme measure, if you invested the entire $10,000 in the stock, the expected return is $10,000(1 + 0.15)^{25} = $329,190$, which is the lot more than the return expected from the diversification. Therefore, it is commonly suggested that if you are a long-term investor, you may increase your exposure to the more risky assets such as stocks.

6.3 Investing in Stocks

As we have illustrated in Chapter 2, stocks are ownership shares in a corporation. You can buy stock in more than 9,000 publicly traded companies, though chances are your portfolio will have only a tiny fraction of what is available. When you buy stock in a corporation, you become one of its owners. If the company does well, you may receive part of its profits as dividends and see the price of your stock increase. However, if the company's earnings drop or its projected earnings do not meet investors' expectation, the value of your investment can drop dramatically.

6.3.1 The Value of Stock

Investors value stock based on one critical question: If I put my money into this company, what are the chances I will get a better return than if I invested in something else? It is a matter of risk and reward. Investors who buy a stock believe other people will buy as well, and that the share price is going to increase. Investing is a kind of gamble, but it is not like betting on horses. A long shot can always win the race even if everyone bets the favorite. In the stock market, the betting itself influences the outcome. If many investors bet on America Online stock, America Online's price will go up. The stock becomes more valuable because investors want it. The reverse is also true: If investors sell the same stock, it will fall in value. The more it falls, the more investors will sell.

Most people buy stocks to make money through capital gains or the profit from selling stock at a higher price than they paid for it. If you buy 100 shares of Coca-Cola at $60 a share and sell it for $80 a share, you have realized a capital gain of $20 a share, or $2,000. If you have held the stock for more than a year, your profits are long-term capital gains. Of course, it does not all go in your pocket. You owe taxes on the gain as well as commissions to your stockbroker for buying and selling the stock.

Some people invest in stocks to get quarterly dividend payments. Dividends are the portion of the company's profit paid out to its shareholders. For example, if IBM declares an annual dividend of $4 a share, and you own 100 shares; you will earn $400 a year, or $100 paid each quarter. A company's board of directors decides how large a dividend the company will pay, or whether it will pay one at all.

Thus, in summary, when investors are contemplating buying stock, they have two things in mind: (1) cash dividends and (2) gains (share appreciation) at the time of sale.[3] From a conceptual standpoint, investors determine market values of stocks by discounting expected future dividends and capital gains at a rate that takes into account any future growth. Since investors seek growth companies, a desired growth factor for future dividends is usually included in the calculation.

Example 6.3 Stock Valuation

Suppose investors in the common stock of IBM Corporation expect to receive a dividend of $1 by the end of 2002. The future annual dividends will grow at an annual rate of 13% for the next 5 years. You will hold the stock for 3 years and expect the market price of the stock rise to $230 by the end of the third year. If your required rate of return were about 10% before any tax consideration, what would be your fair value of the stock in today's market (end of 2001)?

Solution

Given: Dividend for the first year (D_1), required rate of return (i), dividend growth rate (g), and the target price at the end of year 3

Find: Expected market value (P)

We may answer this question by solving the following equation for P:

$$P = \frac{\$1}{(1 + 0.10)} + \frac{\$1(1 + 0.13)}{(1 + 0.10)^2} + \frac{\$1(1 + 0.13)^2 + \$230}{(1 + 0.10)^3}$$

$$= \$175.61$$

The fair market value to you would be $175.61. If the stock were traded on the market less than $175.61, you would be interested in investing in the stock.

---

[3] Investments you own for 6 months or longer are considered as capital assets. When you sell them at a profit, you have a capital gain. If you sell for less than you paid, you have a capital loss. First, you can use your capital losses to offset, or reduce, capital gains. If you do not have any capital gains to offset, you can still offset your loss up to $3,000 per year from your ordinary income. Any remaining losses can be carried over to the future tax years. The 1921 tax act first established a lower rate for capital gains from the sale of assets than for ordinary income. The lower rate for capital gains reflects the strong belief that tax law should encourage investment. Currently, the capital gain tax rate is 20% for any assets held more than 12 months.

Comments: Under the assumption that the future dividend payments and stock price will materialize as expected, the valuation scheme presented above will be meaningful. However, in practice, it is rather difficult to predict how much the dividend payments and future price will fluctuate from the expected values. Therefore, there is always some sort of risk premium built in stock valuation. If there is more upside potential in future earnings, the stock would be traded at a higher than $175. If there is more downside potential, the stock will be traded at a lower price.

Current Practices: A stock has no absolute value. At any given time, its value depends on the supply and demand of the stock. In other words, its value depends on whether the shareholders want to hold it or sell it and on what other investors are willing to pay for it. For any given period, some stocks are undervalued, which means they sell for less than analysts think they are worth, while others are overvalued. Many factors determine investors' attitudes. One of the major factors is whether they expect to make money with the stock under current stock market conditions and the overall state of the economy. Certainly, past performance is no guarantee of future profits. Investing is not about guarantees, but it is about balancing risk with reasonable expectations of reward. With that in mind, the following table summarizes your chances of making or losing money in the leading U.S. stocks over various holding periods based on the historical record. *Newsweek*[4] reported the odds of making a return on investment as a function of holding period. Note that, near term, stocks carry much more risk than you think.

If you hold	Your chance of losing money	Your chance of making return on your investment per year		
		0–10%	10–20%	20+ %
1 year	26%	18%	20%	37%
3 years	14%	28%	39%	19%
5 years	10%	31%	49%	10%
10 years	4%	42%	53%	1%
20 years	0	37%	63%	0

6.3.2 Financial Options

One away to protect the value of financial assets from a significant downside risk is to buy option contracts. Buying an option contract is equivalent to purchasing home or car insurance policies. Let us start with an ordinary "option," which is just the *opportunity* to buy or sell something at a specific price within a certain period. Think of purchasing a car insurance policy that allows you the right to cancel, if you so choose. With the policy, you are paying the right to essentially "cancel" if it is advantageous to you, while your car is protected in case of an accident. If it is not advantageous to cancel, then you will hold that policy and never use your option to cancel. If you sold the

---

[4] "What are your odds? Standard & Poor's 500-stock average, annualized returns since 1926. All dividends reinvested." Source: *Newsweek,* November 10, 1997.

car during the contract period, you may exercise your option—cancel the policy and you are entitled to receive the unused portion of premium back.

Definition

- **Buy Put Options—When you expect price to go down:** When investors purchase a "put," they are betting that the underlying investment is going to decrease in value within a certain amount of time. That means if the value does go down, they will *exercise* the *put*. That is kind of like your homeowner's insurance. You pay an annual premium for protection if, say, your house burns to the ground. If that happens, you will exercise the put and get some money. If it never happens during the period, your loss is just limited to the entire premium.

- **Buy Call Options—When you expect price to go up:** With a "call," investors purchase the right to buy something at a specific price within a certain period. Consider when you buy your Sunday paper, which includes a bunch of coupons. You pay a certain amount for the paper, but in turn, you receive the right to use those coupons within a specific period to buy something at a specific price. Take those coupons to the grocer, and you have exercised a call.

  An investor purchases a call option when he or she is bullish on a company and thinks the stock price will rise within a specific amount of time. For example, you have the right to buy the underlying security at a fixed price before a specified date in the future—usually three, six, or nine months. For this right, you pay the call option seller (called the "writer"), a fee (called a "premium"), which is forfeited if you do not exercise the option before the agreed-upon expiration date. Figure 6.2 illustrates under what situations you would make

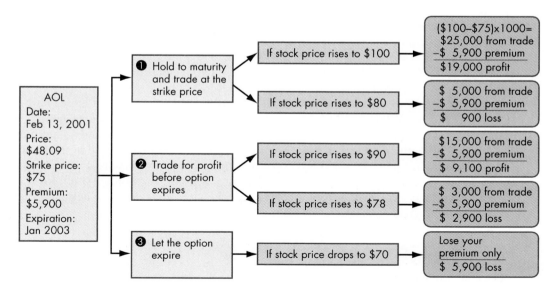

Figure 6.2 How call options work: Investor buys ten Call options (1,000 shares) on America Online (AOL) on Feb. 13, 2001: Price, $48.09 per share; strike price, $75; Expiration date, January 2003; Premium, $5.90 per share or $5,900 for the entire contract.

money by buying call options on America Online (AOL) stocks. You as a call buyer have three options: (1) hold the option to maturity and trade at the strike price, (2) trade for profit before your option expires (also known as exercising options), and (3) let the option expire if it is advantageous to you. For our AOL example in Figure 6.2, the stock is trading at $48.09. Instead of spending $48,090 for 1000 shares (or known as 10 call options), you purchase a Jan' 2003 call with a $75 strike price for $5.90 (or total premium of $5,900). Depending upon the stock price at a given moment and your trading option, your profit and loss statement would be as shown in Figure 6.2.

6.4 Investing in Bonds

Bonds are loans that investors make to corporations and governments. As shown in Figure 6.3, the borrowers get the cash they need while the lenders earn interest. Americans have more money invested in bonds than in stocks, mutual funds, or other types of securities. One of the major appeals is that bonds pay a set amount of interest on a regular basis. That is why they are called *fixed-income securities*. Another attraction is that the issuer promises to repay the loan in full and on time.

6.4.1 Bond versus Loan

A bond is something similar to a loan. For example, say you lend out $1,000 for 10 years in return for a yearly payment of 7% interest. Here is how that arrangement translates into bond terminology. You did not make a loan; you bought a bond. The $1,000 of principal is the **face value** of the bond, the yearly interest payment is its **coupon**, and the length of the loan, 10 years, is the bond's **maturity**. If you buy a bond at face value, or **par**, and hold it until it matures, you will earn interest at the stated, or coupon rate. For example, if you buy a 20-year $1,000 bond paying 8%, you will earn $80 a year for 20 years. The yield, or your return on your investment, will also be 8%, and you get your $1,000 back.

You can also buy and sell bonds through a broker after their date of issue. This is known as the **secondary market**. There the price fluctuates, with a bond sometimes selling at more than par value, at a premium price (premium bonds), and sometimes below, at a discount. Changes in price are directly tied to the interest rate of the bond. If its rate is higher than the rate being paid on similar bonds, buyers are willing to pay more to get the higher interest. However, if its rate is lower, the bond will sell for less to attract buyers. However, as the price goes up, the yield goes down. When the price goes down, the yield goes up.

6.4.2 Types of Bonds

You can choose different types of bonds to fit your financial needs—whether it is investing for college, finding tax-free income, or a range of other possibilities. That is why it is important to have a sense of how the various types work. Basically there are several types of bonds available in the financial market:

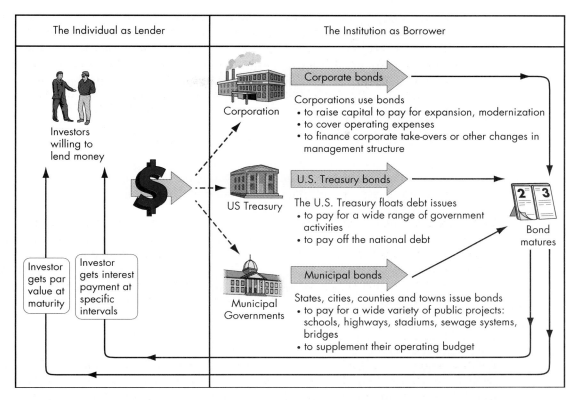

Figure 6.3   Types of bonds and how they are issued in the financial market. (Source: "The Wall Street Journal Guide to Understanding Money & Investing," Kenneth M. Morris and Alan M. Siegel, © 1993 by Lightbulb Press, Inc.)

- **Corporate bonds**—Bonds are the major source of corporate borrowing. *Debentures* are backed by the general credit of the corporation. *Specific corporate assets like property or equipment back asset-backed bonds.*

- **Municipal bonds**—State and local governments issue bonds to finance many projects that could not be funded through tax revenues. *General obligation* bonds are backed by the full faith and credit of the issuer, and *revenue bonds* by the earnings of the particular project being financed.

- **Treasury notes and bonds**—The U.S. Treasury floats debt issues to pay for a wide range of government activities including national debts. Intermediate (2 to 10 years) and long-term (10 to 30 years) government bonds are a major source of government funding.

- **Treasury bills**—Largest components of the money market, where short-term (13 weeks to 52 weeks) securities are bought and sold. Investors use T-bills for part of their cash reserve or as an interim holding place. Interest is the difference between buying price and the amount paid at maturity.

- **Zero-coupon bonds**—Corporations and governments sell zero-coupon bonds at deep discount. Investors do not collect interest; instead, the value of the bond increases to its full value when it matures. In this way, zero-coupon bonds are similar to old Series savings bonds that you bought for $37.50 and could cash in for $50 after 7 years. Organizations like to issue zeros because they can continue to use the loan money without paying periodic interest. Investors like zeros because they can buy more bonds for their money and then time the maturities to coincide with anticipated expenses.

- **Callable bonds**—Callable bonds do not run their full term. The issuer may call the bond—pay off the debt—before the maturity date. It is a process called redemption. Issuers will sometimes call bonds when interest rates drop, so they can reduce their debt. If they pay off their outstanding bonds, they can float another bond at the lower rate. It is the same idea of refinancing a mortgage to get a lower interest and make lower monthly payments. Callable bonds are more risky for investors than non-callable ones because they are often faced with reinvesting the money at a lower, less attractive rate. To protect bondholders expecting long-term steady income, call provisions usually specify that a bond cannot be called before a certain number of years, usually 5 or 10 years.
- **Floating-rate bonds**—These bonds promise periodic adjustments of the interest rate to persuade investors that they are not locked into what seems like an unattractively low rate.

Under the U.S. Constitution, any type of government bond will include a tax break. Because federal and local governments cannot interfere with each other's affairs, income from local government bonds (municipals, or simply **munis**) is immune from federal taxes, and income from U.S. Treasury bonds is free from local taxes. Table 6.3 summarizes major types of bonds and their investment features.

6.4.3 Understanding Bond Prices

Corporate bond prices are quoted in increments of points and eight fractions of a point, with a par of $1,000 as the base. The value of each point is $10, and of each fraction, $1.25, as Table 6.4 shows. For, a bond quoted at 86 1/2 would be selling for $865, and one quoted at $100 3/4 would be selling for $1,007.50. Treasury bonds are measured in 32nds rather than in 100ths of a point. Each 1/32 equals 31.25 cents, and we normally drop the fractional part of the cent when stating a price. For example, if a bond is quoted at 100.2 or 100+2/32, the price translates to $1,000.62.

6.4.4 Are Bonds Safe?

Just because bonds have a reputation as a conservative investment does not mean they are always safe. There are sources of risk in holding or trading bonds.

- To begin with, do not forget that not all loans are paid back. Companies, cities, and counties occasionally do go bankrupt. U.S. Treasury bonds alone are considered rock solid.
- Another source of risk for certain bonds is that your loan may be paid back early, or called. While that is certainly better than not being paid back at all, it forces you to find another, possibly less lucrative, place to put your money.

Type of Bond	Par value	Maturity period	Tax status	Call provisions	Characteristics
Corporate Bonds	$1,000	Short-term: 1–5 years Intermediate-term: 5–10 years Long-term: 10–20 years	Taxable	Callable	• More risky than government bonds • Potentially higher yields than government bonds • Usually large minimum investment required
Municipal Bonds	$5,000 and up	From 1 month to 40 years	Exempt from federal taxes	Sometimes callable	• Lower interest rates than comparable corporate bonds, because of tax exemption • Especially attractive to high tax-bracket investors, who benefit from tax-exemption feature
T-Bonds and T-Notes	$1,000, $5,000, $10,000, $100,000 and $1 million	Bonds—over 10 years Notes—2–10 years	Exempt from state and local taxes	Usually not callable	• Maximum safety, since backed by federal government, but relatively low interest rates
T-Bills	$10,000	3 months 6 months 1 year	Exempt from state and local taxes	Not callable	• Short-term investments—no specific interest payments; instead, interest consists of the difference between a discounted buying price and the par amount at maturity

Table 6.3
Types of U.S. Bonds and Their Characteristics

See more on other types of bond in "The Wall Street Journal Guide to Understanding Money & Investing," Kenneth M. Morris and Alan M. Siegel, © 1993 by Lightbulb Press, Inc.

Table 6.4
Bond Price
Notation
Used in
Financial
Markets

Corporate Bonds		Treasury Bonds	
1/8 = $1.25	5/8 = $6.25	1/32 = $0.3125	17/32 = $5.3125
		2/32 = $0.6250	18/32 = $5.6250
		3/32 = $0.9375	19/32 = $5.9375
		4/32 = $1.25	20/32 = $6.25
1/4 = $2.50	3/4 = $7.50	5/32 = $1.5625	21/32 = $6.5625
		6/32 = $1.8750	22/32 = $6.8750
		7/32 = $2.1875	23/32 = $7.1875
		8/32 = $2.50	24/32 = $7.50
3/8 = $3.75	7/8 = $8.75	9/32 = $2.8125	25/32 = $7.8125
		10/32 = $3.1250	26/32 = $8.1250
		11/32 = $3.4375	27/32 = $8.4375
		12/32 = $3.75	28/32 = $8.75
1/2 = $5.00	1 = $10	13/32 = $4.0625	29/32 = $9.0625
		14/32 = $4.375	30/32 = $9.3750
		15/32 = $4.6875	31/32 = $9.6875
		16/32 = $5.00	32/32 = $10

- The main danger for buy-and-hold investors, however, is a rising inflation rate. Since the dollar amount they earn on a bond investment does not change, the value of that money can be eroded by inflation. Worse yet, with your money locked away in that bond, you will not be able to take advantage of the higher interest rates that are usually available in an inflationary economy. Now you know why bond investors cringe at cheerful headlines about full employment and strong economic growth: These traditional signs of inflation hint that bondholders may soon lose their shirts.

6.4.5 How Do Prices and Yields Work?

You can trade bonds on the market just like stocks. Once purchased, you can keep a bond may for its maturity or for a variable number of interest periods before being sold. You can purchase or sell bonds at prices other than face value, depending on the economic environment. Furthermore, bond prices change over time because of the risk of nonpayment of interest or par value, supply and demand, and the outlook for economic conditions. These factors affect the **yield to maturity**[5](or **return on investment**).

---

[5] The SmartMoney.com (*http://www.smartmoney.com*) provides you with a bond calculator by computing the yield to maturity with inputs of the current price, coupon rate, and maturity date. Begin by entering the bond's coupon rate and maturity. If you then enter a price, the calculator will display the yield of the bond to maturity; if you enter a yield, the calculator will show you the corresponding price. You can also use the sliders to see the effects of price and yield changes. (This is your chance to watch prices and yields literally move in opposite directions.) Note that prices are written in the standard bond format, where 100 means face value, 90 means 90% of face value, and so on. Read our bond primer for more details.

- The **yield to maturity** represents the actual interest earned from a bond over the holding period. In other words, the yield to maturity on a bond is the interest rate that establishes the equivalence between all future interest and face value receipts and the market price of the bond.

- The **current yield** of a bond is the annual interest earned as a percentage of the current market price. This current yield provides an indication of the annual return realized from the bond investment. To illustrate the point, we will explain these values with numerical examples in the following section.

Bond quotes follow a few unique conventions. As an example, we have included two U.S. corporate bonds, AT&T and Amresco, respectively.

Bonds	Current Yield	Volume	Closes	Net Change
ATT 7s05	6.5%	5 million	108 1/4	+1/2
Amresco 10s03	11.7%	34 million	85 3/8	+3/8

- ATT 7s05 means a bond maturing at year 2005 with a coupon rate of 7%. The symbol "s" has no meaning except that it separates the coupon rate (7%) from the maturity (2005).

- Prices are given as percentages of face value, with the last digits not decimals, but in terms of 1/8. The AT&T 7s05 bond on the list, for instance, has just risen to sell for 108 1/4. In other words, an investor who bought the bond when it was issued (at 100) could now sell it for a little more than 8.25% profit. This premium over face value is explained by examining its coupon rate—7%, and its current yield is 6.5%.

- The AT&T bond paying 7% interest is currently yielding only 6.5% because the price is $1,082.50 ($108 \frac{1}{4}$), or $82.50 ($8 \frac{1}{4}$) above par. In contrast, Amresco's 10s03 bond, a diversified financial services company, paying 10% interest is currently yielding 11.7% because the bond is selling for $853.75 (85 3/8), or $146.25 below par.

If you buy a bond at face value, its rate of return, or yield, is just the coupon rate. However, a glance at a table of bond quotes (like the one above) will tell you that after they are first issued, bonds rarely sell for exactly face value. So how much is the yield then? Take the $1,000 ATT 7s05 bond with a 7% ($70) coupon that matures in the year 2005. If you manage to buy it for $900, you are getting two bonuses. First, you have effectively bought a bond with a 7.78% coupon, since the $70 coupon is 7.78% of your $900 purchase price. (Recall that the coupon rate adjusted for the current price is the current yield of the bond.) However, wait, there's more: although you paid $900, in 2005 you will receive the full $1,000 face value. A more accurate calculation of return, yield to maturity considers the resultant $100 capital gain. Example 6.4 will illustrate how you calculate this yield to maturity considering both the purchase price and the capital gain.

Example 6.4 Yield to Maturity and Current Yield

Consider buying a $1,000 denomination AGLC (Atlanta Gas & Light Company) bond at the market price of $996.25. The interest will be paid semi-annually, the interest rate per payment period will be simply 4.8125%, and 20 interest payments over 10 years are required. We show the resulting cash flow to the investor in Figure 6.4. Find (a) the yield to maturity and (b) current yield.

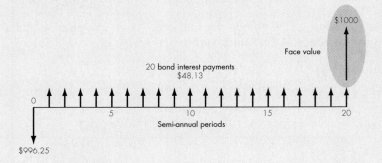

Figure 6.4   A typical cash flow transaction associated with an investment in the AGCL's corporate bond.

Discussion

- **Debenture bond:** The Atlanta Gas & Light Company (AGLC) wants to issue bonds totaling $100 million, and the company will back the bonds with specific pieces of property, such as buildings.
- **Par value:** The AGCL's bond has a par value of $1,000.
- **Maturity date:** The AGLC's bonds, which were issued on January 30, 1997, will mature on January 31, 2007; thus, they have a 10-year maturity at time of issue.
- **Coupon rate:** The AGLC's coupon rate is 9.625%, and interest is payable semi-annually. For example, the AGLC's bonds have a $1000 par value, and they pay $96.25 in simple interest (9 5/8%) each year ($48.13 every 6 months).
- **Discount bond:** The AGLC's bonds are offered at less than the par value at 99.625%, or 0.375% discount. For example, an AGLC bond, with a face value of $1000, can be purchased for just $996.25, which is known as the **market price** (or value) of the bond.

Solution

Given: Initial purchase price =$996.25, coupon rate = 9.625% per year paid semi-annually, 10-year maturity with a par value of $1000

Find: (a) Yield to maturity and (b) current yield

(a) Yield to maturity:

We find the yield to maturity by determining the interest rate that makes the present worth of the receipts equal to the market price of the bond:

$$\$996.25 = \$48.13(P/A, i, 20) + \$1000(P/F, i, 20).$$

The value of i that makes the present worth of the receipts equal to \$996.25 lies between 4.5% and 5%. Solving for i by interpolation yields $i = 4.84\%$:

Present Worth of Receipts	i
\$1,040.71	4.5%
996.25	?
976.70	5%

$$i = 4.5\% + (0.5\%)\left[\frac{1,040.71 - 996.25}{1,040.71 - 976.70}\right] = 4.84\%$$

Note that this is 4.84% yield to maturity per semiannual period. The nominal (annual) yield is 2(4.84) = 9.68%, compounded semiannually. When compared with the coupon rate of 9 5/8% (or 9.625%), purchasing the bond with the price discounted at 0.375% brings about an additional 0.055% yield. The effective annual interest rate is then

$$i_a = (1 + 0.0484)^2 - 1 = 9.91\%.$$

This 9.91% represents the **effective annual yield** to maturity on the bond. Notice that when you purchase a bond at par value and sell at par value, the yield to maturity will be the *same* as the coupon rate of the bond.

Until now, we have observed differences in nominal and effective interest rates because of the frequency of compounding. In the case of bonds, the reason is very different: The stated (par) value of the bond and the actual price for which it is sold are not the same. We normally state the nominal interest as a percentage of par value. However, when the bond is sold at a discount, the same nominal interest on a smaller initial investment is earned; hence, your effective interest earnings are greater than the stated nominal rate.

(b) Current yield:

For our example of AGLC, we compute the current yield as follows:

$$\$48.13/996.25 = 4.83\% \text{ per semiannual period}$$

$$4.83\% \times 2 = 9.66\% \text{ per year (nominal current yield)}$$

$$i_a = (1 + 0.0483)^2 - 1 = 9.90\%.$$

This effective current yield is 0.01% lower than the yield to maturity computed above (9.91%). If the bond is selling at a discount, the current yield is smaller than the yield to maturity. If the bond were selling at a premium, the current yield would be larger than the yield to maturity. A significant difference between the yield to maturity and the current yield of a bond can exist, because the market price of a bond may be more or less than its face value. Moreover, both the current yield and the yield to maturity may differ considerably from the stated coupon value of the bond.

Example 6.5 Bond Value Over Time

Reconsider the AGLC bond investment introduced earlier. If the yield to maturity remains constant at 9.68%, (a) what will be the value of the bond 1 year after it was purchased? (b) If the market interest rate drops to 9% a year later, what would be the market price of the bond?

Solution

Given: Using the same data in Example 6.4

Find: (a) The value of bond one-year later, (b) market price of the bond a year later with the going rate of 9% interest

(a) We can find this value using the same valuation procedure, but now the term to maturity is only 9 years.

$$\$48.13(P/A, 4.84\%, 18) + \$1,000(P/F, 4.84\%, 18) = \$996.80.$$

The value of the bond will remain at $996.80 as long as the yield to maturity remains constant at 9.68% over 9 years.

(b) Now suppose interest rates in the economy have fallen since the AGLC bonds were issued, and consequently, the going rate of interest is 9%. Both the coupon interest payments and the maturity value would remain constant, but now, 9% values would have to be used to calculate the value of the bond. The value of the bond at the end of the first year would be

$$\$48.13(P/A, 4.5\%, 18) + \$1,000(P/F, 4.5\%, 18) = \$1,038.06.$$

Thus, the bond would sell at a premium over its par value.

Comments: The arithmetic of the bond price increase should be clear, but what is the logic behind it? We can explain the reason for the increase as follows: Because the going market interest rate for the bond has fallen to 9%, if we had $1,000 to invest, we could buy new bonds with a coupon rate of 9%. These would pay $90 of interest each year rather than $96.80. We would prefer $96.80 to $90, and therefore we would be willing to pay more than $1,000 for AGLC's bonds to obtain higher coupons. All investors would recognize these facts, and therefore, the AGLC's bonds would be bid up in price to $1,038.06. At that point, they would provide the same yield to maturity (rate of return) to a potential investor, as would the new bonds, 9%.

Summary

- The three basic investment objectives are growth, income, and liquidity.
- Our objectives can change as time passes and new priorities and influences shape the way we view the future.
- There is logical connection between the three objectives and the basic asset classes:

Objective	Asset Class
Growth	Equity (stock)
Income	Debt (bond)
Liquidity	Cash (CDs, T-Bills)

- Once you set your risk tolerance, you are establishing an upper-bound limit on the portfolio's long-term expected rate of return.
- There is no such thing as a risk-free investment. The challenge is to decide what level of risk you are willing to assume and then, having decided on your risk tolerance, to understand the implications of that choice.
- The two greatest risks investors face are inflation and volatility.
- The three basic asset classes are cash, debt, and equities. Cash is the least risky with the lowest return. Debt is moderately risky with moderate returns. Equities are the most risky but offer the greatest payoff.
- Over the long-term, equities outperform debt and cash-type investments.
- Portfolios with long-time horizons need equities to offset inflation while short time frames require debt and/or cash investments to reduce volatility.
- Dollar-cost averaging is a planned transfer, over a period, of equal amounts from one asset class to another.
- There is far more to the power of **diversification** than simply spreading your assets over a number of investments to reduce risk. By combining assets with different patterns of return, it is possible to achieve a higher rate of return without increasing significant risk.
- **Options** give you the right to buy or the obligation to sell. If you buy options, you are buying opportunities. If you sell them, you might have to meet an obligation. Buying **call options** means the right to buy the underlying investment at a fixed price until the expiration date. Buying **put options** means the right to sell the underlying investment at a fixed price until the expiration date.
- The asset allocation is simply a matter of answering the following question: "Given my personal risk tolerance and my investment objectives, what percentage of my assets should be allocated for **growth**, what percentage for **income**, and what percentage for **liquidity**?"
- You can determine the **expected rate of return** for a portfolio by computing the weighted average of the returns for each investment.
- You can determine the **expected risk** of a portfolio by computing the weighted average of the volatility of each investment.
- All other things being equal, if the expected returns are approximately the same choose the portfolio with the lowest expected risk.
- All other things being equal, if the expected risk is about the same choose the portfolio with the highest expected return.
- **Asset-backed bonds:** If a company backs the bonds with specific pieces of property, such as buildings, we call these types of bonds **mortgage bonds**,

which indicate the terms of repayment and the particular assets pledged to the bondholders in case of default. It is much more common, however, for a corporation simply to pledge its overall assets. A **debenture bond** represents such a promise.

- **Par value:** Individual bonds are normally issued in even denominations for $1,000 or multiples of $1,000. The stated face value of an individual bond is termed the **par value**.

- **Maturity date:** Bonds generally have a specified **maturity** date on which the par value is to be repaid.

- **Coupon rate:** We call the interest paid on the par value of a bond the **annual coupon rate**. The time interval between interest payments could be of any duration, but a semi-annual period is the most common.

- **Discount or premium bond:** A bond that sells below its par value is called a **discount bond**. When a bond sells above its par value, it is called a **premium bond**.

Self-Test Questions

6s.1 Which of the following statement is incorrect?

(a) With proper diversification, we can increase our return on investment without increasing significant risk.

(b) Buying U.S. treasury bills would be one of the most risk-free investments if we do not anticipate any inflation in the investment period.

(c) Under inflationary environment, one of the better investment strategies would be to buy a piece of real estates.

(d) Under inflationary environment, holding cash is the best way to protect eroding purchasing power.

6s.2 Dollar-cost averaging is:

(a) a forced payroll deduction plan

(b) an investment principle to reduce market risk due to fluctuation

(c) a specific computational rule used in mutual funds to calculate your return

(d) an investment rule to beat inflation.

6s.3 Which of the following are zero-risk investments?

(a) bank CDs (certificate of deposits)

(b) cash

(c) long-term corporate bonds

(d) U.S. Treasury bills

(e) None of these.

6s.4 The beta is one way to measure an investment's:

(a) Perceived market risk

(b) stock performance

(c) return on investment (or yield)

(d) relative historical risk in reference to other investments in the same business category.

6s.5 You are considering purchasing a 6% bond with a face value of $1,000 with interest paid semiannually. You want to earn a 9% annual return on your investment. Assume that the bond will mature to its face value 5 years from now. What is the required purchasing price of the bond?

(a) $P = \$60(P/A, 6\%, 5) + \$1,000(P/F, 6\%, 5)$

(b) $P = \$90(P/A, 9\%, 5) + \$1,000(P/F, 9\%, 5)$

(c) $P = \$30(P/A, 4.5\%, 10) + \$1,000(P/F, 4.5\%, 10)$

(d) $P = \$30(P/A, 3\%, 10) + \$1,000(P/F, 3\%, 10)$

Problems

Investment Basics

6.1 The following represents the stock prices for Nokia, a Finland-based cellular phone maker over 5 years.

> January 1, 1996 $3.0
> January 1, 1997 $4.75
> January 1, 1998 $5
> January 1, 1999 $16
> January 1, 2000 $45
> January 1, 2001 $40

(a) What is the average annual return on the investment?

(b) What is the compound annual return on the investment?

6.2 You purchased 100 shares of each of the following stocks on June 1, 1998. The price of each stock on June 1, 2000 is as follows:

Stock	Purchase Price	Current Price
America Online (AOL)	$18	$52
Motorola (MOT)	17	30
J.P. Morgan (JPM)	135	112
Coca-Cola (KO)	89	59
Nortel Network (NT)	16	68
Wal-Mart (WMT)	30	54
Boeing (BA)	50	40
Qualcomm (QCOM)	11	61

(a) What is the average annual return on your portfolio?

(b) What is the compound annual return on your portfolio?

(c) Your total investment was $36,600. If you only invested in a single stock instead of diversifying, what is the range of your average annual return on your investment?

(d) In (c), what is the range of compound annual return?

(e) Based on (c) and (d), make comments on your diversification strategy.

6.3 The following represents the monthly stock prices for Cisco for a 6-month period. If you make your purchase in $2,000 amounts for each month, the total investment would be $12,000.

> January 2, 2000 $50
> February 1, 2000 $55
> March 1, 2000 $66
> April 1, 2000 $63
> May 1, 2000 $72
> June 1, 2000 $60

(a) How many shares would you have at the end of June 30, 2000, if you did not sell any shares?

(b) What is the average purchased price per share?

(c) If the stock were selling at $62 per share at the end of June 30, 2000, what would be your account balance?

Investments in Stock

6.4 If you purchased 100 shares of each of the following stocks on January 2000: Dow Chemical (DD), Alcoa (AA),

America Online (AOL), Yahoo (YHOO), American Express (AXP), and Wal-Mart (WMT), how much are they worth today? What things would you consider in deciding to keep or sell these stocks?

6.5 The following represents the historical financial data for Intel Corporation:

Year	Stock Price	Annual Dividends
February 1, 1998	$41.50	$0.058 per share
February 1, 1999	68.75	0.050
February 1, 2000	101.50	0.110

(a) Suppose that you purchase 100 shares on February 1, 1998 and held the shares for two years. What is the compound annual return on your investment over the holding period? (You should include the dividends in your calculation.)

(b) If you plan on selling the stock after 3 years from purchase and you desire a compound return of 60% (without considering the dividends), what should be the share price on February 1, 2001?

(c) If the stock price plunges to $89 on February 1, 2001, what would be your compound return on investment over a 3-year holding period? (Do not consider the dividends.)

Investment in Bonds

6.6 The Jimmy Corporation issued a new series of bonds on January 1, 1996. The bonds were sold at par ($1,000), have a 12% coupon rate, and mature in 30 years, on December 31, 2025. Coupon interest payments are made semi-annually (on June 30 and December 31).

(a) What was the yield-to-maturity (YTM) of the bond on January 1, 1996?

(b) Assuming that the level of interest rates had fallen to 9%, what was the price of the bond on January 1, 2001, 5 years later?

(c) On July 1, 2001, the bonds sold for $922.38. What was the YTM at that date? What was the current yield at that date?

6.7 A $1000, 9.50% semiannual bond is purchased for $1,010. If the bond is sold at the end of 3 years and six interest payments, what should be the selling price be to yield a 10% return on your investment?

6.8 Mr. Gonzalez wishes to sell a bond that has a face value of $1,000. The bond bears an interest rate of 8% with bond interests payable semiannually. Four years ago, $920 was paid for the bond. At least a 9% return (yield) in investment is desired. What must be the minimum selling price?

6.9 Suppose you have the choice of investing in (1) a zero-coupon bond, which costs $513.60 today, pays nothing during its life, and then pays $1,000 after 5 years or (2) a bond which costs $1000 today, pays $113 in interest semiannually, and matures at the end of 5 years. Which bond would provide the higher yield?

6.10 Suppose you were offered a 12-year, 15% coupon, $1,000 par value bond at a price of $1,298.68. What rate of interest (yield-to-maturity) would you earn if you bought the bond and held it to maturity (semiannual interest)?

6.11 The Diversified Products Company has two bond issues outstanding. Both bonds pay $100 semiannual interest plus $1,000 at maturity. Bond A has a remaining maturity of 15 years, and bond B a maturity of 1 year. What will be the value of each of these bonds now, when the going rate of interest is 9%?

6.12 The AirJet Service Company's bonds have 4 years remaining to maturity. Interest is paid annually; the bonds have a $1,000 par value; and the coupon interest rate is 8.75%.

(a) What is the yield-to-maturity at a current market price of $1,108?

(b) Would you pay $935 for one of these bonds if you thought that the market rate of interest was 9.5%?

6.13 Suppose Ford sold an issue of bonds with a 15-year maturity, a $1,000 par value, a 12% coupon rate, and semiannual interest payments.

(a) Two years after the bonds were issued, the going rate of interest on bonds such as these fell to 9%. At what price would the bonds sell?

(b) Suppose that, 2 years after the issue, the going interest rate had risen to 13%. At what price would the bonds sell?

(c) Today, the closing price of this bond is $783.58. What is the current yield?

Short Case Studies

6.14 The following chart illustrates the three different methods of computing the capital gains when you sell stocks. They are (1) first-in, first-out method, (2) specific shares method, and (3) average cost method. Instead of selling the stocks on January 1997, you waited until January 1, 2001 and sold the 200 shares at $25.50 per share. Which method would produce the least capital gains for tax purpose?

Capital gains computation methods

President Clinton proposed that capital gains on all stock and mutual fund sales be figured using the stock or fund shares' average cost. Currently, investors can use one of three methods to figure capital gains and losses, including the average cost method. Here's how the three methods would work, assuming you sell 100 shares at $10.50 a share:

FUND ACCOUNT HISTORY

Date	Activity	Amount
Oct. 1993	Buy 100 shares @ $10 a share	$1,000
April 1994	Buy 50 shares @ $11 a share	$550
Dec. 1994	Reinvested dividends add 5 shares @ $12 a share	$60
March 1995	Buy 150 shares @ $9 a share	$1,350
Jan. 1997	Sell 100 shares @ $10.50 a share	$1,050

FIRST-IN, FIRST-OUT METHOD

The most common way of computing cost basis. As the name implies, the oldest shares are considered to be sold first.

Sale proceeds	$1,050
Cost basis	−$1,000 (100 shrs. @ $10 a share)
Gain	$50

SPECIFIC SHARES METHOD

Lets you choose which shares to sell, giving you control of whether you claim a gain or loss. In this example, you claim a long-term loss:

Sale proceeds	$1,050
Cost basis	−$1,060 (45 shrs. @ $10; 50 shrs. @ $11; 5 shrs. @ $12)
(Loss)	($10)

AVERAGE COST METHOD

Under current law, if you pick this method, you must state it on your tax return and use it for all sales in that fund. And you must have permission from the IRS to revoke this choice. Most mutual funds will figure your average cost for you.

Average cost per share		Cost of shares owned		Number of shares owned
		($1,000+$550+$60+$1,350)		(100+50+5+150)
$9.70	=	$2,960	÷	305

Sale proceeds	$1,050
Cost basis	− $970 (100 shrs. @ $9.70 a share)
Gain	$80

Source: Strong Funds; USA TODAY research

PART THREE

Evaluating Business
and Engineering Assets

CHAPTER 7

Present Worth Analysis

Automating Food Processors with Electronic Sensors: Would you care for a digitally developed doughnut? How about an electronically evolved animal cracker? Nabisco is among the large food producers that have recently installed sensors to help make their products—everything from packaged puddings to cream-filled cookies are now prepared with electronic assistance. Nabisco began to look to computerized process control instead of human inspection to enhance the quality and safety of their products. Before installing the sensor, plant operators ran moisture analyses on just-baked crackers every 15 minutes to make sure they were adequately cooked. The laboratory test takes about 30 minutes to perform, so if the crackers were not right, that meant 30 extra minutes of substandard production before an adjustment could be made. This observation, coupled with a highly competitive marketplace in which profit margins are typically a slim 5% to 6%, has encouraged Nabisco to reduce any inspection cost by automating the entire manufacturing process. By contrast, the sensor makes readings continuously and communicates information to a central process logic controller, which regulates oven temperature. Nabisco engineers believe that the sensor system—which cost $100,000—has already paid for itself in the "reduction of scrap." How would you justify this nature of investment project?

This is a cost reduction project where the main question is "Would there be enough annual savings from reducing waste to recoup the $100,000 investment?" In the past, the trend toward automation in food industry was slowed by the conviction that creating things culinary is a craft. Nametre is one of hundreds of companies that have worked in the last decade to develop more accurate, easily integrated, and durable electronic sensing devices. These devices detect everything from the creaminess of sandwich cookie filling to the piquancy of pasta sauce. However, no matter how sophisticated food taste instrumentation becomes, no one has come up with a replacement for the human tongue. Therefore, if customers do not like electronically prepared cookies, what would be the potential financial risk?

In Chapters 4 and 5, we presented the concept of the time value of money and developed techniques for establishing cash flow equivalence with compound interest factors. This background provides a foundation for accepting or rejecting a capital investment—the economic evaluation of a project's desirability. Forthcoming coverage of investment worth in this chapter will allow us to go a step beyond accepting or rejecting an investment to making comparisons of alternative investments. We will determine how to compare alternatives on an equal basis and select the wisest alternative from an economic standpoint.

The three common measures based on cash flow equivalence are (1) equivalent present worth (PW), (2) equivalent future worth (FW), and (3) equivalent annual worth (AE). Present worth represents a measure of future cash flow relative to the time point "now" with provisions that account for earning opportunities. Future worth is a measure of cash flow at some future planning horizon, which offers a consideration of the earning opportunities of intermediate cash flows. Annual worth is a measure of cash flow in terms of equivalent equal payments made on an annual basis.

Our treatment of measures of investment worth is divided into three chapters. Chapter 7 begins with a consideration of the payback period, a project screening tool that was the first formal method used to evaluate investment projects. Then it introduces two measures based on basic cash flow equivalence techniques: present worth and future worth analysis. Because the annual worth approach has many useful engineering applications related to estimating the unit cost, Chapter 8 is devoted to annual cash flow analysis. Chapter 9 presents measures of investment worth based on yield; these measures are known as rate of return analysis.

We must also recognize that one of the most important parts of the capital budgeting process is the estimation of relevant cash flows. For all examples in this chapter, and those in Chapters 8 and 9, however, net cash flows can be viewed as before-tax values or after-tax values for which tax effects have been recalculated. Since some organizations (e.g., governments and nonprofit organizations) are not subject to tax, the before-tax situation provides a valid base for this type of economic evaluation. Taking this view will allow us to focus on our main area of concern, the economic evaluation of investment projects. The procedures for determining after-tax net cash flows in taxable situations are developed in Chapter 9.

7.1 Describing Project Cash Flows

In Section 1.6, we described many engineering economic decision problems, but we did not provide suggestions on how to actually solve them. What do Examples 1.1 through 1.5 have in common? Note that all these problems involve two dissimilar types of amounts. First, there is the investment, which is usually made in a lump sum at the beginning of the project. Although not literally made "today," the investment is made at a specific point in time that, for analytical purposes, is called today, or time 0. Second, there is a stream of cash benefits that are expected to result from this investment over a period of future years.

7.1.1 Loan versus Project Cash Flows

An investment made in a fixed asset is similar to an investment made by a bank when it lends money. The essential characteristic of both transactions is that funds are committed today in the expectation of their earning a return in the future. In the case of the bank loan, the future return takes the form of interest plus repayment of the principal. This is known as the **loan cash flow**. In the case of the fixed asset, the future return takes the form of profits generated by productive use of the asset. As shown in Figure 7.1, the representation of these future earnings, along with the capital expenditures, and annual expenses (such as wages, raw materials, operating costs, maintenance costs, and income taxes), is the **project cash flow**. This similarity between the loan cash flow and the project cash flow brings us to an important conclusion—i.e., we can use the same equivalence techniques developed in Chapter 4 to measure economic worth. Example 7.1 illustrates a typical procedure for obtaining a project's cash flows.

Example 7.1 Identifying Project Cash Flows

XL Chemicals is considering the installation of a computer process control system in one of its process plants. This plant is used about 40% of the time or 3,500 operating hours per year to produce a proprietary demulsifaction chemical, and during the remaining 60% of the time it is used to produce other specialty chemicals. The annual production of the demulsification chemical amounts to 30,000 kilograms per year and it sells for $15 per kilogram. The proposed computer process control system will cost $650,000 and is expected to provide specific benefits in the production of the demulsification chemical. First, the selling price of the product could be increased by $2 per kilogram because the product will be of higher purity, which translates into better demulsification performance. Secondly, production volumes will increase by 4,000 kilograms per year as a result of higher reaction yields without any increase in raw material quantities or production time. Finally, the number of process operators can be reduced by one per shift which represents a savings of $25 per hour. The new control system would result in additional maintenance costs of $53,000 per year and has an expected useful life of

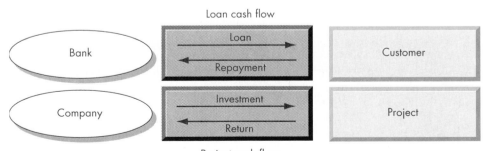

Project cash flow

Figure 7.1 A bank loan versus an investment project

8 years. While the system is likely to provide similar benefits in the production of the other specialty chemicals manufactured in the process plant, these have not been quantified as yet.

Discussion: As in real-world engineering economic analysis, this problem contains a great deal of data from which we must extract and interpret critical cash flows. You would be wise to begin organizing your solution into a table consisting of the following categories.

Year (*n*)	Cash Inflows (Benefits)	Cash Outflows (Costs)	Net Cash Flows
0			
1			
⋮			
8			

Solution

Given: Cost and benefit information as stated above

Find: Net cash flow in each year over the life of the new system

Although we could assume similar benefits are derivable from the production of the other speciality chemicals, let's restrict our consideration to the demulsification chemical and allocate the full initial cost of the control system and the annual maintenance costs to this chemical product. (You could logically argue that only 40% of these costs belong to this production activity.) The gross benefits are the additional revenues realized from the increased selling price and the extra production, as well as the cost savings resulting from having one less operator. Revenues from the price increases are 30,000 kilograms per year × $2/kilogram or $60,000 per year. The added production volume at the new pricing adds revenues of 4,000 kilograms per year × $17 per kilogram or $68,000 per year. The elimination of one operator results in annual savings of 3,500 operating hours per year × $25 per hour or $87,500 per year. The net benefits in each of the 8 years are the gross benefits less the maintenance costs ($60,000 + $68,000 + $87,500 − $53,000) = $162,500 per year. Now we are ready to summarize a cash flow table as follows:

Year (n)	Cash Inflows (Benefits)	Cash Outflows (Costs)	Net Cash Flows
0	0	$650,000	−$650,000
1	215,500	53,000	162,500
2	215,500	53,000	162,500
⋮	⋮	⋮	⋮
8	215,500	53,000	162,500

Comments: In Example 7.1, if the company purchases the computer process control system for $650,000 now, it could expect an annual saving of $162,500 for 8 years. (Note that these savings occur in discrete lumps at the ends of years.) We also considered only the benefits associated with the production of the demulsification chemical. We could quantify some benefits attributable to the production of the other chemicals from this plant. Suppose that the demulsification chemical benefits alone justify the acquisition. Then, it is obvious that, had other benefits deriving from the other chemicals have been considered as well, acquisition would have been even more clearly justified.

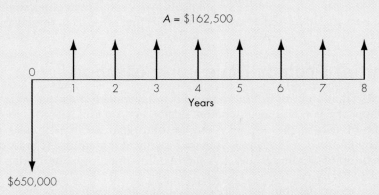

Figure 7.2   Cash flow diagram for the computer process control project described in Example 7.1

We can draw a cash flow diagram of this situation as shown in Figure 7.2. Assuming these cost savings and cash flow estimates are correct, should management give the go-ahead for installation of the system? If management decides not to purchase the computer control system, what should it do with the $650,000 (assuming they have this amount in the first place)? The company could buy $650,000 of Treasury bonds, or it could invest the amount in other cost saving projects. How would the company compare cash flows that differ both in timing and amount for the alternatives it is considering? This is an extremely important question because virtually every engineering investment decision involves a comparison of alternatives. These are the types of questions this chapter is designed to help you answer.

7.1.2   Independent versus Mutually Exclusive Investment Projects

Most firms will have a number of unrelated investment opportunities available. For example, in the case of XL Chemicals, other projects being considered in addition to the computer process control project in Example 7.1 are a new waste heat recovery boiler, a CAD system for the engineering department, and a new warehouse. The economic attractiveness of each of these projects can be measured, and an accept or reject decision made without reference to any of the other projects. The decision on any one project has no effect on the decision made on another project. Such projects are said to be **independent.**

In Section 7.5 we will see that in many engineering situations we are faced with selecting the most economically attractive project from a number of alternative projects, which all resolve the same problem or meet the same need. It is unnecessary to choose more than one project on this situation, and acceptance of one automatically rejects all of the others. Such projects are said to be **mutually exclusive.**

As long as the total cost of all the independent projects found to be economically attractive is less than the investment funds available to a firm, then all of these projects could proceed. However, this is rarely the case. The selection of those projects which should proceed when investment funds are limited is the subject of capital budgeting. Apart from Chapter 16 which deals with capital budgeting, funds availability will not be a consideration in accepting or rejecting projects in this chapter

7.2 Initial Project Screening Method

Let's suppose that you are in the market for a new punch press for your company's machine shop, and you visit an equipment dealer. As you give a serious look at one of punch press models in the display room, an observant equipment sales person approaches you and says, "That press you are looking at is state-of-art in its category. If you buy that top-of-the-line model, it will cost a little bit more, but it will pay for itself in less than 2 years."

Before studying the four measures of investment attractiveness, we will review a simple, but nonrigorous, method commonly used to screen capital investments. One of the primary concerns of most business people is whether, and when, the money invested in a project can be recovered. The **payback method** screens projects on the basis of how long it takes for net receipts to equal investment outlays. This calculation can take one of two forms by either ignoring time value of money considerations or including them. The former case is usually designated as the **conventional payback method,** whereas the latter case is known as the **discounted payback method.**

A common standard used to determine whether or not to pursue a project is that a project does not merit consideration unless its payback period is shorter than some specified period of time. (This time limit is largely determined by management policy. For example, a high-tech firm, such as a computer chip manufacturer, would set a short time limit for any new investment because high-tech products rapidly become obsolete.) If the payback period is within the acceptable range, a formal project evaluation (such as a present worth analysis) may begin. It is important to remember that **payback screening** is not an *end* in itself, but rather a method of screening out certain obvious unacceptable investment alternatives before progressing to an analysis of potentially acceptable ones.

7.2.1 Payback Period—The Time It Takes to Pay Back

Determining the relative worth of new production machinery by calculating the time it will take to pay back what it cost is the single most popular method of project screening. If a company makes investment decisions based solely on the payback period, it considers only those projects with a payback period *shorter* than the maximum acceptable payback period. (However, because of shortcomings of the payback screening method, which we will discuss later, it is rarely used as the only decision criterion.)

What does the payback period tell us? One consequence of insisting that each proposed investment has a short payback period is that investors can assure themselves of being restored to their initial position within a short span of time. By restoring their initial position, investors can take advantage of additional, perhaps better, investment possibilities that may come along.

Example 7.2 Conventional Payback Period for the Computer Process Control System Project

Consider the cash flows given in Example 7.1. Determine the payback period for this computer process control system project.

Solution

Given: Initial cost = $650,000, annual net benefits = $162,500

Find: Conventional payback period

Given a uniform stream of receipts, we can easily calculate the payback period by dividing the initial cash outlay by the annual receipt:

$$\text{Payback period} = \frac{\text{Initial cost}}{\text{Uniform annual benefit}} = \frac{\$650,000}{\$162,500}$$

$$= 4 \text{ years}$$

If the company's policy is to consider only projects with a payback period of 5 years or less, this computer process control system project passes the initial screening.

In Example 7.2, dividing the initial payment by annual receipts to determine payback period is a simplification we can make because the annual receipts are uniform. Wherever the expected cash flows vary from year to year, however, the payback period must be determined by adding the expected cashflows for each year until the sum is equal to, or greater than, zero. The significance of this procedure can be easily explained. The cumulative cash flow equals zero at the point where cash inflows exactly match or pay back the cash outflows; thus, the project has reached the payback point. Similarly, if the cumulative cash flows exceed zero, the cash inflows exceed the cash outflows, and the project has begun to generate a profit, thus exceeding its payback point. To illustrate, consider Example 7.3.

Example 7.3 Conventional Payback Period with Salvage Value

Autonumerics Company has just bought a new spindle machine at a cost of $105,000 to replace one that had a salvage value of $20,000. The projected annual after-tax savings via improved efficiency, which will exceed the investment cost, are as follows:

Period	Cash Flow	Cumulative Cash Flow
0	−$105,000 + $20,000	−$85,000
1	15,000	−70,000
2	25,000	−45,000
3	35,000	−10,000
4	45,000	35,000
5	45,000	80,000
6	35,000	115,000

Solution

Given: Cash flow series as shown in Figure 7.3(a)

Find: Conventional payback period

The salvage value of retired equipment becomes a major consideration in most justification analysis. (In this example, the salvage value of the old machine should be taken into account, as the company already decided to replace the old machine.) When used, the salvage value of the retired equipment is subtracted from the purchase price of new equipment, revealing a closer true cost of the investment. As we see from the cumulative cash flow in Figure 7.3(b), the total investment is recovered during year 3. If the firm's stated maximum payback period is 3 years, the project would not pass the initial screening stage.

Comments: In Example 7.2, we assumed that cash flows only occur in discrete lumps at the ends of years. If cash flows occur continuously throughout the year, the payback period calculation needs adjustment. A negative balance of $10,000 remains at the start of year 4. If the $45,000 is expected to be received as a more or less continuous flow during year 4, the total investment will be recovered two-tenths ($10,000/$45,000) of the way through the 4th year. In this situation, the payback period is thus 3.2 years.

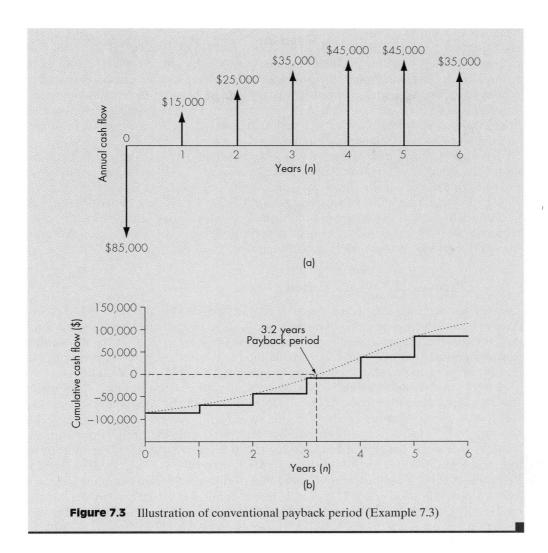

Figure 7.3   Illustration of conventional payback period (Example 7.3)

7.2.2   Benefits and Flaws of Payback Screening

The simplicity of the payback method is one of its most appealing qualities. Initial project screening by the payback method reduces the information search by focusing on that time at which the firm expects to recover the initial investment. The method may also eliminate some alternatives, thus reducing a firm's need to make further analysis efforts.

But the much-used payback method of equipment screening has a number of serious drawbacks. The principal objection to the payback method is that it fails to measure profitability, i.e., no "profit" is made during the payback period. Simply measuring how long it will take to recover the initial investment outlay contributes little to gauging the earning power of a project. (In other words, you already know that the money you borrowed for the drill press is costing you 12% per year. The payback method can't tell you how much your invested money is contributing toward the interest expense.) Because payback period analysis ignores differences in the timing of cash flows, it fails to recognize the difference between the present and future value of money. For example, although the payback for both investments can be the same in terms of numbers of years, the front-loaded investment is better because money available today is worth more than that to be gained later. Because payback screening also ignores all proceeds after the payback period, it does not allow for the possible advantages of a project with a longer economic life.

By way of illustration, consider two investment projects in Table 7.1. Each requires an initial investment outlay of $90,000. Project 1, with expected annual cash proceeds of $30,000 for the first 3 years, has a payback period of 3 years. Project 2 is expected to generate annual cash proceeds of $25,000 for 6 years; hence, its payback period is 3.6 years. If the company's maximum payback period is set to 3 years, then project 1 would pass the initial project screening, whereas project 2 would fail even though it is clearly the more profitable investment.

7.2.3 Discounted Payback Period

To remedy one of the shortcomings of the payback period, we may modify the procedure to consider the time value of money, i.e., the cost of funds (interest) used to support the project. This modified payback period is often referred to as the **discounted payback period.** In other words, we may define the discounted payback period as the number of years required to recover the investment from *discounted* cash flows.

For the project in Example 7.3, suppose the company requires a rate of return of 15%. To determine the period necessary to recover both the capital investment and the cost of funds required to support the investment, we may construct Table 7.2 showing cash flows and costs of funds to be recovered over the project life. To illustrate, let's consider the cost of funds during the first year: With $85,000 committed at the beginning of the year, the interest in year 1 would be $12,750 ($85,000 × 0.15). Therefore, the total commitment grows to $97,750 but the $15,000 cash flow in year 1 leaves a net commitment of $82,750. The cost of funds during the second year would be $12,413 ($82,750 × 0.15). But with the $25,000 receipt from the project, the net commitment reduces to $70,163. When this process repeats for the remaining project years, we find that the net commitment to project ends during year 5. Depending on

Table 7.1
Investment
Cash Flows
for Two
Competing
Projects

n	Project 1	Project 2
0	−$90,000	−$90,000
1	30,000	25,000
2	30,000	25,000
3	30,000	25,000
4	1,000	25,000
5	1,000	25,000
6	1,000	25,000
	$3,000	$60,000

Period	Cash Flow	Cost of Funds (15%)*	Cumulative Cash Flow
0	–$85,000	0	–$85,000
1	15,000	–$85,000(0.15) = –$12,750	82,750
2	25,000	–$82,750(0.15) = –12,413	–70,163
3	35,000	–$70,163(0.15) = –10,524	–45,687
4	45,000	–$45,687(0.15) = –6,853	–7,540
5	45,000	–$7,540(0.15) = –1,131	36,329
6	35,000	$36,329(0.15) = 5,449	76,778

*Cost of funds = Unrecovered beginning balance × interest rate

Table 7.2
Payback Period Calculation Considering the Cost of Funds (Example 7.3)

the cash flow assumption, the project must remain in use about 4.2 years (continuous cash flows) or 5 years (year end cash flows) in order for the company to cover its cost of capital and also recover the funds invested in the project. Figure 7.4 illustrates this relationship.

Inclusion of time value of money effects has increased the payback period calculated for this example by a year. Certainly, this modified measure is an improved one, but it does not show the complete picture of the project profitability, either.

7.2.4 Where Do We Go from Here?

Should we abandon the payback method? Certainly not. But, if you use payback screening exclusively for analysis of capital investments, look again. You may be missing something that another method can help you spot. Therefore, it is illogical to claim that payback is either a good or bad method of justification. Clearly, it is not a measure of profitability. But when it is used to supplement other methods of analysis, it can provide useful information. For example, payback can be useful when a company needs a measure of speed of cash recovery, when the company has a cash flow problem, when a product is built to last only for a short period of time, and when the machine itself is known to have a short market life.

7.3 Present Worth Analysis

Until the 1950s, the payback method was widely used as a means of making investment decisions. As flaws in this method were recognized, however, business people began to search for methods to improve project evaluations. The result was the development of **discounted cash flow techniques (DCF)**, which take into account the time value of money. One of the DCFs is the net present worth (or value) method (NPW or NPV). A capital investment problem is essentially one of determining whether the anticipated cash inflows from a proposed project are sufficiently attractive to invest funds in the project. In developing the NPW criterion, we will use the concept of cash flow equivalence discussed in Chapter 4. As we observed, the most convenient point

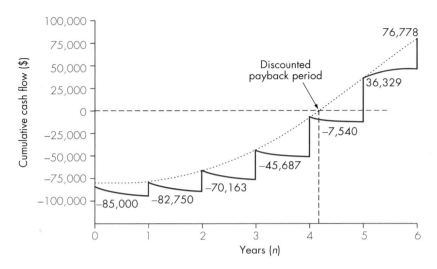

Figure 7.4   Illustration of discounted payback period

at which to calculate the equivalent values is often at time 0. Under the NPW criterion, the present worth of all cash inflows is compared against the present worth of all cash outflows associated with an investment project. The difference between the present worth of these cash flows, referred to as the **net present worth** (NPW), determines whether or not the project is an acceptable investment. When two or more projects are under consideration, NPW analysis further allows us to select the best project by comparing their NPW figures.

7.3.1 Net Present Worth Criterion

We will first summarize the basic procedure for applying the present worth criterion to a typical investment project.

- Determine the interest rate that the firm wishes to earn on its investments. The interest rate you determine represents the rate at which the firm can always invest the money in its **investment pool**. This interest rate is often referred to as either a **required rate of return** or a **minimum attractive rate of return** (MARR). Usually this selection is a policy decision made by top management. It is possible for the MARR to change over the life of a project, as we saw in Section 5.5, but for now we will use a single rate of interest when calculating NPW.

- Estimate the service life of the project.

- Estimate the cash inflow for each period over the service life.

- Estimate the cash outflow over each service period.

- Determine the net cash flows (net cash flow = cash inflow − cash outflow).

- Find the present worth of each net cash flow at the MARR. Add up these present worth figures; their sum is defined as the project's NPW:

$$PW(i) = \frac{A_0}{(1 + i)^0} + \frac{A_1}{(1 + i)^1} + \frac{A_2}{(1 + i)^2} + \cdots + \frac{A_N}{(1 + i)^N}$$

$$= \sum_{n=0}^{N} \frac{A_n}{(1 + i)^n} \qquad (7.1)$$

$$= \sum_{n=0}^{N} A_n(P/F, i, n),$$

where $PW(i)$ = NPW calculated at i,

A_n = Net cash flow at end of period n,

i = MARR (or cost of capital),

n = Service life of the project.

A_n will be positive if the corresponding period has a net cash inflow, or negative if there is a net cash outflow.

- In this context, a positive NPW means the equivalent worth of the inflows is greater than the equivalent worth of outflows, so, the project makes a profit. Therefore, if the $PW(i)$ is positive for a single project, the project should be accepted; if negative, it should be rejected. The decision rule is

    If $PW(i) > 0$, accept the investment.

    If $PW(i) = 0$, remain indifferent.

    If $PW(i) < 0$, reject the investment.

Note that the decision rule is for a single project evaluation[1] where you can estimate the revenues as well as the costs associated with a project. As you will find in Section 7.5, when you compare mutually exclusive alternatives with the same revenues, they are compared based on a cost only basis. In this situation (because you are minimizing costs, rather than maximizing profits), you should accept the project that results in smallest, or least negative, NPW.

Example 7.4 Net Present Worth—Uniform Flows

Consider the investment cash flows associated with the computer process control project in Example 7.1. If the firm's MARR is 15%, compute the NPW of this project. Is this project acceptable?

Solution

Given: Cash flows in Fig. 7.2, MARR = 15% per year

Find: NPW

---

[1] Some projects cannot be avoided, e.g., the installation of pollution control equipment. In a case such as this, the project would be accepted even though its NPW $\leq$ 0. This type of project will be discussed in Chapter 12.

Since the computer process control project requires an initial investment of $650,000 at $n = 0$ followed by the eight equal annual savings of $162,000, we can easily determine the NPW as follows:

$$PW(15\%)_{\text{Outflow}} = \$650,000$$

$$PW(15\%)_{\text{Inflow}} = \$162,500(P/A, 15\%, 8)$$

$$= \$729,190$$

Then, the NPW of the project is

$$PW(15\%) = PW(15\%)_{\text{Inflow}} - PW(15\%)_{\text{Outflow}}$$

$$= \$729,190 - \$650,000$$

$$= \$79,190$$

or, using Eq. (7.1)

$$PW(15\%) = -\$650,000 + \$162,500(P/A, 15\%, 8)$$

$$= \$79,190$$

Since $PW(15\%) > 0$, the project would be acceptable.

Now let's consider an example where the investment cash flows are not uniform over the service life of the project.

Example 7.5 Net Present Worth—Uneven Flows

Tiger Machine Tool Company is considering the acquisition of a new metal cutting machine. The required initial investment of $75,000 and the projected cash benefits[2] over 3-year's project life are as follows.

End of Year	Net Cash Flow
0	-$75,000
1	24,400
2	27,340
3	55,760

---

[2] As we stated at the beginning of this chapter, we treat net cash flows as before tax values or as having their tax effects precalculated. Explaining the process of obtaining cash flows requires an understanding of income taxes and the role of depreciation, which are discussed in Chapters 11 and 12.

You have been asked by the president of the company to evaluate the economic merit of the acquisition. The firm's MARR is known to be 15%.

Solution

Given: Cash flows as tabulated, MARR = 15% per year

Find: NPW

If we bring each flow to its equivalent at time zero we find

$$PW(15\%) = -\$75,000 + \$24,000(P/F, 15\%, 1) + \$27,340(P/F, 15\%, 2)$$
$$+ \$55,760(P/F, 15\%, 3)$$
$$= \$3,553.$$

Since the project results in a surplus of \$3,553, the project is acceptable.

In Example 7.5, we computed the NPW of the project at a fixed interest rate of 15%. If we compute the NPW at varying interest rates, we obtain the data in Table 7.3. Plotting the NPW as a function of interest rate gives Figure 7.5, the present worth profile. (As you will see in Section 7.5, you may use a spreadsheet program such as Excel to generate Table 7.2 or Fig. 7.5.)

Figure 7.5 indicates that the investment project has a positive NPW, if the interest rate is below 17.45% and a negative NPW, if the interest rate is above 17.45%. As we will see in Chapter 9, this **break-even interest rate** is known as the **internal rate of return**. If the firm's MARR is 15%, the project has a NPW of \$3,553 and so may be accepted. The figure of \$3,553 measures the equivalent immediate gain in present worth to the firm following the acceptance of the project. On the other hand, at $i = 20\%$, $PW(20\%) = -\$3,412$, the firm should reject the project. (Note that either accepting or

i (%)	PW(i)	i(%)	PW(i)
0	\$32,500	20	−\$3,412
2	27,743	22	−5,924
4	23,309	24	−8,296
6	19,169	26	−10,539
8	15,296	28	−12,662
10	11,670	30	−14,673
12	8,270	32	−16,580
14	5,077	34	−18,360
16	2,076	36	−20,110
17.45*	0	38	−21,745
18	−750	40	−23,302

Table 7.3
Present Worth Amounts at Varying Interest Rates (Example 7.5)

*Break-even interest rate (also known as the rate of return)

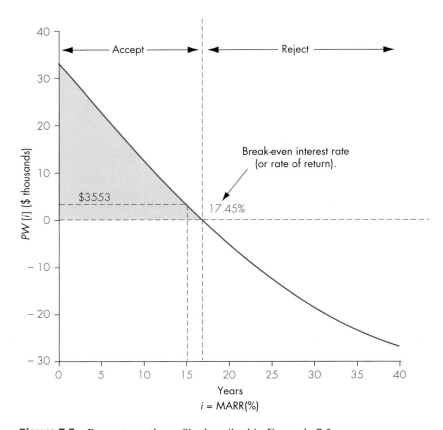

Figure 7.5 Present worth profile described in Example 7.5

rejecting an investment is influenced by the choice of a MARR. So it is crucial to estimate the MARR correctly. We will defer this important issue until Chapter 16. For now we will assume that the firm has an accurate MARR estimate available for use in investment analysis.)

7.3.2 Meaning of Net Present Worth

In present worth analysis, we assume that all the funds in a firm's treasury can be placed in investments that yield a return equal to the MARR. We may view these funds as an **investment pool**. Alternatively, if no funds are available for investment, we assume the firm can borrow them at the MARR from the capital market. In this section, we will examine these two views when explaining the meaning of MARR in NPW calculations.

Investment Pool Concept

An investment pool is equivalent to a firm's treasury where all fund transactions are administered and managed by the firm's comptroller. The firm may withdraw funds from this investment pool for other investment purposes, but if left in the pool, these funds will earn at the MARR. Thus, in investment analysis, net cash flows will be net cash flows relative to this investment pool. To illustrate the investment pool concept, we consider again the project in Example 7.5 that required an investment of $75,000.

If the firm did not invest in the project and left $75,000 in the investment pool for 3 years, these funds would grow as follows:

$$\$75,000(F/P, 15\%, 3) = \$114,066.$$

Suppose the company decided instead to invest $75,000 in the project described in Example 7.5. Then the firm would receive a stream of cash inflows during its project life of 3 years in the amounts as follows:

Period (n)	Net Cash Flow (A_n)
1	$24,400
2	27,340
3	55,760

Since the funds that return to the investment pool earn interest at a rate of 15%, it would be of interest to see how much the firm would benefit from this investment. For this alternative, the returns after reinvestment are

$$\$24,400(F/P, 15\%, 2) \quad = \quad \$32,269$$
$$\$27,340(F/P, 15\%, 1) \quad = \quad \$31,441$$
$$\$55,760(F/P, 15\%, 0) \quad = \quad \$55,760$$
$$\text{Total} \qquad \$119,470.$$

These returns total $119,470. The additional cash accumulation at the end of 3 years from investing in the project is

$$\$119,470 - \$114,066 = \$5,404.$$

If we compute the equivalent present worth of this net cash surplus at time 0, we obtain

$$\$5,404(P/F, 15\%, 3) = \$3,553,$$

which is exactly the same as the NPW of the project computed by Eq. (7.1). Clearly, on the basis of its positive NPW, the alternative of purchasing a new machine should be preferred to that of simply leaving the funds in the investment pool at the MARR. Thus, in PW analysis, any investment is assumed to be returned at the MARR. If a surplus exists at the end of the project, then $PW(MARR) > 0$. Figure 7.6 illustrates the reinvestment concept related to the firm's investment pool.

Borrowed Funds Concept

Suppose that the firm does not have $75,000 at the outset. In fact, the firm doesn't have to have an investment pool at all. Let's further assume that the firm borrows all its capital from a bank at an interest rate of 15%, invests in the project, and uses the

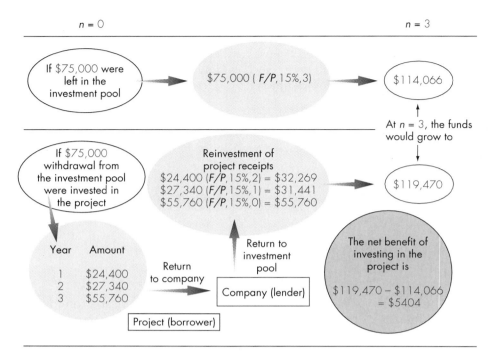

Figure 7.6 The concept of an investment pool with the company as a lender and the project as a borrower

proceeds from the investment to pay off the principal and interest on the bank loan. How much is left over for the firm at the end of the project period?

At the end of first year, the interest on the project's use of the bank loan would be $75,000(0.15) = $11,250$. Therefore, the total loan balance grows to $75,000(1 + 0.15) = $86,250$. Then, the firm receives $24,400 from the project and applies the entire amount to repay the loan portion. This repayment leaves a balance due of

$$\$75,000(1 + 0.15) - \$24,400 = \$61,850.$$

This amount becomes the net amount the project is borrowing at the beginning of year 2, which is also known as **project balance**. At the end of period 2, the bank debt grows to $61,850(1.15) = $71,128$, but with the receipt of $27,340, the project balance reduces to

$$\$61,850(1.1) - \$27,340 = \$43,788.$$

Similarly, at the end of year 3, the project balance becomes

$$\$43,788(1.15) = \$50,356.$$

But, with the receipt of $55,760 from the project, the firm should be able to pay off the remaining balance and come out with a surplus in the amount of $5,404. This terminal project balance is also known as the **net future worth** of the project. In other words, the firm fully repays its initial bank loan and interest at the end of period 3, with a re-

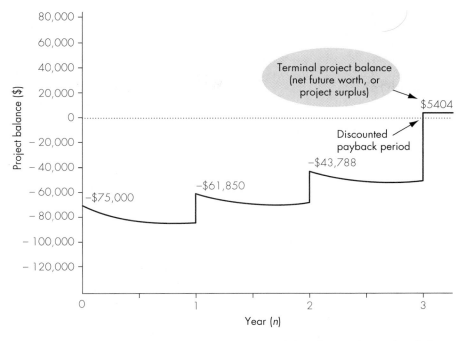

Figure 7.7   Project balance diagram as a function of time (a negative project balance indicates the amount of remaining loan to be paid off, or the amount of investment to be recovered)

sulting profit of $5,404. If we compute the equivalent present worth of this net profit at time 0, we obtain

$$PW(15\%) = \$5,404(P/F, 15\%, 3) = \$3,553.$$

The result is identical to the case where we directly computed the NPW of the project at $i = 15\%$, shown in Example 7.5. Figure 7.7 illustrates the project balance as a function of time.[3]

7.4 Variations of Present Worth Analysis

As variations of present worth analysis, we will consider two additional measures of investment worth. These measures are **future worth analysis** and **capitalized equivalent worth analysis**. (The equivalent annual worth measure is another variation of the present worth measure, but we present it in Chapter 8.) Future worth analysis calculates the future worth of an investment undertaken. Capitalized equivalent worth analysis calculates the present worth of a project with a perpetual life span.

---

[3] Note that the sign of the project balance changes from negative to positive during year 3. The time at which the project balance becomes zero is known as the **discounted payback period**.

7.4.1 Future Worth Analysis

Net present worth measures the surplus in an investment project at time 0. **Net future worth (NFW)** measures this surplus at a time period other than 0. Net future worth analysis is particularly useful in an investment situation where we need to compute the equivalent worth of a project at the end of its investment period, rather than at its beginning. For example, it may take 7 to 10 years to build a nuclear power plant because of the complexities of engineering design and the many time-consuming regulatory procedures that must be satisfied to ensure public safety. In this situation, it is more common to measure the worth of the investment at the time of the project's commercialization, i.e., to conduct an NFW analysis at the end of the investment period.

Net Future Worth Criterion and Calculations

When A_n represents the cash flow at time n for $n = 0, 1, 2, ..., N$ for a typical investment project that extends over N periods, the net future worth (NFW) expression at the end of period N is

$$
\begin{aligned}
FW(i) &= A_0(1 + i)^N + A_1(1 + i)^{N-1} + A_2(1 + i)^{N-2} + ... + A_N \\
&= \sum_{n=0}^{N} A_n(1 + i)^{N-n} \\
&= \sum_{n=0}^{N} A_n(F/P, i, N - n).
\end{aligned}
\tag{7.2}
$$

As you might expect, the decision rule for the NFW criterion is the same as for the NPW criterion: For a single project evaluation,

If $FW(i) > 0$, accept the investment.

If $FW(i) = 0$, remain indifferent to the investment.

If $FW(i) < 0$, reject the investment.

Example 7.6 Net Future Worth—at the End of the Project

Consider the project cash flows in Example 7.5. Compute the NFW at the end of year 3 at $i = 15\%$.

Solution

Given: Cash flows in Example 4.5, MARR = 15% per year

Find: NFW

As seen in Figure 7.8, the NFW of this project at an interest rate of 15% would be

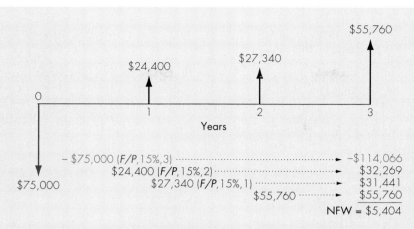

Figure 7.8 Future worth calculation at the end of year 3 (Example 7.6)

$$FW(15\%) = -\$75,000\ (F/P, 15\%, 3) + \$24,400(F/P, 15\%, 2)$$
$$+ \$27,340(F/P, 15\%, 1) + \$55,760$$
$$= \$5,404.$$

Note that the net future worth of this project is equivalent to the terminal project balance as calculated in Section 7.3.2. Since $FW(15\%) > 0$, the project is acceptable. We reach the same conclusion under present worth analysis.

Example 7.7 Future Equivalent—at an Intermediate Time

Higgins Corporation (HC), a Detroit-based robot-manufacturing company, has developed a new advanced-technology robot called Helpmate, which incorporates advanced technology such as vision systems, tactile sensing, and voice recognition. These features allow the robot to roam the corridors of a hospital or office building without following a predetermined track or bumping into objects. HC's marketing department plans to target sales of the robot toward major hospitals to ease nurses' workloads by performing low-level duties such as delivering medicines and meals to patients.

The firm would need a new plant to manufacture the Helpmates; this plant could be built and made ready for production in 2 years. It would require a 30-acre site, which can be purchased for $1.5 million in year 0. Building construction would begin early in year 1 and continue throughout year 2. The building would cost an estimated $10 million, would involve a $4 million payment due to the contractor at the end of year 1, and another $6 million would be payable at the end of year 2. The necessary manufacturing equipment would be installed late in year 2 and would be paid for at the end of year 2. The equipment would cost $13 million, including transportation and installation. When the project terminates, the land is expected to have an after-tax market value of $2 million, the building an after-tax value of $3 million, and the equipment an after-tax value of $3 million.

For capital budgeting purposes, assume that the cash flows occur at the end of each year. Because the plant would begin operations at the beginning of year 3, the first operating cash flows would occur at the end of year 3. The Helpmate plant's estimated economic life is 6 years after completion, with the following expected after-tax operating cash flows in millions:

Calendar Year End of Year	00 0	01 1	02 2	03 3	04 4	05 5	06 6	07 7	08 8
After-tax cash flows									
A. Operating revenue				$6	$8	$13	$18	$14	$8
B. Investment									
Land	−1.5								+2
Building		−4	−6						+3
Equipment			−13						+3
Net cash flow	−$1.5	−$4	−$19	$6	$8	$13	$18	$14	$16

Compute the equivalent worth of this investment at the start of operation. Assume that HC's MARR is 15%.

Solution

Given: Cash flows above, MARR = 15% per year

Find: Equivalent project's worth at the end of calendar year 2

One easily understood method involves calculating the present worth, then transforming this to the equivalent worth at the end of year 2. First, we can compute PW(15%) at time 0 of this project.

$$PW(15\%) = -\$1.5 - \$4(P/F, 15\%, 1) - \$19(P/F, 15\%, 2)$$
$$+ \$6(P/F, 15\%, 3) + \$8(P/F, 15\%, 4) + \$13(P/F, 15\%, 5)$$
$$+ \$18(P/F, 15\%, 6) + \$14(P/F, 15\%, 7) + \$16(P/F, 15\%, 8)$$
$$= \$13.91 \text{ million.}$$

Then, the equivalent project worth at the start of operation is

$$FW(15\%) = PW(15\%)(F/P, 15\%, 2)$$
$$= \$18.40 \text{ million.}$$

A second method brings all flows prior to year 2 up to that point and discounts future flows back to year 2. The equivalent worth of the earlier investment, when the plant begins full operation, is

$$-\$1.5(F/P, 15\%, 2) - \$4(F/P, 15\%, 1) - \$19 = -\$25.58 \text{ million},$$

which produces an equivalent flow as shown in Figure 7.9. If we discount the future flows to the start of operation, we obtain

$$FW(15\%) = -\$25.58 + \$6(P/F, 15\%, 1) + \$8(P/F, 15\%, 2) + \ldots$$
$$+ \$16(F/P, 15\%, 6)$$
$$= \$18.40 \text{ million}.$$

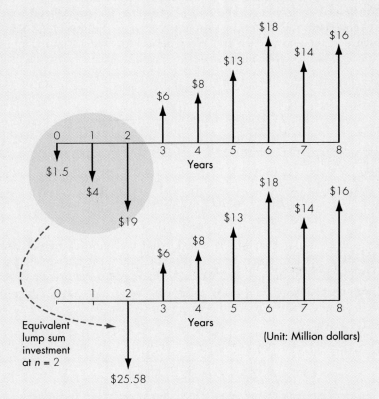

Figure 7.9 Cash flow diagram for the Helpmate project (Example 7.7)

Comments: If another company is willing to purchase the plant and the right to manufacture the robots immediately after completion of the plant (year 2), HC would set the price of the plant at $43.98 million ($18.40 + $25.58) at a minimum.

7.4.2 Capitalized Equivalent Method

Another special case of the PW criterion is useful when the life of a proposed project is **perpetual** or the planning horizon is extremely long (say, 40 years or more). Many public projects such as bridges, waterway constructions, irrigation systems, and hydroelectric dams are expected to generate benefits over an extended period of time (or forever). In this section, we will examine the **capitalized equivalent** $(CE(i))$ method for evaluating such projects.

Perpetual Service Life

Consider the cash flow series shown in Figure 7.10. How do we determine the PW for an infinite (or almost infinite) uniform series of cash flows or a repeated cycle of cash flows? The process of computing the PW cost for this infinite series is referred to as the **capitalization** of project cost. The cost is known as the **capitalized cost**. The capitalized cost represents the amount of money that must be invested today to yield a certain return A at the end of each and every period forever, assuming an interest rate of i. Observe the limit of the uniform series present worth factor as N approaches infinity:

$$\lim_{N \to \infty}(P/A, i, N) = \lim_{N \to \infty}\left[\frac{(1 + i)^N - 1}{i(1 + i)^N}\right] = \frac{1}{i}.$$

Thus, it follows that

$$PW(i) = A(P/A, i, N \to \infty) = \frac{A}{i}. \tag{7.3}$$

Another way of looking at this, is to ask what constant income stream could be generated by $PW(i)$ dollars today in perpetuity. Clearly, the answer is $A = iPW(i)$. If withdrawals were greater than A, you would be eating into the principal, which would eventually reduce it to 0.

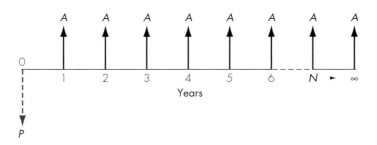

Figure 7.10 Equivalent present worth of an infinite cash flow series

Example 7.8 Capitalized Equivalent Cost

An engineering school has just completed a new engineering complex worth $50 million dollars. A campaign, targeting alumni, is planned to raise funds for future maintenance costs, which are estimated at $2 million per year. Any unforeseen costs above $2 million per year would be obtained by raising tuition. Assuming that the school can create a trust fund that earns 8% interest annually, how much has to be raised now to cover the perpetual string of $2 million annual costs?

Solution

Given: $A = \$2$ million, $i = 8\%$ per year, $N = \infty$

Find: $CE(8\%)$

The capitalized cost equation is

$$CE(i) = \frac{A}{i}$$

$$CE(8\%) = \$2,000,000/0.08$$

$$= \$25,000,000.$$

Comments: It is easy to see that this lump sum amount should be sufficient to pay maintenance expenses for the school forever. Suppose the school deposited $25 million in a bank that paid 8% interest annually. At the end of the first year, the $25 million would earn 8%($25 million) = $2 million interest. If this interest were withdrawn, the $25 million would remain in the account. At the end of the second year, the $25 million balance would again earn 8%($25 million) = $2 million. The annual withdrawal could be continued forever, and the endowment (gift funds) would always remain at $25 million.

Project's Service Life is Extremely Long

The benefits of typical civil engineering projects, such as bridge and highway construction, although not perpetual, can last for many years. In this section, we will examine the use of the $CE(i)$ criterion to approximate the NPW of engineering projects with long lives.

Example 7.9 Comparison of Present Worth for Long Life and Infinite Life

Mr. Gaynor L. Bracewell amassed a small fortune developing real estate in Florida and Georgia over the past 30 years. He sold more than 700 acres of timber and farmland to raise $800,000, with which he built a small hydroelectric plant, known as High Shoals Hydro. The plant was a decade in the making. The design for Mr. Bracewell's plant, which he developed using his Army training as a civil engineer, is relatively simple. A 22-foot-deep canal, blasted out of solid rock just above the higher of two dams on his property, carries water 1,000 feet along the river to a "trash rack," where leaves and other debris are caught. A 6-foot-wide pipeline capable of holding 3 million pounds of liquid then funnels the water into the powerhouse at 7.5 feet per second, thereby creating 33,000 pounds of thrust against the turbines. Under a 1978 federal law designed to encourage alternative power sources, Georgia Power Company is required to purchase any electricity Mr. Bracewell can supply. Mr. Bracewell estimates that his plant can generate 6 million kilowatt-hours per year.[4]

Suppose that, after paying income taxes and operating expenses, Mr. Bracewell's annual income from the hydroelectric plant will be $120,000. With normal maintenance, the plant is expected to provide service for at least 50 years. Figure 7.11 illustrates when, and in what quantities, Mr. Bracewell spent his $800,000 (not considering the time value of money) during the last 10 years. Was Mr. Bracewell's $800,000 investment a wise one? How long will he have to wait to recover his initial investment, and will he ever make a profit? Examine the situation by computing the project worth at varying interest rates.

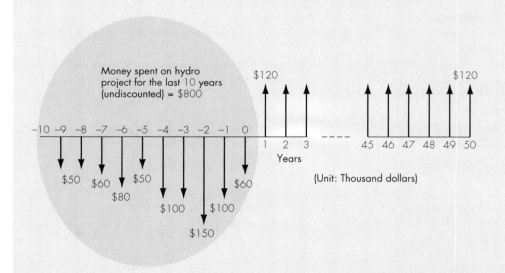

Figure 7.11   Net cash flow diagram for Mr. Bracewell's hydroelectric project (Example 7.9)

---

[4] *The Atlanta Constitution.* May 29, 1989. (Reprinted with permission from *The Atlanta Journal* and *The Atlanta Constitution.*)

(a) If Mr. Bracewell's interest rate is 8%, compute the NPW(at time 0 in Fig. 7.11) of this project with a 50-year service life and infinite service, respectively.

(b) Repeat part (a) assuming an interest rate of 12%.

Solution

Given: Cash flow in Fig. 7.11 (to 50 years or ∞), $i = 8\%$ or 12%

Find: NPW at time 0

One of the main questions was whether Mr. Bracewell's plant would be profitable.

Now we will compute the equivalent total investment and the equivalent worth of receiving future power revenues at the start of generation, i.e., at time 0.

(a) At $i = 8\%$

- With a plant service life of 50 years, we can make use of single payment compound amount factors in the invested cash flow to help us find the equivalent total investment at the start of power generation. Using "K" to indicate thousands,

$$V_1 = -\$50K(F/P, 8\%, 9) - \$50K(F/P, 8\%, 8)$$
$$- \$60K(F/P, 8\%, 7) \ldots -\$100K(F/P, 8\%, 1) - \$60K$$
$$= -\$1101K.$$

The equivalent total benefit at the start of generation is

$$V_2 = \$120K(P/A, 8\%, 50) = \$1468K.$$

Summing, we find the net equivalent worth at the start of power generation:

$$V_1 + V_2 = -\$1,101K + \$1,468K.$$
$$= \$367K.$$

- With an infinite service life, the net equivalent worth is called the capitalized equivalent worth. The investment portion prior to time 0 is identical, so the capitalized equivalent worth is

$$CE(8\%) = -\$1,101K + \$120K/(0.08)$$
$$= \$399K.$$

Note that the difference between the infinite situation and the planning horizon of 50 years is only $32,000.

(b) At $i = 12\%$

- With a service life of 50 years, proceeding as for (a), the equivalent total investment at the start of generation is

$$V_1 = -\$50K(F/P, 12\%, 9) - \$50K(F/P, 12\%, 8)$$

$$-\$60K(F/P, 12\%, 7) \ldots -\$100K(F/P, 12\%, 1) - 60K$$

$$= -\$,299K.$$

Equivalent total benefits at the start of generation:

$$V_2 = \$120K(P/A, 12\%, 50) = \$997K.$$

Net equivalent worth at the start of power generation:

$$V_1 + V_2 = -\$1,299K + \$997K$$

$$= -\$302K.$$

- With infinite cash flows, the capitalized equivalent worth at the current time is

$$CE(12\%) = -\$1,299K + \$120K/(0.12)$$

$$= -\$299K.$$

Note that the difference between the infinite situation and a planning horizon of 50 years is merely $3,000, which demonstrates that we may closely approximate the present worth of long cash flows (i.e., 50 years or more) by using the capitalized equivalent value. The accuracy of the approximation improves as the interest rate increases (or number of years is greater).

Comments: At $i = 12\%$, Mr. Bracewell's investment is not a profitable one, but at 8% it is. The outcome indicates the importance of using the appropriate i in investment analysis. The issue of selecting an appropriate i will be presented once again in Chapter 16.

7.5 Comparing Mutually Exclusive Alternatives

Until now, we have considered situations involving only a single project, or projects that were independent of each other. In both these cases, we made the decision to reject or accept each project individually based on whether it met the MARR requirements, evaluated using the PW or FW criterion.

In the real world of engineering practice, however, it is more typical for us to have two or more choices of projects for accomplishing a business objective. (As we shall see, even when it appears that we have only one project to consider, the implicit "do-nothing" alternative must be factored into the decision-making process.) In this section, we extend our evaluation techniques to multiple projects, which are mutually exclusive. Other dependencies between projects will be considered in Chapter 16.

Often various projects or investments under consideration do not have the same duration, or do not match the desired study period. Adjustments must be made when comparing multiple options to properly account for such differences. In this section, we explain the concepts of an analysis period and the process of accommodating for different lifetimes as important considerations for selecting among several alternatives. In the first few sections of the chapter, all available options in a decision problem are assumed to have equal lifetimes. In Section 7.5.5, this restriction is relaxed.

> **Current Practices—Identifying Mutually Exclusive Investment Alternatives:** The proliferation of computers into all aspects of business has created an ever-increasing need for data capture systems that are fast, reliable, and cost-effective. One technology that has been adopted by many manufactures, distributors, and retailers is a bar-coding system. Hermes Electronics, a leading manufacturer of underwater surveillance equipment, evaluated the economic benefits of installing an automated data acquisition system into their plant. They could use the system on a limited scale, such as for tracking parts and assemblies for inventory management, or opt for a broader implementation by recording information useful for quality control, operator efficiency, attendance, and other functions. All these aspects are currently monitored, but although computers are used to manage the information, the recording is primarily conducted manually. The advantages of an automated data collection system, which include faster and more accurate data capture, quicker analysis of and response to production changes, and savings due to tighter control over operations, could easily outweigh the cost of the new system.
>
> Two alternative systems from competing suppliers are under consideration. System 1 relies on handheld bar-code scanners that transmit frequencies (RF). The hub of this wireless network can then be connected to their existing LAN, and integrated with their current MRP II system and other management software. System 2 consists primarily of specialized data terminals installed at every collection point, with connected bar-code scanners where required. This system configured in such a way as to facilitate phasing in the components over two stages or installing the system all at once. The former would allow Hermes to defer some of the capital investment, while becoming thoroughly familiar with the functions introduced in the first stage.
>
> Either of these systems would satisfy Hermes' data collection needs. They each have some unique elements, and the company needs to compare the relative benefits of the features offered by each system. From the point of view of engineering economics, the two systems have different capital costs, and their operating and maintenance costs are not identical. One system may be also rated to last longer than the other system before replacement is required, particularly if the option of acquiring System 2 in phases is selected. There are many issues to be considered before making the best choice.

7.5.1 Meaning of Mutually Exclusive and "Do-Nothing"

As we briefly mentioned in Section 7.1.2, **mutually exclusive** means that any one of several alternatives will fulfill the same need and that the selection of one alternative implies that the others will be excluded. Take, for example, buying versus leasing an automobile for business use—when one alternative is accepted, the other is excluded. We use the terms **alternative** and **project** interchangeably to mean decision option.

"Do-Nothing" Is a Decision Option

When considering an investment, we are in one of two situations. The project is either aimed at replacing an existing asset or system, or it is a new endeavor. In either case, a do-nothing alternative may exist. If a process or system already in place to accomplish our business objectives is still adequate, then we must determine which, if any, new proposals are economical replacements. If none are feasible, then we do nothing. On the other hand, if the existing system has terminally failed, the choice among proposed alternatives is mandatory (i.e., do-nothing is not an option).

New endeavors occur as alternatives to the "green fields" do-nothing situation, which has zero revenues and zero costs (i.e., nothing currently exists). For most new endeavors, do-nothing is generally an alternative, as we won't proceed unless at least one of the proposed alternatives is economically sound. In fact, undertaking even a single project entails making a decision between two alternatives when it is optional, because the do-nothing alternative is implicitly included. Occasionally a new initiative must be undertaken, cost notwithstanding, and in this case the goal is to choose the most economical alternative since do-nothing is not an option.

When the option of retaining an existing asset or system is available, there are two ways to incorporate it into the evaluation of the new proposals. One way is to treat the do-nothing option a distinct alternative, but this approach will be primarily covered in Chapter 15 where methodologies specific to replacement analysis are presented. The second approach, mostly used in this chapter, is to generate the cash flows of the new proposals relative to that of the do-nothing alternative. That is, for each new alternative, the **incremental costs** (and incremental savings or revenues if applicable) relative to do-nothing are used for the economic evaluation. For a replacement type problem, this is calculated by subtracting the do-nothing cash flows from those of each new alternative. For new endeavors, the incremental cash flows are the same as the absolute amounts associated with each alternative, since the do-nothing values are all zero.

Because the main purpose of this chapter is to illustrate how to choose among mutually exclusive alternatives, most of the problems are structured so that one of the options presented must be selected. Therefore, unless otherwise stated, it is assumed that do-nothing is not an option, and costs and revenues can be viewed as incremental to do-nothing.

Service Projects Versus Revenue Projects

When comparing mutually exclusive alternatives, we need to classify investment projects into either service or revenue projects. **Service projects** are those whose revenues do not depend on the choice of project, but *must produce the same amount of output (revenue)*. In this situation, we certainly want to choose an alternative with the least input (or cost). For example, suppose an electric utility is considering building a new power plant to meet the peak load demand during either hot summer or cold winter days. Two alternative service projects could meet this peak-load demand: a combustion turbine plant or a fuel-cell power plant. No matter which type of plant is selected, the firm will generate the same amount of revenue from its customers. The only difference is how much it will cost to generate electricity from each plant. If we were to compare these service projects, we would be interested in knowing which plant could provide the cheaper power (lower production cost). Further, if we were to use the NPW criterion to compare these alternatives to minimize expenditures, we would choose the alternative with the **lower present value** production cost over the service life.

On the other hand, **revenue projects** are those whose revenues depend on the choice of alternative. For revenue projects, we are not limiting the amount of input going into the project or the amount of output that the project would generate. Then, our decision is to select the alternative with the largest net gains (output-input). For example, a TV manufacturer is considering marketing two types of high-resolution monitors. With its present production capacity, the firm can market only one of them. Distinct production processes for the two models could incur very different manufacturing costs, so the revenues from each model would be expected to differ due to divergent market prices and potentially different sales volumes. In this situation, if we were to use the NPW criterion, we would select the model that promises to bring in the higher net present worth.

Total Investment Approach

Applying an evaluation criterion to each mutually exclusive alternative individually, and then comparing the results to make a decision, is referred to as the total investment approach. We compute the PW for each individual alternative, as we did in Section 7.2, and select the one with the highest PW. This approach only guarantees valid results when using PW, FW, and AE criteria. (As you will see in Chapters 9 and 17, this total investment approach does not work for any decision criterion based on either percentage [rate of return] or ratio [benefit-cost ratio]. You need to use the incremental investment approach, which also works with any decision criteria, including PW, FW, or AE.) The incremental investment approach will be discussed in Chapter 9 in detail.

Considering Scale of Investment

Frequently, mutually exclusive investment projects may require different levels of investments. At first, it seems unfair to compare a project requiring the smaller investment with the one requiring a larger initial investment. However, the disparity in scale of investment should not be of concern in comparing mutually exclusive alternatives, as long as you understand the basic assumptions—*funds not invested in the project will continue to earn interest at the rate of MARR.* We will look at both situations for service projects or revenue projects:

- Service Projects: Typically, what you are asking yourself is to decide if the higher initial investment can be justified to additional savings to occur in the future. More efficient machines are usually more expensive to acquire initially, but it will reduce the future operating costs, thereby generating more savings.

- Revenue projects: If two mutually exclusive revenue projects require different levels of investments with varying future revenue streams, then your question is what to do with the difference in investment if you decide to go with an alternative with the smaller investment. To illustrate, consider the two revenue projects in Figure 7.12.

Suppose you have exactly $4,000 to invest. If you choose Project B, you do not have any leftover funds. However, if you go with Project A, you will have $3,000 unused funds. Our assumption is that this unused fund will continue to earn an interest rate of MARR. Therefore, the unused fund will grow at 10%, or $3,933, at the end of the project term, or 3 years from now. Consequently, selecting Project A is equivalent to having a modified project cash flow as shown in Figure 7.12.

Let's calculate the net present worth for each option at 10%:

- Project A:

$$PW(10\%)A = -\$1,000 + \$450(P/F, 10\%, 1) + \$600(P/F, 10\%, 2)$$
$$+ \$500(P/F, 10\%, 3)$$
$$= \$283$$

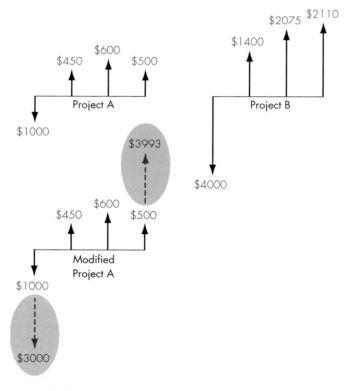

Figure 7.12   Comparing mutually exclusive revenue projects requiring different levels of investment

- Project B:

$$PW(10\%)B = -\$4,000 + \$1,400(P/F, 10\%, 1)$$
$$+ \$2,075(P/F, 10\%, 2) + \$2,110(P/F, 10\%, 3)$$
$$= \$579$$

Clearly Project B is a better choice. How about the modified Project A? If we calculate the present worth for the modified Project A, we will have

- Modified Project A:

$$PW(10\%)A = -\$4,000 + \$450(P/F, 10\%, 1) + \$600(P/F, 10\%, 2)$$
$$+ \$4,493(P/F, 10\%, 3)$$
$$= \$283,$$

which is exactly the same as the net present worth without including the investment consequence of the unused fund. This is not a surprising result, as the return on investment in the unused fund will be exactly 10%. If we also discount them at 10%, there will be no surplus. What is the conclusion? If there is any disparity in investment scale for mutually exclusive revenue projects, go ahead and calculate the net present worth for each option without worrying about the investment differentials.

7.5.2 Analysis Period

The **analysis period** is the time span over which the economic effects of an investment will be evaluated. The analysis period may also be called the **study period** or **planning horizon**. The length of the analysis period may be determined in several ways: It may be a predetermined amount of time set by company policy, or it may be either implied or explicit in the need the company is trying to fulfill—for example, a diaper manufacturer sees the need to dramatically increase production over a 10-year period in response to an anticipated "baby boom." In either of these situations, we consider the analysis period to be the **required service period**. When the required service period is not stated at the outset, the analyst must choose an appropriate analysis period over which to study the alternative investment projects. In such a case, one convenient choice of analysis period is the period of the useful life of the investment project.

When the useful life of an investment project does not match the analysis or required service period, we must make adjustments in our analysis. A further complication, in a consideration of two or more mutually exclusive projects, is that the investments themselves may have differing useful lives. We must compare projects with different useful lives over an **equal time span**, which may require further adjustments in our analysis. (Figure 7.13 is a flow chart showing the possible combinations of the analysis period and the useful life of an investment.) In the sections that follow, we will explore in more detail how to handle situations in which project lives differ from the analysis period and from each other. But we begin with the most straightforward situation, when the project lives and the analysis period coincide.

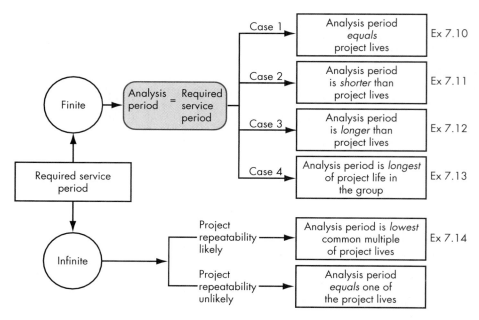

Figure 7.13   Analysis period implied in comparing mutually exclusive alternatives

7.5.3 Analysis Period Equals Project Lives

Let's begin our analysis with the simplest situation where the project lives equal the analysis period. In this case, we compute the NPW for each project and select the one with the highest NPW. Example 7.10 will illustrate this point.

Example 7.10 Present Worth Comparison (Revenue Projects with Equal Lives)—Three Alternatives

Bullard Company (BC) is considering expanding its range of industrial machinery products by manufacturing machine tables, saddles, machine bases, and other similar parts. Several combinations of new equipment and manpower could serve to fulfill this new function:

- Method 1 (M1): new machining center, with three operators
- Method 2 (M2): new machining center with an automatic pallet changer, and with three operators
- Method 3 (M3): new machining center with an automatic pallet changer, and with two task-sharing operators

Each of these arrangements incurs different costs and revenues. The time for loading and unloading parts is reduced in the pallet-changer cases. It costs more to acquire, install, and tool-fit a pallet-changer, but because it is more efficient and versatile, it can generate larger annual revenues. Although saving on labor costs, task-sharing operators take longer to train and are more inefficient initially. As operators become more experienced at their tasks and used to collaborating, it is expected that the annual benefits will increase by 13% per year over the 5-year study period. BC has estimated the investment costs and additional revenues as follows:

	Machining Center Methods		
	M1	M2	M3
Investment:			
Machine tool purchase	$121,000	$121,000	$121,000
Automatic pallet changer		$66,600	$66,600
Installation	$30,000	$42,000	$42,000
Tooling expense	$58,000	$65,000	$65,000
Total Investment	$209,000	$294,600	$294,600
Annual benefits: Year 1			
Additional revenues	$65,000	$69,300	$36,000
Direct labor savings			$17,300
Setup savings		$4,700	$4,700
Year 1: Net revenues	$65,000	$74,000	$58,000
Year 2–5: Net revenues	constant	constant	$g = +13\%/\text{year}$
Salvage value in year 5	$90,000	$200,000	$200,000

All cash flows include any tax effects. Do-nothing is obviously an option, since BC will not undertake this expansion if none of the proposed methods is economically viable. If a method is chosen, BC expects to operate the machining center over the next 5 years. Which option would be selected based on the use of the PW measure at $i = 12\%$?

Solution

Given: Cash flows for three revenue projects, $i = 12\%$ per year

Find: NPW for each project, which to select

For these revenue projects, the net present worth figures at $i = 12\%$ would be as follows:

- For Option M1:

$$PW(12\%)_{M1} = -\$209{,}000 + \$65{,}000(P/A, 12\%, 5) +$$
$$\$90{,}000(P/F, 12\%, 5)$$
$$= \$76{,}379$$

- For Option M2:

$$PW(12\%)_{M2} = -\$294{,}600 + \$74{,}000(P/A, 12\%, 5) +$$
$$\$200{,}000(P/F, 12\%, 5)$$
$$= \$85{,}639$$

- For Option M3:

$$PW(12\%)_{M3} = -\$294{,}600 + \$58{,}000(P/A_1, 13\%, 12\%, 5) +$$
$$\$200{,}000(P/F, 12\%, 5)$$
$$= \$82{,}479$$

Clearly, option M2 is the most profitable. Given the nature of BC parts and shop orders, management has decided that the best way to expand would be with an automatic pallet changer but without task sharing.

7.5.4   Analysis Period Differs from Project Lives

In Example 7.10, we assumed the simplest scenario possible when analyzing mutually exclusive projects: The projects had useful lives equal to each other and to the required service period. In practice, this is seldom the case. Often project lives do not match the required analysis period and/or do not match each other. For example, two machines may perform exactly the same function, but one lasts longer than the other, and both of them last longer than the analysis period for which they are being considered. In the following sections and examples, we will develop some techniques for dealing with these complications.

Project's Life Is Longer than Analysis Period

Project lives rarely conveniently coincide with a firm's predetermined required analysis period; they are often too long or too short. The case of project lives that are too long is the easier one to address.

Consider the case of a firm that undertakes a 5-year production project when all of the alternative equipment choices have useful lives of 7 years. In such a case, we analyze each project for only as long as the required service period (in this case, 5 years). We are then left with some unused portion of the equipment (in this case, 2 years' worth), which we include as salvage value in our analysis. **Salvage value** is the amount of money for which the equipment could be sold after its service to the project has been rendered or the dollar measure of its remaining usefulness.

A common instance of project lives that are longer than the analysis period occurs in the construction industry, where a building project may have a relatively short completion time, but the equipment purchased has a much longer useful life.

Example 7.11 Present Worth Comparison—Project Lives Longer than the Analysis Period

Waste Management Company (WMC) has won a contract that requires the firm to remove radioactive material from government-owned property and transport it to a designated dumping site. This task requires a specially made ripper-bulldozer to dig and load the material onto a transportation vehicle. Approximately 400,000 tons of waste must be moved in a period of 2 years. Model A costs $150,000 and has a life of 6,000 hours before it will require any major overhaul. Two units of model A would be required to remove the material within 2 years, and the operating cost for each unit would run to $40,000/year for 2,000 hours of operation. At this operational rate, the model would be operable for 3 years, and at the end of that time, it is estimated that the salvage value will be $25,000 for each machine.

A more efficient model B costs $240,000 each, has a life of 12,000 hours without any major overhaul, and costs $22,500 to operate for 2,000 hours per year to complete the job within 2 years. The estimated salvage value of model B at the end of 6 years is $30,000. Once again, two units of model B would be required.

Since the lifetime of either model exceeds the required service period of 2 years (Figure 7.14), WMC has to assume some things about the used equipment at the end of that time. Therefore, the engineers at WMC estimate that, after 2 years, the model A units could be sold for $45,000 each, and the model B units for $125,000 each. After considering all tax effects, WMC summarized the resulting cash flows (in thousand of dollars) for each project as follows:

Period	Model A		Model B	
0	-$300		-$480	
1	-80		-45	
2	-80	+90	-45	+250
3	-80	+50	-45	
4			-45	
5			-45	
6			-45	+60

Here, the figures in the boxes represent the estimated salvage values at the end of the analysis period (end of year 2). Assuming that the firm's MARR is 15%, which option would be acceptable?

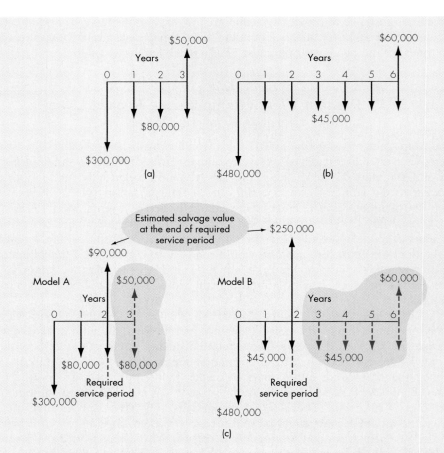

Figure 7.14 (a) Cash flow for model A; (b) cash flow for model B; (c) comparison of unequal-lived service projects when the required service period is shorter than the individual project life (Example 7.11)

Solution

Given: Cash flows for two alternatives as shown in Fig. 7.14, $i = 15\%$ per year

Find: NPW for each alternative, and which is preferred

First, note that these are service projects as we can assume the same revenues for both configurations. Since the firm explicitly estimated the market values of the assets at the end of the analysis period (2 years), we can compare the two models directly. Since the benefits (removal of the wastes) are equal, we can concentrate on the costs:

$$PW(15\%)_A = -\$300 - \$80(P/A, 15\%, 2) + \$90(P/F, 15\%, 2)$$

$$= -\$362$$

$$PW(15\%)_B = -\$480 - \$45(P/A, 15\%, 2) + \$250(P/F, 15\%, 2)$$

$$= -\$364.$$

Model A has the least negative PW costs and thus would be preferred.

Project's Life Is Shorter than Analysis Period

When project lives are shorter than the required service period, we must consider how, at the end of the project lives, we will satisfy the rest of the required service period. Replacement projects—additional projects to be implemented when the initial project has reached the limits of its useful life—are needed in such a case. Sufficient replacement projects must be analyzed to match or exceed the required service period.

To simplify our analysis, we could assume that the replacement project will be exactly the same as the initial project, with the same corresponding costs and benefits. However, this assumption is not necessary. For example, depending on our forecasting skills, we may decide that a different kind of technology—in the form of equipment, materials, or processes—is a preferable replacement. Whether we select exactly the same alternative, or a new technology as the replacement project, we are ultimately likely to have some unused portion of the equipment to consider as salvage value, just as in the case when project lives are longer than the analysis period. On the other hand, we may decide to lease the necessary equipment or subcontract the remaining work for the duration of the analysis period. In this case, we can probably exactly match our analysis period and not worry about salvage values.

In any event, we must make some initial guess concerning the method of completing the analysis period at its outset. Later, when the initial project life is closer to its expiration, we may revise our analysis with a different replacement project. This is only reasonable since economic analysis is an ongoing activity in the life of a company and an investment project, and we should always use the most reliable, up-to-date data we can reasonably acquire.

Example 7.12 Present Worth Comparison—Project Lives Shorter than Analysis Period

The Smith Novelty Company, a mail-order firm, wants to install an automatic mailing system to handle product announcements and invoices. The firm has a choice between two different types of machines. The two machines are designed differently but have identical capacities and do exactly the same job. The $12,500 semi-automatic model A will last 3 years, while the fully automatic model B will cost $15,000 and last 4 years. The expected cash flows for the two machines including maintenance, salvage value, and tax effects are as follows:

n	Model A	Model B
0	-$12,500	-$15,000
1	-5,000	-4,000
2	-5,000	-4,000
3	-5,000 + 2,000	-4,000
4		-4,000 + 1,500
5		

As business grows to a certain level, neither of the models may be able to handle the expanded volume at the end of year 5. If that happens, a fully computerized mail-order system will need to be installed to handle the increased business volume. With this scenario, which model should the firm select at MARR = 15%?

Solution

Given: Cash flows for two alternatives as shown in Figure 7.15, analysis period of 5 years, $i = 15\%$

Find: NPW of each alternative and which to select

Since both models have a shorter life than the required service period (5 years), we need to make an explicit assumption of how the service requirement is to be met. Suppose that the company considers leasing comparable equipment (Model A) that has an annual lease payment of $6,000 (after taxes) with an annual operating cost of $8,000 for the remaining required service period. In this case, the cash flow would look like Figure 7.15.

n	Model A	Model B
0	-$12,500	-$15,000
1	-5,000	-4,000
2	-5,000	-4,000
3	-5,000 + 2,000	-4,000
4	-5,000 - 6,000	-4,000 + 1,500
5	-5,000 - 6,000	-5,000 - 6,000

Here, the boxed figures represent the annual lease payments. (It costs $6,000 to lease the equipment and $5,000 to operate it annually. Other maintenance costs will be paid by the leasing company.) Note that both alternatives now have the same required service period of 5 years. Therefore, we can use NPW analysis:

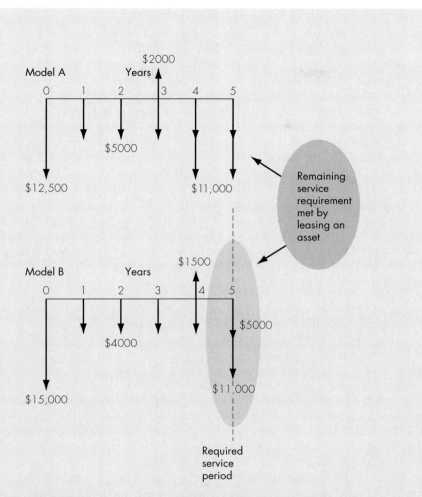

Figure 7.15   Comparison for service projects with unequal lives when the required service period is longer than the individual project life (Example 7.12)

$$PW(15\%)_A = -\$12{,}500 - \$5{,}000(P/A, 15\%, 2) - \$3{,}000(P/F, 15\%, 3)$$

$$-\$11{,}000(P/A, 15\%, 2)(P/F, 15\%, 3)$$

$$= -\$34{,}359.$$

$$PW(15\%)_B = -\$15{,}000 - \$4000(P/A, 15\%, 3) - \$2{,}500(P/F, 15\%, 4)$$

$$-\$11{,}000(P/F, 15\%, 5)$$

$$-\$31{,}031.$$

Since these are service projects, model B is the better choice.

Analysis Period Coincides with Longest Project Life

As seen in the preceding pages, equal future time periods are generally necessary to achieve comparability of alternatives. In some situations, however, revenue projects with different lives can be compared if they require only a one-time investment because the task or need within the firm is a one-time task or need. We may find an example of this situation in the extraction of a fixed quantity of a natural resource such as oil or coal. Consider two mutually exclusive processes: One that requires 10 years to recover some coal and a second that can accomplish the task in only 8 years. There is no need to continue the project if the short-lived process is used and all the coal has been retrieved. In this example, the two processes can be compared over an analysis period of 10 years (longest project life in the group), assuming no cash flows after 8 years for the shorter-lived project. The revenues must be included in the analysis even if the price of coal is constant because of the time value of money. Even if the total revenue (undiscounted) is equal for either process, that for the faster process has a larger present worth. Therefore, the two projects could be compared using the NPW of each over its own life. Note that in this case the analysis period is determined by, and coincides with, the longest project life in the group. (Here we are still, in effect, assuming an analysis period of 10 years.)

Example 7.13 Present Worth Comparison—A Case Where the Analysis Period Coincides with the Project with the Longest Life in the Mutually Exclusive Group

The family-operated Foothills Ranching Company (FRC) owns the mineral rights for land used for growing grain and grazing cattle. Recently, oil was discovered on this property. The family has decided to extract the oil, sell the land, and retire. The company can either lease the necessary equipment and extract and sell the oil itself, or it can lease the land to an oil-drilling company. If the company chooses the former, it will require $300,000 leasing expenses up-front, but the net annual cash flow after taxes from drilling operations will be $600,000 at the end of each year for the next 5 years. The company can sell the land for a net cash flow of $1,000,000 in 5 years when the oil is depleted. If the company chooses the latter, the drilling company can extract all the oil in only 3 years, and it can sell the land for a net cash flow of $800,000 in 3 years. (The difference in resale value of the land is due to the increasing rate of land appreciation anticipated for this property.) The net cash flow from the lease payments to FRC will be $630,000 at the *beginning* of each of the next 3 years. All benefits and costs associated with the two alternatives have been accounted for in the figures listed above. Which option should the firm select at $i = 15\%$?

Solution

Given: Cash flows shown in Figure 7.16, $i = 15\%$ per year

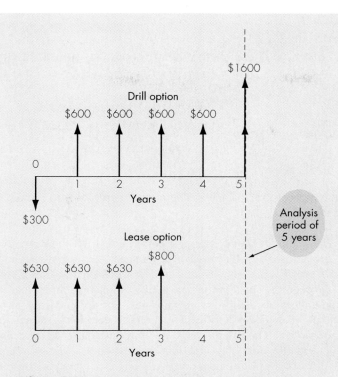

Figure 7.16 Comparison of revenue projects with unequal lives where the analysis period coincides with the project with the longest life in the mutually exclusive group (Example 7.14). In our example, the analysis period is 5 years, assuming zero cash flows in years 4 and 5 for the lease option

Find: NPW of each alternative and which to select

As illustrated in Figure 7.16, the cash flows associated with each option look like this:

n	Drill	Lease
0	−$300,000	$630,000
1	600,000	630,000
2	600,000	630,000
3	600,000	800,000
4	600,000	
5	1,600,000	

After depletion of the oil, the project will terminate.

$$PW(15\%)_{Drill} = -\$300,000 + \$600,000(P/A, 15\%, 4)$$
$$+ \$1,600,000(P/F, 15\%, 5)$$
$$= \$2,208,470.$$
$$PW(15\%)_{Lease} = \$630,000 + \$630,000(P/A, 15\%, 2)$$
$$+ \$800,000(P/F, 15\%, 3)$$
$$= \$2,180,210.$$

Note that these are revenue projects. Therefore, the drill option appears to be the marginally better option.

Comments: The relatively small difference between the two NPW amounts ($28,457) suggests that the actual decision between drilling and leasing might be decided on noneconomic issues. Even if the drilling option were slightly better, the company might prefer to forego the small amount of additional income and select the lease option rather than undertake an entirely new business venture and do their own drilling. A variable that might also have a critical effect on this decision is the sales value of the land in each alternative. The value of land is often difficult to forecast over any long period of time, and the firm may feel some uncertainty about the accuracy of its guesses. In Chapter 14, we will discuss sensitivity analysis, which is a method by which we can factor uncertainty about the accuracy of project cash flows into our analysis.

7.5.5 Analysis Period Is Not Specified

Our coverage so far has focused on situations in which an analysis period is known. When an analysis period is not specified, either explicitly by company policy or practice or implicitly by the projected life of the investment project, it is up to the analyst to choose an appropriate one. In such a case, the most convenient procedure is to choose one based on the useful lives of the alternatives. When the alternatives have equal lives, this is an easy selection. When the lives of alternatives differ, we must select an analysis period that allows us to compare different-lived projects on an equal time basis, i.e., a **common service period.**

Lowest Common Multiple of Project Lives

A required service period of infinity may be assumed if we anticipate an investment project will be ongoing at roughly the same level of production for some indefinite period. It is certainly possible to do so mathematically, though the analysis is likely to be complicated and tedious. Therefore, in the case of an indefinitely ongoing investment project, we typically select a finite analysis period by using the **lowest common multiple** of project lives. For example, if alternative A has a 3-year useful life and alternative B has a 4-year life, we may select 12 years as the analysis or common service period. We would consider alternative A through 4 life cycles and alternative B through 3 life cycles; in each case, we would use the alternatives completely. We then accept the finite model's results as a good prediction of what will be the economically wisest course of action for the foreseeable future. The following example is a case in which we conveniently use the lowest common multiple of project lives as our analysis period.

Example 7.14 Present Worth Comparison—Unequal Lives— Lowest Common Multiple Method

Consider Example 7.12. Suppose that both models A and B can handle the increased future volume and that the system is not going to be phased out at the end of 5 years. Instead, the current mode of operation is expected to continue for an indefinite period of time. We also assume that these two models will be available in the future without significant changes in price and operating costs. At MARR = 15%, which model should the firm select?

Solution

Given: Cash flows for two alternatives as shown in Figure 7.17, $i = 15\%$ per year, indefinite period of need

Find: NPW of each alternative and which to select

Recall that the two mutually exclusive alternatives have different lives, but provide identical annual benefits. In such a case, we ignore the common benefits and can make the decision based solely on costs, as long as a common analysis period is used for both alternatives.

To make the two projects comparable, let's assume that, after either the 3- or 4-year period, the system would be reinstalled repeatedly using the same model and that the same costs would apply. The lowest common multiple of 3 and 4 is 12, so we will use 12 years as the common analysis period. Note that any cash flow difference between the alternatives will be revealed during the first 12 years. After that, the same cash flow pattern repeats every 12 years for an indefinite period. The replacement cycles and cash flows are shown in Figure 7.17.

- Model A: Four replacements occur in a 12-year period. The *PW* for the first investment cycle is

$$PW(15\%) = -\$12,500 - \$5,000(P/A, 15\%, 3)$$
$$+ \$2,000(P/F, 15\%, 3)$$
$$= -\$22,601.$$

  With four replacement cycles, the total *PW* is

$$PW(15\%) = -\$22,601 \left[1 + (P/F, 15\%, 3)\right.$$
$$\left. + (P/F, 15\%, 6) + (P/F, 15\%, 9)\right]$$
$$= -\$53,657.$$

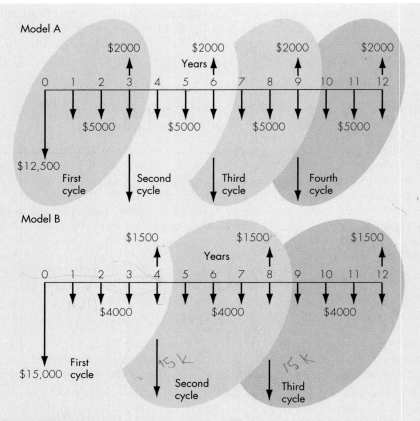

Figure 7.17 Comparison of projects with unequal lives when the required service period is infinite and project repeatability is likely with the same investment and O&M costs in the future replacement (Example 7.14)

- Model B: Three replacements occur in a 12-year period. The *PW* for the first investment cycle is

$$PW(15\%) = -\$15{,}000 - \$4{,}000(P/A, 15\%, 4)$$
$$+ \$1{,}500(P/F, 15\%, 4)$$
$$= -\$25{,}562.$$

With three replacement cycles in 12 years, the total *PW* is

$$PW(15\%) = -\$25{,}562\,[1 + (P/F, 15\%, 4) + (P/F, 15\%, 8)]$$
$$= -\$48{,}534.$$

Comments: In Example 7.14, an analysis period of 12 years seems reasonable. The number of actual reinvestment cycles needed with each type of system will depend on the technology of the future system, so we may or may not actually need the four (model A) or three (model B) reinvestment cycles we used in our analysis. The validity of the analysis also depends on the costs of system and labor remaining constant. If we assume **constant dollar prices** (see Chapter 13), this analysis would provide us with a reasonable result. (As you will see in Example 8.9, the annual worth approach will make it mathematically easier to solve this type of comparison.) If we cannot assume the constant dollar prices in future replacements, we need to estimate the costs for each replacement over the analysis period. This will certainly complicate the problem significantly.

Other Common Analysis Periods

In some cases, the lowest common multiple of project lives is an unwieldy analysis period to consider. Suppose, for example, that you were considering alternatives with lives of 7 and 12 years, respectively. Besides making for tedious calculations, an 84-year analysis period may lead to inaccuracies, since over a long period of time we can be less and less confident about the ability to install identical replacement projects with identical costs and benefits. In a case like this, it would be reasonable to use the useful life of one of the alternatives by either factoring in a replacement project or salvaging the remaining useful life as the case may be. The important rule is to compare both projects on the same time basis.

7.6 Summary

In this chapter, we presented the concept of present worth analysis based on cash flow equivalence along with the payback period. We observed the following important results:

- Present worth is an equivalence method of analysis in which a project's cash flows are discounted to a single present value. It is perhaps the most efficient analysis method we can use for determining project acceptability on an economic basis. Other analysis methods, which we will study in Chapters 5 and 6, are built on a sound understanding of present worth. Table 7.4 summarizes present worth and its related equivalence methods.

- The MARR or minimum attractive rate of return is the interest rate at which a firm can always earn or borrow money. It is generally dictated by management and is the rate at which NPW analysis should be conducted.

- Revenue projects are those for which the income generated depends on the choice of project. Service projects are those for which income remains the same, regardless of which project is selected.

- The term **mutually exclusive** means that, when one of several alternatives that meet the same need is selected, the others will be rejected.

- When not specified by management or company policy, the analysis period to use in a comparison of mutually exclusive projects may be chosen by an indi-

Table 7.4
Summary of Project Analysis Methods

Analysis Method	Equation Form(s)	Description	Comments
Payback period		A method for determining *when* in a project's history it breaks even (the time value of money is not considered).	Should be used as a screening method only—reflects liquidity, not profitability of project.
Net present worth	$PW(i) = \displaystyle\sum_{n=0}^{N} \frac{A_n}{(1 + i)^n}$	An equivalence method that translates a project's cash flows into a net present worth value.	For investments involving disbursements and receipts (revenue projects), the decision rule as follows: If $PW(i) > 0$, accept If $PW(i) = 0$, remain indifferent If $PW(i) < 0$, reject When comparing multiple alternatives, select the one with the greatest NPW value. For investments involving only disbursements (service projects) this is the project with the smallest negative NPW value.
Net future worth	$FW(i) = PW(i)(F/P,i,N)$	An equivalence method variation of NPW; a project's cash flows are translated into a net future worth value.	Most common real-world applications occur when we wish to determine a project's value at commercialization (a future date) not its value when we begin investing (the "present").
Capitalized equivalent	$CE(i) = \dfrac{A}{i}$	An Equivalence method variation of NPW; calculates the NPW of a perpetual or very long-lived project that generates a constant annual net cash flow.	Most common realworld applications are civil engineering projects with lengthy service lives (i.e., $N > 50$) and equal annual costs (or income).

vidual analyst. Several efficiencies can be applied when selecting an analysis period. In general, the analysis period should be chosen to cover the required service period as highlighted in Fig. 7.13.

Self-Test Questions

7s.1 An investment project costs P. It is expected to have an annual net cash flow of $0.125P$ for 20 years. What is the project's payback period?

    (a) 8 years   (b) 0 years
    (c) 17 years   (d) 20 years.

7s.2 You are given the following financial data:

- Investment cost at $n = 0$: $10,000
- Investment cost at $n = 1$: $15,000
- Useful life: 10 years
- Salvage value (at the end of 11 years): $5,000
- Annual revenues: $12,000 per year
- Annual expenses: $4,000 per year
- MARR: 10%.

(**Note:** *The first revenue and expenses will occur at the end of year 2.*)

Which of the following ranges indicates the actual time interval when payback will occur after inception of the project?

    (a) 3 years < payback ≤ 4 years
    (b) 4 years < payback ≤ 5 years
    (c) 5 years < payback ≤ 6 years
    (d) 6 years < payback ≤ 7 years.

7s.3 Which of the following statements is incorrect?

    (a) If two investors are considering the same project, the payback period will be longer for the investor with the higher MARR.

    (b) If you were to consider the cost of funds in a payback period calculation, you would have to wait longer as you increase the interest rate.

    (c) Considering the cost of funds in a payback calculation is equivalent to finding the time period when the project balance becomes zero.

    (d) The simplicity of the payback period method is one of its most appealing qualities even though it fails to measure project profitability.

7s.4 Find the net present worth of the following cash flow series at an interest rate of 9%.

    (a) $670 < PW(9\%) \leq \$700$

End of Period	Cash Flow
0	−$100
1	−200
2	−300
3	−400
4	−500
5 – 8	900

    (b) $770 < PW(9\%) \leq \$800$
    (c) $860 < PW(9\%) \leq \$890$
    (d) $900 < PW(9\%) \leq \$930$.

7s.5 Find the net present worth of the following cash flow series at an interest rate of 9%.

End of Period	Cash Flow	End of Period	Cash Flow
0	−$100	5	−$300
1	−150	6	−250
2	−200	7	−200
3	−250	8	−150
4	−300	9	−100

    (a) $P = -\$1,387$ (b) $P = -\$1,246$
    (c) $P = -\$1,127$ (d) $P = -\$1,027$.

7s.6 You are considering buying an old house that you will convert into an office building for rental. Assuming that you will own the property for 10 years, how much would you be willing to pay for the old house now given the following financial data?

- Remodeling cost at period 0 = $20,000
- Annual rental income = $25,000
- Annual upkeep costs (including taxes) = $5,000
- Estimated net property value (after taxes) at the end of 10 years = $225,000
- The time value of your money (interest rate) = 8% per year.

(a) $201,205   (b) $218,420
(c) $232,316   (d) $250,100.

7s.7 Your R&D group has developed and tested a computer software package that assists engineers to control the proper chemical mix for the various process-manufacturing industries. If you decide to market the software, your first year operating net cash flow is estimated to be $1,000,000. Because of market competition, product life will be about 4 years, and the product's market share will decrease by 25% each year over the previous year's share. You are approached by a big software house which wants to purchase the right to manufacture and distribute the product. Assuming that your interest rate is 15%, for what minimum price would you be willing to sell the software?

(a) $2,507,621   (b) $2,887,776
(c) $2,766,344   (d) $2,047,734.

7s.8 What is the capitalized equivalent amount, at 10% annual interest, for a series of annual receipts of $400 for the first 10 years, which will increase to $500 per year after 10 years, and which will remain constant thereafter?

(a) $4,621   (b) $4,386
(c) $4,452   (d) $9,854.

7s.9 Find the capitalized equivalent worth for the project cash flow series at an interest rate of 10%.

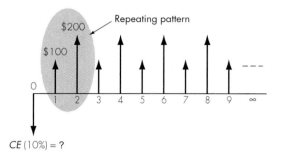

$CE(10\%) = ?$

(a) $CE(10\%) = \$1,476$
(b) $CE(10\%) = \$1,500$
(c) $CE(10\%) = \$3,000$
(d) $CE(10\%) = \$1,753$.

7s.10 The following table contains a summary of how a project's balance is expected to change over its 5-year service life at 10% interest.

End of Period	Project Balance
0	−$1,000
1	−1,500
2	600
3	900
4	1,500
5	2,000

Which of the following statements is incorrect?

(a) The required additional investment at the end of period 1 is $500.
(b) The net present worth of the project at 10% interest is $1,242.
(c) The net future of the project at 10% interest is $2,000.

(d) Within 2 years, the company will recover all its investments and the cost of funds (interest) from the project.

7s.11 NasTech Corporation purchased a vibratory finishing machine for $20,000 in year 0. The useful life of the machine is 10 years, at the end of which, the machine is estimated to have a zero salvage value. The machine generates net annual revenues of $6,000. The annual operating and maintenance expenses are estimated to be $1,000. If NasTech's MARR is 15%, how many years does it take before this machine becomes profitable?

(a) 3 years $< n \leq 4$ years
(b) 4 years $< n \leq 5$ years
(c) 5 years $< n \leq 6$ years
(d) 6 years $< n \leq 7$ years.

7s.12 A newly constructed water-treatment facility cost $2 million. It is estimated that the facility will need renovating every 30 years at a cost of $1 million. Annual repairs and maintenance are estimated to be $100,000 per year. At an interest rate of 6%, determine the capitalized cost of the facility.

(a) $3,360,343   (b) $3,579,806
(c) $3,877,482   (d) $4,301,205.

7s.13 Consider the following two mutually exclusive projects:

End of Year	Net Cash Flow Project A	Net Cash Flow Project B
0	−$1,000	−$2,000
1	475	915
2	475	915
3	475	915

At an interest rate of 12%, what would you recommend?

(a) Select A
(b) Select B

(c) Select either one
(d) Select neither.

7s.14 Two 150-horsepower (HP) motors are being considered for installation at a municipal sewage-treatment plant. The first motor costs $4,500 and has an operating efficiency of 83% at full load. The second motor costs $3,600 and has an efficiency rating of 80% at full load. Both motors are projected to have zero salvage value after a life of 10 years. The annual operating and maintenance cost (excepting power cost) amounts to 15% of the original cost of each motor. The power costs are a flat 5 cents per kilowatt-hour. The minimum number of hours of full-load operation per year necessary to justify the purchase of the more expensive motor at $i = 6\%$ falls in which one of the following ranges?

(a) 800 hours to 1,000 hours
(b) 1,001 hours to 1,200 hours
(c) 1,201 hours to 1,400 hours
(d) 1,401 hours to 1,600 hours
(e) 1,601 hours to 1,800 hours.

7s.15 Alpha Company is planning to invest in a machine, the use of which will result in the following:

- Annual revenues of $10,000 in the first year and increases of $5,000 each year, up to year 9. From year 10, the revenues will remain constant ($52,000) for an indefinite period.

- The machine is to be overhauled every 10 years. The expense for each overhaul is $40,000.

If Alpha expects a present worth of at least $100,000 at a MARR of 10% for this project, what is the maximum investment that Alpha should be prepared to make?

(a) $250,140   (b) $674,697
(c) $350,100   (d) $509,600.

7s.16 Consider the following two investment alternatives:

	Net Cash Flow	
End of Year	Machine A	Machine B
0	−$1,000	−$2,000
1	900	2,500
2	800	800 + 200
3	700	

Suppose that your firm needs either machine for only 2 years. The net proceeds from the sale of machine B is estimated to be $200. What should be the required net proceeds from the sale of machine A so that both machines could be considered economically indifferent at an interest rate of 10%?

(a) $700   (b) $750
(c) $800   (d) $850.

7s.17 Consider the following two mutually exclusive service projects with project lives of 3 years and 2 years, respectively. (The mutually exclusive service projects will have identical revenues for each year of service.) The interest rate is known to be 12%.

	Net Cash Flow	
End of Year	Project A	Project B
0	−$1,000	−$800
1	−400	−200
2	−400	−200 + 0
3	−400 + 200	(salvage)
	(salvage)	

If the required service period is 6 years, and both projects can be repeated with the given costs and better service projects are unavailable in the future, which project is better and why? Choose from the following options:

(a) Select project B because it will save you $344 in present worth over the required service period.

(b) Select project A because it will cost $1,818 in NPW each cycle, with only one replacement, whereas project B will cost $1,138 in NPW each cycle, with two replacements.

(c) Select project B because its NPW exceeds that of project A by $680.

(d) None of the above.

7s.18 Gene Research, Inc., just finished a 4-year R&D and clinical trials successfully and expects a quick approval from the Food and Drug Administration. If the company markets the product on their own, it requires $30 million immediately ($n = 0$) to build a new manufacturing facility, and it is expected to have a 10 year product life. The R&D expenditure in the previous years and the anticipated revenues that the company can generate over the next 10 years is summarized as follows:

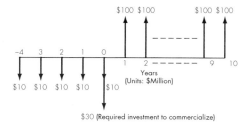

Merck, a large drug company is interested in purchasing the R&D project and the right to commercialize the product from Gene Research, Inc., immediately ($n = 0$). What would be a starting negotiating price for the project from Merck? Assume that Gene's MARR = 20%.

(a) $105 million
(b) $420 million
(c) $494 million
(d) $524 million

7s.19 The following table contains the summary information of how a project's balance changes over its 5-year service life at 10% interest (MARR).

End of Period	Project Balance
0	–$1,000
1	–$1,500
2	$600
3	$900
4	$1,500
5	$2,000

Which of the following statements is *correct?*

(a) The project is not profitable at $i = 10\%$.

(b) The conventional payback period is 1.7 years.

(c) The cash flow in period 3 is $240.

(d) The Net Present Worth of the project is $2,000.

7s.20 A manufacturing company is considering the purchase of a new CNC lathe, which will cost $60,000 and has an annual maintenance cost of $8,000. A few parts in the lathe need to be replaced once every 3 years to enable smooth running of the lathe. This would cost an additional $20,000 (once every 3 years). Assuming that the lathe would last indefinitely under these conditions, what is the capitalized equivalent cost of this investment at an interest rate of 10%.

(a) $140,000   (c) $200,000
(b) $140,159   (d) $200,400

7s.21 A manufacturing company is considering two mutually exclusive machines E1 and E2 with the following cash flow information. Which machine would you recommend if the company needs either machine for only 3 years. Assume a MARR of 12%.

(a) Project E1
(b) Project E2
(c) Indifferent

	Machine E1		Machine E2	
Year	Cash-Flow	Salvage Value	Cash-Flow	Salvage Value
0	–$400		–$600	
1	–$300	$200	–$250	$400
2	–$300	$175	–$250	$360
3	–$300	$150	–$250	$320
4			–$250	$280
5			–$250	$240

(d) Cannot compare without knowing the year-end salvage values over their service lives.

7s.22 Consider the following information for a typical investment project with a service life of 5 years.

n	Cash Flow	Project Balance
0	–$1,000	–$1,000
1	$200	–$900
2	$490	–$500
3	$550	0
4	–$100	–$100
5	$200	$90

Determine the interest rate used in the project balance calculation.

(a) 10%   (c) 20%
(b) 15%   (d) 25%

7s.23 Which of the following investment options would maximize your future wealth at the end of 5 years if you plan to invest $500 today?

(a) 12%, compounded annually
(b) 11.75%, compounded semiannually
(c) 11.5%, compounded quarterly
(d) 11.25%, compounded monthly

Problems

Note: *Unless otherwise stated, all cash flows represent after-tax cash flows. The interest rate (MARR) is also given on an after-tax basis.*

Identifying Cash Inflows and Outflows

7.1 Camptown Togs, Inc., a children's clothing manufacturer, has always found payroll processing to be costly because it must be done by a clerk so that the number of piece-goods coupons collected for each employee can be collected and the types of tasks performed by each employee can be calculated. Recently an industrial engineer has designed a system that partially automates the process by means of a scanner that reads the piece-goods coupons. Management is enthusiastic about this system because it utilizes some personal computer systems that were purchased recently. It is expected that this new automated system will save $45,000 per year in labor. The new system will cost about $30,000 to build and test prior to operation. It is expected that operating costs, including income taxes, will be about $5,000 per year. The system will have a 5-year useful life. The expected net salvage value of the system is estimated to be $3,000.

(a) Identify the cash inflows over the life of the project.

(b) Identify the cash outflows over the life of the project.

(c) Determine the net cash flows over the life of the project.

Payback Period

7.2 Refer to Problem 7.1:

(a) How long does it take to recover the investment?

(b) If the firm's interest rate is 15% after taxes, what would be the discounted payback period for this project?

7.3 For each of the following cash flows:

	Project's Cash Flow ($)			
n	A	B	C	D
0	−$1,500	−$4,000	−$4,500	−$3,000
1	300	2,000	2,000	5,000
2	300	1,500	2,000	3,000
3	300	1,500	2,000	−2,000
4	300	500	5,000	1,000
5	300	500	5,000	1,000
6	300	1,500		2,000
7	300			3,000
8	300			

(a) Calculate the payback period for each project.

(b) Determine whether it is meaningful to calculate a payback period for project D.

(c) Assuming $i = 10\%$, calculate the discounted payback period for each project.

NPW Criterion

7.4 Consider the following sets of investment projects. All projects have a 3-year investment life:

	Project's Cash Flow ($)			
n	A	B	C	D
0	−$1,000	−$1,000	−$1,000	−$1,000
1	0	600	−1,200	900
2	0	800	800	900
3	3,000	1,500	1,500	1,800

(a) Compute the net present worth of each project at $i = 10\%$.

(b) Plot the present worth as function of interest rate (from 0% to 30%) for project B.

7.5 You need to know if the building of a new warehouse is justified under the following conditions:

The proposal is for a warehouse costing $100,000. The warehouse has an expected useful life of 35 years and a net salvage value (net proceeds from sale after tax adjustments) of $25,000. Annual receipts of $17,000 are expected, annual maintenance and administrative costs will be $4,000 year, and annual income taxes are $2,000.

Given these data, which of the following statements are correct?

(a) The proposal is justified for a MARR of 9%.

(b) The proposal has a net present worth of $62,730.50, when 6% is used as the interest rate.

(c) The proposal is acceptable as long as MARR $\leq 10.77\%$.

(d) All of the above are correct.

7.6 Your firm is considering the purchase of an old office building with an estimated remaining service life of 25 years. The tenants have recently signed long-term leases, which leads you to believe that the current rental income of $150,000 per year will remain constant for the first 5 years. Then the rental income will increase by 10% for every 5-year interval over the remaining asset life. For example, the annual rental income would be $165,000 for year 6 through 10, $181,500 for year 11 through 15, $199,650 for year 16 through 20, and $219,615 for year 21 through 25. You estimate that operating expenses, including income taxes, will be $45,000 for the first year, and that they will increase by $3,000 each year thereafter. You estimate that razing the building and selling

the lot on which it stands will realize a net amount of $50,000 at the end of the 25-year period. If you had the opportunity to invest your money elsewhere, and thereby earn interest at the rate of 12% per annum, what would be the maximum amount you would be willing to pay for the building and lot at the present time?

7.7 Consider the following investment project:

n	A_n	i
0	−$2,000	10%
1	2,400	12
2	3,400	14
3	2,500	15
4	2,500	13
5	3,000	10

Suppose the company's reinvestment opportunities change over the life of the project as shown above (i.e., the firm's MARR changes over the life of the project). For example, the company can invest funds available now at 10% for the first year, 12% for the second year, and so forth. Calculate the net present worth of this investment and determine the acceptability of the investment.

7.8 Cable television companies and their equipment suppliers are on the verge of installing new technology that will pack many more channels into cable networks, thereby creating a potential programming revolution with implications for broadcasters, telephone companies, and the consumer electronics industry.

Digital compression uses computer techniques to squeeze 3 to 10 programs into a single channel. A cable system fully using digital compression technology would be able to offer well over 100 channels, compared with about 35 for the average cable television system now used. If the new technology is combined with the increased use of optical fibers it might be possible to offer as many as 300 channels.

A cable company is considering installing this new technology to increase subscription sales and save on satellite time. The company estimates that the installation will take place over two years. The system is expected to have an 8-year service life and have the following savings and expenditures:

Digital Compression	
Investment	
Now	$500,000
First year	$3,200,000
Second year	$4,000,000
Annual savings in satellite time	$2,000,000
Incremental annual revenues due to new subscriptions	$4,000,000
Incremental annual expenses	$1,500,000
Incremental annual income taxes	$1,300,000
Economic service life 8 years	
Net salvage values	$1,200,000

Note that the project has a 2-year investment period which is followed by an 8-year service life (a total 10-year life project). This implies that the first annual savings will occur at the end of year 3 and the last savings will occur at the end of year 10. If the firm's MARR is 15%, justify the economic worth of the project based on the NPW method.

7.9 A large food-processing corporation is considering using laser technology to speed up and eliminate waste in the potato-peeling process. To implement the system, the company anticipates needing $3 million to purchase the industrial strength lasers. The system will save $1,200,000 per year in labor and materials. However, it will require an additional operating and maintenance cost of $250,000. Annual income taxes will also increase by $150,000. The system is expected to have a 10-year service life and will have a salvage value of about $200,000. If the company's MARR is 18%, justify the economics of the project based on the NPW method.

Future Worth and Project Balance

7.10 Consider the following sets of investment projects. All projects have a 3-year investment life.

Period (n)	Project's Cash Flow A	B	C	D
0	−$2,500	−1,000	2,500	−3,000
1	5,400	−3,000	−7,000	1,500
2	14,400	1,000	2,000	5,500
3	7,200	3,000	4,000	6,500

(a) Compute the net present worth of each project at $i = 13\%$.

(b) Compute the net future worth of each project at $i = 13\%$.

Which project(s) are acceptable?

7.11 Consider the following project balances for a typical investment project with a service life of 4 years.

n	A_n	Project Balance
0	−$1,000	−$1,000
1	( )	−1,100
2	( )	−800
3	460	−500
4	( )	0

(a) Construct the original cash flows of the project.

(b) Determine the interest rate used in computing the project balance.

7.12 In Problem 7.11, would this project be acceptable at $i = 15\%$?

7.13 Consider the project balance diagram for a typical investment project with a service life of 5 years. The numbers in the figure indicate the beginning project balances.

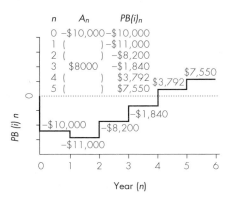

(a) From the project balance diagram, construct the project's original cash flows.

(b) What is the project's conventional payback period (without interest)?

7.14 Consider the following cash flows and present worth profile.

| | Net Cash Flows ($) | |
Year	Project 1	Project 2
0	−$100	−$100
1	40	30
2	80	Y
3	X	80

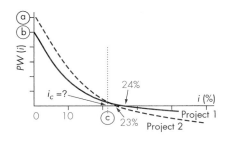

(a) Determine the values for X and Y.

(b) Calculate the terminal project balance of project 1 at MARR=24%.

(c) Find the values for ⓐ,ⓑ, and ⓒ in the NPW plot.

7.15 Consider the project balances for a typical investment project with a service life of 5 years.

n	A_n	Project Balance
0	−$1,000	−$1,000
1	( )	−900
2	490	−500
3	( )	0
4	( )	−100
5	200	( )

(a) Construct the original cash flows of the project and the terminal balance and fill in the blanks above.

(b) Determine the interest rate used in the project balance calculation, and compute the present worth of this project at the computed interest rate.

7.16 Refer to Problem 7.3:

(a) Graph the project balances (at $i = 10\%$) of each project as a function of n.

(b) By examining the graphical results in part (a), determine which project appears to be the safest to undertake if there is some possibility of premature termination of the projects at the end of year 2.

7.17 Consider the following sets of investment projects:

| | Project's Cash Flow | | | | |
n	A	B	C	D	E
0	−$1,500	−$5,000	−$3,500	−$3,000	−$5,500
1	500	2,000	0	500	1,000
2	900	−3,000	0	2,000	3,000
3	1,000	5,000	3,000	3,000	2,000
4	2,000	5,000	7,000	4,000	
5	−500	3,500	13,000	1,250	

(a) Compute the future worth at the end of life for each project, at $i = 15\%$.

(b) Determine the acceptability of each project.

7.18 Refer to Problem 7.17.

(a) Plot the future worth for each project as a function of interest rate $(0\% - 50\%)$.

(b) Compute the project balance of each project at $i = 15\%$.

(c) Compare the terminal project balances calculated in (b) with the results obtained in Problem 7.17 (a). Without using the interest factor tables, compute the future worth based on the project balance concept.

7.19 Refer to Problem 7.3 and compute the future worth for each project at $i = 10\%$.

7.20 Refer to Problem 7.4 and compute the future worth for each project at $i = 12\%$.

7.21 Consider the following set of independent investment projects:

	Project Cash Flows		
n	A	B	C
0	−$100	−$100	$100
1	50	40	−40
2	50	40	−40
3	50	40	−40
4	−100	10	
5	400	10	
6	400		

Assume MARR=10% for the following questions:

(a) Compute the net present worth for each project and determine the acceptability of each project.

(b) Compute the net future worth of each project at the end of each pro-

ject period and determine the acceptability of each project.

(c) Compute the project worth of each project at the end of 6 years with variable MARRs as follows: 10% for $n = 0$ to $n = 3$ and 15% for $n = 4$ to $n = 6$.

7.22 Consider the following project balance profiles for proposed investment projects.

	Project Balances		
n	A	B	C
0	−$1,000	−$1,000	−$1,000
1	−1,000	−650	−1,200
2	−900	−348	−1,440
3	−690	−100	−1,328
4	−359	85	−1,194
5	105	198	−1,000
Interest rate used	10%	?	20%
NPW	?	$79.57	?

Project balance figures are rounded to nearest dollars.

(a) Compute the net present worth of projects A and C, respectively.

(b) Determine the cash flows for project A.

(c) Identify the net future worth of project C.

(d) What interest rate would be used in the project balance calculations for project B?

7.23 Consider the following project balance profiles for proposed investment projects:

Project balance figures are rounded to nearest dollars.

(a) Compute the net present worth of each investment.

	Project Balances		
n	A	B	C
0	−$1,000	−$1,000	−$1,000
1	−800	−680	−530
2	−600	−302	X
3	−400	−57	−211
4	−200	233	−89
5	0	575	0
Interest rate used	0%	18%	12%

(b) Determine the project balance at the end of period 2 for project C, if $A_2 = \$500$.

(c) Determine the cash flows for each project.

(d) Identify the net future worth of each project.

Capitalized Equivalent Worth

7.24 Maintenance money for a new building has been sought. Mr. Kendall would like to make a donation to cover all future expected maintenance costs for the building. These maintenance costs are expected to be $40,000 each year for the first 5 years, $50,000 for each year 6 through 10, and $60,000 each year after that. (The building has an indefinite service life.)

(a) If the money is placed in an account that will pay 13% interest, compounded annually, how large should the gift be?

(b) What is the equivalent annual maintenance cost over the infinite service life?

7.25 Consider an investment project, the cash flow pattern of which repeats itself every 5 years forever as shown. At an interest rate of 14%, compute the capitalized equivalent amount for this project.

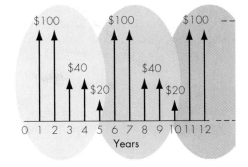

7.26 A group of concerned citizens has established a trust fund that pays 6% interest, compounded monthly, to preserve a historical building by providing annual maintenance funds of $12,000 forever. Compute the capitalized equivalent amount for these building maintenance expenses.

7.27 A newly constructed bridge costs $5,000,000. The same bridge is estimated to need renovation every 15 years at a cost of $1,000,000. Annual repairs and maintenance are estimated to be $100,000 per year.

(a) If the interest rate is 5%, determine the capitalized cost of the bridge.

(b) Suppose that, in (a), the bridge must be renovated every 20 years, not every 15 years. What is the capitalized cost of the bridge?

(c) Repeat (a) and (b) with an interest rate of 10%. What have you to say about the effect of interest on the results?

7.28 To decrease the costs of operating a lock in a large river, a new system of operation is proposed. The system will cost $650,000 to design and build. It is estimated that it will have to be reworked every 10 years at a cost of $100,000. In addition, an expenditure of $50,000 will have to be made at the end of the fifth year for a new type of gear that will not be available until then. Annual operating costs are expected to be $30,000 for the first 15 years and $35,000 a year thereafter. Compute the capitalized cost of perpetual service at $i = 8\%$.

Mutually Exclusive Alternatives

7.29 Consider the following cash flow data for two competing investment projects.

	Cash Flow Data (Unit: $ thousand)	
n	Project A	Project B
0	−$800	−$2,635
1	−1,500	−565
2	−435	820
3	775	820
4	775	1,080
5	1,275	1,880
6	1,275	1,500
7	975	980
8	675	580
9	375	380
10	660	840

At $i = 12\%$, which of the two projects would be a better choice?

7.30 Consider the cash flows for the following investment projects.

Assume MARR=15%.

	Project's Cash Flow				
n	A	B	C	D	E
0	−$1,500	−$1,500	−$3,000	1,500	−$1,800
1	1,350	1,000	1,000	−450	600
2	800	800	X	−450	600
3	200	800	1,500	−450	600
4	100	150	X	−450	600

(a) Suppose Projects A and B are mutually exclusive. Which project would be selected based on the NPW criterion?

(b) Repeat (a) using the NFW criterion.

(c) Find the minimum value of X that makes project C acceptable.

(d) Would you accept D at $i = 18\%$?

(e) Assume that projects D and E are mutually exclusive. Which project would you select based on the NPW criterion?

7.31 Consider the following two mutually exclusive investment projects. Assume that the MARR=12%.

(a) Which alternative would be selected using the NPW criterion?

	Project's Cash Flow	
n	A	B
0	−$4,500	−$2,900
1	2,610	1,210
2	2,930	1,720
3	2,300	1,500

(b) Which alternative would be selected using the NFW criterion?

7.32 Consider the following two mutually exclusive investment projects. Assume MARR=15%.

(a) Using the NPW criterion, which project would be selected?

	Project's Cash Flow	
n	A	B
0	−$3,000	−$8,000
1	400	11,500
2	7,000	400

(b) Sketch the $PW(i)$ function for each alternative on the same chart between 0% and 50%. For what range of i would you prefer project B?

7.33 Two methods of carrying away surface run-off water from a new subdivision are being evaluated:

Method A: Dig a ditch. The first cost would be $30,000, and $10,000 of redigging and shaping would be required at 5-year intervals forever.

Method B: Lay concrete pipe. The first cost would be $75,000, and a replacement would be required at 50-year intervals at a net cost of $90,000 forever.

At $i = 12\%$, which method is the better one?

7.34 A local car dealer is advertising a standard 24-month lease of $1,150 per month for its new XT 3000 series sports car. The standard lease requires a down payment of $4,500 plus a $1,000 refundable initial deposit *now*. The first lease payment is due at the end of month 1. In addition, they also offer a 24-month lease plan that has a single up-front payment of $30,500 plus a refundable initial deposit of $1,000. Under both options, the initial deposit will be refunded at the end of month 24. Assume an interest rate of 6%, compounded monthly. With the present worth criterion, which option is preferred?

7.35 Two alternative machines are being considered for a manufacturing process. Machine A has a first cost of $75,200, and its salvage value at the end of 6 years of estimated service life is $21,000. The operating costs of this machine are estimated to be $6,800 per year. Extra income taxes are estimated at $2,400 per year. Machine B has a first cost of $44,000, and its estimated salvage value at the end of 6 years' service is estimated to be negligible. The annual operating costs will be $11,500. Compare these two alternatives by the present worth method at $i = 13\%$.

7.36 An electric motor is rated at 10 horsepower (HP) and costs $800. Its full load efficiency is specified to be 85%. A newly designed, high-efficiency motor of the same size has an efficiency of 90%, but costs $1,200. It is estimated that the motors will operate at a rated 10 HP output for 1,500 hours a year, and the cost of energy will be $0.07 per kilowatt-hour. Each motor is expected

to have a 15-year life. At the end of 15 years, the first motor will have a salvage value of $50, and the second motor will have a salvage value of $100. Consider the MARR to be 8%. (Note: 1 HP = 0.7457 kW.)

(a) Determine which motor should be installed based on the NPW criterion.

(b) In (a), what if the motors operated 2500 hours a year instead of 1500 hours a year? Would the same motor in (a) be the choice?

7.37 Consider the following two mutually exclusive investment projects.

	Project's Cash Flow	
n	A	B
0	−$10,000	−$22,000
1	7,500	15,500
2	7,000	18,000
3	5,000	

Which project would be selected if you use the infinite planning horizon with project repeatability likely (same costs and benefits) based on the NPW criterion? Assume that $i = 12\%$.

7.38 Consider the following two mutually exclusive investment projects, which have unequal service lives.

	Project's Cash Flow	
n	A1	A2
0	−$900	−$1,800
1	−400	−300
2	−400	−300
3	−400 + 200	−300
4		−300
5		−300
6		−300
7		−300
8		−300 + 500

(a) What assumption(s) do you need to compare a set of mutually exclusive investments with unequal service lives?

(b) With the assumption(s) defined in (a) and using $i = 10\%$, determine which project should be selected.

(c) If your analysis period (study period) is just 3 years, what should be the salvage value of project A2 at the end of year 3 to make the two alternatives economically indifferent?

7.39 Consider the following two mutually exclusive projects B1 and B2.

| | B1 | | B2 | |
| | Cash Flow | Salvage Value | Cash Flow | Salvage Value |
n				
0	−$12,000		−$10,000	
1	−2,000	6,000	−2,100	6,000
2	−2,000	4,000	−2,100	3,000
3	−2,000	3,000	−2,100	1,000
4	−2,000	2,000		
5	−2,000	2,000		

Salvage values represent the net proceeds (after tax) from disposal of the assets if they are sold at the end of each year. Both B1 and B2 will be available (or can be repeated) with the same costs and salvage values for an indefinite period.

(a) With the infinite planning horizon assumption, which project is a better choice at MARR=12%?

(b) With a 10-year planning horizon, which project is a better choice at MARR=12%?

7.40 Consider the following cash flows for two types of models.

Both models will have no salvage value upon their disposal (at the end of their respective service lives). The firm's MARR is known to be 15%.

(a) Notice that both models have a different service life. However, model

| n | Project's Cash Flow | |
	Model A	Model B
0	−$6,000	−$15,000
1	3,500	10,000
2	3,500	10,000
3	3,500	

A will be available in the future with the same cash flows. Model B is available at one time only. If you select model B now, you will have to replace it with model A at the end of year 2. If your firm uses the present worth as a decision criterion, which model should be selected, assuming that your firm will need either model for an indefinite period?

(b) Suppose that your firm will need either model for only 2 years. Determine the salvage value of model A at the end of year 2 that makes both models indifferent (equally likely).

7.41 An electric utility is taking bids on the purchase, installation, and operation of microwave towers.

| | Cost per Tower | |
	Bid A	Bid B
Equipment cost	$65,000	$58,000
Installation cost	$15,000	$20,000
Annual maintenance and inspection fee	$1,000	$1,250
Annual extra income taxes		$500
Life	40 years	35 years
Salvage value	$0	$0

Which is the most economical bid, if the interest rate is considered to be 11%? Either tower will have no salvage value after 20 years of use.

7.42 Refer to Problem 4.61 Consider the two different payment plans for Troy Aikman's contract. Compare these two mutually exclusive plans based on the NPW method.

7.43 Consider the following two investment alternatives:

n	Project's Cash Flow A1	A2
0	−$15,000	−$25,000
1	9,500	0
2	12,500	X
3	7,500	X
PW(15%)	?	9,300

The firm's MARR (minimum attractive rate of return) is known to be 15%.

(a) Compute the PW(15%) for A1.

(b) Compute the unknown cash flow X in years 2 and 3 for A2.

(c) Compute the project balance (at 15%) of A1 at the end of period 3.

(d) If these two projects are mutually exclusive alternatives, which project would you select?

7.44 For each of the following after-tax cash flows:

n	A	B	C	D
0	−$2,500	−$7,000	−$5,000	−$5,000
1	650	−2,500	−2,000	−500
2	650	−2,000	−2,000	−500
3	650	−1,500	−2,000	4,000
4	600	−1,500	−2,000	3,000
5	600	−1,500	−2,000	3,000
6	600	−1,500	−2,000	2,000
7	300		−2,000	3,000
8	300			

(a) Compute the project balances for projects A and D as a function of project year at i = 10%.

(b) Compute the net future worth values for projects A and D at i = 10%.

(c) Suppose that projects B and C are mutually exclusive. Assume also that the required service period is 8

years, and the company is considering leasing comparable equipment that has an annual lease expense of $3,000 for the remaining years of the required service period. Which project is a better choice?

7.45 A bi-level mall is under construction. It is planned to install only nine escalators at the start, although the ultimate design calls for 16. The question arises in the design whether to provide necessary facilities (stair supports, wiring conduits, motor foundations, etc.), which would permit the installation of the additional escalators at the mere cost of their purchase and installation, or to defer investment in these facilities until the escalators need to be installed.

Option 1: Provide these facilities now for all seven future escalators at $200,000.

Option 2: Defer the facility investment as needed. It is planned to install two more escalators in 2 years, three more in 5 years, and the last two in 8 years. The installation of these facilities at the time they are required is estimated to cost $100,000 in year 2, $160,000 in year 5, and $140,000 in 8 years. Additional annual expenses are estimated at $3,000 for each escalator facility installed.

At an interest rate of 12%, compare the net present worth of each option over 8 years.

7.46 An electrical utility is experiencing a sharp power demand, which continues to grow at a high rate in a certain local area.

Two alternatives are under consideration. Each alternative is designed to provide enough capacity during the next 25 years. Both alternatives will consume the same amount of fuel, so fuel cost is not considered in the analysis.

Alternative A: Increase the generating capacity now so that the ultimate demand can be met with additional expenditures later. An initial investment of $30 million would be required, and it is estimated that this plant facility would be in service for 25 years and have a salvage value of $0.85 million. The annual operating and maintenance costs (including income taxes) would be $0.4 million.

Alternative B: Spend $10 million now and follow this expenditure with future additions during the 10th year and the 15th year. These additions would cost $18 million and $12 million, respectively. The facility would be sold 25 years from now with a salvage value of $1.5 million. The annual operating and maintenance costs (including income taxes) initially will be $250,000, increasing to $0.35 million after the second addition (from 11th year to the 15th year) and to $0.45 million during the final 10 years. (Assume that these costs begin 1 year subsequent to the actual addition.)

If the firm uses 15% as a MARR, which alternative should be undertaken based on the present worth criterion?

7.47 A large refinery-petrochemical complex is planning to manufacture caustic soda, which will use feed water of 10,000 gallons per day. Two types of feeder-water storage installation are being considered over 40 years of useful life.

Option 1: Build a 20,000-gallon tank on a tower. The cost of installing the tank and tower is estimated to be $164,000. The salvage value is estimated to be negligible.

Option 2: Place a tank of equal capacity on a hill, which is 150 yards away

from the refinery. The cost of installing the tank on the hill, including the extra length of service lines is estimated to be $120,000 with negligible salvage value. Because of its hill location, an additional investment of $12,000 in pumping equipment is required. The pumping equipment is expected to have a service life of 20 years with a salvage value of $1,000 at the end of that time. The annual operating and maintenance cost (including any income tax effects) for the pumping operation is estimated at $1,000.

If the firm's MARR is known to be 12%, which option is better on the basis of present worth criterion?

Short Case Studies

7.48 Apex Corporation requires a chemical finishing process for a product under contract for a period of 6 years. Three options are available. Neither Option 1 nor Option 2 can be repeated after its process life. However, Option 3 will always be available from H&H Chemical Corporation at the same cost during the contract period.

- Option 1: Process device A, which costs $100,000 has annual operating and labor costs of $60,000, and a useful service life of 4 years with an estimated salvage value of $10,000.

- Option 2: Process device B, which costs $150,000, has annual operating and labor costs of $50,000, and a useful service life of 6 years with an estimated salvage value of $30,000.

- Option 3: Subcontract out the process at a cost of $100,000 per year.

According to present worth criterion, which option would you recommend at $i = 12\%$?

7.49 Tampa Electric Company, an investor-owned electric utility serving approximately 2,000 square miles in west central Florida, was faced with providing electricity to a newly developed industrial park complex. The distribution engineering department needs to develop guidelines for design of the distribution circuit. The "main feeder," which is the backbone of each 13 kV distribution circuit, represents a substantial investment by the company.[5]

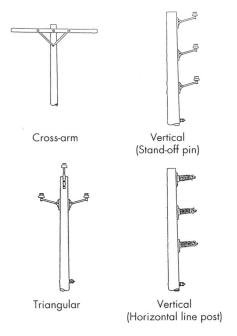

Cross-arm

Vertical (Stand-off pin)

Triangular

Vertical (Horizontal line post)

Tampa Electric has four approved main feeder construction configurations. They are: (1) cross-arm, (2) vertical (horizontal line post), (3) vertical (stand-off pin), and (4) triangular, as illustrated. The width of the easement sought depends on the planned construction configuration. If cross-arm construction is planned, a 15-foot easement is sought. A 10-foot wide easement is sought for vertical and triangular configurations. Once the required easements are obtained, the line clearance department clears any foliage that would impede the construction of the line. The clearance cost is dictated by the typical tree densities along road rights-of-way. The average cost to trim one tree is estimated at $20, and the average tree density in the service area is estimated to be 75 trees per mile. The costs of each construction type are as follows:

| Factors | Design Configurations | | | |
	Cross Arm	Triangular	Horizontal Line	Stand Off
Easements	$487,000	$388,000	$388,000	$388,000
Line clearance	$613	$1,188	$1,188	$1,188
Line construction	$7,630	$7,625	$12,828	$8,812

Additional factors to consider in selecting the best main feeder configuration are as follows:

In certain sections of Tampa Electric's service territory osprey often nest on transmission and distribution poles. These osprey nests reduce the structural and electrical integrity of the pole on which the nest is built. Cross-arm construction is most vulnerable to osprey nesting, since the cross arm and braces provide a secure area for nest construction. Vertical and triangular construction do not provide such spaces. In areas where osprey are known to nest, vertical and triangular configuration have added advantages. The insulation strength of a construction configuration may favorably or adversely affect the reliability of the line for which the configuration is used. A common measure of line insulation strength is the critical flashover (CFO) voltage. The greater the value of CFO, the less susceptible the line to suffer from nuisance flashovers from lightning and other electrical phenomena.

---

[5]Example provided by Andrew Hanson.

The existing inventory of cross arms is used primarily for main feeder construction and maintenance. Use of another configuration for main feeder construction would result in a substantial reduction of cross-arm inventory. The line crews complain that line spacing on vertical and triangular construction is too restrictive for safe live line work. Each accident would cost $65,000 in lost work and other medical expenses. The average cost of each flashover repair would be $3,000.

| Factors | Cross Arm | Design Configurations | | Stand Off |
		Triangular	Horizontal Line	
Nesting	Severe	None	None	None
Insulation strength				
CFO (kV)	387	474	476	462
Annual Flashover occurrence (n)	2	1	1	1
Annual inventory savings		$4,521	$4,521	$4,521
Safety	OK	Problem	Problem	Problem

All configurations would last about 20 years with no salvage values. It appears that non-cross-arm designs are better, but engineers need to consider other design factors, such as safety, rather than just monetary factors when implementing the project. It is true that the line spacing on triangular construction is restrictive. However, with a better clearance design between phases for vertical construction, the safety issue would be minimized. In the utility industry, the typical opposition to new construction types is caused by the confidence acquired from constructing lines in the cross-arm configuration for many years. As more vertical and triangular lines are built, the opposition to these configurations should decrease. Which of the four designs described in the table would you recommend to the management? Assume Tampa Electric's interest rate to be 12%.

B	C	D	E	F Equivalent Worth at r	O
		Interest (%)	Option 1	Option 1	$
Option 1	Option 2	0	$700,000		$
		1	$721,354		$
$100,000		2	$743,428		$
$100,000		3	$766,246		
$100,000		4	$789,829		
$100,000		5	$814,201		
$100,000		6	$839,384		
$100,000		7	$865,402		
$100,000		8	$892,280		
	$100,000	9	$920,043		
	$100,000	10	$948,717		
	$100,000	11	$978,327		
	$100,000	12	$1,008,901		
	$100,000	13	$1,040,466		
	$100,000	14	$1,073,049		
	$100,000				

Chart1 \ Sheet1 \ Sheet2 \ Sheet3 \ Sheet4 \ Sheet5

Annual Equivalent Worth Analysis

Here Come the Pint-size Power Plants: Capstone Turbine Corporation, a new start-up company, is now in the process of manufacturing a micro turbine, a miniaturized, cheaply produced cousin of the turbine-powered electric generators used in airliners. Capstone was able to raise $250 million from various venture capitalists. Much of the money was invested in plant facilities including buildings and equipment. Weighing in at a mere 165 pounds, Capstone's micro turbine, no noisier than a vacuum cleaner, is simplicity itself. Just one moving part, a spinning shaft with 96,000-rpm speed, serves simultaneously as compressor, turbine, and electric generator rotor. The micro turbine is designed to operate on demand for months without maintenance. The generator is cooled by airflow into the gas turbine, thus eliminating the need for liquid cooling. It can make electricity from a variety of fuels—natural gas, kerosene, diesel oil, and even gasoline. Capstone's turbine generates 24 kilowatts, enough to run the central air conditioner of a big house. Capstone's engineer says the 24-kilowatt size was picked with an eye to a second and potentially gigantic market: low-polluting electric automobiles; air-pollution levels would be low enough that California authorities might classify them as "zero emission" vehicles under proposed future standards. Capstone is tooling up a production line scheduled to begin operating later 1999. One of the major questions among the Capstone executives is "How low does the micro turbine's production cost need to go?" A quick feasibility study indicates that it starts to make sense in some utility operations at a capital cost of $500 per kilowatt, somewhat higher than that of the bigger gas turbines in widespread use. That works out to $12,000 for a 24-kilowatt machine. If the micro turbine were produced for hybrid cars in volumes of more than 100,000 a year, the price would drop to less than $2,000.

(*Source:* Capstone Turbine Corporation)

How does Capstone come up with the capital cost of $500 per kilowatt? Suppose you plan to purchase the 24-kilowatt micro turbine and expect to operate it continuously for 10 years. How would you calculate the operating cost per kilowatt-hour?

Suppose you are considering buying a new car. If you expect to drive 12,000 miles per year, can you figure out how much it costs per mile? You would have good reason to want to know cost if you were being reimbursed by your employer on a per-mile basis for the business use of your car. Or consider a real estate developer who is planning to build a shopping center of 500,000 square feet. What would be the minimum annual rental fee per square foot required to recover the initial investment?

Annual equivalence analysis is the method by which these and other unit costs are calculated. As its name suggests, annual equivalent worth analysis, or AE, is also a method by which we can determine the equivalent annual, rather than overall (e.g., present or future), worth of a project. Annual equivalence analysis, along with present worth analysis, is the second major equivalence technique for putting alternatives on a common basis of comparison. In this chapter, we develop the annual equivalent criterion and demonstrate a number of situations in which annual equivalence analysis is preferable to other methods of comparison.

8.1 Annual Equivalent Criterion

The **annual equivalent worth (AE)** criterion provides a basis for measuring investment worth by determining equal payments on an annual basis. Knowing that any lump-sum cash amount can be converted into a series of equal annual payments, we may first find the net present worth (NPW) of the original series and then multiply this amount by the capital recovery factor:

$$AE(i) = PW(i)(A/P, i, N). \tag{8.1}$$

The accept-reject decision rule for a single **revenue** project is as follows:

If $AE(i) \quad > \quad 0$, accept the investment.

If $AE(i) \quad = \quad 0$, remain indifferent to the investment.

If $AE(i) \quad < \quad 0$, reject the investment.

Notice that the factor $(A/P, i, N)$ in Eq. (8.1) is positive for $-1 < i < \infty$, which indicates that the $AE(i)$ value will be positive if, and only if, $PW(i)$ is positive. In other words, accepting a project that has a positive $AE(i)$ value is equivalent to accepting a project that has a positive $PW(i)$ value. Therefore, the AE criterion provides a basis for evaluating a project that is consistent with the NPW criterion.

As with present worth analysis, when you compare mutually exclusive **service** projects whose revenues are the same, you may compare them on a **cost**-only basis. In this situation, the alternative with the minimum annual equivalent cost (or least negative annual equivalent worth) is selected.

Example 8.1 Annual Equivalent Worth by Conversion from Present Worth (PW)

Singapore Airlines is planning to equip some of their Boeing 747 aircrafts with in-flight email and Internet service on transoceanic flights. Passengers can send and receive email no matter where they are in the skies. As email has become a communications staple, airlines have been under increasing pressure to offer access to it. But the rollout has been slow because airlines are hesitant to invest in systems that could quickly become outdated. Hoping to provide value to their passengers, the airline has decided to offer the service through telephone modems for the first 10 Boeing 747s in 2001. As Boeing unveils a broadband email and Internet system during 2002, Singapore will upgrade the system accordingly. If the project turns out to be a financial success, Singapore will introduce the service to all remaining 56 Boeing 747s on its fleet. The service will be free during the first year. After the promotional period, a nominal charge will be instituted for each email message, about $10. Singapore has estimated the projected cash flows (in million dollars) to install these 10 aircrafts as follows:

n	A_n
0	−$15.0
1	−3.5
2	5.0
3	9.0
4	12.0
5	10.0
6	8.0

First, the airline wants to determine whether this project can be justified at MARR = 15%. Then it wants to know what would be the annual benefit (or loss) that could be generated after installation of these systems.

Discussion: When a cash flow has no special pattern, it is easiest to find AE in two steps: (1) Find the NPW (or NFW) of the flow; and (2) find the AE of the NPW or (NFW). This method is presented below. You might want to try another method with this type of cash flow to demonstrate how difficult this can be.

Solution

Given: Cash flow in Figure 8.1, $i = 15\%$
Find: AE

We first compute the NPW at $i = 15\%$:

$$PW(15\%) = -\$15 - \$3.5(P/F, 15\%, 1) + \$5(P/F, 15\%, 2) + \ldots$$
$$+ \$10(P/F, 15\%, 5) + \$8(P/F, 15\%, 6)$$
$$= \$6.946 \text{ million}$$

Since $PW(15\%) > 0$, the project would be acceptable under the NPW analysis.

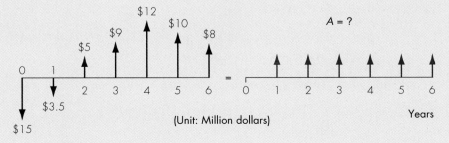

(Unit: Million dollars)

Figure 8.1   Cash flow diagram (Example 8.1)

Now, spreading the NPW over the project life gives

$$AE(15\%) = \$6.946(A/P, 15\%, 6) = \$1.835 \text{ million.}$$

Since $AE(15\%) > 0$, the project is worth undertaking. The positive AE value indicates that the project is expected to bring in a net annual benefit of $1.835 million over the project life.

In some situations a **cyclic cash flow pattern** may be observed over the project life. Unlike the situation in Example 8.1, where we first computed the NPW of the entire cash flow and then calculated the AE value from this NPW, we can compute the AE by examining the first cash flow cycle. Then the NPW for the first cash flow cycle can be computed, and the AE over the first cash flow cycle can be derived. This short-cut method provides the same solution when the NPW of the entire project is calculated, and then the AE can be computed from this NPW.

Example 8.2 Annual Equivalent Worth—Repeating Cash Flow Cycles

SOLEX Company is producing electricity directly from a solar source by using a large array of solar cells and selling the power to the local utility company. SOLEX decided to use amorphous silicon cells because of their low initial cost, but these cells degrade over time, thereby resulting in lower conversion efficiency and power output. The cells must be replaced every 4 years, which results in a particular cash flow pattern that repeats itself as shown in Figure 8.2. Determine the annual equivalent cash flows at $i = 12\%$.

Solution

Given: Cash flows in Figure 8.2, $i = 12\%$
Find: Annual equivalent benefit

To calculate the AE, we need only consider one cycle over its 4-year period. For $i = 12\%$, we first obtain the NPW for the first cycle as follows:

$$
\begin{aligned}
PW(12\%) &= -\$1,000,000 \\
&\quad + [(\$800,000 - \$100,000(A/G, 12\%, 4)](P/A, 12\%, 4) \\
&= -\$1,000,000 + \$2,017,150 \\
&= \$1,017,150.
\end{aligned}
$$

Then, we calculate the AE value over the 4-year life cycle:

$$
\begin{aligned}
AE(12\%) &= \$1,017,150(A/P, 12\%, 4) \\
&= \$334,880.
\end{aligned}
$$

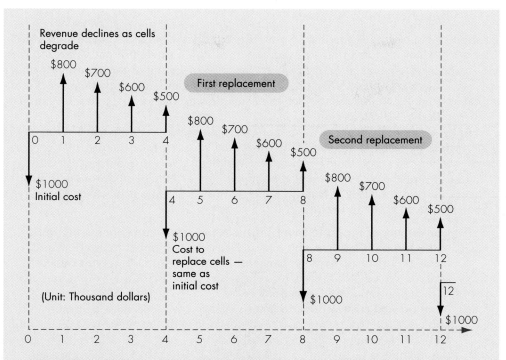

Figure 8.2   Conversion of repeating cash flow cycles into an equivalent annual payment (Example 8.2)

We can now say that the two cash flow series are equivalent:

Original Cash Flows			Annual Equivalent Flows	
n	A_n		n	A_n
0	−$1,000,000	≡	0	0
1	800,000		1	$334,880
2	700,000		2	334,880
3	600,000		3	334,880
4	500,000		4	334,880

We can extend this cash flow equivalency over the remaining cycles of the cash flow. The reasoning is that each similar set of five values (one disbursement and four receipts) is equivalent to four annual receipts of $334,880 each.

8.1.1 Benefits of AE Analysis

Example 8.1 should look familiar to you: It is exactly the situation we encountered in Chapter 4 when we converted a mixed cash flow into a single present value and then into a series of equivalent cash flows. In the case of Example 8.1, you may wonder why we bother to convert NPW to AE at all, since we already know the project is acceptable from NPW analysis. In fact, the example was mainly an exercise to familiarize you with the AE calculation.

However, in the real world, a number of situations can occur in which AE analysis is preferred, or demanded, over NPW analysis. Consider, for example, that corporations issue annual reports and develop yearly budgets. For these purposes, a company may find it more useful to present the annual cost or benefit of an ongoing project rather than its overall cost or benefit. Some additional situations in which AE analysis is preferred include the following:

1. **Consistency of report formats.** Financial managers more commonly work with annual rather than with overall costs in any number of internal and external reports. Engineering managers may be required to submit project analyses on an annual basis for consistency and ease of use by other members of the corporation and stockholders.
2. **Need for unit costs/profits.** In many situations, projects must be broken into unit costs (or profits) for ease of comparison with alternatives. Make-or-buy and reimbursement analyses are key examples, and these will be discussed in this chapter.
3. **Unequal project lives.** As we saw in Chapter 7, comparing projects with unequal service lives is complicated by the need to determine the lowest common multiple life. For the special situation of an indefinite service period and replacement with identical projects, we can avoid this complication by use of AE analysis. This situation will also be discussed in more detail in this chapter.

8.1.2 Capital Costs versus Operating Costs

When only costs are involved, the AE method is sometimes called the **annual equivalent cost** method. In this case, revenues must cover two kinds of costs: **Operating costs** and **capital costs**. Operating costs are incurred by the operation of physical plant or equipment needed to provide service—examples include items such as labor and raw materials. Capital costs are incurred by purchasing assets to be used in production and service. Normally, capital costs are nonrecurring (i.e., one-time costs), whereas operating costs recur for as long as an asset is owned.

Because operating costs recur over the life of a project, they tend to be estimated on an annual basis anyway, so for the purposes of annual equivalent cost analysis, no special calculation is required. However, because capital costs tend to be one-time costs, in conducting an annual equivalent cost analysis, we must translate this one-time cost into its annual equivalent over the life of the project. The annual equivalent of a capital cost is given a special name: **capital recovery cost**, designated $CR(i)$.

Two general monetary transactions are associated with the purchase and eventual retirement of a capital asset: its initial cost (I) and its salvage value (S). Taking into account these sums, we calculate the capital recovery factor as follows:

$$CR(i) = I(A/P, i, N) - S(A/F, i, N). \tag{8.2}$$

Recall algebraic relationships between factors in Table 4.3, and notice that the $(A/F, i, N)$ factor can be expressed as

$$(A/F, i, N) = (A/P, i, N) - i.$$

Then, we may rewrite the $CR(i)$ as

$$\begin{aligned} CR(i) &= i(A/P, i, N) - S[(A/P, i, N) - i] \\ &= (i - S)(A/P, i, N) + iS. \end{aligned} \tag{8.3}$$

We may interpret this situation thus: To obtain the machine, one borrows a total of I dollars, S dollars of which are returned at the end of the Nth year. The first term $(I - S)(A/P, i, N)$ implies that the balance $(I - S)$ will be paid back in equal installments over the N-year period at a rate of i, and the second term iS implies that simple interest in the amount iS is paid on S until it is repaid (Figure 8.3). Thus, the amount to be financed is $I - S(P/F, i, N)$, and the installments of this loan over the N-period are

$$\begin{aligned} AE(i) &= -[I - S(P/F, i, N)](A/P, i, N) \\ &= -I(A/P, i, N) + S(P/F, i, N)(A/P, i, N) \\ &= -[I(A/P, i, N) - S(A/F, i, N)] \\ &= -CR(i). \end{aligned} \tag{8.4}$$

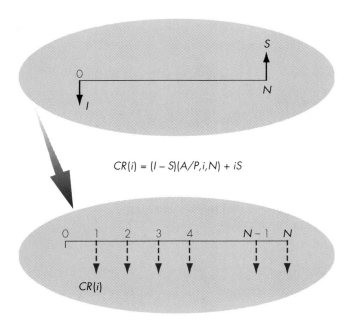

$CR(i) = (I - S)(A/P, i, N) + iS$

Figure 8.3   Capital recovery (ownership) cost calculation

Therefore, the $CR(i)$ tells us what the bank would charge each year. Many auto leases are based on this arrangement in that most require a guarantee of S dollars in salvage. From an industry viewpoint, $CR(i)$ is the annual cost to the firm of owning the asset.

With this information, the amount of annual savings required to recover the capital and operating costs associated with a project can be determined. As an illustration, consider Example 8.3.

Example 8.3 Annual Equivalent Worth—Capital Recovery Cost

Consider a machine that costs $20,000 and has a 5-year useful life. At the end of the 5 years, it can be sold for $4,000 after tax adjustment. If the firm could earn an after-tax revenue of $4,400 per year with this machine, should it be purchased at an interest rate of 10%? (All benefits and costs associated with the machine are accounted for in these figures.)

Solution

Given: $I = \$20,000$, $S = \$4,000$, $A = \$4,400$, $N = 5$ years, $i = 10\%$ per year

Find: AE, and determine whether to purchase the machine or not

We will compute the capital costs in two different ways:

Method 1: First compute the NPW of the cash flows and then compute the AE from the calculated NPW:

$$PW(10\%) = -\$20,000 + \$4,400(P/A, 10\%, 5)$$
$$+ \$4,000(P/F, 10\%, 5)$$
$$= -\$20,000 + \$4,400(3.7908) + \$4,000(0.6209)$$
$$= -\$836.88$$
$$AE(10\%) = -\$836.88(A/P, 10\%, 5) = -\$220.76$$

This negative AE value indicates that the machine does not generate sufficient revenue to recover the original investment so we may reject the project. In fact, there will be an equivalent loss of $220.76 per year over the machine's life (Figure 8.4a).

Method 2: The second method is to separate cash flows associated with the asset acquisition and disposal from the normal operating cash flows. Since the operating cash flows—the $4,400 yearly income—are already given in equivalent annual flows $(AE(i)_2)$, we only need to convert the cash flows associated with asset acquisition and disposal into equivalent annual flows $(AE(i)_1)$ (Fig. 8.4b). Using Eq. (8.3),

$$CR(i) = (I - S)(A/P, i, N) + iS$$
$$AE(i)_1 = -CR(i)$$
$$= -[(\$20,000 - \$4,000)(A/P, 10\%, 5) + (0.10)\$4,000]$$
$$= -\$4,620.76$$

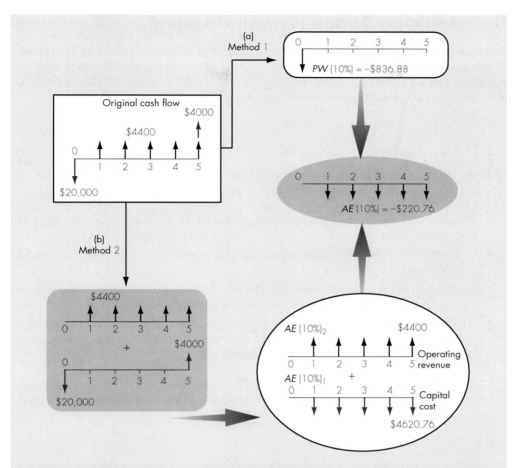

Figure 8.4 Alternative ways of computing capital recovery cost for an investment (Example 8.3)

$$AE(i)_2 = \$4,400$$

$$AE(10\%) = AE(i)_1 + AE(i)_2$$

$$= -\$4,620.76 + \$4,400$$

$$= -\$220.76.$$

Comments: Obviously, Method 2 saves a calculation step, so we may prefer it over Method 1. We may interpret Method 2 as determining that the annual operating benefits must be at least $4,620.76 to recover the asset cost. However, the annual operating benefits actually amount to only $4,400, resulting in a loss of $220.76 per year. Therefore, the project is not worth undertaking.

8.2 Applying Annual Worth Analysis

In general, most engineering economic analysis problems can be solved by the present worth methods that were introduced in Chapter 7. However, some economic analysis problems can be solved more efficiently by annual worth analysis. In this section, we introduce several applications that call for annual worth analysis.

8.2.1   Unit Profit/Cost Calculation

In many situations, we need to know the unit profit (or cost) of operating an asset. To obtain a unit profit (or cost), we may proceed as follows:

- Determine the number of units to be produced (or serviced) each year over the life of the asset.
- Identify the cash flow series associated with production or service over the life of the asset.
- Calculate the net present worth of the project cash flow series at a given interest rate and then determine the equivalent annual worth.
- Divide the equivalent annual worth by the number of units to be produced or serviced during each year. When the number of units varies each year, you may need to convert them into equivalent annual units.

To illustrate the procedure, we will consider Example 8.4 where the annual equivalent concept is useful in estimating the savings per machine hour for a proposed machine acquisition.

Example 8.4 Unit Profit per Machine Hour—Where Annual Operating Hours Remain Constant

Consider the investment in the metal-cutting machine in Example 7.5. Recall that this 3-year investment was expected to generate a NPW of $3,553. Suppose that the machine will be operated for 2000 hours per year. Compute the equivalent savings per machine hour at $i = 15\%$.

Solution

Given: NPW = $3,553, $N = 3$ years, $i = 15\%$ per year, 2,000 machine hours per year

Find: Equivalent savings per machine hour

We first compute the annual equivalent savings from the use of the machine. Since we already know the NPW of the project, we obtain the AE by

$$AE(15\%) = \$3,553(A/P, 15\%, 3) = \$1,556.$$

With an annual usage of 2,000 hours, the equivalent savings per machine hour would be

$$\text{Savings per machine hour} = \$1,556/2,000 \text{ hours} = \$0.78/\text{hour}.$$

Comments: Note that we cannot simply divide the NPW amount ($3,553) by the total number of machine hours over the 3-year period (6,000 hours), or $0.59/hour. This $0.59 figure represents the instant savings in present worth for each hourly use of the equipment, but does not consider the time over which the savings occur. Once we have the annual equivalent worth, we can divide by the desired time unit if the compounding period is 1 year. If the compounding period is shorter, then the equivalent worth should be calculated for the compounding period.

Example 8.5 Unit Profit per Machine Hour—Where Annual Operating Hours Fluctuate

Reconsider Example 8.4 and suppose that the metal cutting machine will be operated according to varying hours: the first year—1,500 hours; the second year—2,500 hours; and third year—2,000 hours. The total operating hours still remain at 6,000 over 3 years. Compute the equivalent savings per machine hour at $i = 15\%$.

Solution

Given: NPW = $3,553, N = 3 years, i = 15% per year, operating hours—1,500 hours first year; 2,500 hours second year; and 2,000 hours third year

Find: Equivalent savings per machine hour

As calculated in Example 8.4, the annual equivalent savings is $1,556. Let C denote the equivalent annual savings per machine hour that needs to be determined. Now, with varying annual usages of the machine, we can set up the equivalent annual savings as a function of C:

$$
\begin{aligned}
\text{Equivalent annual savings} &= [C(1,500)(P/F, 15\%, 1) \\
&\quad + C(2,500)(P/F, 15\%, 2) \\
&\quad + C(2,000)(P/F, 15\%, 3)](A/P, 15\%, 3) \\
&= 1,975.16C.
\end{aligned}
$$

We can equate this amount to $1,556 in Example 8.4 and solve for C. This gives us

$$C = \$1,556/1,975.16 = \$0.79/\text{hour}$$

which is about a penny more than the situation in Example 8.4.

8.2.2 Make-or-Buy Decision

Make-or-buy problems are among the most common of business decisions. At any given time, a firm may have the option of either buying an item or producing it. Unlike the make-or-buy situation we looked at in Section 3.4.1, *if either the "make" or the "buy" alternative requires the acquisition of machinery and/or equipment, then it becomes an investment decision.* Since the cost of an outside service (the "buy" alternative) is usually quoted in terms of dollars per unit, it is easier to compare the two alternatives if the differential costs of the "make" alternative are also given in dollars per unit. This unit cost comparison requires the use of annual worth analysis. The specific procedure is as follows:

Step 1: Determine the time span (planning horizon) for which the part (or product) will be needed.

Step 2: Determine the annual quantity of the part (or product).

Step 3: Obtain the unit cost of purchasing the part (or product) from the outside firm.

Step 4: Determine the equipment, manpower, and all other resources required to make the part (or product).

Step 5: Estimate the net cash flows associated with the "make" option over the planning horizon.

Step 6: Compute the annual equivalent cost of producing the part (or product).

Step 7: Compute the unit cost of making the part (or product) by dividing the annual equivalent cost by the required annual volume.

Step 8: Choose the option with the minimum unit cost.

Example 8.6 Equivalent Worth—Make or Buy

Ampex Corporation currently produces both videocassette cases and metal particle magnetic tape for commercial use. An increased demand for metal particle tapes is projected, and Ampex is deciding between increasing the internal production of empty cassette cases and magnetic tape or purchasing empty cassette cases from an outside vendor. If Ampex purchases the cases from a vendor, the company must also buy specialized equipment to load the magnetic tapes, since its current loading machine is not compatible with the cassette cases produced by the vendor under consideration. The projected production rate of cassettes is 79,815 units per week for 48 weeks of operation per year. The planning horizon is 7 years. After considering the effects of income taxes, the accounting department has itemized the annual costs associated with each option as follows:

- Make Option (annual costs):

Labor	$1,445,633
Materials	$2,048,511
Incremental overhead	$1,088,110
Total annual cost	$4,582,254

- Buy Option:

Capital expenditure	
Acquisition of a new loading machine	$ 405,000
Salvage value at end of 7 years	$ 45,000
Annual Operating Costs	
Labor	$ 251,956
Purchasing empty cassette ($0.85/unit)	$3,256,452
Incremental overhead	$ 822,719
Total annual operating costs	$4,331,127

(Note the conventional assumption that cash flows occur in discrete lumps at the ends of years, as shown in Figure 8.5) Assuming that Ampex's MARR is 14%, calculate the unit cost under each option.

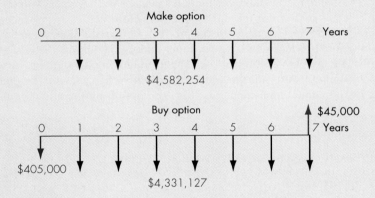

Figure 8.5   Make-or-buy analysis (Example 8.6)

Solution

Given: Cash flows for two options, $i = 14\%$

Find: Unit cost for each option, and which one is preferred

The required annual production volume is

$$79,815 \text{ units/week} \times 48 \text{ weeks} = 3,831,120 \text{ units per year.}$$

We now need to calculate the annual equivalent cost under each option.

- Make option: Since the "make option" is already given on an annual basis, the equivalent annual cost will be

$$AE(14\%)_{\text{Make}} = -\$4,582,254.$$

- Buy option: The two cost components are capital cost and operating cost.

  Capital cost:

  $$
  \begin{aligned}
  CR(14\%) &= (\$405,000 - \$45,000)(A/P, 14\%, 7) \\
  &\quad + (0.14)(\$45,000) \\
  &= \$90,249 \\
  AE(14\%)_1 &= -CR(14\%) = -\$90,249.
  \end{aligned}
  $$

  Operating cost:

  $$
  AE(14\%)_2 = -\$4,331,127.
  $$

  Total annual equivalent cost:

  $$
  AE(14\%)_{\text{Buy}} = AE(14\%)_1 + AE(14\%)_2 = -\$4,421,376.
  $$

Obviously, this annual equivalent calculation indicates that Ampex would be better off buying cassette cases from the outside vendor. However, Ampex wants to know the unit costs in order to set a price for the product. In this situation, we need to calculate the unit cost of producing the cassette tapes under each option. (The negative sign indicates that AE is a cash outflow or cost. When comparing the options, we are interested in the magnitude costs and can ignore the negative sign.) Do this by dividing the magnitude of the annual equivalent cost for each option by the annual quantity required:

- Make Option:

  Unit cost = $\$4,582,254/3,831,120 = \$1.20/\text{unit}$.

- Buy Option:

  Unit cost = $\$4,421,376/3,831,120 = \$1.15/\text{unit}$.

Buying the empty cassette cases from the outside vendor and loading the tape in-house will save Ampex 5 cents per cassette before any tax consideration.

Comments: Two important noneconomic factors should also be considered. The first is the question of whether the quality of the supplier's component is better than (or equal to) or worse, than the component the firm is presently manufacturing. The second is the reliability of the supplier in terms of providing the needed quantities of the cassette cases on a timely basis. A reduction in quality or reliability should virtually always rule out a switch from making to buying.

8.2.3 Break-Even Point: Cost Reimbursement

Companies often need to calculate the cost of equipment that corresponds to a **unit of use** of that equipment. A familiar example is an employer's reimbursement of costs for the use of an employee's personal car for business purposes. If an employee's job depends on obtaining and using a personal vehicle on the employer's behalf, reimbursement on the basis of the employee's overall costs per mile seems fair. Although many car owners think of costs in terms of outlays for gasoline, oil, tires, and tolls, a careful examination shows that, in addition to these operating costs, which are directly related to the use of the car, there are also **ownership costs**, which occur whether or not the vehicle is driven.

Ownership costs include depreciation, insurance, finance charges, registration and titling fees, scheduled maintenance, accessory costs, and garaging (storage). Even if the vehicle is permanently garaged, a portion of each of these costs occurs. **Depreciation** is the loss in value of the vehicle during the time it is owned due to (1) the passage of time, (2) its mechanical and physical condition, and (3) the number of miles it is driven. This type of depreciation, known as **economic depreciation** (as opposed to **accounting depreciation**), is discussed in Chapter 10.

Operating costs include nonscheduled repairs and maintenance, gasoline, oil, tires, parking and tolls, and taxes on gasoline and oil. Certainly, the more a car is used the greater these costs become.

Once the cost of owning and operating a personal vehicle is determined, you may wonder what the minimum reimbursement rate per mile allows you to break even. Here we are looking at the reimbursement cost equation, which is solely determined by a decision variable, reimbursement rate. The reimbursement rate exactly equal to the cost of owning and operating the vehicle is known as the **break-even point**. Example 8.7 illustrates how you obtain the cost of reimbursement for the use of an employee's vehicle for business purposes.

Example 8.7 Break-Even Point—per Unit of Equipment Use

Sam Tucker is a sales engineer at Buford Chemical Engineering Company. Sam owns two vehicles, and one of them is entirely dedicated to business use. His business car is a used small pick-up truck, which he purchased with personal savings for $11,000. Based on his own records, and data compiled by the U.S. Department of Transportation, Sam has estimated the costs of owning and operating his business vehicle for the first 3 years as follows in the table at the top of the next page. Sam expects to drive 14,500, 13,000, and 11,500 business miles, respectively, for the next 3 years. If his interest rate is 6%, what should be Sam's reimbursement rate per mile so that he can break even?

Discussion: You may wonder why the initial cost of the vehicle ($11,000) is not explicitly considered in the costs Sam has estimated. The answer is in the depreciation listings. In economic analysis, capital costs are not considered all at once in the year in which they are incurred. Rather, the costs are spread out over the useful life of the

	First Year	Second Year	Third Year
Depreciation	$2,879	$1,776	$1,545
Scheduled maintenance	100	153	220
Insurance	635	635	635
Registration and taxes	78	57	50
Total ownership cost	$3,692	$2,621	$2,450
Nonscheduled repairs	35	85	200
Replacement tires	35	30	27
Accessories	15	13	12
Gasoline and taxes	688	650	522
Oil	80	100	100
Parking and tolls	135	125	110
Total operating costs	$988	$1,003	$971
Total of all costs	$4,680	$3,624	$3,421
Expected miles driven	14,500 miles	13,000 miles	11,500 miles

asset, based on how much of the cost of the asset is used up each year. Thus, the depreciation amount of $2,879 during the first year represents the fact that, if the pickup truck were bought for $11,000 and then sold at the end of the first year, when it had been driven 14,500 miles, Sam could expect the sale price to be $2879 less than the original purchase price. In Chapter 10, we will explore depreciation in more detail and learn conventions for its calculation. For this example, depreciation amounts more closely represent the major cost of owning the truck.

Solution

Given: Yearly costs and mileages, $i = 6\%$ per year

Find: Equivalent cost per mile

Suppose Buford pays Sam X per mile for his personal car. Assuming that Sam expects to travel 14,500 miles the first year, 13,000 miles the second year, and 11,500 miles the third year, his annual reimbursements would be

Year	Total Miles Driven	Reimbursement ($)
1	14,500	$(X)(14,500) = 14,500X$
2	13,000	$(X)(13,000) = 13,000X$
3	11,500	$(X)(11,500) = 11,500X$

As depicted in Figure 8.6, the annual equivalent reimbursement would be

$$[14{,}500X(P/F, 6\%, 1) + 13{,}000X(P/F, 6\%, 2)$$
$$+ 11{,}500X(P/F, 6\%, 3)] (A/P, 6\%, 3) = 13{,}058X.$$

The annual equivalent costs of owning and operating would be

$$[\$4{,}680(P/F, 6\%, 1) + \$3{,}624(P/F, 6\%, 2)$$
$$+ \$3{,}421(P/F, 6\%, 3)] (A/P, 6\%, 3) = \$3{,}933.$$

Then, the minimum reimbursement rate should be

$$13{,}058X = \$3{,}933$$
$$X = 30.12 \text{ cents per mile.}$$

If Buford pays him 30.12 cents per mile or more, Sam's decision to use his car for business makes sense economically.

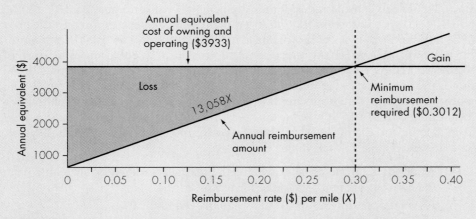

Figure 8.6   Annual equivalent reimbursement as a function of cost per mile (Example 8.7)

8.3 Mutually Exclusive Projects

In this section, we will consider a situation where two or more mutually exclusive alternatives need to be compared based on annual equivalent worth. In Section 7.5, we discussed the general principles that should be applied when mutually exclusive alternatives with unequal service lives were compared. The same general principles should be applied when comparing mutually exclusive alternatives based on annual equivalent worth—mutually exclusive alternatives in equal time spans must be compared. Therefore, we must give a careful consideration of the time period covered by the analysis, the **analysis period**. We will consider two situations: (1) The analysis period equals project lives, and (2) the analysis period differs from project lives.

8.3.1 Analysis Period Equals Project Lives

Let's begin our comparison with a simple situation where project lives equal the analysis period. In this situation, we compute the AE value for each project and select the one that has the least negative AE for service projects (or the largest AE value for revenue projects).

In many situations, we need to compare a set of different design alternatives where each design would produce the same number of units (constant revenues), but would require different amounts of investment and operating costs (because of different degrees of mechanization). Example 8.8 illustrates the use of the annual equivalent cost concept to compare the cost of operating a conventional electric motor with that of a premium-efficiency motor in a strip-processing mill.

Example 8.8 How Premium Efficiency Motors Can Cut Your Electric Costs

Birmingham Steel Inc. is considering replacing 20 conventional, 25 HP, 230 V, 60 Hz, 1800 rpm induction motors in its plant with modern premium efficiency (PE) motors. Both types of motors have a power outputs of 18.650 kW per motor (25 HP × 0.746 kW/HP). Conventional motors have a published efficiency of 89.5% while the PE motors are 93% efficient. The initial cost of the conventional motor is $13,000 while the initial cost of the proposed PE motors is $15,600. The motors are operated 12 hours per day, 5 days per week, 52 weeks per year, with a local utility cost of $0.07 per kilowatt-hour (kWh). The motors are to be operated at 75% load, and the life cycle of both the conventional and PE motor is 20 years, with no appreciable salvage value.

(a) At an interest rate of 13%, what is the savings per kWh by switching from the conventional motors to the PE motors?

(b) At what operating hours are they equally economical?

Discussion: Whenever we compare machines with different efficiency ratings, we need to determine the input powers required to operate the machines. Since percent efficiency is equal to the ratio of output power to input power, we can determine the input power by dividing the output power by the motor's percent efficiency:

$$\text{Input power} = \frac{\text{output power}}{\text{percent efficiency}}$$

For example, a 30 HP motor with 90% efficiency will require an input power of

$$\text{Input power} = \frac{(30 \text{ HP} \times 0.746 \text{ kW/HP})}{0.90}$$

$$= 24.87 \text{ kW.}$$

Once we determine the input power requirement and the number of operating hours, we can convert it into equivalent energy cost (power cost). Although the company needs 20 motors, we can compare them based on a single unit.

Solution

Given: $P = \$15,600$, $S = 0$, $N = 20$ years, $i = 13\%$, rated power output = 18.650 kW, efficiency rating = 93%, utility rate = $0.07/kWh, operating hours = 3,120 hours per year, 20 motors required

Find: (a) Cost of operating the PE motor per hour and (b) break-even operating hours

(a) Operating cost per kWh per unit
- Determine total input power for both motor types:
  Conventional motor:

$$\text{Input power} = \frac{18.650 \text{ kW}}{0.895} = 20.838 \text{ kW.}$$

  PE motor:

$$\text{Input power} = \frac{18.650 \text{ kW}}{0.93} = 20.054 \text{ kW.}$$

  Note that each PE motor requires 0.784 kW less input power (or 15.68 kW for 20 motors), which results in energy savings.
- Determine total kWh per year. We assumed a total of 3,120 hours per year in motor operation.
  Conventional motor:

$$3,120 \text{ hrs/year} \times 20.838 \text{ kW} = 65,018 \text{ kWh/year.}$$

  PE motor:

$$3,120 \text{ hrs/year} \times 20.054 \text{ kW} = 62,568 \text{ kWh/year.}$$

- Determine annual energy costs for both motor types. Since the utility rate is $0.07/kWh, the annual energy cost for
  Conventional motor:

$$\$0.07/\text{kWh} \times 65,018 \text{ kWh/year} = \$4,551/\text{year.}$$

  PE motor:

$$\$0.07/\text{kWh} \times 62,568 \text{ kWh/year} = \$4,380/\text{year.}$$

- Determine capital costs for both types of motors. Recall that we assumed that the useful life for both motor types is 20 years. To determine the annualized capital cost at 13%, we use the capital recovery factor.

Conventional motor:

$$(\$13,000)(A/P, 13\%, 20) = \$1,851.$$

PE motor:

$$(\$15,600)(A/P, 13\%, 20) = \$2,221.$$

- Determine the total equivalent annual cost which is equal to the capital cost plus the annual energy cost. Then calculate the unit cost per kWh based on output power. Note that the total output power is 58,188 kWh per year (25 HP × 0.746 kW/HP × 3120 hours/year).

  Conventional motor:

$$AE(13\%) = \$4,551 + \$1,851 = \$6,402$$
$$\text{Cost per kWh} = \$6,402/58,188 \text{ kWh} = \boxed{11 \text{ cents/kWh.}}$$

  PE motor:

$$AE(13\%) = \$4,380 + \$2,221 = \$6,601$$
$$\text{Cost per kWh} = \$6,601/58,188 \text{ kWh} = \boxed{11.34 \text{ cents/kWh.}}$$

  Clearly, conventional motors are cheaper to operate if the motors are expected to run only 3120 hours per year.

- Determine the savings (losses) per operating hour from switching from conventional to PE motors.

  Additional capital cost required from switching from conventional to PE:

$$\text{Incremental capital cost} = \$2,221 - \$1,851 = \$370.$$

  Additional energy cost savings from switching from conventional to PE:

$$\text{Incremental energy savings} = \$4,551 - \$4,380 = \$171.$$

  At 3,120 annual operating hours, it will cost the company an additional $370, but energy savings are only $171, which results in a $199 loss from each motor. In other words, for each operating hour, you lose about 6.38 cents.

(b) Break-even operating hours
  - Determine the minimum running time required to make the PE motor economical. It is always important to determine the number of hours the motor will be running. For example, it would not be economical to install a PE motor that runs only 3,120 hours per year. Would that result change if the same motor were to operate 5,000 hours per year? Clearly, if a motor is to run around-the-clock, the savings in kWh would result in a substantial annual savings in electricity bills, which are an operating cost. As we calculate the annual equivalent cost by varying the operating hours, we obtain the situation in Figure 8.7.[1] Observe that, if Birmingham Steel uses the PE motors for more than 6,742 hours annually, replacing the conventional motors with the PE motors would be justified.

---

[1] Figure 8.7 is also known as a **sensitivity graph**. These graphics will be discussed in detail in Chapter 14.

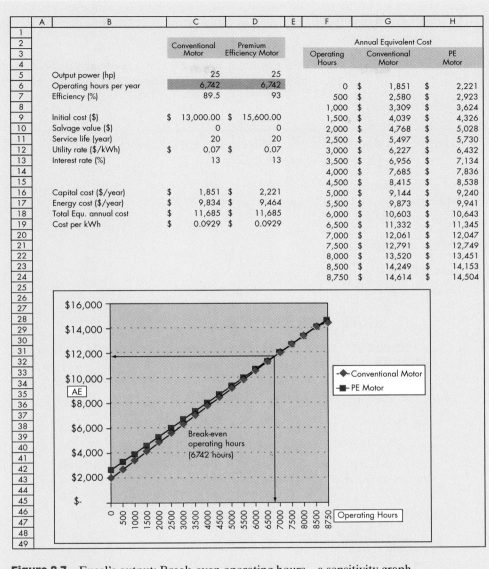

The spreadsheet shown in the figure contains the following data:

		Conventional Motor	Premium Efficiency Motor		Operating Hours	Conventional Motor	PE Motor
						Annual Equivalent Cost	
Output power (hp)		25	25		0	$ 1,851	$ 2,221
Operating hours per year		6,742	6,742		500	$ 2,580	$ 2,923
Efficiency (%)		89.5	93		1,000	$ 3,309	$ 3,624
					1,500	$ 4,039	$ 4,326
Initial cost ($)	$	13,000.00	$ 15,600.00		2,000	$ 4,768	$ 5,028
Salvage value ($)		0	0		2,500	$ 5,497	$ 5,730
Service life (year)		20	20		3,000	$ 6,227	$ 6,432
Utility rate ($/kWh)	$	0.07	$ 0.07		3,500	$ 6,956	$ 7,134
Interest rate (%)		13	13		4,000	$ 7,685	$ 7,836
					4,500	$ 8,415	$ 8,538
Capital cost ($/year)	$	1,851	$ 2,221		5,000	$ 9,144	$ 9,240
Energy cost ($/year)	$	9,834	$ 9,464		5,500	$ 9,873	$ 9,941
Total Equ. annual cost	$	11,685	$ 11,685		6,000	$ 10,603	$ 10,643
Cost per kWh	$	0.0929	$ 0.0929		6,500	$ 11,332	$ 11,345
					7,000	$ 12,061	$ 12,047
					7,500	$ 12,791	$ 12,749
					8,000	$ 13,520	$ 13,451
					8,500	$ 14,249	$ 14,153
					8,750	$ 14,614	$ 14,504

Figure 8.7   Excel's output: Break-even operating hours—a sensitivity graph

8.3.2   Analysis Period Differs from Project Lives

In Section 7.5.3, we learned that, in present worth analysis, we must have a common analysis period (least common multiple periods) when mutually exclusive alternatives are compared. Annual worth analysis also requires establishing common analyses periods, but AE offers some computational advantages as opposed to present worth analysis, provided the following criteria are met:

1. The service of the selected alternative is required on a continuous basis.
2. Each alternative will be replaced by an *identical* asset that has the same costs and performance.

When these two criteria are in effect, we may solve for the AE of each project based on its initial life span, rather than on the lowest common multiple of the projects' lives.

Example 8.9 Annual Equivalent Cost Comparison— Unequal Project Lives

Reconsider Example 7.14 where we compared two types of equipment with unequal service lives. Apply the annual equivalent approach to select the most economical equipment.

Solution

Given: Cost cash flows shown in Figure 8.8, $i = 15\%$ per year

Find: AE cost, and which is the preferred alternative

An alternative procedure for solving Example 7.14 is to compute the annual equivalent cost of an outlay of \$12,500 for model A every 3 years and the annual equivalent cost of an outlay of \$15,000 for model B every 4 years. Notice that the AE of each 12-year cash flow is the same as that of the corresponding 3- or 4-year cash flow, (Figure 8.8). From Example 8.14, we calculate

- Model A:
  For a 3-year life:

$$PW(15\%) = -\$22,601$$
$$AE(15\%) = -22,601(A/P, 15\%, 3)$$
$$= -\$9,899.$$

  For 12-year period (computed for the entire analysis period):

$$PW(15\%) = -\$53,657$$
$$AE(15\%) = -53,657(A/P, 15\%, 12)$$
$$= -\$9,899.$$

- Model B:
  For a 4-year life:

$$PW(15\%) = -\$25,562$$
$$AE(15\%) = -\$25,562(A/P, 15\%, 4)$$
$$= -\$8,954.$$

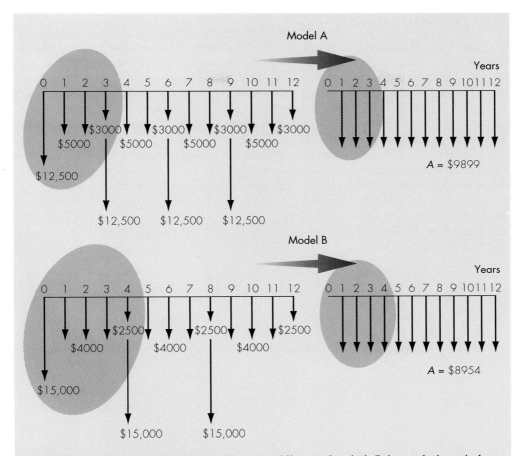

Figure 8.8 Comparison of projects with unequal lives and an indefinite analysis period using the equivalent annual worth criterion (Example 8.9)

For a 12-year period (computed for the entire analysis period):

$$PW(15\%) = -\$48,534$$
$$AE(15\%) = -\$48,534(A/P, 15\%, 12)$$
$$= -\$8,954.$$

Notice that the annual equivalent values that were calculated based on the common service period are the same as those that were obtained over their initial life spans. Thus, for alternatives with unequal lives, we will obtain the same selection by comparing NPW over a common service period using repeated projects or comparing AE for initial lives.

8.4 Design Economics

Engineers are frequently involved in making design decisions that provide the required functional quality at the lowest cost. For example, GM engineers used the "design for manufacturability" discipline to trim part counts for the front suspension design on the 1992 Cadillac Seville. This redesign eliminates two parts and 68 seconds of assembly time. For the overall real suspension, 50 parts were trimmed, and assembly time was cut from nearly 19 minutes to under 6 minutes! Redesigns in the suspension alone yielded savings of over $2 million.

Another valuable extension of AE analysis is minimum-cost analysis, which is used to determine optimal engineering designs. The AE analysis method is useful when two or more cost components are affected differently by the same design element, i.e., for a single design variable, some costs may increase while others decrease. When the equivalent annual total cost of a design variable is a function of increasing and decreasing cost components, we can usually find the optimal value that will minimize its cost as follows:

$$AE(i) = a + bx + \frac{c}{x}, \tag{8.5}$$

where x is a common design variable, and a, b, and c are constants.

To find the value of the common design variable that minimizes the $AE(i)$, we need to take the first derivative, equate the result to zero, and solve for x:

$$\frac{dAE(i)}{dx} = b - \frac{c}{x^2}$$

$$= 0 \tag{8.6}$$

$$x^* = \sqrt{\frac{c}{b}}.$$

The logic of the first-order requirement is that an activity should, if possible, be carried to a point where its **marginal yield** ($dAE(i)/dx$) is zero. However, to be sure whether we have found a maximum or a minimum when we have located a point whose marginal yield is zero, we must examine it in terms of what are called the **second-order conditions**. Second-order conditions are equivalent to the usual requirements that the second derivative be negative in the case of a maximum, and positive for a minimum. In our situation,

$$\frac{d^2AE(i)}{dx^2} = \frac{2C}{x^3}.$$

As long as $C > 0$, the second derivative will be positive, indicating that the value x^* is the minimum cost point for the design alternative. To illustrate the optimization concept, two examples are provided. The first example is about designing the optimal cross-sectional area of a conductor, and the second example is about selecting an optimal size for a pipe.

Example 8.10 Optimal Cross-Sectional Area

A constant electric current of 5,000 amps is to be transmitted a distance of 1,000 feet from a power plant to a substation for 24 hours a day, 365 days a year. A copper conductor can be installed for $8.25 per pound. The conductor will have an estimated life of 25 years and a salvage value of $0.75 per pound. Power loss from the conductors is inversely proportional to the cross-sectional area (A) of the conductors. It is known that the resistance of conductors is 0.8145×10^{-5} ohms for one square inch per foot, cross section. The cost of energy is $0.05 per kilowatt-hour, the interest rate is 9%, and the density of copper is 555 lb/ft^3. For the given data, calculate the optimum cross-sectional area (A) of the conductor.

Discussion: The resistance of transmission-line conductors is the most important cause of power loss in a transmission line. The resistance of an electrical conductor varies with its length and inversely with its cross-sectional area:

$$R = \rho(L/A), \tag{8.7}$$

where R = resistance of the conductor, L = length of the conductor, A = cross-sectional area of the conductor, r = resistivity of the conductor material.

   Any consistent set of units may be used. In power work in the United States, L is usually given in feet, A in circular mils (cmil), and ρ in ohm-circular mils per foot.

   A circular mil is the area of a circle that has a diameter of 1 mil. A mil is equal to 1×10^{-3} in. The cross-sectional area of a wire in square inches equals its area in circular mils multiplied by 0.7854×10^{-6}. More specifically, one unit relates to another as follows:

1 linear mil	=	0.001 in.
	=	0.0254 milimeter.
1 circular mil	=	area of circle 1 linear mil in diameter
	=	$(0.5)^2 \, \pi \, \text{mils}^2$
	=	$0.7854 \times 10^{-6} \, \text{in.}^2$
1 in.^{2}	=	$1/(0.7854 \times 10^{-6})$
	=	1.27324×10^6 cmil.

In SI units (the official designation for the Systèm International d'unités system of units), L is in meters, A in square meters, and ρ in ohm-meters. In terms of the SI units, the copper conductor has a ρ value of 1.7241×10^{-8} Ω-meter. We can easily convert this value in the unit of Ω-in.^{2} per foot. From Eq. (8.7), solving for ρ yields

$$\rho = R.A/L = 1.7241 \times 10^{-8} \, \Omega (1\text{meter})^2/1 \text{ meter}$$

$$= 1.7241 \times 10^{-8} \, \Omega (39.37 \text{ in})^2/3.2808 \text{ ft}$$

$$= 0.8145421 \times 10^{-5} \, \Omega \text{in}^2/\text{ft}.$$

When current flows through a circuit, power is used to overcome the resistance. The unit of electrical work is the kilowatt hour (kWh), which is equal to the power in kilowatts multiplied by the hours during which work is performed. If the current (I) is steady, the charge passing through the wire in time T (or power so used converts to heat, which is known as energy loss) and which is equal to

$$\text{Power} = I^2RT/1000 \text{ kWh}, \tag{8.8}$$

where I = current in amperes, R = resistance in ohms, and T = time duration in hours.

Solution

Given: Cost components as a function of cross-sectional area (A), $N = 25$ years, $i = 9\%$
Find: Optimal A value

Step 1: This classic minimum-cost example, the design of the cross-sectional area of an electrical conductor, involves increasing and decreasing cost components. Since resistance is inversely proportional to the size of the conductor, energy loss will decrease as conductor size increases. More specifically, the energy loss in kilowatt-hours (kWh) in a conductor due to resistance is equal to

$$
\begin{aligned}
\text{Energy loss in kilowatt hours} \;&=\; \frac{I^2R}{1,000A}T \\[2mm]
&=\; \frac{(5,000^2)(0.008145)}{1,000A}(24 \times 365) \\[2mm]
&=\; \frac{1,783,755}{A} \text{ kWh.}
\end{aligned}
$$

Step 2: Since electrical resistance is inversely proportional to the area of the cross section (A), the total energy loss in dollars per year for a specified conductor material is

$$
\begin{aligned}
\text{Energy loss cost} \;&=\; \frac{1,783,755}{A}(\phi) \\[2mm]
&=\; \frac{1,783,755}{A}(\$0.05) \\[2mm]
&=\; \frac{\$89,188}{A},
\end{aligned}
$$

where ϕ = cost of energy in dollars per kWh.

Step 3: As we increase the size of the conductor, however, it costs more to build. First, we need to calculate the total amount of conductor material in pounds. Since the cross-sectional area is given in square inches, we need to convert the total length in feet to inches before finding the material weight:

$$\text{Material weight in pounds} = \frac{(1,000)(12)(555)A}{12^3}$$

$$= 3,854(A)$$

$$\text{Total material cost} = 3,854(A)(\$8.25)$$

$$= \$31,797(A).$$

Here, we are looking for the trade-off between the cost of installation and the cost of energy loss.

Step 4: Since the copper material will be salvaged at the rate of $0.75 per pound at the end of 25 years, we can compute the capital recovery cost as follows:

$$CR(9\%) = [31,797\ A - 0.75(3,854\ A)](A/P, 9\%, 25) + 0.75(3,854\ A)(0.09)$$

$$= 2,943\ A + 260\ A$$

$$= 3,203\ A.$$

Step 5: Using Eq. (8.5), we express the total annual equivalent cost as a function of a design variable (A) as follows:

$$\overset{\text{Capital cost}}{AE(9\%) = 3,203A + \underset{\text{Operating cost}}{\frac{89,188}{A}}}.$$

To find the minimum annual equivalent cost, we use the result of Eq. (8.6):

$$\frac{dAE(9\%)}{dA} = 3,203 - \frac{89,188}{A^2} = 0.$$

$$A^* = \sqrt{\frac{89,188}{3,203}}$$

$$= 5.276\ \text{in.}^2$$

The minimum annual equivalent total cost is

$$AE(9\%) = 3,203(5.276) + \frac{89,188}{5.276}$$

$$= \$33,802.$$

Figure 8.9 illustrates the nature of this design trade-off problem.

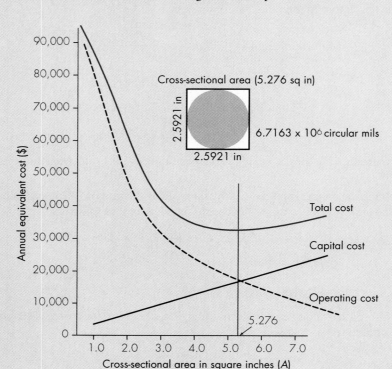

Figure 8.9 Optimal cross-sectional area for a copper conductor. Note that the minimum point almost coincides with the crossing point of the capital cost and operating cost lines. This is not always true. Since the cost components can have a variety of cost patterns, in general, the minimum point does not occur at the crossing point (Example 8.10).

Example 8.11 Economical Pipe Size

As a result of the 1990 conflict in the Persian Gulf, Kuwait is studying the feasibility of running a steel pipeline across the Arabian Peninsula to the Red Sea. The pipeline will be designed to handle 3 million barrels of crude oil per day under optimum conditions. The length of the line will be 600 miles. Calculate the optimum pipeline diameter that will be used for 20 years for the following data at $i = 10\%$:

$$\text{Pumping power} = \frac{1.333 \, Q \Delta P}{1,980,000} \, \text{HP}$$

Q = volume flow rate, cubic ft/hour

$$\Delta P = \frac{128Q\mu L}{g\pi D^4}, \text{ pressure drop lb/sq ft}$$

L = pipe length, ft

D = inside pipe diameter, ft

$t = 0.01\,D$, pipeline wall thickness, ft

μ = 8,500 lb/hour ft, oil viscosity

$g = 32.2 \times 12,960,000 \text{ ft/hour}^2$

Power cost, $0.015 per HP hour

Oil cost, $18 per barrel

Pipeline cost, $1.00 per pound of steel

Pump and motor costs, $195 per HP

The salvage value of the steel after 20 years is assumed to be zero because removal costs exhaust scrap profits from steel. (See Figure 8.10 for relationship between D and t.)

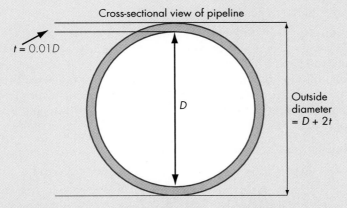

Figure 8.10   Designing economical pipe size to handle 3 million barrels of crude oil per day (Example 8.11)

Discussion: In general, when a progressively larger size pipe is used to carry a given fluid at a given volume flow rate, the energy required to move the fluid will progressively decrease. However, as we increase the pipe size, the cost of its construction will increase. In practice, to obtain the best pipe size for a particular situation, you may choose a reasonable, but small, starting size. Compute the energy cost of pumping fluid through this size and the total construction cost and then compare the difference in energy cost with the difference in construction cost. When the savings in energy cost exceeds the added construction cost, the process may be repeated with progressively larger pipe sizes until the added construction cost exceeds the savings in energy cost. As soon as this happens, the best pipe size to use in the particular application is identified. However, this search process can be simplified by using the minimum cost concept as explained by Eqs. (8.5) and (8.6).

Solution

Given: Cost components as a function of pipe diameter (D), $N = 20$ years, $i = 10\%$

Find: Optimal pipe size (D)

Step 1: Several different units are introduced; however, we need to work with common units. We will assume the following conversion units.

1 mile	=	5,280 ft
600 miles	=	$600 \times 5,280 = 3,168,000$ ft
1 barrel	=	42 U.S. gallons
1 barrel	=	42 gallons $\times$ 231 in^3/gallon = 9,702 in^3
1 barrel	=	9,702 in^3/12^3 = 5.6146 ft^3
Density of steel	=	490.75 lb/ft^3

Step 2: Power required to pump oil:

For any given set of operating conditions involving the flow of a noncompressible fluid, such as oil, through a pipe of constant diameter, it is well known that a small-diameter pipe will have a high fluid velocity and a high fluid pressure drop through the pipe. This will require a pump that will deliver a high discharge pressure and a motor with large energy consumption. To determine the pumping power to boost the pressure drop, we need to determine both the volume flow rate and the amount of pressure drop. Then, we can calculate the cost of the power required to pump oil.

- Volume flow rate per hour:

$$
\begin{aligned}
Q &= 3,000,000 \text{ barrels/day} \times 5.6146 \text{ ft}^3/\text{barrel} \\
&= 16,843,800 \text{ ft}^3/\text{day} \\
&= 701,825 \text{ ft}^3/\text{hour.}
\end{aligned}
$$

- Pressure drop:

$$
\begin{aligned}
\Delta P &= \frac{128 Q \mu L}{g \pi D^4} \\
&= \frac{128 \times 701,825 \times 8,500 \times 3,168,000}{32.3 \times 12,960,000 \times 3.14159 D^4} \\
&= \frac{1,845,153,595}{D^4} \text{ lb/ft}^2.
\end{aligned}
$$

- Pumping power required to boost the pressure drop:

$$\text{Power} = \frac{1.333Q\Delta P}{1,980,000}$$

$$= \frac{1.333 \times 701,825 \times \dfrac{1,845,153,595}{D^4}}{1,980,000}$$

$$= \frac{871,818,975}{D^4} \text{ HP.}$$

- Power cost to pump oil:

$$\text{Power Cost} = \frac{871,818,975}{D^4} \text{ HP} \times \$0.015/\text{HP.hour}$$

$$\times 24 \text{ hours/day} \times 365 \text{ days/year}$$

$$= \frac{\$114,557,013,315}{D^4} /\text{year.}$$

Step 3: Pump and motor cost calculation:

Once we determine the required pumping power, we can determine the size of the pump and the motor costs. This is because the capacity of the pump and motor is proportional to the required pumping power.

$$\text{Pump and motor cost} = \frac{871,818,975}{D^4} \times \$195/\text{HP}$$

$$= \frac{\$170,004,700,125}{D^4}.$$

Step 4: Required amount and cost of steel:

The pumping cost will be counterbalanced by the lower costs for the smaller pipe, valves, and fittings. If the pipe diameter is made larger, the fluid velocity drops markedly and the pumping costs become substantially lower. Conversely, the capital cost for the larger pipe, fittings, and valves becomes greater. For a given cross-sectional area of the pipe, we can determine the total volume of the pipe as well as the weight. Once the total weight of the pipe is determined, we can easily convert it into the required investment cost.

$$\text{Cross-sectional area} = 3.14159[(0.51D)^2 - (0.50D)^2]$$

$$= 0.032 \, D^2 \text{ ft}^2$$

$$\text{Total volume of pipe} = 0.032 \, D^2 \text{ ft}^2 \times 3,168,000 \text{ ft}$$

$$= 101,376 \, D^2 \text{ ft}^3$$

$$\text{Total weight of steel} = 101{,}376 \, D^2 \, \text{ft}^3 \times 490.75 \, \text{lb/ft}^3$$

$$= 49{,}750{,}272 \, D^2 \, \text{lbs}$$

$$\text{Total pipeline cost} = \$1.00/\text{lb} \times 49{,}750{,}272 \, D^2 \, \text{lbs}$$

$$= \$49{,}750{,}272 \, D^2.$$

Step 5: Annual equivalent cost calculation:

For a given total pipline cost and its salvage value at the end of 20 years of service life, we can find the equivalent annual capital cost by using the capital recovery factor with return formula.

$$\text{Capital cost} = \left(\$49{,}750{,}272 \, D^2 + \frac{\$170{,}004{,}700{,}125}{D^4} \right)(A/P, 10\%, 20)$$

$$= 5{,}843{,}648 \, D^2 + \frac{19{,}968{,}752{,}076}{D^4}$$

$$\text{Annual power cost} = \frac{\$114{,}557{,}013{,}315}{D^4}.$$

Step 6: Economical pipe size:

Now we have determined the annual pumping and motor costs and the equivalent annual capital cost, we can express the total equivalent annual cost as a fraction of the pipe diameter (D).

$$AE(10\%) = 5{,}843{,}648 D^2 + \frac{19{,}968{,}752{,}076}{D^4} + \frac{114{,}557{,}013{,}315}{D^4}.$$

To find the optimal pipe size (D) that results in the minimum annual equivalent cost, we take the first derivative of $AE(10\%)$ with respect to D, equate the result to zero, and solve for D:

$$\frac{d\,AE(10\%)}{dD} = 11{,}687{,}297 D - \frac{538{,}103{,}061{,}567}{D^5}$$

$$= 0.$$

$$11{,}687{,}297 D^6 = 538{,}103{,}061{,}567.$$

$$D^6 = 46{,}041.70.$$

$$D^* = 5.9868 \, \text{ft}.$$

Note that velocity in a pipe should ideally be no more than approximately 10 ft/sec, because friction wears the pipe. To check whether the answer is reasonable, we may compute

$$Q = \text{velocity} \times \text{pipe inner area:}$$

$$701{,}825 \text{ ft}^3/\text{hr} \times \frac{1}{3{,}600} \text{ hr/sec} = V\frac{3.14159(5.9868)^2}{4},$$

$$V = 6.93 \text{ ft/sec}$$

which is less than 10 ft/sec. Therefore, the optimal answer as calculated is practical.

Step 7: Equivalent annual cost at optimal pipe size:

$$\text{Capital cost} = [\$49{,}750{,}272(5.9868)^2$$

$$+ \frac{\$170.004{,}700{,}125}{5.9868^4}](A/P, 10\%, 20)$$

$$= 5{,}843{,}648(5.9868)^2 + \frac{19{,}968{,}752{,}076}{5.9868^4}$$

$$= \$224{,}991{,}039.$$

$$\text{Annual power cost} = \frac{114{,}557{,}013{,}315}{5.9868^4}$$

$$= \$89{,}174{,}911.$$

$$\text{Total annual cost} = \$224{,}991{,}039 + \$89{,}174{,}911$$

$$= \$314{,}165{,}950.$$

Step 8: Total annual oil revenue:

$$\text{Annual oil revenue} = \$18/\text{bbl} \times 3{,}000{,}000 \text{ bbls/day}$$

$$\times 365 \text{ days/year}$$

$$= \$19{,}710{,}000{,}000 \text{ year.}$$

Enough revenues are available to offset the capital as well as the operating cost.

Comments: A variety of other criteria exist for choosing pipe size for a particular fluid transfer application. For example, low velocity may be required where erosion or corrosion concerns must be considered. Alternatively, higher velocities may be desirable for slurries where settling is a concern. Constructional ease may also weigh significantly when choosing pipe size. A small pipe size may not accommodate the head and flow requirements efficiently, whereas space limitations may prohibit the selection of large pipe sizes.

8.5 Summary

- Annual equivalent worth analysis, or AE, is—along with present worth analysis—one of the two main analysis techniques based on the concept of equivalence. The equation for AE is

$$AE(i) = PW(i)(A/P, i, N).$$

  AE analysis yields the same decision result as PW analysis.

- The capital recovery cost factor, or $CR(i)$, is one of the most important applications of AE analysis in that it allows managers to calculate an annual equivalent cost of capital for ease of itemization with annual operating costs. The equation for $CR(i)$ is

$$CR(i) = (I - S)(A/P, i, N) + iS,$$

  where I = initial cost and S = salvage value.

- AE analysis is recommended over NPW analysis in many key real-world situations for the following reasons:

  1. In many financial reports, an annual equivalent value is preferred to a present worth value.

  2. Calculation of unit costs is often required to determine reasonable pricing for sale items.

  3. Calculation of cost per unit of use is required to reimburse employees for business use of personal cars.

  4. Make-or-buy decisions usually require the development of unit costs for the various alternatives.

  5. Minimum cost analysis is easy to do when based on annual equivalent worth.

Self-Test Questions

8s.1 The annual equivalent of an income stream of $1,000 per year, to be received at the end of each of the next 3 years, at an interest rate of 12% is

(a) $1,000 per year

(b) less than $1,000 per year

(c) greater than $1,000 per year

(d) $1,000($A/P$, 12%, 3).

8s.2 Two options are available for painting your house: (1) Oil-based painting, which costs $5,000 and (2) water-based painting, which costs $3,000. The estimated lives are 10 years and 5 years respectively. For either option, no salvage value will remain at the end of respective service lives. Assume that you will keep and maintain the house for 10 years. If your personal interest rate is 10% per year, which of the following statement is correct?

(a) On an annual basis, Option 1 will cost about $850

(b) On an annual basis, Option 2 is about $22 cheaper than Option 1

(c) On an annual basis, both options cost about the same

(d) On an annual basis, Option 2 will cost about $820.

8s.3 Find the annual equivalent worth for the following infinite cash flow series at an interest rate of 10%:

n	Net Cash Flow
0	0
1 – 10	$400
11 – ∞	$500

(a) $461.20   (b) $438.60

(c) $445.20   (d) $985.40.

8s.4 Your firm has purchased an injection molding machine at a cost of $100,000. The machine's useful life is estimated at 8 years. Your accounting department has estimated the capital cost for this machine at about $25,455 per year. If your firm's MARR is 20%, how much salvage value do you think the accounting department assumed at the end of 8 years?

(a) $11,000   (b) $12,000

(c) $10,000   (d) $9,000.

8s.5 You purchased a CNC machine for $18,000. It is expected to have a useful life of 10 years and a salvage value of $3,000. At $i = 15\%$, what is the annual capital cost of this machine?

(a) $3,900   (b) $2,990

(c) $3,740   (d) $3,440.

8s.6 What is the annual equivalent amount at $i = 10\%$ for an infinite series of annual receipts of $500 for the first 10 years, which will increase to $1,000 per year after 10 years, and which will remain constant thereafter?

(a) $750   (b) $735

(c) $713   (d) $693.

8s.7 You just purchased a pin inserting machine to relieve some bottleneck problems that have been created in manufacturing a PC board. The machine cost $56,000 and has an estimated service life of 5 years. At that time, the estimated salvage value would be $5,000. The machine is expected to operate 2,500 hours per year. The expected annual operating and maintenance cost would be $6,000. If your firm's interest rate is 15%, what would be the machine cost per hour *without* considering income tax?

(a) $8.79   (b) $5.89

(c) $11.85   (d) $7.85.

8s.8 The City of Greenville has decided to build a softball complex on land donated by one of the city residents. The city council has already voted to fund the project at a level of $800,000 (initial capital investment). The city engineer has collected the following financial information for the project.

- Annual upkeep costs: $120,000
- Annual utility costs: $13,000
- Renovation costs: $50,000 for every 5 years
- Annual team user fees (revenues): $32,000
- Useful life: Infinite
- Interest rate: 5%.

If the city expects 40,000 visitors to the complex each year, what should be the minimum ticket price per person, so that the city can break-even?

(a) $2.50 < price ≤ $3.00
(b) $3.00 < price ≤ $3.50
(c) $3.50 < price ≤ $4.00
(d) $4.00 < price ≤ $4.50.

8s.9 Consider manufacturing equipment that has an installed cost of $100K. The equipment is expected to generate $30K of annual energy savings during its first year of installation. The value of these annual savings is expected to increase by 3% per year because of increased fuel costs. Assume that the equipment has a service life of 5 years (or 3,000 operating hours per year) with no appreciable salvage value. Determine the equivalent dollar savings per each operating hour at $i = 14\%$.

(a) $1.300 per hour
(b) $1.765 per hour
(c) $0.827 per hour
(d) $0.568 per hour.

8s.10 Colgate Printing Co. (CPC) has the book binding contract for the Ralph Brown library. The library pays $25 per book to CPC. CPC binds 1,000 books every year for the library. Ralph Brown library is considering the option of binding the books in-house in the basement of the library complex. In order to do this, the library would have to invest in a binding machine and other printing equipment at a cost of $100,000. The useful life of the machine is 8 years, at the end of which time, the machine is estimated to have a salvage value of $12,000. The annual operating and maintenance costs of the machine are estimated to be $10,000. Assuming an interest rate of 12%, what is the cost of binding per book for the in-house option?

(a) $26.27 per book
(b) $30.13 per book
(c) $22.50 per book
(d) $29.15 per book.

8s.11 In Problem 8s.10, what annual volume of books in need of binding would make both the options (in-house versus subcontracting) equivalent?

(a) $1,050 copies   (b) $900 copies
(c) $1,166 copies   (d) $1,096 copies.

8s.12 A small machine shop with an available electrical load of 37 HP purchases its electricity at the following rates:

kWh/month	@ $/kWh
First 1,500	0.025
Next 1,250	0.015
Next 3,000	0.009
All over 5,750	0.008

The current monthly consumption of electric power averages 3,200 kWh. A bid is to be made on some new business which would require an additional 3,500 kWh per month. If the new business lasted 2 years, what would be the equivalent monthly power cost that should be allocated to the new business? Assume that the shop's interest rate is 9%, compounded monthly.

(a) $75.58   (b) $30.55
(c) $63.00   (d) $90.85.

8s.13 The following infinite cash flow series has a rate of return of 10%. Determine the unknown value of X.

(a) $150   (c) $250
(b) $200   (d) $300

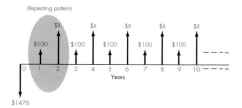

(Repeating pattern)

$1476

8s.14 The owner of a workshop is planning to purchase a special machine for $50,000. The annual operating cost (such as fuel, labor, power) is expected to be $8,000 per year. If the machine has a useful life of 12 years, what is the minimum required annual equivalent revenue needed to break even with an 8% annual interest rate? Assume that the machine would have an estimated market value of $5,000 at the end of its useful life.

(a) $5,972   (c) $8,000
(b) $6,372   (d) $14,371

8s.15 You are considering two types of electric motors for your paint shop. Their financial information and operating characteristics are summarized below. If you plan to operate the motor for 2,000 hours annually, what is the total cost savings per operating hour associated with the more efficient brand (Brand X) at an interest rate of 12%. The motor will be needed for 10 years. Assume that power costs are 5 cents per kilowatt-hour. (1 H.P. $= 0.746$ kW)

Summary Info and Characteristics	Brand X	Brand Y
Price	$4,500	$3,600
Operating expenses per year	$300	$500
Salvage value after 10 years	$250	$100
Capacity	150 H.P.	150 H.P.
Efficiency	83%	80%

(a) Less than 10 cents/hr
(b) Between 10 cents/hr and 20 cents/hr
(c) Between 20.1 cents/hr and 30 cents/hr
(d) Greater than 30.1 cents/hr

8s.16 Suppose that a young investor is considering investing $100 in an interest bearing account that pays 10% interest compounded annually. The investor's plan is to leave the money in the account for one year and then withdraw the principal plus all accrued interest. If the investor's MARR is known to be 10%, does the investor:

(a) Increase his/her wealth by an additional $10
(b) Break even on the investment
(c) The investor earns an additional $10 and also breaks even on the investment
(d) None of the above

8s.17 What is the annual equivalent amount at $i = 10\%$ for the infinite series shown below with annual receipts of $500 for the first 10 years, which will increase to $1,000 per year after 10 years and which will remain constant thereafter?

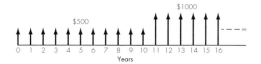

(a) $750   (c) $713
(b) $735   (d) $693

8s.18 The Gillette company is considering introducing a new shaving system called MACH3 in the market. The company plans to manufacture 75 million units of MACH3 a year. The investment at time 0 that is required for building the manufacturing plant is estimated as $500 million, and the economic life of the project is assumed to be 10 years. The annual total operating expenses, including manufacturing costs and overheads, are estimated as $175 million. The salvage value that can be realized from the project is estimated as $120 million. If the company's MARR is 25%, determine the minimum price that Gillette should charge for a MACH3 shaving system. (Do not consider any income tax effect.)

(a) $3.15

(b) $4.15

(c) $5.15

(d) $2.80

Problems

Note: *Unless otherwise stated, all cash flows given in the problems represent after-tax cash flows.*

8.1 Consider the following cash flows and compute the equivalent annual worth at $i = 12\%$.

		A_n
n	Investment	Revenue
0	−$10,000	
1		$2,000
2		2,000
3		3,000
4		3,000
5		1,000
6	+ 2,000	500

8.2 Consider the cash flow diagram. Compute the equivalent annual worth at $i = 12\%$.

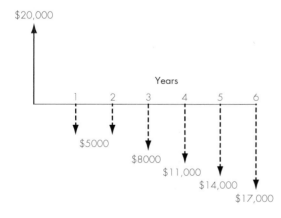

8.3 Consider the cash flow diagram. Compute the equivalent annual worth at $i = 10\%$.

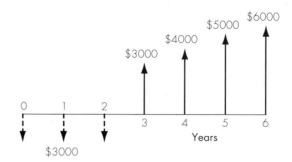

8.4 Consider the cash flow diagram. Compute the equivalent annual worth at $i = 13\%$.

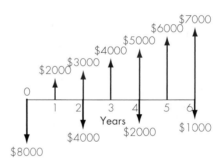

8.5 Consider the cash flow diagram. Compute the equivalent annual worth at $i = 8\%$.

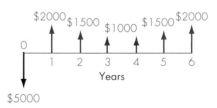

8.6 Consider the following sets of investment projects:

	Project's Cash Flow ($)			
n	A	B	C	D
0	−$2,000	−$4,000	−$3,000	−$9,000
1	400	3,000	−2,000	2,000
2	500	2,000	4,000	4,000
3	600	1,000	2,000	8,000
4	700	500	4,000	8,000
5	800	500	2,000	4,000

Compute the equivalent annual worth of each project at $i = 10\%$ and determine the acceptability of each project.

8.7 In Problem 8.6, plot the equivalent annual worth of each project as a function of interest rate (0%–40%).

8.8 Consider the following sets of investment projects:

Period	Project's Cash Flow			
(n)	A	B	C	D
0	−$3,500	−$3,000	−3,000	−$3,600
1	0	1,500	3,000	1,800
2	0	1,800	2,000	1,800
3	5,500	2,100	1,000	1,800

Compute the equivalent annual worth of each project at $i = 13\%$ and determine the acceptability of each project.

8.9 Consider the following project's cash flows:

	Net Cash Flow	
n	Investment	Operating Income
0	−$600	
1		$500
2		400
3	−600	300
4		500
5		400
6	−600	300
7		500
8		400
9		300

Find the equivalent annual worth for this project at $i = 10\%$ and determine the acceptability of the project.

8.10 The owner of a business is considering investing $55,000 in new equipment. He estimates that the net cash flows during the first year will be $5,000, but these will increase by $2,500 per year the next year, and each year thereafter. The equipment is estimated to have a 10-year service life and a net salvage value at this time of $6,000. The firm's interest rate is 12%.

(a) Determine the annual capital cost (ownership cost) for the equipment.

(b) Determine the equivalent annual savings (revenues).

(c) Determine if this is a wise investment?

8.11 Nelson Electronics Company just purchased a soldering machine to be used in its assembly cell for flexible disk drives. This soldering machine cost $250,000. Because of the specialized function it performs, its useful life is estimated to be 5 years. At that time its salvage value is also estimated to be $40,000. What is the capital cost for this investment if the firm's interest rate is 18%?

8.12 Beginning next year, a foundation will support an annual seminar on campus by the earnings of a $100,000 gift it received this year. It is felt that 8% interest will be realized for the first 10 years, but that plans should be made to anticipate an interest rate of 6% after that time. What amount should be added to the foundation now to fund the seminar at the $10,000 level into infinity?

8.13 The present price (year 0) of kerosene is $2.50 per gallon, and its cost is expected to increase by $.30 per year. (Kerosene at the end of year 1 will cost $2.80 per gallon.) Mr. Garcia uses about 800 gallons of kerosene during a winter season for space heating. He has an opportunity to buy a storage tank for $700, and at the end of 4 years, he can sell the storage tank for $100. The tank has a capacity to supply 4 years of Mr. Garcia's heating needs, so he can buy 4 years of kerosene at its present price ($2.50). He can invest his money elsewhere at 8%. Should he purchase the storage tank? Assume that kerosene purchased on a pay-as-you-go basis is paid for at the end of the year. (However, kerosene purchased for the storage tank is purchased now.)

8.14 Consider the following advertisement that appeared in a local paper.

Pools-Spas-Hot Tubs—Pure Water without Toxic Chemicals: The IONETICS water purification system has proven highly effective in killing algae and bacteria in pools and spas. Here is how it works: The "Ion Chamber" installed in the return water line contains copper/silver electrodes. A safe, low-voltage current is sent through those electrodes from a "Computerized Controller." Copper and silver ions enter the water stream and the pool or spa where they attack and kill the algae and bacteria. These charged, dead microorganisms mutually attract, forming larger particles easily removed by the existing filtration system. The IONETICS system can make your pool water pure enough to drink without the use of chlorine or other toxic chemicals. The ion level need only be tested about once per week and ion output is easily adjusted. The comparative costs between the conventional chemical system (chlorine) and the IONETICS systems are as follows:

Item	Conventional System	IONETICS System
Annual costs		
Chemical	$471	
IONETICS		$85
Pump	$576	$100
($0.667/kWh)		
Capital investment		$1,200

Note that the IONETICS system pays for itself less than 2 years.

Assume that the IONETICS system has a 12-year service life, and the interest rate is 6%. What is the equivalent monthly cost of operating the IONETICS system?

8.15 A construction firm is considering establishing an engineering computing center. This center will be equipped with three engineering workstations that would cost $25,000 each, and each has a service life of 5 years. The expected salvage value of each workstation is $2,000. The annual operating and maintenance cost would be $15,000 for each workstation. At a MARR of 15%, determine the equivalent annual cost for operating the engineering center.

8.16 Consider the cash flows for the following investment projects:

n	Project's Cash Flow A	B
0	−$4,000	$5,500
1	1,000	−1,400
2	X	−1,400
3	1,000	−1,400
4	1,000	−1,400

(a) For project A, find the value of X that makes the equivalent annual receipts equal the equivalent annual disbursement at $i = 13\%$.

(b) Would you accept B at $i = 15\%$ based on AE criterion?

8.17 An industrial firm can purchase a special machine for $40,000. A down payment of $4,000 is required, and the balance can be paid in five equal year-end installments at 7% interest on the unpaid balance. As an alternative, the machine can be purchased for $36,000 in cash. If the firm's MARR is 10%, determine which alternative should be accepted, based on the annual equivalent method.

8.18 An industrial firm is considering purchasing several programmable controllers and automating their manufacturing operations. It is estimated that the equipment will initially cost $100,000 and the labor to install it will cost $35,000. A service contract to maintain the equipment will cost $5,000 per year. Trained service personnel will have to be hired at an annual salary of $30,000. Also estimated is an approximate $10,000 annual income-tax savings (cash inflow). How much will this investment in equipment and services have to increase the annual revenues after taxes to break even? The equipment is estimated to have an operating life of 10 years with no salvage value because of obsolescence. The firm's MARR is 10%.

Unit Cost Profit Calculation

8.19 The engineering department of a large firm is overly crowded. In many cases, several engineers share one office. It is evident that the distraction caused by crowded conditions considerably reduces the productive capacity of the engineers. Management is considering the possibility of new facilities for the department, which could result in fewer engineers per office and a private office for some. For an office presently occupied by five engineers, what minimum individual increase in effectiveness must result to warrant the assignment of only three engineers to an office if the following data apply?

- Office size of 16×20 feet.
- Average annual salary of $60,000 per each engineer.
- Cost of building per square foot is $65.

- Estimated life of building of 25 years.
- Estimated salvage value of the building is 10% of the first cost.
- Annual taxes, insurance, and maintenance are 6% of the first cost.
- Cost of janitor service, heating and illumination, etc. per square foot per year is $3.00.
- Interest rate is 12%.

Assume that engineers reassigned to other office space will maintain their present productive capability as a minimum.

8.20 Two 150-horsepower (HP) motors are being considered for installation at a municipal sewage-treatment plant. The first costs $4,500 and has an operating efficiency of 83%. The second costs $3,600 and has an efficiency of 80%. Both motors are projected to have zero salvage value after a life of 10 years. If all the annual charges such as insurance, maintenance, etc., amount to a total of 15% of the original cost of each motor, and if power costs are a flat 5 cents per kilowatt-hour, how many minimum hours of full-load operation per year are necessary to justify purchase of the more expensive motor at $i = 6\%$? (A conversion factor you might find useful is 1 HP = 746 watts = 0.746 kilowatts.)

8.21 Danford Company, a farm-equipment manufacturer, currently produces 20,000 units of gas filters for use in its lawnmower production annually. The following costs are reported based on the previous year's production:

It is anticipated that gas-filter production will last 5 years. If the company continues to produce the product in-house, the annual direct material costs will increase at the rate of 5%.

Item	Expense ($)
Direct materials	$60,000
Direct labor	180,000
Variable overhead (power and water)	135,000
Fixed overhead (light and heat)	70,000
Total cost	$445,000

(For example, the annual material costs during the first production year will be $63,000.) The direct labor will also increase at the rate of 6% per year. However, the variable overhead costs would increase at the rate of 3%, but the fixed overhead would remain at the current level over the next 5 years. Tompkins Company has offered to sell Danford 20,000 units of gas filters for $25 per unit. If Danford accepts the offer, some of the manufacturing facilities currently used to manufacture the gas filter could be rented to a third party at annual rental of $35,000. Additionally, $3.5 per unit of the fixed overheard applied to gas-filter production would be eliminated. The firm's interest rate is known to be 15%. What is the unit cost of buying the gas filter from the outside source? Should Danford accept Tompkins' offer, and why?

8.22 Southern Environmental Consulting (SEC), Inc., designs plans and specifications for asbestos abatement (removal) projects. These projects involve public, private, and governmental buildings. Currently, SEC must also conduct an air test before allowing the re-occupancy of a building from which asbestos has been removed. SEC subcontracts air-test samples to a laboratory for analysis by transmission electron microscopy (TEM). To offset the cost of TEM analysis, SEC charges its clients $100 more than the subcontractor's fee. The only expenses in this system are the costs of shipping the air-test samples to the subcontractor and the labor involved in this shipping. As business grows, SEC needs to consider either continuing to subcontract the TEM analysis to outside companies or developing its own TEM laboratory. Because of the passage of the Asbestos Hazard Emergency Response ACT (AHERA) by the U.S. Congress, SEC expects about 1000 airsample testings per year, over 8 years. The firm's MARR is known to be 15%.

Subcontract option: The client is charged $400 per sample, which is $100 above the subcontracting fee of $300. Labor expenses are $1,500 per year, and shipping expenses are estimated to be $0.50 per sample.

TEM purchase option: The purchase and installation cost for the TEM is $415,000. The equipment would last for 8 years when it should have no salvage value. The design and renovation cost is estimated to be $9,500. The client is charged $300 per sample based on the current market price. One full-time manager and two part-time technicians are needed to operate the laboratory. Their combined annual salaries will be $50,000. Material required to operate the lab includes carbon rods, copper grids, filter equipment, and acetone. These material costs are estimated at $6,000 per year. Utility costs, operating and maintenance costs, and the indirect labor needed to maintain the lab are estimated at $18,000 per year. The extra income tax expenses would be $20,000.

(a) Determine the cost of air-sample test by the TEM (in-house).

(b) What is the required number of air samples per year to make two options equivalent?

8.23 A company is currently paying its employees $0.33 per mile to drive their own cars when on company business. The company is considering supplying employees with cars, which would involve the following: car purchase at $22,000, with an estimated 3-year life; a net salvage value of $5,000; taxes and insurance at a cost of $700 per year; and operating and maintenance expenses of $0.15 per mile. If the interest rate is 10%, and the company anticipates an employee's annual travel to be 12,000 miles, what is the equivalent cost per mile (without considering income tax)?

8.24 An electrical automobile can be purchased for $25,000. The automobile is estimated to have a life of 12 years, with annual travel of 20,000 miles. Every 3 years, a new set of batteries will have to be purchased at a cost of $3,000. Annual maintenance of the vehicle is estimated to cost $700 per year. The cost of recharging the batteries is estimated at $0.015 per mile. The salvage value of the batteries and the vehicle at the end of 12 years is estimated at $2,000. Consider the MARR to be 7%. What is the cost per mile to own and operate this vehicle, based on the above estimates? The $3,000 cost of the batteries is a net value with the old batteries traded in for the new ones.

8.25 The estimated cost of a completely installed and ready-to-operate 40-kilowatt generator is $30,000. Its annual maintenance costs are estimated at $500. The energy that can be generated annually, at full load, is estimated to be 100,000 kilowatt-hours. If the value of the energy generated is considered to be $0.08 per kilowatt-hour, how long will it take before this machine becomes profitable? Consider the MARR to be 9% and the salvage value

of the machine to be $2,000 at the end of its estimated life of 15 years.

8.26 A large, land grant university, currently facing severe parking problems on its campus, is considering constructing parking decks off campus. A shuttle service could pick up students at the off-campus parking deck, and quickly transport them to various locations on campus. The university would charge a small fee for each shuttle ride, and the students could be quickly and economically transported to their classes. The funds raised by the shuttle would be used to pay for the trolleys, which cost about $150,000 each. Each trolley has a 12-year service life with an estimated salvage value of $3,000. To operate each trolley, the following additional expenses must be considered:

Item	Annual Expenses ($)
Driver	$40,000
Maintenance	7,000
Insurance	2,000

If students pay 10 cents for each ride, determine the annual ridership per trolley (number of shuttle rides per year) required to justify the shuttle project, assuming an interest rate of 6%.

8.27 The following cash flows represent the potential annual savings associated with two different types of production processes, each of which requires an investment of $12,000.

n	Process A	Process B
0	−$12,000	−$12,000
1	9,120	6,350
2	6,840	6,350
3	4,560	6,350
4	2,280	6,350

Assuming an interest rate of 15%,

(a) Determine the equivalent annual savings for each process.

(b) Determine the hourly savings for each process assuming 2,000 hours operation per year.

(c) Which process should be selected?

8.28 Eradicator Food Prep, Inc., has invested $7 million to construct a food irradiation plant. This technology destroys organisms that cause spoilage and disease, thus extending the shelf life of fresh foods and the distances over which it can be shipped. The plant can handle about 200,000 pounds of produce in an hour, and it will be operated for 3,600 hours a year. The net expected operating and maintenance costs (considering any income-tax effects) would be $4 million per year. The plant is expected to have a useful life of 15 years, with a net salvage value of $700,000. The firm's interest rate is 15%.

(a) If investors in the company want to recover the plant investment within 6 years of operation (rather than 15 years), what would be the equivalent after-tax annual revenues that must be generated?

(b) To generate annual revenues determined in part (a), what minimum processing fee per pound should the company charge to their producers?

8.29 The local government of Santa Catalina Island, off the coast near Long Beach, California, is completing plans to build a desalination plant to help ease a critical drought on the island. The drought and new construction on Catalina have combined to leave the island with an urgent need for a new water source. A modern desalination plant could produce fresh water from seawater for $1,000 an acre foot. An acre foot is 326,000 gallons, or enough to supply two households for 1 year. On Catalina, the cost from natural sources is about the same as for desalting. The $3 million plant, with a daily desalting capacity of 0.4 acre foot, can produce 132,000 gallons of fresh water a day (enough to supply 295 households daily), more than a quarter of the island's total needs. The desalination plant has an estimated service life of 20 years with no appreciable salvage value. The annual operating and maintenance costs would be about $250,000. Assuming an interest rate of 10%, what should be the minimum monthly water bill for each household?

8.30 A California utility firm is considering building a 50-megawatt geothermal plant that generates electricity from naturally occurring underground heat. The binary geothermal system will cost $85 million to build and $6 million (including any income-tax effect) to operate per year. (Virtually no fuel cost will be needed as compared to fuel costs related to a conventional fossil fuel plant.) The geothermal plant is to last for 25 years. At that time, the expected salvage value will be about the same as the cost to remove the plant. The plant will be in operation for 70% (plant utilization factor) of the year (or 70% of 8760 hours per year). If the firm's MARR is 14% per year, determine the cost of generating electricity per kilowatt-hour.

8.31 A corporate executive jet with a seating capacity of 20 has the following cost factors:

The company flies three round trips from Boston to London per week, a distance of 3,280 miles one way. How many passengers must be carried on an average trip in order to justify the use

Item	Cost
Initial cost	$12,000,000
Service life	15 years
Salvage value	$2,000,000
Crew costs per year	$225,000
Fuel cost per mile	$1.10
Landing fee	$250
Maintenance per year	$237,500
Insurance cost per year	$166,000
Catering per passenger trip	$75

of the jet if the first-class round-trip fare is $3,400? The firm's MARR is 15%. (Ignore income-tax consequences.)

Comparing Mutually Exclusive Alternatives by the AE Method

8.32 A certain factory building has an old lighting system, and lighting this building costs, on average, $20,000 a year. A lighting consultant tells the factory supervisor that the lighting bill can be reduced to $8,000 a year if $50,000 were invested in relighting the factory building. If the new lighting system is installed, an incremental maintenance cost of $3,000 per year must be considered. If the old lighting system has zero salvage value, and the new lighting system is estimated to have a life of 20 years, what is the net annual benefit for this investment in new lighting? Consider the MARR to be 12%. Also consider that the new lighting system has zero salvage value at the end of its life.

8.33 Travis Wenzel has $2,000 to invest. Usually, he would deposit the money in his savings account, which earns 6% interest, compounded monthly. However, he is considering three alternative investment opportunities:

Option 1: Purchasing a bond for $2,000. The bond has a face value of $2,000 and pays $100 every 6 months for 3 years. The bond matures in 3 years.

Option 2: Buying and holding a growth stock that grows 11% per year for 3 years.

Option 3: Making a personal loan of $2,000 to a friend and receiving $150 per year for 3 years.

Determine the equivalent annual cash flows for each option, and select the best option.

8.34 A chemical company is considering two types of incinerators to burn solid waste generated by a chemical operation. Both incinerators have a burning capacity of 20 tons per day. The following data have been compiled for comparison:

	Incinerator A	Incinerator B
Installed cost	$1,200,000	$750,000
Annual O&M costs	$50,000	$80,000
Service life	20 years	10 years
Salvage value	$60,000	$30,000
Income taxes	$40,000	$30,000

If the firm's MARR is known to be 13%, determine the processing cost per ton of solid waste by each incinerator. Assume that incinerator B will be available in the future at the same cost.

8.35 Consider the cash flows for the following investment projects (assume MARR = 15%):

	Project's Cash Flow		
n	A	B	C
0	−$2,500	−$4,000	−$5,000
1	1,000	1,600	1,800
2	1,800	1,500	1,800
3	1,000	1,500	2,000
4	400	1,500	2,000

(a) Suppose that projects A and B are mutually exclusive. Which project would you select based on the AE criterion?

(b) Assume that projects B and C are mutually exclusive. Which project would you select based on the AE criterion?

8.36 An airline is considering two types of engine systems for use in its planes. Each has the same life and the same maintenance and repair record.

- System A costs $100,000 and uses 40,000 gallons per 1,000 hours of operation at the average load encountered in passenger service.

- System B costs $200,000 and uses 32,000 gallons per 1,000 hours of operation at the same level.

Both engine systems have 3-year lives before any major overhaul. Based on the initial investment, the systems have 10% salvage values. If jet fuel costs $1.80 a gallon currently, and fuel consumption is expected to increase at the rate of 6% because of degrading engine efficiency (each year), which engine system should the firm install? Assume 2,000 hours of operation per year, and a MARR of 10%. Use the AE criterion. What is the equivalent operating cost per hour for each engine?

8.37 Norton Auto-Parts, Inc., is considering one of two forklift trucks for their assembly plant. Truck A costs $15,000

and requires $3,000 annually in operating expenses. It will have a $5,000 salvage value at the end of its 3-year service life. Truck B costs $20,000, but requires only $2,000 annually in operating expenses; its service life is 4 years, at which time, truck B's expected salvage value will be $8,000. The firm's MARR is 12%. Assuming that the trucks are needed for 12 years and that no significant changes are expected in the future price and functional capacity of both trucks, select the most economical truck based on AE analysis.

8.38 A small manufacturing firm is considering the purchase of a new machine to modernize one of its current production lines. Two types of machines are available on the market. The lives of machine A and machine B are 4 years and 6 years, respectively, but the firm does not expect to need the service of either machine for more than 5 years. The machines have the following expected receipts and disbursements.

Item	Machine A	Machine B
First cost	$6,500	$8,500
Service life	4 years	6 years
Estimated salvage value	$600	$1,000
Annual O&M costs	$800	$520
Change oil filter every other year	$100	None
Engine overhaul	$200 (every 3 years)	$280 (every 4 years)

The firm always has another option: To lease a machine at $3,000 year, fully maintained by the leasing company. After 4 years of use, the salvage value for machine B will remain at $1,000.

(a) How many decision alternatives are there?

(b) Which decision appears to be the best at $i = 10\%$?

8.39 A plastic manufacturing company owns and operates a polypropylene production facility that converts the propylene from one of its cracking facilities to polypropylene plastics for outside sale. The polypropylene production facility is currently forced to operate at less than capacity due to lack of enough propylene production capacity in its hydrocarbon cracking facility. The chemical engineers are considering alternatives for supplying additional propylene to the polypropylene production facility. Some of the feasible alternatives are as follows: Option 1: Build a pipeline to the nearest outside supply source; Option 2: Provide additional propylene by truck from an outside source. The engineers also gathered the following projected cost estimates:

- Future costs for purchased propylene excluding delivery: $0.215 per lb
- Cost of pipeline construction: $200,000 per pipeline mile
- Estimated length of pipeline: 180 miles
- Transportation costs by tank truck: $0.05 per lbs, utilizing common carrier
- Pipeline operating costs: $0.005 per lb, excluding capital costs
- Projected additional propylene needs: 180 million lbs per year
- Projected project life: 20 years
- Estimated salvage value of the pipeline: 8% of the installed costs

Determine the propylene cost per pound under each option, if the firm's

MARR is 18%. Which option is more economical?

8.40 The City of Prattsville is comparing two plans for supplying water to a newly developed subdivision.

- Plan A will take care of requirements for the next 15 years; at the end of that period, the initial cost of $400,000 will have to be duplicated to meet the requirements of subsequent years. The facilities installed at dates 0 and 15 may be considered permanent; however, certain supporting equipment will have to be replaced every 30 years from the installation dates, at a cost of $75,000. Operating costs are $31,000 a year for the first 15 years and $62,000 thereafter, although they are expected to increase by $1,000 a year beginning in the 21st year.

- Plan B will supply all requirements for water indefinitely into the future, although it will only be operated at half capacity for the first 15 years. Annual costs over this period will be $35,000 and will increase to $55,000 beginning in the 16th year. The initial cost of Plan B is $550,000; the facilities can be considered permanent, although it will be necessary to replace $150,000 of equipment every 30 years after initial installation.

The city will charge the subdivision the use of water based on the equivalent annual cost. At an interest rate of 10%, determine the equivalent annual cost for each plan, and make a recommendation to the city.

Minimum-Cost Analysis

8.41 A continuous electric current of 2000 amps is to be transmitted from a generator to a transformer located 200 feet away. A copper conductor can be installed for $6 per pound, will have an estimated life of 25 years, and can be salvaged for $1 per pound. Power loss from the conductor will be inversely proportional to the cross-sectional area of the conductor and may be expressed as $6.516/A$ kilowatt, where A is given in square inches. The cost of energy is $0.0825 per kilowatt-hour, the interest rate is 11%, and the density of copper is 555 pounds per cubic foot.

(a) Calculate the optimum cross-sectional area of the conductors.

(b) Calculate the annual equivalent total cost for the value obtained in part (a).

(c) Graph the two individual cost factors (capital cost and power loss cost) and the total cost as a function of cross-sectional area A and discuss the impact of increasing energy cost on the optimum obtained in part (a).

Short Case Studies

8.42 Automotive engineers at Ford are considering the laser blank welding (LBW) technique to produce a windshield frame rail blank (see below). The engineers believe that the LBW as compared with the conventional sheet metal blanks would result in a significant savings as follows:

   1. Scrap reduction through more efficient blank nesting on coil.
   2. Scrap reclamation (weld scrap offal into a larger usable blank).

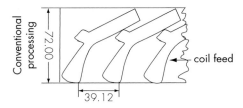

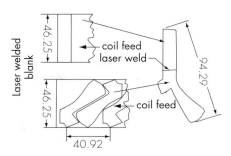

Use of a laser welded blank to provide a reduction in engineered scrap for production of a window frame rail blank. (Problem 5.42)

Based on an annual volume of 3,000 blanks, Ford engineers have estimated the following financial data:

| | Blanking Method | |
| | | Laser Blank |
Description	Conventional	Welding
Weight per blank (lbs/part)	63.764	34.870
Steel cost/part	$14.98	$8.19
Transportation/part	$0.67	$0.42
Blanking/part	$0.50	$0.40
Die investment	$106,480	$83,000

The LBW technique appears to achieve significant savings, so Ford's engineers are leaning toward adopting the LBW technique. Since Ford engineers have had no prior experience with the LBW, they are not sure if producing the windshield frames in-house at this time is a good strategy. For this windshield frame, it may be cheaper to use the services of a supplier, which has both the experience and the machinery for laser blanking. Ford's lack of skill in laser blanking, may require that it take 6 months to get up to the required production volume. On the other hand, if Ford relies on a supplier, it can only assume that supplier labor problems will not halt production of Ford's parts. The make-or-buy decision depends on two factors: The amount of new investment that is required in laser welding, and whether additional machinery will be required for future products. Assuming an analysis of 10 years and an interest rate of 16%, recommend the best course of action. Assume also that the salvage value at the end of 10 years is estimated to be nonsignificant for either system. If Ford is considering the subcontracting option, what would be the acceptable range of contract bid (unit cost per part)?

8.43 A Veterans Administration (VA) hospital is to decide which type of boiler fuel system will most efficiently provide the required steam energy output for heating, laundry, and sterilization purposes. The present boilers were installed in the early 1930s and are now obsolete. Much of the auxiliary equipment is also old and in need of repair. Because of these general conditions, an engineering recommendation was made to replace the entire plant with a new boiler plant building that would house modern equipment. The cost of demolishing the old boiler plant would be almost a complete loss as the salvage value of the scrap steel and used brick was estimated to be only about $1,000. The VA hospital's engineer finally selected two alternative proposals as being worthy of more intensive analysis. The hospital's annual energy requirement, measured in terms of steam output, is approximately 145,000,000 pounds of steam. As a rule of thumb for analysis, one pound of steam is approximately 1,000 BTUs, and one cubic foot of natural gas is approximately 1,000 BTUs. The two alternatives were as follows:

Proposal 1: A new, coal-fired boiler plant. This boiler plant would cost $1,770,300. To meet the requirements for particulate emission as set by the Environmental Protection Agency (EPA), this coal-fired boiler, even if it burned low sulfur coal, would need an electrostatic precipitator which would cost approximately $100,000. This plant would last for 20 years. One pound of dry coal yields about 14,300 BTUs. To convert the 145,000,000 pounds of steam energy to the common denominator of BTUs, it is necessary to multiply by 1000. To find BTU input requirements it is necessary to divide by the relative boiler efficiency for type of fuel. The boiler efficiency for coal is 0.75. The coal price is estimated to be $35.50 per ton.

Proposal 2: A gas-fired boiler plant with No.2 fuel oil as standby. This system would cost $889,200 with an expected service life of 20 years. Since small household or commercial gas users entirely dependent on gas have priority, large plants must have oil switch-over capability. It has been estimated that 6% of 145,000,000 pounds of steam energy (or 8,700,000 pounds) would come about as a result of the oil switch. The boiler efficiency with each fuel would be 0.78 for gas and 0.81 for oil, respectively. The heat value of natural gas is approximately 1,000,000 BTU/MCF (thousand cubic feet), and for No. 2 fuel oil it is 139,400 BTU/gal. The estimated gas price is $2.50/MCF, and the fuel oil (No. 2) price is $0.82 per gallon.

(a) Calculate the annual fuel costs for each proposal.

(b) Determine the unit cost per steam pound for each proposal. Assume $i = 10\%$.

(c) Which proposal is the more economical?

8.44 The following is a letter that I received from a local city engineer. After reading the letter, recommend a reasonable amount of compensation to the citizen for the replaced drainage pipe.

Dear Professor Park:

Thank you for taking the time to assist with this problem. I'm really embarassed at not being able to solve it myself, since it seems rather straightforward. The situation is as follows:

A citizen of Opelika paid for concrete drainage pipe approximately 20 years ago to be installed on his property. (We have a policy that if drainage trouble exists on private property and the owner agrees to pay for the material, City crews will install it.) That was the case in this problem. Therefore, we are dealing with only material costs, disregarding labor.

However, this past year, we removed the pipe purchased by the citizen, due to a larger area drainage project. He thinks, and we agree, that he is due some refund for salvage value of the pipe due to its remaining life.

Problem:
- Known: 80′ of 48″ pipe purchased 20 years ago, Original purchase price
- Unknown: Current quoted price of 48″ pipe = $52.60/foot, times 80 feet = $4,208 total cost in today's dollars.
- Assumptions: 50-year life, therefore assume 30 years of life remaining at removal after 20 years. A 4% price increase per year, average, over twenty years

Thus, we wish to calculate the cost of the pipe 20 years ago. Then, calculate present salvage value after 20 years use, with 30 years of life remaining in today's dollars. Thank you again for your help. We look forward to your reply.

Charlie Thomas, P.E.

Director of Engineering

City of Opelika

CHAPTER 9

Rate of Return Analysis

McDonald Is Testing New Technology on Its Menu: One of McDonald's franchisees in Chicago is experimenting with new technology known as a "digital menu board" that could change the way you look at the old menu board you've been used to. Picture this: higher-priced sandwiches are not moving in the afternoon, so an ad pops up on the restaurant's menu board enticing customers to try the more expensive sandwich. It's called point-of-purchase promotion, the attempt by retailers to grab consumers' attention at the moment they are making up their mind what to buy and are pulling out their money. But the technology is not cheap. The initial outlay for equipment can cost anywhere from $5,000 to $20,000 per store location. Then comes the expense of tailoring promotions to the store, which can represent hundreds of dollars per promotion or more—roughly the same price as traditional cardboard promotions. Will the franchisee make more money with this creative menu advertising? If the franchisees want to make a 15% return on their investment, how much additional revenue should be generated from the digital menu board?

Source: *Chicago Tribune*, December 26, 1997, by Jim Kirk.

In competitive industries like fast food, more retailers are finding it's an advantage to be able to change promotions quickly in response to a rival's activity or other local market conditions. In the fast food industry, a third of the people buying food in a restaurant on any given day are not regular customers. This means they tend to spend more time looking at the choices on the menu board than do regular patrons. These consumers usually cannot make decisions fast enough; Retailers hope to try to grab a more profitable transaction by promoting a more profitable sandwich, while they are in waiting. Any franchisee's main question is how many of these customers would switch to more expensive menu items by seeing the ad on the digital menu.

What does the 15% rate of return figure for the franchisee really represent? How do we compute the figure from the projected cash flow series? And once computed, how do we use the figure when evaluating an investment project? Our consideration of the concept of rate of return in this chapter will answer these and other questions.

Along with the NPW and the AE criteria, the third primary measure of investment worth is **rate of return**. As shown in Chapter 7, the NPW measure is easy to calculate and apply. Nevertheless, many engineers and financial managers prefer rate of return analysis to the NPW method because they find it intuitively more appealing to analyze investments in terms of percentage rates of return rather than in dollars of NPW. Consider the following statements regarding an investment's profitability:

- This project will bring in a 15% rate of return on the investment.
- This project will result in a net surplus of $10,000 in the NPW.

Neither statement describes the nature of an investment project in any complete sense. However, the rate of return figure is somewhat easier to understand because many of us are so familiar with savings and loan interest rates, which are in fact rates of return.

In this chapter, we will examine four aspects of rate of return analysis: (1) the concept of return on investment, (2) calculation of a rate of return, (3) development of an internal rate of return criterion, and (4) the comparison of mutually exclusive alternatives based on rate of return.

9.1 Rate of Return

Many different terms are used to refer to **rate of return**, including **yield** (i.e., the yield to maturity, commonly used in bond valuation), **internal rate of return**, and **marginal efficiency of capital**. We will first review three common definitions of rate of return. Then we will use the definition of internal rate of return as a measure of profitability for a single investment project throughout the text.

9.1.1   Return on Investment

There are several ways of defining the concept of rate of return on investment. The first is based on a typical loan transaction, and the second on the mathematical expression of the present worth function.

Definition 1

Rate of return is the interest rate earned on the unpaid balance of an amortized loan.

Suppose that a bank lends $10,000, and it is repaid $4,021 at the end of each year for 3 years. How would you determine the interest rate that the bank charges on this transaction? As we learned in Chapter 4, you would set up the following equivalence equation:

$$\$10,000 = \$4,021(P/A, i, 3)$$

and solve for *i*. It turns out that $i = 10\%$. In this situation, the bank will earn a return of 10% on its investment of $10,000. The bank calculates the loan balances over the life of the loan as follows:

Year	Unpaid Balance at Beginning Year	Return on Unpaid Balance (10%)	Payment Received	Unpaid Balance at End of Year
0	−$10,000	$0	$0	−$10,000
1	−10,000	−1,000	4,021	−6,979
2	−6,979	−698	4,021	−3,656
3	−3,656	−366	4,021	0

A negative balance indicates an unpaid balance.

Observe that for the repayment schedule shown above, the 10% interest is calculated only for each year's outstanding balance. In this situation, only part of the $4,021 annual payment represents interest; the remainder goes toward repaying the principal. Namely, the three annual payments repay the loan itself and additionally provide a return of 10% on the *amount still outstanding each year*.

Note that, when the last payment is made, the outstanding principal is eventually reduced to zero.[1] If we calculate the NPW of the loan transaction at its rate of return (10%), we see

$$PW(10\%) = -\$10,000 + \$4,021(P/A, 10\%, 3) = 0,$$

which indicates that the bank can break even at a 10% rate of interest. In other words, the rate of return becomes the rate of interest that equates the present value of future cash repayments to the amount of the loan. This observation prompts the second definition of rate of return.

Definition 2

Rate of return is the break-even interest rate, i*, which equates the present worth of a project's cash outflows to the present worth of its cash inflows, or

$$PW(i^*) = PW_{\text{Cash inflows}} - PW_{\text{Cash outflows}}$$
$$= 0.$$

Note that the NPW expression is equivalent to

$$PW(i^*) = \frac{A_0}{(1 + i^*)^0} + \frac{A_1}{(1 + i^*)^1} + \cdots + \frac{A_N}{(1 + i^*)^N} = 0. \tag{9.1}$$

Here we know the value of A_n for each period, but not the value of i^*. Since it is the only unknown, we can solve for i^*. [Inevitably, there will be N values of i^* that satisfy this

---

[1] As we learned in Section 7.3.2, this terminal balance is equivalent to the net future worth of the investment. If the net future worth of the investment is zero, its NPW should also be zero.

equation. In most project cash flows, you would be able to find a unique positive $i*$ that satisfies Eq. (9.1). However, you may encounter some cash flows that cannot be solved for a single rate of return greater than -100%. By the nature of the NPW function in Eq. (9.1), it is certainly possible to have more than one rate of return for certain types of cash flows. For some cash flows, we may not find a specific rate of return at all.][2]

Note that the $i*$ formula in Eq. (9.1) is simply the NPW formula, Eq. (7.1), solved for the particular interest rate $(i*)$ at which $PW(i)$ is equal to zero. By multiplying both sides of Eq. (9.1) by $(1 + i*)^N$, we obtain

$$PW(i*)(1 + i*)^N = FW(i*) = 0.$$

If we multiply both sides of Eq. (9.1) by the capital recovery factor, $(A/P, i*, N)$, we obtain the relationship $AE(i*) = 0$. Therefore, the $i*$ of a project may be defined as the rate of interest that equates the present worth, future worth, and annual equivalent worth of the entire series of cash flows to zero.

9.1.2 Return on Invested Capital

Investment projects can be viewed as analogous to bank loans. We will now introduce the concept of rate of return based on the return on invested capital in terms of a project investment. A project's return is referred to as the internal rate of return (IRR) or the **yield** promised by an **investment project** over its **useful life**.

Definition 3

Internal rate of return is the interest rate charged on the unrecovered project balance of the investment such that, when the project terminates, the unrecovered project balance will be zero.

Suppose a company invests $10,000 in a computer with a 3-year useful life and equivalent annual labor savings of $4,021. Here, we may view the investing firm as the lender and the project as the borrower. The cash flow transaction between them would be identical to the amortized loan transaction described under Definition 1.

n	Beginning Project Balance	Return on Invested Capital	Ending Cash Payment	Project Balance
0	$0	$0	−$10,000	−$10,000
1	−10,000	−1,000	4,021	−6,979
2	−6,979	−697	4,021	−3,656
3	−3,656	−365	4,021	0

[2] You will always have N of them. The issue is whether these are real or imaginary. If they are real, the question are they in the $(-100\%, \infty)$ interval or not should be asked. A **negative rate of return** implies that you never recover your initial investment.

In our project balance calculation, we see that 10% is earned (or charged) on $10,000 during year 1, 10% is earned on $6979 during year 2, and 10% is earned on $3,656 during year 3. This indicates that the firm earns a 10% rate of return on funds that remain *internally* invested in the project. Since it is a return *internal* to the project, we refer to it as the **internal rate of return**, or IRR. This means that the computer project under consideration brings in enough cash to pay for itself in 3 years and also to provide the firm with a return of 10% on its invested capital. To put it differently, if the computer is financed with funds costing 10% annually, the cash generated by the investment will be exactly sufficient to repay the principal and the annual interest charge on the fund in 3 years.

Notice also that only one cash outflow occurs at time 0, and the present worth of this outflow is simply $10,000. There are three equal receipts, and the present worth of these inflows is $4,021(P/A, 10\%, 3) = \$10,000$. Since the NPW $= PW_{\text{Inflow}} - PW_{\text{Outflow}} = \$10,000 - \$10,000 = 0$, 10% also satisfies Definition 2 for rate of return. Even though the above simple example implies that i^* coincides with IRR, only Definitions 1 and 3 correctly describe the true meaning of internal rate of return. As we will see later, if the cash expenditures of an investment are not restricted to the initial period, several break-even interest rates (i^*s) may exist that satisfy Eq. (9.1). However, there may not be a rate of return *internal* to the project.

Current Practices—Some Rate of Return Numbers: One of the highest prices ever paid for a painting at a public auction was $53.9 million for Vincent Van Gogh's *Irises,* at Sotheby's in New York, on November 11, 1987. The seller was art collector John Whitney Payson, who had purchased the painting for $80,000 in 1947. Mr. Payson's investment in art brought him a rate of return of 17.68%.

In 1970, Wal-Mart offered 300,000 shares of its common stock to the public at a price of $16.50 per share. Since that time, Wal-Mart has had eleven (11) two for one (2/1) stock splits. On a purchase of 100 shares at $16.50 per share on our first offering, the number of shares has grown to 204,800 shares or worth $13,312,000 on January 2000. This is equivalent to a 34.97% rate of return on the investment over 30 years, even if the dividends were not reinvested.

Johnson Controls spent more than $2.5 million retrofitting a government complex and installing a computerized energy-management system for the State of Massachusetts. As a result, the state's energy bill dropped from an average of $6 million a year to $3.5 million. Moreover, both parties will benefit from the 10-year life contract. Johnson recovers half the money it saved in reduced utility costs (about $1.2 million a year over 10 years); Massachusetts has the other half to spend elsewhere. It is estimated that Johnson Controls will receive a 46.98% rate of return on its investment in this energy-control system.

9.2 Methods for Finding Rate of Return

We may find i^* by several procedures, each of which has its advantages and disadvantages. To facilitate the process of finding the rate of return for an investment project, we will first classify types of investment cash flow.

9.2.1 Simple versus Nonsimple Investments

We can classify an investment project by counting the number of sign changes in its net cash flow sequence. A change from either "+" to "−" or "−" to "+" is counted as one sign change. (We ignore a zero cash flow.) Then,

- a **simple (or conventional) investment** is one in which the initial cash flows are negative, and only one sign change occurs in the net cash flow series. If the initial flows are positive and only one sign change occurs in the subsequent net cash flows, they are referred to as **simple borrowing** cash flows.

- a **nonsimple (or nonconventional) investment** is one in which more than one sign change occurs in the cash flow series.

Multiple $i*$s, as we will see later, occur only in nonsimple investments. The different types of investment possibilities may be illustrated as follows:

Investment Type	Cash Flow Sign at Period					
	0	1	2	3	4	5
Simple	−	+	+	+	+	+
Simple	−	−	+	+	0	+
Nonsimple	−	+	−	+	+	−
Nonsimple	−	+	+	−	0	+

Example 9.1 Investment Classification

Consider the three cash flow series shown below and classify them into either simple or nonsimple investments.

Period n	Net Cash Flow		
	Project A	Project B	Project C
0	−$1,000	−$1,000	+$1,000
1	−500	3,900	−450
2	800	−5,030	−450
3	1,500	2,145	−450
4	2,000		

Solution

Given: Cash flow sequences above

Find: Classify into either simple or nonsimple investments

- Project A represents many common simple investments. This type of investment reveals the NPW profile shown in Figure 9.1(a). The curve only crosses the i-axis once.
- Project B represents a nonsimple investment. The NPW profile for this investment has the shape shown in Fig. 9.1(b). The i-axis is crossed at 10%, 30%, and 50%.
- Project C represents neither a simple nor a nonsimple investment, even though only one sign change occurs in the cash flow sequence. Since the first cash flow is positive, this is a **simple borrowing** cash flow, not an investment flow. The NPW profile for this type of investment looks like the one in Fig. 9.1(c).

Comments: Not all NPW profiles for nonsimple investments have multiple crossings of the i-axis. In Appendix A, we illustrate when to expect such multiple crossings by examining types of cash flows.

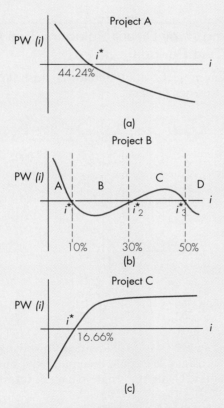

Figure 9.1   Present worth profiles: (a) Simple investment, (b) nonsimple investment with multiple rates of return, and (c) simple borrowing cash flows

9.2.2   Computational Methods

Once we identify the type of an investment cash flow, several ways to determine its rate of return are available. We will discuss some of the most practical methods here. They are as follows:

- Direct solution method,
- Trial-and-error method, and
- Computer solution method.

Direct Solution Method

For the very special case of a project with only a two-flow transaction (an investment followed by a single future payment) or a project with a service life of 2 years of return, we can seek a direct mathematical solution for determining the rate of return. These two cases will be examined in Example 9.2.

Example 9.2   Finding i^* by Direct Solution: Two Flows and Two Periods

Consider two investment projects with the following cash flow transactions. Compute the rate of return for each project.

n	Project 1	Project 2
0	−$1,000	−$2,000
1	0	1,300
2	0	1,500
3	0	
4	1,500	

Solution

Given: Cash flows for two projects

Find: i^* for each

Project 1: Solving for i^* in $PW(i^*) = 0$ is identical to solving $FW(i^*) = 0$ because FW equals PW times a constant. We could do either here, but we will set $FW(i^*) = 0$ to demonstrate the latter. Using the single-payment future worth relationship, we obtain

$$FW(i^*) = -\$1{,}000(F/P, i^*, 4) + \$1{,}500 = 0$$

$$\$1{,}500 = \$1{,}000(F/P, i^*, 4) = \$1{,}000(1 + i^*)^4$$

$$1.5 = (1 + i^*)^4.$$

Solving for $i*$ yields

$$i* = \sqrt[4]{1.5} - 1$$

$$= 0.1067 \text{ or } 10.67\%.$$

Project 2: We may write the NPW expression for this project as follows:

$$PW(i) = -\$2,000 + \frac{\$1,300}{(1 + i)} + \frac{\$1,500}{(1 + i)^2} = 0.$$

Let $x = \dfrac{1}{(1 + i)}$. We may then rewrite the $PW(i)$ as a function of X as follows:

$$PW(i) = -\$2,000 + \$1,300X + \$1,500X^2 = 0.$$

This is a quadratic equation that has the following solution:[3]

$$X = \frac{-1,300 \pm \sqrt{1,300^2 - 4(1,500)(-2,000)}}{2(1,500)}$$

$$= \frac{-1,300 \pm 3,700}{3,000}$$

$$= 0.8 \text{ or } -1.667$$

Replacing X values and solving for i gives us

$$0.8 = \frac{1}{(1 + i)} \rightarrow i = 25\%$$

$$-1.667 = \frac{1}{(1 + i)} \rightarrow i = -160\%.$$

Since an interest rate less than −100% has no economic significance, we find that the project's $i*$ is 25%.

Comments: In both projects, one sign change occurred in the net cash flow series, so we expected a unique $i*$. Also these projects had very simple cash flows. When cash flows are more complex, generally we must use a trial-and-error method or a computer to find $i*$.

---

[3]Given $aX^2 + bX + c = 0$, the solution of the quadratic equation is $X = \dfrac{-b \pm \sqrt{b^2 - 4ac}}{2a}$

Trial-and-Error Method

The first step in the trial-and-error method is to make an estimated ***guess***[4] at the value of i^*. For a simple investment, we compute the present worth of net cash flows using the "guessed" interest rate and observe whether it is positive, negative, or zero. Suppose the $PW(i)$ is negative. Since we are aiming for a value of i that makes $PW(i) = 0$, we must raise the present worth of the cash flow. To do this, we lower the interest rate and repeat the process. If $PW(i)$ is positive, however, we raise the interest rate in order to lower $PW(i)$. The process is continued until $PW(i)$ is approximately equal to zero. Whenever we reach the point where $PW(i)$ is bounded by one negative and one positive value, we use **linear interpolation** to approximate the i^*. This process is somewhat tedious and inefficient. (The trial-and-error method does not work for nonsimple investments in which the NPW function is not, in general, a monotonically decreasing function of interest rate.)

Example 9.3 Finding *i** by Trial and Error

Agdist Corporation distributes agricultural equipment. The board of directors is considering a proposal to establish a facility to manufacture an electronically controlled "intelligent" crop sprayer invented by a professor at a local university. This crop sprayer project would require an investment of $10 million in assets and would produce an annual after-tax net benefit of $1.8 million over a service life of 8 years. All costs and benefits are included in these figures. When the project terminates, the net proceeds from the sale of the assets would be $1 million (Figure 9.2). Compute the rate of return of this project.

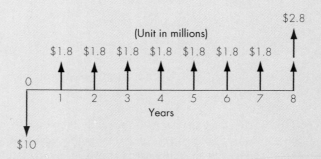

Figure 9.2   Cash flow diagram for a simple investment (Example 9.3)

Solution

Given: Initial investment $(I) = \$10$ million, $A = \$1.8$ million, $S = 1$ million, $N = 8$ years

---

[4] As we see later in this chapter, the ultimate objective of finding the i^* is to compare it to the MARR. Therefore, it is a good idea to use the MARR as the initial guess value.

Find: $i*$

We start with a guessed interest rate of 8%. The present worth of the cash flows in millions of dollars is

$$PW(8\%) = -\$10 + \$1.8(P/A, 8\%, 8) + \$1(P/F, 8\%, 8) = \$0.88.$$

Since this present worth is positive, we must raise the interest rate to bring this value toward zero. When we use an interest rate of 12%, we find that

$$PW(12\%) = -\$10 + \$1.8(P/A, 12\%, 8) + \$1(P/F, 12\%, 8) = -\$0.65.$$

We have bracketed the solution: $PW(i)$ will be zero at i somewhere between 8% and 12%. Using straight-line interpolation, we approximate

$$i* \cong 8\% + (12\% - 8\%)\left[\frac{0.88 - 0}{0.88 - (-0.65)}\right]$$

$$= 8\% + 4\%(0.5752)$$

$$= 10.30\%.$$

Now we will check to see how close this value is to the precise value of $i*$. If we compute the present worth at this interpolated value, we obtain

$$PW(10.30\%) = -\$10 + \$1.8(P/A, 10.30\%, 8) + \$1(P/F, 10.30\%, 8)$$

$$= -\$0.045.$$

As this is not zero, we may recompute the $i*$ at a lower interest rate, say 10%:

$$PW(10\%) = -\$10 + \$1.8(P/A, 10\%, 8) + \$1(P/A, 10\%, 8) = \$0.069.$$

With another round of linear interpolation, we approximate

$$i* \cong 10\% + (10.30\% - 10\%)\left[\frac{0.069 - 0}{0.069 - (-0.045)}\right]$$

$$= 10\% + 0.30\%(0.6053)$$

$$= 10.18\%.$$

At this interest rate,

$$PW(10.18\%) = -\$10 + \$1.8(P/A, 10.18\%, 8) + \$1(P/F, 10.18\%, 8)$$

$$= \$0.0007,$$

which is practically zero, so we may stop here. In fact, there is no need to be more precise about these interpolations because the final result can be no more accurate than the basic data, which ordinarily are only rough estimates. Computing the $i*$ for this problem on computer, incidentally, gives us 10.1819%.

Comment: With Excel, you can evaluate the IRR for the project as follows: = IRR(range, guess), where you specify the cell range for the cash flow (e.g., C0:C7) and the initial guess such as 10%.

Computer Solution Method

We don't need to do laborious manual calculations to find i^*. Many financial calculators have built-in functions for calculating i^*. It is worth noting that many spreadsheet packages have i^* functions, which solve Eq. (9.1) very rapidly.[5] This is usually done by entering the cash flows via a computer keyboard or by reading a cash flow data file. For example, Microsoft Excel has an IRR financial function that analyzes investment cash flows as will be illustrated in Appendix C.

The most easily generated and understandable graphic method of solving for i^* is to create the **NPW profile** on computer. In the graph, the horizontal axis indicates interest rate, and the vertical axis indicates NPW. For a given project's cash flows, the NPW is calculated at an interest rate of zero (which gives the vertical axis intercept) and several other interest rates. Points are plotted, and a curve sketched. Since i^* is defined as the interest rate at which $PW(i^*) = 0$, the point at which the curve crosses the horizontal axis closely approximates the i^*. The graphical approach works for both simple and nonsimple investments.

Example 9.4 Graphical Approach to Estimate i^*

Consider the cash flow series shown in Figure 9.3(a). Estimate the rate of return by generating the NPW profile on a computer.

Solution

Given: Cash flow series in Fig. 9.3(a)

Find: i^* by plotting the NPW profile

The present worth function for the project cash flow series is

$$PW(i) = -\$10,000 + \$20,000(P/A, i, 2) - \$25,000(P/F, i, 3)$$

We first use $i = 0$ in this equation, to obtain NPW = $5000, which is the vertical axis intercept. Substitute several other interest rates—10%, 20%, . . ., 140%—and plot these values of $PW(i)$ as well. The result is Fig. 9.3(b), which shows the curve crossing the horizontal axis at roughly 140%. This value can be verified by other methods if we desire substitution. Note that, in addition to establishing the interest rate that makes NPW = 0, the NPW profile indicates where positive and negative NPW values fall, thus giving us a broad picture of those interest rates for which the project is acceptable or unacceptable. (Note that, a trial-and-error method would lead to some confusion—as you increase the interest rate from 0% to 20%, the NPW value also keeps increasing, instead of decreasing.) Even though the project is a nonsimple investment, the curve crosses the horizontal axis only once. As mentioned in the previous section, however, most nonsimple pro-

---

[5] An alternative method of solving i^* problems is to use a computer-aided economic analysis program. **Ez-Cash** finds i^* visually by specifying the lower and upper bounds of the interest search limit and generates NPW profiles when given a cash flow series. In addition to the savings of calculation time, the advantage of computer-generated profiles is their precision. EzCash can be downloaded from the book's web site.

jects have more than one value of i^*, which makes NPW = 0; i.e., more than one i^* for a project. In such a case, the NPW profile would cross the horizontal axis more than once.[6]

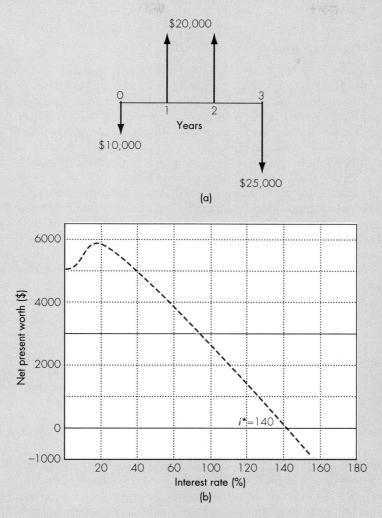

Figure 9.3 Graphical solution to rate of return for a typical non-simple investment (Example 9.4)

---

[6] In Appendix A, we discuss methods of predicting the number of i^* values by looking at cash flows. However, generating an NPW profile to discover multiple i^*s is as practical and informative as any other method.

9.3 Internal Rate of Return Criterion

Now that we have classified investment projects and learned methods to determine the i^* value for a given project's cash flows, our objective is to develop an accept/reject decision rule that gives results consistent with those obtained from NPW analysis.

9.3.1 Relationship to the PW Analysis

As we already observed in Chapter 7, NPW analysis is dependent on the rate of interest used for the NPW computation. A different rate may change a project from being considered acceptable to being unacceptable, or it may change the ranking of several projects. Consider again the NPW profile as drawn for the simple project in Fig. 9.1(a). For interest rates below i^*, this project should be accepted as NPW > 0; for interest rates above i^* it should be rejected.

On the other hand, for certain nonsimple projects, the NPW may look like the one shown in Fig. 9.1(b). Use of NPW analysis would lead you to accept the projects in regions **A** and **C**, but reject those in regions **B** and **D**. Of course, this result goes against intuition—a higher interest rate would change an unacceptable project into an acceptable one. The situation graphed in Fig. 9.1(b) is one of the cases of multiple i^*s mentioned in Definition 2. Therefore, for the simple investment situation in Fig. 9.1(a), the i^* can serve as an appropriate index for either accepting or rejecting the investment. However, for the nonsimple investment in Fig. 9.1(b), it is not clear which i^* to use to make an accept/reject decision. Therefore, the i^* value fails to provide an appropriate measure of profitability for an investment project with multiple rates of return.

9.3.2 Decision Rule for Simple Investments

Suppose we have a simple investment. Why are we interested in finding the particular interest rate that equates a project's cost with the present worth of its receipts? Again, we may easily answer this by examining Fig. 9.1(a). In this figure, we notice two important characteristics of the NPW profile. First, as we compute the project's $PW(i)$ at a varying interest rate (i), we see that the NPW is positive for $i < i^*$, indicating that the project would be acceptable under the PW analysis for those values of i. Second, the NPW is negative for $i > i^*$, indicating that the project is unacceptable for those values of i. Therefore, the i^* serves as a **bench mark** interest rate. By knowing this bench mark rate, we will be able to make an accept/reject decision consistent with the NPW analysis.

Note that, for a simple investment, i^* is indeed the IRR of the investment (see Section 9.1.2). Merely knowing the i^* is not enough to apply this method, however. Because firms typically wish to do better than break even (recall that at NPW = 0 we were indifferent to the project), a minimum acceptable rate of return (MARR) is indicated by company policy, management, or the project decision maker. If the IRR exceeds this MARR, we are assured that the company will more than break-even. Thus, the IRR becomes a useful gauge against which to judge project acceptability, and the decision rule for a simple project is as follows:

If IRR $>$ MARR, accept the project.

If IRR $=$ MARR, remain indifferent.

If IRR $<$ MARR, reject the project.

   Note that this decision rule is designed to be applied for a single project evaluation. When we have to compare mutually exclusive investment projects, we need to apply the **incremental analysis approach**, as we shall see in Section 9.4.2.

Example 9.5 Investment Decision for a Simple Investment

Merco Inc, a machinery builder in Louisville, Kentucky, is considering making an investment of $1,250,000 in a complete, structural beam-fabrication system. The increased productivity resulting from the installation of the drilling system is central to the project's justification. Merco estimates the following figures as a basis for calculating productivity:

- Increased fabricated steel production: 2,000 tons/year
- Average sales price/ton fabricated steel: $2,566.50/ton
- Labor rate : $10.50/hour
- Tons of steel produced in a year: 15,000 tons
- Cost of steel per ton (2,205 lb): $1,950/ton
- Number of workers on layout, hole making, sawing, and material handling: 17
- Additional maintenance cost: $128,500/year

With the cost of steel at $1,950 per ton and the direct labor cost of fabricating 1 lb at 10 cents, the cost of producing a ton of fabricated steel is about $2,170.50. With a selling price of $2,566.50 per ton, the resulting contribution to overhead and profit becomes $396 per ton. Assuming that Merco will be able to sustain an increased production of 2,000 tons per year by purchasing the system, the projected additional contribution has been estimated to be 2,000 tons $\times$ $396 = $792,000.

   Since the drilling system has the capacity to fabricate the full range of structural steel, two workers can run the system, one on the saw and the other on the drill. A third operator is required to operate a crane for loading and unloading materials. Merco estimates that, to do the equivalent work of these three workers with conventional manufacture, would require, on the average, an additional 14 people for center punching, hole making with a radial or magnetic drill, and material handling. This translates into a labor savings in the amount of $294,000 per year (14 $\times$ $10.50 $\times$ 40 hours/week $\times$ 50 weeks/year). The system can last for 15 years with an estimated after-tax salvage value of $80,000. However, after an annual deduction of $226,000 in corporate income taxes, the net investment cost as well as savings are as follows:

- Project investment cost : $1,250,000
- Projected annual net savings:

($792,000 + $294,000) − $128,500 − $226,000 = $731,500

- Projected after-tax salvage value at the end of year 15: $80,000
- (a) What is the projected IRR on this fabrication investment?
- (b) If Merco's MARR is known to be 18%, is this investment justifiable?

Solution

Given: Projected cash flows as shown in Figure 9.4, MARR = 18%

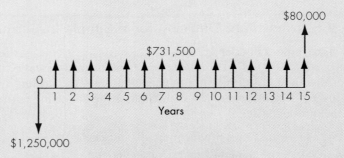

Figure 9.4   Cash flow diagram (Example 9.5)

Find: (a) IRR and (b) whether to accept or reject the investment

(a) Since only one sign change occurs in the net cash flow series, the fabrication project is a simple investment. This indicates that there will be a unique rate of return that is internal to the project:

$$PW(i) = -\$1,250,000 + \$731,500(P/A, i, 15)$$
$$+ \$80,000(P/F, i, 15)$$
$$= 0.$$

Using the trial-and-error approach outlined in Section 9.2.2, let's calculate the present values at two interest rates:

At $i = 50\%$:

$$PW(50\%) = -\$1,250,000 + \$731,500(P/A, 50\%, 15)$$
$$+ 80,000(P/F, 50\%, 15)$$
$$= \$209,842.$$

At $i = 60\%$:

$$PW(60\%) = -\$1,250,000 + \$731,500(P/A, 60\%, 15)$$
$$+ \$80,000(P/F, 60\%, 15)$$
$$= -\$31,822.$$

After making further iterative calculations, you will find that the IRR is about 58.71% for the net investment of $1,250,000.

(b) The IRR figure far exceeds Merco's MARR, indicating that the fabrication system project is an economically attractive one. Merco's management believes that, over a broad base of structural products, there is no doubt that the installation of the fabricating system would result in a significant savings, even after considering some potential deviations from the estimates used in the above analysis.

9.3.3 Decision Rule for Nonsimple Investments

When applied to simple projects, the i^* provides an unambiguous criterion for measuring profitability. However, when multiple rates of return occur, none of them is an accurate portrayal of project acceptability or profitability. Clearly, then, we should place a high priority on discovering this situation early in our analysis of a project's

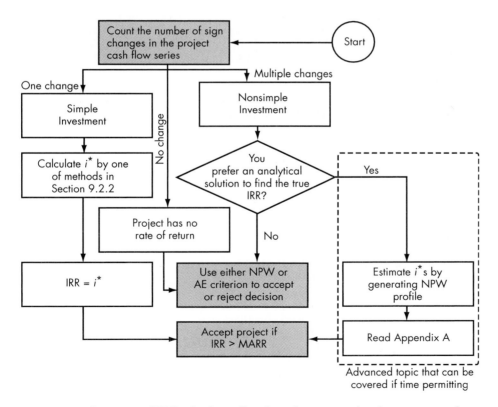

Figure 9.5 Summary of IRR criterion—flowchart that summarizes how to proceed when applying the net cash flow sign rule

cash flows. The quickest way to predict multiple i^*s is to generate a NPW profile on computer and to check if it crosses the horizontal axis more than once.

In addition to the NPW profile, there are good—although somewhat more complex—analytical methods for predicting multiple i^*s. Perhaps more importantly, there is a good method, which uses an **external interest rate**, for refining our analysis when we do discover multiple i^*s. An external rate of return allows us to calculate a single accurate rate of return; it is discussed in Appendix A.

If you choose to avoid these more complex applications of rate of return techniques, you must be able to predict multiple i^*s via the NPW profile and, when they occur, select an alternative method such as NPW or AE analysis for determining project acceptability. Figure 9.5 summarizes the decision logic that should be followed in making investment decisions based on the IRR criterion.

Example 9.6 Investment Decision for a Nonsimple Project

By outbidding its competitors Trane Image Processing (TIP), a defense contractor, has received a contract worth $7,300,000 to build navy flight simulators for U.S. Navy pilot training over 2 years. For some defense contracts, the U.S. government makes an advance payment when the contract is signed, but in this case, the government will make two progressive payments; $4,300,000 at the end of the first year, and the $3,000,000 balance at the end of the second year. The expected cash outflows required to produce these simulators are estimated to be $1,000,000 now, $2,000,000 during the first year, and $4,320,000 during the second year. The expected net cash flows from this project are summarized as follows:

Year	Cash Inflow	Cash Outflow	Net Cash Flow
0		$1,000,000	–$1,000,000
1	$4,300,000	2,000,000	2,300,000
2	3,000,000	4,320,000	–1,320,000

In normal situations, TIP would not even consider a marginal project such as this one in the first place. However, hoping that it can establish itself as a technology leader in the field, management felt that it was worth outbidding its competitors by providing the lowest bid. Financially, what is the economic worth of outbidding the competitors for this project?

(a) Compute the values of i^*s for this project.
(b) Make an accept/reject decision based on the results in part (a). Assume that the contractor's MARR is 15%.

Solution

Given: Cash flow shown above, MARR = 15%

Find: (a) i^* and (b) determine whether to accept the project

(a) Since this project has a 2-year life, we may solve the net present worth equation directly via the quadratic formula method:

$$-\$1,000,000 + \$2,300,000/(1 + i^*) - \$1,320,000/(1 + i^*)^2 = 0$$

If we let $X = 1/(1 + i^*)$, we can rewrite the expression

$$-1,000,000 + 2,300,000X - 1,320,000X^2 = 0.$$

Solving for X gives $X = 10/11$ and $10/12$, or $i^* = 10\%$ and 20%. As shown in Figure 9.6, the NPW profile intersects the horizontal axis twice, once at 10% and again at 20%. The investment is obviously not a simple one, and thus neither 10% nor 20% represents the true internal rate of return of this government project.

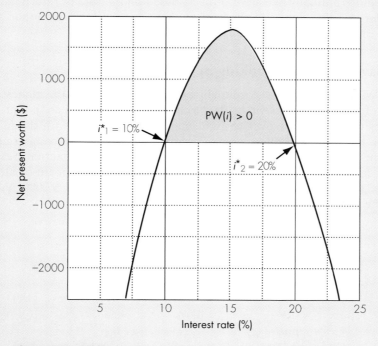

Figure 9.6   NPW plot for a nonsimple investment with multiple rates of return (Example 9.6)

(b) Since the project is a nonsimple project, we may abandon the IRR criterion for practical purposes and use the NPW criterion. If we use the present worth method at MARR = 15%, we obtain

$$PW(15\%) = -\$1,000,000 + \$2,300,000(P/F, 15\%, 1)$$
$$-\$1,320,000(P/F, 15\%, 2)$$
$$= \$1,890 > 0,$$

which verifies that the project is marginally acceptable, and it is not as bad as initially believed.

Comments: Example A.4 in Appendix A illustrates how you go about finding the true rate of return for this nonsimple investment project.

9.4 Mutually Exclusive Alternatives

In this section, we present the decision procedures that should be used when comparing two or more mutually exclusive projects based on the rate of return measure. We will consider two situations: (1) alternatives that have the same economic service life and (2) alternatives that have unequal service lives.

9.4.1 Flaws in Project Ranking by IRR

Under NPW or AE analysis, the mutually exclusive project with the highest worth figure was preferred (this is known as the "total investment approach"). Unfortunately, the analogy does not carry over to IRR analysis. The project with the highest IRR may *not* be the preferred alternative. To illustrate the flaws of comparing IRRs to choose from mutually exclusive projects, suppose you have two mutually exclusive alternatives, each with a 1-year service life: One requires an investment of $1,000 with a return of $2,000, and the other requires $5,000 with a return of $7,000. You already obtained the IRRs and NPWs at MARR = 10% as follows:

n	A1	A2
0	−$1,000	−$5,000
1	2,000	7,000
IRR	100%	40%
$PW(10\%)$	$818	$1,364

Assuming that you have enough money in your investment pool to select either one, would you prefer the first project simply because you expect a higher rate of return?

We can see that A2 is preferred over A1, by the NPW measure. On the other hand, the IRR measure gives a numerically higher rating for A1. This inconsistency in ranking occurs because the NPW, NFW, and AE are **absolute (dollar)** measures of investment worth, whereas the IRR is a **relative (percentage)** measure and cannot be applied in the same way. That is, the IRR measure ignores the **scale** of the investment. Therefore, the answer is no; instead, you would prefer the second project with the lower rate of return, but higher NPW. Either the NPW or the AE measure would lead to that choice, but comparison of IRRs would rank the smaller project higher. Another approach, referred to as **incremental analysis**, is needed.

9.4.2   Incremental Investment Analysis

In our previous ranking example, the more costly option requires an incremental investment of $4,000 at an incremental return of $5,000.

- If you decide to invest in option A1, you will need to withdraw only $1,000 from your investment pool. The remaining $4,000 will continue to earn 10% interest. One year later, you will have $2,000 from the outside investment and $4,400 from the investment pool. With an investment of $5,000, in 1 year you will have $6,400. The equivalent present worth of this wealth change is $PW(10\%) = -\$5,000 + \$6,400(P/F, 10\%, 1) = \$818$.

- If you decide to invest in option A2, you will need to withdraw $5,000 from your investment pool, leaving no money in the pool, but you will have $7,000 from your outside investment. Your total wealth changes from $5,000 to $7,000 in a year. The equivalent present worth of this wealth change is $PW(10\%) = -\$5,000 + \$7,000(P/F, 10\%, 1) = \$1,364$.

In other words, if you decide to take the more costly option, certainly you would be interested in knowing that this additional investment can be justified at the MARR. The 10% of MARR value implies that you can always earn that rate from other investment sources (i.e., $4,400 at the end of 1 year for a $4,000 investment). However, in the second option, by investing the additional $4,000, you would make an additional $5,000, which is equivalent to earning at the rate of 25%. Therefore, the incremental investment can be justified.

Now we can generalize the decision rule for comparing mutually exclusive projects. For a pair of mutually exclusive projects (*A* and *B*, with *B* defined as a more costly option), we may rewrite *B* as

$$B = A + (B - A).$$

In other words, *B* has two cash flow components: (1) The same cash flow as *A* and (2) the incremental component (*B* − *A*). Therefore, the only situation in which *B* is preferred to *A* is when the rate of return on the incremental component (*B* − *A*) exceeds the MARR. Therefore, for two mutually exclusive projects, rate of return analysis is done by computing the *internal rate of return on incremental investment* (*IRR*Δ) between the projects. Since we want to consider increments of investment, we compute the cash flow for the difference between the projects by subtracting the cash flow for the lower investment-cost project (*A*) from that of the higher investment-cost project (*B*). Then, the decision rule is

$$\text{If IRR}_{B-A} > \text{MARR, select B,}$$

$$\text{If IRR}_{B-A} = \text{MARR, select either one,}$$

$$\text{If IRR}_{B-A} < \text{MARR, select A,}$$

where $B - A$ is an investment increment (negative cash flow). *If a "do-nothing" alternative is allowed, the smaller cost option must be profitable (its IRR must be greater than MARR) at first.* This means that you compute the rate of return for each alternative in the mutually exclusive group, and then eliminate the alternatives whose IRRs are less than MARR before applying the incremental analysis.

It may seem odd to you how this simple rule allows us to select the right project. Example 9.7 will illustrate the incremental investment decision rule for you.

Example 9.7 IRR on Incremental Investment: Two Alternatives

John Covington, a college student, wants to start a small-scale painting business during his off-school hours. To economize the start-up business, he decides to purchase some used painting equipment. He has two mutually exclusive options: Do most of the painting by himself by limiting his business to only residential painting jobs (B1) or purchase more painting equipment and hire some helpers to do both residential and commercial painting jobs that he expects will have a higher equipment cost, but provide higher revenues as well (B2). In either case, he expects to fold up the business in 3 years when he graduates from college.

The cash flows for the two mutually exclusive alternatives are given as follows:

n	B1	B2	B2 – B1
0	–$3,000	–$12,000	–$9,000
1	1,350	4,200	2,850
2	1,800	6,225	4,425
3	1,500	6,330	4,830
IRR	25%	17.43%	

Knowing that both alternatives are revenue projects, which project would he select at MARR =10%? (Note that both projects are profitable at 10%.)

Solution

Given: Incremental cash flow between two alternatives, MARR = 10%

Find: (a) IRR on the increment and (b) which alternative is preferable

(a) To choose the best project, we compute the incremental cash flow for $B2 - B1$. Then we compute the IRR on this increment of investment by solving

$$-\$9,000 + \$2,850(P/F, i, 1) + \$4,425(P/F, i, 2) + \$4,830(P/F, i, 3) = 0.$$

(b) We obtain $i^*_{B2-B1} = 15\%$, as plotted in Figure 9.7. By inspection of the incremental cash flow, we know it is a simple investment, so $IRR_{B2-B1} = i^*_{B2-B1}$. Since $IRR_{B2-B1} > MARR$, we select $B2$, which is consistent with the NPW analysis. Note that, at $MARR > 25\%$, neither project would be acceptable

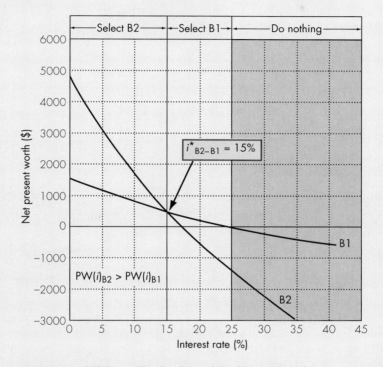

Figure 9.7   NPW profiles for $B1$ and $B2$ (Example 9.7)

Comments: Why did we choose to look at the increment $B2 - B1$ instead of $B1 - B2$? We want the first flow of the incremental cash flow series to be negative (investment flow) so that we can calculate an IRR. By subtracting the lower initial investment project from the higher, we guarantee that the first increment will be an investment flow. If we ignore the investment ranking, we might end up with an increment that involves borrowing cash flow and has no internal rate of return. This is the case for $B1 - B2$. (i^*_{B1-B2} is also 15%, not −15%.) If, erroneously, we had compared this i^* with the MARR, we might have accepted project $B1$ over $B2$. This undoubtedly would have damaged our credibility with management! We will revisit this problem in Example 9.11.

The next example indicates that the ranking inconsistency between NPW and IRR can also occur when differences in the timing of a project's future cash flows exist, even if their initial investments are the same.

Example 9.8 IRR on Incremental Investment when Initial Flows are Equal

Consider the following two mutually exclusive investment projects that require the same amount of investment:

n	C1	C2
0	-$9,000	-$9,000
1	480	5,800
2	3,700	3,250
3	6,550	2,000
4	3,780	1,561
IRR	18%	20%

Which project would you select based on rate of return on incremental invest-ment, assuming that MARR = 12%?(Once again both projects are profitable at 12%.)

Solution

Given: Cash flows for two mutually exclusive alternatives as shown, MARR = 12%

Find: (a) IRR on incremental investment and (b) which alternative is preferable

(a) When initial investments are equal, we progress through the cash flows until we find the first difference, then set up the increment so that this first non-zero flow is negative (i.e., an investment). Thus, we set up the incre-mental investment by taking (C1– C2):

n	C1 – C2
0	$ 0
1	−5,320
2	450
3	4,550
4	2,219

We next set the PW equation equal to zero, as follows:

$$- \$5{,}320 + \$450(P/F, i, 1) + \$4{,}550(P/F, i, 2) + \$2{,}219(P/F, i, 3) = 0.$$

(b) Solving for i yields $i* = 14.71\%$, which is also an IRR since the increment is a simple investment. Since the $\text{IRR}_{C1-C2} = 14.71\% > \text{MARR}$, we would select C1. If we used NPW analysis, we would obtain $PW(12\%)_{C1} = \$1,443$ and $PW(12\%)_{C2} = \$1,185$, indicating the preference of C1 over C2.

When you have more than two mutually exclusive alternatives, they can be compared in pairs by successive examination. Example 9.9 will illustrate how to compare three alternative problems. (In Chapter 16, we will examine some multiple-alternative problems in the context of capital budgeting.)

Example 9.9 IRR on Increment Investment: Three Alternatives

Consider the following three sets of mutually exclusive alternatives:

n	D1	D2	D3
0	−$2,000	−$1,000	−$3,000
1	1,500	800	1,500
2	1,000	500	2,000
3	800	500	1,000
IRR	34.37%	40.76%	24.81%

Which project would you select based on rate of return on incremental investment, assuming that MARR $= 15\%$?

Solution

Given: Cash flows given above, MARR $= 15\%$

Find: IRR on incremental investment and which alternative is preferable

Step 1: Examine the IRR for each alternative. At this point, we can eliminate any alternative that fails to meet the MARR. In this example, all three alternatives exceed the MARR.

Step 2: Compare D1 and D2 in pairs.[7] Because D2 has a lower initial cost, compute the rate of return on the increment (D1 − D2), which represents an increment of investment.

---

[7] When faced with many alternatives, you may arrange them in order of increasing initial cost. This is not a required step, but it makes the comparison more tractable.

n	D1 – D2
0	–$1,000
1	700
2	500
3	300

The incremental cash flow represents a simple investment. To find the incremental rate of return, we set up

$$-\$1,000 + \$700(P/F, i, 1) + \$500(P/F, i, 2) + \$300(P/F, i, 3) = 0.$$

Solving for i^*_{D1-D2} yields 27.61%, which exceeds the MARR; therefore, D1 is preferred over D2.

Step 3: Compare D1 and D3. Once again, D1 has a lower initial cost. Examine the (D3 – D1) increment.

n	D3 – D1
0	–$1,000
1	0
2	1,000
3	200

Here, the incremental cash flow represents another simple investment. The (D3 – D1) increment has an unsatisfactory 8.8% rate of return; therefore, D1 is preferred over D3. In summary, we conclude that D1 is the best alternative.

Example 9.10 Incremental Analysis for Cost-Only Projects

Falk Corporation is considering two types of manufacturing systems to produce its shaft couplings over 6 years: (1) A cellular manufacturing system (CMS) and (2) a flexible manufacturing system (FMS). The average number of pieces to be produced on either system would be 544,000 per year. Operating cost, initial investment, and salvage value for each alternative are estimated as follows:

Items	CMS Option	FMS Option
Annual operating and maintenance costs		
Annual labor cost	$1,169,600	$707,200
Annual material cost	832,320	598,400
Annual overhead cost	3,150,000	1,950,000
Annual tooling cost	470,000	300,000
Annual inventory cost	141,000	31,500
Annual income taxes	1,650,000	1,917,000
Total annual costs	$7,412,920	$5,504,100
Investment	$4,500,000	$12,500,000
Net salvage value	$500,000	$1,000,000

Figure 9.8 illustrates the cash flows *associated* with each *alternative*. The firm's MARR is 15%. Which alternative would be a better choice based on the IRR criterion?

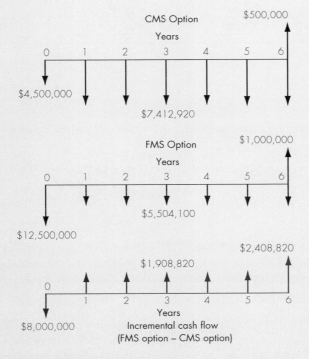

Figure 9.8   Comparison of mutually exclusive alternatives with equal revenues (cost only) (Example 9.10)

Discussion: Since we can assume that both manufacturing systems would provide the same level of revenues over the analysis period, we can compare these alternatives based on cost only. (These are service projects.) Although we cannot compute the IRR for each option without knowing the revenue figures, we can still calculate the IRR on incremental cash flows. Since the FMS option requires a higher initial investment than that for the CMS, the incremental cash flow is the difference (FMS –CMS).

n	CMS Option	FMS Option	Incremental (FMS – CMS)
0	−$4,500,000	−$12,500,000	−$8,000,000
1	−7,412,920	−5,504,100	1,908,820
2	−7,412,920	−5,504,100	1,908,820
3	−7,412,920	−5,504,100	1,908,820
4	−7,412,920	−5,504,100	1,908,820
5	−7,412,920	−5,504,100	1,908,820
6	−7,412,920 + $500,000	−5,504,100 + $1,000,000	$2,408,820

Solution

Given: Cash flows shown in Fig. 9.8, $i = 15\%$ per year

Find: Incremental cash flows, and select the better alternative based on the IRR criterion

$$PW(i)_{FMS-CMS} = -\$8,000,000 + \$1,908,820(P/A, i, 5)$$
$$+ \$2,408,820(P/F, i, 6)$$
$$= 0.$$

Solving for i by trial and error yields 12.43%. Since $IRR_{FMS-CMS} = 12.43\% < 15\%$, we would select CMS. Although the FMS would provide an incremental annual savings of $1,908,820 in operating costs, the savings do not justify the incremental investment of $8,000,000.

Comments: Note that the CMS option was marginally preferred to the FMS option. However, there are dangers in relying solely on the easily quantified savings in input factors—such as labor, energy, and materials—from FMS and in not considering gains from improved manufacturing performance that are more difficult and subjective to quantify. Factors such as improved product quality, increased manufacturing flexibility (rapid response to customer demand), reduced inventory levels, and increased capacity for product innovation are frequently ignored because we have inadequate means for quantifying their benefits. If these intangible benefits were considered, however, the FMS option could come out better than the CMS option.

9.4.3   Incremental Borrowing Analysis

Subtracting the less costly alternative from the more costly one is not absolutely necessary in incremental analysis. In fact, we can examine the difference between two projects A and B as either an (A − B) increment or a (B − A) increment. If the difference in flow (B − A) represents an increment of **investment**, then (A − B) is an increment of **borrowing**. When looking at increments of investment, we accepted the increment when its rate of return exceeded the MARR. When considering an increment of borrowing, however, the rate we calculated (that is, i^* for borrowing) was essentially the rate we paid to borrow money from the increment. We will call this the **borrowing rate of return (BRR)**. Conceptually, we would prefer to get a loan from the increment rather than from our initial investment pool, if the loan rate is less than the MARR. In other words, it is cheaper to borrow than to use your own money. Therefore, the decision rule is reversed:

$$\text{If } BRR_{B-A} < \text{MARR, select B,}$$

$$\text{If } BRR_{B-A} = \text{MARR, select either one,}$$

$$\text{If } BRR_{B-A} > \text{MARR, select A,}$$

where B − A is a borrowing increment (positive first cash flow).

Example 6.11 Borrowing Rate of Return on Incremental Projects

Consider Example 9.7 again, but this time compute the rate of return on the increment B1 − B2.

n	B1	B2	B1 − B2
0	−$3,000	−$12,000	+$9,000
1	1,350	4,200	−2,850
2	1,800	6,225	−4,425
3	1,500	6,330	−4,830

Note that the first incremental cash flow is positive, and all others are negative, which indicates that the cash flow difference is an increment of borrowing. What is the rate of return on this increment of borrowing, and which project should we prefer?

Solution

Given: Incremental borrowing cash flow, MARR = 10%

Find: BRR on increment, and determine whether it is acceptable

Note that the cash flow signs are simply reversed from Example 9.7. The rate of return on this increment of borrowing will be the same as that on the increment of investment, $i^*_{B1-B2} = i^*_{B2-B1} = 15\%$. However, because we are *borrowing* $9,000, the 15% interest rate is what we are losing, or paying, by not investing the $9,000 in B2. If we invest in B1, we are saving $9,000 now, but we are giving up the opportunity of making $2,850 at the end of first year, $4,428 at the end of second year, and $4,830 at the end of third year. This is equivalent to a situation where we are borrowing $9,000, with the repayment series of ($2,850, $4,428, $4,830). In

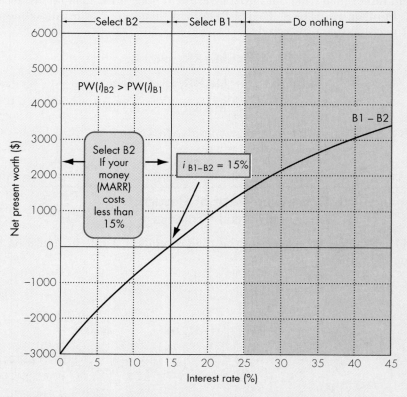

Figure 9.9   Incremental NPW profile for B1 − B2 (Example 9.11)

effect, we are paying 15% interest on a borrowed sum. Is this an acceptable rate for a loan, considering the cost of using our own money is only 10%? Since the firm's MARR is 10%, we can assume that our maximum interest rate for borrowing should also be 10%. Certainly, borrowing under these circumstances is undesirable. Since 15% > 10%, this borrowing situation is not desirable, and we should reject the lower cost alternative—B1—which produced the increment of borrowing (Figure 9.9). We choose instead B2, the same result as in Example 9.7.

Since the incremental investment and incremental borrowing methods yield the same results, if you find one intuitively easier to understand, you should feel free to set up project comparisons consistently by that method. Remember that a negative increment indicates investment and a positive increment indicates borrowing. However, we recommend that alternatives be compared on the basis of incremental *investment* rather than *borrowing*. This strategy allows you to avoid having to remember two decision rules. It also circumvents the ranking problem of deciding which rule to apply to mixed increments. (The first rule always applies to nonsimple investments: If the IRR on the increment is greater than MARR, the increment is acceptable.)

9.4.4   Handling Unequal Service Lives

In Chapters 7 and 8, we discussed the use of the NPW and AE criteria as bases for comparing projects with unequal lives. The IRR measure can also be used to compare projects with unequal lives, as long as we can establish a common analysis period. The decision procedure is then exactly the same as for projects with equal lives. It is likely, however, that we will have a multiple-root problem, which creates a substantial computational burden. For example, suppose we apply the IRR measure to a case in which one project has a 5-year life and the other project has an 8-year life, resulting in a least common multiple of 40 years. When we determine the incremental cash flows over the analysis period, we are bound to observe many sign changes. This leads to the possibility of having many $i*$s. The following two examples use $i*$ to compare mutually exclusive projects, one with only one sign change in the incremental cash flows, and one with several sign changes. (Our purpose is not to encourage you to use the IRR approach to compare projects with unequal lives. Rather, it is to show the correct way to compare them if the IRR approach must be used.)

Example 9.12   IRR Analysis for Projects with Unequal Lives in which the Increment is a Simple Investment

Consider the following mutually exclusive investment projects (E1, E2):

n	E1	E2
0	−$2,000	−$3,000
1	1,000	4,000
2	1,000	
3	1,000	

Project E1 has a service life of 3 years, whereas project E2 has only 1 year of service life. Assume that project E2 can be repeated with the same investment costs and benefits over the analysis period of 3 years. Assume that the firm's MARR is 10%. Determine which project should be selected.

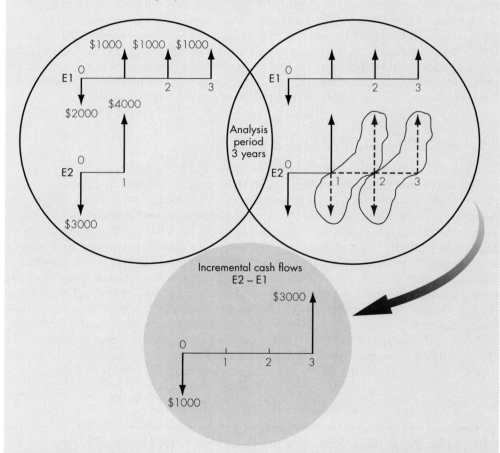

Figure 9.10 Comparison of projects with unequal lives when the required service period is indefinite (Example 9.12)

Solution

Given: Two alternatives with unequal lives, cash flows as shown, MARR = 10%

Find: IRR on incremental investment, and determine which alternative is preferable

By assuming three repetitions of project E2 over the analysis period (as shown in Figure 9.10) and taking the incremental cash flows (E2 – E1), we obtain

n	E1	E2		E2 – E1
0	–$2,000	–$3,000 =	–$3,000	–$1,000
1	1,000	4,000 – 3,000 =	1,000	0
2	1,000	4,000 – 3,000 =	1,000	0
3	1,000	4000 =	4,000	3,000

By inspection, the increment in this case is an investment. To compute i^*_{E2-E1}, we evaluate

$$- \$1,000 + \frac{\$3,000}{(1+i)^3} = 0.$$

Solving for i yields

$$\text{IRR}_{E2-E1} = 44.22\% > 10\%.$$

Therefore, we select project E2.

Example 9.13 IRR Analysis for Projects with Different Lives in which the Increment is a Nonsimple Investment

Consider Example 7.14, in which a mail-order firm wants to install an automatic mailing system to handle product announcements and invoices. The firm had a choice between two different types of machines. Using the IRR as a decision criterion, select the best machine. Assume a MARR of 15% as before.

Solution

Given: Cash flows for two projects with unequal lives, as shown in Figure 9.11, MARR = 15%

Find: The alternative that is preferable

Since the analysis period is equal to the least common multiple of 12 years, we may compute the incremental cash flow over this 12-year period. As shown in Fig. 9.11, we subtract cash flows of model A from those of model B to form the increment of investment. (We want the first cash flow difference to be a negative value.)

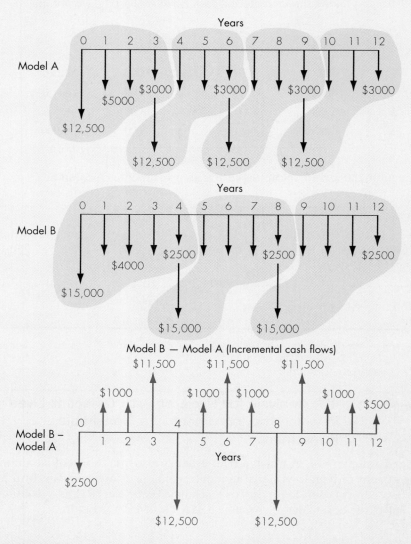

Figure 9.11 Comparison of projects with unequal lives (Example 9.13)

Five sign changes occur in the incremental cash flows, indicating a nonsimple incremental investment. As suggested in Section 9.3.3, we may abandon the rate of

n	Model A		Model B		Model B – Model A
0	–$12,500		–$15,000		–$2,500
1		–5,000		–4,000	1,000
2		–5,000		–4,000	1,000
3	–12,500	–3,000		–4,000	11,500
4		–5,000	–15,000	–2,500	–12,500
5		–5,000		–4,000	1,000
6	–12,500	–3,000		–4,000	11,500
7		–5,000		–4,000	1,000
8		–5,000	–15,000	–2,500	–12,500
9	–12,500	–3,000		–4,000	11,500
10		–5,000		–4,000	1,000
11		–5,000		–4,000	1,000
12		–3,000		–2,500	500

return analysis and use the NPW criterion. Otherwise, we need to find out the true rate of return on incremental investment using the technique explained in Appendix A (See Example A6).

9.5 Summary

- **Rate of return (ROR)** is the interest rate earned on unrecovered project balances such that an investment's cash receipts make the terminal project balance equal to zero. Rate of return is an intuitively familiar and understandable measure of project profitability that many managers prefer to NPW or other equivalence measures.

- Mathematically we can determine the rate of return for a given project cash flow series by locating an interest rate that equates the net present worth of its cash flows to zero. This break-even interest rate is denoted by the symbol i^*.

- **Internal rate of return (IRR)** is another term for ROR that stresses the fact that we are concerned with the interest earned on the portion of the project that is internally invested, not those portions that are released by (borrowed from) the project.

- To apply rate of return analysis correctly, we need to classify an investment into either a simple or a nonsimple investment. A **simple investment** is defined as one in which the initial cash flows are negative and only one sign change in the net cash flow occurs, whereas a **nonsimple investment** is one for which more than one sign change in the cash flow series occurs. Multiple i^*s occur only in nonsimple investments. However, not all nonsimple investments will have multiple i^*s.

- For a simple investment, the solving rate of return (i^*) is the rate of return internal to the project; so the decision rule is:

$$\text{If IRR} > MARR, \text{accept the project.}$$
$$\text{If IRR} = MARR, \text{remain indifferent.}$$
$$\text{If IRR} < MARR, \text{reject the project.}$$

IRR analysis yields results consistent with NPW and other equivalence methods.

- For a nonsimple investment, because of the possibility of having multiple rates of return, it is recommended that the IRR analysis be abandoned and either the NPW or AE analysis be used to make an accept/reject decision. Procedures are outlined in Appendix A for determining the rate of return internal to nonsimple investments. Once you find the IRR (or return on invested capital), you can use the same decision rule for simple investments.

- When properly selecting among alternative projects by IRR analysis, **incremental investment** must be used.

Self-Test Questions

9s.1 You are considering an investment that costs $2,000. It is expected to have a useful life of 3 years. You are very confident about the revenues during the first and the third year, but you are unsure about the revenue in year 2. If you hope to make at least a 10% rate of return on your investment ($2,000), what should be the minimum revenue in year 2?

Year	Cash Flow
0	−$2,000
1	1,000
2	X
3	1,200

(a) $X = \$290$ (b) $X = \$260$
(c) $X = \$230$ (d) $X = \$190$.

9s.2 You are considering a CNC machine that costs $150,000. This machine will have an estimated service life of 10 years with a net after-tax salvage value of $15,000. Its annual after-tax operating and maintenance costs are estimated to be $50,000. To expect an 18% rate of return on investment after-tax, what would be the required minimum annual after-tax revenues?

(a) $63,500   (b) $82,740
(c) $92,435   (d) $94,568.

9s.3 Find the rate of return for the following infinite cash flow series.

Year	Cash Flow
0	−$15,459
1	3,000
2	3,000
⋮	⋮

(a) 15%   (b) 515.30%
(c) 19.41%   (d) 17.83%.

9s.4 Consider the following two investment situations:

- In 1970, when Wal-Mart Stores, Inc. went public, an investment of 100 shares cost $1,650. That investment would have been worth $2,991,080 after 25 years. The Wal-Mart investors' rate of return would be around 35%.
- In 1980, if you bought 100 shares of Fidelity Mutual Funds, it would have cost $5,245. That investment would have been worth $80,810 after 15 years.

Which of the following statement is correct?

(a) If you bought only 50 shares of the Wal-Mart stocks in 1970 and kept it for 25 years, your rate of return would be 0.5 times 35%.

(b) The investors in Fidelity Mutual Funds would have made profit at the annual rate of 30% on the funds remaining invested.

(c) If you bought 100 shares of Wal-Mart in 1970 but sold them after 10 years. (Assume that the Wal-Mart stocks grew at the annual rate of 35% for the first 10 years.) Then immediately, you put all the proceeds into Fidelity Mutual Funds. After 15 years, the total worth of your investment would be around $511,140.

(d) None of the above.

9s.5 A manufacturing company is considering two types of industrial projects that require the same level of initial investment but provide different levels of operating cash flows over the project life. The in-house engineer has compiled the following financial data related to both projects, including the rate of return figures.

The company has been using the internal rate of return as a project justification tool,

	Net Cash Flow		Incremental
n	Project A	Project B	(B – A)
0	–$18,000	–$18,000	0
1	960	11,600	$10,640
2	7,400	6,500	–900
3	13,100	4,000	–9,100
4	7,560	3,122	–4,438
$i*$	18%	20%	14.72%

and these projects will be evaluated based on the principle of rate of return. The firm's minimum required rate of return is known to be 12%. Which project should you select?

(a) Select A, because its increment of borrowing exceeds 12%.

(b) Select B, because its increment of investment exceeds 12%.

(c) Select B, because its increment of borrowing exceeds 12%.

(d) Select B, because it can generate 20% profit (as opposed to 18% for project A) for every dollar invested.

9s.6 The following information on two revenue projects is given:

- IRR of project A = 16%.
- IRR of project B = 16%.
- Both the projects have a service life of 6 years and need the same initial investment of $23,000.
- IRR on incremental cash flows (A – B) = 14%.

If your MARR is 20%, which of the following statements is correct?

(a) Select A.

(b) Select B.

(c) Select either one of the projects.

(d) Select neither project.

(e) Information is insufficient to make a decision.

9s.7 Consider a project that costs $14,762, with an indefinite life. If the cash flow in years 1, 3, 5, . . . (i.e., every odd year) is $1,000 and the cash flow in years 2, 4, 6,. . . (i.e., every even year) is $2,000, find the rate of return of the investment.

(a) 9%   (b) 10%

(c) 11%   (d) 12%.

9s.8 The following information on three mutually exclusive projects is given:

Project	Investment in Year 0	IRR(%)
A	−$1,250	43%
B	−$1,000	56
C	−$1,200	67
B − C		146
B − A		144
A − C		155

All three projects have the same service life and require investment in year 0 only. Incremental rate of return on (B − C) is 146%; the incremental rate of return on (B − A) is 144%; and so forth. Which project would you choose based on the IRR criterion at a MARR of 29%?

(a) Project C   (b) Project B

(c) Project A   (d) None.

9s.9 The following data contains a summary of how a project's balance changes over its 3-year service life for two different interest rates of $i = 10\%$ and $i = 15\%$.

End of Period	Project Balance $i = 10\%$	$i = 15\%$
0	−$1,000	−$1,000
1	−$850	−$900
2	−$400	−$500
3	0	−$1,350

Which of the following statements is correct?

(a) The IRR of the project should be 15%.

(b) The IRR of the project should be greater than 10%.

(c) The IRR of the project should be equal to 10%.

(d) It is not possible to determine the IRR from the given data.

9s.10 Consider the investment project with the following Net Cash Flows.

Year	Net Cash Flow
0	−$1,500
1	$X
2	$650
3	$X

What would be the value of X if the project's IRR is 10%?

(a) $425   (c) $580

(b) $1,045 (d) $635.

9s.11 The following information on three mutually exclusive projects is given below:

Project	Investment at Year 0	IRR
A	$1,000	41%
B	$1,200	65%
C	$1,250	61%

All three projects have the same service life and require investment in year 0 only. The following additional information about incremental rate of returns is available.

$$IRR_{B-A} = 30\%$$

$$BRR_{B-C} = 45\%$$

$$BRR_{A-C} = 40\%$$

Which project would you choose based on the rate of return criterion at MARR of 30%?

(a) Project A

(b) Project B

(c) Project C

(d) Not sufficient information available to decide.

9s.12 Consider the following cash flow series. Assume that the firm's MARR is 10%.

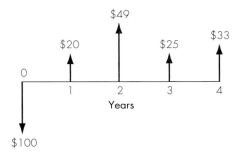

Which of the following statements is *incorrect?*

(a) The project's net future worth is zero.

(b) The project's net present worth is zero.

(c) The project's IRR is zero.

(d) The project's annual equivalent is zero.

9s.13 If you purchase stock for $100 now, what is the internal rate of return on your investment if the stock is worth $337.50 at the end of 3 years?

(a) 75.80%   (c) 33.75%

(b) 25%   (d) 50%

9s.14 The following information on two mutually exclusive projects is given below:

n	Project A	Project B
0	−3,000	−5,000
1	1,350	1,350
2	1,800	1,800
3	1,500	5,406.25
IRR	25%	25%

Which of the following statements is *correct?*

(a) Since the two projects have the same rate of return, they are indifferent.

(b) Project A would be a better choice as the required investment is smaller with the same rate of return.

(c) Project B would be a better choice as long as the investor's MARR is less than 25%.

(d) Project B is a better choice regardless of the investor's MARR.

9s.15 The following infinite cash flow has a rate of return of 15%. Compute the unknown value of X.

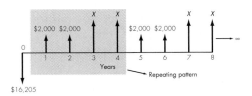

(a) $2,542   (c) $3,640

(b) $3,000   (d) $5,644

Problems

Note: *The symbol i* represents the interest rate that makes the net present value of the project equal to zero. The symbol IRR represents the **internal** rate of return of the investment. For a simple investment, IRR = i*. For a nonsimple investment, generally i* is not equal to IRR.*

Concept of Rate of Return

9.1 Assume that you are going to buy a new car worth $22,000. You will be able to make a down payment of $7,000. The remaining $15,000 will be financed by the dealer. The dealer computes your monthly payment to be $512 for 48 months financing. What is the dealer's rate of return on this loan transaction?

9.2 Mr. Smith wishes to sell a bond that has a face value of $1,000. The bond bears an interest rate of 8% with bond interests payable semiannually. Four years ago, the bond was purchased at $900. At least 9% annual return on investment is desired. What must be the minimum selling price now for the bond in order to make the desired return on the investment?

9.3 Reconsider the art auction story (Vincent Van Gogh) presented in the chapter opening. John Whitney Payson, who purchased the painting for $80,000 in 1947, sold it for $53.9 million in 1988. If Mr. Payson invested his $80,000 in another investment vehicle (such as stock), how much interest would he need to earn to accumulate the same wealth as from the painting investment? Assume for simplicity that the investment period is 40 years, and the interest is compounded annually.

Investment Classification and Calculation i*

9.4 Consider four investments with the following sequences of cash flows:

		Net Cash Flow		
n	Project A	Project B	Project C	Project D
0	−$18,000	−$20,000	+$34,578	−$56,500
1	10,000	32,000	−18,000	−2,500
2	20,000	32,000	−18,000	−6,459
3	30,000	−22,000	−18,000	−78,345

(a) Identify all the simple investments.
(b) Identify all the nonsimple investments.
(c) Compute i* for each investment.
(d) Which project has no rate of return?

9.5 Consider the following infinite cash flow series with repeated cash flow patterns.

n	A_n
0	−$1000
1	400
2	800
3	500
4	500
5	400
6	800
7	500
8	500
⋮	⋮

Determine the i* for this infinite cash flow series.

9.6 An investor bought 100 shares of stock at a cost of $10 per share. He held the stock for 15 years and then sold it for a total of $4,000. For the first 3 years, he received no dividends. For each of the next 7 years, he received total dividends of $50 per year. For the remaining period he received total dividends of $100 per year. What rate of return did he make on the investment?

9.7 Consider the following sets of investment projects:

	Project Cash Flow				
n	A	B	C	D	E
0	−$100	−$100		−$200	−$50
1	60	70	$20	120	−100
2	900	70	10	40	−50
3		40	5	40	0
4		40	−180	−20	150
5			60	40	150
6			50	30	100
7			40		100
8			30		
9			20		
10			10		

(a) Classify each project as either simple or nonsimple.

(b) Compute the $i*$ for A using the quadratic equation.

(c) Obtain the rate(s) of return for each project by plotting the NPW as a function of interest rate.

9.8 Consider the following sets of projects:

	Net Cash Flow			
n	A	B	C	D
0	−$1,000	−$1,000	−$1,700	−$1,000
1	500	800	5,600	360
2	100	600	4,900	4,675
3	100	500	−3,500	2,288
4	1,000	700	−7,000	
5			−1,400	
6			2,100	
7			900	

(a) Classify each project as either simple or nonsimple.

(b) Identify all positive $i*$s for each project.

(c) Plot the present worth as a function of interest rate (i) for each project.

9.9 Consider the following financial data for a project:

Initial investment	$10,000
Project life	8 years
Salvage value	$0
Annual revenue	$5,229
Annual expenses (including income taxes)	$3,000

(a) What is the $i*$ for this project?

(b) If annual expense increases at a 7% rate over the previous year's expenses but annual income is unchanged, what is the new $i*$?

(c) In part (b), at what annual rate will annual income have to increase to maintain the same $i*$ obtained in part (a)?

9.10 Consider two investments, A and B, with the following sequences of cash flows:

	Net Cash Flow	
n	Project A	Project B
0	−$25,000	−$25,000
1	2,000	10,000
2	6,000	10,000
3	12,000	10,000
4	24,000	10,000
5	28,000	5,000

(a) Compute the $i*$ for each investment.

(b) Plot the present worth curve for each project on the same chart, and find the interest rate that makes the two projects equivalent.

IRR Analysis

9.11 Consider an investment project with the following cash flows:

n	Cash Flow
0	−$5000
1	0
2	4840
3	1331

Compute the IRR for this investment. Is this project acceptable at MARR = 10 %?

9.12 Consider the following project's cash flow:

n	Net Cash Flow
0	−$2000
1	800
2	900
3	X

If the project's IRR is 10%,

(a) Find the value of X.

(b) Is this project acceptable at MARR = 8%?

9.13 The InterCell Company wants to participate in the World Fair to be held in Mexico. To participate, the firm needs to spend $1 million at year 0 to develop a showcase. The showcase will produce a cash flow of $2.5 million at the end of year 1. Then, at the end of year 2, $1.54 million must be expended to restore the land to its original condition. Therefore, the project's expected net cash flows are as follows (in thousands of dollars):

n	Net Cash Flow
0	−$1000
1	2500
2	−1540

(a) Plot the present worth of this investment as a function of i.

(b) Compute the i*s for this investment.

(c) Would you accept this investment at MARR = 14%?

9.14 Champion Chemical Corporation is planning to expand one of its propylene manufacturing facilities. At n = 0, a piece of property costing $1.5 million must be purchased to build a plant. The building, which needs to be expanded during the first year, costs $3 million. At the end of the first year, the company needs to spend about $4 million on equipment and other start-up costs. Once the plan becomes operational, it will generate revenue in the amount of $3.5 million during the first operating year. This will increase at the annual rate of 5% over the previous year's revenue, for next 9 years. After 10 years, the sales revenue will stay constant for another 3 years before the operation is phased out. (It will have a project life of 13 years after construction.) The expected salvage value of the land at the end of the project life would be about $2 million, the building about $1.4 million, and the equipment about $500,000. The annual operating and maintenance costs are estimated to be about 40% of the sales revenue each year. What is the IRR for this investment? If the company's MARR is 15%, determine if this is a good investment. (Assume that all figures represent the effect of the income tax.)

9.15 Recent technology has made possible a computerized vending machine that can grind coffee beans and brew fresh coffee on demand. The computer also makes possible such complicated functions as changing $5 and $10 bills and tracking the age of an item, moving the oldest stock to the front of the line, thus cutting down on spoilage. With a price tag of $4,500 for each unit, Easy Snack has estimated the cash flows in millions of dollars over the product's 6-year useful life, including the initial investment, as follows:

n	Net Cash Flow
0	−$20
1	8
2	17
3	19
4	18
5	10
6	3

(a) If the firm's MARR is 18%, is this product worth marketing based on the IRR criterion?

(b) If the required investment remains unchanged, but the future cash flows are expected to be 10% higher than the original estimates, how much increase in IRR do you expect?

(c) If the required investment has increased from $20 million to $22 million, but the expected future cash flows are projected to be 10% smaller than the original estimates, how much decrease in IRR do you expect?

Comparing Alternatives

9.16 Consider two investments A and B with the following sequences of cash flows:

n	Net Cash Flow	
	Project A	Project B
0	−$120,000	−$100,000
1	20,000	15,000
2	20,000	15,000
3	120,000	130,000

(a) Compute the IRR for each investment.

(b) At a MARR = 15%, determine the acceptability of each project.

(c) If A and B are mutually exclusive projects, which project would you select based on the rate of return on incremental investment?

9.17 With $10,000 available, you have two investment options. The first option is to buy a certificate of deposit from a bank at an interest rate of 10% annually for 5 years. The second choice is to purchase a bond for $10,000 and invest the bond's interests in the bank at an interest rate of 9%. The bond pays 10% interest annually and will mature to its face value of $10,000 in 5 years. Which option is better? Assume your MARR is 9% per year.

9.18 A manufacturing firm is considering the following mutually exclusive alternatives:

n	Net Cash Flow	
	Project A	Project B
0	−$2,000	−$3,000
1	1,400	2,400
2	1,640	2,000

Determine which project is a better choice at a MARR = 15%, based on the IRR criterion.

9.19 Consider the following two mutually exclusive alternatives:

n	Net Cash Flow	
	Project A1	Project A2
0	−$10,000	−$12,000
1	5,000	6,100
2	5,000	6,100
3	5,000	6,100

(a) Determine the IRR on the incremental investment in the amount of $2000.

(b) If the firm's MARR is 10%, which alternative is the better choice?

9.20 Consider the following two mutually exclusive investment alternatives:

	Net Cash Flow	
n	Project A1	Project A2
0	−$15,000	−$20,000
1	7,500	8,000
2	7,500	15,000
3	7,500	5,000
IRR	23.5%	20%

(a) Determine the IRR on the incremental investment in the amount of $5000. (Assume that MARR = 10%.)

(b) If the firm's MARR is 10%, which alternative is the better choice?

9.21 You are considering two types of automobiles. Model A costs $18,000 and model B costs $15,624. Although the two models are essentially the same, model A can be sold for $9,000 while model B can be sold for $6,500 after 4 years of use. Model A commands a better resale value because its styling is popular among young college students. Determine the rate of return on the incremental investment of $2,376. For what range of values of your MARR is model A preferable?

9.22 A plant engineer is considering two types of solar water heating system:

	Model A	Model B
Initial cost	$7,000	$10,000
Annual savings	$700	$1,000
Annual maintenance	$100	$50
Expected life	20 years	20 years
Salvage value	$400	$500

The firm's MARR is 12%. Based on the IRR criterion, which system is the better choice?

9.23 Consider the following sets of investment projects. Assume that MARR = 15%.

			Net Cash Flow			
n	A	B	C	D	E	F
0	−$100	−$200	−$4,000	−$2,000	−$2,000	−$3,000
1	60	120	2,410	1,400	3,700	2,500
2	50	150	2,930	1,720	1,640	1,500
3	50					
i*	28.89%	21.65%	21.86%	31.10%	121.95%	23.74%

(a) Projects A and B are mutually exclusive projects. Assuming that both projects can be repeated for an indefinite period, which project would you select based on the IRR criterion?

(b) Suppose projects C and D are mutually exclusive. Using the IRR criterion, which project would be selected?

(c) Suppose projects E and F are mutually exclusive. Which project is better based on the IRR criterion?

9.24 Fulton National Hospital is reviewing ways of cutting the stocking costs of medical supplies. Two new stockless systems are being considered, to lower the hospital's holding and handling costs. The hospital's industrial engineer has compiled the relevant financial data for each system as follows. Dollar values are in millions.

	Current Practice	Just-in-Time System	Stockless Supply System
Start-up cost	$0	$2.5	$5
Annual stock holding cost	$3	$1.4	$0.2
Annual operating cost	$2	$1.5	$1.2
System life	8 years	8 years	8 years

The system life of 8 years represents the contract period with the medical suppliers. If the hospital's MARR is 10%, which system is more economical?

9.25 Consider the cash flows for the following investment projects.

n	A	B	C	D	E
			Project Cash Flow		
0	-$1,000	-$1,000	-$2,000	$1,000	-$1,200
1	900	600	900	-300	400
2	500	500	900	-300	400
3	100	500	900	-300	400
4	50	100	900	-300	400

Assume that the MARR = 12%.

(a) Suppose A, B, and C are mutually exclusive projects. Which project would be selected based on the IRR criterion?

(b) What is the BRR (borrowing rate of return) for D?

(c) Would you accept D at MARR = 20%?

(d) Assume that projects C and E are mutually exclusive. Using the IRR criterion, which project would you select?

9.26 Consider the following investment projects:

n	Project 1	Project 2	Project 3
		Net Cash Flow	
0	-$1,000	-$5,000	-$2,000
1	500	7,500	1,500
2	2,500	600	2,000

Assume the MARR = 15%.

(a) Compute the IRR for each project.

(b) If the three projects are mutually exclusive investments, which project should be selected based on the IRR criterion?

9.27 Consider the following two investment alternatives:

n	Project A	Project B
	Net Cash Flow	
0	-$10,000	-$20,000
1	5,500	0
2	5,500	0
3	5,500	40,000
IRR	30%	?
PW(15%)	?	6300

The firm's MARR is known to be 15%.

(a) Compute the IRR of project B.

(b) Compute the NPW of project A.

(c) Suppose that projects A and B are mutually exclusive. Using the IRR, which project would you select?

9.28 The E. F. Fedele Company is considering the acquisition of an automatic screwing machine for its assembly operation of a personal computer. Three different models with varying automatic features are under consideration. The required investments are $360,000 for model A, $380,000 for model B, and $405,000 for model C, respectively. All three models are expected to have the same service life of 8 years. The following financial information is available. In the following, model (B − A) represents the incremental cash flow determined by subtracting model A's cash flow from model B's.

Model	IRR (%)
A	30%
B	15
C	25

Model	Incremental IRR (%)
(B − A)	5%
(C − B)	40
(C − A)	15

If the firm's MARR is known to be 12%, which model should be selected?

9.29 The GeoStar Company, a leading wireless communication device manufacturer, is considering three cost-reduction proposals in its batch job shop manufacturing operations. The company already calculated rates of return for the three projects along with some incremental rates of return. A_0 denotes the do-nothing alternative. The required investments are $420,000 for A_1, $550,000 for A_2, and $720,000 for A_3. If the MARR is 15%, what system should be selected?

Incremental Investment	Incremental Rate of Return (%)
$A_1 - A_0$	18%
$A_2 - A_0$	20
$A_3 - A_0$	25
$A_2 - A_1$	10
$A_3 - A_1$	18
$A_3 - A_2$	23

9.30 A electronic circuit board manufacturer is considering six mutually exclusive cost reduction projects for its PC-board manufacturing plant. All have lives of 10 years and zero salvage values. The required investment and the estimated after-tax reduction in annual disbursements are given for each alternative. Along with these gross rates of return, rates of return on incremental investments are also computed.

Proposal A_j	Required Investment	After-Tax Savings	Rate of Return (%)
A_1	$60,000	$22,000	35.0%
A_2	100,000	28,200	25.2
A_3	110,000	32,600	27.0
A_4	120,000	33,600	25.0
A_5	140,000	38,400	24.0
A_6	150,000	42,200	25.1

Incremental Investment	Incremental Rate of Return (%)
$A_2 - A_1$	9.0%
$A_3 - A_2$	42.8
$A_4 - A_3$	0.0
$A_5 - A_4$	20.2
$A_6 - A_5$	36.3

Which project would you select based on the rate of return on incremental investment if it is stated that the MARR is 15%?

9.31 Addison Wesley Longman published a nonfiction mystery book after paying $350,000 to the author, who had an earlier best-seller on his resume. The cover price was $18.95. Of that amount $9.85 went to the publisher and the rest to the booksellers. Of the 75,000 copies printed, 32,250 were sold. The sequence of cash flow transactions was as follows:

- June 1999: A $350,000 non-refundable advance against royalties paid to author as a guaranteed minimum.
- March 2000: Book manuscripts were delivered, and book production began.
- June 2000: 75,000 copies were printed at $179,455 (paper, printing, binding, and plant cost).
- June 2000–March 2001: 32,250 copies sold at $317,663 (= $32,250 * $9.85).

Other expenses incurred included the market cost of $150,000 and the overhead cost (salaries, rent, etc.) of $95,337. Because of poor sales, the company sold the paperback rights to another publisher for $170,000. The remaining unsold copies had to be sold on the bargain tables as remainders or at $91,058 (= 42,750 * $2.13). The company's fiscal year ends June 30.

(a) Ignoring any tax implications and using the end-of-year convention, can you determine the rate of return for this book project?

(b) If MacMillan were to make a before-tax 20% rate of return on this book, how many copies had to be sold? (Assume that all other figures remain unchanged.)

9.32 Baby Doll Shop is currently employed manufacturing wooden parts for doll houses. The worker is paid $8.10 an hour, and using a hand saw, can produce a year's required production (1,600 parts) in just 8 weeks of 40 hours per week. That is, the worker averages five parts per hour when working by hand. The shop is considering the purchase of a power band saw with associated fixtures, to improve the productivity of this operation. Three models of power saw could be purchased: Model A (economy version), model B (high-powered version), and model C (deluxe high-end version). The major operating difference between these models is their speed of operation. The investment costs, including the required fixtures and other operating characteristics, are summarized as follows:

Category	By Hand	Model A	Model B	Model C
Production rate (parts/hour)	5	10	15	20
Labor hours required (hours/year)	320	160	107	80
Annual labor cost (@ $8.10/hour)	2,592	1,296	867	648
Annual power cost ($)		400	420	480
Initial investment ($)		4,000	6,000	7,000
Salvage value ($)		400	600	700
Service life (years)		20	20	20

Assume that MARR = 10%. Are there enough savings to purchase any of the power band saws? Which model is most economical based on the rate of return principle? (Assume that any effect of income tax has been already considered in the dollar estimates.) (Source: This problem is adapted with the permission of Professor Peter Jackson of Cornell University.)

Unequal Service Lives

9.33 Consider the following two mutually exclusive investment projects. Assume that the MARR = 15%.

	Net Cash Flow	
n	Project A	Project B
0	−$100	−$200
1	60	120
2	50	150
3	50	
IRR	28.89%	21.65%

Which project would be selected under infinite planning horizon with project repeatability likely, based on the IRR criterion?

9.34 Consider the following two mutually exclusive investment projects:

	Net Cash Flow	
n	Project A1	Project A2
0	−$10,000	−$15,000
1	5,000	20,000
2	5,000	
3	5,000	

(a) To use the IRR criterion, what assumption must be made to compare a set of mutually exclusive investments with unequal service lives?

(b) With the assumption defined above, determine the range of MARR that will indicate the selection of project A1.

Short Case Studies

9.35 Critics have charged that the commercial nuclear power industry does not consider the cost of "decommissioning," or "mothballing," a nuclear power plant when doing an economic analysis, and that the analysis is therefore unduly optimistic. As an example, consider the Tennessee Valley Authority's "Bellefont" twin nuclear generating facility under construction at Scottsboro, in northern Alabama: The first cost is $1.5 billion (present worth at start of operations), the estimated life is 40 years, the annual operating and maintenance costs the first year are assumed to be 4.6% of the first cost and are expected to increase at the fixed rate of 0.05% of the first cost each year, and annual revenues have been estimated to be three times the annual operating and maintenance costs throughout plant life.

(a) The criticism of over-optimism in the economic analysis caused by omitting "mothballing" costs is not justified since the addition of a cost to "mothball" the plant equal to 50% of the first cost only decreases the 10% rate of return to approximately 9.9%.

(b) If the estimated life of the plants is more realistically taken as 25 years instead of 40 years, then the criticism is justified. By reducing the life to 25 years, the rate of return of approximately 9% without a "mothballing" cost drops to approximately 7.7% when a cost to "mothball" the plant equal to 50% of the first cost is added to the analysis.

Comment on these statements.

9.36 The B&E Cooling Technology Company, a maker of automobile air-conditioners, faces an uncertain but impending deadline to phase out the traditional chilling technique, which uses chlorofluorocarbons, (CFCs) a family of refrigerant chemicals believed to attack the earth's protective ozone layer. B&E has been pursuing other means of cooling and refrigeration. As a near-term solution, the engineers recommend a cold technology known as absorption chiller, which uses plain water as a refrigerant and semiconductors that cool down when charged with electricity. B&E is considering two options:

- Option 1: Retrofitting the plant now to adapt the absorption chiller and continuing to be a market leader in cooling technology. Because of untested technology in the large scale, it may cost more to operate the new facility while learning the new system takes place.

- Option 2: Deferring the retrofitting until the federal deadline, which is 3 years away. With expected improvement in cooling technology and technical know-how, the retrofitting cost will be cheaper, but there will be tough market competition, and the revenue would be less than that of Option 1.

The financial data for the two options are as follows:

	Option 1	Option 2
Investment timing	Now	3 years from now
Initial investment	$6 million	$5 million
System life	8 years	8 years
Salvage value	$1 million	$2 million
Annual revenue	$15 million	11 million
Annual O&M costs	$6 million	$7 million

(a) What assumptions must be made to compare these two options?

(b) If B&E's MARR is 15%, which option is the better choice, based on the IRR criterion?

9.37 An oil company is considering changing the size of a current pump (small pump) that is operational in wells in an oil field. If the current smaller pump is kept, it will extract 50% of the known crude oil reserve in the first year of operation and the remaining 50% in the second year. A pump larger than the current pump will cost $1.6 million, but it will extract 100% of the known reserve in the first year. The total oil revenues over the 2 years is the same for both pumps, $20 million. The advantage of the large pump is that it allows 50% of the revenues to be realized a year earlier than with the small pump.

	Current Pump	Larger Pump
Investment, year 0	0	$1.6 million
Revenue, year 1	$10 million	$20 million
Revenue, year 2	$10 million	0

If the firm's MARR is known to be 20%, what do you recommend based on IRR criterion?

9.38 You have been asked by the president of the company to evaluate the proposed acquisition of a new injection molding machine for the firm's manufacturing plant. Two types of injection molding machines have been identified with the following estimated cash flows:

	Net Cash Flow	
n	Project A	Project B
0	−$30,000	−$40,000
1	20,000	43,000
2	18,200	5,000
IRR	18.1%	18.1%

You return to your office and quickly retrieve your old engineering economics text, then begin to smile: Aha—this is a classic rate of return problem! Now, using a calculator, you find out that both projects have about the same rate of return, 18.1%. This rate of return figure seems to be high enough for project justification, but you recall that the ultimate justification should be done in reference to the firm's MARR. You call the accounting department to find out the current MARR the firm should use for project justification. "Oh boy, I wish I could tell you, but my boss will be back next week, and he can tell you what to use," said the accounting clerk.

A fellow engineer approaches you and says, "I couldn't help overhearing you talking to the clerk. I think I can help you. You see, both projects have the same IRR and on top of that, project 1 requires less investment but returns more cash flows (−$3,000 + $2,000 + $1,820 = $820, and −$4,000 + 4,00 + $500 = $800), thus project 1 dominates project 2. For this type of decision problem, you don't need to know a MARR!"

(a) Comment on your fellow engineer's statement.

(b) At what range of MARR would you recommend the selection of project 2?

Development of Project Cash Flows

CHAPTER 10

Depreciation

Know What It Costs to Own a Piece of Equipment: Peter Chong is a design engineer employed by Hicom Diecasting, a Malaysian firm that manufactures die-cast automobile parts. To enhance the firm's competitive position in the marketplace, management has decided to purchase a computer-aided design system that features 3-D solid modeling and full integration with sophisticated simulation and analysis capabilities. As a part of the design team, Peter is excited at the prospect that the design of die-cast molds, the testing of product variations, and the simulation of processing and service conditions can be made highly efficient by use of this state-of-the-art system. In fact, the more Peter thinks about it, the more he wonders why this purchase wasn't made earlier.

Now ask yourself, how does the cost of this system affect the financial position of the firm? In the long run, the system promises to create greater wealth for the organization by improving design productivity, increasing product quality, and cutting down design lead-time. In the short run, however, the high cost of this system will negatively impact the organization's "bottom line," because it involves high initial costs that are only gradually rewarded by the benefits of the system.

Another consideration should come to mind. This state-of-the-art equipment must inevitably wear out over time, and even if its productive service extends over many years, the cost of maintaining its high level of functioning will increase as the individual pieces of hardware wear out and need to be replaced. Of even greater concern is the question of how long this system will be "state-of-the-art." When will the competitive advantage the firm has just acquired become a competitive disadvantage through obsolescence?

One of the facts of life that organizations must deal with and account for is that fixed assets lose their value—even as they continue to function and contribute to the engineering projects that use them. This loss of value, called **depreciation,** can involve deterioration and obsolescence.

The main function of **depreciation accounting** is to account for the cost of fixed assets in a pattern that matches their decline in value over time. The cost of the CAD system we have just described, for example, will be allocated over several years in the firm's financial statements so that its pattern of costs roughly matches its pattern of service. In this way, as we shall see, depreciation accounting enables the firm to stabilize the statements of financial position that it distributes to stockholders and the outside world.

On a project level, engineers must be able to assess how the practice of depreciating fixed assets influences the investment value of a given project. To do this, they need to estimate the allocation of capital costs over the life of the project, which requires understanding the conventions and techniques, that accountants use to depreciate assets. In this chapter, we will review the conventions and techniques of asset depreciation.

We begin by discussing the nature and significance of depreciation, distinguishing its general economic definition from the related but different accounting view of depreciation. We then focus our attention almost exclusively on the rules and laws that govern asset depreciation and the methods that accountants use to allocate depreciation expenses. Knowledge of these rules will prepare you to apply them in assessing the depreciation of assets acquired in engineering projects. Finally, we turn our attention to the subject of depletion, which utilizes similar ideas, but specialized techniques, to allocate the cost of the depletion of natural resource assets.

10.1 Asset Depreciation

Fixed assets such as equipment and real estate are economic resources that are acquired to provide future cash flows. Generally, **depreciation** can be defined as the gradual decrease in utility of fixed assets with use and time. While this general definition does not adequately capture the subtleties inherent in a more specific definition of depreciation, it does provide us with a starting point for examining the variety of underlying ideas and practices that are discussed in this chapter. Figure 10.1 will serve as a road map for understanding the distinctions inherent in the meaning of depreciation we will explore in this chapter.

We can classify depreciation into the categories of physical or functional depreciation. **Physical depreciation** can be defined as a reduction in an asset's capacity to perform its intended service due to physical impairment. Physical depreciation can occur in any fixed asset in the form of (1) deterioration from interaction with the environment, including such agents as corrosion, rotting, and other chemical changes; and (2) wear and tear from use. Physical depreciation leads to a decline in performance and high maintenance costs.

Functional depreciation occurs as a result of changes in the organization or in technology that decrease or eliminate the need for an asset. Examples of functional depreciation include obsolescence attributable to advances in technology, a declining need for the services performed by an asset, or the inability to meet increased quantity and/or quality demands.

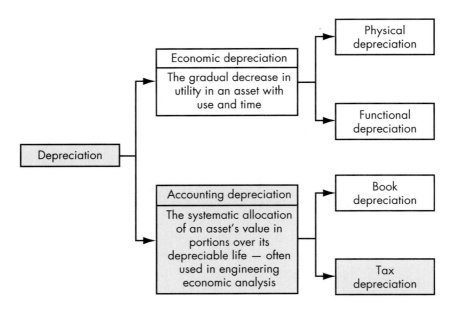

Figure 10.1 Classification of depreciation

10.1.1   Economic Depreciation

This chapter is primarily concerned with accounting depreciation, which is the form of depreciation that provides an organization with the information it uses to assess its financial position. It would also be useful, however, to discuss briefly the economic ideas upon which accounting depreciation is based. In the course of the discussion, we will develop a precise definition of economic depreciation that will help us distinguish between various conceptions of depreciation.

If you have ever owned a car, you are probably familiar with the term depreciation as it is used to describe the decreasing value of your vehicle (see Example 8.7). Because a car's reliability and appearance usually decline with age, the vehicle is worth less with each passing year. You can calculate the economic depreciation accumulated for your car by subtracting the current market value or "blue book" value of the car from the price you originally paid for it. We can define **economic depreciation** as follows:

$$\text{Economic depreciation} = \text{Purchase price} - \text{market value}$$

Physical and functional depreciation are categories of economic depreciation.

The measurement of economic depreciation does not require that an asset be sold: the market value of an asset can be closely estimated without actually testing its value in the marketplace. The need to have a precise scheme for recording the ongoing decline in the value of an asset as a part of the accounting process leads us to an exploration of how organizations account for depreciation.

10.1.2   Accounting Depreciation

The acquisition of fixed assets is an important activity for a business organization. This is true whether the organization is starting up or acquiring new assets to remain competitive. Like other disbursements, the cost of these fixed assets must be recorded as expenses on a firm's balance sheet and income statement. However, unlike costs such as maintenance, material, and labor, the costs of fixed assets are not treated simply as expenses to be accounted for in the year that they are acquired, rather, these assets are **capitalized;** that is, their costs are distributed by subtracting them as expenses from gross income—one part at a time over a number of periods. The systematic allocation of the initial cost of an asset in parts over a time, known as its depreciable life, is what we mean by **accounting depreciation.** Because accounting depreciation is the standard of the business world, we sometimes refer to it more generally as **asset depreciation.**

Accounting depreciation is based on the **matching concept:** A fraction of the cost of the asset is chargeable as an expense in each of the accounting periods in which the asset provides service to the firm, and each charge is meant to be a percentage of the whole cost which "matches" the percentage of the value utilized in the given period. The matching concept suggests that the accounting depreciation allowance generally reflects to some extent the actual economic depreciation of the asset.

In engineering economic analysis, we exclusively use the concept of accounting depreciation. This is because accounting depreciation provides a basis for determining the income taxes associated with any project undertaken. As we will see in Chapter

11, depreciation has a significant influence on the income and cash position of a firm. With a clear understanding of the concept of depreciation, we will be able to appreciate fully the importance of utilizing depreciation as a means to maximize the value both of engineering projects and of the organization as a whole.

10.2 Factors Inherent to Asset Depreciation

The process of depreciating an asset requires that we make several preliminary determinations: (1) what is the cost of the asset? (2) what is the asset's value at the end of its useful life? (3) what is the depreciable life of the asset? and, finally, (4) what method of depreciation do we choose? In this section, we will discuss each of these factors.

10.2.1 Depreciable Property

As a starting point, it is important to recognize what constitutes a **depreciable asset,** that is, a property for which a firm may take depreciation deductions against income. For the purposes of U.S. tax law, any depreciable property has the following characteristics:

1. It must be used in business or held for production of income.
2. It must have a definite service life, and that life must be longer than 1 year.
3. It must be something that wears out, decays, gets used up, becomes obsolete, or loses value from natural causes.

Depreciable property includes buildings, machinery, equipment, and vehicles. Inventories are not depreciable property because these are held primarily for sale to customers in the ordinary course of business. If an asset has no definite service life, the asset cannot be depreciated. For example, *you can never depreciate land.*[1]

As a side note, we should add that while we have been focusing on depreciation within firms, individuals may also depreciate assets as long as they meet the conditions listed above. For example, an individual may depreciate an automobile if the vehicle is used exclusively for business purposes.

10.2.2 Cost Basis

The **cost basis** of an asset represents the total cost that is claimed as an expense over an asset's life, i.e., the sum of the annual depreciation expenses.

Cost basis generally includes the actual cost of an asset and all other incidental expenses, such as freight, site preparation, and installation. This total cost, rather than the cost of the asset only, must be the depreciation basis charged as an expense over an asset's life.

---

[1] This also means that you cannot depreciate the cost of clearing, grading, planting, and landscaping. All four expenses are considered part of the cost of the land.

Besides being used in figuring depreciation deductions, an asset's cost basis is used in calculating the gain or loss to the firm if the asset is ever sold or salvaged. (We will discuss these topics in Chapter 11.)

Example 10.1 Cost Basis

Lanier Corporation purchased an automatic hole-punching machine priced at $62,500. The vendor's invoice included a sales tax of $3,263. Lanier also paid the inbound transportation charges of $725 on the new machine as well as labor cost of $2,150 to install the machine in the factory. Lanier also had to prepare the site at the cost of $3,500 before installation. Determine the cost basis for the new machine for depreciation purpose.

Solution

Given: Invoice price = $62,500, freight = $725, installation cost = $2,150, and site preparation = $3,500

Find: The cost basis

The cost of machine that is applicable for depreciation is computed as follows:

Cost of new hole-punching machine	$62,500
Freight	725
Installation labor	2,150
Site preparation	3,500
Cost of machine (cost basis)	$68,875

Comments: Why do we include all the incidental charges relating to the acquisition of a machine in its cost? Why not treat these incidental charges as expenses of the period in which the machine is acquired? The matching of costs and revenue is the basic accounting principle. Consequently, the total costs of the machine should be viewed as an asset and allocated against the future revenue that the machine will generate. All costs incurred in acquiring the machine are costs of the services to be received from using the machine.

If the asset is purchased by trading in a similar asset, the difference between the book value (cost basis minus the total accumulated depreciation) and trade-in allowance must be considered in determining the cost basis for the new asset. If the trade-in allowance exceeds the book value, the difference (known as **unrecognized gain**) needs to be subtracted from the cost basis of the new asset. If the opposite is true (**unrecognized loss**), it should be added to the cost basis for the new asset.

Example 10.2 Cost Basis with Trade-In Allowance

In Example 10.1, suppose Lanier purchased the hole-punching press by trading in a similar machine and paying cash for the remainder. The trade-in allowance is $5,000, and the book value of the hole-punching machine that was traded in is $4,000.

Old hole-punching machine (book value)	$4,000
Less: Trade-in allowance	5,000
Unrecognized gains	$1,000
Cost of new hole-punching machine	$62,500
Less: Unrecognized gains	(1,000)
Freight	725
Installation labor	2,150
Site preparation	3,500
Cost of machine (cost basis)	$67,875

10.2.3   Useful Life and Salvage Value

Over how many periods will an asset be useful to the company? What do published statutes allow you to choose as the life of an asset? These are the central questions to be answered when determining an asset's depreciable life, i.e., the number of years over which an asset is to be depreciated.

Historically, depreciation accounting included choosing a depreciable life that was based on the service life of an asset. Determining the service life of an asset, however, was often very difficult, and the uncertainty of these estimates often led to disputes between taxpayers and the IRS. To alleviate the problems, the IRS published guidelines on lives for categories of assets known as **Asset Depreciation Ranges,** or **ADRs.** These guidelines specified a range of lives for classes of assets based on historical data, and taxpayers were free to choose a depreciable life within the specified range for a given asset. An example of ADRs for some assets is given in Table 10.1.

The salvage value is an asset's estimated value at the end of its life; it is the amount eventually recovered through sale, trade-in, or salvage. The eventual salvage value of an asset must be estimated when the depreciation schedule for the asset is established. If this estimate subsequently proves to be inaccurate, then an adjustment must be made. We will discuss these specific issues in Section 10.6.

10.2.4   Depreciation Methods: Book and Tax Depreciation

One important distinction within the general definition of accounting depreciation should be introduced. Most firms calculate depreciation in two different ways, depending on whether the calculation is (1) intended for financial reports (**book depreciation method**), such as for the balance sheet or income statement or (2) for the Internal Revenue Service (IRS), for the purpose of calculating taxes (**tax depreciation**

Assets Used	Asset Depreciation Range (Years)		
	Lower Limit	Midpoint Life	Upper Limit
Office furniture, fixtures, and equipment	8	10	12
Information systems (computers)	5	6	7
Airplanes	5	6	7
Automobiles, taxis	2.5	3	3.5
Buses	7	9	11
Light trucks	3	4	5
Heavy trucks (concrete ready-mixer)	5	6	7
Railroad cars and locomotives	12	15	18
Tractor units	5	6	7
Vessels, barges, tugs, and water transportation systems	14.5	18	21.5
Industrial steam and electrical generation and or distribution systems	17.5	22	26.5
Manufacturer of electrical and nonelectrical machinery	8	10	12
Manufacturer of electronic components, products, and systems	5	6	7
Manufacturer of motor vehicles	9.5	12	14.5
Telephone distribution plant	28	35	42

Table 10.1
Some Selected Asset Guideline Classes—Asset Depreciation

Source: IRS Publication 534. *Depreciation*. U.S. Government Printing Office: Washington, DC, 1995.

method). In the United States, this distinction is totally legitimate under IRS regulations, as it is in many other countries. Calculating depreciation differently for financial reports and for tax purposes allows for the following benefits:

- It enables firms to report depreciation to stockholders and other significant outsiders based on the matching concept. Therefore the actual loss in value of the assets is generally reflected.

- It allows firms to benefit from the tax advantages of depreciating assets more quickly than would be possible using the matching concept. In many cases, tax depreciation allows firms to defer paying income taxes. This does not mean that they pay less tax overall, because the total depreciation expense accounted for over time is the same in either case. However, because tax depreciation methods generally permit a higher depreciation in earlier years than do book depreciation methods, the tax benefit of depreciation is enjoyed earlier, and firms generally pay lower taxes in the initial years of an investment project. Typically this leads to a better cash position in early years, the added cash leading to greater future wealth because of the time value of the funds.

As we proceed through the chapter, we will make increasing use of the distinction between depreciation accounting for financial reporting and depreciation accounting used for income tax calculation. Now that we have established the context for our interest in both tax and book depreciation, we can survey the different methods with an accurate perspective.

10.3 Book Depreciation Methods

Three different methods can be used to calculate the periodic depreciation allowances. These are the (1) straight-line method, (2) accelerated methods, and (3) the unit-of-production method. In engineering economic analysis, we are primarily interested in depreciation in the context of income tax computation. Nonetheless, a number of reasons make the study of book depreciation methods useful. First, tax depreciation methods are based largely on the same principles that are used in book depreciation methods. Second, firms continue to use book depreciation methods for financial reporting to stockholders and outside parties. Third, book depreciation methods are still used for state income tax purposes in many states and even for federal income tax purposes for assets that were put into service before 1981. Finally, our discussion of depletion in Section 10.5 is based largely on one of these three book depreciation methods.

10.3.1 Straight-Line Method

The **straight-line method (SL)** of depreciation interprets a fixed asset as one that provides its services in a uniform fashion. The asset provides an equal amount of service in each year of its useful life.

The straight-line method charges, as an expense, an equal fraction of the net cost of the asset each year, as expressed by the relation

$$D_n = \frac{(I - S)}{N},$$ (10.1)

where D_n = Depreciation charge during year n

I = Cost of the asset including installation expenses

S = Salvage value at the end of useful life

N = Useful life.

The book value of the asset at the end of n years is then defined as

Book value in a given year = Cost basis − total depreciation charges made to date or

$$B_n = I - (D_1 + D_2 + D_3 + ... + D_n).$$ (10.2)

Example 10.3 Straight-Line Depreciation

Consider the following automobile data:

Cost basis of the asset, I = $10,000

Useful life, N = 5 years

Estimated salvage value, S = $2,000

Compute the annual depreciation allowances and the resulting book values using the straight-line depreciation method.

Solution

Given: $I = \$10,000$, $S = \$2,000$, $N = 5$ years

Find: D_n and B_n for $n = 1$ to 5

The straight-line depreciation rate is 1/5 or 20%. Therefore, the annual depreciation charge is

$$D_n = (0.20)(\$10,000 - \$2,000) = \$1,600.$$

Then, the asset would have the following book values during its useful life:

n	B_{n-1}	D_n	B_n
1	$10,000	$1,600	$8,400
2	8,400	1,600	6,800
3	6,800	1,600	5,200
4	5,200	1,600	3,600
5	3,600	1,600	2,000

where B_{n-1} represents the book value before the depreciation charge for year n. This situation is illustrated in Figure 10.2.

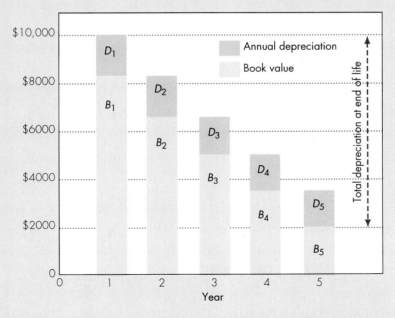

Figure 10.2   Straight-line depreciation method (Example 10.3)

10.3.2 Accelerated Methods

The second concept inherent in depreciation recognizes that the stream of services provided by a fixed asset may decrease over time; in other words, the stream may be greatest in the first year of an asset's service life and least in its last year. This pattern may occur because the mechanical efficiency of an asset tends to decline with age, because maintenance costs tend to increase with age, or because of the increasing likelihood that better equipment will become available and make the original asset obsolete. This reasoning leads to a method that charges a larger fraction of the cost as an expense of the early years than of the later years. Any such method is called an **accelerated method.** The two most widely used accelerated methods are the **declining balance method** and the **sum-of-the-years'-digits method.**

(The SOYD method is rarely used for depreciating an individual asset in the U.S., but may be used in the depreciation of multiple-asset accounts.)

Declining Balance Method (DB)

The **declining balance method** of calculating depreciation allocates a fixed fraction of the beginning book balance each year. The fraction, α, is obtained as follows:

$$\alpha = (1/N)(\text{multiplier}) \tag{10.3}$$

The most commonly used multipliers in the United States are 1.5 (called 150% DB) and 2.0 (called 200%, or double declining balance, DDB). As N increases, α decreases, thereby resulting in a situation in which depreciation is highest in the first year and then decreases over the asset's depreciable life.

The fractional factor can be utilized to determine depreciation charges for a given year, D_n, as follows:

$$D_1 = \alpha I,$$
$$D_2 = \alpha(I - D_1) = \alpha I(1 - \alpha),$$
$$D_3 = \alpha(I - D_1 - D_2) = \alpha I(1 - \alpha)^2,$$

and thus for any year, n, we have a depreciation charge, D_n, of

$$D_n = \alpha I(1 - \alpha)^{n-1} \tag{10.4}$$

We can also compute the total DB depreciation (TDB) at the end of n years as follows:

$$\begin{aligned} TDB &= D_1 + D_2 + \dots + D_n \\ &= \alpha I + \alpha I(1 - \alpha) + \alpha I(1 - \alpha)^2 + \dots + \alpha I(1 - \alpha)^{n-1} \\ &= \alpha I[1 + (1 - \alpha) + (1 - \alpha)^2 + \dots + (1 - \alpha)^{n-1}]. \end{aligned} \tag{10.5}$$

Multiplying Eq. (10.5) by $(1 - a)$, we obtain

$$TDB(1 - \alpha) = \alpha I[(1 - \alpha) + (1 - \alpha)^2 + (1 - \alpha)^3 + \dots + (1 - \alpha)^n]. \tag{10.6}$$

Subtracting Eq. (10.5) from Eq. (10.6) and dividing by α gives

$$TDB = I[1 - (1 - \alpha)^n]. \tag{10.7}$$

The book value, B_n, at the end of n years will be the cost of the asset I minus the total depreciation at the end of N years:

$$B_n = I - TDB$$
$$= I - I[1 - (1 - \alpha)^n] \qquad (10.8)$$
$$B_n = I(1 - \alpha)^n.$$

Example 10.4 Declining Balance Depreciation

Consider the following accounting information for a computer system:

$$\text{Cost basis of the asset, } I = \$10,000$$
$$\text{Useful life, } N = 5 \text{ years}$$
$$\text{Estimated salvage value, } S = \$778$$

Compute the annual depreciation allowances and the resulting book values using the double declining depreciation method (Figure 10.3).

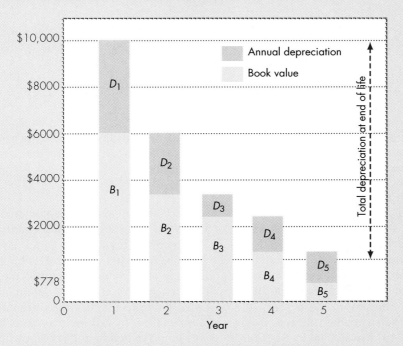

Figure 10.3 Double declining balance method (Example 10.4)

Solution

Given: $I = \$10,000$, $S = \$778$, $N = 5$ years

Find: D_n and B_n for $n = 1$ to 5

The book value at the beginning of the first year is $10,000, and the declining balance rate (α) is $(1/5)(2) = 40\%$. Then, the depreciation deduction for the first year will be $4,000 ($40\% \times \$10,000 = \$4,000$). To figure the depreciation deduction in the second year, we must first adjust the book value for the amount of depreciation we deducted in the first year. The first year's depreciation from the beginning book value is subtracted ($10,000 - \$4,000 = \$6,000$). This amount is multiplied by the rate of depreciation ($6,000 \times 40\% = \$2,400$). By continuing the process, we obtain

n	B_{n-1}	D_n	B_n
1	10,000	4,000	6,000
2	6,000	2,400	3,600
3	3,600	1,440	2,160
4	2,160	864	1,296
5	1,296	518	778

The declining balance is illustrated in terms of the book value of time in Figure 10.3.

Salvage value (S) must be estimated at the outset of depreciation analysis. In Example 10.4, the final book value (B_N) conveniently equals the estimated salvage value of $778, an occurrence that is rather unusual in the real world. When $B_N \neq S$, we would want to make adjustments in our depreciation methods.

• **Case 1: $B_N > S$**
When $B_N > S$, we are faced with a situation in which we have not depreciated the entire cost of the asset and thus have not taken full advantage of depreciation's tax-deferring benefits. If you would prefer to reduce the book value of an asset to its salvage value as quickly as possible, it can be done by switching from DB to SL whenever SL depreciation results in larger depreciation charges and therefore a more rapid reduction in the book value of the asset. The switch from DB to SL depreciation can take place in any of the n years, the objective being to identify the optimal year to switch. The switching rule is as follows: If depreciation by DB in any year is less than (or equal to) it would be by SL; therefore we should switch to and remain with the SL method for the duration of the project's depreciable life. The straight-line depreciation in any year n is calculated by

$$D_n = \frac{\text{Book value at beginning of year } n - \text{salvage value}}{\text{Remaining useful life at beginning of year } n} \qquad (10.9)$$

Example 10.5 Declining Balance with Conversion to Straight-Line Depreciation ($B_N > S$)

Suppose the asset given in Example 10.4 has a zero salvage value instead of $778, i.e.,

Cost basis of the asset, I	=	$10,000
Useful life, N	=	5 years
Salvage value, S	=	$0
$\alpha = (1/5)(2)$	=	40%

Determine the optimal time to switch from DB to SL depreciation and the resulting depreciation schedule.

Solution

Given: $I = \$10,000$, $S = 0$, $N = 5$ years, $\alpha = 40\%$

Find: Optimal conversion time, D_n and B_n for $n = 1$ to 5

We will first proceed by computing the DDB depreciation for each year as before.

Year	D_n	B_n
1	$4,000	$6,000
2	2,400	3,600
3	1,440	2,160
4	864	1,296
5	518	778

Then, we compute the SL depreciation for each year using Eq. (10.9). We compare SL to DDB depreciation for each year and use the decision rule for when to change:

If Switch to SL at Beginning Year	SL Depreciation	DDB Depreciation	Decision
2	($6,000 – 0)/4 = $1,500	< $2,400	Do not switch
3	(3,600 – 0)/3 = 1,200	< 1,440	Do not switch
4	(2,160 – 0)/2 = 1,080	> 864	Switch to SL

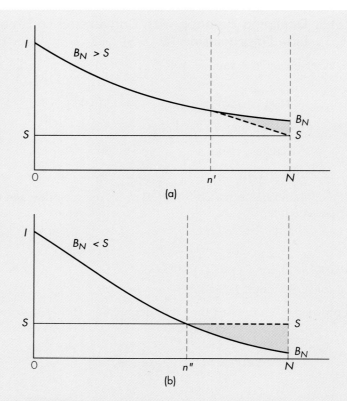

Figure 10.4 Adjustments to the declining balance method: (a) Switch from DB to the SL after n', (b) no further depreciation allowances are available after n'' (Examples 10.5 and 10.6)

The optimal time (year 4) in this situation corresponds to n' in Figure 10.4(a). The resulting depreciation schedule is

Year	DDB with Switching to SL	End-of-year Book Value
1	$4,000	$6,000
2	2,400	3,600
3	1,440	2,160
4	1,080	1,080
5	1,080	0
	$10,000	

• **Case 2: $B_N < S$**

With a relatively high salvage value, it is possible that the book value of the asset could decline below the estimated salvage value. When $B_n < S$, we must readjust our analysis because tax law does not permit us to depreciate assets below their salvage value. To avoid deducting depreciation charges that would drop the book value below the salvage value, you simply stop depreciating the asset whenever you get down to $B_n = S$. In other words, if the implied book value is lower than S, at any period, then the depreciation amounts are adjusted so that $B_n = S$.

Example 10.6 Declining Balance, $B_N < S$

Compute the double declining balance (DDB) depreciation schedule for the data used in Example 10.5, with a salvage value of $2,000.

$$\text{Cost basis of the asset, } I = \$10,000$$
$$\text{Useful life, } N = 5 \text{ years}$$
$$\text{Salvage value, } S = \$2,000$$
$$\alpha = (1/5)(2) = 40\%$$

Solution

Given: $I = \$10,000$, $S = \$2,000$, $N = 5$ years, $\alpha = 40\%$

Find: D_n and B_n for $n = 1$ to 5

End of Year	D_n	B_n
1	0.4($10,000) = $4,000	$10,000 − $4,000 = $6,000
2	0.4(6,000) = 2,400	6,000 − 2,400 = 3,600
3	0.4(3,600) = 1,440	3,600 − 1,440 = 2,160
4	0.4(2,160) = 864 > 160	2,160 − 160 = 2,000
5	0	2,000 − 0 = 2,000
	Total = $8,000	

Note that B_4 would be less than $S = \$2,000$, if the full deduction ($864) had been taken. Therefore, we adjust D_4 to $160, making $B_4 = \$2,000$. D_5 is zero and B_5 remains at $2,000. Year 4 is equivalent to n'' in Figure 10.4(b).

Sum-of-Years'-Digits (SOYD) Method

Another accelerated method for allocating the cost of an asset is called **sum-of-years'-digits** (SOYD) depreciation. Compared with SL depreciation, SOYD results in larger depreciation charges during the early years of an asset's life and smaller charges as the asset reaches the end of its estimated useful life.

In the SOYD method, the numbers 1, 2, 3, ..., N are summed, where N is the estimated years of useful life. We find this sum by the equation[2]

$$SOYD = 1 + 2 + 3 + ... + N = \frac{N(N + 1)}{2}. \tag{10.10}$$

The depreciation rate each year is a fraction in which the denominator is the SOYD and the numerator is, for the first year, N; for the second year, $N - 1$; for the third year, $N - 2$; and so on. Each year the depreciation charge is computed by dividing the remaining useful life by the SOYD and by multiplying this ratio by the total amount to be depreciated $(I - S)$.

$$D_n = \frac{N - n + 1}{SOYD}(I - S). \tag{10.11}$$

Example 10.7 SOYD Depreciation

Compute the SOYD depreciation schedule for Example 10.3.

Cost basis of the asset, I	=	$10,000
Useful life, N	=	5 years
Salvage value, S	=	$2,000

Solution

Given: $I = \$10,000$, $S = \$2,000$, $N = 5$ years

Find: D_n and B_n for $n = 1$ to 5

We first compute the sum-of-years' digits:

$$SOYD = 1 + 2 + 3 + 4 + 5 = 5(5 + 1)/2 = 15$$

---

[2]You may derive this sum equation by writing the SOYD expression in two ways:

$$
\begin{aligned}
SOYD &= 1 + 2 + 3 + ... + N \\
SOYD &= N + (N - 1) + ... + 1.
\end{aligned}
$$

Then you add these two expressions and solve for SOYD

$$
\begin{aligned}
2SOYD &= (N + 1) + (N + 1) + ... + (N + 1) \\
&= N(N + 1) \\
SOYD &= N(N + 1)/2.
\end{aligned}
$$

Year	D_n		B_n
1	(5/15) ($10,000 − $2,000) =	$2,667	$7,333
2	(4/15) (10,000 − 2,000) =	2,133	5,200
3	(3/15) (10,000 − 2,000) =	1,600	3,600
4	(2/15) (10,000 − 2,000) =	1,067	2,533
5	(1/15) (10,000 − 2,000) =	533	2,000

This situation is illustrated in Figure 10.5.

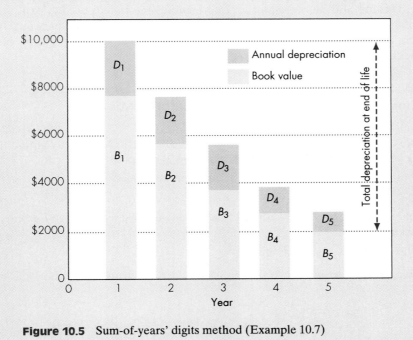

Figure 10.5   Sum-of-years' digits method (Example 10.7)

10.3.3 Units-of-Production Method

Straight-line depreciation can only be defended if the machine is used for exactly the same amount of time each year. What happens when a punch press machine is run 1,670 hours one year and 780 the next, or when some of its output is shifted to a new machining center? This leads us to a consideration of another depreciation concept that views the asset as consisting of a bundle of service units; unlike the SL and accelerated methods, however, this concept does not assume that the service units will be consumed in a time-phased pattern. The cost of each service unit is the net cost of the asset divided by the total number of such units. The depreciation charge for a period is then related to the number of service units consumed in that period. This leads to the **units-of-production method.** By this method, the depreciation in any year is given by

$$D_n = \frac{\text{Service units consumed during year } n}{\text{Total service units}} (I - S) \tag{10.12}$$

When using the units-of-production method, depreciation charges are made proportional to the ratio of actual output to the total expected output, usually this ratio is figured in machine hours. Advantages of using this method include the fact that depreciation varies with production volume, and therefore the method gives a more accurate picture of machine usage. A disadvantage of the units-of-production method is that the collecting of data on machine use and the accounting methods are somewhat tedious. This method can be useful for depreciating equipment used to exploit natural resources, if the resources will be depleted before the equipment wears out. It is not, however, considered a practical method for general use in depreciating industrial equipment.

Example 10.8 Units-of-Production Depreciation

A truck for hauling coal has an estimated net cost of $55,000 and is expected to give service for 250,000 miles, resulting in a $5,000 salvage value. Compute the allowed depreciation amount for truck usage of 30,000 miles.

Solution

Given: $I = \$55,000$, $S = \$5,000$, total service units = 250,000 miles, usage for this year = 30,000 miles

Find: Depreciation amount in this year

The depreciation expense in a year in which the truck traveled 30,000 miles would be

$$\frac{30,000 \text{ miles}}{250,000 \text{ miles}}(\$55,000 - \$5,000) = \left(\frac{3}{25}\right)(\$50,000)$$

$$= \$6,000.$$

10.4 Tax Depreciation Methods

Prior to the Economic Recovery Act of 1981, taxpayers could choose among several methods when depreciating assets for tax purposes. The most widely used methods were the straight-line method, the declining balance method, and the sum-of-years'-digit method. The subsequent imposition of the Accelerated Cost Recovery System (ACRS) and the Modified Accelerated Cost Recovery System (MACRS) superseded these methods for use in tax purposes.

10.4.1 MACRS Depreciation

From 1954 to 1981, congressional changes in tax law have evolved fairly consistently toward simpler, more rapid depreciation methods. Prior to 1954, the straight-line method was required for tax purposes, but in 1954 accelerated methods such as double declining balance and sum-of-years'-digits were permitted. In 1981, these conventional accelerated methods were replaced by the simpler procedure known as the Accelerated Cost Recovery System (ACRS). In 1986, Congress again modified the ACRS (MACRS) by sharply reducing depreciation allowances that were enacted in the Economic Recovery Tax Act of 1981. This section will present some of the primary features of MACRS tax depreciation.

MACRS Recovery Periods

Historically, for tax purposes as well as for accounting, an asset's depreciable life was determined by its estimated useful life; it was intended that an asset would be fully depreciated at approximately the end of its useful life. The MACRS scheme, however, totally abandoned this practice, and simpler guidelines were set which created several classes of assets, each with a more or less arbitrary life called a **recovery period. Note:** *These recovery periods do not necessarily bear a relationship to expected useful lives.*

A major effect of the original ACRS method of 1981 was to shorten the depreciable lives of assets, thus giving businesses larger depreciation deductions that resulted in lower taxes in early years and increased cash flows available for reinvestment. As shown in Table 10.2, the MACRS method of 1986 reclassified certain assets based on midpoint lives under the ADR system. The MACRS scheme includes eight categories of assets: 3, 5, 7, 10, 15, 20, 27.5, and 39 years. Under the MACRS, *the salvage value of property is always treated as zero.*

Recovery Period	ADR* Midpoint Class	Applicable Property
3-year	ADR ≤ 4	Special tools for manufacture of plastic products, fabricated metal products, and motor vehicles
5-year	4 < ADR ≤ 10	Automobiles,† light trucks, high-tech equipment, equipment used for R&D, computerized telephone switching systems
7-year	10 < ADR ≤16	Manufacturing equipment, office furniture, fixtures
10-year	16 < ADR ≤ 20	Vessels, barges, tugs, railroad cars
15-year	20 < ADR ≤ 25	Waste-water plants, telephone- distribution plants, or similar utility property
20-year	25 ≤ ADR	Municipal sewers, electrical power plant
27.5-year		Residential rental property
39-year		Nonresidential real property including elevators and escalators

Table 10.2 MACRS Property Classifications

*ADR = Asset Depreciation Range: Guidelines are published by the IRS.

†Automobiles have a midpoint life of 3 years in the ADR Guidelines, but are classified into a 5-year property class.

- Investments in some short-lived assets are depreciated over 3 years by using 200% DB and switching to SL depreciation.

- Computers, automobiles, and light trucks are written off over 5 years by using 200% DB and then switching to SL depreciation.

- Most types of manufacturing equipment are depreciated over 7 years, but some long-lived assets are written off over 10 years. Most equipment write-offs are calculated by the 200% DB method and switching to SL depreciation, which allows faster write-offs in the first few years after an investment is made.

- Sewage-treatment plants and telephone-distribution plants are written off over 15 years by using 150% DB and switching to SL depreciation.

- Sewer pipes and certain other very long-lived equipment are written off over 20 years by using 150% DB and switching to SL depreciation.

- Investments in residential rental property are written off in straight-line fashion over 27.5 years. On the other hand, nonresidential real estate (commercial buildings) is written off by the SL method over 39 years.

10.4.2 MACRS Depreciation Rules

Under earlier depreciation methods, the rate at which the value of an asset actually declined was estimated, and this rate was then used as the basis for tax depreciation. Thus, different assets were depreciated along different paths over time. The MACRS method, however, establishes prescribed depreciation rates, called **recovery allowance percentages,** for all assets within each class. These rates, as set forth in 1986 and 1993, are shown in Table 10.3. The yearly recovery, or depreciation expense, is determined by multiplying the asset's depreciation base by the applicable recovery allowance percentage.

Half-Year Convention

The MACRS recovery percentages as shown in Table 10.3 use the **half-year convention,** i.e., it is assumed that all assets are placed in service at mid-year and that they will have *zero* salvage value. As a result, only a half-year of depreciation is allowed for the first year that property is placed in service. With half of one year's depreciation being taken in the first year, a full year's depreciation is allowed in each of the remaining years of the asset's recovery period, and the remaining half-year's depreciation in the year following the end of the recovery period. A half-year of depreciation is also allowed for the year in which property is disposed of, or is otherwise retired from service, any time before the end of the recovery period.

Switching from DB to the Straight-Line Method

The MACRS asset is depreciated initially by the DB method and then by SL depreciation. Consequently, the MACRS adopts the switching convention illustrated in Section 10.3.2. To demonstrate how the MACRS depreciation percentages were calculated by the IRS using the half-year convention, let's consider Example 10.9.

Year	Class Depreciation Rate	3 200% DB	5 200% DB	7 200% DB	10 200% DB	15 150% DB	20 150% DB
1		33.33	20.00	14.29	10.00	5.00	3.750
2		44.45	32.00	24.49	18.00	9.50	7.219
3		14.81*	19.20	17.49	14.40	8.55	6.677
4		7.41	11.52*	12.49	11.52	7.70	6.177
5			11.52	8.93*	9.22	6.93	5.713
6			5.76	8.92	7.37	6.23	5.285
7				8.93	6.55*	5.90*	4.888
8				4.46	6.55	5.90	4.522
9					6.56	5.91	4.462*
10					6.55	5.90	4.461
11					3.28	5.91	4.462
12						5.90	4.461
13						5.91	4.462
14						5.90	4.461
15						5.91	4.462
16						2.95	4.461
17							4.462
18							4.461
19							4.462
20							4.461
21							2.231

Table 10.3 MACRS Depreciation Schedules for Personal Properties with Half-Year Convention

*Year to switch from declining balance to straight line. Source: IRS Publication 534. *Depreciation*. U.S. Government Printing Office: Washington, DC, December, 1995.

Example 10.9 MACRS Depreciation: Personal Property

A taxpayer wants to place in service a $10,000 asset that is assigned to the 5-year class. Compute the MACRS percentages and the depreciation amounts for the asset.

Solution

Given: 5-year asset, half-year convention, $\alpha = 40\%$, $S = 0$

Find: MACRS depreciation percentages, D_n for $10,000 asset

MACRS deduction percentages, beginning with the first taxable year and ending with the sixth year, are computed as follows:

Straight $\qquad$ = line rate = 1/5 = 0.20

200% declining balance rate = 2(0.20) = 40%

Under MACRS, salvage (S) = 0

Year	Calculation (%)		MACRS Percentage
1	$\frac{1}{2}$ year DDB dep. = 0.5(0.40)(100%)	=	20%
2	DDB dep. = (0.40)(100% − 20%)	=	32%
	SL dep. = (1/4.5)(100% − 20%)	=	17.78%
3	DDB dep. = (0.40)(100% − 52%)	=	19.20%
	SL dep. = (1/3.5)(100% − 52%)	=	13.71%
4	DDB dep. = (0.40)(100% − 71.20%)	=	11.52%
	SL dep. = (1/2.5)(100% − 71.20%)	=	11.52%
5	SL dep. = (1/1.5)(100% − 82.72%)	=	11.52%
6	$\frac{1}{2}$ year SL dep. = (0.5)(11.52%)	=	5.76%

In year 2, we check to see what the SL depreciation would be. Since 4.5 years are left to depreciate, SL depreciation = (1/4.5)(100% − 20%) = 17.78%. The DDB depreciation is greater than the SL depreciation, so DDB applies. Note that SL depreciation $\geq$ DDB depreciation in year 4 and we switch to SL.

We can calculate the depreciation amounts from the percentages we determined above. In practice, the percentages are taken directly from Table 10.3 supplied by the IRS.

Year n	MACRS Percentage (%)		Depreciation Basis		Depreciation Amount (D_n)
1	20	×	$10,000	=	$2,000
2	32	×	10,000	=	3,200
3	19.20	×	10,000	=	1,920
4	11.52	×	10,000	=	1,152
5	11.52	×	10,000	=	1,152
6	5.76	×	10,000	=	576

The results are also shown in Figure 10.6.

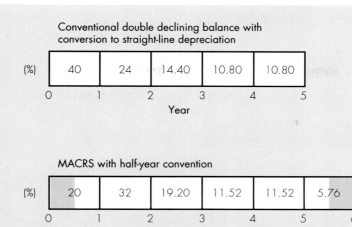

Figure 10.6   MACRS with a 5-year recovery period (Example 10.9)

Note that when an asset is disposed of before the end of recovery period, only half of the normal depreciation is allowed. If, for example, the $10,000 asset were to be disposed of in year 2, the MACRS deduction for that year would be $1,600.

Mid-Month Convention for Real Property

Table 10.4 shows the useful lives and depreciation percentages for real property. When depreciating such property, the straight-line method and the mid-month convention are used. For example, a property placed in service in March would be allowed 9.5 months depreciation for year 1. If it is disposed of before the end of the recovery period, the depreciation percentage must take into account the number of months the property was in service during the year of its disposal.

Example 10.10   MACRS for Real Property

On May 1, Jack Costanza paid $100,000 for a residential rental property. This purchase price represents $80,000 for the cost of the building and $20,000 for the cost of the land. Three years and 5 months later, on October 1, he sold the property for $130,000. Compute the MACRS depreciation for each of the 4 calendar years during which he owned the property.

Solution

Given: Residential real property, cost basis = $80,000; the building was put into service on May 1

Table 10.4
MACRS Percentages for Real Property

Month Property Placed in Service

(a) Residential Rental Property—Straight Line over 27.5 Years with Mid-Month Convention

Year	1	2	3	4	5	6	7	8	9	10	11	12
1	3.485%	3.182%	2.879%	2.576%	2.273%	1.970%	1.667%	1.364%	1.061%	0.758%	0.455%	0.152%
2–9	3.636	3.636	3.636	3.636	3.636	3.636	3.636	3.636	3.636	3.636	3.636	3.636
10	3.637	3.637	3.637	3.637	3.637	3.637	3.636	3.636	3.636	3.636	3.636	3.636
11	3.636	3.636	3.636	3.636	3.636	3.636	3.637	3.637	3.637	3.637	3.637	3.637
12	3.637	3.637	3.637	3.637	3.637	3.637	3.636	3.636	3.636	3.636	3.636	3.636
13	3.636	3.636	3.636	3.636	3.636	3.636	3.637	3.637	3.637	3.637	3.637	3.637
14	3.637	3.637	3.637	3.637	3.637	3.637	3.636	3.636	3.636	3.636	3.636	3.636
15	3.636	3.636	3.636	3.636	3.636	3.636	3.637	3.637	3.637	3.637	3.637	3.637
16	3.637	3.637	3.637	3.637	3.637	3.637	3.636	3.636	3.636	3.636	3.636	3.636
17	3.636	3.636	3.636	3.636	3.636	3.636	3.637	3.637	3.637	3.637	3.637	3.637
18	3.637	3.637	3.637	3.637	3.637	3.637	3.636	3.636	3.636	3.636	3.636	3.636
19	3.636	3.636	3.636	3.636	3.636	3.636	3.637	3.637	3.637	3.637	3.637	3.637
20	3.637	3.637	3.637	3.637	3.637	3.637	3.636	3.636	3.636	3.636	3.636	3.636
21	3.636	3.636	3.636	3.636	3.636	3.636	3.637	3.637	3.637	3.637	3.637	3.637
22	3.637	3.637	3.637	3.637	3.637	3.637	3.636	3.636	3.636	3.636	3.636	3.636
23	3.636	3.636	3.636	3.636	3.6363	3.636	3.637	3.637	3.637	3.637	3.637	3.637
24	3.637	3.637	3.637	3.637	3.637	3.637	3.636	3.636	3.636	3.636	3.636	3.636
25	3.636	3.636	3.636	3.636	3.636	3.636	3.637	3.637	3.637	3.637	3.637	3.637
26	3.637	3.637	3.637	3.637	3.637	3.637	3.636	3.636	3.636	3.636	3.636	3.636
27	3.636	3.636	3.636	3.636	3.636	3.636	3.637	3.637	3.637	3.637	3.637	3.637
28	1.97	2.273	2.576	2.879	3.182	3.485	3.636	3.636	3.636	3.636	3.636	3.636
29							0.152	0.455	0.758	1.061	1.364	1.667

(b) Nonresidential Real Property—Straight Line over 39 Years with Mid-Month Convention

Year	1	2	3	4	5	6	7	8	9	10	11	12
1	2.4573	2.2436	2.0299	1.8162	1.6026	1.3889	1.1752	0.9615	0.7479	0.5342	0.3205	0.1068
2–39	2.5641	2.5641	2.5641	2.5641	2.5641	2.5641	2.5641	2.5641	2.5641	2.5641	2.5641	2.5641
40	0.1068	0.3205	0.5342	0.7479	0.9615	1.1752	1.3889	1.6026	1.8162	2.0299	2.2436	2.4573

Source: IRS Publication No. 534. *Depreciation*. U.S. Government Printing Office: Washington, DC, 2000.

Find: The depreciation in each of 4 tax years property in service

In this example, the mid-month convention assumes that the property is placed in service on May 15, which gives 7.5 months of depreciation in the first year. Remembering that only the building (not the land) may be depreciated, we compute the depreciation over a 27.5 year recovery period using the SL method:

Year	Calculation		D_n	Recovery Percentages
1	$\left(\dfrac{7.5}{12}\right)\dfrac{80{,}000 - 0}{27.5}$	=	$1,818	2.273%
2	$\dfrac{80{,}000 - 0}{27.5}$	=	$2,909	3.636%
3	$\dfrac{80{,}000 - 0}{27.5}$	=	$2,909	3.636%
4	$\left(\dfrac{9.5}{12}\right)\dfrac{80{,}000 - 0}{27.5}$	=	$2,303	2.879%

Notice that the mid-month convention also applies to the year of disposal. Now compare the percentages with those in Table 10.4. As for personal property, calculations for real property generally use the precalculated table percentages.

10.5 Depletion

If you own mineral property (distinguished from personal and real properties), such as oil, gas, a geothermal well, or standing timber, you may be able to claim a deduction as you deplete the resource. A capital investment in natural resources needs to be recovered as the natural resources are being removed and sold. The process of amortizing the cost of natural resources in the accounting periods is called **depletion.** The objective of depletion is the same as that for depreciation: to amortize the cost in a systematic manner over the asset's useful life.

There are two ways of figuring depletion: **cost depletion** and **percentage depletion.** These depletion methods are used for book as well as tax purposes. In most instances, depletion is calculated by both methods, and the larger value is taken as the depletion allowance for the year. For standing timber, and most oil and gas wells, only cost depletion is permissible.

10.5.1 Cost Depletion

The cost depletion method is based on the same concept as the units-of-production method. To determine the amount of cost depletion, the adjusted basis of the mineral property is divided by the total number of recoverable units in the deposit and the resulting rate is multiplied by the number of units sold:

$$\text{Cost depletion} = \frac{(\text{Adjusted basis of mineral property})}{\text{Total number of recoverable units}}$$
$$\times (\text{number of units sold}). \tag{10.13}$$

The **adjusted basis** represents all the depletion allowed (or allowable cost on the property). Estimating the number of recoverable units in a natural deposit is largely an engineering problem.

Example 10.11 Cost Depletion for a Timber Tract

You bought a timber tract for $200,000, and the land was worth $80,000. The basis for the timber is therefore $120,000. The tract has an estimated 1.5 million board feet (1.5 MBF) of standing timber. If you cut 0.5 MBF of timber, determine your depletion allowance.

Solution

Given: Basis = $120,000, total recoverable volume = 1.5 MBF, amount sold this year = 0.5 MBF

Find: The depletion allowance this year

Timber depletion may be figured only by the cost method. Percentage depletion does not apply to timber. Your depletion basis does not include any part of the cost of the land. Because depletion takes place when standing timber is cut, you may figure your depletion deduction only after the timber is cut, and the quantity is first accurately measured.

$$\text{Depletion allowance per MBF} = \$120,000/1.5 \text{ MBF}$$
$$= \$80,000/\text{MBF}$$
$$\text{Depletion allowance for the year} = 0.5 \text{ MBF} \times \$80,000/\text{MBF}$$
$$= \$40,000.$$

10.5.2 Percentage Depletion

Percentage depletion is an alternative method of calculating the depletion allowance for certain mineral properties. For a given mineral property, the depletion allowance calculation is based on a prescribed percentage of the gross income from the property during the tax year. Notice the distinction between depreciation and depletion: **Depreciation** is the allocation of cost over a useful life, whereas **percentage depletion** is an annual allowance of a percentage of the gross income from the property.

Since percentage depletion is computed based on the income rather than the cost of the property, the total depletion on a property may exceed the cost of the property. To prevent this from happening, the annual allowance under the percentage method cannot be more than 50% of the taxable income from the property (figured without the deduction for depletion). Table 10.5 shows the allowed percentages for selected mining properties.

Deposits	Percentage
Oil and gas wells (only for certain domestic and gas production)	15
Sulfur and uranium and, if from deposits in the United States, asbestos, lead, zinc, nickel, mica, and certain other ores and minerals	22
Gold, silver, copper, iron ore, and oil shale, if from deposits in the United States	15
Coal, lignite, and sodium chloride	10
Clay and shale to be used in making sewer pipe or bricks	7.5
Clay (used for roofing tile), gravel, sand, and stone	5
Most other minerals; includes carbon dioxide produced from a well and metallic ores	14

Table 10.5 Percentage Depletion Allowances for Mineral Properties

---

Example 10.12 Percentage Depletion versus Cost Depletion

A gold mine with an estimated deposit of 300,000 ounces of gold has a basis of $30 million (cost minus land value). The mine has a gross income of $16,425,000 for the year from selling 45,000 ounces of gold (unit price of $365 per ounce). Mining expenses before depletion equal $12,250,000. Compute the percentage depletion allowance. Would it be advantageous to apply cost depletion rather than percentage depletion?

Solution

Given: Basis = $30 million, total recoverable volume = 300,000 ounces of gold, amount sold this year = 45,000 ounces, gross income = $16,425,000, this year's expenses before depletion = $12,250,000

Find: Maximum depletion allowance (cost or percentage)

Percentage depletion: Table 10.5 indicates that gold has a 15% depletion allowance. The percentage depletion allowance is computed from gross income:

Gross income from sale of 45,000 ounces	$16,425,000
Depletion percentage	× 15%
Computed percentage depletion	$ 2,463,750

Next, we need to compute the taxable income. The percentage depletion allowance is limited to the computed percentage depletion or 50% of taxable income, whichever is smaller:

Gross income from sale of 45,000 ounces	$16,425,000
Less mining expenses	12,250,000
Taxable income from mine	4,175,000
Deduction limitation	× 50%
Maximum depletion deduction	$ 2,088,000

Since the maximum depletion deduction ($2,088,000) is less than the computed percentage depletion ($2,463,750), the allowable percentage deduction is $2,088,000.

Cost depletion: It is worth computing the depletion allowance using the cost depletion method:

$$\text{Cost depletion} = (\$30,000,000/300,000)(45,000)$$
$$= \$4,500,000.$$

Note that percentage depletion is less than cost depletion. Since the law allows the taxpayer to take whichever deduction is larger in any one year, in this situation it would be advantageous to apply the cost depletion method.

Figure 10.7 illustrates the steps to be taken to apply the depletion method.

10.6 Repairs or Improvements to Depreciable Assets

If any major repairs (engine overhaul) or improvements (additions) are made during the life of the asset, we need to determine whether these actions will extend the life of the asset or will increase the originally estimated salvage value. When either of these situations arises, a revised estimate of useful life should be made, and the periodic depreciation expense should be updated accordingly. We will examine how repairs or improvements affect both book and tax depreciations.

10.6.1   Revision of Book Depreciation

Recall that book depreciation rates are based on estimates of the useful life of assets. Estimates of useful life are seldom precise. Therefore, after a few years of use you may find that the asset could last for a considerably longer or shorter period than was originally estimated. If this happens, the annual depreciation expense, based on the estimated useful life, may be either excessive or inadequate. (If the repairs or improvements do not extend the life or increase the salvage value of the asset, these costs may be treated as maintenance expenses during that year.) The procedure for

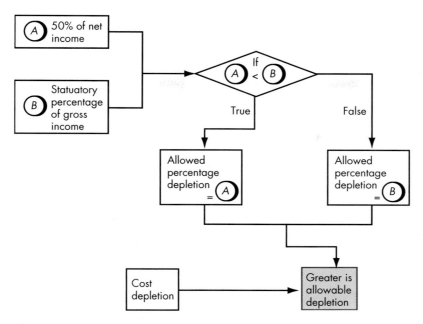

Figure 10.7   Calculating the allowable depletion deduction for federal tax purposes

correcting the book depreciation schedule is to revise the current book value and to allocate this cost over the remaining years of useful life.

10.6.2   Revision of Tax Depreciation

For tax depreciation, repairs or improvements made to any property are treated as *separate* property items. The recovery period for a repair or improvement to the initial property normally begins on the date the repair or improvement is placed in service. The recovery class of the repair or improvement is the recovery class that would apply to the property if it were placed in service at the same time as the repair or improvement. Example 10.13 illustrates the procedure for correcting the depreciation schedule for an asset that had repairs or improvements made to it during its depreciable life.

Example 10.13 Depreciation Adjustment for an Overhauled Asset

In January 1998, Kendall Manufacturing Company purchased a new numerical control machine at a cost of $60,000. The machine had an expected life of 10 years at the time of purchase and a zero expected salvage value at the end of the 10 years. For book depreciation purposes, no major overhauls had been planned for that period and the machine was being depreciated using the straight-line method toward a zero salvage value, or $6,000 per year. For tax purposes, the machine was classified as a 7-year MACRS property. In December 2000, however, the machine was thoroughly overhauled and rebuilt at a cost of $15,000. It was estimated that the overhaul would extend the machine's useful life by 5 years.

(a) Calculate the book depreciation for the year 2003 on a straight-line basis.

(b) Calculate the tax depreciation for the year 2003 for this machine.

Solution

Given: I = $60,000, S = $0, N = 10 years, machine overhaul = $15,000, extended life = 15 years from the original purchase

Find: D_6 for book depreciation, D_6 for tax depreciation

(a) Since an improvement was made at the end of year 2000, the book value of the asset at that time consists of the original book value plus the cost added to the asset. First, the original book value at the end of 2000 is calculated:

$$B_3 \text{ (before improvement)} = \$60,000 - 3(\$6000) = \$42,000.$$

After adding the improvement cost of $15,000, the revised book value is

$$B_3 \text{ (after improvement)} = \$42,000 + \$15,000 = \$57,000.$$

To calculate the book depreciation in the year 2003, which is 3 years after the improvement, we need to calculate the annual straight-line depreciation amount with the extended useful life. The remaining useful life before the improvement was made was 7 years. Therefore, the revised remaining useful life should be 12 years. The revised annual depreciation is then $57,000/12 = $4,750. Using the straight-line depreciation method, we compute the depreciation amount for 2003 as follows:

$$D_6 = \$4,750.$$

(b) For tax depreciation purposes, the improvement made is viewed as a separate property with the same recovery period with the initial asset. Thus, we need to calculate both the tax depreciation under the original asset and that of the new asset. For the 7-year MACRS property, the 6th-year depreciation allowance is 8.92% of $60,000, or $5,352. The third year depreciation for the improvement asset is 17.49% of $15,000 or $2,623. Therefore, the total tax depreciation in year 2003 is

$$D_6 = \$5,352 + \$2,623 = \$7,975.$$

Figure 10.8 illustrates how the additions or improvements are treated in revising depreciation amounts for book and tax purposes.

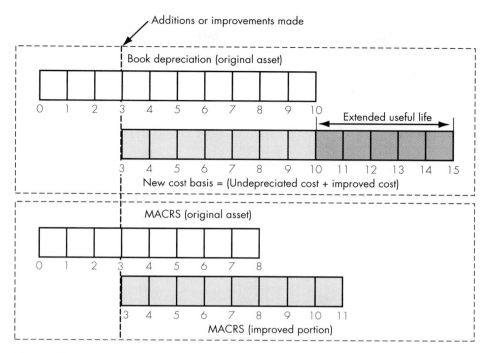

Figure 10.8   Revision of depreciation amount as additions or improvements are made as described in Example 10.13

10.7 Summary

- Machine tools and other manufacturing equipment, and even the factory buildings themselves, are subject to wear over time. However, it is not always obvious how to account for the cost of their replacement. Clearly, the choice of estimated service life for a machine, and the method used to calculate the cost of operating it, can have significant effects on an asset's management.

- The entire cost of replacing a machine cannot be properly charged to any 1 year's production; rather, the cost should be spread (or capitalized) over the years in which the machine is in service. The cost charged to operations during a particular year is called **depreciation.** Several different meanings and applications of depreciation have been presented in this chapter. From an engineering economics point of view, our primary concern is with **accounting depreciation**: The systematic allocation of an asset's value over its depreciable life.

- Accounting depreciation can be broken into two categories:

  1. **Book depreciation**—the method of depreciation used for financial reports and pricing products;

  2. **Tax depreciation**—the method of depreciation used for calculating taxable income and income taxes; it is governed by tax legislation.

- The four components of information required to calculate depreciation are:

  1. The cost basis of the asset,

  2. The salvage value of the asset,

  3. The depreciable life of the asset, and

  4. The method of its depreciation.

  Table 10.6 summarizes the differences in the way these components are treated for purposes of book and tax depreciation.

- Because it employs accelerated methods of depreciation and shorter-than-actual depreciable lives, the **MACRS (Modified Accelerated Cost Recovery System)** gives taxpayers a break: It allows them to take earlier and faster advantage of the tax-deferring benefits of depreciation.

- Many firms select straight-line depreciation for book depreciation because of its relative ease of calculation.

Table 10.6
Summary of Book versus Tax Depreciation

Component of Depreciation	Book Depreciation	Tax Depreciation (MACRS)
Cost basis	Based on the actual cost of the asset, plus all incidental costs such as freight, site preparation, installation, etc.	Same as for book depreciation
Salvage value	Estimated at the outset of depreciation analysis. If the final book value does not equal the estimated salvage value, we may need to make adjustments in our depreciation calculations	Salvage value is zero for all depreciable assets
Depreciable life	Firms may select their own estimated useful lives or follow government guidelines for asset depreciation ranges (ADRs)	Eight recovery periods—3, 5, 7, 10, 15, 20, 27.5, or 39 years—have been established; all depreciable assets fall into one of these eight categories.
Method of depreciation	Firms may select from the following: • straight-line • accelerated methods (declining balance, double declining balance, and sum-of-years'-digits) • units-of-production	Exact depreciation percentages are mandated by tax legislation but are based largely on DDB and straight-line methods. The SOYD method is rarely used in the U.S. except for some cost analysis in engineering valuation.

Function Description	Excel	Table 10.7
Straight-line depreciation	SLN (cost, salvage, life)	Excel's Built-In Depreciation Functions
Double declining balance depreciation	DDB (cost, salvage, life, period, factor)	
Sum-of-years'-digits depreciation	SYD (cost, salvage, life, period)	
Declining balance depreciation (other than DDB; assumes no switching)	VDB (cost, salvage, life, life start, end period, factor, no switch)	
MACRS depreciation command	No built-in commands available	

- Depletion is a cost allocation method used particularly for natural resources. **Cost depletion** is based on the units-of-production method of depreciation. **Percentage depletion** is based on a prescribed percentage of the gross income of a property during a tax year.
- Given the frequently changing nature of depreciation and tax law, we must use whatever percentages, depreciable lives, and salvage values mandated *at the time an asset is acquired*.
- Table 10.7 summarizes the built-in depreciation functions for Excel.

Self-Test Questions

10s.1 A machine, purchased for $45,000, has a depreciable life of 4 years. It will have an expected salvage value of $5,000 at the end of the depreciable life. Using the straight-line method, what is the book value at the end of year 2?

(a) $27,500   (b) $20,000
(c) $35,000   (d) $25,000.

10s.2 Consider problem 10s.1. If the double declining balance (200% DB) method is used, what is the depreciation amount for year 2?

(a) $10,000   (b) $11,250
(c) $20,000   (d) $17,500.

10s.3 Consider problem 10s.1. If the sum-of-the years'-digits (SOYD) method is used, what is the depreciation amount for year 2?

(a) $11,000   (b) $10,000
(c) $12,000   (d) $18,000.

10s.4 Your accounting records indicate that an asset in use has a book value of $8,640. The asset cost $30,000 when it was purchased, and it has been depreciated under the 5 MACRS method. Based on the information available, determine how many years the asset has been in service.

(a) 3 years   (b) 4 years
(c) 5 years   (d) 6 years.

10s.5 Which of the following statements is correct?

(a) The reason why the U.S. Congress allows business to use MACRS depreciation as opposed to conventional methods is to reduce the business tax burden over the project life.

(b) Under the declining balance depreciation system, it is always desirable to switch to straight-line depreciation.

(c) When determining the cost basis for an asset's depreciation, you must include all the costs that were incurred to keep the asset in operable condition.

(d) The main reason why a typical firm may use a straight-line depreciation method in reporting an income to outside investors (as opposed to any other accelerated tax depreciation methods) is to abide by the accounting principle—that is to report the true cost of doing business.

10s.6 A company purchased a drill press priced at $170,000 in year 0. The company additionally incurred $30,000 for site preparation and labor to install the machine. The drill press is classified as a 7-year MACRS class property. The company is considering selling the drill press for $70,000 at the end of year 4. Compute the book value at the end of year 4 that should be used in calculating the taxable gains.

(a) $62,480

(b) $53,108

(c) $63,725

(d) $74,970.

Problems

Economic Depreciation

10.1 A machine now in use was purchased 4 years ago at a cost of $5,000. It has a book value of $1,300. It can be sold for $2,300, but could be used for 3 more years, at the end of which time it would have no salvage value. What is the amount of economic depreciation now for this asset?

Cost Basis

10.2 General Service Contractor Company paid $120,000 for a house and lot. The value of the land was appraised at $65,000, and the value of the house at $55,000. The house was then torn down at an additional cost of $5,000 so that a warehouse could be built on the lot at a cost of $50,000. What is the total value of the property with the warehouse? For depreciation purposes, what is the cost basis for the warehouse?

10.3 To automate one of their production processes, Milwaukee Corporation bought three flexible manufacturing cells at a price of $500,000 each. When they were delivered, Milwaukee paid freight charges of $25,000 and handling fees of $12,000. Site preparation for these cells cost $35,000. Six foremen, each earning $15 an hour, worked five 40-hour weeks to set up and test the manufacturing cells. Special wiring and other materials applicable to the new manufacturing cells cost $1,500. Determine the cost basis (amount to be capitalized) for these cells?

10.4 A new drill press was purchased for $95,000 by trading in a similar ma-

chine that had a book value of $25,000. Assuming that the trade-in allowance is $20,000 and that $75,000 cash is to be paid for the new asset, what is the cost basis of the new asset for depreciation purposes?

10.5 A lift truck priced at $35,000 is acquired by trading in a similar lift truck and paying cash for the remaining balance. Assuming that the trade-in allowance is $10,000, and the book value of the asset traded in is $6,000, what is the cost basis of the new asset for the computation of depreciation for tax purposes?

Book Depreciation Methods

10.6 Consider the following data on an asset:

Cost of the asset, I	$100,000
Useful life, N	5 years
Salvage value, S	$10,000

Compute the annual depreciation allowances and the resulting book values, using

(a) The straight-line depreciation method,

(b) Double declining balance method, and

(c) Sum-of-the-years' digits method.

10.7 A firm is trying to decide whether to keep an item of construction equipment another year. The firm is using DDB for book purposes, and this is the fourth year of ownership. The item cost $150,000 new. What is the depreciation in year 3?

10.8 Consider the following data on an asset:

Cost of the asset, I	$30,000
Useful life, N	7 years
Salvage value, S	$0

Compute the annual depreciation allowances and the resulting book values using the DDB and switching to SL.

10.9 The double declining balance method is to be used for an asset with a cost of $80,000, estimated salvage value of $22,000, and estimated useful life of 6 years.

(a) What is the depreciation for the first 3 fiscal years, assuming that the asset was placed in service at the beginning of the year?

(b) If switching to the straight-line method is allowed, when is the optimal time to switch?

10.10 Compute the double declining balance (DDB) depreciation schedule for the following asset:

Cost of the asset, I	$60,000
Useful life, N	8 years
Salvage value, S	$5,000

10.11 Compute the SOYD depreciation schedule for the following asset:

Cost of the asset, I	$12,000
Useful life, N	5 years
Salvage value, S	$2,000

(a) What is the denominator of the depreciation fraction?

(b) What is the amount of depreciation for the first full year of use?

(c) What is the book value of the asset at the end of the fourth year?

10.12 Upjohn Company purchased new packaging equipment with an estimated useful life of 5 years. Cost of the equipment was $20,000 and the salvage value was estimated to be $3,000 at the end of year 5. Compute the annual depreciation expenses through the 5-year life if the equipment under each of the following methods of book depreciation:

(a) Straight-line method.

(b) Double declining balance method (limit the depreciation expense in the fifth year to an amount that will cause the book value of the equipment at year end to equal the $3,000 estimated salvage value.)

(c) Sum-of-years'-digits method.

10.13 A second-hand bulldozer acquired at the beginning of the fiscal year at a cost of $58,000 has an estimated salvage value of $8,000 and an estimated useful life of 12 years. Determine the following:

(a) The amount of annual depreciation by the straight-line method,

(b) The amount of depreciation for the third year computed by the double declining balance method,

(c) The amount of depreciation for the second year computed by the sum-of-years'-digits method.

Units-of-Production Method

10.14 If a truck for hauling coal has an estimated net cost of $85,000 and is expected to give service for 250,000 miles, resulting in a salvage value of $5,000, depreciation would be charged at a rate of 32 cents per mile. Compute the allowed depreciation amount for the truck usage amounting to 55,000 miles.

10.15 A diesel-powered generator with a cost of $60,000 is expected to have a useful operating life of 50,000 hours. The expected salvage value of this generator is $8,000. In its first operating year, the generator was operated 5,000 hours. Determine the depreciation for the year.

10.16 Ingot Land Company owned four trucks dedicated primarily for its landfills business. The company's accounting record indicates the following:

| Description | Truck Type | | | |
	A	B	C	D
Purchase cost ($)	50,000	25,000	18,500	35,600
Salvage value ($)	5,000	2,500	1,500	3,500
Useful life (miles)	200,000	120,000	100,000	200,000
Accumulated depreciation as year begins ($)	0	1,500	8,925	24,075
Miles driven during year	25,000	12,000	15,000	20,000

Determine the amount of depreciation for each truck during the year.

Tax Depreciation

10.17 Zerex Paving Company purchased a hauling truck on January 1, 2000, at a cost of $32,000. The truck has a useful life of 8 years with an estimated salvage value of $5,000. The straight-line method is used for book purposes. For tax purposes, the truck would be depreciated using MACRS over its 5-year class life. Determine the annual depreciation amount to be taken over the useful life of the hauling truck for both book and tax purposes.

10.18 The Harris Foundry Company purchased new casting equipment in 2000 at a cost of $180,000. Harris also paid $35,000 to have the equipment delivered and installed. The casting machine has an estimated useful life of 12 years, but it will be depreciated using MACRS over its 7-year class life.

(a) What is the cost basis of the casting equipment?

(b) What will be the depreciation allowance in each year on the 7-year class life casting equipment?

10.19 An item of equipment uses the 7-year MACRS recovery period. Compute the book value for tax purposes at the end of 3 years. The cost basis is $100,000.

10.20 A piece of machinery purchased at a cost of $68,000 has an estimated salvage value of $9,000 and an estimated useful life of 5 years. It was placed in service on May 1 of the current fiscal year, which ends on December 31. The asset falls into a 7-year MACRS property. Determine the depreciation amounts over the useful life.

10.21 Suppose that a taxpayer places in service a $10,000 asset that is assigned to the 6-year class (say, a new property class) with half-year convention. Develop the MACRS deductions assuming a 200% declining balance rate switching to straight line.

10.22 On April 1st, Leo Smith paid $170,000 for a residential rental property. This purchase price represents $130,000 for the building and $40,000 for the land. Five years later, on November 1, he sold the property for $200,000. Compute the MACRS depreciation for each of the 5 calendar years during which he had the property.

10.23 In 2000, you purchased a spindle machine (7-year MACRS property) for $26,000, which you placed in service in January. Use the calendar year as your tax year. Compute the depreciation allowances.

10.24 In 2000, three assets were purchased and placed in service.

Asset Type	Date Placed in Service	Cost Base	MACRS Property Class
Car	Feb. 17	$15,000	5-year
Furniture	Mar. 25	$5,000	7-year
Copy machine	Apr. 3	$10,000	5-year

Compute the depreciation allowances for each asset.

10.25 On October 1, you purchased a residential home in which to locate your professional office for $150,000. The appraisal is divided into $30,000 for the land and $120,000 for the building.

(a) In your first year of ownership, how much can you deduct for depreciation for tax purpose?

(b) Suppose that the property was sold at $187,000 at the end of 4th year of ownership. What is the book value of the property?

10.26 Given the data below, identify the depreciation method used for each depreciation schedule as one of the following:

First cost	$80,000
Book depreciation life	7 years
MACRS property class	7-year
Salvage value	$24,000

	Depreciation Schedule			
n	A	B	C	D
1	$14,000	22,857	11,429	22,857
2	12,000	16,327	19,592	16,327
3	10,000	11,661	13,994	11,661
4	8,000	5,154	9,996	8,330
5	6,000	0	7,140	6,942
6	4,000	0	7,140	6,942
7	2,000	0	7,140	6,942
8	0	0	3,570	0

- Double declining balance (DDB) depreciation
- Sum-of-years'-digits depreciation
- DDB with conversion to straight-line, assuming a zero salvage value
- MACRS 7-year with half-year convention

10.27 A manufacturing company has purchased four assets:

Item	Lathe	Asset Type Truck	Building	Computer
Initial cost ($)	45,000	25,000	800,000	40,000
Book life	12 yr	200,000 mi	50 yr	5 yr
MACRS class	7	5	39	5
Salvage value ($)	3,000	2,000	100,000	0
Book depreciation	DDB	UP	SL	SOYD

The truck was depreciated using the unit production method. Usage of the truck was 22,000 miles and 25,000 miles during the first 2 years, respectively.

(a) Calculate the book depreciation for each asset of the first 2 years.

(b) Calculate the tax depreciation for each asset of the first 2 years.

(c) If the lathe is to be depreciated over the early portion of its life using DDB and then by switching to SL for the remainder of the asset's life, when should the switch occur?

10.28 For each of five assets in the following table, determine the missing amounts. For asset type IV, annual usage is 15,000 miles.

Types of Asset	I	II	III	IV	V
Depreciating methods	SL	DDB	SOYD	UP	MACRS
End of year	7	4	3	3	4
Initial cost ($)	10,000	18,000	☐	30,000	$8,000
Salvage value($)	2,000	2,000	7,000	0	$1,000
Book value ($)	3,000	2,320	☐	h	$1,382
Depreciable life	8 yrs	5 yr	5 yr	90,000 mi	☐
Depreciating amount ($)	☐	☐	16,600	☐	☐
Accumulated depreciation ($)	☐	15,680	66,400	☐	☐

10.29 Flint Metal Shop purchased a stamping machine for $147,000 on March 1, 2000. It is expected to have a useful life of 10 years, salvage value of $27,000, production of 250,000 units, and working hours of 30,000. During 2000, Flint used the stamping machine for 2,450 hours to produce 23,450 units. From the information given, compute the book depreciation expense for 2000 under each of the following methods:

(a) Straight-line

(b) Units of production method

(c) Working hours

(d) Sum-of-years'-digits

(e) Double declining balance (without conversion to straight-line)

(f) Double declining balance (with conversion to straight-line).

Depletion

10.30 Early in 2000, Atlantic Mining Company began operation at its West Virginia Mine. The mine had been acquired at a cost of $6,900,000 in 1998. The mine is expected to contain 3 million tons of silver and to have a residual value of $1,500,000 (once the mine is depleted). Before beginning mining operations, the company installed equipment at a cost of $2,700,000. This equipment will have no economic usefulness once the mine is depleted. Therefore, depreciation of the equipment is based upon the estimated number of tons of ore produced each year. Ore removed from the West Virginia Mine amounted to 500,000 tons in 2000 and 682,000 tons in 2001.

(a) Compute the per-ton depletion rate of the mine and the per-ton depreciation rate of mining equipment.

(b) Determine the depletion expense for the mine and the depreciation expense for the mining equipment.

10.31 You bought a timber tract for $450,000, and the land was worth as much as $180,000. Four point eight million board feet (4.8 MBF) of standing timber was estimated to be in the timber tract. If you cut 1.5 MBF of timber, determine your depletion allowance.

10.32 A gold mine with an estimated deposit of 500,000 ounces of gold is purchased for $40 million. The mine has a gross income of $22,623,000 for the year obtained from selling 52,000 ounces of gold. Mining expenses before depletion equal $12,250,000. Compute the percentage depletion allowance. Would it be advantageous to apply cost depletion rather than percentage depletion?

10.33 Oklahoma Oil Company incurred acquisition, exploration, and development costs during 2000 as follows:

Items*	Site Parcel A	Parcel B	Total
Acquisition costs	6	4	10
Exploration costs	13	9	22
Development costs	20	11	31
Recoverable oil (Mbbls)	9	5	14

*Units are in millions of dollars, except recoverable oil.

The market price of oil during 2000 was $16 per bbl.

(a) Determine the cost basis for depletion on each parcel.

(b) During 2000, 1,200,000 barrels were extracted from parcel A at a production cost of $3,600,000. Determine the depletion charge allowed for parcel A.

(c) In (b), if Oklahoma Oil Company sold 1,000,000 of the 1,200,000 barrels extracted during 2000 at a price of $17 per barrel, the sales revenue would be $17,000,000. If it qualified for use of percentage depletion (15%), what would be the allowed depletion amount for 2000?

(d) During 2000, 800,000 barrels were extracted from parcel B at a production cost of $3,000,000. Assume that during 2001 it is ascertained that the remaining proved reserves on parcel B total only 4,000,000 barrels instead of the originally estimated 5,000,000. This revision in proved reserves is considered a change in an accounting estimate that must be corrected during the current and future years. (A correction of prior years' depletion amounts is not permitted.) If 1,000,000 barrels are extracted during 2001, what is the total depletion charge allowed using the unit cost method?

10.34 A coal mine expected to contain 6.5 million tons of coal was purchased at a cost of $30 million. One million tons of coal are produced this year. The gross income for this coal is $600,000, and operating costs (excluding depletion expense) are $450,000. Knowing that coal has a 10% depletion allowance, what will be the depletion allowance for

(a) Cost depletion, and

(b) Percentage depletion?

Revision of Depreciation Rates

10.35 Perkins Construction Company bought a building for $800,000; it is to be used as a warehouse. A number of major structural repairs, completed at the beginning of the current year at a cost of $125,000, are expected to extend the life of the building 10 years beyond the original estimate. The building has been depreciated by the straight-line method for 25 years. Salvage value is expected to be negligible and has been ignored. The book value of the building before the structural repairs is $400,000.

(a) What has the amount of annual depreciation been in past years?

(b) What is the book value of the building after the repairs have been recorded?

(c) What is the amount of depreciation for the current year, using the straight-line method? (Assume that the repairs were completed at the very beginning of the year.)

10.36 The Dow Ceramic Company purchased a glass molding machine in 1995 for $140,000. The company has been depreciating the machine over an estimated useful life of 10 years, assuming no salvage value, by the straight-line method of depreciation.

For tax purposes, the machine has been depreciated under 7-year MACRS property. At the beginning of 1998, Dow overhauled the machine at a cost of $25,000. As a result of the overhaul, Dow estimated that the useful life of the machine would extend 5 years beyond the original estimate.

(a) Calculate the book depreciation for year 2000.

(b) Calculate the tax depreciation for year 2000.

10.37 On January 2, 1998, Hines Food Processing Company purchased a machine priced at $75,000 that dispenses a premeasured amount of tomato juice into a can. The estimated useful life of the machine is estimated at 12 years with a salvage value of $4,500. At the time of purchase, Hines incurred the following additional expenses:

Freight-in	$800
Installation cost	2,500
Testing costs prior to regular operation	1,200

Book depreciation was calculated by the straight-line method but, for tax purpose, the machine was classified into a 7-year MACRS property. In January 2000, accessories costing $5,000 were added to the machine to reduce its operating costs. These accessories neither prolonged the machine's life nor provided any additional salvage value.

(a) Calculate the book depreciation expense for 2001.

(b) Calculate the tax depreciation expense for 2001.

Short Case Studies

10.38 On January 2, 1997, Allen Flour Company purchased a new machine at a cost of $63,000. Installation costs for the machine are $2,000. The machine was expected to have a useful life of 10 years with a salvage value of $4,000. The company uses straight-line depreciation for financial reporting. On January 3, 1999, the machine broke down, and an extraordinary repair had to be made to the machine at a cost of $6,000. The repair resulted in extending the machine's life to 13 years, but left the salvage value unchanged. On January 2, 2000, an improvement was made to the machine in the amount of $3,000, which increases the machine's productivity and increases the salvage value to $6,000, but does not affect the remaining useful life. Determine depreciation expenses for the years December 31, 1997, 1999, and 2000.

10.39 On March 17, 1997, Wildcat Oil Company began operations at its Louisiana Oil Field. The oil field had been acquired several years earlier at a cost of $11.6 million. The field is estimated to contain 4 million barrels of oil and to have a salvage value of $2 million before, and after, all of the oil has been pumped out. Equipment costing $480,000 was purchased for use at the oil field. The equipment will have no economic usefulness once Louisiana Field is depleted; therefore, it is depreciated on a units-of-production method. Additionally, Wildcat Oil built a pipeline at a cost of $2,880,000 to serve the Louisiana Field. Although this pipeline is physically capable of being used for many years, its economic usefulness is limited to the productive life of Louisiana Field and, therefore, has no salvage value. Depreciation of the pipeline is based on the estimated number of barrels of oil to be produced. Production at the Louisiana Oil Field amounted to 420,000 barrels in 2000 and 510,000 barrels in 2001.

(a) Compute the per-barrel depletion rate of the oil field during years 1998 and 2001.

(b) Compute the per-barrel depreciation rates of the equipment and the pipeline during years 2000 and 2001.

10.40 At the beginning of the fiscal year, Borland Company acquired new equipment at a cost of $65,000. The equipment has an estimated life of 5 years and an estimated salvage value of $5,000.

(a) Determine the annual depreciation (for financial reporting) for each of the 5 years of estimated useful life of the equipment, the accumulated depreciation at the end of each year, and the book value of the equipment at the end of each year by (1) straight-line method, (2) the double declining balance method, and (3) the sum-of-years'-digits method.

(b) Determine the annual depreciation for tax purpose, assuming that the equipment falls into 7-year MACRS property class.

(c) Assume that the equipment was depreciated under 7-year MACRS. In the first month of the fourth year, the equipment was traded in for similar equipment priced at $82,000. The trade-in allowance on the old equipment was $10,000, and cash was paid for the balance. What is the cost basis of the new equipment for computing the amount of depreciation for income tax purposes?

10.41 A vessel purchased by AT&T at a cost of $50 million will be used, along with several like it, to lay 20,000 miles of underwater fiber-optic cable over the next several years.

(a) What would be the reasonable depreciable life of this vessel for book depreciation purposes?

(b) What category of MACRS recovery period could be used for tax depreciation purposes?

(c) Determine the allowed annual depreciation amounts over the depreciable life for tax purposes.

CHAPTER 11

Corporate Income Taxes

Government is a silent corporate partner: The total revenues and income taxes paid by five U.S industries during 2000 are summarized below (dollars in millions).(All figures are from the company's 2000 annual reports.)

Company	Gross Income	Taxable Income	Income Taxes	Net Income	Average Tax Rate
Intel	33,726	15,141	4,606	10,535	30.42%
Cisco	18,920	4,343	1,675	2,668	38.57%
Amazon	2,762	(1,107)	0	(1,411)	0%
Broadcom	1,132	339	68	271	20.00%
Oracle	17,173	10,123	3,827	6,297	37.80%

What do these companies have in common? All are among some of the most well-known Internet players with extensive foreign operations. But how do we explain the apparent discrepancy of Intel, which makes a considerably larger gross income than Oracle, a database software company, but which pays a smaller percentage of taxes? On the other hand, Amazon.com generated 2.4 times more revenue than Broadcom, but the company incurred losses and ended up paying no income taxes at all. Depending on how a firm is structured, and the types of actions it takes, a firm can pay federal tax rates of up to 35%. Including the state and local taxes, the combined rate could go as high as 43%. Most of these companies' effective tax rates also reflect the tax benefits derived from having significant operations outside the United States that are taxed at rates lower than the U.S. statutory rate of 35%.

lthough tax law is subject to frequent changes, the analytical procedures presented in this chapter provide a basis for tax analysis that can be adapted to reflect future changes in tax law. Thus, while we present many examples based on current tax rates, in a larger context we present a general approach to the analysis of any tax law.

There are many forms of government taxation, including sales taxes, property taxes, user taxes, and state and federal income taxes. In this chapter, we will focus on federal income taxes. When you are operating a business, any profits or losses you incur are subject to income tax consequences. Therefore, we cannot ignore the impact of income taxes in project evaluation. This chapter will give you a good idea of how the U.S. tax system operates and of how federal income taxes affect economic analysis.

11.1 After-Tax Cash Flow

Once again, we make our investment decisions based on the net project cash flows. The net project cash flows mean cash flows after taxes. In order to calculate the amount of taxes involved in project evaluation, we need to understand how business determines the taxable income and thereby the net income (profit).

Firms invest in a project because they expect it to increase their wealth. If the project does this—if project revenues exceed project costs—we say it has generated a **profit**, or **income**. If the project reduces a firm's wealth—if project costs exceed project revenues—we say that the project has resulted in a **loss**. One of the most important roles of the accounting function within an organization is to measure the amount of profit or loss a project generates each year, or in any other relevant time period. Any profit generated will be taxed. The accounting measure of a project's after-tax profit during a particular time period is known as **net income**.

11.1.1 Calculation of Net Income

Accountants measure the net income of a specified operating period by subtracting expenses from revenues for that period. These terms can be defined as follows:

1. The **project revenue** is the income earned[1] by a business as a result of providing products or services to customers. Revenue comes from sales of merchandise to customers and from fees earned by services performed for clients or others.
2. The **project expenses** that are incurred[2] are the cost of doing business to generate the revenues of the specified operating period. Some common expenses are the cost of the goods sold (labor, material, inventory, and supplies), depreciation, the cost of employees' salaries, the operating costs (such as the cost of renting buildings and the cost of insurance coverage), and income taxes.

The business expenses listed above are accounted for in a straightforward fashion on a company's income statement and balance sheet: The amount paid by the organization for each item would translate dollar for dollar into expenses in financial reports for the period. One additional category of expenses, the purchase of new assets, is treated by depreciating the total cost gradually over time. Because capital goods are given this unique accounting treatment, depreciation is accounted for as a separate expense in financial reports. In the following section, we will discuss how depreciation accounting is reflected in net income calculations.

11.1.2 Treatment of Depreciation Expenses

Whether you are starting or maintaining a business, you will probably need to acquire assets (such as buildings and equipment). The cost of this property becomes part of your business expenses. The accounting treatment of capital expenditures differs from

---

[1] Note that the cash may be received in a different accounting period.
[2] Note that the *cash* may be paid in a different accounting period.

the treatment of manufacturing and operating expenses, such as cost of goods sold and business operating expenses. As you recall from Chapter 10, **capital expenditures must be capitalized,** i.e., they must be systematically allocated as expenses over their depreciable lives. Therefore, when you acquire a piece of property that has a productive life extending over several years, you cannot deduct the total costs from profits in the year the asset was purchased. Instead, a depreciation allowance[3] is established over the life of the asset, and an appropriate portion of that allowance is included in the company's deductions from profit each year. Because it plays a role in reducing taxable income, depreciation accounting is of special concern to a company. In the next section, we will investigate the relationship between depreciation and net income.

11.1.3 Taxable Income and Income Taxes

Corporate taxable income is defined as follows:

$$\text{Taxable income} = \text{Gross income (revenues)} - \text{expenses}.$$

Once taxable income is calculated, income taxes are determined as follows:

$$\text{Income taxes} = (\text{Tax rate}) \times (\text{taxable income}).$$

(We will discuss how we determine the applicable tax rate in Section 11.2.) We then calculate net income as follows:

$$\text{Net income} = \text{Taxable income} - \text{income taxes}.$$

A more common format is to present the net income in the following tabular income statement:

Item
Gross income
Expenses
Cost of goods sold (revenues)
Depreciation
Operating expenses
Taxable income
Income taxes
Net income

Our first example illustrates this relationship using numerical values.

---

[3] This allowance is based on the total cost basis of the property.

Example 11.1 Net Income Within a Year

A company buys a numerically controlled (NC) machine for $28,000 (year 0) and uses it for 5 years, after which it is scrapped. The allowed depreciation deduction during the first year is $4,000 as the equipment falls into 7-year MACRS property. (The first year depreciation rate is 14.29%.) The cost of the goods produced by this NC machine should include a charge for the depreciation of the machine. Suppose the company estimates the following revenues and expenses including the depreciation for the first operating year.

Gross income	=	$50,000
Cost of goods sold	=	$20,000
Depreciation on NC machine	=	$4,000
Operating expenses	=	$6,000

If the company pays taxes at the rate of 40% on its taxable income, what is its net income during the first year from the project?

Solution

Given: Gross income and expenses as stated, income tax rate = 40%

Find: Net income

At this point, we will defer the discussion of how the tax rate (40%) is determined and treat it as given. We consider the purchase of the machine to have been made in year 0, which is also the beginning of year 1. (Note that our example explicitly assumes that the only depreciation charges for year 1 are those for the NC machine, a situation that may not be typical.)

Item	Amount
Gross income (revenues)	$50,000
Expenses	
Cost of goods sold	20,000
Depreciation	4,000
Operating expenses	6,000
Taxable income	20,000
Taxes (40%)	8,000
Net income	$12,000

Comments: In this example, the inclusion of a depreciation expense reflects the true cost of doing business. This expense is meant to match the amount of the $28,000 total cost of the machine that has been put to use or "used up" during the first year. This example also highlights some of the reasons why income tax laws govern the depreciation of assets. If the company were allowed to claim the entire $28,000 as a year 1 expense, a discrepancy would exist between the one-time cash outlay for the machine's cost and the gradual benefits of its productive use. This discrepancy would lead to dramatic variations in the firm's net income, and net income would become a less accurate measure of the organization's performance. On the other hand, failing to account for this cost would lead to increased reported profit during the accounting period. In this situation, the profit would be a "false profit" in that it would not accurately account for the usage of the machine. Depreciating the cost over time allows the company a logical distribution of costs that matches the utilization of the machine's value.

11.1.4   Cash Flow versus Net Income

Traditional accounting stresses net income as a means of measuring a firm's profitability, but it is desirable to discuss why cash flows are relevant data to be used in project evaluation. As seen in section 11.1.1, net income is an accounting measure based, in part, on the **matching concept**. Costs become expenses as they are matched against revenue. The actual timing of cash inflows and outflows is ignored.

Over the life of a firm, net incomes and net cash inflows will usually be the same. However, the timing of incomes and cash inflows can differ substantially. Given the time value of money, it is better to receive cash now rather than later, because cash can be invested to earn more cash. (You cannot invest net income.) For example, consider two firms and their income and cash flow schedules over 2 years as follows:

		Company A	Company B
Year 1	Net income	$1,000,000	$1000,000
	Cash flow	1,000,000	0
Year 2	Net income	1,000,000	1,000,000
	Cash flow	1,000,000	2,000,000

Both companies have the same amount of net income and cash sum over 2 years, but Company A returns $1 million cash yearly, while Company B returns $2 million at the end of the second year. If you received $1 million at the end of the first year from Company A, you could invest it at 10%, for example. While you would receive only $2 million in total from Company B at the end of the second year, you would receive in total $2.1 million from Company A.

Apart from the concept of the time value of money, certain expenses do not even require a cash outflow. Depreciation and amortization are the best examples of this type of expense. Even though depreciation (or amortization expense) is deducted from revenue on a daily basis, no cash is paid to anyone.

In Example 11.1, we have just seen that the annual depreciation allowance has an important impact on both taxable and net income. However, although depreciation has a direct impact on net income, it is *not* a cash outlay; as such, it is important to distinguish between annual income in the presence of depreciation and annual cash flow.

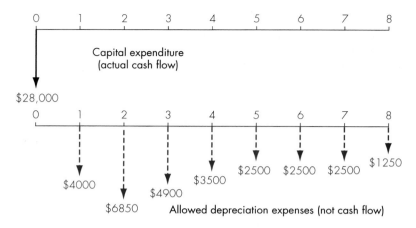

Figure 11.1   Cash flow versus depreciation expenses for an asset with a cost basis of $28,000

The situation described in Example 11.1 serves as a good vehicle to demonstrate the difference between depreciation costs as expenses and the cash flow generated by the purchase of a fixed asset. In this example, cash in the amount of $28,000 was expended in year 0, but the $4,000 depreciation charged against the income in year 1 is not a cash outlay. Figure 11.1 summarizes the difference.

Net income (**accounting profit**) is important for accounting purposes, but **cash flows** are more important for project evaluation purposes. However, as we will now demonstrate, net income can provide us with a starting point to estimate the cash flow of a project.

The procedure for calculating net income is identical to that used for obtaining net cash flow (after-tax) from operations, with the exception of depreciation, which is excluded from the net cash flow computation (it is needed only for computing income taxes). Assuming that revenues are received and expenses are paid in cash, we can obtain the net cash flow by adding the **non-cash expense** (depreciation) to net income, which cancels the operation of subtracting it from revenues.

$$\text{Cash flows} = \text{Net income} + \text{non-cash expense (depreciation)}$$

Example 11.2 illustrates this relationship.

Example 11.2 Cash Flow versus Net Income

Using the situation described in Example 11.1, assume that (1) all sales are cash sales, and (2) all expenses except depreciation were paid during year 1. How much cash would be generated from operations?

Solution

Given: Net income components
Find: Cash flow

We can generate a cash flow statement by simply examining each item in the income statement and determining which items actually represent receipts or disbursements. Some of the assumptions listed in the problem statement make this process simpler.

Item	Income	Cash Flow
Gross income (revenues)	$50,000	$50,000
Expenses		
Cost of goods sold	20,000	−20,000
Depreciation	4,000	
Operating expenses	6,000	−6,000
Taxable income	20,000	
Taxes (40%)	8,000	−8,000
Net income	$12,000	
Net cash flow		$16,000

Column 2 shows the income statement, while Column 3 shows the statement on a cash flow basis. The sales of $50,000 are all cash sales. Costs other than depreciation were $26,000; these were paid in cash, leaving $24,000. Depreciation is not a cash flow—the firm did not pay out $4,000 in depreciation expenses. Taxes, however, are paid in cash, so the $8,000 for taxes must be deducted from the $24,000, leaving a net cash flow from operations of $16,000.

Check: As shown in Figure. 11.2, this $16,000 is exactly equal to net income plus depreciation: $12,000 + $4,000 = $16,000.

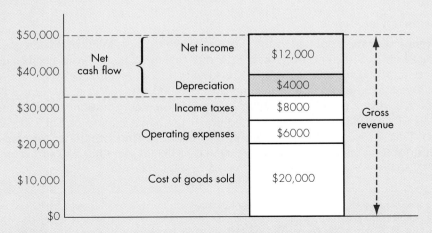

Figure 11.2   Net income versus net cash flow (Example 11.2)

As we've just seen, depreciation has an important impact on annual cash flow in its role as an accounting expense that reduces taxable income and thus taxes. (Although depreciation expenses are not actual cash flows, depreciation has a positive impact on the after-tax cash flow of the firm.) Of course, during the year in which an asset is actually acquired, the cash disbursed to purchase it creates a significant negative cash flow and, during the depreciable life of the asset, the depreciation charges will affect the taxes paid and, therefore, cash flows.

As shown in Example 11.2, we clearly see that depreciation, through its influence on taxes, plays a critical role in project cash flow analysis, which we will explore further in Chapter 12.

11.2 Corporate Taxes

Now that we have learned what elements constitute taxable income, we turn our attention to the process of computing income taxes. The corporate tax rate is applied to the taxable income of a corporation, which is defined as its gross income minus allowable deductions. As we briefly discussed in Section 11.1, the allowable deductions include the cost of goods sold, salaries and wages, rent, interest, advertising, depreciation, amortization,[4] depletion, and various tax payments other than federal income tax.

11.2.1   Income Taxes on Operating Income

The corporate tax rate structure for 2000 is relatively simple. As shown in Table 11.1, there are four basic rate brackets (15%, 25%, 34%, and 35%) plus two surtax rates (5% and 3%) based on taxable incomes. U.S. tax rates are progressive; that is, businesses with lower taxable incomes are taxed at lower rates than those with higher taxable incomes.

Marginal Tax Rate

Marginal tax rate is defined as the rate applied to the last dollar of income earned. Income of up to $50,000 is taxed at a 15% rate (meaning that, if your taxable income is less than $50,000, your marginal tax rate is 15%); income between $50,000 and $75,000 is taxed at 25%; and income over $75,000 is taxed at a 34% rate. An additional 5% surtax (resulting in 39%) is imposed on a corporation's taxable income in excess of $100,000, the maximum additional tax being limited to $11,750 (235,000 × 0.05). This surtax provision phases out the benefit of graduated rates for corporations with taxable incomes between $100,000 and $335,000. Another 3% surtax rate is im-

---

[4] The **amortization expense** is a special form of depreciation for an intangible asset, such as patents, goodwill, and franchises. More precisely, the amortization expense is the systematic write-off to expenses of the cost of an intangible asset over the periods of its economic usefulness. Normally a straight-line method is used to calculate the amortization expense.

Taxable Income (X)	Tax Rate	Tax Computation Formula
$0 – $50,000	15%	$0 $\quad$ + 0.15 X
50,001 – 75,000	25%	7,500 $\quad$ + 0.25(X – $50,000)
75,001 – 100,000	34%	13,750 $\quad$ + 0.34(X – 75,000)
100,001 – 335,000	34% + 5%	22,250 $\quad$ + 0.39(X – 100,000)
335,001 – 10,000,000	34%	113,900 $\quad$ + 0.34(X – 335,000)
10,000,001 – 15,000,000	35%	3,400,000 + 0.35(X – 10,000,000)
15,000,001 – 18,333,333	35% + 3%	5,150,000 + 0.38(X – 15,000,000)
18,333,334 and up	35%	6,416,666 + 0.35 (X – 18,333,333)

Table 11.1
Corporate Tax Schedule for 2001

posed on a corporate taxable income, in the range $15,000,001 to $18,333,333. Corporations with incomes in excess of $18,333,333, in effect, pay a flat tax of 35%. As shown in Table 11.1, the corporate tax is also progressive up to $18,333,333 in taxable income, but essentially it is constant thereafter.

Effective (Average) Tax Rate

Effective tax rates can be calculated from the data in Table 11.1. For example, if your corporation had a taxable income of $16,000,000 in 2001, then the income tax owed by the corporation would be

Taxable Income	Tax Rate	Taxes	Cumulative Taxes
First $50,000	15%	$7,500	$7,500
Next 25,000	25%	6,250	13,750
Next 25,000	34%	8,500	22,250
Next 235,000	39%	91,650	113,900
Next 9,665,000	34%	3,286,100	3,400,000
Next 5,000,000	35%	1,750,000	5,150,000
Remaining 1,000,000	38%	380,000	$5,530,000

or using the tax formulas in Table 11.1, we obtain

$$\$5,150,000 + 0.38(\$16,000,000 - \$15,000,000) = \$5,530,000.$$

The effective tax rate would be

$$\$5,530,000/\$16,000,000 = 0.3456 \text{ or } 34,56\%,$$

as opposed to the marginal rate of 38%. In other words, on the average, the company paid 34.56 cents for each taxable dollar it generated during the accounting period.

Example 11.3 Corporate Taxes

A mail-order computer company sells personal computers and peripherals. The company leased showroom space and a warehouse for $20,000 a year and installed $290,000 worth of inventory checking and packaging equipment. The allowed depreciation expense for this capital expenditure ($290,000) amounted to $58,000. The store was completed and operations begun on January 1. The company had a gross income of $1,250,000 for the calendar year. Supplies and all operating expenses other than the lease expense were itemized as follows:

Merchandise sold in the year	$600,000
Employee salaries and benefits	150,000
Other supplies and expenses	90,000
	$840,000

Compute the taxable income for this company. How much will the company pay in federal income taxes for the year?

Solution

Given: Income, cost information above, and depreciation

Find: Taxable income, federal income taxes

First we compute the taxable income as follows:

Gross revenues	$1,250,000
Expenses	840,000
Lease expense	20,000
Depreciation	58,000
Taxable income	$332,000

Note that capital expenditures are not deductible expenses. Since the company is in the 39% marginal tax bracket, the income tax can be calculated by using the formula given in Table 11.1, $22,250 + 0.39 (X - 100,000)$:

$$\text{Income tax} = \$22,250 + 0.39(\$332,000 - \$100,000)$$
$$= \$112,730.$$

The firm's current marginal tax rate is 39%, but its average corporate tax rate is

$$\$112,730/\$332,000 = 33.95\%.$$

Comments: Instead of using the tax formula in Table 11.1, we can also compute the federal income tax in the following fashion:

First $50,000 at 15%	$7,500
Next 25,000 at 25%	6,250
Next 25,000 at 34%	8,500
Next 232,000 at 39%	90,480
Income Tax	$112,730

11.2.2   Income Taxes on Corporate Capital Gains and Losses

Corporate capital gains are taxed at ordinary corporate tax rates. A company might realize capital gains from sources such as the sale of nondepreciable assets (e.g., land) or the sale of financial assets (e.g., stocks and bonds). **Capital gains** in financial assets occur when an asset is sold for more than its cost basis. **Capital losses** occur when an asset is sold for less than its cost basis. For a nondepreciable or financial asset, the book value is normally the same as its cost basis.

Corporate capital losses can be subtracted from any capital gains during the tax year. The net remaining losses may be carried back or over. In general, capital losses, when carried back or over, are deducted from capital gains in each year, but cannot be used to offset ordinary operating income.

11.3 Tax Treatment of Gains or Losses on Depreciable Assets

As in the disposal of capital assets, generally gains or losses are associated with the sale (or exchange) of depreciable assets. To calculate a gain or loss, we first need to determine the book value of the depreciable asset at the time of disposal.

11.3.1   Disposal of a MACRS Property

For a MACRS property, one important consideration at the time of disposal is whether the property is disposed of *during* or *before* its specified recovery period. Moreover, with the half-year convention, which is now mandated by all MACRS depreciation methods, the year of disposal is charged one half of that year's annual depreciation amount, if it should occur during the recovery period.

Example 11.4 Book Value in the Year of Disposal

Consider a 5-year MACRS asset purchased for $10,000. Note that property belonging to the 5-year MACRS class is depreciated over 6 years due to the half-year convention. The applicable depreciation percentages, shown in Table 10.3, are 20%, 32%, 19.20%, 11.52%, 11.52%, and 5.76%. Compute the allowed depreciation amounts and the book value when the asset is disposed of (a) in year 3, (b) in year 5, and (c) in year 6, respectively.

Solution

Given: 5-year MACRS asset, cost basis = $10,000, MACRS depreciation percentages as shown in Figure 11.3

Find: Total depreciation and book value at disposal if sold in year 3, 5, or 6

(a) If the asset is disposed of in year 3 (or at the end of year 3), the total accumulated depreciation amount and the book value would be

$$\text{Total depreciation} = \$10,000(0.20 + 0.32 + 0.192/2)$$
$$= \$6,160.$$
$$\text{Book value} = \$10,000 - \$6,160$$
$$= \$3,840.$$

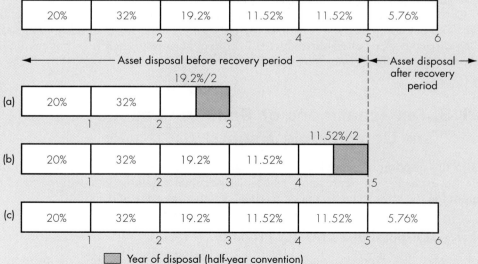

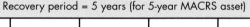

Figure 11.3 Disposal of a MACRS property and its effect on depreciation allowances (Example 11.4)

(b) If the asset is disposed of during[5] year 5, the depreciation and book values will be

$$\text{Total depreciation} = \$10,000(0.20 + 0.32 + 0.192 + 0.1152 + 0.1152/2)$$
$$= \$8,848.$$
$$\text{Book value} = \$10,000 - \$8,848$$
$$= \$1,152.$$

(c) If the asset is disposed of after the recovery period, there will be no penalty due to early disposal. Since the asset is depreciated fully, we have

$$\text{Total depreciation} = \$10,000$$
$$\text{Book value} = \$0.$$

11.3.2 Calculations of Gains and Losses on MACRS Property

When a depreciable asset used in business is sold for an amount that differs from its book value, the gain or loss has an important effect on income taxes. The gain or loss is found as follows:

$$\text{Gains (losses)} = \text{Salvage value} - \text{book value},$$

where the salvage value represents the proceeds from the sale (selling price) less any selling expense or removal cost.

These gains, commonly known as **depreciation recapture**, are taxed as ordinary income under current tax law. In the unlikely event that an asset is sold for an amount greater than its cost basis, the gains (salvage value – book value) are divided into two parts for tax purposes:

$$\text{Gains} = \text{Salvage value} - \text{book value}$$
$$= \underbrace{(\text{Salvage value} - \text{cost basis})}_{\text{Capital gains}}$$
$$+ \underbrace{(\text{Cost basis} - \text{book value})}_{\text{Ordinary gains}}$$

Recall from Section 10.2.2 that cost basis is the purchase cost of an asset plus any incidental costs, such as freight and installation. As illustrated in Figure 11.4:

---

[5] Note that even though you dispose of the asset at the *end* of the recovery period, the half-year convention still applies.

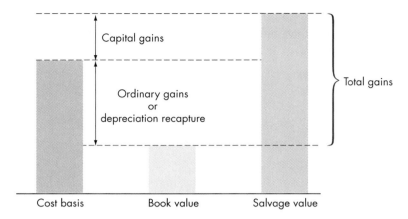

Figure 11.4   Capital gains and ordinary gains (or depreciation recapture) when the salvage value exceeds the cost basis

$$\text{Capital gains} \quad = \quad \text{Salvage value} - \text{cost basis}$$
$$\text{Ordinary gains} \quad = \quad \text{Cost basis} - \text{book value}$$

This distinction is only necessary when capital gains are taxed at the capital gain tax rate and ordinary gains (or depreciation recapture) at the ordinary income tax rate. Current tax law does not provide a special low rate of taxation for capital gains. Currently, capital gains are treated as ordinary income, but the maximum tax rate is set at the U.S. statutory rate of 35%. Nevertheless, the statutory structure for capital gains has been retained in the tax code. This provision could allow Congress to restore preferential treatment for capital gains at some future time.

Example 11.5 Gains and Losses on Depreciable Assets

A company purchased a drill press costing $230,000 in year 0. The drill press, classified as 7-year recovery property, has been depreciated by the MACRS method. If it is sold at the end of 3 years, compute the gains (losses) for the following four salvage values: (a) $150,000, (b) $120,693, (c) $100,000, and (d) $250,000. Assume that both capital gains and ordinary income are taxed at 34%.

Solution

Given: 7-year MACRS asset, cost basis = $230,000, sold 3 years after purchase

Find: Gains or losses, tax effects and net proceeds from the sale if sold for $150,000, $120,693, $100,000, or $250,000

In this example, we first compute the current book value of the machine. From the MACRS depreciation schedule in Table 11.3, the allowed annual depreciation percentages for the first 3 years of a 7-year MACRS property are 14.29%, 24.49%, and 17.49%. Since the asset is disposed of before the end of its recovery period, the depreciation amount in year 3 will be reduced by half. The total depreciation and final book value will be:

$$\text{Total allowed depreciation} = \$230,000 \, (0.1429 + 0.2449 + 0.1749/2)$$
$$= \$109,308.$$
$$\text{Book value} = \$230,000 - \$109,308$$
$$= \$120,693.$$

(a) Case 1: Book value < Salvage value < Cost basis

In this case, there are no capital gains to consider. All gains are ordinary gains.

$$\text{Ordinary gains} = \text{Salvage value} - \text{book value}$$
$$= \$150,000 - \$120,693$$
$$= \$29,308$$
$$\text{Gains tax } (34\%) = 0.34 \, (\$29,308)$$
$$= \$9,965.$$
$$\text{Net proceeds from sale} = \text{Salvage value} - \text{gains tax}$$
$$= \$150,000 - \$9,965$$
$$= \$140,035.$$

This situation (salvage value exceeds book value) is denoted as Case 1 in Figure 11.5.

(b) Case 2: Salvage value = Book value

In Case 2, the book value is again $120,693. Thus, if the drill press's salvage value equals $120,693, the book value, no taxes are levied on that salvage value. Therefore, the net proceeds equal the salvage value.

(c) Case 3: Salvage value < Book value

Case 3 illustrates a loss when the salvage value (say, $100,000) is less than the book value. We compute the net salvage value after tax as follows:

$$\text{Gain (loss)} = \text{Salvage value} - \text{book value}$$
$$= \$100,000 - \$120,693$$
$$= (\$20,693).$$
$$\text{Tax savings} = 0.34(\$20,693)$$
$$= \$7,036.$$
$$\text{Net proceeds from sale} = \$100,000 + \$7,036$$
$$= \$107,036.$$

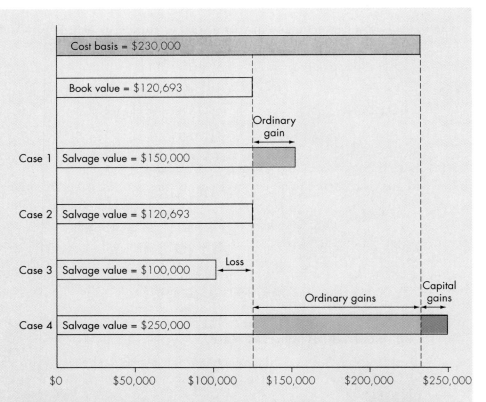

Figure 11.5 Calculations of gains or losses on MACRS property (Example 11.5)

(d) Case 4: Salvage value > Cost basis

This situation is not likely for most depreciable assets (except for real property). But this is the only situation where both capital gains and ordinary gains can be observed. Nevertheless, the tax treatment on this gain is as follows:

$$\text{Capital gains} = \text{Salvage value} - \text{cost basis}$$
$$= \$250,000 - \$230,000$$
$$= \$20,000.$$
$$\text{Capital gains tax} = \$20,00(0.34)$$
$$= \$6,800.$$
$$\text{Ordinary gains} = \$230,000 - \$120,693$$
$$= \$109,307.$$
$$\text{Gains tax} = \$109,307(0.34)$$
$$= \$37,164.$$

$$\text{Net proceeds from sale} = \$250,000 - (\$6,800 + \$37,164)$$
$$= \$206,036.$$

Comments: Note that in (c) the reduction in tax, due to the loss, actually increases the net proceeds. This is realistic when the incremental tax rate (34% in this case) is positive, indicating the corporation is still paying tax, but less than if the asset had not been sold at a loss. The incremental tax rate will be discussed in Section 11.4.

11.4 Income Tax Rate to be Used in Economic Analysis

As we have seen in the earlier sections, average income tax rates for corporations vary with the level of taxable income from 0 to 35%. Suppose that a company now paying a tax rate of 25% on its current operating income is considering a profitable investment. What tax rate should be used in calculating the taxes on the investment's projected income?

11.4.1 Incremental Income Tax Rate

As we will explain, the choice of the rate depends on the incremental effect on taxable income because of undertaking the investment. In other words, the tax rate to use is the rate that applies to the additional taxable income projected in the economic analysis.

To illustrate, consider ABC Corporation, whose taxable income from operations is expected to be $70,000 for the current tax year. ABC management wishes to evaluate the incremental tax impact of undertaking a project during the same tax year. The revenues, expenses, and taxable incomes before and after the project are estimated as follows:

	Before	After	Incremental
Gross revenue	$200,000	$240,000	$40,000
Salaries	100,000	110,000	10,000
Wages	30,000	40,000	10,000
Taxable income	$70,000	$90,000	$20,000

Because the income tax rate is progressive, the tax effect of the project cannot be isolated from the company's overall tax obligations. The base operations of ABC without the project are projected to yield a taxable income of $70,000. With the new project, the taxable income increases to $90,000. Using the tax computation formula in Table 11.1, the corporate income taxes with, and without, the project are as follows:

$$\text{Income tax without the project} = \$7{,}500 + 0.25(\$70{,}000 - \$50{,}000)$$
$$= \$12{,}500.$$

$$\text{Income tax with the project} = \$13{,}750 + 0.34(\$90{,}000 - \$75{,}000)$$
$$= \$18{,}850.$$

The additional income tax is then $18,850 – $12,500 = $6,350. The $6,350 tax on the additional $20,000 of taxable income, a rate of 31.75%, is an incremental rate. This is the rate we should use in evaluating the project in isolation from the rest of ABC's operations. As shown in Figure 11.6, the 31.75% is not an arbitrary figure, but a weighted average of two distinct marginal rates. Because the new project pushes ABC into a higher tax bracket, the first $5,000 it generates is taxed at 25%; the remaining $15,000 it generates is taxed in the higher bracket, at 34%. Thus, we could have calculated the incremental tax rate by

$$0.25(\$5{,}000)/\$20{,}000) + 0.34(\$15{,}000/\$20{,}000) = 31.75\%.$$

The average tax rates before and after the new project being considered would be

	Before	After	Incremental
Taxable income	$70,000	$90,000	$20,000
Income taxes	12,500	18,850	6,350
Average tax rate	17.86%	20.94%	
Incremental tax rate			31.75%

However, in conducting an economic analysis of an individual project, neither one of the company wide average rates is appropriate—we want the incremental rate applicable to just the new project for use when generating its cash flows.

A corporation with continuing base operations that place it consistently in the highest tax bracket will have both marginal and average federal tax rates of 35%. For such firms, the tax rate on an additional investment project is, naturally, 35%. But for corporations in lower tax brackets, and those that fluctuate between losses and profits, marginal and average tax rates are likely to vary. For such corporations, estimating

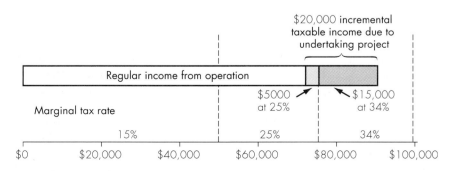

Figure 11.6   Illustration of incremental tax rate

a prospective incremental tax rate for a new investment project may be difficult. The only solution may be to perform scenario analysis, which is to examine how much the income tax fluctuates due to undertaking the project (in other words, calculate the total taxes and the incremental taxes for each scenario). A typical scenario example is presented in Example 11.6.

Example 11.6 Scenario Analysis for a Small Company

EverGreen Grass Company expects to have an annual taxable income of $320,000 from its regular grass-sodding business over the next 2 years. EverGreen has just won a contract to sod grasses for a new golf complex over the next 2 years. This 2-year project requires a purchase of new equipment costing $50,000. The equipment falls into the MACRS 5-year class with depreciation allowances of 20%, 32%, 19.2%, 11.52%, 11.52%, and 5.76% in each of the 6 years, respectively. After the 2-year term of the contract, equipment will be retained for future use (instead of being sold), indicating no salvage cash flow, gain, or loss on this property. The project will bring in an additional annual revenue of $150,000, but it is expected to incur additional annual operating costs of $90,000. Compute the incremental (marginal) tax rates applicable to the project's operating profits for years 1 and 2.

Solution

Given: Base taxable income = $320,000 per year, incremental income, expenses, and depreciation amounts as stated

Find: Incremental tax rate for this new project in years 1 and 2

First, we compute the additional taxable income from the golf course project over the next 2 years:

Year	1	2
Gross Revenue	$150,000	$150,000
Expenses	90,000	90,000
Depreciation	10,000	16,000
Taxable income	$ 50,000	$ 44,000

Next, we compute the income taxes. To do this, we need to determine the applicable marginal tax rate, but because of the progressive income tax rate, the project cannot be isolated from the other operations of EverGreen.

We can solve this problem much more efficiently using the incremental tax rate concept discussed in the beginning of this section. Because the golf course project pushes EverGreen into the 34% tax bracket from the 39%, we want to know what proportions of the incremental taxable income of $50,000 in year 1 are taxed at 39% and 34%, respectively. Without the project, the firm's taxable income is $320,000 and its marginal tax rate is 39%. With the additional taxable income of $50,000, EverGreen's tax bracket reverts to 34% as it's combined taxable income changes from $320,000 to $370,000. Since the rate changes at $335,000, the first $15,000 of the $50,000 taxable income will still be in the 39% bracket, and the remaining $35,000 will be in the 34% bracket. In year 2, we can divide the additional taxable income of $44,000 in a similar fashion. Then, we can calculate the incremental tax rates for the first 2 years as follows:

$$0.39(\$15,000/\$50,000) + 0.34(\$35,000/\$50,000) = 0.3550.$$

$$0.39(\$15,000/\$44,000) + 0.34(\$29,000/\$44,000) = 0.3570.$$

Note that these incremental tax rates vary slightly from year to year. Much larger changes could occur if a company's taxable income fluctuates drastically from its continuing base operation.

11.4.2   Consideration of State Income Taxes

For large corporations, the top federal marginal tax rate is 35%. In addition to federal income taxes, state income taxes are levied on corporations in most states. State income taxes are an allowable deduction in computing federal taxable income, and two ways are available to consider explicitly the effects of state income taxes in an economic analysis.

The first approach is to estimate explicitly the amount of state income taxes before calculating the federal taxable income. We then reduce the federal taxable income by the amount of the state taxes, and then apply the marginal tax rate to the resulting federal taxes. The total taxes would be the sum of the state taxes plus the federal taxes.

The second approach is to calculate a single tax rate that reflects both state and federal income taxes. This single rate is then applied to the federal taxable income without subtracting state income taxes. Taxes computed in this fashion represent total taxes. If state income taxes are considered, the combined state and federal marginal tax rate may be higher than 35%. Since state income taxes are deductible as expenses when determining federal taxes, the marginal rate for combined federal and state taxes can be calculated with the following expression:

$$t_m = t_f + t_s - (t_f)(t_s), \tag{11.1}$$

where t_m = combined marginal tax rate,

   t_f = federal marginal tax rate,

   t_s = state marginal tax rate.

This second approach provides a more convenient and efficient way to handle taxes in an economic analysis where the incremental tax rates are known. Therefore, incremental tax rates will be stated as combined marginal tax rates unless indicated otherwise. (For large corporations, these would be about 40%, but they vary from state to state.)

Example 11.7 Combined State and Federal Income Taxes

Consider a corporation whose revenues and expenses before income taxes are:

Gross revenue	$1,000,000
All expenses	400,000

If the marginal federal tax rate is 35% and the marginal state rate is 7%, compute the combined state and federal taxes using the two methods described above.

Solution

Given: Gross income = $1,000,000, deductible expenses = $400,000, $t_f = 35\%$, $t_s = 7\%$

Find: Combined income taxes, t_m

(a) Explicit calculation of state income taxes:

Let's define FT as federal taxes and ST as state taxes. Then, the state income taxes for this corporation are

$$\text{State taxable income} = \$1,000,000 - \$400,000$$
$$ST = (0.07)(\$600,000)$$
$$= \$42,000.$$
$$\text{Federal taxable income} = \$1,000,000 - \$400,000 - ST$$
$$FT = (0.35)(\$600,000 - \$42,000)$$
$$= (0.35)(\$558,000)$$
$$= \$195,300.$$
$$\text{Combined taxes} = FT + ST$$
$$= \$237,300.$$

(b) Tax calculation based on the combined tax rate:

Compute the combined tax rate directly from the formula:

$$\text{Combined tax rate}(t_m) = 0.35 + 0.07 - (0.35)(0.07)$$
$$= 39.55\%.$$
$$\text{Combined taxes} = \$600,000(0.3955)$$
$$= \$237,300.$$

As expected, these two methods always produce exactly the same results.

11.4.3   Consideration of Investment Tax Credits

For many years, Congress has used income tax laws to encourage corporations to invest in new productive assets. The two primary mechanisms that have been employed to encourage capital investment are depreciation allowances and investment tax credits.

In most years since 1962, tax laws have permitted a reduction in a year's income taxes equal to a percentage of the cost of any business machinery and equipment acquired during the year. This is called the **investment tax credit**. Businesses were able to deduct from 4% to 10% of their new business equipment purchases as a tax credit. The tax savings are equivalent to reducing the net cost of the equipment by the amount of the investment tax credit, yet at the same time the cost basis for computing depreciation remained the full cost of the equipment. The Tax Reform Act of 1986 repealed the investment credit for property placed in service after 1986. It is likely, however, that this tax credit will reappear at some future time.

11.4.4   Electing to Expense or Capitalize

As a benefit to very small companies, current tax law permits companies to "expense," which is equivalent to depreciating over 1 year up to $20,000 of equipment in year 2000. After year 2000, the maximum deduction limit increases according to the following schedule:

Year	Allowed Maximum Section 179 Deduction
2000	$20,000
2001–2002	24,000
After 2002	25,000

Thus, if a small company bought assets worth up to $24,000 in year 2001, it could write off the asset in the year it was placed in service. This is called Section 179 expensing. This allowance is phased out for taxpayers investing more than $200,000—if the investment cost is over $200,000, reduce the maximum dollar limit from each dollar over $200,000 (but not below zero). For example, in 2001, Hilton Plumbing Supplies placed in service machinery costing $207,000. Because this cost is $7,000 more than $200,000, the company must reduce its maximum dollar limit of $24,000 by $7,000. If Hilton's taxable income is $17,000 or more, Hilton can claim a $17,000 Section 179 deduction in year 2001.

Given that a firm has the choice between capitalizing an asset or writing it off as an expense, is there any reason why it would prefer to capitalize the asset and thus depreciate it over a period of years? In general, depreciation does result in a "smoothing" of the firm's income statement, while writing off large expenses in the year in which they occur may lead to large fluctuations in the company's reported earnings. However, the amount allowed in Section 179 expensing is relatively small, so the effect on the income statement for large companies would be insignificant.

11.5 Summary

- Explicit consideration of taxes is a necessary aspect of any complete economic study of an investment project.

- Since we are interested primarily in the measurable financial aspects of depreciation, we consider the effects of depreciation on two important measures of an organization's financial position, **net income** and **cash flow**. Once we understand that depreciation has a significant influence on the income and cash position of a firm, we will be able to appreciate fully the importance of utilizing depreciation as a means to maximize the value both of engineering projects and of the organization as a whole.

- For corporations, the U.S. tax system has the following characteristics:

  1. Tax rates are progressive: The more you earn, the more you pay.

  2. Tax rates increase in stair-step fashion: four brackets for corporations and two additional surtax brackets, giving a total of six brackets.

  3. Allowable exemptions and deductions may reduce the overall tax assessment.

- Three distinct terms to describe taxes were used in this chapter: **marginal tax rate**, which is the rate applied to the last dollar of income earned; **effective (average) tax rate**, which is the ratio of income tax paid to net income; and **incremental tax rate**, which is the average rate applied to the incremental income generated by a new investment project.

- **Capital gains** are currently taxed as ordinary income, and the maximum rate is capped at 35%. **Capital losses** are deducted from capital gains; net remaining losses may be carried backward and forward for consideration in years other than the current tax year.

- An **investment tax credit** is a direct reduction of income taxes payable, arising from the acquisition of depreciable assets. Government uses the investment tax credit to stimulate investments in specific assets or in specific industries.

Self-Test Questions

11s.1 Which of the following statements is correct?

(a) Over a project's life, a typical business will generate a greater amount of total project cash flows (undiscounted) if a faster depreciation method is adopted.

(b) No matter which depreciation method you adopt, total tax obligations over a project's life remain unchanged.

(c) Depreciation recapture equals cost basis minus an asset's book value at the time of disposal, that is, if the salvage value is less than the asset's cost basis.

(d) Cash flows normally include depreciation expenses since they represent a cost of doing business.

11s.2 You purchased a computer system which cost $50,000 5 years ago. At that time, the system was estimated to have a service life of 5 years with salvage value of $5,000. These estimates are still good. The property has been depreciated according to a 5-year MACRS property class. Now (at the end of year 5 from purchase) you are considering selling the computer at $10,000. What book value should you use in determining the taxable gains?

(a) $8,640   (b) $10,368

(c) $5,760   (d) $11,520.

11s.3 Omar Shipping Company bought a tugboat for $75,000 (year 0) and expected to use it for 5 years, after which it will be sold for $12,000. Suppose the company estimates the following revenues and expenses for the first operating year.

Operating revenue	$200,000
Operating expenses	$84,000
Depreciation	$4,000

If the company pays taxes at the rate of 30% on its taxable income, what is the net income during the first year?

(a) $28,700   (b) $81,200

(c) $78,400   (d) $25,900.

11s.4 In Problem 11s.3, assume for the moment that (1) all sales are for cash, and (2) all costs, except depreciation, were paid during year 1. How much cash would have been generated from operations?

(a) $82,400   (b) $32,700

(c) $85,200   (d) $3,400.

11s.5 Gilbert Corporation had a gross income of $500,000 in tax year 1, $150,000 in salaries, $30,000 in wages, $20,000 in interest, and $60,000 in depreciation expenses for an asset purchased 3 years ago. Ajax Corporation has a gross income of $500,000 in tax year 1, and $150,000 in salaries, $90,000 in wages, and $20,000 in interest expenses. Apply the tax rates in Table 11.1 and determine which of the following statements is correct.

(a) Both corporations will pay the same amount of income taxes in year 1.

(b) Both corporations will have the same amount of net cash flows in year 1.

(c) Ajax Corporation will have a larger net cash flow than Gilbert in year 1.

(d) Gilbert Corporation will have a larger taxable income than Ajax Corporation in year 1.

11s.6 A company purchased an industrial fork-lift for $75,000 in year 0. The company expects to use it for the next 7 years after which it plans to sell it for $10,000. The estimated gross income and expenses excluding depreciation for the first year are given below. The fork-lift will be depreciated according to a 5-year MACRS.

	Year 1
Gross revenue	$120,000
Expenses	$40,000
(Depreciation not included)	

Determine the average tax rate applicable in the first year of operation, using the corporate tax rate schedule in Table 11.1.

(a) 15%   (c) 18.75%

(b) 17.31%   (d) 25%.

11s.7 Minolta Machine Shop purchased a computer-controlled vertical drill press for $100,000. The drill press is classified as a 3-year MACRS property. Minolta is planning to use the press for 5 years. Then Minolta will sell the press at the end of service life at $20,000. The annual revenues are estimated to be $110,000. If the estimated net cash flow at the end of year 5 is $30,000, what are the estimated operating and maintenance expenses in year 5? Minolta's income tax rate is 40%.

(a) $60,000   (c) $80,000

(b) $65,000   (d) $88,333.

Problems

Note: *Unless otherwise specified, use current tax rates for corporate taxes. Check the Web site (described in Preface) for the most current tax rates for corporations.*

Corporate Tax Systems

11.1 Tiger Construction Company had a gross income of $20,000,000 in tax year 1, $3,000,000 in salaries, $4,000,000 in wages, $800,000 in depreciation expenses, a loan principal payment of $200,000, and a loan interest payment of $210,000. Determine the net income of the company in tax year 1.

11.2 A consumer electronics company was formed to sell a portable handset system that allows people with cellular car phones to receive calls up to 1,000 feet from their vehicles. The company purchased a warehouse and converted it into a manufacturing plant for $2,000,000 (including the warehouse). It completed installation of assembly equipment worth $1,500,000 on December 31. The plant began operation on January 1. The company had a gross income of $2,500,000 for the calendar year. Manufacturing costs and all operating expenses, excluding the capital expenditures, were $1,280,000. The depreciation expenses for capital expenditures amounted to $128,000.

(a) Compute the taxable income of this company.

(b) How much will the company pay in federal income taxes for the year?

11.3 Quick Printing Company had a sales revenue of $1,250,000 from operations during tax year 1. Here are some operating data on the company:

(a) What is Quick's taxable income?

(b) What is Quick's taxable gains?

Labor expenses	$550,000
Materials costs	185,000
Depreciation expenses	32,500
Interest income on time deposit	6,250
Bond interest income on Apple Computer	4,500
Stock dividend income from Sears	3,900
Interest expense	12,200
Rental expenses	45,000
Dividend payment to Quick's shareholders	40,000
Proceeds from sale of old equipment that had a book value of $20,000	23,000

(c) What is Quick's marginal and effective (average) tax rate?

(d) What is Quick's net cash flow after tax?

(Note: Interest income received by a corporation is taxed as ordinary income at regular corporate tax rates. However, 70% of dividends received by one corporation from another is excluded from taxable income, while the remaining 30% is taxed at the ordinary tax rate.)

11.4 Elway Aerospace Company had gross revenues of $1,200,000 from operations. The following financial transactions were posted during the year:

Manufacturing expenses (including depreciation)	$450,000
Operating expenses (excluding interest expenses)	120,000
A new short-term loan from a bank	50,000
Interest expenses on borrowed funds (old and new)	40,000
Dividends paid to common stockholders	80,000
Old equipment sold	60,000

The old equipment had a book value of $75,000 at the time of sale.

(a) What is Elway's income tax liability?

(b) What is Elway's operating income?

Gains or Losses

11.5 Consider a 5-year MACRS asset, which was purchased at $60,000.

(Note that a 5-year MACRS property class is depreciated over 6 years due to the half-year convention. The applicable salvage values would be $20,000 in year 3, $10,000 in year 5, and $5,000 in year 6, respectively.) Compute the gain or loss amounts when the asset is disposed of

(a) in year 3,   (b) in year 5, and

(c) in year 6, respectively.

11.6 An electrical appliance company purchased an industrial robot costing $300,000 in year 0. The industrial robot to be used for welding operations, classified as a 7-year recovery property, has been depreciated by the MACRS method. If the robot is to be sold after 5 years, compute the amounts of gains (losses) for the following three salvage values (assume that both capital gains and ordinary incomes are taxed at 34%).

(a) $10,000   (b) $125,460

(c) $200,000.

11.7 LaserMaster, Inc., a laser-printing service company, had a sales revenue of $1,250,000 during tax year 2001. The following represents the other financial information relating to the tax year.

(The printers had a combined book value of $20,000 at the time of sale.) Consult the corporate tax rates for 2001, and

Labor expenses	$550,000
Material costs	$185,000
Depreciation	$32,500
Interest income	$6,250
Interest expenses	$12,200
Rental expenses	$45,000
Proceeds from the sale of old printers	$23,000

(a) Determine the taxable income for the tax year.

(b) Determine the taxable gains for the tax year.

(c) Determine the amount of income taxes and gain taxes (or loss credits) for the tax year.

11.8 Valdez Corporation will commence operations on January 1, 2001. The company projects the following financial performance during its first year of operation.

- Sales revenues are estimated at $1,500,000.

- Labor, material, and overhead costs are projected at $600,000.

- The company will purchase a warehouse worth $500,000 in February. To finance this warehouse, the company will issue $500,000 of long-term bonds on January 1, which carry an interest rate of 10%. The first interest payment would occur on December 31.

- For depreciation purposes, the purchase cost of the warehouse is divided into $100,000 in land and $400,000 in building. The building will be classified as 39-year MACRS real property class and will be depreciated accordingly.

- On January 5, the company purchases $200,000 of equipment, which has a 5-year MACRS class life.

(a) Determine the total depreciation expenses allowed in 2001.

(b) Determine Valdez's tax liability in 2001.

11.9 AmSouth Inc., bought a machine for $50,000 on January 2, 2000. Management expects to use the machine for 10 years, at the end which time it will have a $1,000 salvage value. Consider the following questions independently.

(a) If AmSouth uses straight-line depreciation, what is the book value of the machine on December 31, 2002?

(b) If AmSouth uses double declining balance depreciation, what is the depreciation expense for 2002?

(c) If AmSouth uses double declining balance depreciation switching to straight-line depreciation, when is the optimal time to switch?

(d) If AmSouth uses sum-of-years'-digits depreciation, what is the total depreciation on December 31, 2005?

(e) If Amsouth uses 7-year MACRS and sells the machine on April 1, 2003, at a price of $30,000, what is the taxable gains?

11.10 A machine now in use that was purchased 3 years ago at a cost of $4,000 has a book value of $1,800. It can be sold for $2,500, but could be used for 3 more years, at the end of which time it would have no salvage value. The annual O&M costs amount to $10,000 for the old machine. A new machine can be purchased at an invoice price of $14,000 to replace the present equipment. Freight-in will amount to $800, and the installation cost will be $200. The new machine has an expected service life of 5 years and will have no salvage value at the end of that time. With the new machine, the expected direct cash savings amount to $8,000 the first year and $7,000 in

O&M for each of the next 2 years. Corporate income taxes are at an annual rate of 40%, and the net capital gain is taxed at the ordinary income tax rate. The present machine has been depreciated according to a straight-line method, and the proposed machine would be depreciated on a 7-year MACRS. (Note: Each question should be considered independently.)

(a) If the old asset is to be sold now, what would be the amount of its equivalent economic depreciation?

(b) For depreciation purposes, what would be the first cost of the new machine (depreciation base)?

(c) If the old machine is to be sold now, what would be the amount of taxable gains and gains taxes?

(d) If the old machine can be sold for $5,000 now instead of $2,500, then what would be the gains tax?

(e) If the old machine had been depreciated using 175% DB and then switching to SL depreciation, what would be the current book value?

(f) If the machine were not replaced by the new one, when would be the time to switch from DB to SL?

Marginal Tax Rate in Project Evaluation

11.11 Buffalo Ecology Corporation expects to generate a taxable income of $250,000 from its regular business in 2001. The company is considering a new venture: cleaning up oil spills made by fishing boats in lakes. This new venture is expected to generate an additional taxable income of $150,000.

(a) Determine the firm's marginal tax rates before and after the venture.

(b) Determine the firm's average tax rates before and after the venture.

11.12 Boston Machine Shop expects to have annual taxable incomes of $270,000 from its regular business over the next 6 years. The company is considering the proposed acquisition of a new milling machine during year 0. The machine's installed price is $200,000. The machine falls into the MACRS 5-year class, and it will have an estimated salvage value of $30,000 at the end of 6 years. The machine is expected to generate an additional before-tax revenue of $80,000 per year.

(a) What is the total amount of economic depreciation for the milling machine, if the asset is sold at $30,000 at the end of 6 years?

(b) Determine the company's marginal tax rates over the next 6 years with the machine.

(c) Determine the company's average tax rates over the next 6 years with the machine.

11.13 Major Electrical Company expects to have an annual taxable income of $450,000 from its residential accounts over the next 2 years. The company is bidding on a 2-year wiring service for a large apartment complex. This commercial service requires the purchase of a new truck equipped with wire-pulling tools at a cost of $50,000. The equipment falls into the MACRS 5-year class and will be retained for future use (instead of selling) after 2 years, indicating no gain or loss on this property. The project will bring in an additional annual revenue of $200,000, but it is expected to incur additional annual operating costs of $100,000. Compute the marginal tax rates applicable to the project's operating profits for the next 2 years.

11.14 Florida Citrus Corporation estimates its taxable income for next year at $2,000,000. The company is consider-

ing expanding its product line by introducing pineapple-orange juice for the next year. The market responses could be (1) good, (2) fair, or (3) poor. Depending on the market response, the expected additional taxable incomes are (1) $2,000,000 for a good response, (2) $500,000 for fair response, and (3) $100,000 (loss) for a poor response.

(a) Determine the marginal tax rate applicable to each situation.

(b) Determine the average tax rate that results from each situation.

11.15 A small manufacturing company has an estimated annual taxable income of $95,000. Owing to an increase in business, the company is considering purchasing a new machine that will generate an additional (before-tax) annual revenue of $50,000 over the next 5 years. The new machine requires an investment of $100,000, which will be depreciated under the 5-year MACRS method.

(a) What is the increment in income tax due to the purchase of the new machine in tax year 1?

(b) What is the incremental tax rate due to the purchase of the new equipment in year 1?

11.16 Simon Machine Tools Company is considering the purchase of a new set of machine tools to process special orders. The following financial information is available.

• Without the project: The company expects to have taxable incomes of $300,000 each year from its regular business over the next 3 years.

• With the project: This 3-year project requires the purchase of a new set of machine tools at a cost of $50,000. The equipment falls into the MACRS 3-year class. The tools will be sold at the end of project life for $10,000. The project will be bringing in an additional annual revenue of $80,000, but it is expected to incur additional annual operating costs of $20,000.

(a) What are the additional taxable incomes (due to undertaking the project) during years 1 through 3, respectively?

(b) What are the additional income taxes (due to undertaking the new orders) during years 1 through 3, respectively?

(c) Compute the gain taxes when the asset is disposed of at the end of year 3.

Combined Marginal Income Tax Rate

11.17 Consider a corporation whose taxable income without state income tax is as follows:

Gross revenue	$2,000,000
All expenses	1,200,000

If the marginal federal tax rate is 34% and the marginal state rate is 6%, compute the combined state and federal taxes using the two methods described in the text.

11.18 A corporation has the following financial information for a typical operating year:

Gross revenue	$4,500,000
Cost of goods sold	2,450,000
Operating costs	630,000
Federal taxes	352,000
State taxes	193,120

(a) Based on this financial information, determine both federal and state marginal tax rates.

(b) Determine the combined marginal tax rate for this corporation.

11.19 Denver Sewer Service Company expects to have taxable incomes of $300,000 from its regular residential sewer line installations over the next 2 years. The company is considering a new commercial account for a proposed shopping-mall complex during year 0. This 2-year project requires the purchase of new digging equipment at a cost of $55,000. The equipment falls into the MACRS 5-year class. The equipment will be sold for $35,000 at the end of year 2. The project will bring in an additional annual revenue of $100,000, but it is expected to incur additional annual operating costs of $50,000.

(a) What incremental tax rate would you use in evaluating the acquisition of the equipment during years 1 and 2, respectively?

(b) With the project, what is the effective (average) tax rate of the firm during project year 1?

(c) What is the incremental project cash flow over the 2-year period?

11.20 Van-Line Company, a small electronics repair firm, expects an annual income of $70,000 from its regular business. The company is considering expanding its repair business to include personal computers. This expansion would bring in an additional annual income of $30,000, but will require an additional expense of $10,000 each year over the next 3 years. Using applicable current tax rates, answer the following:

(a) What is the marginal tax rate in tax year 1?

(b) What is the average tax rate in tax year 1?

(c) Suppose that the new business expansion requires a capital investment of $20,000 (a 3-year MACRS property). What is the PW of the total income taxes to be paid over the project life, at $i = 10\%$?

11.21 A company purchased a new forging machine to manufacture disks for airplane turbine engines. The new press cost $3,500,000, and it falls into a 7-year MACRS property class. The company has to pay property taxes for ownership of this forging machine at a rate of 1.2% on the beginning book value of each year to the local township.

(a) Determine the book value of the asset at the beginning of each tax year.

(b) Determine the amount of property taxes over the machine's depreciable life.

Short Case Studies

11.22 Chuck Robbins owns and operates a small electrical service business, Robbins Electrical Service (RES). Chuck is married with two children, so he claims four exemptions on his tax return. As business grows steadily, tax considerations are important to him. Therefore, Chuck is considering whether to incorporate the business. Under either form, the family will initially own 100% of the firm. Chuck plans to finance the firm's expected growth by drawing a salary just sufficient for his family living expenses and by retaining all other income in the business. He estimates the expected income and expenses over the next 3 years to be as follows (see table top of page).

Which form of business (corporation or sole ownership) will allow

	Year 1	Year 2	Year 3
Gross income	$80,000	$95,000	$110,000
Expenses			
Salary	40,000	45,000	50,000
Business expenses	15,000	20,000	30,000
Personal exemptions	10,000	10,000	10,000
Itemized deductions	6,000	8,000	10,000

Chuck to pay the lowest taxes (and retain the most income) during the period 1 to 3? Personal income tax brackets and amount of personal exemption are updated yearly, so you need to consult the IRS tax manual for the tax rates as well as for the exemptions that are applicable to the tax years.

11.23 Electronic Measurement and Control Company (EMCC) has developed a laser speed detector that emits infrared light invisible to humans and radar detectors alike. For full-scale commercial marketing, EMCC needs to invest $5 million in new manufacturing facilities. The system is priced at $3,000 per unit. The company expects to sell 5,000 units annually over the next 5 years. The new manufacturing facilities will be depreciated according to a 7-year MACRS property class. The expected salvage value of the manufacturing facilities at the end of 5 years is $1.6 million. The manufacturing cost for the detector is $1,200 per unit, excluding depreciation expenses. The operating and maintenance costs are expected to run to $1.2 million per year. EMCC has a combined federal and state income tax rate of 35%, and undertaking this project will not change this current marginal tax rate.

(a) Determine the incremental taxable income, income taxes, and net income due to undertaking this new product for the next 5 years.

(b) Determine the gains or losses associated with the disposal of the manufacturing facilities at the end of 5 years.

11.24 Diamonid is a start-up diamond coating company planning to manufacture a microwave plasma reactor that synthesizes diamonds. Diamonid anticipates that the industry demand for diamonds will skyrocket over the next decade for use in industrial drills, high-performance microchips, and artificial human joints, among other things. Diamonid has decided to raise $50 million through issuing common stocks for investment in plant ($10 million) and equipment ($40 million). Each reactor can be sold at a price of $100,000 per unit. Diamonid can expect to sell 300 units per year during the next 8 years. The unit manufacturing cost is estimated at $30,000, excluding depreciation. The operating and maintenance cost for the plant is estimated at $12 million per year. Diamonid expects to phase out the operation at the end of 8 years and revamp the plant and equipment to adopt a new diamond manufacturing technology. At that time, Diamonid estimates that the salvage values for the plant and equipment will be about 60% and 10% of the original investments, respectively. The plant and equipment will be depreciated according to 39-year real property (placed in service in January) and 7-year MACRS, respectively. Diamonid pays 5% of state and local income taxes on their taxable income.

(a) If the 2000 corporate tax system continues over the project life, determine the combined state and federal income tax rate each year.

(b) Determine the gains or losses at the time of plant revamping.

(c) Determine the net income each year over the plant life.

11.25 Julie Magnolia has $50,000 cash to invest for 3 years. Two types of bond are available for consideration. She can buy a tax-exempt Arizona State bond that pays interest of 9.5% per year. Julie's marginal tax rate is 25% for both ordinary income and capital gains. In the following questions, assume that any investment decision considered will not change her marginal tax bracket.

(a) If Julie were looking for a corporate bond that was just as safe as the state bond, what interest rate on the corporate bond is required so that Julie would be indifferent between the two bonds? There are no capital gains or losses at the time of trading the bond.

(b) In (a), suppose at the time of trading (year 3) that the corporate bond is expected to be sold at a price 5% higher than the face value. What interest rate on the corporate bond is required so that Julie would be indifferent between the two bonds?

(c) Julie can invest the amount in a tract of land that could be sold at $75,000 (after paying the real estate commission) at the end of year 3. Is this investment better than the state bond?

CHAPTER 12

Developing Project Cash Flows

Installing Cooling Fans at McCook Plant: Alcoa Aluminum's McCook plant (formerly known as Reynolds Metals before merging with Alcoa) produces aluminum coils, sheets, and plates. Its annual production runs at 400 million pounds. In an effort to improve McCook's current production system, an engineering team, led by the divisional vice president, went on a fact-finding tour of Japanese aluminum and steel companies to observe their production systems and methods. The large fans, which the Japanese companies used to reduce the time that coils need to cool down after various processing operations, were cited among the observations. Cooling the hot process coils with the fans was estimated to significantly reduce the queue or work-in-process (WIP) inventory buildup allowed for cooling. The process also reduced production lead time and improved delivery performance. The possibility of reducing production time, and, as a consequence, the WIP inventory, excited the team members. Particularly excited was the vice president. After the trip, Neal Donaldson, the plant engineer, was asked to investigate the economic feasibility of installing cooling fans at the McCook plant. Neal's job is to justify the purchase of cooling fans for his plant. He was given 1 week to prove the idea was a good one. Essentially, all he knew were these brief background details: The number of fans, their locations, and the project cost. Everything else was left to Neal's devices. He alone was to determine any effects the purchase might have; what products would be involved; what would be the number of days by which the cooling queue would be reduced; and how much money would be saved. Can Neal logically explain how it all works?

rojecting cash flows is the most important—and the most difficult—step in the analysis of a capital project. Typically, a capital project initially requires investment outlays and only later produces annual net cash inflows. A great many variables are involved in forecasting cash flows, and many individuals, ranging from engineers to cost accountants and marketing people, participate in the process. This chapter provides the general principles on which determining a project's cash flows are based.

To help us imagine the range of activities that are typically initiated by project proposals, we begin this chapter with an overview of how firms classify projects. A whole variety of types of projects exists, each having its own characteristic set of economic concerns. We will next provide an overview of the typical cash flow elements of engineering projects in Section 12.2. Once we have defined these elements, we will examine how to develop cash flow statements that are used to analyze the economic value of projects. In Section 12.3, we will use several examples to demonstrate the development of after-tax project cash flow statements. Then, in Section 12.4, we will present some alternative techniques for developing a cash flow statement based on generalized cash flow approach. By the time you have finished this chapter, you should be prepared not only to understand the format and significance of after-tax cash flow statements, but also how to develop them yourself.

12.1   Estimating Cost/Benefit for Engineering Projects

Before the economics of an engineering project can be evaluated, it is necessary to reasonably estimate the various cost and revenue components that describe the project. Engineering projects may range from something as simple as the purchase of a new milling machine to the design and construction of a multibillion dollar process or resource recovery complex.

The engineering projects appearing in this book as examples and problems already include the necessary cost and revenue estimates. Developing adequate estimates for these quantities is extremely important and can be a time-consuming activity. Cost estimating techniques are not the focus of this book. However, it is worthwhile to mention some of the possible approaches in the context of simple projects which are straightforward and involve little or no engineering design, and complex projects which tend to be large and may involve many thousands of hours of engineering design. Obviously, there are projects that fall between these extremes, and some combination of approaches may be appropriate in these cases.

12.1.1   Simple Projects

Projects in this category usually involve a single "off-the-shelf" component or a series of such components that are integrated in a simple manner. The acquisition of a new milling machine is an example.

The installed cost is the price of the equipment as determined from catalogues or supplier quotations, shipping and handling charges, and the cost of building modifications and changes in utility requirements. The latter may require some design effort to define the scope of the work which is the basis for contractor quotations.

Project benefits are in the form of new revenue and or cost reduction. Estimating new revenue requires agreement on the total units produced and the selling price per unit. These quantities are related through supply and demand considerations in the market place. In highly competitive product markets, sophisticated marketing studies are often required to establish price–volume relationships. These studies are undertaken as one of the first steps in an effort to define the appropriate scale for the project.

The ongoing costs to operate and maintain the equipment can be estimated at various levels of detail and accuracy. Familiarity with the cost–volume relationships for similar facilities would allow the engineer to establish a "ball park" cost. Some of this information may be used in conjunction with more detailed estimates in other areas. For example, maintenance costs may be estimated as a percentage of the installed cost value. Such percentages are derived from historical data and are frequently available from equipment suppliers. Other costs such as manpower, energy, etc., may be estimated in detail to reflect specific local considerations. The most comprehensive and time-consuming type of estimate involves a detailed estimate around each type of cost associated with the project.

12.1.2   Complex Projects

The estimates developed for complex projects involve the same general considerations as discussed for simple projects. However, such projects usually include specialized equipment which is not "off the shelf" and must be fabricated from detailed engineering drawings. For certain projects such drawings are not even available until after a commitment has been made to proceed with the project. The typical phases of a project are:

- Development
- Conceptual Design
- Preliminary Design
- Detailed Design

Depending upon the specific project, some phases may be combined. Project economics are performed during each of these phases to confirm the project attractiveness and the incentive to continue. The types of estimates and estimating techniques are a function of the stage of the project.

Benefits are usually well known at the outset in terms of pricing–volume relationships and/or cost reduction potential when the project involves the total or partial replacement of an existing facility. In the case of natural resource projects, oil and gas and mineral price forecasts are subject to considerable uncertainty.

At the development phase, work is undertaken to identify potential technologies and confirm the technical viability of these. Installed cost estimates and operating estimates are based on similar existing facilities or parts of these.

Conceptual design examines issues of project scale and technology alternatives. Again estimates tend to be based on large pieces of or processes within the overall project which correspond to similar facilities already in operation elsewhere.

Preliminary design takes the most attractive alternative from the conceptual design phase to a level of detail which provides specific sizing and layout for the actual equipment and associated infrastructure. Estimates at this stage tend to be based on individual pieces of equipment. The estimating basis is similar pieces of equipment in use elsewhere.

For very large projects, the cost of undertaking the detailed engineering is prohibitive unless the project is going forward. At this stage, detailed fabrication and construction drawings which would provide the basis for an actual vendor quotation become available.

The accuracy of the estimates available improves with each phase as the project becomes defined in greater detail. Normally, the installed cost estimate from each phase includes a contingency which is some fraction of the actual estimate calculated. The contingency at the development phase can be 50 to 100% and decreases to something in the order of 10% after preliminary design.

More information on cost estimating techniques can be found in reference books on project management and cost engineering. Industry specific data books are also available for some sectors where costs are summarized on some normalized basis such as dollars per square foot of building space or dollars per tonne of material moved in operating mines. In these cases, the data are categorized by type of building and type of mine. Engineering design companies maintain extensive databases of such cost information.

12.2 Incremental Cash Flows

When a company purchases a fixed asset such as equipment, it makes an investment. The company commits funds today in the expectation of earning a return on those funds in the future. Such investments are similar to those made by a bank when it lends money. In the case of a bank loan, the future cash flow consists of interest plus repayment of the principal. For a fixed asset, the future return is in the form of cash flows generated by the profitable use of the asset. In evaluating a capital investment, we are concerned only with those cash flows that result directly from the investment. These cash flows, called **differential** or **incremental cash flows,** represent the change in the firm's total cash flow that occurs as a direct result of the investment. In this section, we will look into some of the cash flow elements common to most investments.

12.2.1   Elements of Cash Outflows

We first consider the potential uses of cash in undertaking an investment project.

Purchase of New Equipment

A typical project usually involves a cash outflow in the form of an initial investment in equipment. The relevant investment costs are incremental costs, such as the cost of the asset, shipping and installation costs, and the cost of training employees to use the new asset.

   If the purchase of a new asset results in the sale of an existing asset, the net proceeds from this sale reduce the amount of the incremental investment. In other words, the incremental investment represents the total amount of additional funds that must be committed to the investment project. When existing equipment is sold, the transaction results in either an accounting gain or loss. The gain or loss is dependent on whether the amount realized from the sale is greater or less than the equipment's book value. In any event, when existing assets are disposed of, the relevant amount by which the new investment is reduced consists of the proceeds of the sale, adjusted for tax effects.

Investments in Working Capital

Some projects require investment in nondepreciable assets. If a project increases a firm's revenues, for example, more funds will be needed to support the higher level of operations. Investment in nondepreciable assets is often called **investment in working capital.** In accounting, working capital means the amount carried in cash, accounts receivable, and inventory that is available to meet day-to-day operating needs. For example, additional working capital may be needed to meet the greater volume of business that will be generated by a project. Part of this increase in current assets may be supplied from increased accounts payable, but the remainder must come from permanent capital. This additional working capital is as much a part of the initial investment as the equipment itself. (We explain the amount of working capital required for a typical investment project in Section 12.3.2.)

Manufacturing, Operating, and Maintenance Costs

The costs associated with manufacturing a new product need to be determined. Typical manufacturing costs include labor, materials, and overhead costs. Overhead costs cover items such as power, water, and indirect labor. Investments in fixed assets normally require periodic outlays for repairs and maintenance and for additional operating costs, all of which must be considered in investment analysis.

Leasing Expenses

When a piece of equipment or a building is leased (instead of purchased) for business use, leasing expenses become cash outflows. Many firms lease computers, automobiles, and industrial equipment subject to technological obsolescence.

Interest and Repayment of Borrowed Funds

When we borrow money to finance a project, we need to make interest payments as well as principal payments. Proceeds from both short-term borrowing (bank loans) and long-term borrowing (bonds) are treated as cash inflows, but repayments of debts (both principal and interest) are classified as cash outflows.

Income Taxes and Tax Credits

Any income tax payments following profitable operations should be treated as cash outflows for capital budgeting purposes. As we learned in Chapter 11, when an investment is made in depreciable assets, depreciation is an expense that offsets part of what would otherwise be additional taxable income. This is called a **tax shield,** or **tax savings,** and we must take it into account when calculating income taxes. If any investment **tax credit** is allowed, this tax credit will directly reduce income taxes, resulting in cash inflows.

12.2.2 Elements of Cash Inflows

The potential sources of cash inflow over the project life may include such items as (1) borrowed funds to finance the investment, (2) operating revenues from sale of goods and services, (3) cost savings due to better efficiency, (4) sale of any asset, and (5) release of working capital.

Borrowed Funds

If you finance your investment by borrowing, the borrowed funds will appear as cash inflow to the project at the time of borrowing. From these borrowed funds, the purchase of new equipment or any other investment will be paid out.

Operating Revenues

If the primary purpose of undertaking the project is to increase the production or service capacity to meet the increased demand, the new project will be bringing in additional revenues.

Cost Savings (or Cost Reduction)

If the primary purpose of undertaking a new project is to reduce operating costs, the amount involved should be treated as a cash inflow for capital budgeting purposes. A reduction in costs is equivalent to an increase in revenues, even though the actual sales revenues may remain unchanged.

Salvage Value

In many cases, the estimated salvage value of a proposed asset is so small, and occurs so far in the future that it may have no significant effect on the decision. Furthermore, any salvage value that is realized may be offset by removal and dismantling costs. In situations where the estimated salvage value is significant, the net salvage value is viewed as a cash inflow at the time of disposal. The **net salvage value** of the existing asset is its selling price minus any costs incurred in selling, dismantling, and removing it, and this value is subject to taxable gain or loss.

Working Capital Release

As a project approaches termination, inventories are sold off and receivables are collected, that is, at the end of the project, these items can be liquidated at their cost. As this occurs, the company experiences an end-of-project cash flow that is equal to the net working capital investment that was made when the project began. This recovery of working capital is not taxable income, since it merely represents a return of investment funds to the company.

In summary, the following types of cash flows, depicted in Figure 12.1, are common in engineering investment projects.

12.2.3 Classification of Cash Flow Elements

Once the cash flow elements are determined (both inflows or outflows), we may group them into three areas: (1) cash flow elements associated with operations, (2) cash flow elements associated with investment activities (capital expenditures), and (3) cash flow elements associated with project financing (such as borrowing). The main purpose of grouping cash flows this way is to provide information about the operating, investing, and financing activities of a project.

Operating Activities

In general, cash flows from operations include current sales revenues, cost of goods sold, operating expenses, and income taxes. Cash flows from operations should generally reflect the cash effects of transactions entering into the determination of net income. The interest portion of a loan repayment is a deductible expense allowed when determining net income, and it is included in the operating activities. Since we usually look only at yearly flows, it is logical to express all cash flows on a yearly basis.

As we discussed in Chapter 11, we can determine the net cash flow from operations either (1) based on the net income or (2) based on cash flow by computing income taxes in a separate step. When we use net income as the starting point for cash flow determination, we should add any noncash expenses (mainly, depreciation and amortization expenses) to net income to estimate the net cash flow from the operation. Recall that depreciation (or amortization) is not a cash flow but is deducted,

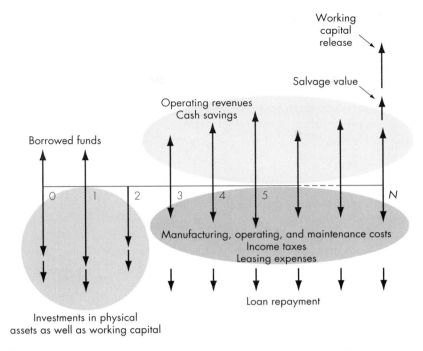

Figure 12.1   Types of cash flow elements used in project analysis

along with operating expenses and lease costs, from gross income to find taxable income and therefore taxes. Accountants calculate net income by subtracting taxes from taxable income. But depreciation—which is not a cash flow—was subtracted to find taxable income, so it must be added back on to taxable income if we wish to use the net income figure as an intermediate step along the path to after-tax cash flow. Mathematically, it is easy to show that the two approaches are identical. Thus,

Cash flow from operation = Net income +

(Depreciation or amortization)

Approach 1 Income Statement Approach	Approach 2 Direct Cash Flow Approach
Operating revenues	Operating revenues
Cost of goods sold	Cost of goods sold
Depreciation	
Operating expenses	Operating expenses
Interest expenses	Interest expenses
Taxable income	
Income taxes	Income taxes
Net income	Cash flow from operation
+ Depreciation	

Note that to use Approach 2, you still need to find out the taxable income before computing the income taxes, indicating that you must revisit most steps (boxed in shade) in Approach 1. Therefore, Approach 1 could be more intuitive to many practitioners.

In business practice, accountants usually prepare cash flow statements based on net income, namely using Approach 1, whereas Approach 2 is commonly used in many traditional engineering economic texts. If you learn only Approach 2, it is more than likely that you need to be retrained to learn Approach 1 to communicate with the financing and accounting professionals within your organization. Therefore, we will use the income statement approach (Approach 1) whenever possible throughout the text.

Investing Activities

In general, three types of investment flows are associated with buying a piece of equipment: The original investment, salvage value at the end of its useful life, and the working capital investment or recovery. We will assume that our outflow for both capital investment and working capital investment take place in year 0. It is possible, however, that both investments will not occur instantaneously but, rather, over a few months as the project gets into gear; we could then use year 1 as an investment year. (Capital expenditures may occur over several years before a large investment project becomes fully operational. In this case, we should enter all expenditures as they occur.) For a small project, either method of timing these flows is satisfactory, because the numerical differences are likely to be insignificant.

Financing Activities

Cash flows classified as financing activities include (1) the amount of borrowing and (2) the repayment of principal. Recall that interest payments are tax deductible expenses so that they are classified as operating, not financing, activities.

Net cash flow for a given year is simply the sum of the net cash flows from operating, investing, and financing activities. Table 12.1 can be used as a checklist when you set up a cash flow statement, because it groups each type of cash flow element into operating, investing, or financing activities.

12.3 Developing Cash Flow Statements

In this section, we will illustrate through a series of numerical examples how we actually prepare a project's cash flow statement; a generic version is shown in Figure 12.2 where we first determine the net income from operations and then adjust the net income by adding any noncash expenses, mainly depreciation (or amortization). We will also consider a case in which a project generates a negative taxable income for an operating year.

12.3.1 When Projects Require Only Operating and Investing Activities

We will start with the simple case of generating after-tax cash flows for an investment project with only operating and investment activities. In the sections ahead, we will add complexities to this problem by including working capital investments (Section 12.3.2) and borrowing activities (Section 12.3.3).

Cash Flow Element	Other Terms Used in Business
Operating activities:	
Gross income	Gross revenue, Sales revenue, Gross Profit, Operating revenue
Cost savings	Cost reduction
Manufacturing expenses	Cost of goods sold, Cost of revenue
O&M cost	Operating expenses
Operating income	Operating profit, Gross margin
Interest expenses	Interest payments, Debt cost
Income taxes	Income taxes owed
Investing activities:	
Capital investment	Purchase of new equipment, Capital expenditure
Salvage value	Net selling price, Disposal value, Resale value
Investment in working Capital	Working capital requirement
Working capital release	Working capital recovery
Gains taxes	Capital gains taxes, Ordinary gains taxes
Financing activities:	
Borrowed funds	Borrowed amounts, Loan amount
Principal repayments	Loan repayment

Table 12.1 Classifying Cash Flow Elements and Their Equivalent Terms Practiced in Business

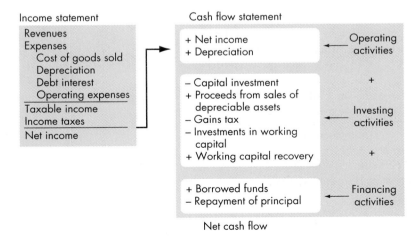

Figure 12.2 A popular format used for presenting a cash flow statement

Example **12.1** **Cash Flow Statement—Operating and Investing Activities for an Expansion Project**

A computerized machining center has been proposed for a small tool manufacturing company. If the new system, which costs $125,000, is installed, it will generate annual revenues of $100,000 and will require $20,000 in annual labor, $12,000 in annual material expenses, and another $8,000 in annual overhead (power and utility) expenses. The automation facility would be classified as a 7-year MACRS property. The company expects to phase out the facility at the end of 5 years, at which time it will be sold for $50,000. Find the year-by-year after-tax net cash flow for the project at a 40% marginal tax rate based on the net income (Approach 1), and determine the after-tax net present worth of the project at the company's MARR of 15%.

Discussion: We can approach the problem in two steps by using the format shown in Fig. 12.2 to generate an income statement and then a cash flow statement. We will follow this form in our listing of givens and unknowns below. In year 0 (that is, at present) we have an investment cost of $125,000 for the equipment.[1] This cost will be depreciated in years 1 to 5. The revenues and costs are uniform annual flows in years 1 to 5. We can see that once we find depreciation allowances for each year, we can easily compute the results for years 1 to 4, which have fixed revenue and expense entries along with the variable depreciation charges. In year 5, we will need to incorporate the salvage value and any gains tax from the asset's disposal.

We will use the business convention that no signs (positive or negative) be used in preparing the income statement, except in the situation where we have a negative taxable income or tax savings. In this situation we will use ( ) to denote a negative entry. However, in preparing the cash flow statement, we will observe explicitly the sign convention: A positive sign indicates a cash inflow; a negative sign or ( ) indicates a cash outflow.

Solution

Given: Cash flow information stated above

Find: After-tax cash flow

Before presenting the cash flow table, we need to do some preliminary calculations. The following notes explain the essential items in Table 12.2.

- Depreciation calculation
  1. If it is held for all 8 years, we can depreciate a 7-year property in respective percentages of 14.29%, 24.49%, 17.49%, 12.49%, 8.93%, 8.92%, 8.93%, and 4.46% (see Table 10.3).

---

[1] We will assume that the asset is purchased and placed in service at the beginning of year 1 (or end-of-year 0), and the first year's depreciation will be claimed at the end of year 1.

Income Statement (Year *n*)

Revenues	$100,000
Expenses	
Labor	$20,000
Material	$12,000
Overhead	$8,000
Depreciation	D_n
Taxable Income	TI_n
Income tax (40%)	T_n
Net income	NI_n

In years 1–5

Cash Flow Statement (Year *n*)

Operating activities		
Net income	NI_n	
Depreciation	$+D_n$	
Investing activities		
Machining center	$-P_0$	Only in year 0
Salvage value	$+S_5$	Only in year 5
Gains tax	$-G_5$	May come in year 5
Net cash flow	A_n	

2. If the asset is sold at the end of the fifth tax year (during the recovery period), the applicable depreciation amounts would be $17,863, $30,613, $21,863, $15,613, and $5,581. Since the asset is disposed of in the fifth tax year, the last year's depreciation, which would ordinarily be $11,163, is halved due to the half-year convention.

We now have a value for our unknown D_n, which will enable us to complete the statement for years 1 to 5. The results of these simple calculations appear in Table 12.2.

- Salvage value and gain taxes

  In year 5, we must deal with two aspects of the asset's disposal, salvage value and gains (both ordinary as well as capital). We list the estimated salvage value as a positive cash flow. Taxable gains are calculated as follows:

  1. The total depreciation in years 1 to 5 is $17,863 + $30,613 + $21,863 + $15,613 + $5,581 = $91,533.
  2. The book value at the end of period 5 is the cost basis minus the total depreciation, or $125,000 − $91,533 = $33,467.

Table 12.2
Cash Flow
Statement
for the
Automated
Machining
Center
Project
Using
Approach 1
(Example
12.1)

Year	0	1	2	3	4	5
Income Statement						
Revenues		$100,000	$100,000	$100,000	$100,000	$100,000
Expenses						
Labor		20,000	20,000	20,000	20,000	20,000
Material		12,000	12,000	12,000	12,000	12,000
Overhead		8,000	8,000	8,000	8,000	8,000
Depreciation		17,863	30,613	21,863	15,613	5,581
Taxable income		$ 42,137	$ 29,387	$ 38,137	$ 44,387	$ 54,419
Income taxes (40%)		16,855	11,755	15,255	17,755	21,768
Net income		$ 25,282	$ 17,632	$ 22,882	$ 26,632	$ 32,651
Cash Flow Statement[2]						
Operating activities						
Net income		25,282	17,632	22,882	26,632	32,651
Depreciation		17,863	30,613	21,863	15,613	5,581
Investment activities						
Investment	(125,000)					
Salvage						50,000
Gains tax						(6,613)
Net cash flow	$(125,000)	$ 43,145	$ 48,245	$ 44,745	$ 42,245	$ 81,619

3. The gains on the sale are the salvage value minus the book value, or $50,000 – $33,467 = $16,533. (The salvage value is less than the cost basis, so all the gain is ordinary.)
4. The tax on the ordinary gains is $16,533 × 40% = $6,613. This is the amount placed in the table under "gains tax."

See Table 12.2 for the summary cash flow profile.[2]

- Investment analysis

  Once we obtain the project's after-tax net cash flows, we can determine their equivalent present worth at the firm's interest rate. The after-tax cash flow series from the cash flow statement is shown in Figure 12.3. Since this series does not contain any patterns to simplify our calculations, we must find the net present worth of each payment. Using $i = 15\%$, we have

  $$PW(15\%) = -\$125,000 + \$43,145(P/F, 15\%, 1)$$
  $$+ \$48,245(P/F, 15\%, 2) + \$44,745(P/F, 15\%, 3)$$

---

[2] Even though gains from equipment disposal have an effect on income tax calculations, they should not be viewed as ordinary operating income. Therefore, in preparing the income statement, the capital expenditures or related items such as gains tax and salvage value are not included. Nevertheless, these items represent actual cash flows in the year they occur and must be shown in the cash flow statement.

Figure 12.3   Cash flow diagram (Example 12.1)

$$+ \ \$42{,}245(P/F, 15\%, 4) \ + \ \$81{,}619(P/F, 15\%, 5)$$
$$= \ \$43{,}152.$$

This means that investing \$125,000 in this automated facility would bring in enough revenue to recover the initial investment and the cost of funds, with a surplus of \$43,152.

Comment: As a variation to Table 12.2 which uses Approach 1, a tabular format widely used in traditional engineering economics texts is shown in Table 12.3. The approach taken in developing Table 12.3 is based on Approach 2, which computes income taxes directly. Without seeing the footnotes in Table 12.3, however, it is not intuitively clear how the last column (net cash flow) is obtained. Therefore, we will use the cash flow statement based on the income (Approach 1) whenever possible throughout this text.

Table 12.3
Net Cash Flow Table Generated by Traditional Method Using Approach 2 (Example 12.1)

A	B	C	D	E	F	G	H	I	J
Year End	Investment and Salvage Value	Revenue	Labor	Expenses Materials	Overhead	Depreciation	Taxable Income	Income Taxes	Net Cash Flow
0	\$(125,000)								\$(125,000)
1		\$100,000	\$(20,000)	\$(12,000)	\$(8,000)	\$(17,863)	\$42,137	\$(16,855)	\$ 43,145
2		100,000	(20,000)	(12,000)	(8,000)	\$30,613)	29,387	(11,755)	\$ 48,245
3		100,000	(20,000)	(12,000)	(8,000)	(21,863)	38,137	(15,255)	\$ 44,745
4		100,000	(20,000)	(12,000)	(8,000)	(15,613)	44,387	(17,755)	\$ 42,245
5		100,000	(20,000)	(12,000)	(8,000)	(5,581)	54,419	(21,768)	$\begin{cases} \$ 38{,}232 \\ \$ 43{,}387 \end{cases}$
	50,000*						16,533	(6,613)	

* Salvage value. Note that col. H = col. C + col. D + col. E + col. F + col. G except for salvage value;
col. I = 0.40 × col. H; and col. J = col. B + col. C + col. D + col. E + col. F + col. I

1

Information required to calculate income taxes

12.3.2 When Projects Require Working Capital Investments

In many cases, changing a production process by replacing old equipment or by adding a new product line will have an impact on cash balances, accounts receivable, inventory, and accounts payable. For example, if a company is going to market a new product, inventories of the product and larger inventories of raw materials will be needed. Accounts receivable from sales will increase, and management might also decide to carry more cash because of the higher volume of activities. These investments in working capital are investments just as are those in depreciable assets (except that they have no tax effects. The flows always sum to zero over the life of a project, but the inflows and outflows are shifted in time so they do affect net present worth).

Consider the case of a firm that is planning a new product line. The new product will require a 2-month's supply of raw materials at a cost of $40,000. The firm could provide $40,000 in cash on hand to pay them. Alternatively, the firm could finance these raw materials via a $30,000 increase in accounts payable (60-day purchases) by buying on credit. The balance of $10,000 represents the amount of net working capital that must be invested.

Working capital requirements differ according to the nature of the investment project. For example, larger projects may require greater average investments in inventories and accounts receivable than would smaller ones. Projects involving the acquisition of improved equipment entail different considerations. If the new equipment produces more rapidly than the old equipment, the firm may be able to decrease its average inventory holdings because new orders can be filled faster as a result of using the new equipment. (One of the main advantages cited in installing advanced manufacturing systems, such as flexible manufacturing systems, is the reduction in inventory made possible by the ability to respond to market demand more quickly.) Therefore, it is also possible for working capital needs to decrease because of an investment. If inventory levels were to decrease at the start of a project, the decrease would be considered a cash inflow, since the cash freed up from inventory could be put to use in other places. (See Example 12.5.)

Two examples will be provided to illustrate the effects of working capital on a project's cash flows. Example 12.2 will show how the net working capital requirement is computed, and Example 12.3 will examine the effects of working capital on the automated machining center project discussed in Example 12.1.

Example 12.2 Working Capital Requirements

Consider Example 12.1. Suppose that the tool manufacturing company's annual revenue projection of $100,000 is based on an annual volume of 10,000 units (or 833 units per month). Assume the following accounting information:

Price (revenue) per unit	$10
Unit variable manufacturing costs	
Labor	$2
Material	$1.20
Overhead	$0.80
Monthly volume	833 units
Finished goods inventory to maintain	2-month supply
Raw materials inventory to maintain	1-month supply
Accounts payable	30 days
Accounts receivable	60 days

The accounts receivable period of 60 days means that revenues from the current month's sales will be collected 2 months later. Similarly, accounts payable of 30 days indicates that payment for materials will be made approximately 1 month after the materials are received. Determine the working capital requirement for this operation.

Solution

Given: Information stated above

Find: Working capital requirement

Figure 12.4 illustrates the working capital requirements for the first 12-month period. Accounts receivable of $16,666 (2 months' sales) means that in year 1 the company will have cash inflows of $83,333, which is less than the projected sales of $100,000 ($8,333 × 12). In years 2 to 5, collections will be $100,000, equal to sales, because beginning and ending accounts receivable will be $16,666, with sales of $100,000. Collections of the final accounts receivable of $16,666 would occur in the first 2 months of year 6, but can be added to the year 5 revenue to simplify the calculations. The important point is that cash inflow lags sales by $16,666 in the first year.

Assuming the company wishes to build up 2 months' inventory during the first year, it must produce 833 × 2 = 1,666 more units than are sold the first year. The extra cost of these goods in the first year is 1,666 units ($4 variable cost per unit), or $6,665. The finished goods inventory of $6,665 represents the variable cost incurred to produce 1,666 more units than are sold in the first year. In years 2 to 4, the company will produce and sell 10,000 units per year, while maintaining its 1,666 units supply of finished goods. In the final year of operations, the company will produce only 8,334 units (for 10 months) and will use up the finished goods inventory. As 1,666 units of the finished goods inventory get liquidated during the last year, a working capital release in the amount of $6,665 will occur. Along with the collections of the final accounts re-

During year 1			
	Income/Expense reported	Actual cash received/paid	Difference
Sales	$100,000 (10,000 units)	$83,333	−$16,666
Expenses	$40,000 (10,000 units)	$46,665 (11,667 units)	+$6665
Income taxes	$16,855	$16,855	0
Net amount	$43,145	$19,814	−$23,333

This differential amount must be invested at the beginning of the year

Figure 12.4 Illustration of working capital requirements (Example 12.2)

ceivable of $16,666, a total working capital release of $23,331 will remain when the project terminates. Now we can calculate the working capital requirements as follows:

Accounts receivables (833 units/month × 2 months × $10)	$16,666
Finished good inventory (833 units/month × 2 months × $4)	6,665
Raw materials inventory (833 units/month × 1 month × $1.20)	1,000
Accounts payable (raw material purchase) (833 units/month × 1 month × $1.20)	(1,000)
Net change in working capital	$23,331

Comments: In our example, during the first year, the company produces 11,666 units to maintain 2 months' finished good inventory, but it sells only 10,000 units. On what basis should the company calculate the net income during the first year (use 10,000 or 11,666 units)? Any increases in inventory expenses will reduce the taxable income, therefore, this calculation is based on 10,000 units. The reason is that accounting measure of net income is based on the **matching concept**. If we report revenue when it is *earned* (whether it is actually received or not), and we report expenses when they are *incurred* (whether they are paid or not), we are using the *accrual method* of accounting. By tax law, this accrual method must be used for purchases and sales whenever business transactions involve an inventory. Therefore, most manufacturing and merchandising businesses use the accrual basis in recording revenues and expenses. Any cash inventory expenses not accounted for in the net income calculation will be reflected in changes in working capital.

Example 12.3 Cash Flow Statement—Including Working Capital

Reconsider Example 12.1. Update the after-tax cash flows for the automated machining center project by including a working capital requirement of $23,331 in year 0 and full recovery of the working capital at the end of year 5.

Solution

Given: Flows as in Example 12.1, with the addition of a working capital requirement = $23,331

Find: Net after-tax cash flows with working capital and present worth

Using the procedure outlined above, the net after-tax cash flows for this machining center project are grouped as shown in Table 12.4. As the table indicates, investments in working capital are cash outflows when they are expected to occur, and recoveries are treated as cash inflows at the times they are expected to materialize. In this example, we assume that the investment in working capital made at period 0 will be recovered at the end of the project's life.[3] Moreover, we also assume a full recovery of the initial working capital. However, many situations occur in which the investment in working capital may not be fully recovered (e.g., inventories may deteriorate in value or become obsolete). The equivalent net present worth of the after-tax cash flows including the effects of working capital is calculated as

$$PW(15\%) = -\$148,331 + \$43,145(P/F, 15\%, 1) \ldots$$
$$+ \$104,950(P/F, 15\%, 5)$$
$$= \$31,420.$$

This present worth value is $11,732 less than in the situation with no working capital requirement (Example 12.1). This example demonstrates that working capital requirements must be considered when properly assessing a project's worth.

Comment: The $11,732 reduction in present worth is just the present worth of an annual series of 15% interest payments on the working capital, which is borrowed by the project at time 0 and repaid at the end of year 5:

$$\$23,331(15\%)(P/A, 15\%, 5) = \$11,732.$$

The investment tied up in working capital results in lost earnings.

---

[3] In fact, we could assume that the investment made in working capital would be recovered at the end of the first operating cycle (say, year 1). However, the same amount of investment in working capital has to be made again at the beginning of year 2 for the second operating cycle, and the process repeats until the project terminates. Therefore, the net cash flow transaction looks as though the initial working capital will be recovered at the end of project life (Figure 12.5.)

Period	0	1	2	3	4	5
Investment	−$23,331	−$23,331	−$23,331	−$23,331	−$23,331	0
Recovery	0	23,331	23,331	23,331	23,331	23,331
Net flow	−$23,331	0	0	0	0	$23,331

Table 12.4
Cash Flow
Statement
for Auto-
mated
Machining
Center
Project with
Working
Capital
Require-
ment
(Example
12.3)

Year	0	1	2	3	4	5
Income Statement						
Revenues		$100,000	$100,000	$100,000	$100,000	$100,000
Expenses						
Labor		20,000	20,000	20,000	20,000	20,000
Material		12,000	12,000	12,000	12,000	12,000
Overhead		8,000	8,000	8,000	8,000	8,000
Depreciation		17,863	30,613	21,863	15,613	5,581
Taxable income		$ 42,137	$ 29,387	$ 38,137	$ 44,387	$ 54,419
Income taxes (40%)		16,855	11,755	15,255	17,755	21,768
Net income		$ 25,282	$ 17,632	$ 22,882	$ 26,632	$ 32,651
Cash Flow Statement						
Operating activities						
Net income		25,282	17,632	22,882	26,632	32,651
Depreciation		17,863	30,613	21,863	15,613	5,581
Investment activities						
Investment	(125,000)					
Salvage						50,000
Gains tax						(6,613)
Working capital	(23,331)					23,331
Net cash flow	$(148,331)	$ 43,145	$ 48,245	$ 44,745	$ 42,245	$104,950

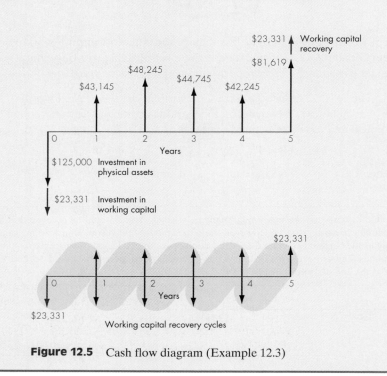

Figure 12.5   Cash flow diagram (Example 12.3)

12.3.3 When Projects Are Financed with Borrowed Funds

Many companies use a mixture of debt and equity to finance their physical plant and equipment. The ratio of total debt to total investment, generally called the **debt ratio,** represents the percentage of the total initial investment provided by borrowed funds. For example, a debt ratio of 0.3 indicates that 30% of the initial investment is borrowed, and the rest is provided from the company's earnings (also known as **equity**). Since interest is a tax-deductible expense, companies in high tax brackets may incur lower after-tax financing costs by financing through debt. (Along with the effect on taxes, the method of loan repayment can also have a significant impact. We will discuss the issue of project financing in Chapter 16.)

Example 12.4 Cash Flow Statement— with Financing (Borrowing)

Rework Example 12.3, assuming that $62,500 of the $125,000 paid for the investment is obtained through debt financing (debt ratio = 0.5). The loan is to be repaid in equal annual installments at 10% interest over 5 years. The remaining $62,500 will be provided by equity (e.g., from retained earnings).

Solution

Given: Same as in Example 12.3, but $62,500 is borrowed, repaid in equal installments over 5 years at 10%

Find: Net after-tax cash flows in each year

We first need to compute the size of the annual loan repayment installments:

$$\$62,500(A/P, 10\%, 5) = \$16,487.$$

Next, we determine the repayment schedule of the loan by itemizing both the interest and principal represented in each annual repayment.

Year	Beginning Balance	Interest Payment	Principal Payment	Ending Balance
1	$62,500	$6,250	$10,237	$52,263
2	52,263	5,226	11,261	41,002
3	41,002	4,100	12,387	28,615
4	28,615	2,861	13,626	14,989
5	14,989	1,499	14,988	0

The resulting after-tax cash flow is detailed in Table 12.5. The present value equivalent of the after-tax cash flow series is

Table 12.5
Cash Flow
Statement
for Auto-
mated
Machining
Center
Project with
Debt
Financing
(Example
12.4)

Year	0	1	2	3	4	5
Income Statement						
Revenues		$100,000	$100,000	$100,000	$100,000	$100,000
Expenses						
Labor		20,000	20,000	20,000	20,000	20,000
Material		12,000	12,000	12,000	12,000	12,000
Overhead		8,000	8,000	8,000	8,000	8,000
Depreciation		17,863	30,613	21,863	15,613	5,581
Debt interest		6,250	5,226	4,100	2,861	1,499
Taxable income		$ 35,887	$ 24,161	$ 34,037	$ 41,526	$ 52.920
Income taxes (40%)		14,355	12.664	13,615	16,610	21,168
Net income		$ 21,532	$ 14,497	$ 20,422	$ 24,916	$ 31,752
Cash Flow Statement						
Operating activities						
Net income		21,532	14,497	20,422	24,916	31,752
Depreciation		17,863	30,613	21,863	15,613	5,581
Investment activities						
Investment	(125,000)					
Salvage						50,000
Gains tax						(6,613)
Working capital	(23,331)					23,331
Financing activities						
Borrowed funds	62,500					
Principal repayment		(10,237)	(11,261)	(12,387)	(13,626)	(14,988)
Net cash flow	$ (85,831)	$ 29,158	$ 33,849	$ 29,898	$ 26,903	$ 89,063

$$PW(15\%) = -\$85,351 + \$29,158(P/F, 15\%, 1) + \ldots$$
$$+ \$89,063(P/F, 15\%, 5)$$
$$= \$44,439.$$

When this amount is compared with the amount found in the case that involved no
borrowing ($31,420), we see that debt financing actually increases the present worth
by $13,019. This surprising result is largely caused by the firm being able to borrow the
funds at a cheaper rate (10%) than its MARR (opportunity cost rate) of 15%. We
should be careful in interpreting the result. It is true, to some extent, that firms can
usually borrow money at lower rates than their MARR. However, if the firm can borrow money at a significantly lower rate, it also affects its MARR because the borrowing rate is one of the elements in determining the MARR. Therefore, a significant
difference in present values between "with borrowings" and "without borrowings" is
not expected in practice. We will address this important issue in Chapter 16.

12.3.4   When Projects Result in Negative Taxable Income

In a typical project year, revenues may not be large enough to offset expenses, thereby resulting in a negative taxable income. A negative taxable income does *not* mean that a firm does not need to pay income tax: Rather, the negative figure can be used to reduce the taxable incomes generated by other business operations.[4] Therefore, a negative taxable income usually results in a **tax savings**. When we evaluate an investment project using an incremental tax rate, we also assume that the firm has sufficient taxable income from other activities so that changes due to the project under consideration will not change the incremental tax rate.

When we compare **cost-only** mutually exclusive projects (service projects), we have no revenues to consider in their cash flow analysis. In this situation, we typically assume no revenue (zero), but proceed as before when constructing the after-tax cash flow statement. With no revenue to match expenses, we have a negative taxable income, resulting in tax savings as before. Example 12.5 illustrates how we may develop an after-tax cash flow statement for this type of project.

Example 12.5   After-Tax Cash Flow Analysis for a Cost-Only Project

Reconsider the McCook plant's cooling fan project that was introduced in the opening of this chapter. Suppose that Mr. Donaldson compiled the following financial data:

- The project will require an investment of $536,000 in cooling fans now.
- The cooling fans would provide 16 years of service with no appreciable salvage values considering the removal costs.
- It is expected that the amount of time required between hot rolling and the next operation would be reduced from 5 days to 2 days. Cold rolling queue time would be reduced from 2 days to 1 day for each cold roll pass. The net effect of these changes would be a reduction of WIP inventory at a value of $2,121,000. Because of the lead time involved in installing the fans, as well as the consumption of stock-piled WIP inventory, this working capital release will be realized 1 year after the fans are installed.
- The cooling fans will be depreciated according to 7-year MACRS.
- Annual electricity costs are estimated to be increased by $86,000.
- Reynolds' after-tax required rate of return is known to be 20% for this type of cost reduction project.
- Develop the project cash flows over the service period, and determine if the investment is a wise one, based on 20% interest.

---

[4] Even if the firm does not have any other taxable income to offset in the current tax year, the operating loss can be carried back to each of the preceding 3 years and forward for the following 15 years to offset taxable income in those years.

Solution

Given: Required investment = \$536,000; service period = 16 years; salvage value = \$0; depreciation method for cooling fans = 7-year MACRS; working capital release = \$2,121,000 1 year later; annual operating cost (electricity) = \$86,000

Find: (a) annual after-tax cash flows, (b) make the investment decision based on the NPW, and (c) make the investment decision based on the IRR.

(a) Because we can assume that the annual revenues would stay the same as before and after the installation of the cooling fans, we can treat these unknown revenue figures as zero. Table 12.6 summarizes the cash flow statement for the cooling fan project. With no revenue to offset the expenses, the taxable income will be negative, resulting in tax savings. Note that the working capital recovery (as opposed to working capital investment for a typical investment project) is shown in year 1. Note also that there is no gains tax because the cooling fans are fully depreciated with a zero salvage value.

(b) At $i = 20\%$, the NPW of this investment would be

$$PW(20\%) = -\$536,000 + \$2,100,038(P/F, 20\%, 1)$$
$$+ \$906(P/F, 20\%, 2) + \dots$$
$$-\$2,172,600(P/F, 20\%, 16)$$
$$= \$991,008$$

Even with only one time savings in WIP, this cooling fan project is economically justifiable.

(c) If we had to justify the investment based on the internal rate of return, we need to first determine to see if the investment is a simple or nonsimple one. Since there are more than one sign changes in the cash flow series, this is not a simple investment, indicating the possibility of multiple rates of return. In fact, as shown in Figure 12.6 the project has two rates of return: one at 4.24% and the other at 291.56%. Neither is a true rate of return as we learned from Section 9.3.3, so we may proceed to abandon the rate of return approach and use the NPW criterion. If you desire to find the true IRR for this project, you need to follow the procedures outlined in Appendix A. At a MARR of 20%, the true IRR (or return on invested capital) is 241.87%, which is significantly larger than the MARR and indicates the acceptance of the investment. Note that this is the same conclusion that we reached in (b).

Comments: As the cooling fans reach the end of their respective service lives, we need to add the working capital investment (\$2,120,000) at the end of year 16 working with the assumption that the plant will return to the system without the cooling fans and thus will require the additional investment in working capital. If the system has

Table 12.6 Cash Flow Statement for the Cooling Fan Project without Revenue (Example 12.5)

Year	0	1	2	3	4	5	6	7	8	9-15	16
Income Statement											
Revenues											
Expenses											
Depreciation		$76,594	$131,266	$93,746	$66,946	$47,865	$47,811	$47,865	$23,906		
Electricity cost		86,000	86,000	86,000	86,000	86,000	86,000	86,000	86,000	86,000	86,000
Taxable income		(162,594)	(217,266)	(179,746)	(152,946)	(133,865)	(133,811)	(133,865)	(109,906)	(86,000)	(86,000)
Income taxes		(65,038)	(86,906)	(71,898)	(61,178)	(53,546)	(53,524)	(53,546)	(43,962)	(34,400)	(34,400)
Net income		$ (97,556)	$(130,360)	$(107,848)	$ (91,768)	$ (80,319)	$ (80,287)	$ (80,319)	$ (65,944)	$ (51,600)	$ (51,600)
Cash Flow Statement											
Operating activities											
Net income		(97,556)	(130,360)	(107,848)	(91,768)	(80,319)	(80,287)	(80,319)	(65,944)	(51,600)	(51,600)
Depreciation		76,594	131,266	93,746	66,946	47,865	47,811	47,865	23,906	0	0
Investment activities											
Cooling fans	(536,000)										
Salvage value											
Gains tax											
Working capital		2,121,000									(2,121,000)
Net cash flow	$ (536,000)	$2,100,038	$ 906	$ (14,102)	$ (24,822)	$ (32,454)	$ (32,476)	$ (32,454)	$ (42,038)	$ (51,600)	$ (2,172,600)

Note: The working capital release attributable to reduction in work-in-process inventories will be materialized at the end of year 1.

547

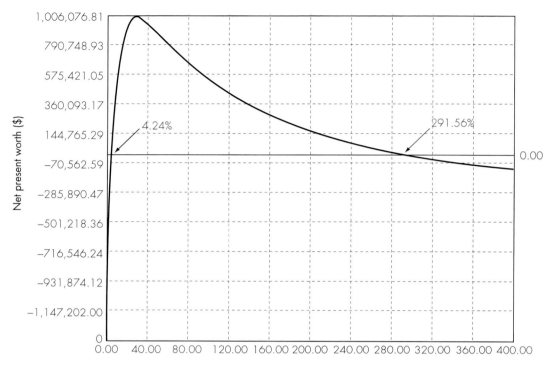

Figure 12.6   NPW plot for the cooling fan project (Example 12.5)

> proven to be effective, and the plant will remain in definite service, we need to make another investment to purchase a new set of cooling fans at the end of year 16. As a consequence of this new investment, there will be a working capital release in the amount of $2,120,000. However, this investment should bring benefits to the second cycle of the operation, so that it should be charged against the cash flows for the second cycle, not the first cycle.

12.3.5 When Projects Require Multiple Assets

Up to this point, our examples have been limited to situations where only one asset was employed in a project. In many situations, however, a project may require purchase of multiple assets with different property classes. For example, a typical engineering project may involve more than just the purchase of equipment—it may need a building in which to carry out manufacturing operations. Even the various assets may be placed in service at different points in time. What we have to do is itemize the timing of the investment requirement and the depreciation allowances according to the asset placement. Example 12.6 illustrates the development of project cash flows that require multiple assets.

Example 12.6 A Project Requiring Multiple Assets

Langley Manufacturing Company (LMC), a manufacturer of fabricated metal products, is considering the purchase of a new computer-controlled milling machine for $90,000 to produce a custom-ordered metal product. The costs for installation of the machine, site preparation, and wiring are expected to be $10,000. The machine also needs special jigs and dies, which will cost $12,000. The milling machine is expected to last 10 years, and the jigs and dies to last 5 years. The machine will have a $10,000 salvage value at the end of its life. The special jigs and dies are worth only $1,000 as scrap metal at any time in their lives. The milling machine is classified as a 7-year MACRS property and the jigs and dies as a 3-year MACRS property. LMC needs to either purchase or build an 8,000 ft^2 warehouse in which to store the product before shipping to the customer. LMC has decided to purchase a building near the plant at a cost of $160,000. For depreciation purposes, the warehouse cost of $160,000 is divided into $120,000 for the building (39-year real property) and $40,000 for land. At the end of 10 years, the building will have a salvage value of $80,000, but the value of the land will have appreciated to $110,000. The revenue from increased production is expected to be $150,000 per year. The additional annual production costs are estimated as follows: materials, $22,000, labor, $32,000, energy $3,500, and other miscellaneous costs, $2,500. For the analysis, a 10-year life will be used. LMC has a marginal tax rate of 40% and a MARR of 18%, respectively. No money is borrowed to finance the project. Capital gains will also be taxed at 40%.[5]

Discussion: Three types of assets are to be considered in this problem: The milling machine, the jigs and dies, and the warehouse. The first two assets are personal properties and the last is a real property. The cost basis for each asset has to be determined separately. For the milling machine, we need to add the site-preparation expense to the cost basis, whereas we need to subtract the land cost from the warehouse cost to establish the correct cost basis for the real property.

- The milling machine: $90,000 + $10,000 = $100,000.
- Jigs and dies: $12,000.
- Warehouse (building): $120,000.
- Warehouse (land): $40,000.

Since the jigs and dies last only 5 years, we need to make a specific assumption regarding the replacement cost at the end of 5 years. In this problem, we will assume that the replacement cost would be approximately equal to the cost of the initial purchase. We will also assume that the warehouse property will be placed in service in January, which indicates that the first year's depreciation percentage will be 2.4573% (see Table 10.4).

---

[5] Capital gains for corporation are taxed at a maximum rate of 35%. However, capital gains are also subject to state taxes, so that the combined tax rate will be approximately 40%.

Solution

Given: Cash flow elements provided above, $t_m = 40\%$, MARR = 18%

Find: Net after-tax cash flow, NPW

Table 12.7 and Figure 12.7 summarize the net after-tax cash flows associated with the multiple-asset investment. In this table, we see that the milling machine, as well as the jigs and dies, are fully depreciated during the project life, whereas the building is not. We need to calculate the book value of the building at the end of the project life. We assume that the building will be disposed of December 31 on the tenth year, so that the mid-month convention also applies to the book value calculation.

$$B_{10} = \$120,000 - (\$2,949 + \$3,077 \times 8 + \$2,949) = \$89,486.$$

Then, gains (losses) are

$$\text{Salvage value} - \text{book value} = \$80,000 - \$89,486 = (\$9,486).$$

We then can calculate the gains or losses associated with disposal of each asset as follows:

Property (Asset)	Cost Basis	Salvage Value	Book Value	Gains (Losses)	Gains Taxes
Land	$40,000	$110,000	$40,000	$70,000	$28,000
Building	120,000	80,000	89,486	(9,486)	(3,794)
Milling machine	100,000	10,000	0	10,000	4,000
Jigs and dies	12,000	1,000	0	1,000	400

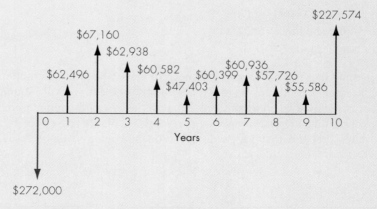

Figure 12.7   Cash flow diagram (Example 12.6)

Table 12.7 Cash Flow Statement for the LMC's Machining Center Project with Multiple Assets (Example 12.6)

Year	0	1	2	3	4	5	6	7	8	9	10
Income Statement											
Revenues		$150,000	$150,000	$150,000	$150,000	$150,000	$150,000	$150,000	$150,000	$150,000	$150,000
Expenses											
Materials		22,000	22,000	22,000	22,000	22,000	22,000	22,000	22,000	22,000	22,000
Labor		32,000	32,000	32,000	32,000	32,000	32,000	32,000	32,000	32,000	32,000
Energy		3,500	3,500	3,500	3,500	3,500	3,500	3,500	3,500	3,500	3,500
Others		2,500	2,500	2,500	2,500	2,500	2,500	2,500	2,500	2,500	2,500
Depreciation											
Building		2,949	3,077	3,077	3,077	3,077	3,077	3,077	3,077	3,077	2,949
Machines		14,290	24,490	17,490	12,490	8,930	8,920	8,930	4,460		
Tools		4,000	5,333	1,778	889		4,000	5,333	1,778	889	
Taxable income		68,761	57,100	67,655	73,544	77,993	74,003	72,660	80,685	86,034	87,051
Income taxes		27,504	22,840	27,062	29,418	31,197	29,601	29,064	32,274	34,414	34,820
Net income		$41,257	$34,260	$40,593	$44,126	$46,796	$44,402	$43,596	$48,411	$51,620	$52,231
Cash Flow Statement											
Operating activities:											
Net income		41,257	34,260	40,593	44,126	46,796	44,402	43,596	48,411	51,620	52,231
Depreciation		21,239	32,900	22,345	16,456	12,007	15,997	17,340	9,315	3,966	2,949
Investment activities:											
Land	(40,000)										110,000
Building	(120,000)										80,000
Machines	(100,000)										10,000
Tools (1st cycle)	(12,000)					1,000					
Tools (2nd cycle)						(12,000)					1,000
Gains tax:											
Land											(28,000)
Building											3,794
Machines											(4,000)
Tools						(400)					(400)
Net cash flow	$ (272,000)	$ 62,496	$ 67,160	$ 62,938	$ 60,582	$ 47,403	$ 60,399	$ 60,936	$ 57,726	$ 55,586	$227,574

Note: Investment in tools (jigs and dies) will be repeated at the end of year 5 with the same cost of the initial purchase.

The NPW of the project is

$$PW(18\%) = -\$272,000 + \$62,496(P/F, 18\%, 1) + \$67,160(P/F, 18\%, 2)$$
$$+ \ldots + \$227,574(P/F, 18\%, 10)$$
$$= \$32,343 > 0,$$

and the IRR for this investment is about 21%, which exceeds the MARR. Therefore, this project is acceptable.

Comment: Note that the gains (losses) posted in the table above can be classified into two types: ordinary gains (losses) and capital gains. Only $70,000 for land represents a true sense of capital gains, whereas others represent ordinary gains (losses).

12.3.6 When Investment Tax Credits Are Allowed

The Economic Recovery Act of 1981 created an investment tax credit (ITC) of 10% applicable to all qualifying property with lives of more than 3 years. However, this ITC was repealed by Congress in the Tax Reform Act of 1986. The reasons given for abandoning the ITC were twofold: The tax credit resulted in a misallocation of resources, and it did not provide a fiscal stimulus to the economy. However, particularly during recessionary periods, Congress may consider reinstating the ITC to stimulate an ailing economy.

How do we adjust our after-tax cash flow analysis to consider a tax credit? The answer is rather straightforward—assuming that the ITC is taken at the end of the first year the asset is in service, we simply reduce the income tax amount by the ITC. This will result in an increase in the NPW of the investment. (Or, you simply add the ITC amount to the first year's cash flow.) Depending on the tax laws in effect, it may require some adjustment in cost basis for the depreciable asset.[6]

Example 12.7 Considering an Investment Tax Credit

Suppose, in Example 12.6, that a 10% ITC was allowed on the purchase of the milling machine and the jigs and dies only. Assume no change occurred in the cost basis of any of these assets. How does the ITC affect the profitability of the investment?

Solution

Given: Net cash flows in Table 12.6, ITC = 10%, no adjustment in cost basis
Find: NPW

---

[6] This ITC is a one-time tax credit and may be subject to being recaptured (returned to the government) if an asset is sold before the end of a specified period.

The ITC amount for each asset is

Milling machine	(0.10)($100,000)	=	$10,000
Jigs and dies	(0.10)($12,000)	=	1,200
Total ITC			$11,200

Assuming that the ITC is claimed at the *end* of first year, we can obtain the NPW of the project by

$$\text{New NPW} = \text{Old NPW} + \text{ITC}(P/F, 18\%, 1)$$
$$= \$32{,}343 + (\$11{,}200)(P/F, 18\%, 1)$$
$$= \$32{,}343 + \$9{,}492$$
$$= \$41{,}835.$$

Note that the ITC increases the profitability of the project. (For a certain investment situation, the ITC could increase the profitability of the project from somewhat marginal to moderate.) This increase will certainly encourage the firm to undertake the investment.

12.4 Generalized Cash Flow Approach

If we analyze project cash flows for a corporation that consistently operates in the highest tax bracket, we can assume that the firm's marginal tax rate will remain the same, whether the project is accepted or rejected. In this situation, we may apply the top marginal tax rate (currently 35%) to each taxable item in the cash profile, thus obtaining the after-tax cash flows. By aggregating individual items, we obtain the project's net cash flows. This approach is referred to as the **generalized cash flow approach**. As we shall see in later chapters, this approach affords several analytical advantages. In particular, when we compare service projects, the generalized cash flow method is computationally efficient. (Examples are provided in Chapters 14 and 15.)

12.4.1 Setting up Net Cash Flow Equations

To produce the generalized cash flow table, we first examine each cash flow element. We can do this as follows, using the scheme for classifying cash flows that we have just developed:

$$A_n = \begin{array}{l} + \text{Revenues at time } n, (R_n) \\ - \text{Expenses (excluding depreciation and debt} \\ \quad \text{interest) at time } n, (E_n) \\ - \text{Debt interest at time } n, (I_n) \\ - \text{Income taxes at time } n, (T_n) \end{array} \left.\vphantom{\begin{array}{l}x\\x\\x\\x\\x\end{array}}\right\} \text{Operating activities}$$

$$\begin{array}{l} - \text{Investment at time } n, (P_n) \\ + \text{Net proceeds from sale at time } n, (S_n - G_n) \\ - \text{Working capital investment at time } n, (W_n) \end{array} \left.\vphantom{\begin{array}{l}x\\x\\x\end{array}}\right\} \text{Investing activities}$$

$$\begin{array}{l} + \text{Proceeds from loan } n, (B_n) \\ - \text{Repayment of principal } n, (PP_n), \end{array} \left.\vphantom{\begin{array}{l}x\\x\end{array}}\right\} \text{Borrowing activities}$$

where A_n is the net after-tax cash flow at the end of period n.

Depreciation (D_n) is *not* a cash flow and is therefore excluded from E_n (although it must be considered when calculating income taxes). Note also that ($S_n - G_n$) represents the net salvage value after adjustments for gains tax or loss credits (G_n). Not all terms are relevant in calculating a cash flow in every year. The term ($S_n - G_n$) appears only when the asset is disposed of.

In terms of symbols, we can express A_n as

$$A_n = R_n - E_n - I_n - T_n \qquad \longleftarrow \text{Operating activities}$$
$$\quad - P_n + (S_n - G_n) - W_n \longleftarrow \text{Investing activities} \qquad (12.1)$$
$$\quad + B_n - PP_n \qquad\qquad \longleftarrow \text{Financing activities.}$$

If we designate T_n as the total income taxes paid at time n and t_m as the marginal tax rate, income taxes on this project are

$$T_n = (\text{Taxable income})(\text{marginal tax rate})$$
$$\quad = (R_n - E_n - I_n - D_n)\, t_m \qquad\qquad (12.2)$$
$$\quad = (R_n - E_n - I_n)t_m - D_n t_m$$

The term ($I_n + D_n$) t_m is known as the tax shield (or tax savings) from financing and asset depreciation. Now substituting the result of Eq. (12.2) into Eq. (12.1) we obtain

$$A_n = (R_n - E_n - I_n)(1 - t_m) + t_m D_n$$
$$\quad - P_n + (S_n - G_n) - W_n \qquad\qquad (12.3)$$
$$\quad + B_n - PP_n.$$

12.4.2 Presenting Cash Flows in Compact Tabular Formats

After-tax cash flow components over the project life can be grouped by type of activities in a compact tabular format as follows:

Cash Flow Elements	End of Period 0 1 2 ... N
Investment activities	
$-P_n$	
$+S_n - G_n$	
$-W_n$	
Operating activities	
$+(1 - t_m)(R_n)$	
$-(1 - t_m)(E_n)$	
$-(1 - t_m)(I_n)$	
$+t_m D_n$	
Financing activities	
$+B_n$	
$-PP_n$	
Net cash flow	
A_n	

However, in preparing their after-tax cash flow, most business firms adopt the income statement approach presented in previous sections because they want to know the accounting income along with the cash flow statement.

Example 12.8 Generalized Cash Flow Approach

Reconsider Example 12.4. Use the generalized cash flow approach to obtain the after-tax cash flows:

Solution

Given:

Investment cost (P_n) = $125,000

Investment in working capital (W_n) = $23,331

Annual revenues (R_n) = $100,000

Annual expenses other than depreciation and debt interest (E_n) = $40,000

Debt interest (I_n), years 1 to 5 = $6,250, $5,226, $4,100, $2,861, $1,499

Principal repayment (PP_n), years 1 to 5 = $10,237; $11,261; $12,387; $13,626; $14,988

Depreciation (D_n), years 1 to 5 = $17,863; $30,613; $21,863; $15,613; $5,581

Marginal tax rate (t_m) = 40%

Salvage value (S_n) = $50,000

Find: Annual after-tax cash flows (A_n)

Step 1: Find the cash flow at year 0.

1. Investment in depreciable asset $(P_0) = -\$125,000$
2. Investment in working capital $(W_0) = -\$23,331$
3. Borrowed funds $(PP_0) = \$62,500$
4. Net cash flow $(A_0) = -\$125,000 - \$23,331 + \$62,500 = -\$85,831.$

Step 2: Find the cash flow at years 1 to 4.

$$A_n = (R_n - E_n - I_n)(1 - t_m) + D_n t_m - PP_n$$

n	Net Operating Cash Flow ($)
1	$(100,000 - 40,000 - 6250)(0.60) + 17,863(0.40) - 10,237 = \$29,158$
2	$(100,000 - 40,000 - 5226)(0.60) + 30,613(0.40) - 11,261 = \$33,849$
3	$(100,000 - 40,000 - 4100)(0.60) + 21,863(0.40) - 12,387 = \$29,898$
4	$(100,000 - 40,000 - 2861)(0.60) + 15,613(0.40) - 13,626 = \$26,903$

Step 3: Find the cash flow for year 5.

1. Operating cash flow:
   $(100,000 - 40,000 - 1499)(0.60) + 5581(0.40) - 14,988 = \$22,345$
2. Net salvage value (as calculated in Example 12.1) $= \$50,000 - \$6,613 = \$43,387$
3. Recovery of working capital, $W_5 = \$23,331$
4. Net cash flow in year 5, $A_5 = \$22,345 + \$43,387 + \$23,331 = \$89,063.$

Our results and overall calculations are summarized in Table 12.8. Checking this versus the results we obtained in Table 12.5 confirms our result.

Table 12.8
Cash Flow Statement for Example 12.4 Using the Generalized Cash Flow Approach (Example 12.8)

	0	1	2	3	4	5
Investment	$(125,000)					
Net proceeds from sale						$43,387
Investment in working capital	(23,331)					
Recovery of working capital						23,331
(1 – 0.40) (Revenue)		$ 60,000	$ 60,000	$ 60,000	$ 60,000	$ 60,000
–(1 – 0.40) (Expenses)		(24,000)	(24,000)	(24,000)	(24,000)	(24,000)
–(1 – 0.40) (Debt interest)		(3,750)	(3,136)	(2,460)	(1,717)	(899)
+(0.40) (Depreciation)		7,145	12,245	8,745	6,245	2,232
Borrowed funds	62,500					
Principal repayment		(10,237)	(11,261)	(12,387)	(13,626)	(14,988)
Net cash flow	$ (85,831)	$ 29,158	$ 33,849	$ 29,898	$ 26,903	$ 89,063

12.4.3   An Application of Generalized Cash Flow Approach—Lease-or-Buy Decision

A lease-or-buy decision begins only after a company has decided that the acquisition of a piece of equipment is necessary to carry out an investment project. Having made this decision, the company may be faced with several alternative methods of financing the acquisition: cash purchase, debt purchase, or acquisition via a lease.

In a situation of debt purchase, the present worth expression is similar to that of a purchase for cash, except that it has additional items—loan repayments and a tax shield on interest payments. The only way that the lessee can evaluate the cost of a lease is to compare it against the best available estimate of what the cost would be if the lessee owned the equipment.

To lay the groundwork for a more general analysis, we shall first consider how to analyze the lease-or-buy decision for a project with a single fixed asset for which the company expects a service life of N periods. Since the net after-tax revenue is the same for both alternatives, we need only consider the incremental cost of owning the asset and the incremental cost of leasing. Using the generalized cash flow approach, the incremental cost of owning the asset by borrowing may be expressed as

$$
\begin{aligned}
PW(i)_{\text{Buy}} = \ & - \text{PW of loan repayment} \\
& - \text{PW of after-tax O\&M costs} \\
& + \text{PW of tax credit on depreciation and interest} \\
& + \text{PW of net proceeds from sale.}
\end{aligned}
\tag{12.4}
$$

Note that the asset acquisition (investment) cost is offset by the same amount of borrowing at time 0, so that we only need to consider the loan repayment series.

Assume that the firm can lease the asset at a constant amount per period. The project's incremental cost of leasing, $PW(i)_{\text{Lease}}$, then becomes

$$
PW(i)_{\text{Lease}} = -\text{PW of after-tax lease expenses.}
$$

If the lease does not provide for the maintenance of the equipment leased, then the lessee must assume this responsibility. In this situation, the maintenance term in Eq. (12.4) can be dropped when calculating the incremental cost of owning the asset.

The criterion for the decision to lease as opposed to purchase thus reduces to a comparison between $PW(i)_{\text{Buy}}$ and $PW(i)_{\text{Lease}}$. In our terms, purchase is preferred if the combined present value of loan repayment series and after-tax O&M expense, reduced by the present values of the depreciation tax shield and the net proceeds from disposal of the asset, is less than the present value of the net lease costs.

Example 12.9 Lease-or-Buy Decision

The Dallas Electronics Company is considering replacing an old, 1,000-pound-capacity industrial forklift truck. The truck has been used primarily to move goods from production machines into storage. The company is working nearly at capacity and is operating on a two-shift basis, 6 days per week. Dallas management is considering the possibility of either owning or leasing the truck. The plant engineer has compiled the following data for the management:

- The initial cost of a gas-powered truck is $20,000. The new truck would use about 8 gallons of gasoline (single shift for 8 hours per day) at a cost of $1.50 per gallon. If the truck is operated 16 hours per day, its expected life will be 4 years, and an engine overhaul at a cost of $1,500 will be required at the end of 2 years.

- The Austin Industrial Truck Company was servicing the old forklift truck, and Dallas would buy the new truck through Austin. Austin offers a service agreement to users of its trucks that provides for a monthly visit by an experienced service representative to lubricate and tune the trucks and costs $120 per month. Insurance and property taxes for the truck are $650 per year.

- The truck is classified as 5-year property under the MACRS and will be depreciated accordingly. Dallas is in the 40% marginal tax bracket; the estimated resale value of the truck at the end of 4 years will be 15% of the original cost.

- Austin also has offered to lease a truck to Dallas. Austin will maintain the equipment and guarantee to keep the truck in serviceable condition at all times. In the event of a major breakdown, Austin will provide a replacement truck, at its expense. The cost of the operating lease plan is $850 per month. The contract term is 3 years' minimum, with the option to cancel on 30 days' notice thereafter.

- Based on recent experience, the company expects that funds committed to new investments should earn at least a 12% rate of return after taxes.

Compare the cost of owning versus leasing the truck.

Discussion: We may calculate the fuel costs for two-shift operations as

$$(8 \text{ gallons/shift})(2 \text{ shifts/day})(\$1.50/\text{gallon}) = \$24.00 \text{ per day.}$$

The truck will operate 300 days per year, so the annual fuel cost will be $5,280. However, both alternatives require the company to supply its own fuel, so the fuel cost is not relevant for our decision making. Therefore, we may drop this common cost item from our calculation.

Solution

Given: Cost information as given, MARR = 12%, marginal tax rate = 40%

Find: Incremental cost of owning versus leasing the truck

(a) Incremental cost of owning the truck

   To compare the incremental cost of ownership with the incremental cost of leasing, we make the following additional estimates and assumptions.

- Step 1: The preventive-maintenance contract, which costs $120 per month (or $1,440 per year) will be adopted. With annual insurance and taxes of $650, the equivalent present worth of the after-tax O&M is

$$P_1 = (\$1,440 + \$650)(1 - 0.40)(P/A, 12\%, 4) = -\$3,809.$$

- Step 2: The engine overhaul is not expected to increase either the salvage value or the service life. Therefore, the overhaul cost ($1,500) will be expensed all at once rather than capitalized. The equivalent present worth of this after-tax overhaul expense is

$$P_2 = \$1,500(1 - 0.40)(P/F, 12\%, 2) = -\$717.$$

- Step 3: If Dallas decided to purchase the truck through debt financing, the first step in determining financing costs would be to compute the annual installments of the debt-repayment schedule. Assuming that the entire investment of $20,000 is financed at a 10% interest rate, the annual payment would be

$$A = \$20,000(A/P, 10\%, 4) = \$6,309.$$

The equivalent present worth of this loan-payment series is

$$P_3 = -\$6,309(P/A, 12\%, 4) = -\$19,163.$$

- Step 4: The interest payment each year (10% of the beginning balance) is calculated as follows:

Year	Beginning Balance	Interest Charged	Annual Payment	Ending Balance
1	$20,000	$2,000	-$6,309	$15,691
2	15,691	1,569	-6,309	10,951
3	10,951	1,095	-6,309	5,737
4	5,737	573	-6,309	0

With the 5-year MACRS depreciation schedule, the combined tax savings due to depreciation expenses and interest payments can be calculated as follows:

n	D_n	I_n	Combined Tax Savings		
1	$4,000	$2,000	$6,000(0.40)	=	$2,400
2	6,400	1,569	7,969(0.40)	=	3,188
3	3,840	1,095	4,935(0.40)	=	1,974
4	1,152	573	1,725(0.40)	=	690

Therefore, the equivalent present worth of the combined tax credit is

$$P_4 = \$2,400(P/F, 12\%, 1) + \$3,188(P/F, 12\%, 2)$$
$$+ \$1,974(P/F, 12\%, 3) + \$690(P/F, 12\%, 4)$$
$$= \$6,527.$$

- Step 5: With the estimated salvage value of 15% of the initial investment ($3,000) and with the 5-year MACRS depreciation schedule above, we compute the net proceeds from the sale of the truck at the end of 4 years as follows:

Book value	=	$20,000 − $15,392 = $4,608.
Gains (losses)	=	$3,000 − $4,608 = ($1,608).
Tax savings	=	0.40($1,608) = $643.
Net proceeds from sale	=	$3,000 + $643 = $3,643.

The present equivalent amount of the net salvage value is

$$P_5 = \$3,643(P/F, 12\%, 4) = \$2,315.$$

- Step 6: Therefore, the net incremental cost of owning the truck through 100% debt financing is thus

$$PW(12\%)_{Buy} = P_1 + P_2 + P_3 + P_4 + P_5$$
$$= -\$14,847.$$

(b) Incremental cost of leasing the truck

How does the cost of acquiring a forklift truck under the lease compare with the cost of owning the truck?

- Step 1: The lease payment is also a tax-deductible expense; however, the net cost of leasing has to be computed explicitly on an after-tax basis. The calculation of the annual incremental leasing costs is as follows.

Annual lease payments (12 months)	=	$10,200
Less 40% taxes	=	4,080
Annual net costs after taxes	=	$ 6,120

- Step 2: The total net present equivalent incremental cost of leasing is

$$PW(12\%)_{Lease} = -\$6,120(P/A, 12\%, 4)$$
$$= -\$18,589.$$

Purchasing the truck with debt financing would save Dallas $3,742 in NPW.

- Step 3: Here, we have assumed that the lease payments occur at the end of the period; however, many leasing contracts require the payments to be made at the beginning of each period. In the latter situation, we can easily modify the present worth expression of the lease expense to reflect this cash-flow timing:

$$PW(12\%)_{Lease} = -\$6,120 - \$6,120(P/A, 12\%, 3)$$
$$= -\$20,819.$$

Comments: In our example, leasing the truck appears to be more expensive than purchasing the truck with debt financing. You should not conclude, however, that leasing is always more expensive than owning. In many cases, analysis favors a lease option. The interest rate, salvage value, lease-payment schedule, and debt financing all have an important effect on decision making.

12.5 Summary

- Identifying and estimating relevant project cash flows is perhaps the most challenging aspect of engineering economic analysis. All cash flows can be organized into one of the following three categories:
  1. Operating activities
  2. Investing activities
  3. Financing activities.
- The following cash flow types account for the most common flows a project may generate:
  1. New investment and disposal of existing assets
  2. Salvage value (or net selling price)
  3. Working capital
  4. Working capital release
  5. Cash revenues/savings
  6. Manufacturing, operating, and maintenance costs
  7. Interest and loan payments
  8. Taxes and tax credits.

  In addition, although not cash flows, the following elements may exist in a project analysis, and they must be accounted for:
  1. Depreciation expenses
  2. Amortization expenses.

  Table 12.1 summarizes these elements and organizes them as investment, financing, or operating elements.
- The **income statement approach** is typically used in organizing project cash flows. This approach groups cash flows according to whether they are operating, investing, or financing functions.
- The **generalized cash flow approach** (shown in Table 12.8) to organizing cash flows can be used when a project does not change a company's marginal tax rate. The cash flows can be generated more quickly and the formatting of the

results is less elaborate than with the income statement approach. There are also analytical advantages, which we will discover in later chapters. However, the generalized approach is less intuitive and not commonly understood by business people.

Self-Test Questions

12s.1 Consider the following financial data for an investment project:

- Required capital investment at $n =$ 0: $100,000

- Project service life: 10 years

- Salvage value at $n = 10$: $15,000

- Annual revenue: $150,000

- Annual O&M costs (not including depreciation): $50,000

- Depreciation method for tax purpose: 7-year MACRS

- Income tax rate: 40%.

Determine the project cash flow at the end of year 10.

(a) $69,000   (b) $73,000
(c) $66,000   (d) $67,000.

12s.2 Suppose in Problem 12s.1 the firm borrowed the entire capital investment at 10% interest over 10 years. If the required principal and interest payments in year 10 are

- Principal payment in year 10: $14,795, and

- Interest payment in year 10: $1,480,

what would be the net cash flow at the end of year 10?

(a) $46,725   (b) $63,000
(c) $62,112   (d) $53,317.

12s.3 You are considering purchasing industrial equipment to expand one of your production lines. The equipment costs $100,000 and has an estimated service life of 6 years. Assuming that the equipment will be financed entirely from your business-retained earnings (equity funds), a fellow engineer has calculated the expected after-tax cash flows, including the salvage value, at the end of its project life are as follows:

End of Year	Net Cash Flow
0	– $100,000
1–5	500,000
6	600,000

Now you are pondering the possibility of financing the entire amount by borrowing from a local bank at 12% interest. You can make an arrangement to pay only the interest each year over the project period by deferring the principal payment until the end of 6 years. Your firm's interest rate is also 12%. The expected marginal income tax rate over the project period is known to be 40%. What is the amount of economic gain (or loss) in present worth by using debt financing over equity financing?

(a) $0   (b) $18,000
(c) $19,735   (d) $22,350.

12s.4 A corporation is considering purchasing a machine that will save $130,000 per year before taxes. The cost of operating the machine, including maintenance, is $20,000 per year. The machine will be needed for 4 years, after which it will have a zero salvage value. MACRS depreciation will be used assuming a 3-year class. The allowed depreciation amount as a percentage of the initial cost will be 33.33%, 44.45%, 14.81%, and 7.41%, respectively. If the firm wants a 12% rate of return after taxes, how much can it afford to pay for this machine? The firm's income tax rate is 40%.

(a) $295,582   (b) $350,800

(c) $400,750   (d) $435,245.

12s.5 Your company needs an air compressor and has narrowed the choice to two alternatives, A and B. The following financial data have been collected:

	Model A	Model B
First cost	$25,000	$35,000
Annual O&M cost	5,600	3,500
Salvage value	0	4,000
Service life	10 years	10 years
Depreciation	5-year MACRS	5-year MACRS

The marginal tax rate is 40%. Which of the following statements is incorrect? Assume the MARR = 20%.

(a) Select Model A because you can save $440 annually.

(b) Select Model A because its incremental rate of return (Model A – Model B) exceeds 20%.

(c) Select Model A because you can save $1,845 in present worth.

(d) Select Model A because its incremental rate of return (Model B – Model A) is 14.12%, which is less than 20%.

12s.6 You are planning to lease an automobile for 36 months from an auto dealer. The negotiated price is $20,000. The re-

quired security deposit is $500 at the time of the lease and will be refunded in full at the end of the lease. The monthly lease payment at the end of each month is calculated to be $300. What kind of residual value (salvage value) was assumed by the dealer in calculating the monthly lease? The dealer's interest rate is known to be 9%, compounded monthly.

(a) $9,200   (c) $13,674

(b) $12,124   (d) $14,481.

12s.7 A corporation is considering purchasing a machine that costs $120,000 and will save $X per year after taxes. The cost of operating the machine, including maintenance and depreciation, is $20,000 per year after taxes. The machine will be needed for 4 years, after which it will have a zero salvage value. If the firm wants a 14% rate of return after taxes, what is the minimum after tax annual savings that must be generated to realize a 14% rate of return after taxes?

(a) $50,000   (c) $91,974

(b) $61,184   (d) $101,974.

12s.8 Cutter Ltd. is planning to invest $150,000 in an automated screw cutting machine to enhance its current operations. Due to a lack of internal funds, its management is planning to borrow 60% of the investment from a local bank at an interest rate of 10% payable in 5 equal payments. What is the principal payment in year 1 that should be included in the cash flow analysis?

(a) $14,740   (c) $18,000

(b) $14,940   (d) $24,570.

12s.9 A corporation is considering purchasing a machine that will save $200,000 per year before taxes. The cost of operating the machine, including maintenance, is $80,000 per year. The machine will be needed for 5 years, after which it will have a zero salvage value. A straight-line depreciation with no half-year convention applies (i.e., 20% each year). If the firm wants 15% rate of return after taxes, how much can they afford to pay for this machine? The firm's income tax rate is 40%.

(a) $600,000 (c) $441,538
(b) $518,313 (d) $329,821.

Consider the following financial data for an investment project:

- Required capital investment at $n = 0$: $200,000
- Project service life: 5 years
- Salvage value at the end of 5 years: $50,000
- Depreciation method for tax purposes: 5-year MACRS
- Annual revenue: $300,000
- Annual O&M expenses (not including depreciation and interest): $180,000
- Required investment in working capital at $n = 0$ (which will be recovered in full at the end of project year): $40,000
- The income tax rate to use: 40%

12s.10 Determine the project cash flow at the end of year 5.

(a) $115,088 (c) $155,824
(b) $115,824 (d) $144,304.

12s.11 In 12s.10, 50% of the capital investment will be financed at 10% interest over 5 years. The contract calls for equal annual interest payment for each period and one lump sum principal payment at the end of project year. What is the cash flow in year 2?

(a) $91,600 (c) $97,600
(b) $85,600 (d) $27,600.

Problems

Generating Net Cash Flows

12.1 Tampa Construction Company builds residential solar homes. Because of anticipated increase in business volume, the company is considering acquisition of a loader at a cost of $54,000. This acquisition cost includes delivery charges and applicable taxes. The firm has estimated that if the loader is acquired, the following additional revenues and operating expenses (excluding depreciation) should be expected:

End of Year	Additional Operating Revenue	Additional Operating Expenses Excluding Depreciation	Allowed Tax Depreciation
1	$66,000	$29,000	$10,800
2	70,000	28,400	17,280
3	74,000	32,000	10,368
4	80,000	38,800	6,221
5	64,000	31,000	6,221
6	50,000	25,000	3,110

The projected revenue is assumed to be in cash in the year indicated, and all the additional operating expenses are expected to be paid in the year in which they are incurred. The estimated salvage value for the loader at the end of sixth year is $8,000. The firm's incremental (marginal) tax rate is 35%. What is the after-tax cash flow if the loader is acquired? Using this data, (a) develop the cash flow statement based on net income (Approach 1 as shown in Table 12.2) and (b) develop the cash flow statement based on the conventional tabular method (Approach 2 as shown in Table 12.3).

12.2 An automobile manufacturing company is considering the purchase of an industrial robot to do spot welding, which is currently done by skilled labor. The initial cost of the robot is $235,000, and the annual labor savings are projected to be $122,000. If purchased, the robot will be depreciated under MACRS as a 5-year recovery property. This robot will be used for 7 years, at the end of which time the firm expects to sell the robot for $50,000. The company's marginal tax rate is 38% over the project period. Determine the net after-tax cash flows for each period over the project life.

12.3 A Los Angeles company is planning to market an answering device for people working alone, who want the prestige that comes with having a secretary, but who cannot afford one. The device, called Tele-Receptionist, is similar to a voice-mail system. It uses digital recording technology to create the illusion that a person is operating the switchboard at a busy office. The company purchased a 40,000-ft^2 building and converted it to an assembly plant for $600,000 ($100,000 of land and $500,000 worth of building). Installation of the assembly equipment worth $500,000 was completed on December 31. The plant will begin operation on January 1. The company expects to have a gross annual income of $2,500,000 over the next 5 years. Annual manufacturing costs, and all other operating expenses (excluding depreciation), are projected to be $1,280,000. For depreciation purposes, the assembly plant building will be classified as 39-year real property and the assembly equipment as a 7-year MACRS property. The property value of the land and the building at the end of year 5 would appreciate as much as 15% over the initial purchase cost. The residual value of the assembly equipment is estimated to be about $50,000 at the end of year 5. The firm's marginal tax rate is expected to be about 40% over the project period. Determine the project's after-tax cash flows over the period of 5 years.

12.4 A highway contractor is considering buying a new trench excavator that costs $200,000 and can dig a 3-foot wide trench at the rate of 16 feet per hour. With adequate maintenance, the production rate will remain constant for the first 1,200 hours of operation, then decrease by 2 feet per hour for each additional 400 hours thereafter. The expected average annual use is 400 hours, and maintenance and operating costs will be $15 per hour. The contractor will depreciate the equipment using a 5-year MACRS. At the end of 5 years, the excavator will be sold for $40,000. Assuming that the contractor's marginal tax rate is 34% per year, determine the annual after-tax cash flow.

12.5 A small children's clothing manufacturer is considering an investment to computerize its management information system for material requirement planning, piece-goods coupon printing, and invoice and payroll. An outside consultant has been asked to estimate the initial hardware requirement and installation costs. He suggests the following:

The expected life of these computer systems is 6 years, with an estimated salvage value of $1,000. The proposed system is classified as a 5-year property class under the MACRS depreci-

PC systems (10 PCs, 4 printers)	$65,000
Local area networking system	15,000
System installation and testing	4,000

ation system. A group of computer consultants need to be hired to develop various customized software packages to run on these systems. Software development costs will be $20,000 and can be expensed during the first tax year. The new system will eliminate two clerks, whose combined annual payroll expenses would be $52,000. Additional annual expenses to run this computerized system are expected to be $12,000. Borrowing is not considered an option for this investment. No tax credit is available for this system. The firm's expected marginal tax rate over the next 6 years will be 40%. The firm's interest rate is 18%. Compute the after-tax cash flows over the investment's life by using the income statement approach.

12.6 The Manufacturing Division of Ohio Vending Machine Company is considering its Toledo Plant's request for a half-inch-capacity automatic screw-cutting machine to be included in the division's 2001 capital budget:

- Name of project : Mazda Automatic Screw Machine
- Project cost : $68,701
- Purpose of project: To reduce the cost of some of the parts that are now being subcontracted by this plant, to cut down on inventory due to a shorter lead time, and to better control the quality of the parts. The proposed equipment includes the following cost basis:

Machine cost	$48,635
Accessory cost	8,766
Tooling	4,321
Freight	2,313
Installation	2,110
Sales tax	2,556
Total Cost	$68,701

- Anticipated savings: As shown in the table below
- Tax depreciation method: 7-year MACRS
- Marginal tax rate: 40%
- MARR: 15%.

(a) Determine the net after-tax cash flows over the project life of 6 years. Assume a salvage value of $3,500.

(b) Is this project acceptable based on the NPW criterion?

(c) Determine the IRR for this investment.

Item	Hours Present M/C labor	Proposed M/C labor	Present Method	Proposed Method
Setup		350		$7,700
Run	2,410	800		17,600
Overhead				
Indirect labor				3,500
Fringe benefits				8,855
Maintenance				1,350
Tooling				6,320
Repair				890
Supplies				4,840
Power				795
Taxes and insurance				763
Other relevant costs				
Floor space				3,210
Subcontracting			$122,468	
Material				27,655
Other				210
Total			$122,468	$83,688
Operating advantage				$38,780

12.7 A firm has been paying a print shop $18,000 annually to print the company's monthly newsletter. The agreement with this print shop has now expired, but it could be renewed for a further 5 years. The new subcontracting charges are expected to be 12% higher than they were in the previous contract. The company is also considering the purchase of a desktop publishing system with a high-quality laser printer driven by a microcomputer. With appropriate word/graphics software, the newsletter can be composed and printed in near typeset quality. A special device is also required to print photos in the newsletter. The following estimates have been quoted by a computer vendor:

Microcomputer	$5,500
Laser printer	8,500
Photo device/scanner	10,000
Software	2,000
Total cost basis	$26,000
Annual O&M costs	10,000

The salvage value of each piece of equipment at the end of 5 years is expected to retain only 10% of the original cost. The company's marginal tax rate is 40%, and the desktop publishing system will be depreciated by MACRS under 5-year property class.

(a) Determine the projected net after-tax cash flows for the investment.

(b) Compute the IRR for this project.

(c) Is this project justifiable at MARR = 12%?

12.8 An asset in the 5-year MACRS property class cost $100,000 and has a zero estimated salvage value after 6 years of use. The asset will generate annual revenues of $300,000 and will require $100,000 in annual labor and $50,000 in annual material expenses. There are no other revenues and expenses. Assume a tax rate of 40%.

(a) Compute the after-tax cash flows over the project life.

(b) Compute the NPW at MARR = 12%. Is this an acceptable investment?

12.9 American Aluminum Company is considering making a major investment of $150 million ($5 million for land, $45 million for buildings, and $100 million for manufacturing equipment and facilities) to develop stronger, lighter materials, called aluminum lithium, which will make aircraft sturdier and more fuel efficient. Aluminum lithium, which has been sold commercially for only a few years as an alternative to composite materials, will likely be the material of choice for the next generation of commercial and military aircraft because it is so much lighter than conventional aluminum alloys, which use a combination of copper, nickel, and magnesium to harden aluminum. Another advantage of aluminum lithium is that it is cheaper than composites. The firm predicts that aluminum lithium will account for about 5% of the structural weight of the average commercial aircraft within 5 years and 10% within 10 years. The proposed plant, which has an estimated service life of 12 years, would have a capacity of about 10 million pounds of aluminum lithium, although domestic consumption of the material is expected to be only 3 million pounds during the first 4 years, 5 million for the next 3 years, and 8 million for the remaining plant life. Aluminum lithium costs $12 a pound to produce, and the firm would expect to sell it at $17 a pound. The building will be depreciated according to the 39-year MACRS real property class with the building placed in service in July 1

during the first year. All manufacturing equipment and facilities will be classified as 7-year MACRS property. At the end of project life, the land will be worth $8 million, the building $30 million, and the equipment $10 million. Assuming that the firm's marginal tax rate is 40%, and its capital gains tax rate is 35%,

(a) Determine the net after-tax cash flows.

(b) Determine the IRR for this investment.

(c) Determine whether the project is acceptable if firm's MARR is 15%.

12.10 An automaker is considering installing a three-dimensional (3-D) computerized car-styling system at a cost of $200,000 (including hardware and software). With the 3-D computer modeling system, designers will have the ability to view their design from many angles and to fully account for the space required for the engine and passengers. The digital information used to create the computer model can be revised in consultation with engineers, and the data can be used to run milling machines that make physical models quickly and precisely. The automaker expects to decrease turnaround time by 22% for designing a new automobile model (from configuration to final design). The expected savings in dollars is $250,000 per year. The training and operating maintenance cost for the new system is expected to be $50,000 per year. The system has a 5-year useful life and can be depreciated according to 5-year MACRS class. The system will have an estimated salvage value of $5,000. The automaker's marginal tax rate is 40%. Determine the annual cash flows for this investment. What is the return on investment for this project?

12.11 A facilities engineer is considering a $50,000 investment in an energy management system (EMS). The system is expected to save $10,000 annually in utility bills for N years. After N years, the EMS will have a zero salvage value. In an after-tax analysis, how many years would N need to be to earn 10% return on the investment? Assume MACRS depreciation with a 3-year class life and a 35% tax rate.

12.12 A corporation is considering purchasing a machine that will save $130,000 per year before taxes. The cost of operating the machine, including maintenance, is $20,000 per year. The machine will be needed for 5 years after which it will have a zero salvage value. MACRS depreciation will be used, assuming a 3-year class life. The marginal income tax rate is 40%. If the firm wants 12% IRR after taxes, how much can it afford to pay for this machine?

12.13 Ampex Corporation produces a wide variety of tape cassettes for commercial and government markets. Due to increased competition in VHS cassette production, Ampex is concerned that it price its product competitively. Currently, Ampex has 18 loaders that load cassette tapes in half-inch VHS cassette shells. Each loader is manned by one operator per shift. Ampex currently produces 25,000 half-inch tapes per week and operates 15 shifts per week, 50 weeks per year.

As a means of reducing the unit cost, Ampex can purchase cassette shells for $0.15 less (each) than it can currently produce them. A supplier has guaranteed a price of $0.77 per cas-

sette for the next 3 years. However, Ampex's current loaders will not be able to load the proposed shells properly. However, to accommodate the vendor's shells, Ampex must purchase eight KING-2500 VHS loaders at a cost of $40,000 each. For these new machines to operate properly, Ampex must also purchase $20,827 worth of conveyor equipment, the cost of which will be included in the overall depreciation base of $340,827. The new machines are much faster and will handle more than the current demand of 25,000 cassettes per week. The new loaders will require two people per machine per shift, three shifts per day, 5 days a week. The new machines will fall into a 7-year MACRS equipment class and will have an approximate life of 8 years. At the end of the project life, Ampex expects the market value for each loader to be $3,000.

The average pay of the needed new employees is $8.27 per hour, and adding 23% for benefits will make it $10.17 per hour. The new loaders are simple to operate; therefore, the training impact of the alternative is minimal. The operating cost, including maintenance, is expected to stay the same for the new loaders. This cost will not be considered in the analysis. The cash inflows from the project will be a material savings per cassette of $0.15, and the labor savings of two employees per shift. This gives an annual savings in materials and labor costs of $187,500 and $122,065, respectively. If the new loaders are purchased, the old machines will be shipped to other plants for standby use. Therefore, no gains will be realized. Ampex's combined marginal tax rate is running at 40%.

(a) Determine the after-tax cash flows over the project life.

(b) Determine the IRR for this investment.

(c) Is this investment profitable at MARR = 15%?

Investment in Working Capital

12.14 The Motch Machinery Company is planning to expand its current spindle product line. The required machinery would cost $500,000. The building to house the new production facility would cost $1.5 million. The land would cost $250,000, and $150,000 working capital would be required. The product is expected to result in additional sales of $675,000 per year for 10 years, at which time the land can be sold for $500,000, the building for $700,000, and the equipment for $50,000. All of the working capital will be recovered. The annual disbursements for labor, materials, and all other expenses are estimated to be $425,000. The firm's income tax rate is 40%, and any capital gains will be taxed at 35%. The building will be depreciated according to a 39-year property class. The manufacturing facility will be classified as a 7-year MACRS. The firm's MARR is also known to be 15% after taxes.

(a) Determine the projected net after-tax cash flows from this investment. Is the expansion justified?

(b) Repeat (a) above using the traditional tabular method (Approach 2).

(c) Compare the IRR of this project with that of a situation with no working capital.

12.15 An industrial engineer at Lagrange Textile Mill proposed the purchase of scanning equipment for the company's warehouse and weave rooms. The engineer felt that the purchase would ensure a better system of locating cartons in the warehouse by recording the location of the cartons and storing this data in the computer. The estimated investment, annual operating and maintenance costs, and expected annual savings are as follows:

- Cost of equipment and installation: $44,500
- Project life: 6 years
- Expected salvage value: $0
- Investment in working capital (fully recoverable at the end of project life): $10,000
- Expected annual labor and materials savings: $62,800
- Expected annual expenses: $8,120
- Depreciation method: 5-year MACRS.

The firm's marginal tax rate is 35%.

(a) Determine the net after-tax cash flows over the project life.

(b) Compute the IRR for this investment.

(c) At MARR = 18%, is this project acceptable?

12.16 Delaware Chemical Corporation is considering investing in a new composite material. R&D engineers are investigating exotic metal-ceramic and ceramic-ceramic composites to develop materials that will withstand high temperatures, such as those to be encountered in the next generation of jet fighter engines. The company expects a 3-year R&D period before these new materials can be applied to commercial products. The following financial information is presented for management review:

- R&D cost: $5 million over a 3-year period. Annual R&D expenditure of $0.5 million at the beginning of year 1, $2.5 million at the beginning of year 2, and $2 million at the beginning of year 3. These R&D expenditures will be expensed rather than amortized for tax purposes.
- Capital investment: $5 million at the beginning of year 4. This investment consists of $2 million in a building, and $3 million in plant equipment. The company already owns a piece of land as the building site.
- Depreciation Method: The building (39-year real property class with the asset placed in service in January) and plant equipment (7-year MACRS recovery class).
- Project life: 10 years after a 3-year R&D period.
- Salvage value: 10% of the initial capital investment for the equipment and 50% for the building (at the end of project life).
- Total sales: $50 million (at the end of year 4) with an annual sales growth rate of 10% per year (compound growth) during the next 5 years (year 5 through year 9) and −10% (negative compound growth) per year for the remaining process life.
- Out-of-pocket expenditures: 80% of annual sales.
- Working capital: 10% of annual sales (considered as investment at the beginning of each production year and investments fully recovered at the end of project life).
- Marginal tax rate: 40%.

(a) Determine the net after-tax cash flows over the project life based on net income (Approach 1) and the traditional tabular method (Approach 2).

(b) Determine the IRR for this investment.

(c) Determine the equivalent annual worth for this investment at MARR = 20%.

Effects of Borrowing

12.17 Refer to the data for the Tampa Construction Company in Problem 12.1. If the firm expects to borrow the initial investment ($54,000) at 10% over 2 years (equal annual payments of $31,114), determine the project's net cash flows.

12.18 In Problem 12.2, to finance the industrial robot, the company will borrow the entire amount from a local bank, and the loan will be paid off at the rate of $50,000 per year, plus 12% on the unpaid balance. Determine the net after-tax cash flows over the project life.

12.19 Refer to the data for the children's clothing company in Problem 12.5. Suppose that the initial investment of $84,000 will be borrowed from a local bank at an interest rate of 11% over 5 years (five equal annual payments). Recompute the after-tax cash flow.

12.20 Ann Arbor Die Casting Company is considering the installation of a new process machine for their manufacturing facility. The machine costs $250,000 installed, will generate additional revenues of $80,000 per year, and will save $50,000 per year in labor and material costs. The machine will be financed by a $150,000 bank loan repayable in three equal, annual principal installments, plus 9% interest on the outstanding balance. The machine will be depreciated using 7-year MACRS. The useful life of this process machine is 10 years, at which time it will be sold for $20,000. The combined marginal tax rate is 40%.

The amount comprises 34% federal taxes plus state and local income taxes.

(a) Find the year-by-year after-tax cash flow for the project using the income statement approach (Approach 1 as shown in Table 12.2).

(b) Repeat (a) above using the conventional tabular method (Approach 2 as shown in Table 12.3).

(c) Compute the IRR for this investment.

(d) At MARR = 18%, is this project economically justifiable?

12.21 Consider the following financial information about a retooling project at a computer manufacturing company:

- The project costs $2 million and has a 5-year service life.
- The retooling project can be classified as a 7-year property under the MACRS rule.
- At the end of fifth year, any assets held for the project will be sold. The expected salvage value will be about 10% of the initial project cost.
- The firm will finance 40% of the project money from an outside financial institution at an interest rate of 10%. The firm is required to repay the loan with five equal annual payments.
- The firm's incremental (marginal) tax rate on this investment is 35%.
- The firm's MARR is 18%.
- With the financial information above,

(a) Determine the after-tax cash flows.

(b) Compute the annual equivalent worth for this project.

12.22 An auto-part manufacturing plant wants to add two delivery trucks to its current fleet system. The trucks would cost $50,000 each, and the firm expects to use these trucks for 5 years. The salvage value of the trucks at the end of year 5 would be about $5,000 each. The system qualifies for the 5-year class property under the MACRS. The 40% of the initial cost ($40,000) will be financed from a local bank at an annual interest rate of 12% over 3 years. The borrowed money will be repaid in three equal annual payments. These trucks will speed up auto-part delivery to customers, and the firm expects additional revenue over the next 5 years. The firm already computed the projected net income and recorded it as shown in the cash flow statement below.

Cash Flow Elements	End of Period					
	0	1	2	3	4	5
Cash flow operation						
Net income		30,000	40,000	42,500	40,000	35,800
Depreciation		20,000	☐	19,200	☐	☐
Investment/ salvage	−100,000					10,000
Gains tax						☐
Loan/ principal repayment	+40,000	☐	☐	☐		
Net cash flow	☐	☐	☐	☐	☐	☐

(a) Complete the remaining cash flow statement based on the information above, assuming a tax rate of 40%.

(b) Compute the NPW at $i = 18\%$.

(c) Compute the IRR for this project.

12.23 A fully automatic chucker and bar machine is to be purchased for $35,000. This amount is to be borrowed with the stipulation that it be repaid by six equal end-of-year payments at 12%, compounded annually. The machine is expected to provide an annual revenue of $10,000 for 6 years and is to be depreciated by the

MACRS 7-year recovery period. The salvage value at the end of 6 years is expected to be $3,000. Assuming a marginal tax rate of 36% and a MARR of 15%.

(a) Determine the after-tax cash flow for this asset through 6 years.

(b) Determine if this project is acceptable based on the IRR criterion.

12.24 A manufacturing company is considering the acquisition of a new injection molding machine at a cost of $100,000. Because of a rapid change in product mix, the need for this particular machine is expected to last only 8 years, after which time the machine is expected to have a salvage value of $10,000. The annual operating cost is estimated to be $5,000. The addition of this machine to the current production facility is expected to generate an annual revenue of $40,000. The firm has only $60,000 available from its equity funds, so it must borrow the additional $40,000 required at an interest rate of 10% per year with repayment of principal and interest in eight equal annual amounts. The applicable marginal income tax rate for the firm is 40%. Assume that the asset qualifies for a 7-year MACRS property class.

(a) Determine the after-tax cash flows.

(b) Determine the NPW of this project at MARR = 14%.

Generalized Cash Flow Method

12.25 Suppose an asset has a first cost of $6,000, a life of 5 years, a salvage value of $2,000 at the end of 5 years, and a net annual before-tax revenue of $1,500. The firm's marginal tax rate is 35%. The asset will be depreciated by 3-year MACRS.

(a) Using the generalized cash flow approach, determine the cash flow after taxes.

(b) Rework part (a) assuming that the entire investment would be financed by a bank loan at an interest rate of 9%.

(c) Given a choice between the financing methods of parts (a) and (b), show calculations to justify your choice of which is the better one at an interest rate of 9%.

12.26 A construction company is considering the proposed acquisition of a new earthmover. The purchase price is $100,000, and an additional $25,000 is required to modify the equipment for special use by the company. The equipment falls into the MACRS 7-year classification (tax life), and it will be sold after 5 years (project life) for $50,000. Purchase of the earthmover will have no effect on revenues, but it is expected to save the firm $60,000 per year in before-tax operating costs, mainly labor. The firm's marginal tax rate is 40%. Assume that the initial investment is to be financed by a bank loan at an interest rate of 10%, payable annually. Determine the after-tax cash flows by using the generalized cash flow approach and the worth of investment for this project if the firm's MARR is known to be 12%.

12.27 Air South, a leading regional airline that is now carrying 54% of all the passengers that pass through the southeast, is considering the possibility of adding a new long-range aircraft to its fleet. The aircraft being considered for purchase is the McDonnell Douglas DC-9-532 "Funjet," which is quoted at $60 million per unit. McDonnell Douglas requires 10% down payment at the time of delivery, and the balance is to be paid over a 10-year period at an interest rate of 12% compounded annually. The actual payment schedule calls for only interest payments over the 10-year period, with the original principal amount to be paid off at the end of the tenth year. Air South expects to generate $35 million per year by adding this aircraft to its current fleet but also estimates an operating and maintenance cost of $20 million per year. The aircraft is expected to have a 15-year service life with a salvage value of 15% of the original purchase price. If the aircraft is bought, it will be depreciated by the 7-year MACRS property classifications. The firm's combined federal and state marginal tax rate is 38%, and its required minimum attractive rate of return is 18%.

(a) Determine the cash flow associated with the debt financing using the generalized cash flow approach.

(b) Is this project acceptable?

Comparing Mutually Exclusive Alternatives

12.28 The Pittsburgh Division of Vermont Machinery Inc. manufactures drill bits. One of production processes of a drill bit requires tipping, where carbide tips are inserted into the bit to make it stronger and more durable. This tipping process usually requires four or five operators depending on the weekly work load. The same operators were assigned to the stamping operation, in which the size of the drill bit and the company's "logo" are imprinted into the bit. Vermont is considering acquiring three automatic tipping machines to replace the manual tipping and stamping operations. If the tipping process is automated,

Vermont engineers will have to redesign the shapes of the carbide tips to be used in the machine. The new design requires less carbide, resulting in material savings. The following financial data has been compiled:

- Project life: 6 years.
- Expected annual savings: reduced labor, $56,000; reduced material, $75,000; other benefits (reduced carpal tunnel syndrome and related problems), $28,000; reduced overhead, $15,000.
- Expected annual O&M costs: $22,000.
- Tipping machines and site preparation: equipment (three machines) costs including delivery, $180,000; site preparation, $20,000.
- Salvage value: $30,000 (three machines) at the end of 6 years.
- Depreciation method: 7-year MACRS.
- Investment in working capital: $25,000 at the beginning of the project year, and that same amount will be fully recovered at the end of project year.
- Other accounting data: Marginal tax rate of 39%, MARR of 18%.

To raise $200,000, Vermont is considering the following financing options:

- Option 1: Finance the tipping machines using their retained earnings.
- Option 2: Secure a 12% term loan over 6 years (six equal annual installments).
- Option 3: Lease the tipping machines. Vermont can obtain a 6-year financial lease on the equipment (no maintenance) for payments of $55,000 at the *beginning* of each year.

(a) Determine the net after-tax cash flows for each financing option.

(b) What is Vermont's present value cost of owning the equipment by borrowing?

(c) What is Vermont's present value cost of leasing the equipment?

(d) Recommend the best course of action for Vermont.

12.29 The headquarters building owned by a rapidly growing company is not large enough for current needs. A search for enlarged quarters revealed two new alternatives that would provide sufficient room, enough parking, and the desired appearance and location.

- Option 1: Lease for $144,000 per year.
- Option 2: Purchase for $800,000, including a $150,000 cost for land.
- Option 3: Remodel the current headquarters building.

It is believed that land values will not decrease over the ownership period, but the value of all structures will decline to 10% of the purchase price in 30 years. Annual property tax payments are expected to be 5% of the purchase price. The present headquarters building is already paid for and is now valued at $300,000. The land it is on is appraised at $60,000. The structure can be remodeled at a cost of $300,000 to make it comparable to other alternatives. However, the remodeling will occupy part of the existing parking lot. An adjacent, privately owned parking lot can be leased for 30 years under an agreement that the first year's rental of $9,000 will increase by $500 each year. The annual property taxes on the remodeled property will again be 5% of the present valuation plus the cost to remodel. The study period for the comparison is 30 years, and the de-

sired rate of return on investments is 12%. Assume that the firm's marginal tax rate is 40% and the new building and remodeled structure will be depreciated under MACRS using a real property recovery period of 39 years. If the annual upkeep costs are the same for all three alternatives, which one is preferable?

12.30 An international manufacturer of prepared food items needs 50,000,000 kWh of electrical energy a year with maximum demand of 10,000 kW. The local utility presently charges $0.085 per kWh, a rate considered high throughout the industry. Because the firm's power consumption is so large, its engineers are considering installing a 10,000 kW steam-turbine plant. Three types of plant have been proposed ($ units in thousand):

	Plant A	Plant B	Plant C
Average station heat rate (BTU/kWh)	16,500	14,500	13,000
Total investment (boiler/turbine/ electrical/ structures)	$8,530	$9,498	$10,546
Annual operating cost			
Fuel	1,128	930	828
Operating labor	616	616	616
Maintenance	150	126	114
Supplies	60	60	60
Insurance and property taxes	10	12	14

The service life of each plant is expected to be 20 years. The plant investment will be subject to a 20-year MACRS property classification. The expected salvage value of the plant at the end of its useful life is about 10% of its original investment. The firm's MARR is known to be 12%. The firm's marginal income tax rate is 39%.

(a) Determine the unit power cost ($/kWh) for each plant.

(b) Which plant would provide the most economical power?

Lease-versus-Buy Decisions

12.31 The Jacob Company needs to acquire a new lift truck for transporting its final product to the warehouse. One alternative is to purchase the lift truck for $40,000, which will be financed by the bank at an interest rate of 12%. The loan must be repaid in four equal installments, payable at the end of each year. Under the borrow-to-purchase arrangement, Jacob would have to maintain the truck at a cost of $1,200 payable at year-end. Alternatively, Jacob could lease the truck on a 4-year contract for a lease payment of $11,000 per year. Each annual lease payment must be made at the beginning of each year. The truck would be maintained by the lessor. The truck falls into the 5-year MACRS classification, and it has a salvage value of $10,000, which is the expected market value after 4 years, at which time Jacob plans to replace the truck irrespective of whether it leases or buys. Jacob has a marginal tax rate of 40% and a MARR of 15%.

(a) What is Jacob's cost of leasing in present worth?

(b) What is Jacob's cost of owning in present worth?

(c) Should the truck be leased or purchased?

This is an operating lease so that the truck would be maintained by the lessor.

12.32 Janet Wigandt, an electrical engineer for Instrument Control, Inc. (ICI), has been asked to perform a lease-buy analysis on a new pin-inserting machine for its pc-board manufacturing.

- Buy Option: The equipment costs $120,000. To purchase it, ICI could obtain a term loan for the full amount at 10% of interest with four equal annual installments (end-of-year payment). The machine falls into a 5-year MACRS property classification. Annual revenues of $200,000 and operating costs of $40,000 are anticipated. The machine requires annual maintenance at a cost of $10,000. Because technology is changing rapidly in pin-inserting machinery, the salvage value of the machine is expected to be only $20,000.

- Lease Option: Business Leasing, Inc. (BLI) is willing to write a 4-year operating lease on the equipment for payments of $44,000, at the beginning of each year. Under this operating-lease arrangement, BLI will maintain the asset, so that the maintenance cost of $10,000 will be saved annually. ICI's marginal tax rate is 40%, and its MARR is 15% during the analysis period.

(a) What is ICI's present value (incremental) cost of owning the equipment?

(b) What is ICI's present value (incremental) cost of leasing the equipment?

(c) Should ICI buy or lease the equipment?

12.33 Consider the following lease versus borrow-and-purchase problem:

- Borrow-and-Purchase Option:

  1. Jensen Manufacturing Company plans to acquire sets of special industrial tools with a 4-year life and a cost of $200,000, delivered and installed. The tools will be depreciated by the MACRS 3-year classification.

  2. Jensen can borrow the required $200,000 at a rate of 10% over 4 years. Four equal annual payments (end-of-year) would be made in the amount of $63,094 = $200,000(A/P, 10%,4). The annual interest and principal payment schedule along with the equivalent present worth of these payments is as follows:

End of Year	Interest	Principal
1	$20,000	$43,094
2	15,961	47,403
3	10,950	52,144
4	5,736	57,358

  3. The estimated salvage value for the tool sets at the end of 4 years is $20,000.

  4. If Jensen borrows and buys, it will have to bear the cost of maintenance, which will be performed by the tool manufacturer at a fixed contract rate of $10,000 per year.

- Lease Option:

  1. Jensen can lease the tools for 4 years at an annual rental charge of $70,000, payable at the end of each year.

  2. The lease contract specifies that the lessor will maintain the tools at no additional charge to Jensen.

Jensen's tax rate is 40%. Any gains will also be taxed at 40%.

(a) What is Jensen's PW of after-tax cash flow of leasing at $i = 15\%$?

(b) What is Jensen's PW of after-tax cash flow of owning at $i = 15\%$?

12.34 Tom Hagstrom has decided to acquire a new car for his business. One alternative is to purchase the car outright for $16,170, financing with a bank loan for the net purchase price. The bank loan calls for 36 equal monthly payments of $541.72 at an interest rate of 12.6% compounded monthly. Payments must be made at the end of each month.

Buy or Lease Your New '01	
Buy	**Lease**
$16,170	$425 per month 36-month open end lease Annual mileage allowed = 15,000 miles

If Tom takes the lease option, he is required to pay $500 for a security deposit refundable at the end of the lease, and $425 a month at the beginning of each month for 36 months. If the car is purchased, it will be depreciated according to a 5-year MACRS property classification. It has a salvage value of $5,800, which is the expected market value after 3 years, at which time Tom plans to replace the car irrespective of whether he leases or buys. Tom's marginal tax rate is 35%. His MARR is known to be 13% per year.

(a) Determine the annual cash flows for each option.

(b) Which option is a better choice?

12.35 The Boggs Machine Tool Company has decided to acquire a pressing machine. One alternative is to lease the machine on a 3-year contract for a lease payment of $15,000 per year, with payments to be made at the beginning of each year. The lease would include maintenance. The second alternative is to purchase the machine outright for $100,000, financing with a bank loan for the net purchase price and amortizing the loan over a 3-year period at an interest rate of 12% per year (annual payment = $41,635).

Under the borrow-to-purchase arrangement, the company would have to maintain the machine at an annual cost of $5,000, payable at year-end. The machine falls into 5-year MACRS classification; and it has a salvage value of $50,000, which is the expected market value at the end of year 3, at which time the company plans to replace the machine irrespective of whether it leases or buys. Boggs has a tax rate of 40% and a MARR of 15%.

(a) What is Boggs's PW cost of leasing?

(b) What is Boggs's PW cost of owning?

(c) From the financing analysis in (a) and (b) above, what are the advantages and disadvantages of leasing and owning?

12.36 An asset is to be purchased for $25,000. The asset is expected to provide revenue of $10,000 a year, and have operating costs of $2500 a year. The asset is considered to be a 7-year MACRS property. The company is planning to sell the asset at the end of year 5 for $5,000. Given that the company's marginal tax rate is 30% and that it has a MARR of 10% for any project undertaken, answer the following questions:

(a) What is the net cash flow for each year given the asset is purchased

with borrowed funds at an interest rate of 12% with repayment in five equal end-of-year payments?

(b) What is the net cash flow for each year, given the asset is leased at a rate of $3,500 a year (financial lease)?

(c) Which method (if either) should be used to obtain the new asset?

12.37 Enterprise Capital Leasing Company is in the business of leasing tractors for construction companies. The firm wants to set a 3-year lease payment schedule for a tractor purchased at $53,000 from the equipment manufacturer. The asset is classified as a 5-year MACRS property. The tractor is expected to have a salvage value of $22,000 at the end of 3-years' rental. Enterprise will require a lessee to make a security deposit in the amount of $1,500 that is refundable at the end of the lease term. Enterprise's marginal tax rate is 35%. If Enterprise wants an after-tax return of 10%, what lease-payment schedule should be set?

Short Case Studies

12.38 Morgantown Mining Company is considering a new mining method at its Blacksville mine. The method, called longwall mining, is carried out by a robot. Coal is removed by the robot, not by tunneling like a worm through an apple, which leaves more of the target coal than is removed, but rather by methodically shuttling back and forth across the width of the deposit and devouring nearly everything. The method can extract about 75% of the available coal, compared with 50% for conventional mining, which is largely done with machines that dig tunnels. Moreover, the coal can be recovered far more inexpensively. Currently, at Blacksville alone, the company mines 5 million tons a year with 2,200 workers. By installing two longwall robot machines, the company can mine 5 million tons with only 860 workers. (A robot miner can dig more than 6 tons a minute.) Despite the loss of employment, the United Mine Workers union generally favors longwall mines for two reasons: The union officials are quoted as saying, (1) "It would be far better to have highly productive operations that were able to pay our folks good wages and benefits than to have 2,200 shovelers living in poverty," and (2) "Longwall mines are inherently safer in their design." The company projects the following financial data upon installation of the longwall mining:

Robot installation (2 units)	$19.3 million
Total amount of coal deposit	50 million tons
Annual mining capacity	5 million tons
Project life	10 years
Estimated salvage value	$0.5 million
Working capital requirement	$2.5 million
Expected additional revenues:	
Labor savings	$6.5 million
Accident prevention	$0.5 million
Productivity gain	$2.5 million
Expected additional expenses:	
O&M costs	$2.4 million

(a) Estimate the firm's net after-tax cash flows over the project life, if the firm uses the unit-production method for asset depreciation. The firm's marginal tax rate is 40%.

(b) Estimate the firm's net after-tax cash flows, if the firm chooses to depreciate the robots based on MACRS (7-year property classification).

12.39 The National Parts Inc., an auto-parts manufacturer, is considering purchasing a rapid prototyping system to reduce prototyping time for form, fit, and function applications in automobile parts manufacturing. An outside consultant has been called in to estimate the initial hardware requirement and installation costs. He suggests the following:

- Prototyping equipment: $187,000
- Posturing apparatus : $10,000
- Software: $15,000
- Maintenance: $36,000 per year by the equipment manufacturer
- Resin: Annual liquid polymer consumption: 400 gallons at $350 per gallon
- Site preparation: Some facility changes are required when installing the rapid prototyping system. (Certain liquid resins contain a toxic substance, so the work area must be well vented.)

The expected life of the system is 6 years with an estimated salvage value of $30,000. The proposed system is classified as a 5-year MACRS property. A group of computer consultants must be hired to develop customized software to run on these systems. These software development costs will be $20,000 and can be expensed during the first tax year. The new system will reduce prototype development time by 75% and the material waste (resin) by 25%. This reduction in development time and material waste will save the firm $114,000 and $35,000 annually, respectively. The firm's expected marginal tax rate over the next 6 years will be 40%. The firm's interest rate is 20%.

(a) Assuming that the entire initial investment will be financed from the firm's retained earnings (equity financing), determine the after-tax cash flows over the investment life. Compute the NPW of this investment.

(b) Assuming the entire initial investment will be financed through a local bank at an interest rate of 13% compounded annually, determine the net after-tax cash flows for the project. Compute the NPW of this investment.

(c) Suppose that a financial lease is available for the prototype system at $62,560 per year, payable at the beginning of each year. Compute the NPW of this investment with lease financing.

(d) Select the best financing option based on the rate of return on incremental investment.

12.40 National Office Automation Inc. (NOAI) is a leading developer of imaging systems, controllers, and related accessories. The company's product line consists of systems for desktop publishing, automatic identification, advanced imaging, and office information markets. The firm's manufacturing plant in Ann Arbor, Michigan, consists of eight different functions: cable assembly, board assembly, mechanical assembly, controller integration, printer integration, production repair, customer repair, and shipping. The process to be considered is the transportation of pallets loaded with eight packaged desktop printers from printer integration to the shipping department. Several alternatives have been examined to minimize operating and maintenance costs. The two most feasible alternatives are:

• Option 1: Use of gas-powered lift trucks to transport pallets of packaged printers from printer integration to shipping. The truck also is used to return printers that must be reworked. The trucks can be leased at a cost of $5,465 per year. With a maintenance contract costing $6,317 per year, the dealer will maintain the trucks. A fuel cost of $1,660 per year is also expected. The truck requires a driver for each of the three shifts, at a total cost of $58,653 per year for labor. It is also estimated that transportation by truck would cause damages to material and equipment totaling $10,000 per year.

• Option 2: Installing an automatic guided vehicle system (AGVS) to transport pallets of packaged printers from printer integration to shipping and to return products that require rework. The AGVS, using an electrical powered cart and embedded wire-guidance system, would do the same job that the truck currently does, but without drivers. The total investment costs, including installation, are itemized as follows:

Vehicle and system installation	$97,255
Staging conveyor	24,000
Power supply lines	5,000
Transformers	2,500
Floor surface repair	6,000
Batteries and charger	10,775
Shipping	6,500
Sales tax	6,970
Total AGVS system cost	$159,000

NOAI could obtain a term loan for the full investment amount ($159,000) at a 10% interest rate. The loan would be amortized over 5 years, with payments made at the end of each year. The AGVS falls into the 7-year MACRS classification, and it has an estimated service life of 10 years and no salvage value. If the AGVS is installed, a maintenance contract would be obtained at a cost of $20,000, payable at the beginning of each year. The firm's marginal tax rate is 35%, and its MARR is 15%.

(a) Determine the net cash flows for each alternative over 10 years.

(b) Compute the incremental cash flows (Option 2 – Option 1) and determine the rate of return on this incremental investment.

(c) Determine the best course of action based on the rate of return criterion.

Note: Assume a zero salvage value for the AGVS.

CHAPTER 13

Inflation and Its Impact on Project Cash Flows

How Much Would It Cost to Send Your Child to College in Year 2005? You may have heard your parents fondly remember the "good old days" of penny candy or 35 cents-per-gallon gasoline. But even a college student in his or her early twenties can relate to the phenomenon of escalating costs. Do you remember when a postage stamp cost a dime? When the price of a first-run movie was under $2?

The table below demonstrates price differences between 1967 and 2000 for some commonly bought items. For example, a loaf of bread cost 22 cents in 1967, whereas it cost $1.84 in 2000. In 2000, the same 22 cents bought only a fraction of the bread it would have bought in 1967—specifically, about 12% of a loaf (0.22/1.84). From 1967 to 2000, the 22-cent sum had lost 88% of its purchasing power.

Item	1967 Price	2000 Price	% Increase
Consumer price index (CPI)	100	512.9	413
Monthly housing expense	$114.31	$943.97	726
Monthly automobile expense	82.69	471.38	470
Loaf of bread	0.22	1.84	736
Pound of hamburger	0.39	2.98	564
Pound of coffee	0.59	4.10	595
Candy bar	0.10	0.90	800
Men's dress shirt	5.00	39.00	680
Postage (first-class)	0.05	0.33	660
Annual public college tuition	294.00	3,960.00	1,247

Note: All prices are average prices compiled by Bureau of Labor Statistics

On a brighter note, we open this chapter with some statistics about engineers' pay rates. The average starting salary of a first-year engineer was $7,500 in 1967. According to a recent survey by the National Society of Professional Engineers,[1] the average starting salary of a first-year engineer in 2000 was $47,350, which is equivalent to an annual increase of 5.74% over 33 years. During the same period, the general inflation rate remained at 5.11%, indicating that engineers' pay rates at least have kept up with the rate of inflation.

Up to this point, we have demonstrated how to develop cash flows in a variety of ways and how to compare them under constant conditions in the general economy. In other words, we have assumed that prices remain relatively unchanged over long periods. As you know from personal experience, this is not a realistic assumption. In this chapter, we define and quantify **inflation** and then go on to apply it in several economic analyses. We will demonstrate inflation's effect on depreciation, borrowed funds, rate of return of a project, and working capital, within the bigger picture of developing projects.

---

[1] Source: 2000 Income and Salarly Survey by National Society of Professional Engineers.

13.1 Meaning and Measure of Inflation

Historically, the general economy has usually fluctuated in such a way as to experience **inflation**, a loss in the purchasing power of money over time. Inflation means that the cost of an item tends to increase over time, or, to put it another way, the same dollar amount buys less of an item over time. **Deflation** is the opposite of inflation in that prices usually decrease over time, and hence, a specified dollar amount gains in purchasing power. Inflation is far more common than deflation in the real world, so our consideration in this chapter will be restricted to accounting for inflation in economic analyses.

13.1.1 Measuring Inflation

Before we can introduce inflation into an engineering economic problem, we need a means of isolating and measuring its effect. Consumers usually have a relative, if not a precise, sense of how their purchasing power is declining. This sense is based on their experience of shopping for food, clothing, transportation, and housing over the years. Economists have developed a measure called the **consumer price index** (CPI), which is based on a typical **market basket** of goods and services required by the average consumer. This market basket normally consists of items from eight major groups: (1) food and alcoholic beverages, (2) housing, (3) apparel, (4) transportation, (5) medical care, (6) entertainment, (7) personal care, and (8) other goods and services.

The CPI compares the cost of the typical market basket of goods and services in a current month with its cost 1 month ago, 1 year ago, or 10 years ago. The point in the past to which current prices are compared is called the **base period**. The index value for this base period is set at $100. The original base period used by the Bureau of Labor Statistics (BLS), U.S. Department of Labor, for the CPI index is 1967. For example, let us say that, in 1967, the prescribed market basket could have been purchased for $100. Suppose the same combination of goods and services costs $512.90 in 2000. We can then compute the CPI for 2000 by multiplying, by 100, the ratio of the current price to the base period price. In our example, the price index is ($512.90/$100)100 = 512.90, which means that the 2000 price of the contents of the market basket is 512.90% of its base period price.

The revised CPI introduced by the BLS in 1987 includes indexes for two populations: urban wage earners and clerical workers (CW), and all urban consumers (CU). As a result of the revision, both the CW and the CU utilize updated expenditure weights based upon data tabulated from the 3 years (1982, 1983, and 1984) of the Consumer Expenditure Survey and incorporate a number of technical improvements. This method of assessing inflation does not imply, however, that consumers actually purchase the same goods and services year after year. Consumers tend to adjust their shopping practices to changes in relative prices and to substitute other items for those whose prices have greatly increased in relative terms. We must understand that the CPI does not take into account this sort of consumer behavior, because it is predicated on the purchase of a fixed market basket of the same goods and services, in the same proportions, month after month. For this reason, the CPI is called a **price index** rather than a **cost-of-living index**, although the general public often refers to it as a cost-of-living index.

Producer Price Index

The consumer price index is a good measure of the general price increase of consumer products. However, it is not a good measure of industrial price increases. When performing engineering economic analysis, the appropriate price indexes must be selected to estimate the price increases of raw materials, finished products, and operating costs. The **Survey of Current Business**, a monthly publication prepared by the BLS, U.S. Department of Labor, provides the industrial product price index for various industrial goods.[2] Table 13.1 lists the CPI together with several price indexes over a number of years.

From Table 13.1, we can easily calculate the inflation rate of gasoline from 2000 to 2001 as follows:

$$\frac{132.8 - 92.6}{92.6} = 0.4341 = 43.41\%.$$

Since the inflation rate calculated is positive, the price of gasoline increased at an annual rate of 43.79% over the year 2000, which is one of the largest jumps in gasoline price in recent years.

Average Inflation Rate (f)

To account for the effect of varying yearly inflation rates over a period of several years, we can compute a single rate that represents an **average inflation rate**. Since each individual year's inflation rate is based on the previous year's rate, these rates have a compounding effect. As an example, suppose we want to calculate the average inflation rate for a 2-year period: The first year's inflation rate is 4%, and the second year's rate is 8%, using a base price of $100.

Year Base Period	New CPI 1982–84	Old CPI 1967	Gasoline 1982	Steel 1982	Passenger Car 1982
1991	135.2	405.1	66.9	110.6	124.2
1992	139.5	417.9	65.6	107.1	127.3
1993	144.0	431.2	67.9	106.7	129.8
1994	147.4	441.4	59.5	111.9	133.3
1995	152.2	455.0	67.7	121.7	134.0
1996	156.6	468.2	76.4	114.9	135.2
1997	160.2	479.7	72.7	116.4	135.2
1998	162.5	487.1	54.0	115.4	132.2
1999	166.2	497.8	64.4	105.3	131.4
2000	171.2	512.9	92.6	109.8	133.4
2001	176.9	529.9	132.8	102.3	142.7

Table 13.1 Selected Price Indexes (Index for Base Year = 100, Calendar Month = April)

---

[2] CPI data are now available on the Internet worldwide computer network through the BLS home page on the World Wide Web: http://stats.bls.gov

- Step 1: To find the price at the end of the second year we use the process of compounding:

$$\underbrace{\overbrace{\$100(1 + 0.04)}^{\text{First year}}(1 + 0.08)}_{\text{Second year}} = \$112.32$$

- Step 2: To find the average inflation rate f, we establish the following equivalence equation:

$$\$100(1 + f)^2 = \$112.32 \text{ or } \$100(F/P, f, 2) = \$112.32$$

Solving for f yields

$$f = 5.98\%.$$

We can say that the price increases in the last 2 years are equivalent to an average annual percentage rate of 5.98% per year. Note that the average is a geometric, not an arithmetic, average over a several-year period. Our computations are simplified by using a single average rate such as this, rather than a different rate for each year's cash flows.

Example 13.1 Average Inflation Rate

Reconsider the price increases for the nine items listed in the table at the opening of this chapter. Determine the average inflation rate for each item over the 33-year period.

Solution

Let's take the first item, the monthly housing expense, for a sample calculation. Since we know the prices during both 1967 and 2000, we can use the appropriate equivalence formula (single-payment compound amount factor or growth formula).

Given: $P = \$114.31$, $F = \$943.97$, and $N = 2000 - 1967 = 33$
Find: f

Equation: $F = P(1 + f)^N$

$$\$943.97 = \$114.31(1 + f)^{33}$$

$$f = \sqrt[33]{8.2580} - 1$$

$$= 0.0661 = 6.61\%$$

In a similar fashion, we can obtain the average inflation rates for the remaining items as follows:

Item	1967 Price	2000 Price	Average Inflation Rate
Consumer price index (CPI)	100	512.9	5.07%
Monthly housing expense	$114.31	$943.97	6.61
Monthly automobile expense	82.69	471.38	5.42
Loaf of bread	0.22	1.84	6.64
Pound of hamburger	0.39	2.98	6.36
Pound of coffee	0.59	4.10	6.05
Candy bar	0.10	0.90	6.88
Men's dress shirt	5.00	39.00	6.42
Postage (first-class)	0.05	0.33	5.89
Annual public college tuition	294.00	3,960.00	8.19

The cost of resident college tuition increased most among the items listed in the table.

General Inflation Rate ($\bar{f}$) versus Specific Inflation Rate (f_j)

When we use the CPI as a base to determine the average inflation rate, we obtain the **general inflation rate**. We need to distinguish carefully between the general inflation rate and the average inflation rate for specific goods:

- **General inflation rate ($\bar{f}$) :** This average inflation rate is calculated based on the CPI for all items in the market basket. The market interest rate is expected to respond to this general inflation rate.

- **Specific inflation rate (f_j):** This rate is based on an index (or the CPI) specific to segment j of the economy. For example, we must often estimate the future cost for an item, e.g., labor, material, housing, or gasoline. (When we refer to the average inflation rate for just one item, we will drop the subscript j for simplicity.)

In terms of CPI, we define the general inflation rate as

$$CPI_n = CPI_0 (1 + \bar{f})^n,$$ (13.1)

or

$$\bar{f} = \left[\frac{CPI_n}{CPI_0} \right]^{1/n} - 1,$$ (13.2)

where $\bar{f}$ = The general inflation rate,

CPI_n = The consumer price index at the end period n,

CPI_0 = The consumer price index for the base period.

Knowing the CPI values for 2 consecutive years, we can calculate the annual general inflation rate:

$$\bar{f}_n = \frac{CPI_n - CPI_{n-1}}{CPI_{n-1}}, \tag{13.3}$$

where $\bar{f}_n$ = the general inflation rate for period n.

As an example, let us calculate the general inflation rate for the year 2000, where $CPI_{1999} = 497.8$ and $CPI_{2000} = 512.9$:

$$\frac{512.9 - 497.8}{497.8} = 0.0303 = 3.03\%,$$

which was an unusually good year for the U.S. economy, when compared with the average general inflation rate of 5.07%[3] over the last 33 years.

Example 13.2 Yearly and Average Inflation Rates

The following table shows a utility company's cost to supply a fixed amount of power to a new housing development; the indices are specific to the utilities industry. Assume that year 0 is the base period.

Year	Cost
0	$504,000
1	538,400
2	577,000
3	629,500

Determine the inflation rate for each period, and calculate the average inflation rate over the 3 years.

Solution

Inflation rate during year 1 (f_1):

---

[3] To calculate the average general inflation rate from the base period (1967) to 2000, we evaluate the following:

$$\bar{f} = \left[\frac{512.9}{100}\right]^{1/33} - 1 = 5.07\%$$

($538,400 − $504,000)/$504,000 = 6.83%.

Inflation rate during year 2 (f_2):

($577,000 − $538,400)/$538,400 = 7.17%.

Inflation rate during year 3 (f_3):

($629,500 − $577,000)/$577,000 = 9.10%.

The average inflation rate over 3 years is

$$f = \left(\frac{\$629,500}{\$504,000}\right)^{1/3} - 1 = 0.0769 = 7.69\%.$$

Note that, although the average inflation rate[4] is 7.69% for the period taken as a whole, none of the years within the period had this rate.

13.1.2   Actual versus Constant Dollars

To introduce the effect of inflation into our economic analysis, we need to define several inflation-related terms.[5]

- **Actual (current) dollars (A_n):** Actual dollars are estimates of future cash flows for year n that take into account any anticipated changes in amount caused by inflationary or deflationary effects. Usually these amounts are determined by applying an inflation rate to base year dollar estimates.

- **Constant (real) dollars (A'_n):** Constant dollars represent constant purchasing power independent of the passage of time. In situations where inflationary effects were assumed when cash flows were estimated, these estimates can be converted to constant dollars (base year dollars) by adjustment using some readily accepted **general inflation rate**. We will assume that the base year is always time 0 unless we specify otherwise.

Conversion from Constant to Actual Dollars

Since constant dollars represent dollar amounts expressed in terms of the purchasing power of the base year, we may find the equivalent dollars in year n using the general inflation rate ($\bar{f}$)

$$A_n = A'_n(1 + \bar{f})^n = A'_n(F/P, \bar{f}, n), \tag{13.4}$$

where A'_n =   Constant-dollar expression for the cash flow occurring at the end of the year n, and

---

[4] Since we obtained this average rate based on costs that are specific to the utility industry, this rate is not the general inflation rate. This is a specific inflation rate for this utility.

[5] Based on the ANSI Z94 Standards Committee on Industrial Engineering Terminology, *The Engineering Economist.* 1988; 33(2): 145–171.

$$A_n = \text{Actual-dollar expression for the cash flow}$$
$$\text{at the end of year } n.$$

If the future price of a specific cash flow element (j) is not expected to follow the general inflation rate, we will need to use the appropriate average inflation rate applicable to this cash flow element, f_j, instead of $\bar{f}$.

Example 13.3 Conversion from Constant to Actual Dollars

Transco Company is considering making and supplying computer-controlled traffic-signal switching boxes to be used throughout Arizona. Transco has estimated the market for its boxes by examining data on new road construction and on deterioration and replacement of existing units. The current price per unit is $550; the before-tax manufacturing cost is $450. The start-up investment cost is $250,000. The projected sales and net before-tax cash flows in constant dollars are as follows:

Period	Unit Sales	Net Cash Flow in Constant $
0		−$250,000
1	1,000	100,000
2	1,100	110,000
3	1,200	120,000
4	1,300	130,000
5	1,200	120,000

Assume that the price per unit as well as the manufacturing cost keeps up with the general inflation rate, which is projected to be 5% annually. Convert the project's before-tax cash flows into the equivalent actual dollars.

Solution

We first convert the constant dollars into actual dollars. Using Eq. (13.4), we obtain the following (note that the cash flow in year 0 is not affected by inflation):

Period	Net Cash Flow in Constant $	Conversion Factor	Cash Flow in Actual $
0	−$250,000	$(1 + 0.05)^0$	−$250,000
1	100,000	$(1 + 0.05)^1$	105,000
2	110,000	$(1 + 0.05)^2$	121,275
3	120,000	$(1 + 0.05)^3$	138,915
4	130,000	$(1 + 0.05)^4$	158,016
5	120,000	$(1 + 0.05)^5$	153,154

Figure 13.1 illustrates this conversion process graphically.

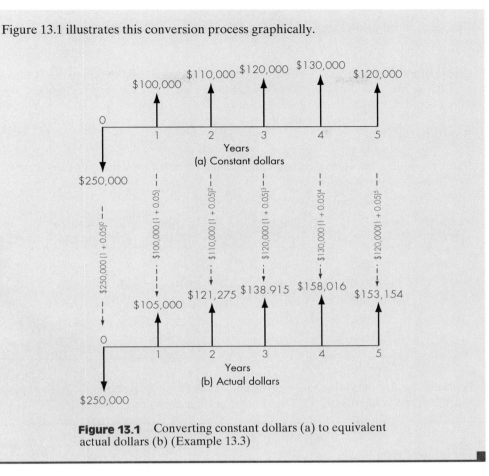

Figure 13.1   Converting constant dollars (a) to equivalent actual dollars (b) (Example 13.3)

Conversion from Actual to Constant Dollars

This is the reverse process of converting from constant to actual dollars. Instead of using the compounding formula, we use a discounting formula (single-payment present worth factor):

$$A'_n = \frac{A_n}{(1 + \bar{f})^n} = A_n(P/F, \bar{f}, n). \tag{13.5}$$

Once again, we may substitute f_j for $\bar{f}$ if future prices are not expected to follow the general inflation rate.

Example 13.4 Conversion from Actual to Constant Dollars

Jagura Creek Fish Company, an aquacultural production firm, has negotiated a 5-year lease on 20 acres of land, which will be used for fish ponds. The annual cost stated in the lease is $20,000, to be paid at the beginning of each of the 5 years.

The general inflation rate $\bar{f} = 5\%$. Find the equivalent cost in constant dollars in each period.

Discussion: Although the $20,000 annual payments are *uniform*, they are not expressed in constant dollars. Unless an inflation clause is built into a contract, any stated amounts refer to *actual dollars*.

Solution

Using Eq. (13.5), the equivalent lease payments in constant dollars can be determined as follows:

End of Period	Cash Flow in Actual $	Conversion at $\bar{f}$	Cash Flow in Constant $	Loss in Purchasing Power
0	–$20,000	$(1 + 0.05)^0$	–$20,000	0%
1	–20,000	$(1 + 0.05)^{-1}$	–19,048	4.76
2	–20,000	$(1 + 0.05)^{-2}$	–18,141	9.30
3	–20,000	$(1 + 0.05)^{-3}$	–17,277	13.62
4	–20,000	$(1 + 0.05)^{-4}$	–16,454	17.73

Note that, under the inflationary environment, the lender's receipt of the lease payment in year 5 is worth only 82.27% of the first lease payment.

13.2 Equivalence Calculations under Inflation

In previous chapters, our equivalence analyses took into consideration changes in the **earning power** of money, i.e., interest effects. To factor in changes in **purchasing power** as well, that is, inflation, we may use either (1) constant dollar analysis or (2) actual dollar analysis. Either method produces the same solution; however, each method requires use of a different interest rate and procedure. Before presenting the two procedures for integrating interest and inflation, we will give a precise definition of the two interest rates.

13.2.1 Market and Inflation-Free Interest Rates

Two types of interest rates are used in equivalence calculations: (1) the market interest rate and (2) the inflation-free interest rate. The rate to apply depends on the assumptions used in estimating the cash flow.

- **Market interest rate (i):** This rate takes into account the combined effects of the earning value of capital (earning power) and any anticipated inflation or deflation (purchasing power). Virtually all interest rates stated by financial institutions for loans and savings accounts are market interest rates. Most firms use a market interest rate (also known as **inflation-adjusted MARR**) in evaluating their investment projects.

- **Inflation-free interest rate (i'):** This rate is an estimate of the true earning power of money when the inflation effects have been removed. This rate is commonly known as **real interest rate**, and it can be computed if the market interest rate and inflation rate are known. In fact, all the interest rates mentioned in previous chapters are inflation-free interest rates. As you will see later in this chapter, in the absence of inflation, the market interest rate is the same as the inflation-free interest rate.

In calculating any cash flow equivalence, we need to identify the nature of project cash flows. The three common cases follow:

Case 1: All cash flow elements are estimated in constant dollars.
Case 2: All cash flow elements are estimated in actual dollars.
Case 3: Some of the cash flow elements are estimated in constant dollars, and others are estimated in actual dollars.

For case 3, we simply convert all cash flow elements into one type—either constant or actual dollars. Then we proceed with either constant-dollar analysis as for case 1 or actual-dollar analysis as for case 2.

13.2.2 Constant Dollar Analysis

Suppose that all cash flow elements are already given in constant dollars, and that you want to compute the equivalent present worth of the constant dollars (A_n') occurring in year n. In the absence of an inflationary effect, we should use i' to account for only the earning power of the money. To find the present worth equivalent of this constant-dollar amount at i', we use

$$P_n = \frac{A'_n}{(1 + i')}. \tag{13.6}$$

Constant dollar analysis is common in the evaluation of many long-term public projects, because governments do not pay income taxes. Typically, income taxes are levied based on taxable incomes in actual dollars.

Example 13.5 Equivalence Calculation when Flows are Stated in Constant Dollars

Consider the constant dollar flows originally given in Example 13.3. If Transco managers want the company to earn a 12% inflation-free rate of return (i') before-tax on any investment, what would be the present worth of this project?

Solution

Since all values are in constant dollars, we can use the inflation-free interest rate. We simply discount the dollar inflows at 12% to obtain the following:

$$PW(12\%) = -\$250,000 + \$100,000(P/A, 12\%, 5)$$
$$+ \$10,000(P/G, 12\%, 4) + \$20,000(P/F, 12\%, 5)$$
$$= \$163,099 \text{ (in year 0 dollars).}$$

Since the equivalent net receipts exceed the investment, the project can be justified before considering any tax effect.

13.2.3 Actual Dollar Analysis

Now let us assume that all cash flow elements are estimated in actual dollars. To find the equivalent present worth of this actual dollar amount (A_n) in year n, we may use either the **deflation method** or the **adjusted-discount method**.

Deflation Method

The deflation method requires two steps to convert actual dollars into equivalent present worth dollars. First we convert actual dollars into equivalent constant dollars by discounting by the general inflation rate, a step that removes the inflationary effect. Now we can find the equivalent present worth using i'.

Example 13.6 Equivalence Calculation when Cash Flows Are in Actual Dollars: Deflation Method

Applied Instrumentation, a small manufacturer of custom electronics, is contemplating an investment to produce sensors and control systems that have been requested by a fruit drying company. The work would be done under a proprietary contract, which would terminate in 5 years. The project is expected to generate the following cash flows in actual dollars.

n	Net Cash Flow in Actual Dollars
0	−$75,000
1	32,000
2	35,700
3	32,800
4	29,000
5	58,000

(a) What are the equivalent year 0 dollars (constant dollars), if the general inflation rate ($\bar{f}$) is 5% per year?

(b) Compute the present worth of these cash flows in constant dollars at $i' = 10\%$.

Solution

The net cash flows in actual dollars can be converted to constant dollars by deflating them, again assuming a 5% yearly deflation factor. The deflated or constant dollar cash flows can then be used to determine the NPW at i'.

(a) We convert the actual dollars into constant dollars as follows:

n	Cash Flows in Actual Dollars	Multiplied by Deflation Factor	Cash Flows in Constant Dollars
0	−$75,000	1	−$75,000
1	32,000	$(1 + 0.05)^{-1}$	30,476
2	35,700	$(1 + 0.05)^{-2}$	32,381
3	32,800	$(1 + 0.05)^{-3}$	28,334
4	29,000	$(1 + 0.05)^{-4}$	23,858
5	58,000	$(1 + 0.05)^{-5}$	45,445

(b) We compute the equivalent present worth of constant dollars using $i' = 10\%$:

n	Cash Flows in Constant Dollars	Multiplied by Discounting Factor	Equivalent Present Worth
0	−$75,000	1	−$75,000
1	30,476	$(1 + 0.10)^{-1}$	27,706
2	32,381	$(1 + 0.10)^{-2}$	26,761
3	28,334	$(1 + 0.10)^{-3}$	21,288
4	23,858	$(1 + 0.10)^{-4}$	16,295
5	45,445	$(1 + 0.10)^{-5}$	28,218
			$45,268

Figure 13.2 shows the conversion from the cash flows stated in year 0 dollars to equivalent present worth.

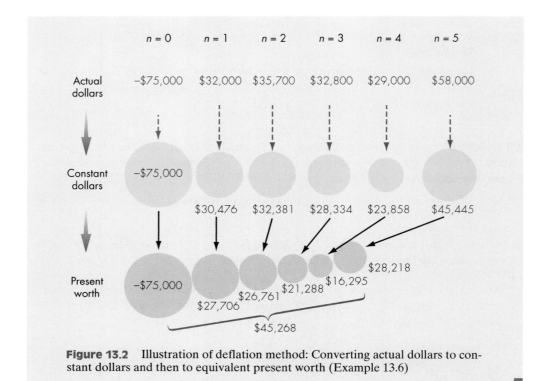

| | $n = 0$ | $n = 1$ | $n = 2$ | $n = 3$ | $n = 4$ | $n = 5$ |

Actual dollars: $-\$75,000$ $\$32,000$ $\$35,700$ $\$32,800$ $\$29,000$ $\$58,000$

Constant dollars: $-\$75,000$ $\$30,476$ $\$32,381$ $\$28,334$ $\$23,858$ $\$45,445$

Present worth: $-\$75,000$ $\$27,706$ $\$26,761$ $\$21,288$ $\$16,295$ $\$28,218$

$\$45,268$

Figure 13.2   Illustration of deflation method: Converting actual dollars to constant dollars and then to equivalent present worth (Example 13.6)

Adjusted-Discount Method

The two-step process shown in Example 13.6 can be greatly streamlined by the efficiency of the **adjusted-discount method**, which performs deflation and discounting in one step. Mathematically we can combine this two-step procedure into one by

$$P_n = \frac{\dfrac{A_n}{(1 + f)^n}}{(1 + i')^n}$$

$$= \frac{A_n}{(1 + \bar{f})^n(1 + i')^n} \tag{13.7}$$

$$= \frac{A_n}{[(1 + \bar{f})^n(1 + i')]^n}.$$

Since the market interest rate (i) reflects both the earning power and the purchasing power, we have the following relationship:

$$P_n = \frac{A_n}{(1 + i)^n}. \tag{13.8}$$

The equivalent present worth values in Eqs. (13.7) and (13.8) must be equal at year 0. Therefore,

$$\frac{A_n}{(1+i)^n} = \frac{A_n}{[(1+\bar{f})(1+i')]^n}$$

This leads to the following relationship among $\bar{f}$, i', and i:

$$(1+i) = (1+\bar{f})(1+i')$$
$$= 1 + i' + \bar{f} + i'\bar{f}.$$

Simplifying the terms yields

$$i = i' + \bar{f} + i'\bar{f}. \tag{13.9}$$

This implies that the market interest rate is a function of two terms, i' and $\bar{f}$. Note that without an inflationary effect, the two interest rates are the same ($\bar{f} = 0 \rightarrow i = i'$). As either i' or $\bar{f}$ increases, i also increases. For example, we can easily observe that, when prices increase due to inflation, bond rates climb, because lenders (i.e., anyone who invests in a money-market fund, a bond, or a certificate of deposit) demand higher rates to protect themselves against erosion in the value of their dollars. If inflation were to remain at 3%, you might be satisfied with an interest rate of 7% on a bond because your return would more than beat inflation. If inflation were running at 10%, however, you would not buy a 7% bond; you might insist instead on a return of at least 14%. On the other hand, when prices are coming down, or at least are stable, lenders do not fear the loss of purchasing power with the loans they make, so they are satisfied to lend at lower interest rates.

In practice, we often approximate the market interest rate (i) by simply adding the inflation rate ($\bar{f}$) to the real interest rate (i'), but ignoring the product term ($i'\bar{f}$). This practice is OK as long as either i' or $\bar{f}$ is relatively small.

Example 13.7 Equivalence Calculation when Flows Are in Actual Dollars: Adjusted-Discounted Method

Consider the cash flows in actual dollars in Example 13.6. Compute the equivalent present worth of these cash flows using the adjusted-discount method.

Solution

First, we need to determine the market interest rate i. With $\bar{f} = 5\%$ and $i = 10\%$, we obtain

$$i = i' + \bar{f} + i'\bar{f}$$
$$= 0.10 + 0.5 + (0.10)(0.05)$$
$$= 15.5\%.$$

n	Cash Flows in Actual Dollars	Multiplied by	Equivalent Present Worth
0	−$75,000	1	−$75,000
1	32,000	$(1 + 0.155)^{-1}$	27,706
2	35,700	$(1 + 0.155)^{-2}$	26,761
3	32,800	$(1 + 0.155)^{-3}$	21,288
4	29,000	$(1 + 0.155)^{-4}$	16,296
5	58,000	$(1 + 0.155)^{-5}$	28,217
			$45,268

The conversion process is shown in Figure 13.3. Note that the equivalent present worth that we obtain using the adjusted-discount method ($i = 15.5\%$) is exactly the same as the result we obtained in Example 13.6.

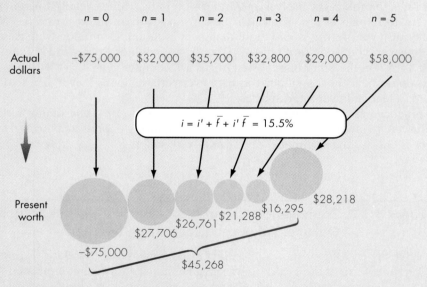

Figure 13.3   Illustration of adjusted-discount method: Converting actual dollars to present worth dollars by applying the market interest rate (Example 13.7)

13.2.4 Mixed Dollar Analysis

We will consider another situation in which some cash flow elements are expressed in constant (or today's) dollars and the other elements in actual dollars. In this situation, we can convert all cash flow elements into the same dollar units (either constant or actual). If the cash flow is converted into actual dollars, the market interest rate (i) should be used in calculating the equivalence value. If the cash flow is converted in terms of constant dollars, the inflation-free interest rate (i') should be used. Example 13.8 illustrates this situation.

Example 13.8 Equivalence Calculation with Composite Cash Flow Elements

Consider establishing a college funds at a bank for a couple with a 5-year old child. The college funds will earn 8% interest, compounded quarterly. Assuming that the child enters college at age 18, the couple estimate that an amount of $30,000 per year, in terms of today's dollars, will be required to support the child's college expenses for 4 years. The college expense is estimated to increase at the annual rate of 6%. Determine the equal quarterly deposits the couple must make until they send their child to college. Assume that the first deposit will be made at the end of the first quarter, and deposits will continue until the child reaches age 17. The child will enter college at age 18, and the annual college expense will be paid at the beginning of each college year. In other words, the first withdrawal will be made at age 18.

Discussion: In this problem, future college expenses are expressed in terms of today's dollars, whereas the quarterly deposits are in actual dollars. Since the interest rate quoted on the college funds is a market interest rate, we may convert the future college expenses into actual dollars.

Age	College expenses (in today's dollars)	College expenses (in actual dollars)
18 (Freshman)	$30,000	$30,000(F/P, 6%,13) = $63,988
19 (Sophomore)	30,000	30,000(F/P, 6%,14) = 67,827
20 (Junior)	30,000	30,000(F/P, 6%,15) = 71,897
21 (Senior)	30,000	30,000(F/P, 6%,16) = 76,211

Solution

College expenses as well as the quarterly deposit series in actual dollars are shown in Figure 13.4.

We first select $n = 12$ or age 17 as the base period for our equivalence calculation. Then, calculate the accumulated total amount at the base period at 2% interest per quarter. Since the deposit period is 12 years, we have a total of 48 quarterly deposits. If the first deposit is made at the end of first quarter, we have a 48-quarter deposit period. Therefore, the total deposit balance at age 17 would be

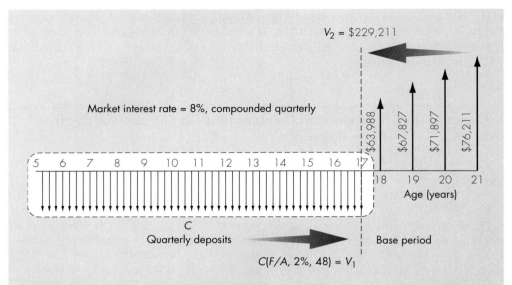

Figure 13.4 Establishing a college fund under an inflationary economy for a 5-year-old child by making 48 quarterly deposits (Example 13.7)

$$V_1 = C(F/A, 2\%, 48)$$
$$= \boxed{79.3535C}.$$

The equivalent lump-sum worth of the total college expenditure at the base period would be

$$V_2 = \$63,988(P/F, 2\%, 4) + \$67,827(P/F, 2\%, 8)$$
$$+ \$71,897(P/F, 2\%, 12) + \$76,211(P/F, 2\%, 16)$$
$$= \$229,211.$$

By setting $V_1 = V_2$ and solving for C, we obtain

$$79.3535C = \$229,211$$
$$C = \$2,888.48 \text{ per quarter.}$$

13.3 Effects of Inflation on Project Cash Flows

We will now introduce inflation into some investment projects. We are especially interested in two elements of project cash flows—depreciation expenses and interest expenses. These two elements are essentially immune to the effects of inflation as they are always given in actual dollars. We will also consider the complication of how to proceed when multiple price indexes have been used to generate various project cash flows.

13.3.1 Depreciation Expenses under Inflation

Because depreciation expenses are calculated on some base-year purchase amount, they do not increase over time to keep pace with inflation. Thus, they lose some of their value to defer taxes because inflation drives up the general price level and hence taxable income. Similarly, the selling prices of depreciable assets can increase with the general inflation rate and, because any gains on salvage values are taxable, they can result in increased taxes. Example 13.9 illustrates how a project's profitability changes under an inflationary economy.

Example 13.9 Effects of Inflation on Projects with Depreciable Assets

Reconsider the automated machining center investment project described in Example 12.1. The summary of the financial facts in the absence of inflation is as follows:

Item	Description or Data
Project	Automated machining center
Required investment	$125,000
Project life	5 years
Salvage value	$50,000
Depreciation method	7-year MACRS
Annual revenues	$100,000 per year
Annual expenses	
Labor	$20,000 per year
Material	$12,000 per year
Overhead	$8,000 per year
Marginal tax rate	40%
Inflation-free interest rate (i')	15%

The after-tax cash flow for the automated machining center project was given in Table 12.2, and the net present worth of the project in the absence of inflation was calculated to be $43,152.

What will happen to this investment project if the general inflation rate during the next 5 years is expected to increase by 5% annually? Sales and operating costs are assumed to increase accordingly. Depreciation will remain unchanged, but taxes, profits, and thus cash flow, will be higher. The firm's inflation-free interest rate (i') is known to be 15%.

(a) Determine the NPW of the project using the deflation method.

(b) Compare the NPW with that in the inflation-free situation.

Discussion: All cash flow elements, except depreciation expenses, are assumed to be in constant dollars. Since income taxes are levied on actual taxable income, we will use the actual-dollar analysis, which requires that all cash flow elements be expressed in actual dollars.

- For the purposes of this illustration, all inflationary calculations are made as of year end.
- Cash flow elements such as sales, labor, material, overhead, and selling price of the asset will be inflated at the same rate as the general inflation rate.[6] For example, whereas annual sales had been estimated at $100,000, under conditions of inflation they become 5% greater in year 1, or $105,000; 10.25% greater in year 2, and so forth.

Period	Sales in Constant $	Conversion $\bar{f}$	Sales in Actual $
1	$100,000	$(1+0.05)^1$	$105,000
2	100,000	$(1+0.05)^2$	110,250
3	100,000	$(1+0.05)^3$	115,763
4	100,000	$(1+0.05)^4$	121,551
5	100,000	$(1+0.05)^5$	127,628

Future cash flows in actual dollars for other elements can be obtained in a similar way.

- No change occurs in the investment in year 0 or in depreciation expenses since these items are unaffected by expected future inflation.
- The selling price of the asset is expected to increase at the general inflation rate. Therefore, the salvage in actual dollars will be

$$\$50,000(1 + 0.05)^5 = \$63,814.$$

This increase in salvage value will also increase the taxable gains as the book value remains unchanged. The calculations for both the book value and gains tax are shown in Table 13.2.

---

[6] This is a simplistic assumption. In practice, these elements may have price indices other than the CPI. Differential price indices will be treated in Example 13.10.

	Inflation Rate	0	1	2	3	4	5
Income Statement							
Revenues	5%		$105,000	$110,250	$115,763	$121,551	$127,628
Expenses							
Labor	5%		21,000	22,050	23,153	24,310	25,526
Material	5%		12,600	13,230	13,892	14,586	15,315
Overhead	5%		8,400	8,820	9,261	9,724	10,210
Depreciation			17,863	30,613	21,863	15,613	5,581
Taxable income			$ 45,137	$ 35,537	$ 47,595	$ 57,317	$ 70,996
Income taxes (40%)			18,055	14,215	19,038	22,927	28,398
Net income			$ 27,082	$ 21,322	$ 28,557	$ 34,390	$ 42,598
Cash Flow Statement							
Operating activities							
Net income			27,082	21,322	28,557	34,390	42,598
Depreciation			17,863	30,613	21,863	15,613	5,581
Investment activities							
Investment		(125,000)					
Salvage	5%						63,814
Gains tax							(12,139)
Net cash flow (in actual dollars)		$(125,000)	$ 44,945	$ 51,935	$ 50,420	$ 50,003	$ 99,854

Table 13.2
Cash Flow Statement for the Automated Machining Center Project under Inflation (Example 13.9)

Solution

Table 13.2 shows after-tax cash flows in actual dollars. Using the deflation method, we convert the cash flows to constant dollars with the same purchasing power as those used to make the initial investment (year 0), assuming a general inflation rate of 5%. Then, we discount these constant-dollar cash flows at i' to determine the NPW.

Year	Net Cash Flow in Actual $	Conversion $\bar{f}$	Net Cash Flow in Constant $	NPW at 15%
0	−$125,000	$(1 + 0.05)^0$	−$125,000	−$125,000
1	44,945	$(1 + 0.05)^{-1}$	42,805	37,222
2	51,935	$(1 + 0.05)^{-2}$	47,107	35,620
3	50,420	$(1 + 0.05)^{-3}$	43,555	28,638
4	50,003	$(1 + 0.05)^{-4}$	41,138	23,521
5	99,854	$(1 + 0.05)^{-5}$	78,238	38,898
				$38,899

Since NPW = $38,899 > 0, the investment is still economically attractive.

Comments: Note that the NPW in the absence of inflation was $43,152 in Example 12.1. The $4,253 decline (known as inflation loss) in the NPW under inflation, illustrated above, is due entirely to income tax considerations. The depreciation expense is a charge against taxable income, which reduces the amount of taxes paid, and as a result, increases the cash flow attributable to an investment by the amount of taxes saved. But the depreciation expense under existing tax laws is based on historic cost. As time goes by, the depreciation expense is charged to taxable income in dollars of declining purchasing power; as a result, the "real" cost of the asset is not totally reflected in the depreciation expense. Depreciation costs are thereby understated, and the taxable income is overstated, resulting in higher taxes. In "real" terms, the amount of this additional income tax is $4,253, which is also known as the **inflation tax**. In general, any investment that, for tax purposes, is expensed over time, rather than immediately, is subject to the inflation tax.

13.3.2 Multiple Inflation Rates

As we noted previously, the inflation rate (f_j) represents a rate applicable to a specific segment j of the economy. For example, if we were estimating the future cost of a piece of machinery, we should use the inflation rate appropriate for that item. Furthermore, we may need to use several rates to accommodate the different costs and revenues in our analysis. The following example introduces the complexity of multiple inflation rates.

Example 13.10 Applying Specific Inflation Rates

We will rework Example 13.9 using different annual indicies (differential inflation rates) in the prices of cash flow components. Suppose that we expect the general rate of inflation $(\bar{f})$ to average 6% during the next 5 years. We also expect that the selling price of the equipment will increase 3% per year, that wages (labor) and overhead will increase 5% per year, and that the cost of material will increase 4% per year. We expect sales revenue to climb at the general inflation rate. Table 13.3 shows the relevant calculations based on the income statement format. For simplicity, all cash flows and inflation effects are assumed to occur at year's end. Determine the net present worth of this investment, using the adjusted-discount method.

Solution

From Table 13.3, the after-tax cash flows in actual dollars are as follows:

	Inflation Rate	0	1	2	3	4	5
Income Statement							
Revenues	6%		$106,000	$112,360	$119,102	$126,248	$133,823
Expenses							
Labor	5%		21,000	22,050	23,153	24,310	25,526
Material	4%		12,480	12,979	13,498	14,038	14,600
Overhead	5%		8,400	8,820	9,261	9,724	10,210
Depreciation			17,863	30,613	21,863	15,613	5,581
Taxable income			$ 46,257	$ 37,898	$ 51,327	$ 62,562	$ 77,906
Income taxes (40%)			18,503	15,159	20,531	25,025	31,162
Net income			$ 27,754	$ 22,739	$ 30,796	$ 37,537	$ 46,744
Cash Flow Statement							
Operating activities							
Net income			27,754	22,739	30,796	37,537	46,744
Depreciation			17,863	30,613	21,863	15,613	5,581
Investment activities							
Investment		(125,000)					
Salvage	3%						57,964
Gains tax							(9,799)
Net cash flow (in actual dollars)		$(125,000)	$ 45,617	$ 53,352	$ 52,659	$ 53,150	$100,490

Table 13.3
Cash Flow Statement for the Automated Machining Center Project under Inflation, with Multiple Price Indices (Example 11.10)

n	Net Cash Flow in Actual Dollars
0	−$125,000
1	45,617
2	53,352
3	52,659
4	53,150
5	100,490

To evaluate the present worth using actual dollars, we must adjust the original discount rate of 15%, which is an inflation-free interest rate, i'. The appropriate interest rate to use is the market interest rate:[7]

$$i = i' + \bar{f} + i'\bar{f}$$
$$= 0.15 + 0.06 + (0.15)(0.06)$$
$$= 21.90\%.$$

The equivalent present worth is obtained as follows:

$$PW(21.90\%) = -\$125,000 + \$45,617(P/F, 21.90\%, 1)$$
$$+ \$53,352(P/F, 21.90\%, 2) + \ldots$$
$$+ \$100,490(P/F, 21.90\%, 5)$$
$$= \$38,801.$$

13.3.3   Effects of Borrowed Funds under Inflation

Loan repayment is based on the historical contract amount; the payment size does not change with inflation. Yet inflation greatly affects the value of these future payments, which are computed in year 0 dollars. First, we shall look at how the values of loan payments change under inflation. Interest expenses are usually already stated in the loan contract in actual dollars and need not be adjusted. Under the effect of inflation, the constant-dollar costs of both interest and debt principal repayments are reduced. Example 13.11 illustrates the effects of inflation on payments with project financing.

Example 13.11   Effects of Inflation on Payments with Financing (Borrowing)

Let us rework Example 13.9 with a debt-to-equity ratio of 0.50, where the debt portion of the initial investment is borrowed at 15.5% annual interest. Assume, for simplicity, that the general inflation rate ($\bar{f}$) of 5% during the project period will affect all revenues, expenses (except depreciation and loan payments), and salvage value. Determine the NPW of this investment. (Note that the borrowing rate of 15.5% reflects the higher cost of debt financing under inflationary environment.)

Solution

For equal future payments, the actual dollar cash flows for the financing activity are represented by the circles in Figure 13.5. If inflation were to occur, the cash flow, measured in year 0 dollars, would be represented by the shaded circles in Figure 13.5.

---

[7] In practice, the market interest rate is usually given and the inflation-free interest rate can be calculated when the general inflation rate is known for years in the past or is estimated for time in the future. In our example, we are considering the opposite situation.

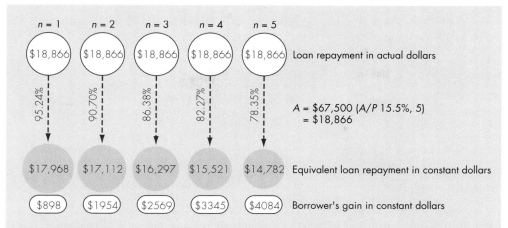

n = 1   n = 2   n = 3   n = 4   n = 5

($18,866) ($18,866) ($18,866) ($18,866) ($18,866) Loan repayment in actual dollars

95.24%   90.70%   86.38%   82.27%   78.35%

$A = \$67,500\,(A/P\ 15.5\%,\ 5)$
$= \$18,866$

$17,968 $17,112 $16,297 $15,521 $14,782 Equivalent loan repayment in constant dollars

($898) ($1954) ($2569) ($3345) ($4084) Borrower's gain in constant dollars

Figure 13.5 Equivalent loan repayment cash flows measured in year 0 dollars and borrower's gain over the loan life (Example 13.11)

Table 13.4 summarizes the after-tax cash flows under this situation. For simplicity, assume that all cash flows and inflation effects occur at year's end. To evaluate the present worth using actual dollars, we must adjust the original discount rate (MARR) of 15%, which is an inflation-free interest rate i'. The appropriate interest rate to use is thus the market interest rate:

$$
\begin{aligned}
i &= i' + \bar{f} + i'\bar{f} \\
&= 0.15 + 0.05 + (0.15)(0.05) \\
&= 20.75\%.
\end{aligned}
$$

Then, from Table 13.4, we compute the equivalent present worth of the after-tax cash flow as follows:

n	Net Cash Flow in Actual Dollars
0	−$62,500
1	29,953
2	36,375
3	34,203
4	35,027
5	82,000

The equivalent present worth is

$$
\begin{aligned}
PW\,(20.75\%) &= -\$62,500 + \$29,953(P/F, 20.75\%, 1) + \ldots \\
&\quad + \$82,000(P/F, 20.75\%, 5) \\
&= \$54,159.
\end{aligned}
$$

Table 13.4
Cash Flow Statement for the Automated Machining Center Project under Inflation, with Borrowed Funds (Example 13.11)

	Inflation Rate	0	1	2	3	4	5
Income Statement							
Revenues	5%		$105,000	$110,250	$115,763	$121,551	$127,628
Expenses:							
Labor	5%		$ 21,000	$ 22,050	$ 23,153	$ 24,310	$ 25,526
Material	5%		$ 12,600	$ 13,230	$ 13,892	$ 14,586	$ 15,315
Overhead	5%		$ 8,400	$ 8,820	$ 9,261	$ 9,724	$ 10,210
Depreciation			$ 17,863	$ 30,613	$ 21,863	$ 15,613	$ 5,581
Debt interest			9,688	8,265	6,622	4,724	2,532
Taxable income			$ 35,449	$ 27,272	$ 40,973	$ 52,593	$ 68,464
Income Taxes (40%)			14,180	10,909	16,389	21,037	27,386
Net Income			$ 21,269	$ 16,363	$ 24,584	$ 31,556	$ 41,078
Cash Flow Statement							
Operating activities:							
Net income			21,269	16,363	24,584	31,556	41,078
Depreciation			$ 17,863	$ 30,613	$ 21,863	$ 15,613	$ 5,581
Investment activities:							
Investment		(125,000)					
Salvage	5%						63,814
Gains tax							(12,139)
Financing activities							
Borrowed funds		62,500					
Princial repayment			(9,179)	(10,601)	(12,244)	(14,142)	(16,334)
Net Cash Flow (in actual dollars)		$ (62,500)	$ 29,953	$ 36,375	$ 34,203	$ 33,027	$ 82,000
Net Cash Flow (in constant dollars)	5%	$ (62,500)	$28,527	$32,993	$29,545	$27,171	$64,249
Equ. Present Worth	15%	$ (62,500)	$24,806	$24,948	$19,427	$15,535	$31,943
Net Present Worth		$54,159					

Comments: In the absence of debt financing, the project would have a net present worth of $38,899 as shown in Example 13.9. When compared with the result of $54,159, the present worth gain due to debt financing is $15,260. This increase in NPW is primarily due to the debt-financing. An inflationary trend decreases the purchasing power of future dollars, which helps long-term borrowers because they can repay a loan with dollars of reduced buying power. That is, the debt-financing cost is reduced in an inflationary environment. In this case, the benefits of financing under inflation have more than offset the *inflation tax* effect on depreciation and salvage value. The amount of gain may vary depending on loan interest rate: The interest rate for borrowing is also generally higher during periods of inflation because it is a market driven rate.

13.4 Rate of Return Analysis under Inflation

In addition to affecting individual aspects of a project's income statement, inflation can have a profound effect on its overall return, i.e., the very acceptability, of an investment project. In this section, we will explore the effects of inflation on return on investments and show several examples.

13.4.1 Effects of Inflation on Return on Investment

The effect of inflation on the rate of return for an investment depends on how future revenues respond to the inflation. Under inflation, a company is usually able to compensate for increasing material and labor prices by raising its selling prices. However, even if future revenues increase to match the inflation rate, the allowable depreciation schedule, as we have seen, does not increase. The result is increased taxable income and higher income-tax payments. This increase reduces the available constant dollar after-tax benefits and, therefore, the inflation-free after-tax rate of return (IRR′). The next example will help us to understand this situation.

Example 13.12 IRR Analysis with Inflation

Hartsfield Company, a manufacturer of autoparts, is considering the purchase of a set of machine tools at a cost of $30,000. The purchase is expected to generate increased sales of $24,500 per year and increased operating costs of $10,000 per year in each of the next 4 years. Additional profits will be taxed at a rate of 40%. The asset falls into the 3-year MACRS property class for tax purposes. The project has a 4-year life with zero salvage value. (All dollar figures represent constant dollars.)

(a) What is the expected internal rate of return?

(b) What is the expected IRR′ if the general inflation is 10% during each of the next 4 years? (Here also assume that $f_j = \bar{f} = 10\%$.)

(c) If this is an independent alternative and the company has an inflation-free MARR (or MARR′) of 20%, should the company invest in the equipment?

Solution

(a) **Rate of Return Analysis without Inflation**

We find the expected rate of return by first computing the after-tax cash flow by the income statement approach, as shown in Table 13.5. The first part of the table shows the calculation of additional sales, operating costs, depreciation, and taxes. The asset will be depreciated fully over 4 years, with no expected salvage value. As we emphasized in Chapter 10, depreciation is not a cash expense, although it affects taxable income, and thus cash flow, indirectly. Therefore, we must add depreciation to net income to determine the net cash flow.

Table 13.5
Rate of Return Calculation without Inflation (Example 13.12)

	0	1	2	3	4
Income Statement					
Revenues		$ 24,500	$ 24,500	$ 24,500	$ 24,500
Expenses					
O&M		10,000	10,000	10,000	10,000
Depreciation		10,000	13,333	4,445	2,222
Taxable income		4,500	1,167	10,055	12,278
Income taxes (40%)		1,800	467	4,022	4,911
Net income		$ 2,700	$ 700	$ 6,033	$ 7,367
Cash Flow Statement					
Operating activities					
Net income		2,700	700	6,033	7,367
Depreciation		10,000	13,333	4,445	2,222
Investment activities					
Machine center	$ (30,000)				
Salvage					0
Gains tax					0
Net cash flow (in actual dollars)	$ (30,000)	$ 12,700	$ 14,033	$ 10,478	$ 9,589

n	Net Cash Flow in Constant Dollars
0	−$30,000
1	12,700
2	14,033
3	10,478
4	9,589

Thus, if the investment is made, we expect to receive additional annual cash flows of $12,700, $14,033, $10,478, and $9,589. This is a simple investment, so we can calculate the IRR for the project as follows:

$$PW(i') = -\$30,000 + \$12,700(P/F, i', 1) + \$14,033(P/F, i', 2)$$
$$+ \$10,478(P/F, i', 3) + \$9,589(P/F, i', 4)$$
$$= 0.$$

Solving for i' yields

$$IRR' = i'^* = 21.88\%.$$

The project has an inflation-free rate of return of 21.88%, i.e., the company will recover its original investment ($30,000) plus interest at 21.88% each year for each dollar still invested in the project. Since the IRR′ > MARR′ of 20%, the company should buy the equipment.

(b) Rate of Return Analysis under Inflation

With inflation, we assume that sales, operating costs, and future selling price of the asset will increase. Depreciation will be unchanged, but taxes, profits, and cash flow will be higher. We might think that higher cash flows will mean an increased rate of return. Unfortunately, this is not the case. We must recognize that cash flows for each year are stated in dollars of declining purchasing power. When the net after-tax cash flows are converted to dollars with the same purchasing power as those used to make the original investment, the resulting rate of return decreases. These calculations, assuming an inflation rate of 10% in sales and operating expenses and a 10% annual decline in the purchasing power of the dollar, are shown in Table 13.6. For example, whereas additional sales had been $24,500 yearly, under conditions of inflation they would be 10% greater in year 1, or $26,950; 21% greater in year 2, and so forth. No change in investment or depreciation expenses will occur since these items are unaffected by expected future inflation. We have restated the after-tax cash flows (actual dollars) in dollars of a common purchasing power (constant dollars) by deflating them, again assuming an annual deflation factor of 10%. The constant-dollar cash flows are then used to determine the real rate of return.

$$PW(i') = -\$30{,}000 + \$12{,}336(P/F, i', 1) + \$13{,}108(P/F, i', 2)$$

$$+ \$10{,}036(P/F, i', 3) + \$9{,}307(P/F, i', 4)$$

$$= 0.$$

Solving for i' yields

$$i'^* = 19.40\%.$$

The rate of return for the project's cash flows in constant dollars (year 0 dollars) is 19.40%, which is less than the 21.88% return in the inflation-free case. Since IRR' < MARR', the investment is no longer acceptable.

Comments: We could also calculate the rate of return by setting the PW of the actual dollar cash flows to 0. This would give a value of IRR = 31.34%, but this is an

Table 13.6
Rate of Return Calculation under Inflation (Example 13.12)

	0	1	2	3	4
Income Statement					
Revenues		$26,950	$29,645	$32,610	$35,870
Expenses					
O&M		11,000	12,100	13,310	14,641
Depreciation		10,000	13,333	4,445	2,222
Taxable income		5,950	4,212	14,855	19,007
Income taxes (40%)		2,380	1,685	5,942	7,603
Net income		$ 3,570	$ 2,527	$ 8,913	$11,404
Cash Flow Statement					
Operating activities					
Net income		3,570	2,527	8,913	11,404
Depreciation		10,000	13,333	4,445	2,222
Investment activities					
Machine center	$(30,000)				
Salvage					0
Gains tax					0
Net cash flow (in actual dollars)	$(30,000)	$13,570	$15,860	$13,358	$13,626
Net cash flow (in constant dollars)	$(30,000)	$12,336	$13,108	$10,036	$ 9,307

PW (20%) = $ (321); IRR (actual dollars) = 31.34%; IRR' (constant dollars) = 19.40%

inflation-adjusted IRR. We could then convert to an IRR' by deducting the amount caused by inflation:

$$i' = \frac{(1 + i)}{1 + \bar{f})} - 1$$

$$= \frac{(1 + 0.3134)}{(1 + 0.10)} - 1$$

$$= 19.40\%,$$

which gives the same final result of IRR' = 19.40%.

13.4.2   Effects of Inflation on Working Capital

The loss of tax savings in depreciation is not the only way that inflation may distort an investment's rate of return. Another source of decrease in a project's rate of return is working-capital drain. Capital projects requiring increased levels of working capital suffer from inflation because additional cash must be invested to maintain new price levels. For example, if the cost of inventory increases, additional outflows of cash are required to maintain appropriate inventory levels over time. A similar phenomenon occurs with funds committed to accounts receivable. These additional working-capital requirements can significantly reduce a project's rate of return. The next example will illustrate the effects of working-capital drain on a project's rate of return.

Example 13.13 Effect of Inflation on Profits with Working Capital

Consider Example 13.12. Suppose that a $1,000 investment in working capital is expected and that all the working capital will be recovered at the end of the project's 4-year life. Determine the rate of return on this investment.

Solution

Using the data in the upper part of Table 13.7, we can calculate the IRR' = IRR of 20.88% in the absence of inflation. The PW (20%) = $499. The lower part of Table 13.7 includes the effect of inflation on the proposed investment. As illustrated in Figure 13.6, working-capital levels can be maintained only by additional infusions of cash—the working-capital drain also appears in the lower part of Table 13.7. For example, the $1,000 investment in working capital made in year 0 will be recovered at the end of the first year assuming a 1-year recovery cycle. However, because of 10% inflation, the required working capital for the second year increases to $1,100. In addi-

Table 13.7
Effects of
Inflation on
Working
Capital and
After-tax
Rate of Return (Example 13.13)

Case 1: Without Inflation	0	1	2	3	4
Income Statement					
Revenue		$ 24,500	$ 24,500	$ 24,500	$ 24,500
Expenses					
O&M		10,000	10,000	10,000	10,000
Depreciation		10,000	13,333	4,445	2,222
Taxable income		4,500	1,167	10,055	12,278
Income taxes (40%)		1,800	467	4,022	4,911
Net income		$ 2,700	$ 700	$ 6,033	$ 7,367
Cash Flow Statement					
Operating activities					
Net income		2,700	700	6,033	7,367
Depreciation		10,000	13,333	4,445	2,222
Investment activities					
Machine center	$ (30,000)				
Working capital	(1,000)				1,000
Salvage					0
Gains tax					0
Net cash flow (in actual dollars)	$ (31,000)	$ 12,700	$ 14,033	$ 10,478	$ 10,589

PW (20%) = $499; IRR′ = 20.88%

Case 2: With Inflation (10%)	0	1	2	3	4
Income Statement					
Revenue		$ 26,950	$ 29,645	$ 32,610	$ 35,870
Expenses					
O&M		11,000	12,100	13,310	14,641
Depreciation		10,000	13,333	4,445	2,222
Taxable income		5,950	4,212	14,855	19,007
Income taxes (40%)		2,380	1,685	5,942	7,603
Net income		$ 3,570	$ 2,527	$ 8,913	$ 11,404
Cash Flow Statement					
Operating activities					
Net income		3,570	2,527	8,913	11,404
Depreciation		10,000	13,333	4,445	2,222
Investment activities					
Machine center	$ (30,000)				
Working capital	(1,000)	(100)	(110)	(121)	$ 1,331
Salvage					0
Gains tax					0
Net cash flow (in actual dollars)	$ (31,000)	$ 13,470	$ 15,750	$ 13,237	$ 14,957
Net cash flow (in constant dollars)	$ (31,000)	$ 12,245	$ 13,017	$ 9,945	$ 10,216

PW (20%) = $(1,074); IRR (actual dollars) = 29.89%; IRR′ (constant dollars) = 18.09%

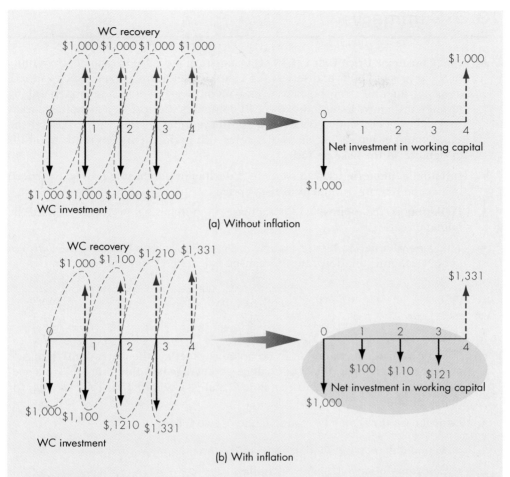

Figure 13.6 Working capital requirements under inflation: (a)Requirements without inflation and (b) requirements with inflation, assuming a 1-year recovery cycle (Example 13.13)

tion to reinvesting the $1,000 revenues, an additional investment of $100 must be made. This $1,100 will be recovered at the end of the second year. However, the project will need a 10% increase, or $1,210 for the third year, and so forth.

As Table 13.7 illustrates, the combined effect of depreciation and working capital is significant. Given an inflationary economy and investment in working capital, the project's IRR′ drops to 18.09%, or *PW* (20%) = −$1,074. By using either IRR analysis or NPW analysis, we end up with the same result (as we must): Alternatives that are attractive when inflation does not exist may not be acceptable when inflation does exist.

13.5 Summary

- The **Consumer Price Index (CPI)** is a statistical measure of change, over time, of the prices of goods and services in major expenditure groups—such as food, housing, apparel, transportation, and medical care—typically purchased by urban consumers. Essentially, the CPI compares the cost of a sample "market basket" of goods and services in a specific period relative to the cost of the same "market basket" in an earlier reference period. This reference period is designated as the **base period**.

- **Inflation** is the term used to describe a **decline in purchasing power** evidenced in an economic environment of rising prices.

- **Deflation** is the opposite: An increase in purchasing power evidenced by falling prices.

- The **general inflation rate** ($\bar{f}$) is an average inflation rate based on the CPI. An annual general inflation rate ($\bar{f}$) can be calculated using the following equation:

$$\bar{f}_n = \frac{CPI_n - CPI_{n-1}}{CPI_{n-1}}.$$

- Specific, individual commodities do not always reflect the general inflation rate in their price changes. We can calculate an **average inflation rate** ($\bar{f}_j$) for a specific commodity (j) if we have an index (that is, a record of historical costs) for that commodity.

- Project cash flows may be stated in one of two forms:

  Actual dollars (A_n): Dollars that reflect the inflation or deflation rate

  Constant dollars (A'_n): Year 0 dollars

- Interest rates for project evaluation may be stated in one of two forms:

  Market interest rate (i): A rate which combines the effects of interest and inflation; used with **actual dollar** analysis

  Inflation-free interest rate (i'): A rate from which the effects of inflation have been removed; this rate is used with constant dollar analysis

- To calculate the present worth of actual dollars, we can use a two-step or a one-step process:

  Deflation method—two steps:
  1. Convert actual dollars by deflating with the general inflation rate of $\bar{f}$.
  2. Calculate the PW of constant dollars by discounting at i' .

  Adjusted-discount method—one step (use the market interest rate):

Item	Effects of Inflation
Depreciation expense	Depreciation expense is charged to taxable income in dollars of declining values; taxable income is overstated, resulting in higher taxes.
Salvage values	Inflated salvage values combined with book values based on historical costs result in higher taxable gains.
Loan repayments	Borrowers repay historical loan amounts with dollars of decreased purchasing power, reducing the debt-financing cost.
Working capital requirement	Known as a *working capital drain*, the cost of working capital increases in an inflationary economy.
Rate of return and NPW	Unless revenues are sufficiently increased to keep pace with inflation, tax effects and/or a working capital drain result in lower rate of return or lower NPW.

Table 13.8
Effects of Inflation on Project Cash Flows and Return

$$P_n = \frac{A_n}{((1 + \bar{f})(1 + i'))^n}$$

$$= \frac{A_n}{(1 + i)^n}.$$

- A number of individual elements of project evaluations can be distorted by inflation. These are summarized in Table 13.8.

Self-Test Questions

13s.1 A series of five constant-dollar (or real-dollar) payments, beginning with $6,000 at the end of the first year, are increasing at the rate of 5% per year. Assume that the average general inflation rate is 4%, and the market (inflation-adjusted) interest rate is 11% during this inflationary period. What is the equivalent present worth of the series?

(a) $24,259   (b) $25,892
(c) $27,211   (d) $29,406.

13s.2 "At a market interest rate of 7% per year and an inflation rate of 5% per year, a series of three equal annual receipts of $100 in constant dollars is equivalent to a series of three annual receipts of $105 in actual dollars." Which of the following statements is correct?

(a) The amount of actual dollars is overstated.
(b) The amount of actual dollars is understated.
(c) The amount of actual dollars is about right.
(d) Sufficient information is not available to make a comparison.

13s.3 The following figures represent the CPI indices (base period 1982–1984 = 100) for urban consumers in U.S. cities. Determine the average general inflation rate between 1991 and 1995. Use the following data:

Year	CPI
1991	136.2
1992	140.3
1993	144.5
1994	148.2
1995	153.6

The average inflation rate between 1991 and 1995 is

(a) 3.07%   (b) 2.43%

(c) 3.05%   (d) 3.64%.

13s.4 How many years will it take for the dollar's purchasing power to be one-half what it is now, if the average inflation rate is expected to continue at the rate of 9% for an indefinite period? (Hint: You may apply the Rule of 72.)

(a) about 6 years   (b) about 8 years

(c) about 10 years   (d) about 12 years.

13s.5 Which of the following statements is incorrect?

(a) A negative inflation rate implies that you are experiencing a deflationary economy.

(b) Under an inflationary economy, debt financing is always a preferred option because you are paying back with cheaper dollars.

(c) If a project requires some investment in working capital under an inflationary economy, its rate of return will decrease when compared with the same project without inflation.

(d) Under an inflationary economy, in general, your tax burden will not increase, as long as inflationary

adjustments are made in tax brackets.

13s.6 You are considering purchasing a $1,000 bond with a coupon rate of 9.5%, interest payable annually. If the current inflation rate is 4% per year, which will continue in the foreseeable future, what would be the real rate of return if you sold the bond at $1,080 after 2 years?

(a) about 8.9%   (b) about 9.5%

(c) about 13.26%   (d) about 9.26%.

13s.7 Vermont Casting has received an order to supply 250 units of casted fireplace inserts each year for Midwestern Builders over a 2-year period. The current sales price is $1,500 per unit, and the current cost per unit is $1,000. Vermont is taxed at a rate of 40%. Both prices and costs are expected to rise at a rate of 6% per year. Vermont will produce these units on fully-depreciated existing machines. Since the orders will be filled at the end of each year, the unit sale price and unit cost during the first year would be $1,590 and $1,060, respectively. Vermont's market interest rate is 15%. Which of the following net present worth calculation is incorrect?

(a) $\text{NPW} = \$500(1 - 0.4)(P/F, 8.49\%, 1) + \$561.8(1 - 0.40)(P/F, 15\%, 2)$

(b) $\text{NPW} = \$530(1 - 0.4)[(P/F, 15\%, 1) + (1.06)(P/F, 15\%, 2)]$

(c) $\text{NPW} = \$500(1 - 0.4)(P/A, 9\%, 2)$

(d) $\text{NPW} = \$500(1 - 0.4)(P/A, 8.49\%, 2)$.

13s.8 A proposed project that requires an investment of $10,000 (now) is expected to generate a series of five equal payments ($6,000 each in constant dollars). Assume that the average inflation rate is 4%, and the market interest rate (i) is 10% during this inflationary period.

What is the equivalent present worth of this investment?

(a) $15,434   (b) $15,274

(c) $12,745   (d) $16,711.

13s.9 A father wants to save in advance for his 8-year old daughter's college expenses. The daughter will enter the college 10 years from now. An annual amount of $20,000 in today's dollars (constant dollars) will be required to support the college for 4 years. Assume that these college payments will be made at the *beginning* of each school year. (The first payment occurs at the end of 10 years). The future general inflation rate is estimated to be 5% per year, and the interest rate on the savings account will be 8% compounded quarterly (market interest rate) during this period. If the father has decided to save only $1,000 (actual dollars) each quarter, how much will the daughter have to borrow to cover her sophomore expenses?

(a) $4,120   (b) $4,314

(c) $4,000   (d) $4,090.

13s.10 Which of the following statements is incorrect under inflationary economy?

(a) Borrowers will always come out ahead as they pay back with cheaper dollars.

(b) In general, you will pay more taxes in real dollars if you have depreciable assets.

(c) In general, there will be more drain in working capital.

(d) Bond interest rates will tend to be higher in the financial market, so that it would cost more to finance a new project.

Problems

Note: In problem statements, the term "market interest rate" represents the "inflation-adjusted MARR" for project evaluation or the "interest rate" quoted by a financial institution for commercial loans.

Measure of Inflation

13.1 The following data indicates the median unleaded gasoline price during the last 5 years for California residents:

Period	Price ($)
1996	$1.10
2000	$1.62

Assuming that the base period (price index = 100) is period 1996, compute the average price index for the unleaded gasoline price for the year 2000.

13.2 The following data indicate the price indices of lumber (base period 1982 = 100) during the last 5 years:

Period	Price (cents)
1996	150.6
1997	155.1
1998	158.3
1999	161.8
2000	165.8
2001	?

(a) Assuming that the base period (price index = 100) is reset to the year 1996 period, compute the average price index for lumber between 1996 and 2000.

(b) If the past trend is expected to continue, how would you estimate the lumber product at time period 2001?

13.3 For prices that are increasing at the annual rate of 5% the first year and 8% the second year, determine the average inflation rate ($\bar{f}$) over these 2 years.

13.4 Because of general price inflation in our economy, the purchasing power of the dollar shrinks with the passage of time. If the average general inflation rate is expected to be 7% per year for the foreseeable future, how many years will it take for the dollar's purchasing power to be one-half of what it is now?

Actual versus Constant Dollars

13.5 An annuity provides for 10 consecutive end-of-year payments of $4,500. The average general inflation rate is estimated to be 5% annually, and the market interest rate is 12% annually. What is the annuity worth in terms of a single equivalent amount of today's dollars?

13.6 A company is considering buying workstation computers to support its engineering staff. In today's dollars, it is estimated that the maintenance costs for the computers (paid at the end of each year) will be $25,000, $30,000, $32,000, $35,000, and $40,000 for years 1 to 5, respectively. The general inflation rate ($\bar{f}$) is estimated to be 8% per year, and the company will receive 15% per year on its invested funds during the inflationary period. The company wants to pay for maintenance expenses in equivalent equal payments (in actual dollars) at the end of each of the 5 years. Find the amount of the company's payment.

13.7 Given the following cash flows in actual dollars, convert to an equivalent cash flow in constant dollars if the base year is time 0. Keep cash flows at same point in time, that is, year 0, 4, 5, and 7. Assume that the market interest rate is 16% and that the general inflation rate ($\bar{f}$) is estimated at 4% per year.

n	Cash Flow (in actual \$)
0	\$1,500
4	2,500
5	3,500
7	4,500

13.8 The purchase of a car requires a \$25,000 loan to be repaid in monthly installments for 4 years at 12% interest, compounded monthly. If the general inflation rate is 6%, compounded monthly, find the actual and constant dollar value of the 20th payment.

13.9 A series of four annual constant-dollar payments beginning with \$7,000 at the end of the first year is growing at the rate of 8% per year [assume that the base year is the current year ($n = 0$)]. If the market interest rate is 13% per year, and the general inflation rate ($\bar{f}$) is 7% per year, find the present worth of this series of payments based on

(a) constant-dollar analysis

(b) actual-dollar analysis.

13.10 Consider the cash flow diagrams, where the equal-payment cash flow in constant dollars (a) is converted from the equal-payment cash flow in actual dollars (b), at an annual general inflation rate of $\bar{f} = 3.8\%$ and $i = 9\%$. What is the amount A in actual dollars equivalent to $A' = \$1,000$ in constant dollars?

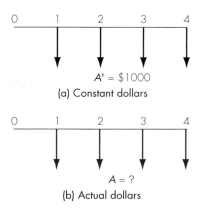

(a) Constant dollars

(b) Actual dollars

13.11 A 10-year \$1,000 bond pays a nominal rate of 9%, compounded semi-annually. If the market interest rate is 12%, compounded annually, and the general inflation rate is 6% per year, find the actual and constant dollar amount (time 0 dollars) of the 16th interest payment on the bond.

Equivalence Calculation under Inflation

13.12 Suppose that you borrow \$20,000 at 12%, compounded monthly, over 5 years. Knowing that the 12% represents the market interest rate, the monthly payment in actual dollars will be \$444.90. If the average monthly general inflation rate is expected to be 0.5%, determine the equivalent equal monthly payment series in constant dollars.

13.13 The annual fuel costs to operate a small solid-waste treatment plant are projected to be \$1.5 million, without considering any future inflation. The best estimates indicate that the annual inflation-free interest rate (i') will be 6% and the general inflation rate ($\bar{f}$) = 5%. If the plant has a remaining useful life of 5 years, what is the present equivalent of its fuel costs using actual dollar analysis?

13.14 Suppose that you just purchased a used car worth $6,000 in today's dollars. Assume also that you borrowed $5,000 from a local bank at 9%, compounded monthly, over 2 years. The bank calculated your monthly payment at $228. Assuming that average general inflation will run at 0.5% per month over the next 2 years,

(a) Determine the annual inflation-free interest rate (i') for the bank.

(b) What equal monthly payments, in terms of constant dollars over the next 2 years, are equivalent to the series of actual payments to be made over the life of the loan?

13.15 A man is planning to retire in 20 years. Money can be deposited at 6%, compounded monthly, and it is also estimated that the future general inflation ($\bar{f}$) rate will be at 5%, compounded annually. What monthly deposit must be made each month until the man retires so that he can make annual withdrawals of $40,000 in terms of today's dollars over the 15 years following his retirement? (Assume that his first withdrawal occurs at the end of the first 6 months after his retirement.)

13.16 A young woman engineer on her 23rd birthday decides to start saving toward building up a retirement fund that pays 8% interest, compounded quarterly (market interest rate). She feels that $600,000 worth of purchasing power in today's dollars will be adequate to see her through her sunset years after her 63rd birthday. Assume a general inflation rate of 6% per year.

(a) If she plans to save by making 160 equal quarterly deposits, what should be the amount of her quarterly deposit in actual dollars?

(b) If she plans to save by making end-of-the-year deposits, increasing by $1,000 over each subsequent year, how much would her first deposit be in actual dollars?

13.17 A father wants to save for his 8-year-old son's college expenses. The son will enter college 10 years from now. An annual amount of $40,000 in constant dollars will be required to support the son's college expenses for 4 years. Assume that these college payments will be made at the beginning of the school year. The future general inflation rate is estimated to be 6% per year, and the market interest rate on the savings account will average 8%, compounded annually. Given this information

(a) What is the amount of the son's freshman-year expense in terms of actual dollars?

(b) What is the equivalent single-sum amount at the present time for these college expenses?

(c) What is the equal amount, in actual dollars, the father must save each year until his son goes to college?

Effects of Inflation on Project Cash Flows

13.18 Consider the following project's after-tax cash flow and the expected annual general inflation rate during the project period:

End of Year	Cash Flow (in actual $)	Expected General Inflation Rate
0	−$45,000	
1	26,000	6.5%
2	26,000	7.7
3	26,000	8.1

(a) Determine the average annual general inflation rate over the project period.

(b) Convert the cash flows in actual dollars into equivalent constant dollars with the base year 0.

(c) If the annual inflation-free interest rate is 5%, what is the present worth of the cash flow? Is this project acceptable?

13.19 Gentry Machines Inc. has just received a special job order from one of its clients. The following financial data has been collected:

- This 2-year project requires purchase of a special purpose equipment of $55,000. The equipment falls into the MACRS 5-year class.

- The machine will be sold at the end of 2 years for $27,000 (today's dollars).

- The project will bring in an additional annual revenue of $114,000 (actual dollars), but it is expected to incur an additional annual operating cost of $53,800 (today's dollars).

- The project requires an investment in working capital in the amount of $12,000 at $n = 0$. In each following year, additional working capital needs to be provided at the rate of general inflation rate. Any investment in working capital will be recovered at the end of project termination.

- To purchase the equipment, the firm expects to borrow $50,000 at 10% over a 2-year period (equal annual payments of $28,810 [actual dollars]). The remaining $5,000 will be taken from the firm's retained earnings.

- The firm expects a general inflation of 5% per year during the project period. The firm's marginal tax rate is 40%, and its market interest rate is 18%.

(a) Compute the after-tax cash flows in actual dollars.

(b) What is the equivalent present value of this amount at time 0?

13.20 Hugh Health Product Corporation is considering the purchase of a computer to control plant packaging for a spectrum of health products. The following data have been collected:

- First cost = $120,000 to be borrowed at 9% interest where only interest is paid each year, and the principal is due in a lump sum at end of year 2.

- Economic service life (project life) = 6 years

- Estimated selling price in year 0 dollars = $15,000

- Depreciation = 5-year MACRS property

- Marginal income tax rate = 40% (remains constant)

- Annual revenue = $145,000 (today's dollars)

- Annual expense (not including depreciation and interest) = $82,000 (today's dollars)

- Market interest rate = 18%

(a) With an average general inflation rate of 5% expected during the project period, which will affect all revenues, expenses, and the salvage value, determine the cash flows in actual dollars.

(b) Compute the net present value of the project under inflation.

(c) Compute the net present value loss (gain) due to inflation.

(d) In (c), how much is the present value loss (or gain) due to borrowing?

13.21 Norcross Textile Company is considering automating their piece-goods screen printing system at a cost of $20,000. The firm expects to phase out this automated printing system at the end of 5 years due to changes in style. At that time, the firm could scrap the system for $2,000 in today's dollars. The expected net savings due to the automation are in today's dollars (constant dollars) as follows:

End of Year	Cash Flow (in constant $)
1	$15,000
2	17,000
3–5	14,000

The system qualifies as a 5-year MACRS property and will be depreciated accordingly. The expected average general inflation rate over the next 5 years is approximately 5% per year. The firm will finance the entire project by borrowing at 10%. The scheduled repayment of the loan will be as follows:

End of Year	Principal Payment	Interest Payment
1	$6,042	$2,000
2	6,647	3,396
3	7,311	731

The firm's market interest rate for project evaluation during this inflation-ridden time is 20%. Assume that the net savings and the selling price will be responsive to this average inflation rate. The firm's marginal tax rate is known to be 40%.

(a) Determine the after-tax cash flows of this project in actual dollars.

(b) Determine the net present value reduction (or gains) in profitability due to inflation.

13.22 The J. F. Manning Metal Co. is considering the purchase of a new milling machine during year 0. The machine's base price is $135,000, and it will cost another $15,000 to modify it for special use by the firm. This results in a $150,000 cost base for depreciation. The machine falls into the MACRS 7-year property class. The machine will be sold after 3 years for $80,000 (actual dollars). Use of the machine will require an increase in net working capital (inventory) of $10,000 at the beginning of the project year. The machine will have no effect on revenues, but it is expected to save the firm $80,000 (today's dollars) per year in before-tax operating costs, mainly labor. The firm's marginal tax rate is 40%, and this rate is expected to remain unchanged over the project's duration. However, the company expects the labor cost will increase at an annual rate of 5%, but that the working capital requirement will grow at an annual rate of 8% caused by inflation. The selling price of the milling machine is not affected by inflation. The general inflation rate is estimated to be 6% per year over the project period. The firm's market interest rate is 20%.

(a) Determine the project cash flows in actual dollars.

(b) Determine the project cash flows in constant (time 0) dollars.

(c) Is this project acceptable?

Rate of Return Analysis under Inflation

13.23 Fuller Ford Company is considering purchasing a vertical drill machine. The machine will cost $50,000 and will have an 8-year service life. The selling price of the machine at the end of 8 years is expected to be $5,000 in today's dollars. The machine will generate annual revenues of $20,000 (today's dollars), but it expects to have an annual expense (excluding depreciation) of $8,000 (today's dollars). The asset is classified as a 7-year MACRS property. The project requires a working capital investment of $10,000 at year 0. The marginal income tax rate for the firm is averaging 35%. The firm's market interest rate is 18%.

(a) Determine the internal rate of return of this investment.

(b) Assume that the firm expects a general inflation rate of 5%, but that it also expects an 8% annual increase in revenue and working capital and a 6% annual increase in expense caused by inflation. Compute the real (inflation-free) internal rate of return. Is this project acceptable?

13.24 Sonja Jensen is considering the purchase of a fast-food franchise. Sonja will be operating on a lot that is to be converted into a parking lot in 6 years, but that may be rented in the interim for $800 per month. The franchise and necessary equipment will have a total first cost of $55,000 and a salvage value of $10,000 (in today's dollars) after 6 years. Sonja is told that the future annual general inflation rate will be 5%. The projected operating revenues and expenses in actual dollars, other than rent and depreciation for the business, are as follows:

End of Year	Revenue	Expenses
1	$30,000	$15,000
2	35,000	21,000
3	55,000	25,000
4	70,000	30,000
5	70,000	30,000
6	60,000	30,000

Assume that the initial investment will be depreciated using the 5-year MACRS, and Sonja's tax rate will be 30%. Sonja can invest her money at a rate of least 10% in other investment activities during this inflation-ridden period.

(a) Determine the cash flows associated with the investment over its investment life.

(b) Compute the projected after-tax rate of return (real) for this investment opportunity.

13.25 You have $10,000 cash, which you want to invest. Normally, you would deposit the money in a savings account that pays an annual interest rate of 6%. However, you are now considering the possibility of investing in a bond. Your alternatives are either a nontaxable municipal bond paying 9% or a taxable corporate bond paying 12%. Your marginal tax rate is 30% for both ordinary income and capital gains. You expect the general inflation to be 3% during the investment period. You can buy a high grade municipal bond costing $10,000, which pays interest of 9% ($900) per year. This interest is not taxable. A comparable high grade corporate bond is also available that is just as safe as the municipal bond, but that pays an interest rate of 12% ($1,200) per year. This interest is taxable as ordinary income. Both bonds mature at the end of year 5.

(a) Determine the real (inflation-free) rate of return for each bond.

(b) Without knowing your MARR, can you make a choice between these two bonds?

13.26 Air Florida is considering two types of engines for use in its planes. Each has the same life, same maintenance, and repair record.

- Engine A costs $100,000 and uses 50,000 gallons per 1,000 hours of operation at the average service load encountered in passenger service.
- Engine B costs $200,000 and uses 32,000 gallons per 1,000 hours of operation at the same service load.

Both engines are estimated to have 10,000 service hours before any major overhaul of the engines is required. If fuel costs $1.25 per gallon currently, and its price is expected to increase at the rate of 8% because of inflation, which engine should the firm install for an expected 2,000 hours of operation per year? The firm's marginal income tax rate is 40%, and the engine will be depreciated based on the unit-of-production method. Assume the firm's market interest rate is 20%. It is estimated that both engines will retain a market value of 40% of their initial cost (actual dollars) if they are sold on market after 10,000 hours of operation.

(a) Using the present worth criterion, which project would you select?

(b) Using the annual equivalent criterion, which project would you select?

(c) Using the future worth criterion, which project would you select?

13.27 Johnson Chemical Company has just received a special subcontracting job from one of its clients. This 2-year project requires purchase of a special-purpose painting sprayer of $60,000. This equipment falls into the MACRS 5-year class. After the subcontracting work is completed, the painting sprayer will be sold at the end of 2 years for $40,000 (actual dollars). The painting system will require an in-

crease in net working capital (spare parts inventory such as spray nozzles) of $5,000. This investment in working capital will be fully recovered at the end of project termination. The project will bring in an additional annual revenue of $120,000 (today's dollars), but it is expected to incur an additional annual operating cost of $60,000 (today's dollars). It is projected that, due to inflation, there will be sales price increases at an annual rate of 5%. (This implies that annual revenues will increase at an annual rate of 5%.) An annual increase of 4% for expenses and working capital requirement is expected. The company has a marginal tax rate of 30% and it uses a market interest rate of 15% for project evaluation during the inflationary period. If the firm expects a general inflation of 8% during the project period:

(a) Compute the after-tax cash flows in actual dollars.

(b) What is the rate of return on this investment (real earnings)?

(c) Is special order profitable?

13.28 Land Development Corporation is considering the purchase of a bulldozer. The bulldozer will cost $100,000 and will have an estimated salvage value of $30,000 at the end of 6 years. The asset will generate annual before-tax revenues of $80,000 over the next 6 years. The asset is classified as a 5-year MACRS property. The marginal tax rate is 40%, and the firm's market interest rate is known to be 18%. All dollar figures represent constant dollars at time 0 and are responsive to general inflation rate $\bar{f}$.

(a) With $\bar{f} = 6\%$, compute the after-tax cash flows in actual dollars.

(b) Determine the real rate of return of this project on after-tax basis.

(c) Suppose that the initial cost of the project will be financed through a local bank at an interest rate of 12%, with an annual payment of $24,323 over 6 years. With this additional condition, answer part (a) above.

(d) In part (a), determine the present value loss due to inflation.

(e) In part (c), determine how much the project has to generate in additional before-tax annual revenue in actual dollars (equal amount) to make up the inflation loss.

13.29 Wilson Machine Tools Inc., a manufacturer of fabricated metal products, is considering the purchase of a high-tech computer-controlled milling machine at a cost of $95,000. The cost of installing the machine, preparing the site, wiring, and rearranging other equipment is expected to be $15,000. This installation cost will be added to the machine cost to determine the total cost basis for depreciation. Special jigs and tool dies for the particular product will also be required at a cost of $10,000. The milling machine is expected to last 10 years, and the jigs and dies only for 5 years. Therefore, another set of jigs and dies has to be purchased at the end of 5 years. The milling machine will have a $10,000 salvage value at the end of its life, and the special jigs and dies are worth only $300 as scrap metal at any time in their lives. The machine is classified as a 7-year MACRS property, and the special jigs and dies are classified as a 3-year MACRS property. With the new milling machine, Wilson expects an additional annual revenue of $80,000 due to increased production. The additional annual production costs are estimated as follows: Materials, $9,000; labor, $15,000; energy, $4,500; and miscellaneous O&M costs, $3,000. Wilson's marginal income tax rate is expected to remain at 35% over the project life of 10 years. All dollar figures represent today's dollars. The firm's market interest rate is 18%, and the expected general inflation rate during the project period is estimated at 6%.

(a) Determine the project cash flows in the absence of inflation.

(b) Determine the internal rate of return for the project in (a).

(c) Suppose that Wilson expects price increases during the project period: material at 4% per year, labor at 5% per year, and energy and other O&M costs at 3% per year. To compensate for these increases in prices, Wilson is planning to increase annual revenue at the rate of 7% per year by charging its customers a higher price. No changes in salvage value are expected for the machine as well as for the jigs and dies. Determine the project cash flows in actual dollars.

(d) In (c), determine the real (inflation-free) rate of return of the project.

(e) Determine the economic loss (or gain) in present worth caused by inflation.

Short Case Studies

13.30 Recent biotechnological research has made possible the development of a sensing device that implants living cells on a silicon chip. The chip is capable of detecting physical and chemical changes in cell processes. Proposed uses include researching the mechanisms of disease on a cellular level, developing new therapeutic drugs, and replacing the use of animals in cosmetic and drug testing. Biotech Device Corporation (BDC) has just perfected a process for mass-producing the chip. The following information has been compiled for the board of directors.

- BDC's marketing department plans to target sales of the device to larger chemical and drug manufacturers. BDC estimates that annual sales would be 2,000 units, if the device were priced at $95,000 per unit (dollars of the first operating year).

- To support this level of sales volume, BDC would need a new manufacturing plant. Once the "go" decision is made, this plant could be built and made ready for production within 1 year. BDC would need a 30-acre tract of land that would cost $1.5 million. If the decision were to be made, the land could be purchased on December 31, 2000. The building would cost $5 million and would be depreciated according to the MACRS 39-year class. The first payment of $1 million would be due to the contractor on December 31, 2001, and the remaining $4 million on December 31, 2002.

- The required manufacturing equipment would be installed late in 1998 and would be paid for on December 31, 2002. BDC would have to purchase the equipment at an estimated cost of $8 million, including transportation, plus a further $500,000 for installation. The equipment would fall into the MACRS 7-year class.

- The project would require an initial investment of $1 million in working capital. This initial working capital investment would be made on December 31, 2002, and on December 31 of each following year, net working capital would be increased by an amount equal to 15% of any sales in-

crease expected during the coming year. The investments in working capital would be fully recovered at the end of project year.

- The project's estimated economic life is 6 years (excluding the 2-year construction period). At that time, the land is expected to have a market value of $2 million, the building a value of $3 million, and the equipment a value of $1.5 million. The estimated variable manufacturing costs would total 60% of the dollar sales. Fixed costs, excluding depreciation, would be $5 million for the first year of operations. Since the plant would begin operations on January 1, 2003, the first operating cash flows would occur on December 31, 2003.

- Sales prices and fixed overhead costs, other than depreciation, are projected to increase with general inflation, which is expected to average 5 percent per year over the 6-year life of the project.

- To date, BDC has spent $5.5 million on research and development (R&D) associated with the cell implanting research. The company has already expensed $4 million R&D costs. The remaining $1.5 million will be amortized over 6 years (i.e., the annual amortization expense would be $250,000). If BDC decides not to proceed with the project, the $1.5 million R&D cost could be written off on December 31, 2000.

- BDC's marginal tax rate is 40%, and its market interest rate is 20%. Any capital gains will also be taxed at 40%.

(a) Determine the after-tax cash flows of the project in actual dollars.

(b) Determine the inflation-free (real) IRR of the investment.

(c) Would you recommend that the firm accept the project?

Special Topics in Engineering Economics

B	C	D	E	F
		Interest (%)	Equivalent Worth at	Option 1
Option 1	Option 2		Option 1	
$100,000		0	$700,000	$
$100,000		1	$721,354	$
$100,000		2	$743,428	$
$100,000		3	$766,246	$
$100,000		4	$789,829	
$100,000		5	$814,201	
$100,000		6	$839,384	
	$100,000	7	$865,402	
	$100,000	8	$892,280	
	$100,000	9	$920,043	
	$100,000	10	$948,717	
	$100,000	11	$978,327	
	$100,000	12	$1,008,901	
	$100,000	13	$1,040,466	
	$100,000	14	$1,073,049	

Project Risk and Uncertainty

Motorola decided to build and operate a small and portable telephone that can be used anywhere on earth. Potential users of the handsets were to include vacationers, business people, and engineers traveling to places where phone service is not available or where an international call could take hours to complete. Because of the sheer size of the project, Motorola decided to share the financial burden by creating a consortium named "Iridium Corporation." Motorola successfully lined up several investment partners, thus raising $5 billion for the project, in which Motorola had a 20% stake. The network of 66 satellites was supposed to provide worldwide cellular, Internet, and paging services, passing calls with ease from anywhere on earth. But the economics of a project like this one are harder to forecast than the technical issues.

Motorola estimated it would need at least 700,000 users to break even. After consulting many international organizations about the proposed venture, company officials forecasted that the system could attract as many as 5 million subscribers worldwide, who would pay at least $100 a month by the year 2000. Despite Motorola management's decision to build the portable telephone, Wall Street investors were uncertain whether the demand for such a portable telephone would be sufficient to justify the risk of undertaking an investment of this magnitude.

During 6 years of construction and testing, the project was plagued with software glitches, failed launches, and costly delays. Finally a full-scale commercial operation was commenced on late October 1999. The 25-ounce handset was priced around $3,000, and Motorola charged calls at $4 to $7 a minute, depending on where the calls were originated. Even after commercial service begun, Motorola and Kyocera, which manufacture the handsets, continued to have production problems. The handset design was too clunky to carry. On the other hand, the cellular industry was working on its own land-based global network and the average call charges were significantly cheaper than Iridium's. But a new handset—a new anything—to make this service more marketable didn't come soon enough. Worse, Iridium continued to miss revenue and customer targets, and was not able to meet mounting debt payments. Finally, Iridium filed for corporate bankruptcy protection and the phone service had to be discontinued eventually on April 17, 2000. It would cost another $57 million to destroy the 66 satellites in the orbit to clear the communication space. What a loss!

Iridium engineers would be recognized as pioneers—they were able to convince many investors including, Motorola management, to undertake such a huge challenging but fascinating engineering project. Sure, they demonstrated the technical feasibility at the expense of $5 billion. However, they were not known for their financial and marketing savvy—they thought that consumers would pay $3,000 for a handset the size of brick. Most of all, they kept believing their vision that people would pay any price for satellite phone service and continued to pay too much attention to the engineering details. Unfortunately, they ignored one important rule in the phone industry, "You don't sell hardware to a telephone user—you sell a service." They couldn't figure out a way to recoup their losses when they realized that there was no market for their product.

What went wrong? Iridium committed so many marketing and sales mistakes that its experience could form the basis of a textbook on how not to sell a product. Its phones started out costing $3,000 and didn't work as promised. They weren't available in stores when Iridium ran a $180 million advertising campaign. And Iridium's prices, which ranged from $3 to $7.50 a call, were much more than that of the average cellphone call. The cellular industry has been working on its own land-based global network, and these players fight for a pretty small market for people in remote locations. Many airplanes and ships have existing systems, so there would be little point in carrying an Iridium phone at such a huge price tag.

In previous chapters, cash flows from projects were assumed to be known with complete certainty; our analysis was concerned with measuring the economic worth of projects and selecting the best investment projects. Although these types of analyses can provide a reasonable decision basis in many investment situations, we should certainly consider the more usual uncertainty. In this type of situation, management rarely has precise expectations about the future cash flows to be derived from a particular project. In fact, the best that a firm can reasonably expect to do is to estimate the range of possible future costs and benefits and the relative chances of achieving a reasonable return on the investment. We use the term **risk** to describe an investment project whose cash flow is not known in advance with absolute certainty, but for which an array of alternative outcomes and their probabilities (odds) are known. We will also use the term **project risk** to refer to variability in a project's NPW. A greater project risk usually means a greater variability in a project's NPW, or simply that the *risk is the potential for loss.* This chapter begins by exploring the origins of project risk.

14.1 Origins of Project Risk

The decision to make a major capital investment such as introducing a new product requires cash flow information over the life of a project. The profitability estimate of an investment depends on cash flow estimations, which are generally uncertain. The factors to be estimated include the total market for the product; the market share that the firm can attain; the growth in the market; the cost of producing the product, including labor and materials; the selling price; the life of the product; the cost and life of the equipment needed; and the effective tax rates. Many of these factors are subject to substantial uncertainty. A common approach is to make single-number "best estimates" for each of the uncertain factors and then to calculate measures of profitability, such as NPW or rate of return for the project. This approach has two drawbacks:

1. No guarantee can ever ensure that the "best estimates" will ever match actual values.
2. No provision is made to measure the risk associated with an investment or the project risk. In particular, managers have no way of determining either the probability that a project will lose money or the probability that it will generate large profits.

Because cash flows can be so difficult to estimate accurately, project managers frequently consider a range of possible values for cash flow elements. If a range of values for individual cash flows is possible, it follows that a range of values for the NPW of a given project is also possible. Clearly, the analyst will want to try to gauge the probability and reliability of individual cash flows occurring and, consequently, the level of certainty about overall project worth.

Quantitative statements about risk are given as numerical probabilities or as values for likelihood (odds) of occurrence. Probabilities are given as decimal fractions in the interval 0.0 to 1.0. An event or outcome that is certain to occur has a probability of 1.0. As the probability of an event approaches 0, the event becomes increasingly less likely to occur. The assignment of probabilities to the various outcomes of an investment project is generally called **risk analysis**. Example 14.1 illustrates some important probability concepts that are easily demonstrated in daily life.

Example 14.1 Improving the Odds—All It Takes Is $7 Million and a Dream

In Virginia's six-number lottery, or lotto, players pick six numbers from 1 to 44. The winning combination is determined by a machine that looks like a popcorn machine, except that it is filled with numbered table-tennis balls. On February 15, 1992, the Virginia lottery drawing offered the following prizes,[1] assuming the first prize is not shared:

Number of Prizes	Prize Category	Total Amount
1	First prize	$27,007,364
228	Second prizes ($899 each)	204,972
10,552	Third prizes ($51 each)	538,152
168,073	Fourth prizes ($1 each)	168,073
	Total winnings	$27,918,561

Common among regular lottery players is this dream: Waiting until the jackpot reaches an astronomical sum and then buying every possible number, thereby guaranteeing a winner. Sure it would cost millions of dollars, but the payoff would be much greater. Is it worth trying? How do the odds of winning the first prize change as you increase the number of ticket purchases?

Discussion: In Virginia, one investment group came tantalizingly close to cornering the market on all possible combinations of six numbers from 1 to 44. State lottery offi-

---

[1] Prizes are based on before-tax values and on the actual number of second- and third-prize winning tickets sold.

cials say that the group bought 5 million of the possible 7 million tickets (precisely 7,059,052). Each ticket cost $1 each. The lottery had a more than $27 million jackpot.[2]

- Economic Logic: If the jackpot is big enough, provided nobody else buys a winning ticket, it makes economic sense to buy one lottery ticket for every possible combination of numbers and be sure to win. A group in Australia apparently tried to do this in the February 15 (1992) Virginia Lottery drawing.
- The Cost: Since 7,059,052 combinations of numbers are possible[3] and each ticket costs $1, it would cost $7,059,052 to cover every combination. The total remains the same regardless of the size of the jackpot.
- The Risk: The first prize jackpot is paid out in 20 equal yearly installments, so the actual payoff on all prizes is $2,261,565 the first year and $1,350,368 per year for the next 19 years. If more than one first prize-winning ticket is sold, the prize is shared so that the maximum payoff depends on an ordinary player not buying a winning ticket. Since Virginia began its lottery in January 1990, 120 of the 170 drawings have not yielded a first-prize winner.

Solution

In the Virginia game, 7,059,052 combinations of numbers are possible. The following table summarizes the winning odds for various prizes for a one ticket-only purchase.

Number of Prizes	Prize Category	Winning Odds
1	First prize	0.0000001416
228	Second prizes	0.0000323
10,552	Third prizes	0.00149
168,073	Fourth prizes	0.02381

So a person who buys one ticket has odds of 1 in slightly more than 7 million. Holding more tickets increases the odds of winning, so that 1,000 tickets have odds of 1 in 7,000 and 1 million tickets have odds of 1 in 7. Since each ticket costs $1, it would receive at least a share in the jackpot and many of the second, third, and fourth place prizes. Together these combined prizes (second through fourth) were worth $911,197 payable in one lump sum. Suppose that the Australia group bought all the tickets (7,059,052). We may consider two separate cases.

---

[2] Source: *The New York Times*, February 25, 1992.
[3] One ticket each for every possible combination of 6 numbers from 1 to 44: Let $C(n,k)$ = the number of combinations of n distinct numbers taken k at a time. Then

$$C(44,6) = \frac{44!}{6!(44-6)!} = 7,059,052.$$

- Case 1: If none of the prizes was shared, the rate of return on this lottery investment, with prizes paid at the end of each year would be

$$PW(i) = -\$7,059,052 + \$911,197(P/F, i, 1)$$
$$+ \$1,350,368(P/A, i, 20)$$
$$= 0.$$
$$i^* = 20.94\%$$

The first-prize payoff over 20 years is equivalent to putting the same $7,059,052 in a more conventional investment that pays a guaranteed 20.94% return before taxes for 20 years, a rate available only from speculative investments with fairly high risk. (If the prizes are paid at the beginning of each year, the rate of return would be 27.88%.)

- Case 2: If the first prize is shared with one other ticket, the rate of return on this lottery investment would be 8.87%. (With the prizes paid at the beginning of each year, the rate of return would be 10.48%.) Certainly, if the first prize is shared by more than one, the rate of return would be far less than 8.87%.

Comments: Only lack of time prevented the group from buying the extra 2 million tickets. On February 15, the winning number, yielding a prize of $1,350,368 a year for 20 years, was pulled. Officials checked their records and found that one winning ticket, with the numbers 8, 11, 13, 15, 19, and 20, had been sold. Several clues pointed to an Australian investment group as being the winner. The investment group from Australia was able to reduce the uncertainty inherent to any lottery game by purchasing the bulk of the lottery tickets. Conceptually, we can entirely reduce the uncertainty (or risk) by purchasing the entire ticket pool. However, ordinarily in a project investment environment, it is not feasible to reduce project risk in the same way that this lottery example illustrated.

14.2   Methods of Describing Project Risk

We may begin analyzing project risk by first determining the uncertainty inherent in a project's cash flows. We can do this analysis in a number of ways, which range from making informal judgments to calculating complex economic and statistical analyses. In this section, we will introduce three methods of describing project risk: (1) sensitivity analysis, (2) break-even analysis, and (3) scenario analysis. Each method will be explained with reference to a single example (Boston Metal Company).

14.2.1   Sensitivity Analysis

One way to glean a sense of the possible outcomes of an investment is to perform a sensitivity analysis. Sensitivity analysis determines the effect on the NPW of variations in the input variables (such as revenues, operating cost, and salvage value) used to estimate after-tax cash flows. A **sensitivity analysis** reveals how much the NPW will change in response to a given change in an input variable. In calculating cash flows, some items have a greater influence on the final result than others. In some problems, the most significant item may be easily identified. For example, the estimate of sales volume is often a major factor in a problem in which the quantity sold varies among the alternatives. In other problems, we may want to locate the items that have an important influence on the final results so that they can be subjected to special scrutiny.

Sensitivity analysis is sometimes called "what-if" analysis because it answers questions such as, What if incremental sales are only 1,000 units, rather than 2,000 units? Then what will the NPW be? Sensitivity analysis begins with a base-case situation, which is developed using the most-likely values for each input. We then change the specific variable of interest by several specified percentages above and below the most-likely value, while holding other variables constant. Next, we calculate a new NPW for each of these values. A convenient and useful way to present the results of a sensitivity analysis is to plot **sensitivity graphs**. The slopes of the lines show how sensitive the NPW is to changes in each of the inputs: The steeper the slope, the more sensitive the NPW is to a change in a particular variable. Sensitivity graphs identify the crucial variables that affect the final outcome most. We will use Example 14.2 to illustrate the concept of sensitivity analysis.

Example 14.2 Sensitivity Analysis

Boston Metal Company (BMC), a small manufacturer of fabricated metal parts, must decide whether to enter the competition to become the supplier of transmission-housings for Gulf Electric. Gulf Electric produces transmission-housings in its own in-house manufacturing facility, but it has almost reached its maximum production capacity. Therefore, Gulf is looking for an outside supplier. To compete, BMC must design a new fixture for the production process and purchase a new forge. The new forge would cost $125,000. This total includes retooling costs for the transmission-housings. If BMC gets the order, it may be able to sell as many as 2,000 units per year to Gulf Electric for $50 each, and variable production costs,[4] such as direct labor and direct material costs, will be $15 per unit. The increase in fixed costs,[5] other than depreciation, will amount to $10,000 per year. The firm expects that the proposed transmission-housings project will have about a 5-year product life. The firm also estimates that the amount ordered by Gulf Electric for the first year will be ordered in each of the subsequent 4 years. (Due to the nature of contracted production, the annual demand and unit price would remain the same over the project after the contract is signed.) The initial investment can be depreciated on a MACRS basis over the 7-year period, and the marginal income tax rate is expected to remain at 40%. At the end of 5 years, the forge is expected to retain a market value of about 32% of the original invest-

---

[4] Expenses that change in direct proportion to the change in volume of sales or production, as defined in Section 3.3.

[5] Expenses that do not vary as the volume of sales or production changes. For example, property taxes, insurance, depreciation, and rent are usually fixed expenses.

ment. Based on this information, the engineering and marketing staffs of BMC
have prepared the cash flow forecasts shown in Table 14.1. Since NPW is positive
($40,168) at the 15% opportunity cost of capital (MARR), the project appears to
be worth undertaking.

However, BMC's managers are uneasy about this project because too many
uncertain elements have not been considered in the analysis. If it decided to take
on the project, BMC must invest in the forging machine to provide Gulf Electric
with some samples as a part of the bidding process. If Gulf Electric does not like
BMC's sample, BMC stands to lose its entire investment in the forging machine.
Another issue is that, if Gulf likes BMC's sample, but it is overpriced, BMC
would be under pressure to bring the price in line with competing firms. Even the
possibility that BMC would get a smaller order must be considered, as Gulf may
utilize their overtime capacity to produce some extra units. It is also not certain
about the variable and fixed cost figures. Recognizing these uncertainties, the
managers want to assess the various potential future outcomes before making a
final decision. Put yourself in BMC's management position and describe how you

	0	1	2	3	4	5
Revenues						
Unit price		$ 50	$ 50	$ 50	$ 50	$ 50
Demand (units)		2,000	2,000	2,000	2,000	2,000
Sales revenue		$100,000	$100,000	$100,000	$100,000	$100,000
Expenses						
Unit variable cost		$ 15	$ 15	$ 15	$ 15	$ 15
Variable cost		30,000	30,000	30,000	30,000	30,000
Fixed cost		10,000	10,000	10,000	10,000	10,000
Depreciation		17,863	30,613	21,863	15,613	5,575
Taxable income		$ 42,137	$ 29,387	$ 38,137	$ 44,387	$ 54,425
Income taxes (40%)		16,855	11,755	15,255	17,755	21,770
Net income		$ 25,282	$ 17,632	$ 22,882	$ 26,632	$ 32,655
Cash Flow Statement						
Operating activities						
Net income		25,282	17,632	22,882	26,632	32,655
Depreciation		17,863	30,613	21,863	15,613	5,575
Investment activities						
Investment	(125,000)					
Salvage						40,000
Gains tax						(2,611)
Net cash flow	$(125,500)	$ 43,145	$ 48,245	$ 44,745	$ 42,245	$ 75,619

Table 14.1
After-Tax
Cash Flow
for BMC's
Transmission-Housings Project
(Example
14.2)

may resolve the uncertainty associated with the project. In doing so, perform a sensitivity analysis for each variable and develop a sensitivity graph.

Discussion: Table 14.1 shows BMC's expected cash flows—but a guarantee that they will indeed materialize cannot be assumed. BMC is not particularly confident in its revenue forecasts. The managers think that, if competing firms enter the market, BMC will lose a substantial portion of the projected revenues by not being able to increase its bidding price. Before undertaking the project described, the company needs to identify the key variables that will determine whether the project will succeed or fail. The marketing department has estimated revenue as follows:

$$\text{Annual revenue} = (\text{Product demand})(\text{unit price})$$
$$= (2,000)(\$50) = \$100,000.$$

The engineering department has estimated variable costs such as labor and material per unit at $15. Since the projected sales volume is 2,000 units per year, the total variable cost is $30,000.

After first defining the unit sales, unit price, unit variable cost, fixed cost, and salvage value, we conduct a sensitivity analysis with respect to these key input variables. This is done by varying each of the estimates by a given percentage and determining what effect the variation in that item will have on the final results. If the effect is large, the result is sensitive to that item. Our objective is to locate the most sensitive item(s).

Solution

Sensitivity analysis: We begin the sensitivity analysis with a consideration of the "base-case" situation, which reflects the best estimate (expected value) for each input variable. In developing Table 14.2, we changed a given variable by 20% in 5% increments, above and below the base-case value and calculated new NPWs, while other variables were held constant. The values for both sales and operating costs were the expected, or base-case, values, and the resulting $40,169 is the base-case NPW. Now we ask a series of "what-if" questions: What if sales are 20% below the expected level? What if operating costs rise? What if the unit price

Table 14.2 Sensitivity Analysis for Five Key Input Variables (Example 14.2)

Deviation	-20%	-15%	-10%	-5%	0% (Base)	5%	10%	15%	20%
Unit price	$ (57)	$ 9,999	$ 20,055	$ 30,111	$ 40,169	$ 50,225	$ 60,281	$ 70,337	$ 80,393
Demand	12,010	19,049	26,088	33,130	40,169	47,208	54,247	61,286	68,325
Variable cost	52,236	49,219	46,202	43,186	40,169	37,152	34,135	31,118	28,101
Fixed cost	44,191	43,185	42,179	41,175	40,169	39,163	38,157	37,151	36,145
Salvage value	37,782	38,378	38,974	39,573	40,169	40,765	41,361	41,957	42,553

drops from $50 to $45? Table 14.2 summarizes the results of varying the values of the key input variables.

Sensitivity graph: Figure 14.1 shows the transmission project's sensitivity graphs for five of the key input variables. The base-case NPW is plotted on the ordinate of the graph at the value 1.0 on the abscissa (or 0% deviation). Next, the value of product demand is reduced to 0.95 of its base-case value, and the NPW is recomputed with all other variables held at their base-case value. We repeat the process by either decreasing or increasing the relative deviation from the base case. The lines for the variable unit price, variable unit cost, fixed cost, and salvage value are obtained in the same manner. In Figure 14.1, we see that the project's NPW is (1) very sensitive to changes in product demand and unit price, (2) fairly sensitive to changes in the variable costs, and (3) relatively insensitive to changes in the fixed cost and the salvage value.

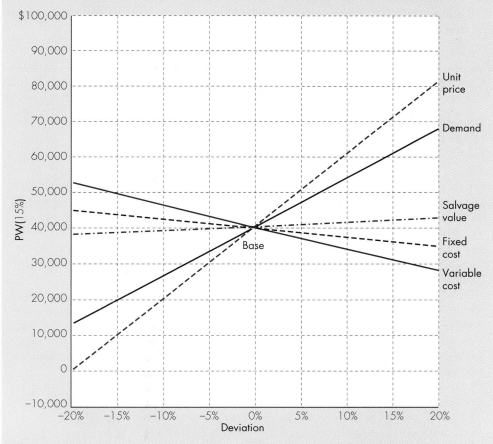

Figure 14.1   Sensitivity graph—BMC's transmission-housings project (Example 14.2)

Graphic displays such as the one in Figure 14.1 provide a useful means to communicate the relative sensitivities of the different variables on the corresponding NPW value. However, sensitivity graphs do not explain any interactions among the variables or the likelihood of realizing any specific deviation from the base case. Certainly, it is conceivable that an answer might not be very sensitive to changes in either of two items, but very sensitive to combined changes in them.

14.2.2   Sensitivity Analysis for Mutually Exclusive Alternatives

In Figure 14.1, each variable is uniformly adjusted by $\pm 20\%$ and all variables are plotted on the same chart. This uniform adjustment can be of too simplistic an assumption; in many situations, each variable can have a different range of uncertainty. Plotting all variables on the same chart could be confusing if there are too many variables to consider. When we perform sensitivity analysis for mutually exclusive alternatives, it may be more effective to plot the PWs (or any other measures, such as AEs) of all alternatives over the range of each variable, basically one plot for each variable, with units of the variable on the horizontal axis. Example 14.3 illustrates this.

Example 14.3 Sensitivity Analysis for Mutually Exclusive Alternatives

A local U.S. Postal Service Office is considering purchasing a 4,000-pound forklift truck, which will be used primarily for processing incoming as well as outgoing postal packages. Forklift trucks traditionally have been fueled by either gasoline, liquid propane gas (LPG), or diesel fuel. Battery-powered electric forklifts, however, are increasingly popular in many industrial sectors due to the economic and environmental benefits that accrue from their use. Therefore, the postal service is interested in comparing the four different types of fuel. The purchase costs as well as annual operating and maintenance costs are provided by a local utility company and the Lead Industries Association. Annual fuel and maintenance costs are measured in terms of number of shifts per year, where one shift is equivalent to 8 hours of operation.

The postal service is unsure of the number of shifts per year, but it expects it should be somewhere between 200 and 260 shifts. Since the U.S. Postal Service does not pay income taxes, no depreciation or tax information is required. The U.S. government uses 10% as an interest rate for any project evaluation of this nature. Develop a sensitivity graph that shows how the choice of alternatives changes as a function of number of shifts per year.

Solution

Two annual cost components are pertinent to this problem: (1) Ownership cost (capital cost) and (2) operating cost (fuel and maintenance cost). Since the operating cost is already given in terms of annual basis, we only need to determine the equivalent annual ownership cost for each alternative.

	Electrical Power	LPG	Gasoline	Diesel Fuel
Life expectancy	7 year	7 years	7 years	7 years
Initial cost	$29,739	$21,200	$20,107	$22,263
Salvage value	$3,000	$2,000	$2,000	$2,200
Maximum shifts				
per year	260	260	260	260
Fuel consumption/shift	31.25 kWh	11 gal	11.1 gal	7.2 gal
Fuel cost/unit	$0.05/kWh	$1.02/gal	$1.20/gal	$1.13/gal
Fuel cost per shift	$1.56	$11.22	$13.32	$8.14
Annual maintenance cost				
Fixed cost	$500	$1,000	$1,000	$1,000
Variable cost/shift	$4.5	$7	$7	$7

(a) Ownership cost (capital cost): Using the capital recovery with return formula developed in Eq. (8.3), we compute

Electrical power: $CR(10\%) = (\$29,739 - \$3,000)(A/P, 10\%, 7) + (0.10)$
$\$3,000 = \$5,792$

LPG: $CR(10\%) = (\$21,200 - \$2,000)(A/P, 10\%, 7) + (0.10)$
$\$2,000 = \$4,144$

Gasoline: $CR(10\%) = (\$20,107 - \$2,000)(A/P, 10\%, 7) + (0.10)$
$\$2,000 = \$3,919$

Diesel fuel: $CR(10\%) = (\$22,263 - \$2,200)(A/P, 10\%, 7) + (0.10)$
$\$2,200 = \$4,341$

(b) Annual operating cost: We can express the annual operating cost as a function of number of shifts per year (M) by combining the variable and fixed cost portions of fuel and maintenance expenditures.

Electrical power: $\$500 + (1.56 + 4.5)M = \$500 + 5.06M$
LPG: $\$1,000 + (11.22 + 7)M = \$1,000 + 18.22M$
Gasoline: $\$1,000 + (13.32 + 7)M = \$1,000 + 20.32M$
Diesel fuel: $\$1,000 + (8.14 + 7)M = \$1,000 + 15.14M$

(c) Total equivalent annual cost: This is the sum of the ownership cost and operating cost.

Electrical power: $AE(10\%) = 6,292 + 5.06M$
LPG: $AE(10\%) = 5,144 + 18.22M$
Gasoline: $AE(10\%) = 4,919 + 20.32M$
Diesel fuel: $AE(10\%) = 5,341 + 15.14M$

In Figure 14.2, these four annual equivalent costs are plotted as a function of number of shifts, M. It appears that the economics of the electric forklift truck can be justified as long as the number of annual shifts exceeds approximately 95.

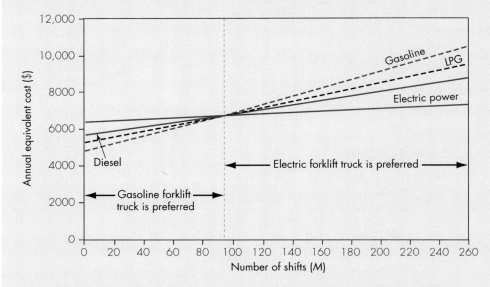

Figure 14.2   Sensitivity analysis for mutually exclusive alternatives (Example 14.3)

14.2.3 Break-Even Analysis

When we perform a sensitivity analysis of a project, we are asking how serious the effect of lower revenues or higher costs will be on the project's profitability. Managers sometimes prefer to ask how much sales can decrease below forecasts before the project begins to lose money. This type of analysis is known as **break-even analysis**. In other words, break-even analysis is a technique for studying the effect of variations in output on a firm's NPW (or other measures). We will present an approach to break-even analysis based on the project's cash flows.

To illustrate the procedure of break-even analysis based on NPW, we use the generalized cash flow approach we discussed in Section 12.4. We compute the PW of cash inflows as a function of an unknown variable (say X)—this variable could be annual sales. For example:

$$\text{PW of cash inflows} = f(x)_1.$$

Next, we compute the PW of cash outflows as a function of X:

$$\text{PW of cash outflows} = f(x)_2.$$

NPW is of course the difference between these two numbers. Then, we look for the break-even value of x that makes

$$f(x)_1 = f(x)_2.$$

Note that this break-even value calculation is similar to that used to calculate the internal rate of return where we want to find the interest rate that makes the NPW equal zero and many other similar "cutoff values" where a choice changes.

Example 14.4 Break-Even Analysis

Through the sensitivity analysis in Example 14.2, BMC's managers are convinced that the NPW is most sensitive to changes in annual sales volumes. Determine the break-even NPW value as a function of that variable.

Solution

The analysis is shown in Table 14.3, where the revenues and costs of the BMC transmission-housings project are set out in terms of an unknown amount of annual sales, X.

We calculate both the PWs of cash inflow and outflows as

- PW of cash inflows

$$
\begin{aligned}
PW(15\%)_{\text{Inflow}} &= (\text{PW of after-tax net revenue}) \\
&\quad + (\text{PW of net salvage value}) \\
&\quad + (\text{PW of tax savings from depreciation}).
\end{aligned}
$$

$$
\begin{aligned}
&= 30X\,(P/A, 15\%, 5) + \$37{,}389(P/F, 15\%, 5) \\
&\quad + \$7{,}145(P/F, 15\%, 1) + \$12{,}245(P/F, 15\%, 2) \\
&\quad + \$8{,}745(P/F, 15\%, 3) + \$6{,}245(P/F, 15\%, 4) \\
&\quad + \$2{,}230(P/F, 15\%, 5) \\
&= 30X(P/A, 15\%, 5) + \$44{,}490 \\
&= 100.5650X + \$44{,}490.
\end{aligned}
$$

	0	1	2	3	4	5
Cash inflow						
Net salvage						37,389
Revenue						
$X(1-0.4)(\$50)$		30X	30X	30X	30X	30X
Depreciation credit						
0.4 (depreciation)		7,145	12,245	8,745	6,245	2,230
Cash outflow						
Investment	−125,000					
Variable cost						
$-X(1-0.4)(\$15)$		−9X	−9X	−9X	−9X	−9X
Fixed cost						
$-(1-0.4)(\$10{,}000)$		−6,000	−6,000	−6,000	−6,000	−6,000
Net cash flow	−125,000	21X + 1,145	21X + 6,245	21X + 2,745	21X + 245	21X + 33,617

Table 14.3
Break-Even Analysis with Unknown Annual Sales (Example 14.4)

- PW of cash outflows:

$$PW(15\%)_{\text{Outflow}} = (\text{PW of capital expenditure})$$

$$+ (\text{PW of after-tax expenses}).$$

$$= \$125{,}000 + (9X + \$6{,}000)(P/A, 15\%, 5)$$

$$= 30.1694X + \$145{,}113.$$

The NPW of cash flows for the BMC is thus

$$PW(15\%) = 100.5650X + \$44{,}490$$

$$- (30.1694X + \$145{,}113)$$

$$= 70.3956X - \$100{,}623.$$

In Table 14.4, we compute the PW of the inflows and the PW of the outflows as a function of demand (X).

Table 14.4
Determination of Break-Even Volume Based on Project's NPW (Example 14.4)

Demand (X)	PW of Inflow (100.5650X + $44,490)	PW of Outflow (30.1694X + $145,113)	NPW (70.3956X − $100,623)
0	$ 44,490	$145,113	(100,623)
500	94,773	160,198	(65,425)
1000	145,055	175,282	(30,227)
1429	188,197	188,225	(28)
1430	188,298	188,255	43
1500	195,338	190,367	4,970
2000	245,620	205,452	40,168
2500	295,903	220,537	75,366

Break-even volume = 1,430 units

The NPW will be just slightly positive if the company sells 1,430 units. Precisely calculated, the zero-NPW point (break-even volume) is 1,429.43 units:

$$PW(15\%) = 70.3956X - \$100{,}623$$

$$= 0$$

$$X_b = 1{,}430 \text{ units}.$$

In Figure 14.3, we have plotted the PWs of the inflows and outflows under various assumptions about annual sales. The two lines cross when sales are 1,430 units, the point at which the project has a zero NPW. Again we see that, as long as sales are greater or equal to 1,430, the project has a positive NPW.

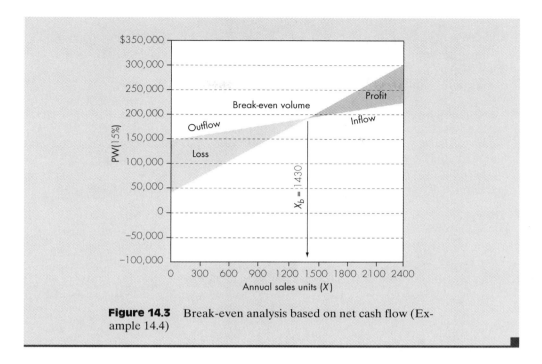

Figure 14.3   Break-even analysis based on net cash flow (Example 14.4)

14.2.4   Scenario Analysis

Although both sensitivity and break-even analyses are useful, they have limitations. Often it is difficult to specify precisely the relationship between a particular variable and the NPW. The relationship is further complicated by interdependencies among the variables. Holding operating costs constant while varying unit sales may ease the analysis, but in reality, operating costs do not behave in this manner. Yet, it may complicate the analysis too much to permit movement in more than one variable at a time.

A scenario analysis is a technique that does consider the sensitivity of NPW to both changes in key variables and to the range of likely variable values. For example, the decision-maker may consider two extreme cases, a "worst-case" scenario (low unit sales, low unit price, high variable cost per unit, high fixed cost, and so on) and a "best-case" scenario. The NPWs under the worst and the best conditions are then calculated and compared to the expected, or base-case, NPW. Example 14.5 will illustrate a plausible scenario analysis for BMC's transmission-housings project.

Example 14.5 Scenario Analysis

Consider again BMC's transmission-housings project in Example 14.2. Assume that the company's managers are fairly confident of their estimates of all the project's cash flow variables, except the estimates for unit sales. Further, assume that they regard a decline in unit sales to below 1,600 or a rise above 2,400 as extremely unlikely. Thus, decremental annual sales of 400 units defines the lower bound, or the worst-case scenario, whereas incremental annual sales of 400 units defines the upper bound, or the best-case scenario. (Remember that the most-likely value was 2,000 in annual unit sales.) Discuss the worst- and best-case scenarios, assuming that the unit sales for all 5 years would be equal.

Discussion: To carry out the scenario analysis, we ask the marketing and engineering staffs to give optimistic (best-case) and pessimistic (worst-case) estimates for the key variables. Then we use the worst-case variable values to obtain the worst-case NPW and the best-case variable values to obtain the best-case NPW.

Solution

The results of our analysis are summarized below. We see that the base-case produces a positive NPW, the worst-case produces a negative NPW, and the best-case produces a large positive NPW.

Variable Considered	Worst-Case Scenario	Most-Likely-Case Scenario	Best-Case Scenario
Unit demand	1,600	2,000	2,400
Unit price ($)	48	50	53
Variable cost ($)	17	15	12
Fixed cost ($)	11,000	10,000	8,000
Salvage value ($)	30,000	40,000	50,000
PW(15%)	−$5,856	$40,169	$104,295

By just looking at the results in the table, it is not easy to interpret the scenario analysis or to make a decision based on it. For example, we could say that there is a chance of losing money on the project, but we do not yet have a specific probability for this possibility. Clearly, we need estimates of the probabilities of occurrence of the worst-case, the best-case, the base-case (most likely), and all the other possibilities. The need to estimate probabilities leads us directly to our next step, developing a probability distribution (or the probability that the variable in question takes on a certain value). If we can predict the effects on the NPW of variations in the parameters, why should we not assign a probability distribution to the possible outcomes of each parameter and combine these distributions in some way to produce a probability distribution for the possible outcomes of the NPW? We shall consider this issue in the next sections.

14.3 Probability Concepts for Investment Decisions

In this section, we shall assume that the analyst has available the probabilities (likelihoods) of future events from either previous experience in a similar project or a market survey. The use of probability information can provide management with a range of possible outcomes and the likelihood of achieving different goals under each investment alternative.

14.3.1   Assessment of Probabilities

We will first define terms related to probability concepts, such as random variable, probability distribution, and cumulative probability distribution.

Random Variables

A **random variable** is a parameter or variable that can have more than one possible value. The value of a random variable at any one time is unknown until the event occurs, but the probability that the random variable will have a specific value is known in advance. In other words, with each possible value of the random variable, associated with it is a likelihood, or probability, of occurrence. For example, when your college team plays a football game, only three events regarding the game outcome are possible: Win, lose, or tie (if allowed). The game outcome is a random variable, which is largely dictated by the strength of your opponent.

To indicate random variables, we will adopt the convention of a capital italic letter (for example, X). To denote the situation where the random variable takes a specific value, we will use a lower-case italic letter (for example, x). Random variables are classified as either discrete or continuous. Any random variables that take on only isolated (countable) values are **discrete random variables**. **Continuous random variables** may have any value in a certain interval. For example, the game outcome described above should be a discrete random variable. But suppose you are interested in the amount of beverage sold on a given game day. The quantity (or volume) of beverage sold will depend on the weather conditions, the number of people attending the game, and other factors. In this case, the quantity is a random variable which takes a continuous value.

Probability Distributions

For a discrete random variable, the probability figure for each random event needs to be assessed. For a continuous random variable, the probability function needs to be assessed as the event takes place over a continuous domain. In either case, a range of possibilities for each feasible outcome exists. These together make up a **probability distribution**.

Probability assessments may be based on past observations or historical data, if the same trends or characteristics of the past are expected to prevail in the future. Forecasting weather or predicting a game outcome in many professional sports is done based on the compiled statistical data. Any probability assessments based on objective data are called **objective probabilities**. We are not restricted to objective probabilities. In many real investment situations, no objective data are available to consider. In these situations, we assign **subjective probabilities** that we think are appropriate to the possible states of nature. As long as we act consistently with our beliefs about the possible events, we may reasonably account for the economic consequences of those events in our profitability analysis.

For a continuous random variable, we usually try to establish a range of values; i.e., we try to determine a **minimum value** (L) and a **maximum value** (H). Next, we determine whether any value within these limits might be *more likely* to occur than

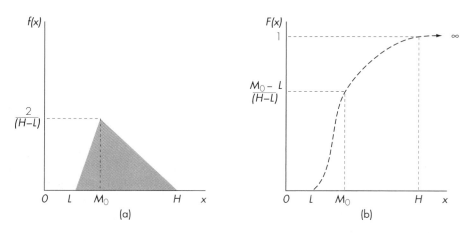

Figure 14.4 A triangular probability distribution: (a) Probability function and (b) cumulative probability distribution

are the other values; i.e., does the distribution have a **mode** (M_o), or a most frequently occurring value?

If the distribution does have a mode, we can represent the variable by a **triangular distribution**, such as that shown in Figure 14.4. If we have no reason to assume that one value is any more likely to occur than any other, perhaps the best we can do is to describe the variable as a **uniform distribution**, as shown in Figure 14.5. These two distributions are frequently used to represent the variability of a random variable when the only information we have is its minimum, its maximum, and whether or not the distribution has a mode. For example, suppose the best judgment of the analyst was that the sales revenue could vary anywhere from $2,000 to $5,000 per day, and any value within the range is equally likely. This judgment about the variability of sales revenue could be represented by a uniform distribution.

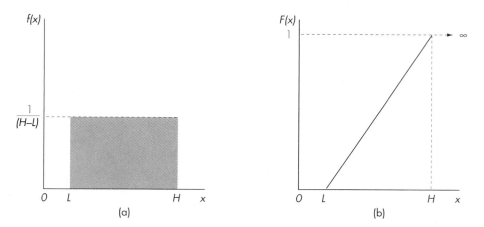

Figure 14.5 A uniform probability distribution: (a) Probability function and (b) cumulative probability distribution

| Product Demand (X) | | Unit Sale Price (Y) | | Table 14.5 |
Units (x)	$P(X = x)$	Unit price (y)	$P(Y = y)$	Probability Distributions
1,600	0.20	$48	0.30	for Unit Demand (X)
2,000	0.60	50	0.50	and Unit
2,400	0.20	53	0.20	Price (Y) for BMC's

X and Y are independent random variables.

Table 14.5 Probability Distributions for Unit Demand (X) and Unit Price (Y) for BMC's Project

For BMC's transmission-housings project, we can think of the discrete random variables (X and Y) as variables whose values cannot be predicted with certainty at the time of decision making. Let us assume the probability distributions in Table 14.5: We see that the product demand with the highest probability is 2,000 units, whereas the unit sales price with the highest probability is $50. These, therefore, are the most-likely values. We also see a substantial probability that a unit demand, other than 2,000 units, will be realized. When we use only the most-likely values in an economic analysis, we are in fact ignoring these other outcomes.

Cumulative Distribution

As we have observed in the previous section, the probability distribution provides information regarding the probability that a random variable will be some value, x. We can use this information, in turn, to define the cumulative distribution function. The **cumulative distribution** function shows the probability that the random variable will attain a value smaller than, or equal to, some value, x. A common notation for the cumulative distribution is

$$F(x) = P(X \le x) = \begin{cases} \sum_{j=1}^{j} p_j & \text{(for a discrete random variable)} \\ \int_L^x f(x)\,dx & \text{(for a continuous random variable)} \end{cases}$$

where p_j is the probability of occurrence of the x_jth value of the discrete random variable, and $f(x)$ is a probability function for a continuous variable. With respect to a continuous random variable, the cumulative distribution rises continuously in a smooth (rather than stairwise) fashion.

In Example 14.6, we will explain the method by which probabilistic information can be incorporated into our analysis. Again, BMC's transmission-housing project will be used. In the next section we will show you how to compute some composite statistics using all the data.

Example 14.6 Cumulative Probability Distributions

Suppose that the only parameters subject to risk are the number of unit sales (X) to Gulf Electric each year and the unit sales price (Y). From experience in the market, BMC assesses the probabilities of outcomes for each variable as shown in Table 14.5. Determine the cumulative probability distributions for these random variables.

Solution

Consider the demand probability distribution (X) given previously in Table 14.5 for BMC's transmission-housings project:

Unit Demand (X)	Probability, $P(X = x)$
1,600	0.2
2,000	0.6
2,400	0.2

If we want to know the probability that demand will be less than, or equal to, any particular value, we can use the following cumulative probability function:

$$F(x) = P(X \leq x) \quad = \begin{cases} 0.2 \ x \leq 1,600 \\ 0.8 \ x \leq 2,000 \\ 1.0 \ x \leq 2,400. \end{cases}$$

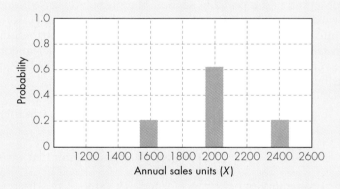

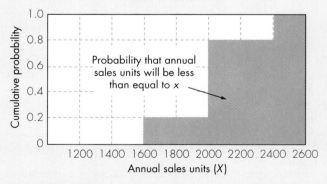

Figure 14.6   Probability and cumulative probability distributions for random variable X (annual sales) (Example 14.6)

For example, if we want to know the probability that the demand will be less than or equal to 2,000, we can examine the appropriate interval ($x \leq 2,000$), and we shall find the probability is 80%.

We can find the cumulative distribution for Y in a similar fashion. Graphic representations of the probability distributions and the cumulative probability distributions for X and Y are given in Figures 14.6 and 14.7, respectively.

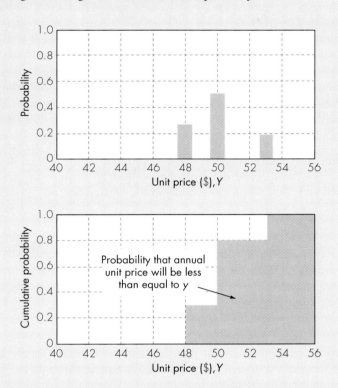

Figure 14.7 Probability and cumulative probability distributions for random variable Y from Table 14.5

14.3.2 Summary of Probabilistic Information

Although knowledge of the probability distribution of a random variable allows us to make a specific probability statement, a single value that may characterize the random variable and its probability distribution is often desirable. Such a quantity is the **expected value** of the random variable. We also want to know something about how the values of the random variable are dispersed about the expected value (i.e., the **variance**). In investment analysis, this dispersion information is interpreted as the degree of project risk. Expected value indicates the weighted average of the random variable, and the variance captures the variability of the random variable.

Measure of Expectation

The **expected value** (also called the **mean**) is a weighted average value of the random variable where the weighting factors are the probabilities of occurrence. All distributions (discrete and continuous) have an expected value. We will use $E[X]$ (or μ) to denote the expected value of random variable X. For a random variable X that has either discrete or continuous values, we compute an expected value with

$$E[X] = \mu = \begin{cases} \sum_{j=1}^{J}(p_j)x_j & \text{(discrete case)} \\ \int_{L}^{H} xf(x)dx & \text{(continuous case)}, \end{cases} \tag{14.1}$$

where J is the number of discrete events and L and H are the lower and upper bounds of the continuous probability distribution.

The expected value of a distribution gives us important information about the "average" value of a random variable, such as the NPW, but it does not tell us anything about the variability on either side of the expected value. Will the range of possible values of the random variable be very small, and will all the values be located at, or near, the expected value? For example, the following represents the temperatures recorded on January 29, 1996, for two cities in the United States.

Location	Low	High	Average
Atlanta	15°F	67°F	41°F
Seattle	36°F	48°F	42°F

Even though both cities had almost identical mean (average) temperatures on that particular day, they had different variations in extreme temperatures. We shall examine this variability issue in the following section.

Measure of Variation

Another measure needed when we are analyzing probabilistic situations is a measure of the risk due to the variability of the outcomes. There are several measures of the variation of a set of numbers that are used in statistical analysis—the **range** and the **variance** (or **standard deviation**), among others. The variance and the standard deviation are used most commonly in the analysis of risk situations. We will use $Var[X]$ or σ_x^2 to denote the variance and σ_x to denote the standard deviation of random variable X. (If there is only one random variable in an analysis, we normally omit the subscript.)

The variance tells us the degree of spread, or dispersion, of the distribution on either side of the mean value. As the variance increases, the spread of the distribution increases; the smaller the variance, the narrower the spread about the expected value.

To determine the variance, we first calculate the deviation of each possible outcome x_j from the expected value $(x_j - \mu)$. Then we raise each result to the second power and multiply it by the probability of x_j occurring (that is, p_j). The summation of all these products serves as a measure of the distribution's variability. For a random variable that has only discrete values, the equation to compute variance[6] is

$$Var[X] = \sigma_x^2 = \sum_{j=1}^{J} (x_j - \mu)^2 (p_j), \qquad (14.2)$$

where p_j is the probability of occurrence of the jth value of the random variable (x_j), and μ is as defined by Eq. (14.1). To be most useful, any measure of risk should have a definite value (unit). One such measure is the **standard deviation**. To calculate the standard deviation, we take the positive square root of $Var[X]$, which is measured in the same units as is X

$$\sigma_x = \sqrt{Var[X]}. \qquad (14.3)$$

The standard deviation is a probability-weighted deviation (more precisely, the square root of the sum of squared deviations) from the expected value. Thus, it gives us an idea of how far above, or below, the expected value the actual value is likely to be. For most probability distributions, the actual value will be observed within the $\pm 3\sigma$ range.

In practice, the actual calculation of the variance is somewhat easier if we use the following formula:

$$\begin{aligned} Var[X] &= \Sigma p_j x_j^2 - (\Sigma p_j x_j)^2 \\ &= E[X^2] - (E[X])^2 \end{aligned} \qquad (14.4)$$

The term $E[X^2]$ in Eq. (14.4) is interpreted as the mean value of the squares of the random variable (i.e., the actual values squared). The second term is simply the mean value squared. Example 14.7 will illustrate how we compute measures of variation.

Example 14.7 Mean and Variance Calculation

Consider BMC's transmission-housings project. Unit sales (X) and unit price (Y) are estimated as in Table 14.5, now compute the means, variances, and standard deviations for the random variables X and Y.

Solution

For the product demand variable (X):

---

[6] For a continuous random variable, we compute the variance as follows:

$$Var[X] = \int_L^H (x - u)^2 f(x) dx.$$

x_j	p_j	$x_j(p_j)$	$(x_j - E[X])^2$	$(x_j - E[X])^2(p_j)$
1,600	0.20	320	$(-400)^2$	32,000
2,000	0.60	1,200	0	0
2,400	0.20	480	$(400)^2$	32,000
		$E[X] = 2,000$		$Var[X] = 64,000$
				$\sigma_x = 252.98$

For the variable unit price (Y):

y_j	p_j	$y_j(p_j)$	$(y_j - E[Y])^2$	$(y_j - E[Y])^2(p_j)$
\$48	0.30	\$14.40	$(-2)^2$	1.20
50	0.50	25.00	$(0)^2$	0
53	0.20	10.60	$(3)^2$	1.80
		$E[Y] = 50.00$		$Var[Y] = 3.00$
				$\sigma_y = 1.73$

14.3.3 Joint and Conditional Probabilities

Thus far, we have not looked at how the values of variables can influence the values of others. It is, however, entirely possible—indeed it is likely—that the values of some parameters will be dependent on the values of others. We commonly express these dependencies in terms of conditional probabilities. An example is product demand, which will probably be influenced by unit price.

We define a **joint probability** as

$$P(x, y) = P(X = x \mid Y = y)P(Y = y), \tag{14.5}$$

where $P(X = x \mid Y = y)$ is the **conditional probability** of observing x, given $Y = y$, and $P(Y = y)$ is the **marginal probability** of observing $Y = y$. Certainly important cases exist where a knowledge of the occurrence of event X does not change the probability of an event Y. That is, if X and Y are **independent**, the joint probability is simply

$$P(x, y) = P(x)P(y). \tag{14.6}$$

The concepts of joint, marginal, and conditional distributions are best illustrated by numerical examples.

Unit Price Y	Probability	Conditional Unit Sales X	Joint Probability	Probability
		1,600	0.10	0.03
$48	0.30	2,000	0.40	0.12
		2,400	0.50	0.15
		1,600	0.10	0.05
50	0.50	2,000	0.64	0.32
		2,400	0.26	0.13
		1,600	0.50	0.10
53	0.20	2,000	0.40	0.08
		2,400	0.10	0.02

Table 14.6
Assessments of Conditional and Joint Probabilities

Suppose that BMC's marketing staff estimates that, for a given unit price of $48, the conditional probability that the company can sell 1,600 units is 0.10. The probability of this joint event (unit sales = 1,600 and unit sales price = $48) is

$$
\begin{aligned}
P(x, y) &= P(1{,}600, \$48) \\
&= P(x = 1{,}600 \mid y = \$48)P(y = \$48) \\
&= (0.10)(0.30) \\
&= 0.03.
\end{aligned}
$$

We can obtain the probabilities for other joint events in a similar fashion; several are presented in Table 14.6.

From Table 14.6, we can see that the unit demand (X) ranges from 1,600 to 2,400 units, the unit price (Y) ranges from $48 to $53, and nine joint events are possible. The sum of these joint probabilities must equal 1.

Joint Event (xy)	P(x, y)
(1,600, $48)	0.03
(2,000, 48)	0.12
(2,400, 48)	0.15
(1,600, 50)	0.05
(2,000, 50)	0.32
(2,400, 50)	0.13
(1,600, 53)	0.10
(2,000, 53)	0.08
(2,400, 53)	0.02
	Sum = 1.00

The marginal distribution for x can be developed from the joint event by fixing x and summing over y:

x_j	$P(x_j) = \Sigma_y\, P(x, y)$
1,600	$P(1,600, \$48) + P(1,600, \$50) + P(1,600, \$53) = 0.18$
2,000	$P(2,000, \$48) + P(2,000, \$50) + P(2,000, \$53) = 0.52$
2,400	$P(2,400, \$48) + P(2,400, \$50) + P(2,400, \$53) = 0.30$

This marginal distribution tells us that 52% of the time we can expect to have a demand of 2,000 units, and that 18% and 30% of the time we can expect to have a demand of 1,600 and 2,400, respectively.

14.4 Probability Distribution of NPW

After we have identified the random variables in a project and assessed the probabilities of the possible events, the next step is to develop the project's NPW distribution.

14.4.1 Procedure for Developing an NPW Distribution

We will consider the situation where all the random variables used in calculating the NPW are independent. To develop the NPW distribution, we may follow these steps:

- Express the NPW as functions of unknown random variables.
- Determine the probability distribution for each random variable.
- Determine the joint events and their probabilities.
- Evaluate the NPW equation at these joint events.
- Order the NPW values in increasing order of NPW.

These steps can best be illustrated by Example 14.8.

Example 14.8 Procedure of Developing an NPW Distribution

Consider BMC's transmission-housings project in Example 14.2. Use the unit demand (X) and price (Y) given in Table 14.5, and develop the NPW distribution for the BMC project. Then, calculate the mean and variance of the NPW distribution.

Solution

Table 14.7 summarizes the after-tax cash flow for the BMC's transmission-housings project as functions of random variables X and Y. From this table, we can compute the PW of cash inflows as follows:

Item	0	1	2	3	4	5
Cash inflow						
Net salvage						37,389
Revenue						
$X(1-0.4)Y$		0.6XY	0.6XY	0.6XY	0.6XY	0.6XY
Depreciation credit						
0.4 (depreciation)		7,145	12,245	8,745	6,245	2,230
Cash outflow						
Investment	−125,000					
Variable cost						
$-X(1-0.4)(\$15)$		−9X	−9X	−9X	−9X	−9X
Fixed cost						
$-(1-0.4)(\$10,000)$		−6,000	−6,000	−6,000	−6,000	−6,000
Net cash flow	−125,000	0.6X(Y−15) +1,145	0.6X(Y−15) +6,245	0.6X(Y−15) +2,745	0.6X(Y−15) +245	0.6X(Y−15) +33,617

Table 14.7
After-Tax Cash Flow as a Function of Unknown Unit Demand (X) and Unit Price (Y) (Example 14.8)

$$PW(15\%) = 0.6XY(P/A, 15\%, 5) + \$44,490$$
$$= 2.0113XY + \$44,490.$$

The PW of cash outflows is

$$PW(15\%) = \$125,000 + (9X + \$6,000)(P/A, 15\%, 5)$$
$$= 30.1694X + \$145,113.$$

Thus, the NPW is

$$PW(15\%) = 2.0113X(Y - \$15) - \$100,623.$$

If the product demand X and the unit price Y are independent random variables, the $PW(15\%)$ will also be a random variable. To determine the NPW distribution, we need to consider all the combinations of possible outcomes.[7] The first possibility is the event where $x = 1,600$ and $y = \$48$. Since X and Y are considered to be independent random variables, the probability of this joint event is

$$P(x = 1,600, y = \$48) = P(x = 1,600)P(y = \$48)$$
$$= (0.20)(0.30)$$
$$= 0.06.$$

With these values as input, we compute the possible NPW outcome as follows:

---

[7] If X and Y are dependent random variables, the joint probabilities developed in Table 14.6 should be used.

$$PW\,(15\%) = 2.0113X\,(Y - \$15) - \$100,623$$
$$= 2.0113(1,600)(\$48 - \$15) - \$100,623$$
$$= \$5,574.$$

Eight other outcomes are possible: they are summarized with their joint probabilities in Table 14.8 and depicted in Figure 14.8.

Table 14.8
The NPW Probability Distribution with Independent Random Variables (Example 14.8)

Event No.	x	y	P(x,y)	Cumulative Joint Probability	NPW
1	1,600	$48.00	0.06	0.06	$ 5,574
2	1,600	50.00	0.10	0.16	12,010
3	1,600	53.00	0.04	0.20	21,664
4	2,000	48.00	0.18	0.38	32,123
5	2,000	50.00	0.30	0.68	40,168
6	2,000	53.00	0.12	0.80	52,236
7	2,400	48.00	0.06	0.86	58,672
8	2,400	50.00	0.10	0.96	68,326
9	2,400	53.00	0.04	1.00	82,808

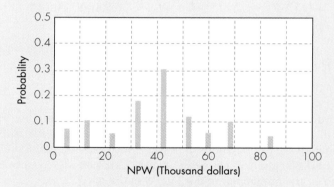

Figure 14.8   NPW probability distributions: When X and Y are independent (Example 14.8)

The NPW probability distribution in Table 14.8 indicates that the project's NPW varies between $5,574 and $82,808, but that no loss under any of the circumstances examined occurs. From the cumulative distribution, we further observe that there is a 0.38 probability that the project would realize a NPW less than that forecast for the base-case situation ($40,168). On the other hand, there is a 0.32 probability that the NPW will be greater than this value. Certainly, the probability distribution provides

Event No.	x	y	P(x,y)	Cumulative Joint Probability	NPW	Weighted NPW
1	1,600	$48.00	0.06	0.06	$ 5,574	$ 334
2	1,600	50.00	0.10	0.16	12,010	1,201
3	1,600	53.00	0.04	0.20	21,664	867
4	2,000	48.00	0.18	0.38	32,123	5,782
5	2,000	50.00	0.30	0.68	40,168	12,050
6	2,000	53.00	0.12	0.80	52,236	6,268
7	2,400	48.00	0.06	0.86	58,672	3,520
8	2,400	50.00	0.10	0.96	68,326	6,833
9	2,400	53.00	0.04	1.00	82,808	3,312
						$E[\text{NPW}] = \$ 40,168$

Table 14.9 Calculation of the Mean of NPW Distribution (Example 14.8)

much more information on the likelihood of each possible event as compared with the scenario analysis presented in Section 14.2.4.

We have developed a probability distribution for the NPW by considering random cash flows. As we observed, a probability distribution helps us to see what the data imply in terms of project risk. Now, we can learn how to summarize the probabilistic information—the mean and the variance. For BMC's transmission-housings project, we compute the expected value of the NPW distribution as shown in Table 14.9. Note that this expected value is the same as the most likely value of the NPW distribution. This equality was expected because both X and Y have a symmetrical probability distribution.

We obtain the variance of the NPW distribution, assuming independence between X and Y and using Eq. (14.2), as shown in Table 14.10. We could obtain the same result more easily by using Eq. (14.4).

Event No.	x	y	P(x,y)	NPW	$(\text{NPW} - E[\text{NPW}])^2$	Weighted $(\text{NPW} - E[\text{NPW}])^2$
1	1,600	$48.00	0.06	$ 5,574	1,196,769,744	$71,806,185
2	1,600	50.00	0.10	12,010	792,884,227	79,288,423
3	1,600	53.00	0.04	21,664	342,396,536	13,695,861
4	2,000	48.00	0.18	32,123	64,725,243	11,650,544
5	2,000	50.00	0.30	40,168	0	0
6	2,000	53.00	0.12	52,236	145,631,797	17,475,816
7	2,400	48.00	0.06	58,672	342,396,536	20,543,792
8	2,400	50.00	0.10	68,326	792,884,227	79,288,423
9	2,400	53.00	0.04	82,808	1,818,132,077	72,725,283
					$Var[\text{NPW}] = $	366,474,326
					$\sigma = $	$19,144

Table 14.10 Calculation of the Variance of NPW Distribution (Example 14.8)

14.4.2 Decision Rules

Once the expected value has been located from the NPW distribution, it can be used to make an accept-reject decision in much the same way that a single NPW is used when a single possible outcome for an investment project is considered. The decision rule is called the **expected value criterion** and using it, we may accept a single project if its expected NPW value is positive. In the case of mutually exclusive alternatives, we select the one with the highest expected NPW. The use of expected NPW has an advantage over the use of a point estimate, such as the likely value, because it includes all possible cash flow events and their probabilities.

The justification for the use of the expected value criterion is based on the **law of large numbers**, which states that, if many repetitions of an experiment are performed, the average outcome will tend toward the expected value. This justification may seem to negate the usefulness of the expected value criterion, since most often in project evaluation we are concerned with a single, non-repeatable "experiment"—i.e., an investment alternative. However, if a firm adopts the expected value criterion as a standard decision rule for *all* its investment alternatives, over the long term, the law of large numbers predicts that accepted projects tend to meet their expected values. Individual projects may succeed or fail, but the average project result tends to meet the firm's standard for economic success.

The expected-value criterion is simple and straightforward to use, but it fails to reflect the variability of investment outcome. Certainly, we can enrich our decision by incorporating the variability information along with the expected value. Since the variance represents the dispersion of the distribution, it is desirable to minimize it. In other words, the smaller the variance, the less the variability (the potential for loss) associated with the NPW. Therefore, when we compare the mutually exclusive projects, we may select the alternative with the smaller variance if its expected value is the same as, or larger than, those of other alternatives. In cases where preferences are not clear cut, the ultimate choice depends on the decision-maker's tradeoffs—how far he or she is willing to take the variability to achieve a higher expected value. In other words, the challenge is to decide what level of risk you are willing to accept and then, having decided on your risk tolerance, to understand the implications of that choice. Example 14.9 illustrates some of the critical issues that need to be considered in evaluating mutually exclusive risky projects.

Example 14.9 Comparing Risky Mutually Exclusive Projects

With ever-growing concerns about air pollution, the greenhouse effect, and increased dependence on oil imports in the United States, Green Engineering has developed a prototype conversion unit that allows a motorist to switch from gasoline to compressed natural gas (CNG) or vice versa. Driving a car equipped with Green's conversion kit is not much different from driving a conventional model. A small dial switch on the dashboard controls which fuel is to be used. Four different configurations are available according to types of vehicles: compact size, midsize, large size, and trucks. In the past, Green has built a few prototype vehicles powered by alternative fuels other than gasoline, but has been reluctant to go into higher-volume production without more evidence of public demand. Therefore, Green Engineering initially would like to target one market segment (one configuration model) in offering the conversion kit. Green Engineering's marketing group has compiled the potential NPW distribution for each different configuration when marketed independently.

Event (NPW) (unit: thousands)	Probabilities			
	Model 1	Model 2	Model 3	Model 4
$1,000	0.35	0.10	0.40	0.20
1,500	0	0.45	0	0.40
2,000	0.40	0	0.25	0
2,500	0	0.35	0	0.30
3,000	0.20	0	0.20	0
3,500	0	0	0	0
4,000	0.05	0	0.15	0
4,500	0	0.10	0	0.10

Evaluate the expected return and risk for each model configuration and recommend which one, if any, should be selected.

Solution

For model 1, we calculate the mean and variance of the NPW distribution as follows:

$$E[NPW]_1 = \$1,000(0.35) + \$2,000(0.40)$$
$$+ \$3,000(0.20) + \$4,000(0.05)$$
$$= \$1,950;$$
$$Var[NPW]_1 = 1,000^2(0.35) + 2,000^2(0.40)$$
$$+ 3,000^2(0.20) + 4,000^2(0.05) - (1,950)^2$$
$$= 747,500.$$

In a similar manner, we compute the other values, with the following results:

Configuration	E [NPW]	Var [NPW]
Model 1	$1,950	747,500
Model 2	2,100	915,000
Model 3	2,100	1,190,000
Model 4	2,000	1,000,000

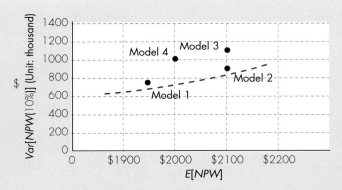

Figure 14.9   Mean-variance chart showing project dominance. Both model 3 and model 4 are dominated by model 2 (Example 14.9)

Results are plotted in Figure 14.9. If we make a choice based solely on the expected value, we may select either model 2 or model 3, because they have the highest expected NPW.

If we consider the variability along with the expected NPW, however, the correct choice is not obvious. We will first eliminate the alternatives that are clearly inferior to other alternatives.

Model 2 versus Model 3: We see that models 2 and 3 have the same mean of $2,100, but model 2 has a much lower variance. In other words, model 2 dominates model 3, so we eliminate model 3 from further consideration.

Model 2 and Model 4: Similarly, we see that model 2 is preferred over model 4, because model 2 has a higher expected value and a smaller variance—model 2 again dominates model 4, so we eliminate model 4. In other words, model 2 dominates both models 3 and 4 as we consider the variability of NPW.

Model 1 and Model 2: Even though the mean and variance rule has enabled us to narrow down our choices to only two models (models 1 and 2), it does not indicate what the ultimate choice should be. In other words, comparing models 1 and 2, we see that $E[\text{NPW}]$ increases from $1,950 to $2,100 at the price of a higher $Var[\text{NPW}]$, increasing from 747,500 to 915,000. That choice will depend on the decision-maker's tradeoffs between the incremental expected return ($150) and the incremental risk (167,500). We cannot choose between the two simply on the basis of mean and variance, so we must resort to other probabilistic information.[8]

Comments: When we need to compare mutually exclusive risky projects where no project dominance is observed, an incremental analysis can provide additional probabilistic information regarding the likelihood that the NPW of one project would ex-

---

[8] As we seek further refinement in our decision under risk, we may consider the expected utility theory, or stochastic dominance rules, which is beyond the scope of our text. See Park, C. S. and Sharp-Bette, G. P. *Advanced Engineering Economics*. New York: John Wiley, 1990 (Chapters 10 and 11).

ceed that of the other project, say $P\{NPW_1 \geq NPW_2\}$. If this probability is 1, model 1 dominates model 2, so model 1 is preferred. If this probability is equal to zero, the reverse is true.

We first consider all the possible joint events that satisfy the statement $P\{NPW_1 > NPW_2\}$. (Note the strict inequality.) For example, when we realize the event of $1,000 from model 1, no events from model 2 can satisfy the condition, because the smallest value we can observe from model 2 is $1,000. But with the $2,000 event observed for model 1, an event of either $1,000 or $1,500 from model 2 will meet the condition. Assuming statistical independence between the two models, we can easily compute the joint probability for such a joint event: For the joint event ($2,000, $1,000), the joint probability is $(0.4)(0.10) = 0.04$; for the event ($2,000, $1,500), it is $(0.40)(0.45) = 0.18$. Similarly, we can enumerate other possible joint events, as follows:

Model 1	Model 2	Joint Probability
$1,000	No event	$(0.35)(0.00) = 0.000$
2,000	1,000	$(0.40)(0.10) = 0.040$
	1,500	$(0.40)(0.45) = 0.180$
3,000	1,000	$(0.20)(0.10) = 0.020$
	1,500	$(0.20)(0.45) = 0.190$
	2,500	$(0.20)(0.35) = 0.070$
4,000	1,000	$(0.05)(0.10) = 0.005$
	1,500	$(0.05)(0.45) = 0.023$
	2,500	$(0.05)(0.35) = 0.018$
		$\boxed{0.445}$

The probability that the NPW of model 1 will exceed that of model 2 is calculated to be only 44.5%. This result implies that there is a 52% probability that the reverse situation may hold true, indicating a possible preference for model 2. (A 0.035 probability exists that both models would result in the identical NPW.) The ultimate decision is again up to the investor, but this type of additional information helps the decision-maker discern the best alternative in a complex decision environment.

14.5 Decision Trees and Sequential Investment Decisions

Most investment problems that we have discussed so far involved only a single decision at the time of investment (accept or reject), or this single decision could entail a different decision option such as, "make a product in house" or "farm out," and so on.

Once the decision is made, there are no later contingencies or decision options to follow up. However, certain classes of investment decisions cannot be analyzed in a single step. As an oil driller, for example, you have an opportunity to bid on an offshore lease. Your first decision is whether to bid. But if you decide to bid and eventually win the contract, you will have another decision regarding whether to drill immediately or run more seismic (drilling) tests. If you drill a test well that turns out to be dry, you have another decision whether to drill another well or drop the entire drilling project. In a sequential decision problem such as the oil drilling in which the actions taken at one stage depend on actions taken in earlier stages, the evaluation of investment alternatives can be quite complex. Certainly all these future options must be considered when evaluating the feasibility of bidding at the initial stage. In this section, we describe a general approach for more complex decisions which is useful both for structuring the decision problems as well as for finding a solution. The approach utilizes a decision tree, a graphic tool for describing the actions available to the decision maker, the events that can occur, and the relationship between the actions and events.

14.5.1   Structuring a Decision Tree Diagram

To illustrate the basic decision tree concept, let's consider an investor, named Bill Henderson, who wants to invest some money in the financial market. He wants to choose between a highly speculative growth stock (d_1) and a very safe U.S. Treasury bond (d_2). Figure 14.10 illustrates the situation where the decision point is represented by a square box ($\square$) or decision node. The alternatives are represented as branches emanating from the decision node. Suppose Bill were to select some particular alternative, say "invest in the stock." There are three chance events that can happen, each event representing a potential return on investment. These are shown in Figure 14.10 as branches emanating from a circle node ($\bigcirc$). Note that these branches represent uncertain events over which Bill has no control. However, Bill can assign probabilities to each chance event and enter beside each branch in the decision tree. At the end of each branch is the conditional monetary transactions associated with the selected action and given event.

Relevant Net After-Tax Cash Flow

Once the structure of the decision tree is determined, the next step is to find the relevant cash flow (monetary value) associated with each of the decision alternatives and the possible chance outcomes. As we have emphasized throughout this book, the decision has to be made on an after-tax basis. Therefore, the relevant monetary value should be on an after-tax basis. Since the costs and revenues occur at different points in time over the study period, we also need to convert the various amounts on the tree's branches to their equivalent lump amounts, say, net present value. The conditional net present value thus represents the profit associated with the decisions and events along the path from the first part of the tree to the end.

Rollback Procedure

To analyze a decision tree, we begin at the end of the tree and work backward—the **rollback** procedure. In other words, starting at the tips of the decision tree's branches and working toward the initial node, we use the following two rules:

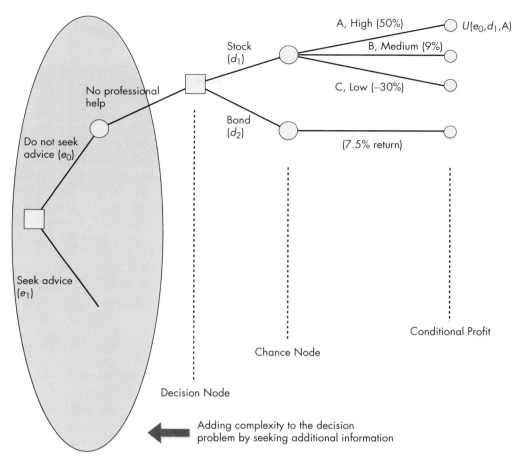

Figure 14.10   Structure of a typical decision tree (Bill's investment problem)

1. For each chance node, we calculate the expected monetary value (EMV).
    This is done by multiplying probabilities by conditional profits associated
    with branches emanating from that chance node and summing these
    conditional profits. We then place the EMV in the node to indicate that
    it is the expected value calculated over all branches emanating from that
    node.
2. For each decision node, we select the one with the highest EMV (or mini-
    mum cost). Then those decision alternatives not selected are eliminated from
    further consideration. On the decision tree diagram, a mark is drawn across
    the non-optimal decision branches, indicating that they are not to be fol-
    lowed.

Example 14.10 will illustrate how Bill could transform his investment problem in a
tree format using numerical values.

Example 14.10 Bill's Investment Problem in a Decision Tree Format

Suppose Bill has $50,000 to invest in the financial market for one year. His choices have been narrowed to two options:

- Option 1: Buy 1,000 shares of ADC Communication (ADCT) @ $50 per share that will be held for one year. Since this is a new initial public offering (IPO), there is not much research information available on the stock. This will entail a brokerage fee of $100 for this size of transaction (for either buying or selling stocks). For simplicity, assume that the stock is expected to provide return at three different levels: high level (A) with a 50% return ($25,000), medium level (B) with a 9% return ($4,500), or low level (C) with a 30% loss (−$15,000), and that the probabilities of these occurrences are assessed at 0.25, 0.40, and 0.35, respectively. It is not anticipated any stock dividend for such a growth oriented company.
- Option 2: Purchase a $50,000 U.S. Treasury bond, which pays interest at an effective annual rate of 7.5% ($3,750). The interest earned from the Treasury bond is non-taxable income. However, there is a $150 transaction fee for either buying or selling the bond.

Bill's question is which alternative to choose to maximize his financial gain. At this point, Bill is not concerned about seeking some professional advice on the stock before making a decision. We will assume that any capital gains will be taxed at 20% for any long-term gains. Bill's minimum attractive rate of return is known to be 5% after taxes. Determine the pay-off amount at the end of each branch tip.

Solution

Figure 14.11(a) shows the costs and revenues on the branches transformed to their present equivalents.

- Option 1:
  1. With a 50% return (real winner) over a one-year holding period:
     - The net cash flow at Period 0 associated with this event will include the amount of investment and the brokerage fee.

$$\text{Period 0: } (-\$50{,}000 - \$100) = -\$50{,}100.$$

  - When you sell stock, you need to pay another brokerage fee. However, any investment expenses such as brokerage commissions must be included in determining the cost basis for investment. The taxable capital gains will be calculated as ($75,000 − $50,000 − $100 − $100) = $24,800. Therefore, the net cash flow at Period 1 would be

$$\text{Period 1: } (+\$75{,}000 - \$100) - 0.20 (\$24{,}800) = \$69{,}940.$$

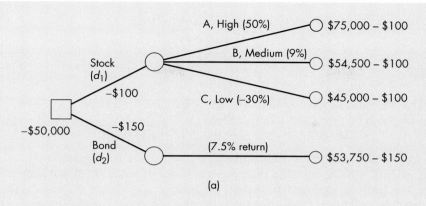

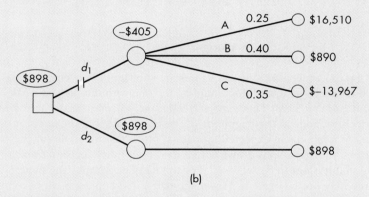

Figure 14.11   Decision tree for Bill's investment problem: (a) Relevant cash flows (before tax), (b) Net present worth for each decision path

- Then, the conditional net present value of this stock transaction is

$$PW(5\%) = -\$50,100 + \$69,940(P/F, 5\%, 1) = \$16,510.$$

This amount of $16,510 is entered at the end of the corresponding branch tip. This procedure is repeated for each possible branch, and the resulting amounts are shown in Figure 14.11.

2. With a 9% return:
   - Period 0: $(-\$50,000 - \$100) = -\$50,100$
   - Period 1: $(+\$54,500 - \$100) - (0.20)(\$4,300) = \$53,540$
   - $PW(5\%) = -50,100 + 53,540(P/F, 5\%, 1) = \$890$
3. With a 30% loss,
   - Period 0: $(-\$50,000 - \$100) = -\$50,100$
   - Period 1: $(+\$35,000 - \$100) - (0.20)(-\$14,800) = \$37,940$
   - $PW(5\%) = -\$50,100 + \$37,940(P/F, 5\%, 1) = -\$13,967$

- For Option 2

  Since the interest income will not be taxed for the U.S. government bond, there will be no capital gains tax. Considering only the brokerage commission, the relevant cash flows would be as follows:

    - Period 0: $(- \$50,000 - \$150) = -\$50,150$
    - Period 1: $(+\$53,750 - \$150) = \$53,600$
    - $PW(5\%) = -\$50,150 + \$53,600 \ (P/F, 5\%, 1) = \$898$

Figure 14.11(b) shows the complete decision tree for the Bill's investment decision problem. Now Bill can calculate the expected monetary value (EMV) at each chance node. The EMV of Option 1 represents the sum of the product of probabilities for high, medium and low returns times the respective conditional profits (or losses):

$$EMV = \$16,510(0.25) + \$890(0.40) - \$13,967(0.35) = -\$405$$

For Option 2, the EMV is simply

$$EMV = \$898$$

In Figure 14.11, the expected monetary values are shown in the event nodes. Bill must choose which action to take, and this would be the one with the highest EMV, namely Option 2 with EMV = $898. This expected value is indicated in the tree by putting $898 in the decision node at the beginning of the tree. Note that the decision tree uses the same idea of maximizing expected monetary value developed in the previous section. In addition, the mark (‖) is drawn across the nonoptimal decision branch (Option 1), indicating that it is not to be followed. For this simple example, the benefit of using a decision tree may not be evident. However, as the decision problem becomes more complex, the decision tree becomes more useful in organizing the information flow needed to make the decision. In particular, this is true if Bill must make a sequence of decisions, rather than a single decision, as the next section illustrates.

14.5.2 Worth of Obtaining Additional Information

In this section, we introduce a general method for evaluating the possibility of obtaining more information. Most of the information that we can obtain is imperfect in the sense that it will not tell us exactly which event will occur. Such imperfect information may have some value if it improves the chances of making a correct decision, that is, if it improves the expected monetary value. The problem is whether the reduced uncertainty is valuable enough to offset its cost. The gain is in the improved efficiency of decisions that may become possible with better information.

We use the term experiment in a broad sense here. An experiment may represent a market survey to predict sales volume for a typical consumer product, statistical sampling of production quality, or a seismic test to give a well-drilling firm some indications of the presence of oil.

The Value of Perfect Information

Let us take the prior decision of "Purchase U.S. Treasury Bonds" as a starting point. How do we determine whether further strategic steps would be profitable? We could do more to obtain additional information about the potential stock return, but such steps cost money. Thus, we have to balance the monetary value of reducing uncertainty with the cost of additional information. In general, we can evaluate the worth of a given experiment only if we can estimate the reliability of the resulting information. In our stock investment problem, an expert's opinion may be helpful in deciding whether to abandon the idea of investing in a stock. This can be of value, however, only if Bill can say beforehand how closely the expert can estimate the potential stock performance.

The best place to start a decision improvement process is with a determination of how much we might improve incremental profit by removing uncertainty. Although we probably couldn't obtain it, the value of perfect information is worth computing as an upper bound to the value of additional information.

We can easily calculate the value of perfect information. Merely note the difference between the incremental profit from an optimal decision based on perfect information and the original decision of "Purchase Treasury Bonds" made without foreknowledge of the actual return on the stock. This difference is called **opportunity loss** and must be computed for each potential level of stock return. Then we compute the expected opportunity loss by weighting each potential loss by the assigned probability associated with that event. What turns out in this expected opportunity loss is exactly the **expected value of perfect information** (EVPI). In other words, if we had perfect information on the level of stock performance, we should have made a correct decision for each situation, resulting in no regrets or no opportunity losses. Example 14.11 will illustrate the opportunity loss concept and how this loss concept is related to the value of perfect information.

Example 14.11   Expected Value of Perfect Information

Bill, reluctant to give up a chance to make a larger profit with the stock option, however, may wonder whether to obtain further information before action. As with the prior decision, Bill needs a single figure to represent the expected value of perfect information (EVPI). Calculate the EVPI based on the financial data in Example 14.10 and Figure 14.11(b).

Solution

For our stock example, the decision may hinge on the return for the stock in the next year. The only unknown, subject to a probability distribution, is the return on the stock. The opportunity loss table for this prior decision is shown in Table 14.11. For example, the conditional net present value of $16,510 is the net profit if the potential return level is high. Recall also that, without receiving any information, the indicated action (prior optimal decision) was to select Option 2 ("Purchase Treasury bond")

Table 14.11
Opportunity
Loss Cost
Associated
with Invest-
ing in Bonds

Potential Return Level	Probability	Decision Option		Optimal Choice with Perfect Information	Opportunity Loss Associated with Investing in Bonds
		Option 1: Invest in Stock	Option 2: Invest in Bonds		
High (A)	0.25	$16,510	$898	Stock	$15,612
Medium (B)	0.40	890	898	Bond	0
Low (C)	0.35	−13,967	898	Bond	0
Expected Monetary Value (EMV)		−$405	$898 ↗ Prior optimal decision		$3,903 ↗ EVPI

with an expected net present value of $898. However, if we know that the stock will be a definite winner (high return), then the prior decision to "Purchase Treasury bond" is inferior to "Invest in stock" to the extent of $16,510 − $898 = $15,612. With either a medium or low return situation, however, there is no opportunity loss, as a decision strategy with and without perfect information is the same.

Again, an average weighted by the assigned chances is used, but this time weights are applied to the set of opportunity losses (regrets) in Table 14.11.

$$EVPI = (0.25)(\$15,612) + (0.40)(\$0) + (0.35)(\$0) = \$3,903$$

This figure represents the maximum expected amount that could be gained in incremental profit from having perfect information. This EVPI places an upper limit on the sum Bill would be willing to pay for additional information.

Updating Conditional Profit (or Loss) after Paying a Fee to the Expert

With a relatively large EVPI in Example 14.11, there is a huge potential that Bill can benefit by obtaining some additional information on the stock. Bill can seek an expert's advice by receiving a detailed research report on the stock under consideration. This stock research report is available at a nominal fee, say, $200. There are two possible outcomes from the report: (1) the report may come up with a "buy" rating on the stock when it perceived a favorable business condition, or (2) the report may not recommend or take a neutral stance when it perceived an unfavorable business condition or the ADCT's business model is not well received in the financial marketplace. Bill's alternative is either to pay for this service before making any further decision or simply to make his own decision without seeing the report.

If Bill seeks advice before acting, he can base his decision upon the stock report. This problem can be expressed in terms of a decision tree, as shown in Figure 14.12. The upper part of the tree shows the decision process if no advice is sought. This is the same as Figure 14.11, with probabilities of 0.25, 0.40, and 0.35 for high, medium, and

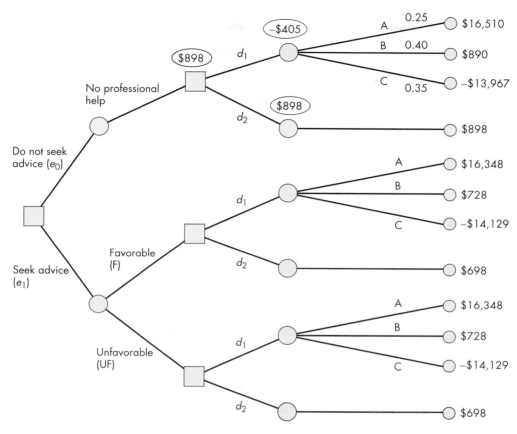

Figure 14.12   Decision tree for Bill's investment problem with an option having a professional advice

low returns, and an expected loss of $405 for stock investment and the expected profit of $898 from purchasing Treasury bonds.

The lower part of the tree, following the branch "Seek Advice," displays the results and the subsequent decision possibilities. Using the analyst's research report, Bill can expect two stages of events—the expert's recommendation and the actual level of stock performance. After each of the two possible report outcomes, Bill must make a decision about whether to invest in the stock. Since Bill has to pay a nominal fee ($200) to get the report, the profit or loss figures at the end of branch tips must also be updated. For the stock investment option, when the actual return proves to be very high (50%), the resulting cash flows would be as follows:

$$\text{Period 0: } (-\$50{,}000 - \$100 - \$200) = -\$50{,}300$$

$$\text{Period 1: } (+\$75{,}000 - \$100) - (0.20)(\$25{,}000 - \$400) = \$69{,}980$$

$$PW(5\%) = -\$50{,}300 + \$69{,}980(P/F, 5\%, 1) = \$16{,}348$$

Since the cost of obtaining the research report ($200) will be a part of the total investment expenses along with the brokerage commissions, the taxable capital gains would be adjusted accordingly. With a medium return, the equivalent net present value would be $728. With a low return, there would be a net loss in the amount of $14,129.

Similarly, Bill can compute the net profit for Option 2 as follows:

$$\text{Period 0: } (-\$50,000 - \$150 - \$200) = -\$50,350$$

$$\text{Period 1: } (+\$53,750 - \$150) = \$53,600$$

$$PW(5\%) = -\$50,350 + \$53,600(P/F, 5\%, 1) = \$698$$

Once again, note that no gains will be taxed for the U.S. government bond.

14.5.3 Decision Making after Having Imperfect Information

In Figure 14.12, the analyst's recommendation precedes the actual investment decision. This allows Bill to change his prior probabilistic assessments on stock performance. In other words, if the analyst's report has any bearing on Bill's reassessments on the stock's performance, he must change his prior probabilistic assessments accordingly. We explain what types of information should be available and how the prior probabilistic assessments should be updated.

**Conditional Probabilities of Expert's Predictions,
Given Potential Stock Return**

Suppose that Bill knows an expert (stock analyst) whom he can provide his research report on the stock described as Option 1. The expert will charge a fee in the amount of $200 to provide Bill with a report on the stock that is either favorable or unfavorable. This research report is not infallible, but it is known to be pretty reliable. From past experience, Bill estimates that, when the stock under consideration is relatively highly profitable (A), there is an 80% chance that the report will be favorable; when the stock return is medium (B), there is a 65% chance that the report will be favorable; and when the stock return is relatively unprofitable (C), there is a 20% chance that the report will be favorable. Bill's estimates on these probabilities are summarized in Table 14.12.

These conditional probabilities can be expressed in terms of a tree format as shown in Figure 14.13. For example, when the stock indeed turned out to be a real winner, the report would have predicted the same (with a favorable recommendation)

Table 14.12
Conditional Probabilities of the Expert's Prediction

What the Report Will Say	Given Level of Stock Performance		
	High (A)	Medium (B)	Low (C)
Favorable (F)	0.80	0.65	0.20
Unfavorable (UF)	0.20	0.35	0.80

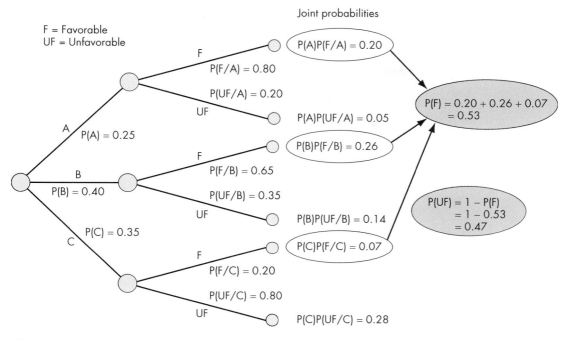

Figure 14.13   Conditional Probabilities and Joint Probabilities as Expressed in Terms of Tree Format

with a probability of 0.80. Or the report would have issued an "unfavorable recommendation" with a probability of 0.20. Such probabilities would reflect past experience with the analyst's service of this type, modified perhaps by Bill's own judgment. In fact, these probabilities represent the reliability or accuracy of the expert opinion. It is a question of counting the number of times the forecast was "on target" or a "complete miss" for each of the three actual levels of stock return. With these estimates, Bill can evaluate the economic worth of the service. Without these reliability estimates, no specific value can be attached to seeking the advice.

Joint and Marginal Probabilities

Now Bill has some critical information on (1) all the probabilities for the three levels of stock performance before seeing the research report (prior probabilities), and (2) with what probability the report will say on the stock (favorable or unfavorable) given the three levels of the future stock performance. A glance at the sequence of events in Figure 14.12, however, shows that these probabilities are not the ones required to find "expected" values of various strategies in the decision path. What are really needed are the chances of what the report will say about the stock—favorable (buy recommendation) or unfavorable (no recommendation or neutral stance), and the conditional probabilities of each of the three potential returns after seeing the report. In other words, the probabilities of Table 14.12 are not directly useful in Figure 14.12.

Rather, we need the marginal probabilities of the "favorable" and "unfavorable" recommendations.

To remedy this, the probabilities must be put in a different form—we must construct a joint probabilities table. To start with, we have the original (prior) probabilities assessed by Bill before seeing the report: a 0.25 chance that the stock will result in a high return (A), a 0.40 chance for a medium return (B), and a 0.35 chance for a loss (C). From these and the conditional probabilities of Table 14.12, the joint probabilities of Figure 14.13 can be calculated. Thus, the joint probability of both a prediction of a favorable stock market (F) and a high level of stock return (A) is calculated by multiplying the conditional probability of a favorable prediction, given a high stock return (which is 0.80 from Table 14.12), by the probability of a high return (A):

$$P(A,F) = P(F/A)P(A) = (0.80)(0.25) = 0.20$$

Similarly,

$$P(A,UF) = P(UF/A)P(A) = (0.20)(0.25) = 0.05,$$

$$P(B,F) = P(F/B)P(B) = (0.65)(0.40) = 0.26,$$

$$P(B,UF) = P(UF/B)P(B) = (0.35)(0.40) = 0.14$$

and so on.

The marginal probabilities of return level in Table 14.13 are obtained by summing the values across the columns. Note that these are precisely the original probabilities for high, medium, and low return, and they are designated prior probabilities because they were assessed before any information from the report was obtained.

In understanding Table 14.13, it is useful to think of it as representing the results of 100 past situations identical to the one under consideration. The probabilities then represent the frequency with which the various outcomes occurred. For example, in 25 of the 100 cases, the actual return for the stock turned out to be high; and in these 25-high return cases, the report predicted a favorable condition in 20 instances [that is $P(A, F) = 0.20$)], and predicted an unfavorable one in five instances.

Table 14.13
Joint as Well
as Marginal
Probabilities

When Potential Level of Return is Given	What Report Will Say		Marginal Probabilities of Return Level
	Joint Probabilities		
	Favorable (F)	Unfavorable (UF)	
High (A)	0.20	0.05	0.25
Medium (B)	0.26	0.14	0.40
Low (C)	0.07	0.28	0.35
Marginal Probabilities	0.53	0.47	1.00

The marginal probabilities of survey prediction in the bottom row of Table 14.13 can then be interpreted as the relative frequency with which the report predicted favorable, and unfavorable conditions. For example, the survey predicted favorable 53 out of 100 times: 20 of these times when stock returns actually were high, 26 times when returns were medium, and 7 times when returns were low. These marginal probabilities of what the report will say are important to our analysis, for they give us the probabilities associated with the information received by Bill before the decision to invest in the stock is made. The probabilities are entered beside the appropriate branches in Figure 14.12.

Determining Revised Probabilities

We still need to calculate the probabilities for the branches labeled "High Return," "Medium Return," and "Low Return" in the lower part of Figure 14.12. We cannot use the values of 0.25, 0.40, and 0.35 for these events as we did in the upper part of the tree, because those probabilities were calculated prior to seeking the advice. At this point on the decision tree, Bill will have received information from the analyst, and the probabilities should reflect this information. (If Bill's judgment has not changed even after seeing the report, then there will be no changes in the probabilities.) The required probabilities are the conditional probabilities for the various levels of stock return given the expert's opinion—for our example $P(A/F)$, the probability of seeing a high return (A) given the expert's buy recommendation (F). This can be computed directly from the definition of conditional probability, using the data from Table 14.13:

$$P(A/F) = P(A,F)/P(F) = 0.20/0.53 = 0.38$$

The probabilities of medium and low return, given a buy recommendation, are

$$P(B/F) = P(B,F)/P(F) = 0.26/0.53 = 0.49$$
$$P(C/F) = P(C,F)/P(F) = 0.07/0.53 = 0.13.$$

These probabilities are called revised (or posterior) probabilities, since they come after the inclusion of the information received from the report. To understand the meaning of the above calculations, think again of Table 14.13 as representing 100 past identical situations. Then, in 53 cases [since $P(F) = 0.53$], 20 actually had a high return result. Hence, the posterior probability for high return is, as calculated, $20/53 = 0.38$.

The posterior probabilities after seeing an unfavorable recommendation of other situations can be calculated similarly:

$$P(A/UF) = P(A,UF)/P(UF) = 0.05/0.47 = 0.11$$
$$P(B/UF) = P(B,UF)/P(UF) = 0.14/0.47 = 0.30$$
$$P(C/UF) = P(C,UF)/P(UF) = 0.28/0.47 = 0.59$$

These values are also shown in Figure 14.14 at the appropriate points in the decision tree.

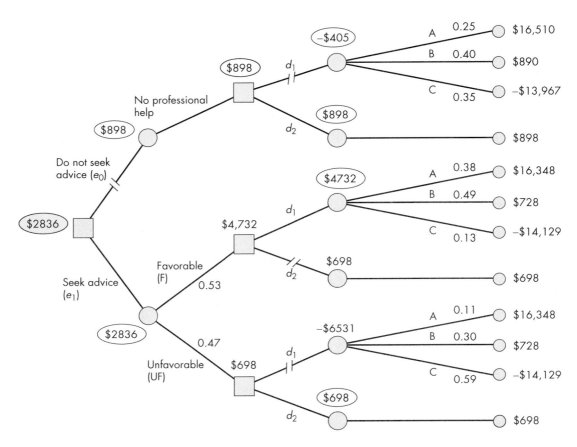

Figure 14.14 Decision tree for Bill's investment problem with an option of having a professional advice

---

Example 14.12 Decision Making after Getting Some Professional Advice

Based on Figure 14.14, determine how much the research report is worth to Bill. Should Bill seek an experts advice (research report)?

Solution

All the necessary information is now available, and Figure 14.14 can be analyzed—starting from the right and working backward. The expected values are shown in the circles. For example, follow the branches "Seek Advice," "Favorable (Buy recommendation)," and "Invest in the Stock." The expected value of $4,732 shown in the circle at the end of these branches is calculated as

$$0.38 (\$16,348) + 0.49(\$728) + 0.13(-\$14,129) = \$4,732.$$

Since Bill can expect a profit of $4,732 from the stock investment after receiving a buy recommendation, the investment in Treasury bonds ($698) becomes a less profitable alternative. Thus, it can be marked out (≠) of the "Purchase Treasury bonds" branch to indicate it is not optimal.

There is an expected loss of $6,531 if the report gives an unfavorable recommendation, but Bill goes ahead with buying the stock. In this situation, the "Treasury bonds" option becomes a better choice, and the "Buy Stock" branch is marked out. The part of the decision tree relating to seeking the advice is now reduced to two chance events ($4,732 and $698). And the expected value in the circle node is calculated as:

$$0.53 (\$4,732) + 0.47(\$698) = \$2,836.$$

Thus, if the survey is taken and Bill acts on the basis of the information received, the expected profit is $2,836. Since this is greater than the $898 profit that would be obtained from not seeking the advice, we conclude that it is worth spending $200 to receive some additional information from the expert.

In summary, the optimal decision should be as follows: (1) Bill should seek professional advice at the outset. (2) If the expert's report indicates a buy recommendation, go ahead and make investment in the stock. (3) If the report says otherwise, invest in the Treasury bonds.

14.6 Summary

- Often, cash flow amounts and other aspects of investment project analysis are uncertain. Whenever such uncertainty exists, we are faced with the difficulty of **project risk**—the possibility that an investment project will not meet our minimum requirements for acceptability and success.

- Three of the most basic tools for assessing project risk are as follows:

  1. **Sensitivity analysis**— a means of identifying the project variables which, when varied, have the greatest effect on project acceptability.

  2. **Break-even analysis**—a means of identifying the value of a particular project variable that causes the project to exactly break even.

  3. **Scenario analysis**—a means of comparing a "base-case" or expected project measurement (such as NPW) to one or more additional scenarios, such as best and worst case, to identify the extreme and most likely project outcomes.

- Sensitivity, break-even, and scenario analyses are reasonably simple to apply, but also somewhat simplistic and imprecise in cases where we must deal with multifaceted project uncertainty. **Probability concepts** allow us to further refine the analysis of project risk by assigning numerical values to the likelihood that project variables will have certain values.

- The end goal of a probabilistic analysis of project variables is to produce a NPW distribution. From the distribution, we can extract such useful information as the **expected NPW value**, the extent to which other NPW values vary from, or are clustered around, the expected value (**variance**), and the best- and worst-case NPWs.

- Our real task is not try to find "risk-free" projects—they don't exist in real life. The challenge is to decide what level of risk we are willing to assume and then, having decided on your risk tolerance, to understand the implications of that choice.

- The decision tree is another technique that can facilitate investment decision making when uncertainty prevails, especially when the problem involves a sequence of decisions. Decision tree analysis involves the choice of a decision criterion—say, to maximize expected profit. If possible and feasible, an experiment is conducted. The prior probabilities of the states of nature are revised, based on the experimental results. The expected profit associated with each possible decision is computed, and the act with the highest expected profit is chosen as the optimum action.

Self-Test Questions

14s.1 For a certain investment project, the net present worth can be expressed as functions of sales price (X) and variable production cost Y of NPW = $10,450(2X - Y) - 7890$. The base values for X and Y are $20 and $10, respectively. If the sales price is increased 10% over the base price, how much change in NPW can be expected?

(a) 10%   (b) 20%
(c) 13.68%   (d) 22.32%.

14s.2 An investor bought 100 shares of stock at a cost of $10 per share. He held the stock for 15 years and wants to sell it now. For the first 3 years, he received no dividends. For each of the next 7 years, he received total dividends of $100 per year. For each of the remaining 5 years, no dividends were paid. In the last 15 years, the investor's marginal tax rate and capital gain tax rate was averaging about 30% and 20%, respectively. What would be the break-even selling price to earn a 15% return on investment after-tax?

(a) $6,579   (b) $7,974
(c) $8,224   (d) $9,398.

14s.3 A company is currently paying a sales representative $0.25 per mile to drive her car for company business. The company is considering supplying the representative with a car, which would involve the following: A car costs $12,000, has a service life of 5 years, and a market value of $3,500 at the end of that time. Monthly storage costs for the car are $80, and the cost of fuel, tires, and maintenance is 15 cents per mile. The car will be depreciated by MACRS, using a recovery period of 5 years (20%, 32%, 19.20%, 11.52%, 11.52%). The firm's marginal tax rate is 40%. What annual mileage must a salesman travel by car for the cost of the two methods of providing transportation to be equal if the interest rate is 15%?

(a) 36,345 miles   (b) 41,235 miles

(c) 45,233 miles   (d) 47,511 miles.

14s.4 Project A has the following probability distribution of expected future returns:

Probability	Net Future Worth
0.1	−$12,000
0.2	4,000
0.4	12,000
0.2	20,000
0.1	30,000

What is the expected future worth for Project A?

(a) $9450   (b) $10,800

(c) $11,400   (d) $12,300.

14s.5 In Problem 14s.4, what is the standard deviation of expected future worth for Project A?

(a) $10,735   (b) $3,686

(c) $11,400   (d) $2,280.

14s.6 In a resort location, you find a slot machine that costs you $1.00 per play. Odds for potential payoffs are as follows:

Payoff	Probability
$25	0.01
$1	0.50
$0	0.49

With 100 plays, what is the expected value of net payoff?

(a) −$25   (b) $0

(c) $25   (d) $75.

Harry Wilson, a mechanical engineer at Lehigh manufacturing, has found that the anticipated profitability of a newly developed motion detector for its popular home security device product line can be estimated as follows:

$$NPW = 40.28 \, V(2X - 11) - 77,860$$

where V is the number of units produced and sold, and X is the sales price per unit. Harry also found that V parameter value could occur anywhere over a range of 1,000 to 6,000 units and the X parameter value anywhere between $20 and $40 per unit.

14s.7 Suppose both V and X are statistically independent continuous random variables with the following means and variances:

$$E[V] = 3,500, \, Var[V] = 2,083,333$$
$$E[X] = 30, \, Var[X] = 33.$$

What is the variance of NPW?

(a) 2.2309×10^{11}

(b) 4.4618×10^{11}

(c) 2.769×10^{9}

(d) 1.118549×10^{13}.

14s.8 In 14s.7, if V and X are mutually independent discrete random variables with the following probabilities:

V		X	
Event	Probability	Event	Probability
1,000	0.4	$20	0.7
6,000	0.6	$40	0.3

What is the probability that the NPW would exceed $6,000,000?

(a) 0.28   (c) 0.60

(b) 0.40   (d) 0.82.

Problems

Sensitivity Analysis

14.1 Ford Construction Company is considering the proposed acquisition of a new earthmover. The mover's basic price is $70,000, and it will cost another $15,000 to modify it for special use by the company. This earthmover falls into the MACRS 5-year class. It will be sold after 4 years for $30,000. The earthmover purchase will have no effect on revenues, but it is expected to save the firm $32,000 per year in before-tax operating costs, mainly labor. The firm's marginal tax rate (federal plus state) is 40%, and its MARR is 15%.

   (a) Is this project acceptable based on the most-likely estimates given in the problem?

   (b) Suppose that the project will require an increase in net working capital (spare parts inventory) of $2,000, which will be recovered at the end of year 5. With consideration of this new requirement, would the project still be acceptable?

   (c) If the firm's MARR is increased to 20%, what would be the required savings in labor so that the project remains profitable?

14.2 Minnesota Metal Forming Company has just invested $500,000 of fixed capital in a manufacturing process, which is estimated to generate an after-tax annual cash flow of $200,000 in each of the next 5 years. At the end of year 5, no further market for the product and no salvage value for the manufacturing process is expected. If a manufacturing problem delays plant start-up for 1 year (leaving only 4 years of process life), what additional after-tax cash flow will be needed to maintain the same internal rate of return as would be experienced if no delay occurred?

14.3 A real estate developer seeks to determine the most economical height for a new office building. The building will be sold after 5 years. The relevant net annual revenues and salvage values are as follows:

	Height			
	2 Floors	3 Floors	4 Floors	5 Floors
First cost (net after-tax)	$500,000	$750,000	$1,250,000	$2,000,000
Lease revenue	199,100	169,200	149,200	378,150
Net resale value (after-tax)	600,000	900,000	2,000,000	3,000,000

   (a) The developer is uncertain about the interest rate (i) to use, but is certain that it is in the range 5% to 30%. For each building height, find the range of values of i for which that building height is the most economical.

   (b) Suppose that the developer's interest rate is known to be 15%. What would be the cost, in terms of net present value, of an error in overestimation in resale value (the true value resulted in a value 10% lower than that of the original estimate)?

14.4 A special purpose milling machine was purchased 4 years ago for $20,000. It was estimated at that time that this machine would have a life of 10 years and a salvage value of $1,000 with a cost of removal of $1,500. These estimates are still good. This machine has annual operating costs of $2,000, and its current book value is $13,000. If the machine is retained over the life of the asset, the remaining annual depreciation schedule would be $2,000 for years 5 through 10. A new machine, which is more efficient, will reduce operating costs to $1,000, but it will require an investment of $12,000. The life of the new machine is estimated to be 6 years with a salvage value of $2,000. The new machine would fall into the 5-year MACRS property class. An offer of $6,000 for the old machine has been made, and the purchaser would pay for removal of the machine. The firm's marginal tax rate is 40%, and its required minimum rate of return is 10%.

(a) What incremental cash flows will occur at the end of Years 0 through 6 as a result of replacing the old machine? Should the old machine be replaced now?

(b) Suppose that the annual operating costs for the old milling machine would increase at an annual rate of 5% over the remaining service life. With this change in future operating costs for the old machine, would the answer in (a) change?

(c) What is the minimum trade-in value for the old machine so that both alternatives are economically equivalent?

14.5 A local telephone company is considering the installation of a new phone line for a new row of apartment complexes. Two types of cables are being considered: conventional copper wire and fiber optics. Transmission by copper wire cables, although cumbersome, involves much less complicated and expensive support hardware than fiber optics. The local company may use five different types of copper wire cables: 100 pairs, 200 pairs, 300 pairs, 600 pairs, and 900 pairs per cable. In calculating the cost per foot cable the following equation is used:

$$\text{Cost per length} = [\text{Cost per foot} + \text{cost per pair (number of pairs)}](\text{length})$$

22 gauge copper
wire = $1.692 per foot
Cost per pair = $0.013 per pair.

The annual cost of the cable as a percent of the first cost is 18.4%. The life of the system is 30 years.

In fiber optics, a cable is referred to as a ribbon. One ribbon contains 12 fibers. The fibers are grouped in fours; therefore, one ribbon contains three groups of four fibers. Each group of four fibers can produce 672 lines (equivalent to 672 pairs of wires) and, since each ribbon contains three groups, the total capacity of the ribbon is 2,016 lines. To transmit signals using fiber optics, many modulators, wave guides, and terminators are needed to convert the signals from electric currents to modulated light waves. Fiber optic ribbon costs $15,000 per mile. At each end of the ribbon three terminators are needed, one for each group of four fibers at a cost of $30,000 per terminator. Twenty-one modulating systems are needed at each end of the ribbon at a cost of $12,092 for a unit in the central office and $21,217 for a unit in the field. Every 22,000 feet, a repeater is required to keep the modulated light waves in the ribbon at an intelligible intensity for detection. The unit cost of this repeater is $15,000. The annual cost, including income taxes for the 21 modulating systems, is 12.5% of the first cost for the units. The annual cost of the ribbon itself is 17.8% for its first cost. The life of the whole system is 30 years. (All figures represent after-tax costs.)

(a) Suppose that the apartments are located 5 miles from the phone company's central switching system and about 2,000 telephones will be required. This would require either 2,000 pairs of copper wire or one fiber optics ribbon and related hardware. If the telephone company's interest rate is 15%, which option is more economical?

(b) In (a), suppose that the apartments are located 10 miles or 25 miles from the phone company's central switching system. Which option is more economically attractive under each scenario?

14.6 A small manufacturing firm is considering the purchase of a new boring machine to modernize one of its production lines. Two types of boring machine are available on the market. The lives of machine A and machine B are 8 years and 10 years, respectively. The machines have the following receipts and disbursements. Use a MARR (after-tax) of 10% and a marginal tax rate of 30%:

Item	Machine A	Machine B
First cost	$6,000	$8,500
Service life	8 years	10 years
Salvage value	500	1,000
Annual O&M costs	700	520
Depreciation (MACRS)	(7-year)	(7-year)

(a) Which machine would be most economical to purchase under the infinite planning horizon? Explain any assumption that you need to make about future alternatives.

(b) Determine the break-even annual O&M costs for machine A so that the present worth of machines A and B is the same.

(c) Suppose that the required service life of the machine is only 5 years. The estimated salvages at the end of the required service period are estimated to be $3,000 for machine A and $3,500 for machine B, respectively. Which machine is more economical?

14.7 The management of Langdale Mill is considering replacing a number of old looms in the mill's weave room. The looms to be replaced are two 86-inch President looms, sixteen 54-inch President looms, and twenty-two 72-inch Draper X-P2 looms. The company may either replace the old looms with new ones of the same kind or buy 21 new shutterless Pignone looms. The first alternative requires the purchase of 40 new President and Draper looms and the scrapping of the old looms. The second alternative involves scrapping the 40 old looms, relocating 12 Picanol looms, and constructing a concrete floor, plus purchasing the 21 Pignone looms and various related equipment.

Description	Alternative 1	Alternative 2
Machinery/related equipment	$2,119,170	$1,071,240
Removal cost of old looms/site preparation	26,866	49,002
Salvage value of old looms	62,000	62,000
Annual sales increase with new looms	7,915,748	7,455,084
Annual labor	261,040	422,080
Annual O&M	1,092,000	1,560,000
Depreciation (MACRS)	7-year	7-year
Project life	8 years	8 years
Salvage value	169,000	54,000

The firm's MARR is 18%. This figure is set by corporate executives. The corporate executives feel that various investment opportunities available for the mills will guarantee a rate of return on investment of at least 18%. The mill's marginal tax rate is 40%.

(a) Perform a sensitivity analysis on the project's data, varying the net operating revenue, labor cost, annual maintenance cost, and the MARR. Assume that each of these variables can deviate from its base case expected value by ±10%, by ±20%, and by ±30%.

(b) From the results of part (a), prepare sensitivity diagrams and interpret the results.

14.8 Mike Lazenby, an industrial engineer at Energy Conservation Service, has found that the anticipated profitability of a newly developed water-heater temperature control device can be measured by present worth as follows:

$$NPW = 4.028V(2X - \$11) - \$77,860,$$

where V is the number of units produced and sold, and X is the sales price per unit. Mike also found that the V parameter value could occur anywhere over the range of 1,000 to 6,000 units and the X parameter value anywhere between $20 to $45 per unit. Develop a sensitivity graph as a function of number of units produced and sales price per unit.

Break-Even Analysis

14.9 Susan Campbell is thinking about going into the motel business near Disney World in Orlando. The cost to build the motel is $2,200,000. The lot costs $600,000. Furniture and furnishings cost $400,000 and should be recovered in 7 years (7-year MACRS property), while the motel building should be recovered in 39 years (39-year MACRS real property placed in service on January 1). The land will appreciate at an annual rate of 5% over the project period, but the building will have a zero salvage value after 25 years. When the motel is full (100% capacity), it takes in (receipts) $4,000 per day for 365 days per year. The motel has fixed operating expenses, exclusive of depreciation, of $230,000 per year. The variable operating expenses are $170,000 at 100% capacity, and these vary directly with percent capacity down to zero at 0% capacity. If the interest is 10%, compounded annually, at what percentage capacity must this motel operate at to break even? (Assume Susan's tax rate is 31%.)

14.10 A plant engineer wishes to know which of two types of light bulbs should be used to light a warehouse. The bulbs currently used cost $45.90 per bulb and last 14,600 hours before burning out. The new bulb ($60 per bulb) provides the same amount of light and consumes the same amount of energy, but lasts twice as long. The labor cost to change a bulb is $16.00. The lights are on 19 hours a day, 365 days a year. If the firm's MARR is 15%, what is the maximum price (per bulb) the engineer should be willing to pay to switch to the new bulb? (Assume the firm's marginal tax rate is 40%.)

14.11 Robert Cooper is considering a piece of business rental property containing stores and offices at a cost of $250,000. Cooper estimates that annual receipts from rentals will be $35,000 and that annual disbursements, other than income taxes, will be about $12,000. The property is expected to appreciate at the annual rate of 5%. Cooper expects to retain the property for 20 years once it is acquired. Then, it will be depreciated based on the 39-year real property class (MACRS), assuming that the property would be placed in service on January 1. Cooper's marginal tax rate is 30%, and his MARR is 15%. What would be the minimum annual total of rental receipts that would make the investment break even?

14.12 The city of Opelika was having a problem locating land for a new sanitary landfill site when the Alabama Energy Extension Service offered the solution of burning the solid waste to generate steam. At the same time, Uniroyal Tire Company seemed to be having a similar problem, disposing of solid waste in the form of rubber tires. It was determined that there would be about 200 tons per day of waste to be burned; this included municipal and industrial waste. The city is considering building a waste-fired steam plant, which would cost $6,688,800. To finance the construction cost, the city will issue resource recovery revenue bonds in the amount of $7,000,000 at an interest rate of 11.5%. Bond interest is payable annually. The differential amount between the actual construction costs and the amount of bond financing ($7,000,000 − $6,688,800 = $311,200) will be used to settle the bond discount and expenses associated with the bond financing. The expected life of the steam plant is 20 years. The expected salvage value is estimated to be about $300,000. The expected labor costs would be $335,000 per year. The annual operating and maintenance costs (including fuel, electricity, maintenance, and water) are expected to be $175,000. This plant would generate 9,360

pounds of waste along with 7,200 pounds of waste after incineration, which will have to be disposed of as land fill. At the present rate of $19.45 per pound, this will cost the city a total of $322,000 per year. The revenues for the steam plant will come from two sources: (1) Steam sales and (2) disposal tipping fees. The city expects 20% down time per year for the waste-fired steam plant. With an input of 200 tons per day and 3.01 pounds of steam per pound refuse, a maximum of 1,327,453 pounds of steam can be produced per day. However, with 20% down time, the actual output would be 1,061,962 pounds of steam per day. The initial steam charge will be approximately $4.00 per thousand pounds. This would bring in $1,550,520 in steam revenue the first year. The tipping fee is used in conjunction with the sale of steam to offset the total plant cost. It is the goal of the Opelika steam plant to phase out the tipping fee as soon as possible. The tipping fee will be $20.85 per ton in the first year of plant operation and will be phased out at the end of the eighth year. The scheduled tipping fee assessment would be as follows:

Year	Tipping Fee
1	$976,114
2	895,723
3	800,275
4	687,153
5	553,301
6	395,161
7	208,585

(a) At an interest rate of 10%, would the steam plant generate sufficient revenue to recover the initial investment?

(b) At an interest rate of 10%, what would be the minimum charge (per thousand pounds) for steam sales to make the project break even?

(c) Perform a sensitivity analysis to determine the input variable of the plant's down time.

14.13 Two different methods of solving a production problem are under consideration. Both methods are expected to be obsolete in 6 years. Method A would cost $80,000 initially and have annual operating costs of $22,000 a year. Method B would cost $52,000 and costs $17,000 a year to operate. The salvage value realized with Method A would be $20,000 and with Method B would be $15,000. Method A would generate $16,000 revenue income a year more than Method B. Investments in both methods are subject to a 5-year MACRS property class. The firm's marginal income tax rate is 40%. The firm's MARR is 20%. What would be the required additional annual revenue for Method A such that both methods would be indifferent?

14.14 Rocky Mountain Publishing Company is considering introducing a new morning newspaper in Denver. Its direct competitor charges $0.25 at retail, with $0.05 going to the retailer. For the level of news coverage the company desires, it determines the fixed cost of editors, reporters, rent, press room expenses, and wire service charges to be $300,000 per month. The variable cost of ink and paper is $0.10 per copy, but advertising revenues of $0.05 per paper will be generated. To print the morning paper, the publisher has to purchase a new printing press, which will cost $600,000. The press machine will be depreciated according to a 7-year MACRS class. The press machine will be used for 10 years, at which time its salvage value would be about $100,000. Assume 25 weekdays in a month, a 40% tax rate, and a 13% MARR. How many copies per day must be sold to break even at a selling price of $0.25 per paper at retail?

Probabilistic Analysis

14.15 A corporation is trying to decide whether or not to buy the patent for a product designed by another company. The decision to buy will mean an investment of $8 million, and the demand for the product is not known. If demand is light, the company expects a return of $1.3 million each year for 3 years. If the demand is moderate, the return will be $2.5 million each year for 4 years, and a high demand means a return of $4 million each year for 4 years. It is estimated the probability of a high demand is 0.4 and the probability of a light demand is 0.2. The firm's interest rate (risk-free) is 12%. Calculate the expected present worth. On this basis should the company make the investment? (All figures represent after-tax values.)

14.16 Juan Carols is considering two investment projects whose present values are described as follows:

- Project 1: $PW(10\%) = 2X(X-Y)$, where X and Y are statistically independent discrete random variables with the following distributions:

	X		Y
Event	Probability	Event	Probability
$20	0.6	$10	0.4
40	0.4	20	0.6

- Project 2:

PW(10%)	Probability
$ 0	0.24
400	0.20
1,600	0.36
2,400	0.20

Note: Cash flows between the two projects are also assumed to be statistically independent.

(a) Develop the NPW distribution for project 1.

(b) Compute the mean and variance of the NPW for project 1.

(c) Compute the mean and variance of the NPW for project 2.

(d) Suppose that Projects 1 and 2 are mutually exclusive. Which project would you select?

14.17 A financial investor has an investment portfolio worth $350,000. A bond in his investment portfolio will mature next month and provide him with $25,000 to reinvest. The choices have been narrowed down to the following two options.

- Option 1: Reinvest in a foreign bond that will mature in one year. This will entail a brokerage fee of $150. For simplicity, assume that the bond will provide interest over the 1-year period of $2,450, $2,000, or $1,675 and that the probabilities of these occurrences are assessed to be 0.25, 0.45, and 0.30, respectively.

- Option 2: Reinvest in a $25,000 certificate with a savings and loan association. Assume this certificate has an effective annual rate of 7.5%.

(a) Which form of reinvestment should the investor choose in order to maximize his expected financial gain?

(b) If the investor can obtain professional investment advice from Solomon and Brother's Inc., what would be the maximum amount the investor should pay for this service?

14.18 Kellog Company is considering the following investment project and has estimated all cost and revenues in constant dollars. The project requires a purchase of $9,000 asset, which will be used for only 2 years (project life).

- The salvage value of this asset at the end of 2 years is expected to be $4,000.

- The project requires an investment of $2,000 in working capital, and this amount will be fully recovered at the end of project year.

- The annual revenue as well as general inflation are discrete random variables, but can be described by the following probability distributions. Both random variables are statistically independent.

Annual Revenue (X)	Probability	General Inflation	Rate (Y)
10,000	0.30	3%	0.25
20,000	0.40	5%	0.50
30,000	0.30	7%	0.25

- The investment will be classified as a 3-MACRS property (tax life).

- It is assumed that the revenues, salvage value, and working capital are responsive to this general inflation rate.

- The revenue and inflation rate dictated during the first year will prevail over the remaining project period.

- The marginal income tax rate for the firm is 40%. The firm's inflation-free interest rate (i') is 10%.

(a) Determine the NPW as a function of X.

(b) In (a), compute the expected NPW of this investment.

(c) In (a), compute the variance of the NPW of this investment.

Comparing Risky Projects

14.19 A manufacturing firm is considering two mutually exclusive projects. Both projects have an economic service life of 1 year with no salvage value. The first cost and the net year end revenue for each project are given as follows:

First Cost	Project 1 ($1,000)		Project 2 ($800)	
	Probability	Revenue	Probability	Revenue
Net revenue given in PW	0.2	$2,000	0.3	$1,000
	0.6	3,000	0.4	2,500
	0.2	3,500	0.3	4,500

We assume that both projects are statistically independent from each other.

(a) If you are an expected value maximizer, which project would you select?

(b) If you also consider the variance of the project, which project would you select?

14.20 A business executive is trying to decide whether to undertake one of two contracts or neither one. He has simplified the situation somewhat and feels that it is sufficient to imagine that the contracts provide alternatives as follows:

Contract A		Contract B	
NPW	Probability	NPW	Probability
$100,000	0.2	$40,000	0.3
50,000	0.4	10,000	0.4
0	0.4	−10,000	0.3

(a) Should the executive undertake either one of the contracts? If so, which one? What would he do if he made decisions by maximizing his expected NPW?

(b) What would be the probability that Contract A would result in a larger profit than that of Contract B?

14.21 Two alternative machines are being considered for a cost reduction project.

• Machine A has a first cost of $60,000 and a salvage value (after-tax) of $22,000 at the end of 6-years' service life. Probabilities of annual after-tax operating costs of this machine are estimated as follows:

Annual O&M Costs	Probability
$ 5,000	0.20
8,000	0.30
10,000	0.30
12,000	0.20

• Machine B has a first cost of $35,000, and its estimated salvage value (after-tax) at the end of 4 years service is to be negligible. The annual after-tax operating costs are estimated to be as follows:

Annual O&M Costs	Probability
$8,000	0.10
10,000	0.30
12,000	0.40
14,000	0.20

The MARR on this project is 10%. The required service period of these machines is estimated to be 12 years, and no technological advance in either machine is expected.

(a) Assuming independence, calculate the mean and variance for the equivalent annual cost of operating each machine.

(b) From the results of part (a), calculate the probability that the annual cost of operating Machine A will exceed the cost of operating Machine B.

14.22 Two mutually exclusive investment projects are under consideration. It is assumed that the cash flows are statistically independent random variables with means and variances estimated as follows:

End of Year	Project A Mean	Project A Variance	Project B Mean	Project B Variance
0	−$5,000	$1,000^2$	−$10,000	$2,000^2$
1	4,000	$1,000^2$	6,000	$1,500^2$
2	4,000	$1,500^2$	8,000	$2,000^2$

(a) For each project, determine the mean and standard deviation for NPW using an interest rate of 15%.

(b) Based on the results of part (a), which project would you recommend?

Decision Tree Analysis

14.23 Delta College's campus police are quite concerned with ever-growing weekend parties taking place at the various dormitories on the campus, where alcohols are commonly served to underage college students. According to reliable information, for a given Saturday night, one may observe the actual party to take place 60% of the time. Police Lieutenant Shark usually receives a tip regarding student drinking to take place in one of the residence halls the following weekend.

According to Officer Shark, this tipster has been correct 40% of the times when the actual party is really planned. On the other hand, this tipster has been also correct 80% of the times when the party does not take place (that is the tipster says there will be no party planned). If Officer Shark does not raid that residence hall at the time of the supposed party, he loses 10 career progress points (the police chief gets the complete information on whether there was a party only after the weekend). If he leads a raid and the tip is indeed false, he loses 50 career progress points; if the tip is correct, he earns 100 points.

(a) What is the probability that no party is actually planned even though the tipster says there will be a party?

(b) If the lieutenant wishes to maximize his expected career progress points, what should he do?

(c) What is the EVPI (in terms of career points)?

14.24 As a plant manager of a firm, you are trying to decide whether to open a new factory outlet store, which would cost about $500,000. Success of the outlet store depends on demand in the new region. If demand is high, you expect to gain $1 million per year; if average, $500,000; and if low, to lose $80,000. From your knowledge of the region and your product, you feel the chances are 0.4 that sales will be average, and equally likely that they will be high or low (0.3, respectively). Assume that the firm's MARR is known to be 15%, and the marginal tax rate will be 40%. Also, assume that the salvage value of the store at the end of 15 years will be about $100,000. The store will be depreciated under a 39-year property class.

(a) If the outlet store will be in business for 15 years, should you open the new outlet store? How much would you be willing to pay to know the true state of nature?

(b) Suppose a market survey is available at $1,000 with the following reliability (values obtained from past experience where actual demand was compared with predictions made by the market survey).

Given Actual Demand	Survey Prediction		
	Low	Medium	High
Low	0.75	0.20	0.05
Medium	0.20	0.60	0.20
High	0.05	0.25	0.70

Determine the strategy that maximizes the expected payoff after taking the market survey. In doing so, compute the EVPI after taking the survey. What is the true worth of the sample information?

Short Case Studies

14.25 Mount Manufacturing Company produces industrial and public safety shirts. As is done in most apparel manufacturing, the cloth must be cut into shirt parts by marking sheets of paper in the way that the particular cloth is to be cut. At present, these sheet markings are done manually and its annual labor cost is running around $103,718. Mount has the option of purchasing one of two automated marking systems. The two systems are the Lectra System 305 and the Tex Corporation Marking System. The comparative characteristics of the two systems are as follows:

	Most Likely Estimates	
	Lectra System	Tex System
Annual labor cost	$51,609	$51,609
Annual material savings	$230,000	$274,000
Investment cost	$136,150	$195,500
Estimated life	6 years	6 years
Salvage value	$20,000	$15,000
Depreciation method (MACRS)	5-year	5-year

The firm's marginal tax rate is 40%, and the interest rate used for project evaluation is 12% after taxes.

(a) Based on the most likely estimates, which alternative is the best?

(b) Suppose that the company estimates the material savings during the first year for each system on the basis of the following probability distribution:

Lectra System	
Material Savings	Probability
$150,000	0.25
230,000	0.40
270,000	0.35

Tex System	
Material Savings	Probability
$200,000	0.30
274,000	0.50
312,000	0.20

Further assume that the annual material savings for both Lectra and Tex are statistically independent. Compute the mean and variance for the equivalent annual value of operating each system.

(c) In part (b), calculate the probability that the annual benefit of operating Lectra will exceed the benefit of operating Tex.

14.26 Burlington Motor Carriers, a trucking company, is considering the installation of a two-way mobile satellite messaging service on their 2,000 trucks. Based on tests done last year on 120 trucks, the company found that satellite messaging could cut 60% from its $5 million bill for long-distance communications with truck drivers. More important, the drivers reduced the number of "deadhead" miles—those driven without paying loads—by 0.5%. Applying that improvement to all 230 million miles covered by the Burlington fleet each year would produce an extra $1.25 million savings.

Equipping all 2,000 trucks with the satellite hook-up will require an investment of $8 million and the construction of a message-relaying system costing $2 million. The equipment and on-board devices will have a service life of 8 years and negligible salvage value; they will be depreciated under the 5-year MACRS class. Burlington's marginal tax rate is about 38%, and its required minimum attractive rate of return is 18%.

(a) Determine the annual net cash flows from the project.

(b) Perform a sensitivity analysis on the project's data, varying savings in telephone bill and savings in deadhead miles. Assume that each of these variables can deviate from its base case expected value by ±10%, ±20%, and ±30%.

(c) Prepare sensitivity diagrams and interpret the results.

14.27 The following is a comparison of the cost structure of a conventional manufacturing technology (CMT) with a flexible manufacturing system (FMS) at one U.S. firm.

| | Most Likely Estimates | |
	CMT	FMS
Number of part types	3,000	3,000
Number of pieces produced/year	544,000	544,000
Variable labor cost/part	$2.15	$1.30
Variable material cost/part	$1.53	$1.10
Total variable cost/part	$3.68	$2.40
Annual overhead	$3.15M	$1.95M
Annual tooling costs	$470,000	$300,000
Annual inventory costs	$141,000	$31,500
Total annual fixed operating costs	$3.76M	$2.28M
Investment	$3.5M	$10M
Salvage value	$0.5M	$1M
Service life	10 years	10 years
Depreciation method (MACRS)	7-year	7-year

(a) The firm's marginal tax rate and MARR is 40% and 15%, respectively. Determine the incremental cash flow (FMS – CMT) based on the most likely estimates.

(b) Management feels confident about all input estimates in CMT. However, the firm does not have any previous experience in operating an FMS. Therefore, many input estimates, except the investment and salvage value, are subject to variations. Perform a sensitivity analysis on the project's data, varying the elements of operating costs. Assume that each of these variables can deviate from

its base case expected value by ±10%, ±20%, and ±30%.

(c) Prepare sensitivity diagrams and interpret the results.

(d) Suppose that probabilities of the variable material cost and the annual inventory cost for the FMS are estimated as follows:

Material Cost	
Cost per Part	Probability
$1.00	0.25
1.10	0.30
1.20	0.20
1.30	0.20
1.40	0.05

Inventory Cost	
Annual Inventory Cost	Probability
$25,000	0.10
31,000	0.30
50,000	0.20
80,000	0.20
100,000	0.20

What are the best and the worst cases of incremental NPW?

(e) In part (d), assuming that the random variables of the cost per part and the annual inventory cost are statistically independent, find the mean and variance of the NPW for the incremental cash flows.

(f) In parts (d) and (e), what is the probability that the FMS would be a more expensive investment option?

CHAPTER 15

Replacement Decisions

In March 2000, Steve Hausmann, a Production Engineer at IKEA-USA, a European-style furniture company, was considering replacing a 1000-pound-capacity industrial forklift truck. The truck was being used to move assembled furniture from the finishing department to the warehouse. Recently the truck had not been dependable and was frequently out of service while awaiting repairs. Its maintenance expenses have been rising steadily. If the truck were not available, the company would have to rent one. Additionally the forklift truck was diesel operated, and workers in the plant were complaining about the air pollution. If retained, it would require an immediate engine overhaul to keep it in operable condition. The overhaul would neither extend the originally estimated service life nor would it increase the value of the truck.

Two types of forklift trucks were recommended as replacements. One was electrically operated, and the other was gasoline-operated. The electric truck would eliminate the air pollution problem entirely, but would necessitate a battery change twice a day, which would significantly increase the operating cost. If the gasoline-operated truck were to be used, it would require more frequent maintenance.

Mr. Hausmann was undecided about whether the company should buy the electric- or the gasoline-operated forklift truck at this point. He felt he should do some homework before approaching upper management for authorization of the replacement. Two questions came to his mind immediately:

1. Should the forklift truck be repaired now or replaced by one of the more advanced and fuel-efficient trucks?
2. If the replacement option were chosen, when should the replacement occur—now or sometime in the future?

The answer to the first question is almost certainly that it should be replaced in the near future. Even if the truck is repaired now, at some point it will fail to meet the company's needs due to physical or functional depreciation. If the replacement option were to be chosen, the company would probably want to take advantage of technological advances that would provide more fuel efficiency and safer lifting capability in newer trucks.

The answer to the second question is less clear. Presumably the current truck can be maintained in serviceable condition for some time—when will be the economically optimal time to make the replacement? Furthermore, when is the technologically optimal time to switch? By waiting for a more technologically advanced truck to evolve and gain commercial acceptance, the company may be able to obtain even better results tomorrow than by just switching to whatever lift-truck is available on the market today.

In Chapters 7 through 9, we presented methods that helped us choose the best investment alternative. The problems we examined in those chapters primarily concerned profit-adding projects. However, economic analysis is also frequently performed on projects with existing facilities or profit-maintaining projects. Profit-maintaining projects are those whose primary purpose is not to increase sales, but rather to simply maintain ongoing operations. In practice, profit-maintaining projects less frequently involve the comparison of new machines; instead, the problem often facing management is whether to buy new and more efficient equipment or to continue to use existing equipment. This class of decision problems is known as the **replacement problem**. In this chapter, we examine the basic concepts and techniques related to replacement analysis.

15.1 Replacement Analysis Fundamentals

In this section and the following two sections, we examine three aspects of the replacement problem: (1) approaches for comparing defender and challenger, (2) determination of economic service life, and (3) replacement analysis when the required service period is long. The impact of income tax regulations will be ignored in these sections. In Section 15.4, we revisit these replacement problems considering income taxes.

15.1.1 Basic Concepts and Terminology

Replacement projects are decision problems involving the replacement of existing obsolete or worn-out assets. The continuation of operations is dependent on these assets. Failure to make an appropriate decision results in a slowdown or shutdown of the operations. The question is when existing equipment should be replaced with more efficient equipment. This situation has given rise to the use of the terms **defender** and **challenger**, terms commonly used in the boxing world. In every boxing class, the current defending champion is constantly faced with a new challenger. In replacement analysis, the defender is the existing machine (or system), and the challenger is the best available replacement equipment.

An existing piece of equipment will be removed at some future time, either when the task it performs is no longer necessary or when the task can be performed more efficiently by newer and better equipment. The question is not whether the existing piece of equipment will be removed, but when it will be removed. A variation of this question is why we should replace existing equipment at this time rather than postponing replacement of the equipment by repairing or overhauling it. Another aspect of the defender-challenger comparison concerns deciding exactly which equipment is the best challenger. If the defender is to be replaced by the challenger, we would generally want to install the very best of the possible alternatives.

Current Market Value

The most common problem encountered in considering the replacement of existing equipment is the determination of what financial information is actually relevant to the analysis. Often a tendency to include irrelevant information in the analysis is apparent. To illustrate this type of decision problem, let us consider Example 15.1.

Example 15.1 Relevant Information for Replacement Analysis

Macintosh Printing, Inc., purchased a $20,000 printing machine 2 years ago. The company expected this machine to have a 5-year life and a salvage value of $5,000. The company spent $5,000 last year on repairs, and current operating costs are now running at the rate of $8,000 per year. Furthermore, the anticipated salvage value has now been reduced to $2,500 at the end of the printer's remaining useful life. In addition, the company has found that the current machine has a market value of $10,000 today. The equipment vendor will allow the company this full amount as a trade-in on a new machine. What value(s) for the defender are relevant in our analysis?

Solution

In this example, three different dollar amounts relating to the defender are presented:

1. Original cost: The printing machine was purchased for $20,000.
2. Market value: The company estimates the old machine's market value at $10,000.
3. Trade-in allowance: It is the same as the market value. (This value, however, could be different from the market value.)

In this example, and in all defender analyses, the relevant cost is the **current market value** of the equipment. The original cost, repair cost, and trade-in value are irrelevant. A common misconception is held that the trade-in value is the same as the current market value of the equipment and thus could be used to assign a suitable current value to the equipment. This is not always true, however. For example, a car dealer typically offers a trade-in value on a customer's old car to reduce the price of a new car. Would the dealer offer the same value on the old car if he were not also selling the new one? This is not generally the case. In many instances, the trade-in allowance is inflated to make the deal look good, and the price of the new car is also inflated to compensate for the dealer's trade-in cost. In this type of situation, the trade-in value does not represent the true value of the item, so we should not use it in economic analysis.[1]

Sunk Costs

As mentioned in Section 3.4.3, a **sunk cost** is any past cost unaffected by any future investment decision. In Example 15.1, the company spent $20,000 to buy the machine two years ago. Last year, $5,000 more was spent on this machine. The total accumulated expenditure on this machine is $25,000. If the machine is sold today, the company can only get $10,000 back (Figure 15.1). It is tempting to think that the company would lose $15,000 in addition to the cost of the new machine if the machine were to be sold and replaced with a new one. This is an incorrect way of doing economic analysis. In a proper engineering economic analysis, only future costs should be considered; past or sunk costs should be ignored. Thus, the value of the defender that should be used in a replacement analysis should be its current market value, not what it cost when it was originally purchased and not the cost of repairs that have already been made on the machine.

Sunk costs are the money that is gone, and no present action can recover it. They represent past actions. They are the results of decisions made in the past. In making

---

[1] If we do make the trade, however, the actual net cash flow at this time, properly used, is certainly relevant.

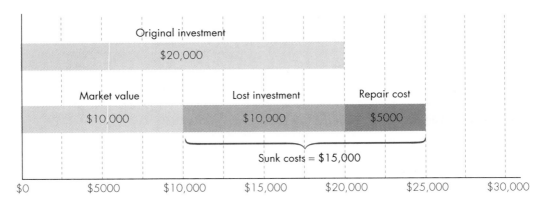

Figure 15.1   Sunk cost associated with an asset's disposal as described in Example 15.1

economic decisions at the present time, one should only consider the possible out-comes of different decisions and pick the one with the best possible future results. Using sunk costs in arguing one option over the other would only lead to more bad decisions.

Operating Costs

The driving force for replacing existing equipment is that it becomes more expensive to operate with time. The total cost of operating a piece of equipment may include re-pair and maintenance costs, wages for the operators, energy consumption costs, and costs of materials. Increases in any one or a combination of these cost items over a pe-riod of time may cause us to find a replacement for the existing asset. The challenger is usually newer than the defender and often incorporates design improvements and newer technology. As a result, some or all of the cost items for the challenger are likely to be less than those for the defender.

We will call the sum of the various cost items related to the operation of an asset the **operating costs**. As is illustrated in the following sections, keeping the defender in-volves a lower initial cost than purchasing the challenger but higher annual operating costs. These operating costs usually increase over time for both the defender and the challenger. In many instances, the labor costs, material costs, and energy costs are the same for the defender and the challenger and do not change with time. It is the repair and maintenance costs that increase and cause the operating costs to increase each year as an asset ages. When repair and maintenance costs are the only cost items that differ between the defender and the challenger on a year by year basis, we only need to include repair and maintenance costs in the operating costs used in the analysis. Regardless of which cost items we choose to include in the operating costs, it is essen-tial that the same items are included for both the defender and the challenger. For ex-ample, if energy costs are included in the operating costs of the defender, they should also be included in the operating costs of the challenger. A more comprehensive dis-cussion of the various types of costs incurred in a complex manufacturing facility is provided in Chapter 3.

15.1.2 Approaches for Comparing Defender and Challenger

Although replacement projects are a subcategory of the mutually exclusive project decisions we studied in Chapter 7, they do possess unique characteristics that allow us to use specialized concepts and analysis techniques in their evaluation. We consider two basic approaches to analyzing replacement problems: the cash flow approach and the opportunity cost approach. We start with a replacement decision problem where both the defender and the challenger have the same useful life, which begins now.

Cash Flow Approach

The cash flow approach can be used as long as the analysis period is the same for all replacement alternatives. In other words, we consider explicitly the actual cash flow consequences for each replacement alternative and compare them based on either PW or AE values.

Example 15.2 Replacement Analysis Using the Cash Flow Approach

Consider Example 15.1. The company has been offered a chance to purchase another printing machine for $15,000. Over its 3-year useful life, the machine will reduce labor and raw materials usage sufficiently to cut operating costs from $8,000 to $6,000. This reduction in costs will allow after-tax profits to rise by $2,000 per year. It is estimated that the new machine can be sold for $5,500 at the end of year 3. If the new machine were purchased, the old machine would be sold to another company rather than traded in for the new machine. Suppose that the firm will need either machine (old or new) for only 3 years and that it does not expect a new, superior machine to become available on the market during this required service period. Assuming that the firm's interest rate is 12%, decide whether replacement is justified now.

Solution

- Option 1: Keep the defender

  If the old machine is kept, there is no additional cash expenditure today. It is in perfect operational condition. The annual operating cost for the next 3 years will be $8,000 per year, and its salvage value 3 years from today will be $2,500. The cash flow diagram for the defender is shown in Figure 15.2(a).

- Option 2: Replace the defender with the challenger

  If this option is taken, the defender (designated D) can be sold for $10,000. The cost of the challenger (designated C) is $15,000. Thus, the initial combined cash flow for this option is a negative cash flow of $15,000 − $10,000 = $5,000. The annual operating cost of the challenger is $6,000. The salvage value of the challenger 3 years later will be $5,500. The actual cash flow diagram for this option is shown in Figure 15.2 (b).

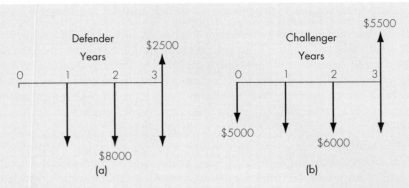

Figure 15.2   Comparison of defender and challenger based on the cash flow approach (Example 15.2)

$$PW(12\%)_D = \$2{,}500(P/F, 12\%, 3) - \$8{,}000(P/A, 12\%, 3)$$
$$= -\$17{,}434.90$$
$$AE(12\%)_D = PW(12\%)_D(A/P, 12\%, 3)$$
$$= -\$7{,}259.10$$
$$PW(12\%)_C = \$5{,}500(P/F, 12\%, 3) - \$5{,}000 - \$6{,}000(P/A, 12\%, 3)$$
$$= \$15{,}495.90$$
$$AE(12\%)_C = PW(12\%)_C(A/P, 12\%, 3)$$
$$= -\$6{,}451.79$$

Because of the annual difference of $807.31 in favor of the challenger, the replacement should be made now.

Comments: If the defender should not be replaced now, we did not address the question of whether the defender should be kept for 1 or 2 years before being replaced with the challenger. This is a valid question, which requires more data on market values over time. We address this situation later, in Section 15.3.

Opportunity Cost Approach

In the previous example, $10,000 in receipts from sale of the old machine was foregone by not selling the defender. Another way to analyze such a problem is to charge the $10,000 as an opportunity cost of keeping the asset. That is, instead of deducting the salvage value from the purchase cost of the challenger, we consider the salvage value as a cash outflow for the defender (or investment required to keep the defender).

Example 15.3 Replacement Analysis Using the Opportunity Cost Approach

Rework Example 15.2 using the opportunity cost approach.

Solution

Recall that the cash flow approach in Example 15.2 credited proceeds in the amount of $10,000 from the sale of the defender toward the $15,000 purchase price of the challenger, and no initial outlay would have been required had the decision been to keep the defender. If the decision to keep the defender had been made, the opportunity cost approach treats the $10,000 current salvage value of the defender as an incurred cost. Figure 15.3 illustrates the cash flows related to these decision options.

Since the lifetimes are the same, we can use either PE or AE analysis as follows:

$$PW(12\%)_D = -\$10,000 - \$8,000(P/A, 12\%, 3) + \$2,500(P/F, 12\%, 3)$$

$$= -\$27,434.90$$

$$AE(12\%)_D = PW(12\%)_D(A/P, 12\%, 3)$$

$$= -\$11,422.64$$

$$PW(12\%)_C = -\$15,000 - \$6,000(P/A, 12\%, 3) + \$5,500(P/F, 12\%, 3)$$

$$= -\$25,495.90$$

$$AE(12\%)_C = PW(12\%)_C(A/P, 12\%, 3)$$

$$= -\$10,615.33$$

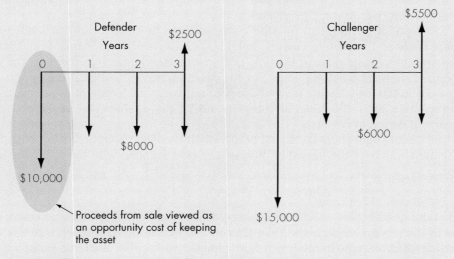

Figure 15.3   Compression of defender and challenger based on the opportunity cost approach (Example 15.3)

The decision outcome is the same as in Example 15.2, that is, the replacement should be made. Since both the challenger and defender cash flows were adjusted by the same amount −$10,000 at time 0, this should not come as a surprise.

Comments: Recall that we assumed the same service life for both the defender and the challenger in Examples 15.2 and 15.3. In general, however, old equipment has a relatively short remaining life compared with new equipment, so this assumption is overly simplistic. In the next section, we discuss how to find the economic service life of equipment.

15.2 Economic Service Life

You have probably seen a 50-year-old automobile still in service. Provided it receives the proper repair and maintenance, almost anything can be kept operating for an extended period of time. If it's possible to keep a car operating for an almost indefinite period, why aren't more old cars spotted on the streets? There are several reasons. Some people may get tired of driving the same old car. Others may want to keep a car as long as it will last, but they realize that repair and maintenance costs will become excessive.

In general, we need to consider explicitly how long an asset should be held, once it is placed in service. For instance, a truck-rental firm that frequently purchases fleets of identical trucks may wish to arrive at a policy decision on how long to keep each vehicle before replacing it. If an appropriate life span is computed, a firm could stagger a schedule of truck purchases and replacements to smooth out annual capital expenditures for overall truck purchases.

The costs of owning and operating an asset can be divided into two categories: **capital costs** and **operating costs.** Capital costs have two components: initial investment and the salvage value at the time of disposal. The initial investment for the challenger is simply its purchase price. For the defender, we should treat the opportunity cost as its initial investment. We will use N to represent the length of time in years the asset will be kept; I, the initial investment; and S_N, the salvage value at the end of the ownership period of N years.

The annual equivalent of capital costs, which is called capital recovery cost (refer to Section 8.2), over the period of N years can be calculated with the following equation:

$$CR(i) = I(A/P, i, N) - S_N(A/F, i, N) \tag{15.1}$$

Generally speaking, as an asset becomes older, its salvage value becomes smaller. As long as the salvage value is less than the initial cost, the capital recovery cost is a decreasing function of N. In other words, the longer we keep an asset, the lower the capital recovery cost becomes. If the salvage value is equal to the initial cost no matter how long the asset is kept, the capital recovery cost is also constant.

As described earlier, the operating costs of an asset include operating and maintenance (O&M) costs, labor costs, material costs, and energy consumption costs. For the same equipment, labor costs, material costs, and energy costs are often constant from year to year if the usage of the equipment remains constant. However, O&M costs tend to increase as a function of the age of the asset. Because of the increasing trend of the O&M costs, the total operating costs of an asset usually increases as the asset ages. We use OC_n to represent the total operating costs in year n of the ownership period and $OC(i)$ to represent the annual equivalent of the operating costs over a lifespan of N years. Then, $OC(i)$ can be expressed as

$$OC(i) = \left(\sum_{n=1}^{N} OC_n (P/F, i, n) \right) (A/P, i, N) \tag{15.2}$$

As long as the annual operating costs increase with the age of the equipment, $OC(i)$ is an increasing function of the life of the asset. If the annual operating costs are the same from year to year, $OC(i)$ is constant and equal to the annual operating costs no matter how long the asset is kept.

The total annual equivalent costs of owning and operating an asset is a summation of the capital recovery costs and the annual equivalent of operating costs of the asset.

$$AE = CR(i) + OC(i) \tag{15.3}$$

The economic service life of an asset is defined to be the period of useful life that minimizes the annual equivalent costs of owning and operating the asset. Based on the foregoing discussions, we need to find the value of N that minimizes AE as expressed in equation (15.3). If $CR(i)$ is a decreasing function of N and $OC(i)$ is an increasing function of N, as this is often the case, AE will be a convex function of N with a unique minimum point (see Figure 15.4). In this book, we assume that AE has a unique minimum point. If the salvage value is constant and equal to the initial cost and the annual operating cost increases with time, AE is an increasing function of N and attains its minimum at $N = 1$. In this case, we should try to replace the asset as

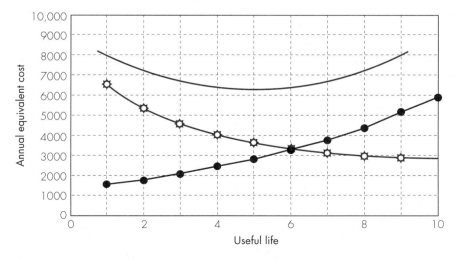

Figure 15.4   A schematic illustrating the trends of capital recovery cost (ownership cost), annual operating cost, and total annual equivalent cost

soon as possible. If the annual operating cost is constant and the salvage value is less than the initial cost and decreases with time, AE is a decreasing function of N. In this case, we would try to delay the replacement of the asset as much as possible. If the salvage value is constant and equal to the initial cost and the annual operating costs are constant, AE will also be constant. In this case, when to replace the asset does not make any economic difference.

If a new asset is purchased and operated for the length of its economic life, the annual equivalent cost is minimized. If we further assume that a new asset of identical price and features can be purchased repeatedly over an indefinite period, we would always replace this kind of asset at its economic life. By replacing perpetually according to an asset's economic life, we obtain the minimum AE cost stream over an indefinite period. However, if the identical replacement assumption cannot be made, we will have to use the methods to be covered in Section 15.3 to make replacement analysis. The next example explains the computational procedure for determining economic service life.

Example 15.4 Economic Service Life for a Lift Truck

As a challenger to the forklift truck described in chapter opening, consider a new electric-lift truck that would cost $18,000, have operating costs of $1,000 in the first year, and have a salvage value of $10,000 at the end of the first year. For the remaining years, operating costs increase each year by 15% over the previous year's operating costs. Similarly, the salvage value declines each year by 25% from the previous year's salvage value. The lift truck has a maximum life of 7 years. An overhaul costing $3,000 and $4,500 will be required during the fifth and seventh year of service, respectively. The firm's required rate of return is 15%. Find the economic service life of this new machine.

Discussion: For an asset whose revenues are either unknown or irrelevant, we compute its economic life based on the costs for the asset and its year-by-year salvage values. To determine an asset's economic service life, we need to compare the options of keeping the asset for one year, two years, three years, and so forth. The option that results in the lowest annual equivalent cost (AE) gives the economic service life of the asset.

- $N = 1$: One-year replacement cycle. In this case, the machine is bought, used for one year, and sold at the end of year 1. The cash flow diagram for this option is shown in Figure 15.5. The annual equivalent cost for this option is:

$$AE(15\%) = \$18,000(A/P, 15\%, 1) + \$1,000 - \$10,000$$

$$= \$11,700$$

Note that $(A/P, 15\%, 1) = (F/P, 15\%, 1)$ and the annual equivalent cost is the equivalent cost at the end of year 1 since $N = 1$. Because we are calculating the annual equivalent costs, we have treated cost items with a positive sign, while the salvage value has a negative sign in the above computation of $AE(15\%)$.

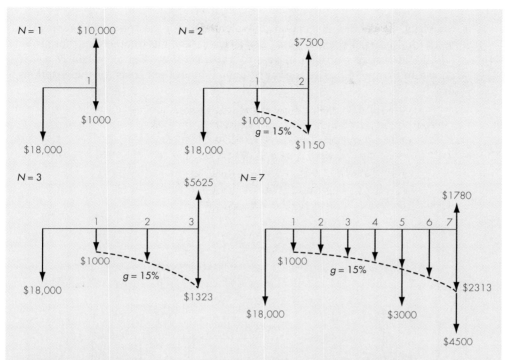

Figure 15.5 Cash flow diagrams for the options of keeping the asset for one year, two years, three years, and seven years (Example 15.4)

- $N = 2$: Two-year replacement cycle. In this case, the truck will be used for two years and disposed of at the end of year 2. The operating cost in year 2 is 15% higher than that in year 1, and the salvage value at the end of year 2 is 25% lower than that at the end of year 1. The cash flow diagram for this option is also shown in Figure 15.5. The annual equivalent cost over the 2-year period is:

$$AE(15\%) = [\$18,000 + \$1,000(P/A_1\, 15\%, 15\%, 2)](A/P, 15\%, 2)$$

$$- \$7,500(A/F, 15\%, 2)$$

$$= \$8,653$$

- $N = 3$: Three-year replacement cycle. In this case, the truck will be used for 3 years and sold at the end of year 3. The salvage value at the end of year 3 is 25% lower than that at the end of year 2, that is, $\$7,500(1 - 25\%) = \$5,625$. The operating cost per year increases at a rate of 15%. The cash flow diagram for this option is also shown in Figure 15.5.

$$AE(15\%) = [\$18,000 + \$1,000(P/A_1\, 15\%, 15\%, 3)](A/P, 15\%, 3)$$

$$- \$5,625(A/F, 15\%, 3)$$

$$= \$7,406$$

- Similarly, we can find the annual equivalent costs for the options of keeping the asset for 4 years, 5 years, 6 years, and 7 years. One has to note that there is an additional cost of overhaul in year 5. The cash flow diagram when $N = 7$ is shown in Figure 15.5. The computed annual equivalent costs for these options are:

$$N = 4, AE(15\%) = \$6,678$$
$$N = 5, AE(15\%) = \$6,642$$
$$N = 6, AE(15\%) = \$6,258$$
$$N = 7, AE(15\%) = \$6,394$$

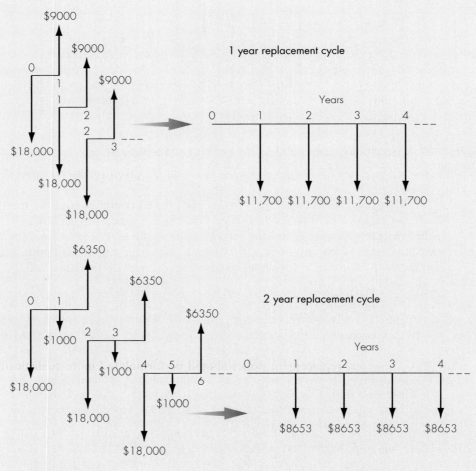

Figure 15.6 Conversion of an infinite number of replacement cycles to infinite AE cost streams (Example 15.4)

From the above calculated AE values for $N = 1, ..., 7$ we find that $AE(15\%)$ is the smallest when $N = 6$. If the truck were to be sold after 6 years, it would have an annual cost of $6,258 per year. If it were to be used for a period other than 6 years, the annual equivalent costs would be higher than $6,258. Thus, a life span of 6 years for this truck results in the lowest annual cost. We conclude that the economic service life of the truck is 6 years. By replacing the assets perpetually according to an economic life of 6 years, we obtain the minimum annual equivalent cost stream. Figure 15.6 illustrates this concept. Of course, we should envision a long period of required service for this kind of asset.

15.3 Replacement Analysis When Required Service Is Long

Now that we understand how the economic service life of an asset is determined, the next question is how to use these pieces of information to decide whether now is the time to replace the defender. If now is not the right time, when is the optimal time to replace the defender? Before presenting an analytical approach to answer this question, we consider several important assumptions.

15.3.1 Required Assumptions and Decision Frameworks

In deciding whether now is the time to replace the defender, we need to consider the following three factors:

- Planning horizon (study period)
- Technology
- Relevant cash flow information

Planning Horizon (Study Period)

By planning horizon, we simply mean the service period required by the defender and a sequence of future challengers. The infinite planning horizon is used when we are simply unable to predict when the activity under consideration will be terminated. In other situations, it may be clear that the project will have a definite and predictable duration. In these cases, replacement policy should be formulated more realistically based on a finite planning horizon.

Technology

Predictions of technological patterns over the planning horizon refer to the development of types of challengers that may replace those under study. A number of possibilities exist in predicting purchase cost, salvage value, and operating cost dictated by the efficiency of the machine over the life of an asset. If we assume that all future machines will be the same as those now in service, we are implicitly saying no technological progress in the area will occur. In other cases, we may explicitly recognize the possibility of machines becoming available in the future which will be significantly more efficient, reliable, or productive than those currently on the market. (Personal computers are a good example.) This situation leads to recognition of technological change or obsolescence. Clearly, if the best available machine gets better and better over time, we should certainly investigate the possibility of delaying replacement for a couple of years, which contrasts with the situation where technological change is unlikely.

Revenue and Cost Patterns over Asset Life

Many varieties of predictions can be used to estimate the patterns of revenue, cost, and salvage value over the life of an asset. Sometimes revenue is constant, but costs increase, while salvage value decreases, over the life of a machine. In other situations, a decline in revenue over equipment life can be expected. The specific situation will determine whether replacement analysis is directed toward cost minimization (with constant revenue) or profit maximization (with varying revenue). We formulate a replacement policy for an asset in which salvage values do not increase with age.

Decision Frameworks

To illustrate how a decision framework is developed, we indicate a replacement sequence of assets by the notation $(j_0, n_0), (j_1, n_1), (j_2, n_2), \ldots, (j_K, n_K)$. Each pair of numbers (j, n) indicates an asset type and the lifetime for which that asset will be retained. The defender, asset 0, is listed first; if the defender is replaced now, $n_0 = 0$. A sequence of pairs may cover a finite period or an infinite period. For example, the sequence $(j_0, 2), (j_1, 5),$ $(j_2, 3)$ indicates retaining the defender for 2 years, replacing the defender with an asset of type j_1, using it for 5 years, replacing j_1 with an asset of type j_2, and using it for 3 years. In this situation, the total planning horizon covers 10 years $(2 + 5 + 3)$. The special case of keeping the defender for n_0 periods, followed by infinitely repeated purchases and the use of an asset of type j for n^* years, is represented by $(j_0, n_0), (j, n_*)_\infty$. This sequence covers an infinite period, and the relationship is illustrated in Figure 15.7.

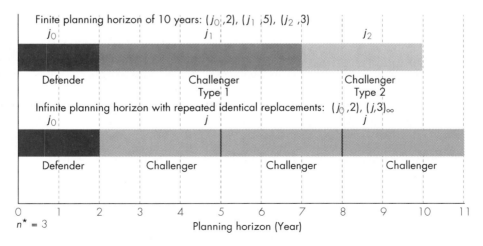

Figure 15.7   Types of typical replacement decision frameworks

Decision Criterion

Although the economic life of the defender is defined as the additional number of years of service that minimizes the annual equivalent cost (or maximizes the annual equivalent revenue), this is *not* necessarily the *optimal time* to replace the defender. The correct replacement time depends on data for the challenger, as well as on data for the defender.

As a decision criterion, the AE method provides a more direct solution when the planning horizon is infinite. When the planning horizon is finite, the PW method is more convenient to use. We will develop the replacement decision procedure for both situations: (1) the infinite planning horizon and (2) the finite planning horizon. We begin by analyzing an infinite planning horizon without technological change. Even though a simplified situation such as this is not likely to occur in real life, the analysis of this replacement situation introduces methods useful for analyzing infinite horizon replacement problems with technological change.

15.3.2 Replacement Strategies under the Infinite Planning Horizon

We consider the situation where a firm has a machine in use in a process. The process is expected to continue for an indefinite period. Presently, a new machine will be on the market that is, in some ways, more effective for the application than the defender. The problem is when, if at all, the defender should be replaced with the challenger.

Under the infinite planning horizon, the service is required for a very long time. Either we continue to use the defender to provide the service, or we replace the defender with the best available challenger for the same service requirement. In this case, the following procedure may be followed in replacement analysis:

1. Compute the economic lives of both defender and challenger. Let's use $N_D{}^*$ and $N_C{}^*$ to indicate the economic lives of the defender and the challenger, respectively. The annual equivalent costs for the defender and the challenger at their respective economic lives are indicated by $AE_D{}^*$ and $AE_C{}^*$.

2. Compare $AE_D{}^*$ and $AE_C{}^*$. If $AE_D{}^*$ is bigger than $AE_C{}^*$, we know that it is more costly to keep the defender than to replace it with the challenger. Thus, the challenger should replace the defender now. If $AE_D{}^*$ is smaller than $AE_C{}^*$, it costs less to keep the defender than to replace it with the challenger. Thus, the defender should not be replaced now. The defender should continue to be used at least for the duration of its economic life if there are no technological changes over the economic life of the defender.

3. If the defender should not be replaced now, when should it be replaced? First we need to continue to use it until its economic life is over. Then, we should calculate the cost of running the defender for one more year after its economic life. If this cost is greater than $AE_C{}^*$, the defender should be replaced at the end of its economic life. Otherwise, we should calculate the cost of running the defender for the second year after its economic life. If this cost is bigger than $AE_C{}^*$, the defender should be replaced one year after its economic

life. This process should be continued until you find the optimal replacement time. This approach is called **marginal analysis**, that is, calculate the incremental cost of operating the defender for just one more year. In other words, we want to see whether the cost of extending the use of the defender for an additional year exceeds the savings resulting from delaying the purchase of the challenger. Here we have assumed the best available challenger does not change.

It should be noted that this procedure might be applied dynamically. It may be performed annually for replacement analysis. Whenever there are updated data on the costs of the defender or new challengers available on the market, these new data should be used in the procedure. Example 15.5 illustrates the above procedure.

Example 15.5 Replacement Analysis Under the Infinite Planning Horizon

Advanced Electrical Insulator Company is considering replacing a broken inspection machine, which has been used to test the mechanical strength of electrical insulators, with a newer and more efficient one. If repaired, the old machine can be used for another 5 years although the firm does not expect to realize any salvage value from scrapping it in 5 years. However, the firm can sell it now to another firm in the industry for $5,000. If the machine is kept, it will require an immediate $1,200 overhaul to restore it to operable condition. The overhaul will neither extend the service life originally estimated nor increase the value of the inspection machine. The operating costs are estimated at $2,000 during the first year, and these are expected to increase by $1,500 per year thereafter. Future market values are expected to decline by $1,000 per year.

The new machine costs $10,000 and will have operating costs of $2,000 in the first year, increasing by $800 per year thereafter. The expected salvage value is $6,000 after 1 year and will decline 15% each year. The company requires a rate of return of 15%. Find the economic life for each option, *and* determine when the defender should be replaced.

Solution

1. Economic Service Life
   - Defender
     If the company retains the inspection machine, it is in effect deciding to overhaul the machine and invest the machine's current market value in that alternative. The opportunity cost of the machine is $5,000. Because an overhaul costing $1,200 is also needed to make the machine operational, the total initial investment of the machine is considered to be $5,000 + $1,200 = $6,200. Other data for the defender are summarized as follows:

n	Overhaul	Forecasted Operating Cost	Market Value If Disposed Of
0	$1,200		$5,000
1	0	$2,000	$4,000
2	0	$3,500	$3,000
3	0	$5,000	$2,000
4	0	$6,500	$1,000
5	0	$8,000	0

We can calculate the annual equivalent costs if the defender is to be kept for 1 year, 2 years, 3 years, and so forth. For example, the cash flow diagram for $N = 4$ years is shown in Figure 15.8.

$$N = 4 \text{ years: } AE(15\%) = \$6,200(A/P, 15\%, 4) + \$2,000$$
$$+ \$1,500(A/G, 15\%, 4) - \$1,000(A/F, 15\%, 4)$$
$$= \$5,961$$

The other AE cost figures can be calculated with the following equation:

$$AE(15\%)_N = \$6,200(A/P, 15\%, N) + \$2,000 + \$1,500(A/G, 15\%, N)$$
$$- \$1,000(5 - N)(A/F, 15\%, N) \text{ for } N = 1, 2, 3, 4, 5$$

$$N = 1: AE(15\%) = \$5,130$$
$$N = 2: AE(15\%) = \$5,116$$
$$N = 3: AE(15\%) = \$5,500$$
$$N = 4: AE(15\%) = \$5,961$$
$$N = 5: AE(15\%) = \$6,434$$

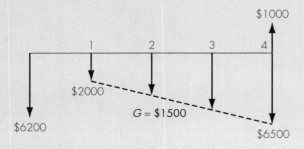

Figure 15.8   Cash flow diagram for defender when $N = 4$ years (Example 15.5)

When $N = 2$ years, we get the lowest AE value. Thus, the defender's economic life is 2 years. Using the notation we have defined in the procedure, we have

$$N_D* = 2 \text{ years}$$

$$AE_D* = \$5,116$$

The AE values as a function of N are plotted in Figure 15.9. Actually, after computing AE for $N = 1, 2,$ and 3, we can stop right there. There is no need to compute AE for $N = 4$ and $N = 5$ because AE is increasing when $N > 2$, and we have assumed that AE has a unique minimum point.

• Challenger
  The economic life of the challenge can be determined using the same procedure shown in this example for the defender and in Example 15.4. A summary of the general equation for AE calculation for the challenger follows. You don't have to summarize such an equation when you need to determine the economic life of an asset, as long as you follow the procedure illustrated in Example 15.4.

$$AE(15\%)_N = \$10,000(A/P, 15\%, N) + \$2000 + \$800(A/G, 15\%, N)$$
$$- \$6000(1 - 15\%)^{N-1}(A/F, 15\%, N)$$

The obtained results are:

$$N = 1 \text{ year: } AE(15\%) = \$7,500$$
$$N = 2 \text{ years: } AE(15\%) = \$6,151$$
$$N = 3 \text{ years: } AE(15\%) = \$5,857$$

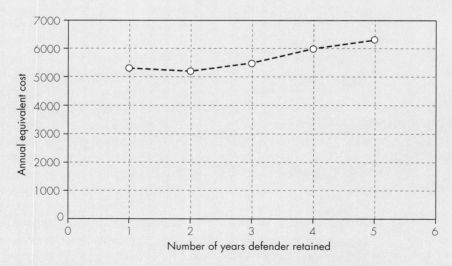

Figure 15.9   AE as a function of the life of the defender (Example 15.5)

$$N = 4 \text{ years: } AE(15\%) = \$5,826$$

$$N = 5 \text{ years: } AE(15\%) = \$5,897$$

The economic life of the challenger is 4 years, that is

$$N_C{}^* = 4 \text{ years}$$

$$AE_C{}^* = \$5,826$$

2. Should the defender be replaced now?

   Since $AE_D{}^* = \$5,116 < AE_C{}^* = \$5,826$, the answer is not to replace the defender now. If there are no technological advances in the next few years, the defender should be used for at least $N_D{}^* = 2$ more years. However, it is not necessarily best to replace the defender right at its economic life.

3. When should the defender be replaced?

   If we need to find the answer to this question today, we have to calculate the cost of keeping and using the defender for the third year from today. That is, what is the cost of not selling the defender at the end of year 2, using it for the third year, and replacing it at the end of year 3? The following cash flows are related to this question:

   (a) Opportunity cost at the end of year 2: Equal to the market value then: $3,000

   (b) Operating cost for the 3rd year: $5,000

   (c) Salvage value of the defender at the end of year 3: $2,000

   The following cash flow diagram represents these cash flows.

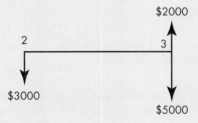

The cost of using the defender for one more year from the end of its economic life is

$$\$3,000 \times 1.15 + \$5,000 - \$2,000 = \$6,450.$$

Now compare this cost with the $AE_C{}^* = \$5,826$ of the challenger. It is greater than $AE_C{}^*$. It is more expensive to keep the defender for the third year than to replace it with the challenger. Thus, the conclusion is to replace the defender at the end of year 2. If this one-year cost is still smaller than $AE_C{}^*$, we need to calculate the cost of using the defender for the fourth year and then compare it with $AE_C{}^*$ of the challenger.

In replacement analysis, it is common for a defender and its challenger to have different economic service lives. The annual equivalent approach is frequently used in replacement analysis, but it is important to know that we use the AE method in replacement analysis, not because we have to deal with the unequal service life problem, but rather because the AE approach provides some computational advantage for a special class of replacement problem.

In Chapter 7, we discussed the general principle for comparing alternatives with unequal service lives. In particular, we pointed out that use of the AE method relies on the concept of repeatability of projects and one of two assumptions: an infinite planning horizon or a common service period. In defender-challenger situations, however, repeatability of the defender cannot be assumed. In fact, by virtue of our problem definition, we are not repeating the defender, but replacing it with its challenger, an asset that in some way constitutes an improvement over the current equipment. Thus, the assumptions we made for using an annual cash flow analysis with unequal service life alternatives are not valid in the usual defender-challenger situation.

The complication—the unequal life problem—can be resolved, however, if we recall that the replacement problem at hand is not whether to replace the defender but when to do so. When the defender is replaced, it will always be by the challenger—the best available equipment. An identical challenger can then replace the challenger repeatedly. In fact, we are comparing the following two options in replacement analysis:

1. Replace the defender now: The cash flows of the challenger will be used from today and will be repeated because an identical challenger will be used if replacement becomes necessary again in the future. This stream of cash flows is equivalent to a cash flow of $AE_C{}^*$ each year for an infinite number of years.
2. Replace the defender, say, x years later: The cash flows of the defender will be used in the first x years. Starting in year $x + 1$, the cash flows of the challenger will be used indefinitely.

The annual equivalent cash flows for the years beyond year x are the same for these two options. We need only to compare the annual equivalent cash flows for the first x years to determine which option is better. This is why we can compare $AE_D{}^*$ with $AE_C{}^*$ to determine whether now is the time to replace the defender.

15.3.3 Replacement Strategies under the Finite Planning Horizon

If the planning period is finite (for example, 8 years), a comparison based on the AE method over a defender's economic service life does not generally apply. The procedure for solving such a problem with a finite planning horizon is to establish all "reasonable" replacement patterns and then use the PW value for the planning period to select the most economical pattern. To illustrate the procedure, let us consider Example 15.6.

Example 15.6 Replacement Analysis under the Finite Planning Horizon (PW Approach)

Reconsider the defender and the challenger in Example 15.5. Suppose that the firm has a contract to perform a given service, using the current defender or the challenger for the next 8 years. After the contract work, neither the defender nor the challenger will be retained. What is the best replacement strategy?

Solution

Recall again the annual equivalent costs for the defender and challenger under the assumed holding periods (a boxed number denotes the minimum AE value at $N_D^* = 2$ and $N_C^* = 4$, respectively).

| | Annual Equivalent Cost ($) | |
n	Defender	Challenger
1	5,130	7,500
2	5,116	6,151
3	5,500	5,857
4	5,961	5,826
5	6,434	5,897

Many ownership options would fulfill an 8-year planning horizon, as shown in Figure 15.10. Of these options, six appear to be the most likely by inspection. These options are listed and the present equivalent cost for each option is calculated as follows:

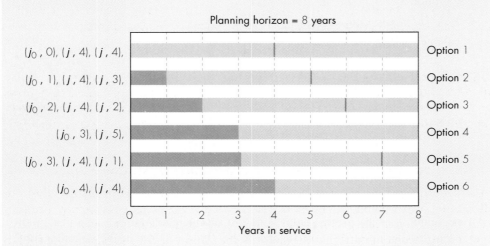

Figure 15.10   Some likely replacement patterns under a finite planning horizon of 8 years (Example 15.6)

- Option 1: $(j_0, 0), (j, 4), (j, 4)$

$$PW(15\%)_1 = \$5,826(P/A, 15\%, 8)$$
$$= \$26,143.$$

- Option 2: $(j_0, 1), (j, 4), (j, 3)$

$$PW(15\%)_2 = \$5,130(P/F, 15\%, 1)$$
$$+ \$5,826(P/A, 15\%, 4)(P/F, 15\%, 1)$$
$$+ \$5,857(P/A, 15\%, 3)(P/F, 15\%, 5)$$
$$= \$25,573.$$

- Option 3: $(j_0, 2), (j, 4), (j, 2)$

$$PW(15\%)_3 = \$5,116(P/A, 15\%, 2)$$
$$+ \$5,826(P/A, 15\%, 4)(P/F, 15\%, 2)$$
$$+ \$6,151(P/A, 15\%, 2)(P/F, 15\%, 6)$$
$$= \$25,217 \leftarrow \text{minimum cost.}$$

- Option 4: $(j_0, 3), (j, 5)$

$$PW(15\%)_4 = \$5,500(P/A, 15\%, 3)$$
$$+ \$5,897(P/A, 15\%, 5)(P/F, 15\%, 3)$$
$$= \$25,555.$$

- Option 5: $(j_0, 3), (j, 4), (j, 1)$

$$PW(15\%)_5 = \$5,500(P/A, 15\%, 3)$$
$$+ \$5,826(P/A, 15\%, 4)(P/F, 15\%, 3)$$
$$+ \$7,500(P/F, 15\%, 8)$$
$$= \$25,946.$$

- Option 6: $(j_0, 4), (j, 4)$

$$PW(15\%)_6 = \$5,961(P/A, 15\%, 4)$$
$$+ \$5,826(P/A, 15\%, 4)(P/F, 15\%, 4)$$
$$= \$26,529.$$

An examination of the present equivalent cost of a planning horizon of 8 years indicates that the least-cost solution appears to be Option 3: Retain the defender for 2 years, purchase the challenger and keep it for 4 years, and purchase another challenger and keep it for 2 years.

Comments: In this example, we examined only six possible decision options likely to lead to the best solution, but it is important to note that several other possibilities have not been looked at. To explain, consider Figure 15.11, which shows a graphical representation of various replacement strategies under a finite planning horizon.

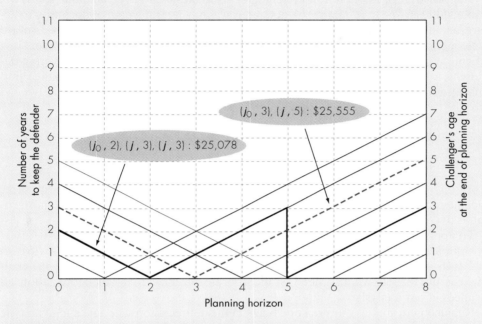

Figure 15.11   Graphical representations of replacement strategies under a finite planning horizon (Example 15.6)

For example, the replacement strategy (shown as a solid line in Figure 15.11), $[(j_0, 2), (j, 3), (j, 3)]$ is certainly feasible, but we did not include it in the previous computation. Naturally, as we extend the planning horizon, the number of possible decision options can easily multiply. To make sure that we indeed find the optimal solution for such a problem, an optimization technique such as dynamic programming can be used.[2]

15.3.4   Consideration of Technological Change

Thus far, we have defined the challenger simply as the best available replacement for the defender. It is more realistic to recognize that often the replacement decision involves an asset now in use versus a candidate for replacement, that is, in some way, an

[2] Hillier, F. S. And Lieberman, G. S. *Introduction To Operations Research,* 6th ed., New York: Mcgraw Hill, 1995.

improvement on the current asset. This, of course, reflects technological progress that is ongoing continually. Future models of a machine are likely to be more effective than a current model. In most areas, technological change appears as a combination of gradual advances in effectiveness; the occasional technological breakthrough, however, can revolutionize the character of a machine.

The prospect of improved future challengers makes a current challenger a less desirable alternative. By retaining the defender, we may later have an opportunity to acquire an improved challenger. If this is the case, the prospect of improved future challengers may affect a current decision between a defender and its challenger. It is difficult to forecast future technological trends in any precise fashion. However, in developing a long-term replacement policy, we need to take technological change into consideration.

15.4 Replacement Analysis with Tax Considerations

Up to this point, we covered various concepts and techniques that are useful in replacement analysis. In this section, we illustrate how to use those concepts and techniques to conduct replacement analysis on an after-tax basis.

To apply the concepts and methods covered in Section 15.1 through 15.3 in after-tax comparison of defender and challenger, we have to incorporate the gains (losses) tax effects whenever an asset is disposed of. Whether the defender is kept or the challenger is purchased, we also need to incorporate the tax effects of depreciation allowances in our analysis.

Replacement studies require knowledge of the depreciation schedule and of taxable gains or losses at disposal. Note that the depreciation schedule is determined at the time of asset acquisition, whereas the relevant tax law determines the gains tax effects at the time of disposal. In this section, we will use the same examples (Example 15.1 through 15.5) to illustrate how to do the following analyses on an after-tax basis:

1. Calculate the net proceeds of defender from disposal (Example 15.7).
2. Use the cash flow approach and the opportunity cost approach in comparison of defender and challenger (Examples 15.8 and 15.9).
3. Calculate the economic life of defender or challenger (Example 15.10).
4. Conduct replacement analysis under the infinite planning horizon (Example 15.11).

Example 15.7 Net Proceeds from Disposal of an Old Machine

Refer to Example 15.1. Suppose that the $20,000 capital expenditure was set up to be depreciated on a 7-year MACRS (allowed annual depreciation: $2,858, $4,898, $3,498, $2,498, $1,786, $1,784, $1,786, and $892). If the firm's marginal income tax rate is 40%, determine the taxable gains (or losses) and the net proceeds from disposal of the old printing machine.

Solution

First we need to find the current book value of the old printing machine. The original cost minus the accumulated depreciation calculated with the half-year convention (if sold now) is

$$\$20,000 - (\$2,858 + \$4,898/2) = \$14,693$$

So, we compute the following:

Allowed book value	=	$14,693
Current market value	=	$10,000
Losses	=	$4,693
Tax savings	=	$4,693(0.40) = $1,877
Net proceeds from the sale	=	$10,000 + $1,877
	=	$11,877

This calculation is illustrated in Figure 15.12.

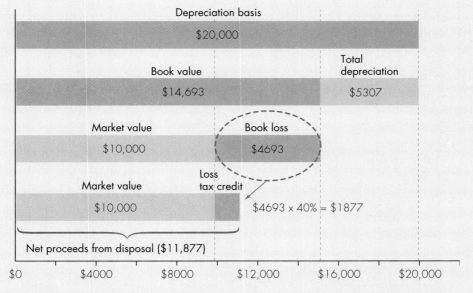

Figure 15.12   Net proceeds from the sale of the old printing machine — defender (Example 15.7)

Example 15.8 Replacement Analysis Using the Cash Flow Approach

Consider Example 15.2. Assume that the new machine would fall into the same 7-year MACRS class. The firm's after-tax interest rate (MARR) is 12%, and the marginal income tax rate is 40%, respectively. Decide whether replacement is justified now using the cash flow approach.

Solution

Tables 15.1 and 15.2 show the worksheet formats the company uses to analyze a typical replacement project on a cash flow basis. Each line is numbered, and a line-by-line description of the table follows.

Table 15.1
Replacement Worksheet: Option 1—Keep the Defender (Example 15.8)

n	-2	-1	0	1	2	3
Financial data (cost information)						
(1) Depreciation		$2,858	$4,898	$3,498	$2,498	$893
(2) Book value	$20,000	17,142	12,244	8,746	6,248	5,355
(3) Salvage value						2,500
(4) Loss from sale						(2,855)
(5) Repair cost		5,000				
(6) O&M costs				8,000	8,000	8,000
Income statement						
(7) Revenue						
(8) Expense						
(9) Depreciation				$3,498	$2,498	$893
(10) O&M costs				8,000	8,000	8,000
(11) Taxable income				$(11,498)	$(10,498)	$(8,893)
(12) Income taxes				(4,599)	(4,199)	(3,557)
(13) Net income				$(6,899)	$(6,299)	$(5,336)
Cash flow statement						
(14) Operating activities						
Net income				$(6,899)	$(6,299)	$(5,336)
Depreciation				3,498	2,498	893
(15) Investment activities						
Investment						
(16) Salvage value						2,500
(17) Gains tax						1,142
(18) Net cash flow				$(3,401)	$(3,801)	$(801)

n	0	1	2	3
Financial data (cost information)				
(1) Net proceeds from sale of old printer	$ 11,877			
(2) Cost of new printer	15,000			
(3) Depreciation		$ 2,144	$ 3,674	$ 1,312
(4) Book value		12,856	9,182	7,870
(5) Salvage value				5,500
(6) Gains from sale of new printer				(2,370)
(7) O&M costs		6,000	6,000	6,000
Income statement				
(8) Revenue				
(9) Expenses				
(10) Depreciation		$ 2,144	$ 3,674	$ 1,312
(11) O&M costs		6,000	6,000	6,000
(12) Taxable income		$ (8,144)	$ (9,674)	$ (7,312)
(13) Income taxes		(3,258)	(3,870)	(2,925)
(14) Net income		$ (4,886)	$ (5,804)	$ (4,387)
Cash flow statement				
(15) Operating activities				
Net income		$ (4,886)	$ (5,804)	$ (4,387)
Depreciation		2,144	3,674	1,312
(16) Investment activities				
Investment	$ (15,000)			
(17) Net proceeds from sale of old printer	11,877			
(18) Salvage value				5,500
(19) Gains tax				948
(20) Net cash flow	$ (3,123)	$ (2,742)	$ (2,130)	$ 3,373

Note: The computational procedure to determine the net proceeds from sale of the old printer is shown in Example 15.7.

Table 15.2
Replacement Worksheet (Challenger): Option 2—Replace the Defender (Example 15.8)

- Option 1: Keep the defender

    Lines 1–4: If the old machine is kept, the depreciation schedule would be ($3,498, $2,498, and $1,786). Following the half-year convention, it is assumed that the asset will be retired at the end of 3 years; thus the depreciation for year 3 is (0.5)($1,786) = $893. This results in total depreciation in the amount of $14,645 and a remaining book value of $20,000 − $14,645 = $5,355.

    Lines 5–6: Repair costs in the amount of $5,000 were already incurred before the replacement decision. This is a sunk cost and should not be considered in the analysis. If a repair in the amount of $5,000 is required to keep the defender in serviceable condition, this will show as an expense in year 0. If the old machine is retained for the next 3 years, the before-tax annual O&M costs are as shown in Line 6.

Line 7: This is a service project, and revenue will remain unchanged regardless of the replacement decision.

Lines 8–10: Two items are listed under the expense category: Depreciation and O&M costs. Because the defender will be disposed of before the end of the recovery period, the depreciation allowance in year 3 is also halved.

Lines 11–13: Without revenues, the net income figures result in negative numbers.

Line 15: With the cash flow approach, no new investment is required to retain the old printer.

Line 17: Since the old equipment would be sold at less than book value, the sale would create a loss, which would reduce the firm's taxable income and, hence, its tax payment. The tax savings should be equal to (book value – market value) (tax rate) = ($5,355 – $2,500)(0.40) = $1,142. This loss is an operating loss, because it reflects the fact that inadequate depreciation was taken on the old machine. Therefore, the net proceeds from sale of old printer at the end of 3 years is $2,500 + $1,142 = $3,642.

Line 18: Since a new investment is not required to keep the old machine, the net operating cash flows and the net proceeds from the sale of the old machine consist of the operating cash flows and the net proceeds from the sale of the old machine at the end of year 3.

- Option 2: Replace the defender

Line 1: The price received from the sale of the old equipment at $n = 0$ is shown. As shown in Example 15.7, the net proceeds from the sale is the sum of the market value and the tax savings.

Line 2: The purchase price of the new machine, including installation and freight charges, is listed.

Lines 3–4: The depreciation schedule along with the book values for the new machine (7-year MACRS) is shown. The depreciation amount of $1,312 in year 3 reflects the half-year convention.

Lines 5–6: With the salvage value estimated at $5,500, we expect a loss ($2,370 = $7,878 – $5,500) on the sale of the new machine at the end of year 3.

Line 7: The O&M costs for the new machine are listed.

Lines 8–14: The net income figures are tabulated here to determine the net cash flows.

Line 15: The net operating cash flows over the project's 3-year life are shown. These flows are determined by adding the non-cash expense (depreciation) to the net income.

Lines 16–17: The total net cash outflow at the time the replacement is made is shown. The company writes a check for $15,000 to pay for the machine. However, this outlay is partially offset by the proceeds from the sale of the old equipment and tax savings in the amount of $11,877.

Lines 18–19: These lines show the cash flows associated with the termination of the new machine. To begin, Line 18 shows the estimated salvage value of the new machine at the end of its 3-year life, $5,500. Since the book value of the new machine at the end of year 3 is $7,870, the company will have a tax credit (or tax savings) of ($2,370)(0.40) = $948 on this book loss.

Line 20: This line represents the net cash flows associated with replacement of the old machine.

The actual cash flow diagrams are shown in Figure 15.13. Since both the defender and challenger have the same service life, we can use either PW or AE analysis.

$$PW(12\%)_{Old} = 0 - \$3,401(P/F, 12\%, 1) - \$3,801(P/F, 12\%, 2)$$
$$- \$801(P/F, 12\%, 3)$$
$$= -\$6,637.$$

$$AE(12\%)_{Old} = -\$6,637(A/P, 12\%, 3)$$
$$= -\$2,763.$$

$$PW(12\%)_{New} = -\$3,123 - \$2,742(P/F, 12\%, 1) - \$2,130(P/F, 12\%, 2)$$
$$+ \$3,373(P/F, 12\%, 3)$$
$$= -\$4,868.$$

$$AE(12\%)_{New} = -\$4,868(A/P, 12\%, 3)$$
$$= -\$2,027.$$

Because of the annual difference of $736 in favor of the challenger, the replacement should be made now.

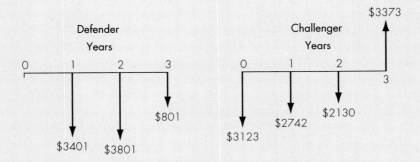

Figure 15.13   Comparison of defender and challenger based on the cash flow approach (Example 15.8)

Example 15.9 Replacement Analysis Using the Opportunity Cost Approach

Rework Example 15.8 using the opportunity cost approach.

Solution

Recall that the cash flow approach in Example 15.8 credited net proceeds in the amount of $11,877 from the sale of the defender toward the $15,000 purchase price of the challenger, and no initial outlay would have been required had the decision been to keep the defender. If the decision to keep the defender had been made, the opportunity cost approach treats the $11,877 current salvage value of the defender as an incurred cost. Figure 15.14 illustrates the cash flows related to these decision options.

Since the lifetimes are the same, we can use either PW or AE analysis as follows:

$$PW(12\%)_{\text{Old}} = -\$11,877 - \$3,401(P/F, 12\%, 1) - \$3,801(P/F, 12\%, 2)$$
$$-\$801(P/F, 12\%, 3)$$
$$= -\$18,514.$$

$$AE(12\%)_{\text{Old}} = -\$18,514(A/P, 12\%, 3)$$
$$= -\$7,708.$$

$$PW(12\%)_{\text{New}} = -\$15,000 - \$2,742(P/F, 12\%, 1)$$
$$-\$2,130(P/F, 12\%, 2) + \$3,373(P/F, 12\%, 3)$$
$$= -\$16,745.$$

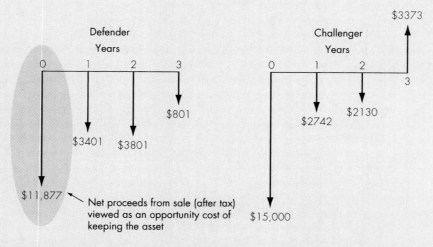

Figure 15.14   Comparison of defender and challenger based on the opportunity cost approach (Example 15.9)

$$AE(12\%)_{\text{New}} = -\$16,745(A/P, 12\%, 3)$$
$$= -\$6,972.$$

The result is the same as in Example 15.8. Since both the challenger and defender cash flows were adjusted by the same amount—$11,877—at time 0, this should not come as a surprise.

Comments: Recall that we assumed the same service life for both the defender and the challenger. In general, however, old equipment has a relatively short remaining life compared to new equipment, so this assumption is overly simplistic. When the defender and challenger have unequal lifetimes, we must make an assumption in order to obtain a common analysis period. A typical assumption is that, after the initial decision, we make **perpetual replacements** with assets similar to the challenger. Certainly, we can still use the PW analysis with actual cash flows, but this requires evaluation of *infinite* cash flow streams.

Example 15.10 Economic Service Life for a Lift Truck

Consider Example 15.4 with the following additional data. The asset belongs to a 5-year MACRS property class with the following annual depreciation allowances: 20%, 32%, 19.20%, 11.52%, 11.52%, and 5.76%. The firm's marginal tax rate is 40%, and its after-tax MARR is 15%. Find the economic service life of this new machine.

Discussion: To determine an asset's economic service life, we will first list the gains or losses that will be realized if the truck were to be disposed of at the end of each operating year. In doing so, we need to compute the book values at the end of each operating year assuming that the asset would be disposed of at that time. Recall that, with the half-year convention, the book value (for the MACRS property) at the end of the year is based on its disposal during the year. As summarized in Table 15.3, these values provide a basis for identifying the relevant after-tax cash flows at the end of an assumed operating period.

Solution

Two approaches may be used to find the economic life of an asset: (a) The generalized cash flow approach and (b) the tabular approach. We will now demonstrate both approaches.

Table 15.3
Forecasted Operating Costs and Net Proceeds from Sale as a Function of Holding Period (Example 15.10)

Holding Period	O&M	Permitted Annual Depreciation Amounts over the Holding Period							Total Depreciation	Book Value	Expected Market Value	Taxable Gains	Gains Tax	Net A/T Salvage Value
		1	2	3	4	5	6	7						
1	$1,000	$3,600							$3,600	$14,400	$10,000	$(4,400)	$(1,760)	$11,760
2	1,150	3,600	$2,880						6,480	11,520	7,500	(4,020)	(1,608)	9,108
3	1,323	3,600	5,760	$1,728					11,088	6,912	5,625	(1,287)	(515)	6,140
4	1,521	3,600	5,760	3,456	$1,037				13,853	4,147	4,219	72	29	4,190
5	4,749	3,600	5,760	3,456	2,074	$1,037			15,927	2,073	3,164	1,091	436	2,728
6	2,011	3,600	5,760	3,456	2,074	2,074	$1,036	$0	18,000	0	2,373	2,373	949	1,424
7	6,813	3,600	5,760	3,456	2,074	2,074	1,036	$0	18,000	0	1,780	1,780	712	1,068

Note: Asset price of $18,000, depreciated under MACRS for 5-year property with the half-year convention; in year 5, normal operating expense ($1,749) + overhaul ($3,000); in year 7, normal operating expense ($2,313) + another engine overhaul ($4500)

(a) Generalized Cash Flow Approach: First we could determine the relevant after-tax cash flows if the lift truck were to be retained for just 1 year by means of the income statement/cash flow statement approach as shown in Table 15.4(a). Since we have only a few cash flow elements (O&M, depreciation, and salvage value), a more efficient way to obtain the after-tax cash flow is to use the generalized cash flow approach discussed in Section 12.4. Recall that the depreciation allowances result in a tax reduction equal to the depreciation amount multiplied by the tax rate. The operating expenses are multiplied by the factor of (1 – the tax rate) to obtain the after-tax O&M. For a situation in which the asset is retained for 1 year, Table 15.4(b) summarizes the cash flows obtained by using the generalized cash flow approach.

If we use the expected operating costs and the salvage values in Table 15.3, we can continue to generate yearly after-tax entries for the asset's remaining physical life. For the first 2 operating years, we compute the equivalent annual costs of owning and operating as follows.

Year	0	1
Income Statement Approach		
Revenue		
Operating expense		
O&M		$ 1,000
Depreciation		3,600
Taxable income		$ (4,600)
Income taxes (40%)		(1,840)
Net income		$ (2,760)
Cash flow statement		
Operating activities		
Net income		$ (2,760)
Depreciation		3,600
Investment activities		
Investment	$ (18,000)	
Salvage value		10,000
Gains tax		1,760
Net cash flow	$ (18,000)	$ 12,600
Generalized Cash Flow Approach		
Investment	$ (18,000)	
+ (0.4) (depreciation)		$ 1,440
– (0.6) (O&M)		(600)
Net proceeds from sale		11,760
Net cash flow	$ (18,000)	$ 12,600

Table 15.4
After-Tax Cash Flow Calculation for Owning and Operating the Asset for 1 Year (Example 15.10)

- $n = 1$: One-year replacement cycle:

$$AE(15\%) = [-\$18,000 + [-(0.6)(\$1,000)$$
$$+ (0.40)(\$3,600)$$
$$+ \$11,760](P/F, 15\%, 1)\}(A/P, 15\%, 1)$$
$$= (-\$7,043)(1.15)$$
$$= -\$8,100.$$

- $n = 2$: Two-year replacement cycle:

$$AE(15\%) = \{-\$18,000 + [-0.6(\$1,000)$$
$$+ 0.4(\$3,600)](P/F, 15\%, 1) + [-0.6(\$1,150)$$
$$+ 0.4(\$2,880) + \$9,108](P/F, 15\%, 2)\}(A/P, 15\%, 2)$$

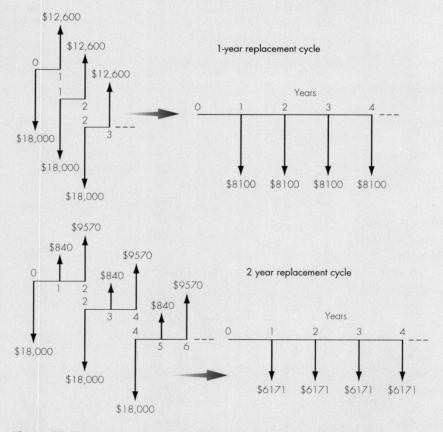

Figure 15.15 Conversion of an infinite number of replacement cycles to infinite AE cost streams (Example 15.10)

$$= (\$10,033)(0.6151)$$

$$= -\$6,171.$$

Similarly, the annual equivalent costs for the subsequent years can be computed as shown in Table 15.5 (column 12). If the truck were to be sold after 6 years, it would have a minimum annual cost of $4,344 per year, and this is the life most favorable for comparison purposes. By replacing the asset perpetually according to an economic life of 6 years, we obtain the minimum infinite AE cost stream. Figure 15.15 illustrates this concept. Of course, we should envision a long period of required service for the asset—this life no doubt being heavily influenced by market values, O&M costs, and depreciation credits.

(b) **Tabular Approach:** The tabular approach separates the annual cost elements into two parts: One associated with the capital recovery of the asset and the other associated with operating the asset. In computing the capital recovery cost, we need to determine the after-tax salvage values at the end of each holding period as calculated previously in Table 15.3. Then we compute the total annual equivalent costs of the asset for any given year's operation:

$$\text{Total equivalent annual costs} = \text{Capital recovery cost}$$
$$+ \text{ equivalent annual}$$
$$\text{operating costs.}$$

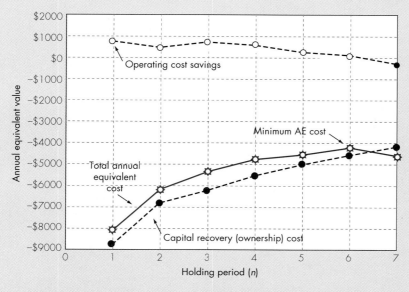

Figure 15.16 Economic service life obtained by finding the minimum AE cost (Example 15.10)

Table 15.5
Tabular Calculation of Economic Service Life (Example 15.10)

[1] Holding Period N	[2] n	[3] Before-Tax Operating Expenses O&M	[4] Depreciation	[5] A/T O&M	[6] A/T Depreciation Credit	[7] Net Operating Cost	[8] Investment and Net Salvage	[9] Net A/T Cash Flow	[10] Capital Cost	[11] Operating Cost	[12] Total Cost
1	0	–	–	–	–	–	$(18,000)	(18,000)	$(8,940)	$ 840	$(8,100)
	1	$1,000	$3,600	$ (600)	$1,440	$ 840	11,760	12,600			
2	0	–	–	–	–	–	(18,000)	(18,000)	(6,835)	664	(6,171)
	1	1,000	3,600	(600)	1,440	840	–	840			
	2	1,150	2,880	(690)	1,152	462	9,108	9,570			
3	0	–	–	–	–	–	(18,000)	(18,000)	(6,115)	825	(5,291)
	1	1,000	3,600	(600)	1,440	840		840			
	2	1,150	5,760	(690)	2,304	1,614		1,614			
	3	1,323	1,728	(794)	691	(103)	6,140	6,037			
4	0	–	–	–	–	–	(18,000)	(18,000)	(5,466)	719	(4,746)
	1	1,000	3,600	(600)	1,440	840		840			
	2	1,150	5,760	(690)	2,304	1,614		1,614			
	3	1,323	3,456	(794)	1,382	589		589			
	4	1,521	1,037	(913)	415	(498)	4,190	3,692			
5	0	–	–	–	–	–	(18,000)	(18,000)	(4,965)	322	(4,643)
	1	1,000	3,600	(600)	1,440	840		840			
	2	1,150	5,760	(690)	2,304	1,614		1,614			
	3	1,323	3,456	(794)	1,382	589		589			
	4	1,521	2,074	(913)	830	(83)		(83)			
	5	4,749	1,037	(2,849)	415	(2,435)	2,728	293			
6	0	–	–	–	–	–	(18,000)	(18,000)	(4,594)	249	(4,344)
	1	1,000	3,600	(600)	1,440	840		840			
	2	1,150	5,760	(690)	2,304	1,614		1,614			
	3	1,323	3,456	(794)	1,382	589		589			
	4	1,521	2,074	(913)	830	(83)		(83)			
	5	4,749	2,074	(2,849)	830	(2,020)		(2,020)			
	6	2,011	1,037	(1,207)	415	(792)	1,424	632			
7	0	–		–	–		(18,000)	(18,000)	(4,230)	(143)	(4,372)
	1	1,000	3,600	(600)	1,440	840		840			
	2	1,150	5,760	(690)	2,304	1,614		1,614			
	3	1,323	3,456	(794)	1,382	589		589			
	4	1,521	2,074	(913)	830	(83)		(83)			
	5	4,749	2,074	(2,849)	830	(2,020)		(2,020)			
	6	2,011	1,037	(1,207)	415	(792)		(792)			
	7	6,813	–	(4,088)	–	(4,088)	1,068	(3,020)			

Economic service life (indicated at N = 6, Total Cost (4,344))

If we examine the equivalent annual costs itemized in Table 15.5 (columns 10 and 11), we see that, as the asset ages, equivalent annual O&M cost savings decrease. At the same time, capital recovery costs decrease with prolonged use of the asset. The combination of decreasing capital recovery costs and increasing annual O&M costs results in the total annual equivalent cost taking on a form similar to that depicted in Figure 15.16. Even though an expensive overhaul is required during the fifth year of service, it is more economical to keep the equipment over a 6-year life.

Example 15.11 Replacement Analysis under the Infinite Planning Horizon

Recall Example 15.5 where Advanced Electrical Insulator Company is considering replacing a broken inspection machine. Let's assume the following additional data:

- The old machine has been fully depreciated, so it has zero book value. The machine could be used for another 5 years, but the firm does not expect to realize any salvage value from scrapping it in 5 years.
- The new machine falls into the 5-year MACRS property class and will be depreciated accordingly.

The marginal income tax rate is 40%, and the after-tax MARR is 15%. Find the useful life for each option, *and* decide whether the defender should be replaced now or later.

Solution

1. Economic Service Life
- Defender

The defender is fully depreciated so that all salvage values can be treated as ordinary gains and taxed at 40%. The after-tax salvage values are thus

n	Current Market Value	After-tax Salvage Value
0	$5,000	$5,000(1 − 0.40) = $3,000
1	4,000	4,000(1 − 0.40) = 2,400
2	3,000	3,000(1 − 0.40) = 1,800
3	2,000	2,000(1 − 0.40) = 1,200
4	1,000	1,000(1 − 0.40) = 600
5	0	0
6	0	0
⋮	⋮	⋮

If the company retains the inspection machine, it is in effect deciding to overhaul the machine and invest the machine's current market value (after taxes) in that alternative. Although no physical investment cash flow transaction will occur, the firm is withholding from the investment the market value of the inspection machine (opportunity cost). Similarly, the after-tax O&M costs are as follows:

n	Overhaul	Forecasted O&M Cost	After-tax O&M Cost
0	$1,200		$1,200(1 − 0.40) = $720
1	0	$2,000	2,000(1 − 0.40) = 1,200
2	0	3,500	3,500(1 − 0.40) = 2,100
3	0	5,000	5,000(1 − 0.40) = 3,000
4	0	6,500	6,500(1 − 0.40) = 3,900
5	0	8,000	8,000(1 − 0.40) = 4,800

Using the current year's market value as the investment required to retain the defender, we obtain the data in Table 15.6, which indicates that the remaining useful life of the defender is 2 years, *in the absence of future challengers*. The overhaul (repair) cost of $1,200 in year 0 can be treated as a deductible operating expense for tax purposes, as long as it does not add value to the property. (Any repair or improvement expenses that increase the value of the property must be capitalized by depreciating them over the estimated service life.)

- Challenger

  Because the challenger will be depreciated over its tax life, we must determine the book value of the asset at the end of each period to compute the after-tax salvage value. This is shown in Table 15.7(a). With the after-tax salvage values computed in Table 15.7(a), we are now ready to find the economic service life of the challenger by generating AE value entries. These calculations are summarized in Table 15.7(b). The economic life of the challenger is 4 years with an AE(15%) value of $4,065.

2. Optimal Time to Replace the Defender

Since the AE value for the defender's remaining useful life (2 years) is $3,070, which is less than $4,065, the decision will be to keep the defender for now. The defender's remaining useful life of 2 years does not necessarily imply that the defender should be kept for 2 years before switching to the challenger. The reason for this is that the defender's remaining useful life of 2 years was calculated without considering what type of challenger would be available in the future. When a challenger's financial data is available, we need to enumerate all replacement timing possibilities. Since the defender can be used for another 5 years, six replacement strategies exist:

- Replace now with the challenger.
- Replace in year 1 with the challenger.

Table 15.6
Economics of Retaining the Defender for N More Years (Example 15.11)

[1] Holding Period N	[2] n	[3] Before-Tax Operating Expenses O&M	[4] Depreciation	[5] After-Tax Cash Flow if the Asset Is Kept for N More Years A/T O&M	[6] Depreciation Credit	[7] Net Operating Cost	[8] Investment and Net Salvage	[9] Net A/T Cash Flow	[10] Equivalent Annual Cost Capital Cost	[11] Operating Cost	[12] Total Cost
1	0	$1,200		$ (720)		$ (720)	$(3,000)	$(3,720)	$(1,050)	$(2,028)	$3,078
	1	2,000		(1,200)		(1,200)	2,400	1,200			
2	0	1,200		(720)		(720)	(3,000)	(3,720)	(1,008)	(2,061)	(3,070)
	1	2,000		(1,200)		(1,200)		(1,200)			
	2	3,500		(2,100)		(2,100)	1,800	(300)			
3	0	1,200		(720)		(720)	(3,000)	(3,720)	(968)	(2,332)	(3,300)
	1	2,000		(1,200)		(1,200)		(1,200)			
	2	3,500		(2,100)		(2,100)		(2,100)			
	3	5,000		(3,000)		(3,000)	1,200	(1,800)			
4	0	1,200		(720)		(720)	(3,000)	(3,720)	(931)	(2,646)	(3,576)
	1	2,000		(1,200)		(1,200)		(1,200)			
	2	3,500		(2,100)		(2,100)		(2,100)			
	3	5,000		(3,000)		(3,000)		(3,000)			
	4	6,500		(3,900)		(3,900)	600	(3,300)			
5	0	1,200		(720)		(720)	(3,000)	(3,720)	(895)	(2,965)	(3,860)
	1	2,000		(1,200)		(1,200)		(1,200)			
	2	3,500		(2,100)		(2,100)		(2,100)			
	3	5,000		(3,000)		(3,000)		(3,000)			
	4	6,500		(3,900)		(3,900)		(3,900)			
	5	8,000		(4,800)		(4,800)	—	(4,800)			

Minimum AE cost → (3,070)

Table 15.7(a) Forecasted Operating Costs and Net Proceeds from Sale as a Function of Holding Period—Challenger (Example 15.11)

Holding Period	O&M	Allowed Depreciation Amounts over the Holding Period							Expected Total Depreciation	Book Value	Market Value	Taxable Gains	Net A/T Gains Tax	Salvage Value
		1	2	3	4	5	6	7						
1	$2,000	$2,000							$2,000	$8,000	$6,000	$(2,000)	$ (800)	$6,800
2	3,000	2,000	$1,600						3,600	6,400	5,100	(1,300)	(520)	5,620
3	4,000	2,000	3,200	$ 960					6,160	3,840	4,335	495	198	4,137
4	5,000	2,000	3,200	1,920	$ 576				7,696	2,304	3,685	1,381	552	3,133
5	6,000	2,000	3,200	1,920	1,152	$ 576			8,848	1,152	3,132	1,980	792	2,340
6	7,000	2,000	3,200	1,920	1,152	1,152	$ 576	$ 0	10,000	0	2,662	2,662	1,065	1,597
7	8,000	2,000	3,200	1,920	1,152	1,152	576	0	10,000	0	2,263	2,263	905	1,358

Note: Asset price of $10,000, depreciated under MACRS for 5-year property with the half-year convention

Table 15.7(b) Economics of Owning and Operating the Challenger for N More Years (Example 15.11)

[1] N Holding Period	[2] n	[3] O&M (Before-Tax Operating Expenses)	[4] Depreciation	[5] A/T O&M	[6] Depreciation Credit	[7] Net Operating Cost	[8] Investment and Net Salvage	[9] Net A/T Cash Flow	[10] Capital Cost	[11] Operating Cost	[12] Total Cost
						(After-Tax Cash Flow if the Asset Is Kept for N More Years)			(Equivalent Annual Cost)		
	0						$(10,000)	$(10,000)			
1	1	$2,000	$2,000	$(1,200)	$800	$(400)	6,800	6,400	$(4,700)	$(400)	$(5,100)
	0						(10,000)	(10,000)			
	1	2,000	2,000	(1,200)	800	(400)		(400)			
2	2	3,000	1,600	(1,800)	640	(1,160)	5,620	4,460	(3,536)	(753)	(4,290)
	0						(10,000)	(10,000)			
	1	2,000	2,000	(1,200)	800	(400)		(400)			
	2	3,000	3,200	(1,800)	1,280	(520)		(520)			
3	3	4,000	960	(2,400)	384	(2,016)	4,137	2,121	(3,188)	(905)	(4,094)
	0						(10,000)	(10,000)			
	1	2,000	2,000	(1,200)	800	(400)		(400)			
	2	3,000	3,200	(1,800)	1,280	(520)		(520)			
	3	4,000	1,920	(2,400)	768	(1,632)		(1,632)			
4	4	5,000	576	(3,000)	230	(2,770)	3,133	363	(2,875)	(1,190)	(4,065)
	0						(10,000)	(10,000)			
	1	2,000	2,000	(1,200)	800	(400)		(400)			
	2	3,000	3,200	(1,800)	1,280	(520)		(520)			
	3	4,000	1,920	(2,400)	768	(1,632)		(1,632)			
	4	5,000	1,152	(3,000)	461	(2,539)		(2,539)			
5	5	6,000	576	(3,600)	230	(3,370)	2,340	1,030	(2,636)	(1,474)	(4,110)

Economic service life → (4,065)

- Replace in year 2 with the challenger.
- Replace in year 3 with the challenger.
- Replace in year 4 with the challenger.
- Replace in year 5 with the challenger.

If the costs and efficiency of the current challenger remain unchanged in the future years, the possible replacement cash patterns associated with each alternative are shown in Figure 15.17. From the figure, we observe that, on an annual basis, the cash flows after the remaining physical life of the defender are the same.

Before we evaluate the economics of various replacement-decision options, recall the AE values for the defender and the challenger under the assumed service lives (a boxed figure denotes the minimum AE value at $n_0 = 2$ and $n^* = 4$, respectively).

n	Annual Equivalent Cost ($) Defender	Challenger
1	−3078	−5100
2	−3070	−4290
3	−3300	−4094
4	−3576	−4065
5	−3860	−4110
6		−4189
7		−4287

Instead of using the marginal analysis in Example 15.5, we will use the PW analysis, which requires evaluation of infinite cash flow streams. (You will have the same result under the marginal analysis.) Immediate replacement of the defender by the challenger, is equivalent to computing the PW for an infinite cash flow stream of −$4065. If we use the capitalized equivalent worth approach in Chapter 7 $(CE(i) = A/i)$, we obtain

- $n = 0$:

$$PW(15\%)_{n_0=0} = (1/0.15)(-\$4,065)$$
$$= -\$27,100.$$

Suppose we retain the old machine n more years and then replace it with the new one. Now we will compute $PW(i)n_0 = n$.

- $n = 1$:

$$PW(15\%)_{n_0=1} = -\$3,078(P/A, 15\%, 1) - \$27,100(P/F, 15\%, 1)$$
$$= -\$26,242.$$

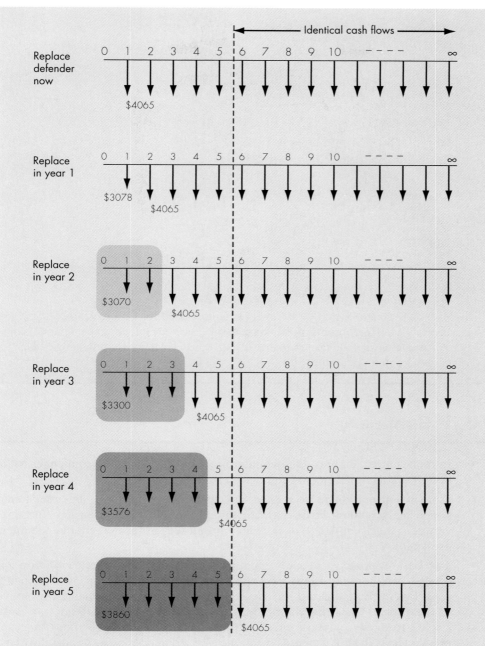

Figure 15.17 Equivalent annual cash flow streams when the defender is kept for *n* years followed by infinitely repeated purchases of the challenger every 4 years (Example 15.11)

- $n = 2$:
$$PW(15\%)_{n_0=2} = -\$3,070(P/A, 15\%, 2) - \$27,100(P/F, 15\%, 2)$$
$$= -\$25,482.$$

- $n = 3$:
$$PW(15\%)_{n_0=3} = -\$3,300(P/A, 15\%, 3) - \$27,100(P/F, 15\%, 3)$$
$$= -\$25,353$$

- $n = 4$:
$$PW(15\%)_{n_0=4} = -\$3,576(P/A, 15\%, 4) - \$27,100(P/F, 15\%, 4)$$
$$= -\$25,704.$$

- $n = 5$:
$$PW(15\%)_{n_0=5} = -\$3,860(P/A, 15\%, 5) - \$27,100(P/F, 15\%, 5)$$
$$= -\$26,413.$$

This leads us to conclude that the defender should be kept for 3 more years. The present worth of $25,353 represents the net cost associated with retaining the defender for 3 years, replacing it with the challenger, and then replacing the challenger every 4 years for an indefinite period.

15.5 Summary

- In replacement analysis, the **defender** is an existing asset; the **challenger** is the best available replacement candidate.

- The **current market value** is the value to use in preparing a defender's economic analysis. **Sunk costs**—past costs that cannot be changed by any future investment decision—should not be considered in a defender's economic analysis.

- Two basic approaches to analyzing replacement problems are the **cash flow approach** and the **opportunity cost approach.** The cash flow approach explicitly considers the actual cash flow consequences for each replacement alternative as they occur. Typically, the net proceeds from sale of the defender are subtracted from the purchase price of the challenger. The opportunity cost approach views the net proceeds from sale of the defender as an opportunity cost of keeping the defender. That is, instead of deducting the salvage value from the purchase cost of the challenger, it is considered as an investment required to keep the asset.

- **Economic service life** is the remaining useful life of a defender, *or* a challenger, that results in the minimum equivalent annual cost or maximum annual equivalent revenue. We should use the respective economic service lives of the defender and the challenger when conducting a replacement analysis.

- Ultimately, in replacement analysis, the question is not *whether* to replace the defender, but *when* to do so. The AE method provides a marginal basis on which to make a year-by-year decision about the best time to replace the defender. As a general decision criterion, the PW method provides a more direct solution to a variety of replacement problems, with either an infinite or a finite planning horizon, or a technological change in a future challenger.

- The role of **technological change** in asset improvement should be weighed in making long-term replacement plans: If a particular item is undergoing rapid, substantial technological improvements, it may be prudent to delay replacement (to the extent where the loss in production does not exceed any savings from improvements in future challengers) until a desired future model is available.

Self-Test Questions

Problem Statement for Questions (15s.1–15s.5)

A machine now in use, which was bought 5 years ago for $4,000, has been fully depreciated. It can be sold for $2,500, but could be used for 3 more years (remaining useful life), at the end of which time it would have no salvage value. The annual operating and maintenance costs for the old machine amount to $10,000. A new machine can be purchased at an invoice price of $14,000 to replace the present equipment. Because of the nature of the manufactured product, the new machine has an expected economic life of 3 years, and it will have no salvage value at the end of that time. The new machine's expected operating and maintenance costs amount to $2,000 for the first year and $3,000 for each of the next 2 years. The income tax rate is 34%. Any gains will also be taxed at 34%. The allowed depreciation amounts for the new machine are $1,400 during the first year, and $2,800 per year for the next 2 years. The firm's interest rate is 15%.

15s.1 If the old machine is to be sold now, what will be the gains tax?

(a) $800 (b) $850
(c) $900 (d) $950.

15s.2 If you decide to retain the old machine for now, what will be the opportunity cost?

(a) $2,500 (b) $4,000
(c) $1,650 (d) $1,500.

15s.3 If the old asset is to be sold now, its sunk cost is

(a) $2,500 (b) $4,000
(c) $1,650 (d) $1,500.

15s.4 For depreciation purposes, the first cost of the new machine will be

(a) $14,000 (b) $11,500
(c) $12,350 (d) $16,500.

15s.5 What is the incremental annual after-tax benefit of replacing the old machine?

(a) $909 (b) $918
(c) $936 (d) $972.

15s.6 A local delivery company has purchased a delivery truck for $15,000. The truck will be depreciated under the MACRS as a 5-year property (for a 5-year property, the MACRS percentages are 20%, 32%, 19.2%, 11.52%, 11.52%, 5.76%). The truck's market value (or selling price) is expected to be $2,500 less each year. The O&M costs are expected to be $3,000 per year. The firm is in a 40% tax bracket, and its MARR is 15%. Compute the annual equivalent cost for retaining the truck for a 2-year period which will be

(a) $5,527   (b) $5,175

(c) $5,362   (d) $5,014.

15s.7 The annual equivalent after-tax costs of retaining a defender over its 3-year remaining life and the annual equivalent operating costs for its challenger over its 4-year physical life are as follows:

Holding Period	Annual Equivalent Cost	
	Defender	Challenger
1	$3,000	$5,000
2	2,500	4,000
3	2,800	3,000
4		4,500

Assume a MARR of 12% and determine the optimal replacement time for the defender. Assume an infinite planning horizon and no technological change (cost) in the challenger. What would be your decision?

(a) Replace now

(b) Replace 1 year later

(c) Replace 2 years later

(d) Replace 3 years later.

"The new machine tool is simply beautiful," exclaimed John Ogletree, industrial engineer for Trent Manufacturing Company. "No wonder the plant manager is so anxious to buy it." "I am not sure it is worth replacing the present machine tool," replied Jerry Hanson, operation manager. "Everyone seems to forget that we purchased our present machine just four years ago at a cost of $12,000. The new tool will cost $15,000. The worst part is that we can only get $2,000 out of old machine tool if we sell it now." "That's quite a loss for the company to absorb." "We can make up the loss very quickly," countered John. "Willow Creek Manufacturing Co. in the northern part of the state says that scrap loss decreased by 20% when they purchased these new tools. I've gathered a lot of information about the new tools, and I will have a recommendation ready for the plant manager tomorrow."

- Option 1: You continue to use an old machine tool that was bought 4 years ago for $12,000. It has been fully depreciated but can be sold for $2,000. If kept, it could be used for 3 more years with proper maintenance and with some extra care. No salvage value is expected at the end of 3 years. The maintenance costs would run $10,000 per year for the old machine tool.

- Option 2: You purchase a brand-new machine tool at a price of $15,000 to replace the present equipment. Because of the nature of the product manufactured, it also has an expected economic life of 3 years, and will have a salvage value of $5,000 at the end of that time. With the new machine tool, the expected operating and maintenance costs (with the scrap savings) amount to $3,000 each year for 3 years. The allowed depreciation amounts for the new machine are $5,000 during the first year, $6,668 during the second year, and $1,110 (with the half-year convention) during the third year.

The income tax rate is 35%. Any gains will also be taxed at 35%.

15s.8 For the old machine tools, what would be the amount of sunk cost that should be recognized in replacement analysis?

(a) $10,000   (c) $12,000

(b) $2,000   (d) 0.

15s.9 What is the opportunity cost of retaining the old machine tool now?

(a) $700   (c) $2,000

(b) $1,300   (d) $10,000.

15s.10 What would be the net proceeds associated with disposing the new machine tool at the end of 3 years?

(a) $4,028   (c) $972

(b) $2,778   (d) $5,000.

15s.11 What is the net incremental benefit or loss in present value associated with replacing the old machine tools at an interest rate of 15%?

(a) $6,648   (c) $421

(b) $2,879   (d) $960.

15s.12 The annual equivalent after-tax revenue of retaining a defender over its 3-year remaining life and the annual equivalent operating revenues for its challenger over its 4-year project life follow. (*Note: All figures in dollars are revenue figures.*)

Holding period	Annual Equivalent Worth (Defender)	Annual Equivalent Worth (Challenger)
1	$3,500	$1,800
2	$2,500	$2,500
3	$1,500	$3,000
4		$2,000

Assume an MARR of 10% and determine the optimal replacement time for the defender. Assume an infinite planning horizon and no technological change (revenue/cost) in the challenger. What would be your decision?

(a) Replace now

(b) Replace 1 year later

(c) Replace 2 years later

(d) Replace 3 years later.

Problems

Sunk Costs, Opportunity Costs, and Cash Flows

15.1 Inland Trucking Company is considering the replacement of a 1,000-pound-capacity forklift truck. The truck was purchased 3 years ago at a cost of $15,000. The diesel-operated forklift truck was originally expected to have a useful life of 8 years and a zero estimated salvage value at the end of that period. The truck has not been dependable and is frequently out of service while awaiting repairs. The maintenance expenses of the truck have been rising steadily and currently amount to about $3,000 per year. The truck could be sold for $6,000. If retained, the truck will require an immediate $1,500 overhaul to keep it in operable condition. This overhaul will neither extend the originally estimated service life nor will it increase the value of the truck. The updated annual operating costs, engine overhaul cost, and market values over the next 5 years are estimated as follows:

n	O&M	Depreciation	Engine Overhaul	Market Value
−3				
−2		$3,000		
−1		4,800		
0		2,880	$1,500	$6,000
1	$3,000	1,728		4,000
2	3,500	1,728		3,000
3	3,800	864		1,500
4	4,500	0		1,000
5	4,800	0	5,000	0

A drastic increase in O&M costs during the fifth year is expected due to another overhaul, which will be required to keep the truck in operating condition. The firm's MARR is 15%.

(a) If the truck is to be sold now, what will be its sunk cost?

(b) What is the opportunity cost of not replacing the truck now?

(c) What is the equivalent annual cost of owning and operating the truck for 2 more years?

(d) What is the equivalent annual cost of owning and operating the truck for 5 years?

15.2 Komatsu Cutting Technologies is considering replacing one of its CNC machines with one that is newer and more efficient. The firm purchased the CNC machine 10 years ago at a cost of $135,000. It had an expected economic life of 12 years at the time of purchase and an expected salvage value of $12,000 at the end of the 12 years. The original salvage estimate is still good, and the machine has a remaining useful life of 2 years. The firm can sell this old machine now to another firm in the industry for $30,000. The new machine can be purchased for $165,000, including installation costs. It has an estimated useful (economic) life of 8 years. The new machine is expected to reduce cash operating expenses by $30,000 per year over its 8-year life. At the end of its useful life, the machine is estimated to be worth only $5,000. The company has a MARR of 12%.

(a) If you decided to retain the old machine, what is the opportunity (investment) cost of retaining the old asset?

(b) Compute the cash flows associated with retaining the old machine in years 1 to 2.

(c) Compute the cash flows associated with purchasing the new machine in years 1 to 8 (use the opportunity cost concept).

(d) If the firm needs the service of these machines for an indefinite period and no technology improvement is expected in future machines, what will be your decision?

15.3 Air Links, a commuter airline company, is considering the replacement of one of its baggage loading/unloading machines with a newer and more efficient one.

- The current book value of the old machine is $50,000, and it has a remaining useful life of 5 years. The salvage value expected from scrapping the old machine at the end of 5 years is zero, but the company can sell the machine now to another firm in the industry for $10,000.

- The new baggage handling machine has a purchase price of $120,000 and an estimated useful life of 7-years. It has an estimated salvage value of $30,000 and is expected to realize economic savings on electric power usage, labor, and repair costs and also to reduce the amount of damaged luggage. In total, an annual savings of $50,000 will be realized if the new machine is installed. The firm uses a 15% of MARR. Using the opportunity cost approach,

(a) What is the initial cash outlay required for the new machine?

(b) What are the cash flows for the defender in years 0 to 5?

(c) Should the airline purchase the new machine?

15.4 Duluth Medico purchased a digital image processing machine 3 years ago at a cost of $50,000. The machine had an expected life of 8 years at the time of purchase and an expected salvage value of $5,000 at the end of the 8 years. The old machine has been slow at handling the increased business volume, so management is considering replacing the machine. A new machine can be purchased for $75,000, including installation costs. Over its 5-year life, the machine will reduce cash operating expenses by $30,000 per year. Sales are not expected to change. At the end of its useful life, the machine is estimated to be worthless. The old machine can be sold today for $10,000. The firm's interest rate for project justification is known to be 15%. The firm does not expect a better machine (other than the current challenger) to be available for the next 5 years. Assuming that the economic service life for the new machine and the remaining useful life for the old machine are 5 years,

(a) Determine the cash flows associated with each option (keeping the defender versus purchasing the challenger).

(b) Should the company replace the defender now?

15.5 The Northwest Manufacturing Company is currently manufacturing one of its products on a hydraulic stamping press machine. The unit cost of the product is $12, and in the past year, 3,000 units were produced and sold for $19 each. It is expected that both the future demand of the product and the unit price will remain steady at 3,000 units per year and $19 per unit. The old machine has a remaining useful life of 3 years. The old machine could be sold on the open market now for $5,500. Three years from now the old machine is expected to have a salvage value of $1,200. The new machine would cost $36,500, and the unit manufacturing cost on the new machine is projected to be $11. The new machine has an expected economic life of 5 years and an expected salvage of $6,300. The appropriate MARR is 12%. The firm does not expect a significant improvement in technology, and it needs the service of either machine for an indefinite period of time.

(a) Compute the cash flows over the remaining useful life, if the firm decides to retain the old machine.

(b) Compute the cash flows over the economic service life, if the firm decides to purchase the machine.

(c) Should the equipment be acquired now?

Economic Service Life

15.6 A firm is considering replacing a machine that has been used for making a certain kind of packaging material. The new improved machine will cost $31,000 installed and will have an estimated economic life of 10 years with a salvage value of $2,500. Operating costs are expected to be $1,000 per year throughout its service life. The old machine in use had an original cost of $25,000 four years ago, and at the time it was purchased, its service life (physical life) was estimated to be 7 years with a salvage value of $5,000. The old machine has a current market value of $7,700. If the firm retains the old machine further, its updated market values and operating costs for the next 4 years will be as follows:

Year End	Market Value	Book Value	Operating Costs
0	$7,700	$7,889	
1	4,300	5,578	$3,200
2	3,300	3,347	3,700
3	1,100	1,116	4,800
4	0	0	5,850

The firm's minimum attractive rate of return is 12%.

(a) Working with the updated estimates of market values and operating costs over the next 4 years, determine the remaining useful life of the old machine.

(b) Determine whether it is economical to make the replacement now.

(c) If the firm's decision is to replace the old machine, then when?

15.7 The University Resume Service has just invested $8,000 in a new desktop publishing system. From past experience, the owner of the company estimates its after-tax cash returns as follows:

$$A_n = \$8,000 - \$4,000(1 + 0.15)^{n-1}$$
$$S_n = \$6,000(1 - 0.3)^n;$$

where A_n stands for net after-tax cash flows from operation during period n, and S_n stands for the after-tax salvage value at the end of period n.

(a) If the company's MARR is 12%, compute the economic service life of the desktop operating system.

(b) Explain how the economic service life varies with the interest rate.

15.8 A special purpose machine is to be purchased at a cost of $15,000. The following table shows the expected annual operating and maintenance cost and the salvage values for each year of service.

Year of Service	O&M Costs	Market Value
1	$2,500	$12,000
2	3,200	8,100
3	5,300	5,200
4	6,500	3,500
5	7,800	0

(a) If the interest rate is 10%, what is the economic service life for this machine?

(b) Repeat (a) above using $i = 15\%$.

Replacement Decisions with an Infinite Planning Horizon and No Technological Change

15.9 A special-purpose turnkey stamping machine was purchased 4 years ago for $20,000. It was estimated at that time that this machine would have a life of 10 years and a salvage value $3,000, with a removal cost of $1,500. These estimates are still good. This machine has annual operating costs of $2,000. A new machine, which is more efficient, will reduce the operating costs to $1,000, but it will require an investment of $20,000 plus $1,000 for installation. The life of the new machine is estimated to be 12 years with a salvage of $2,000 and a removal cost of $1,500. An offer of $6,000 has been made for the old machine, and the purchaser is willing to pay for removal of the machine. Find the economic advantage of replacement or of continuing with the present machine. State any assumptions that you make. (Assume the MARR = 8%).

15.10 A 5-year-old defender has a current market value of $4,000, expects O&M costs of $3,000 this year, increasing by $1,500 per year. Future market values are expected to decline by $1,000 per year. The machine can be used for another 3 years. The challenger costs $6,000 and has O&M costs of $2,000 per year, increasing by $1,000 per year. The machine will be needed for only 3 years, and the salvage value at the end of 3 years is expected to be $2,000. The MARR is 15%.

(a) Determine the annual cash flows for retaining the old machine for 3 years.

(b) Determine if now is the time to replace the old machine. First show the annual cash flows for the challenger.

15.11 Greenleaf Company is considering the purchase of a new set of air-electric quill units to replace an obsolete one. The machine being used for the operation has a market value of zero: however, it is in good working order, and it will last physically for at least an additional 5 years. The new quill units will perform the operation with so much more efficiency, that the firm's engineers estimate that labor, material, and other direct costs will be reduced $3,000 a year if it is installed. The new set of quill units costs $10,000, delivered and installed, and its economic life is estimated to be 5 years with zero salvage value. The firm's MARR is 10%.

(a) What is the investment required to keep the old machine?

(b) Compute the cash flow to use in the analysis for each option.

(c) If the firm uses the internal rate of return criterion, should the firm buy the new machine on that basis?

15.12 Wu Lighting Company is considering the replacement of an old, relatively inefficient vertical drill machine that was purchased 7 years ago at a cost of $10,000. The machine had an original expected life of 12 years and a zero estimated salvage value at the end of that period. The divisional manager reports that a new machine can be bought and installed for $12,000 which, over its 5-year life, will expand sales from $10,000 to $11,500 a year and, furthermore, will reduce labor and raw materials usage sufficiently to cut annual operating costs from $7,000 to $5,000. The new machine, has an estimated salvage value of $2,000 at the end of its 5-year life. The old machine's current market value is $1,000; the firm's MARR is 15%.

(a) Should the new machine be purchased now?

(b) What current market value of the old machine would make the two options equal?

15.13 Advanced Robotics Company is faced with the prospect replacing its old call-switching systems, which have been used in the company's headquarters for 10 years. This particular system was installed at a cost of $100,000, and it was assumed that it would have a 15-year life with no appreciable salvage value. The current annual operating costs are $20,000 for this old system, and these costs would be the same for the rest of its life. A sales representative from North Central Bell is trying to sell this company a computerized switching system. The new system would require an investment of $200,000 for installation. The economic life of this computerized system is estimated to be 10 years with a salvage value of $18,000, and the system will reduce annual operating costs to $5,000. No detailed agreement has been made with the sales representative about the disposal of the old system. Determine the ranges of resale value associated with the old system that would justify installation of the new system at a MARR of 14%.

15.14 A company is currently producing chemical compounds by a process installed 10 years ago at a cost of $100,000. It was assumed that the process would have a 20-year life with

a zero salvage value. The current market value of this equipment, however, is $60,000, and the initial estimate of its economic life is still good. The annual operating costs associated with this process are $18,000. A sales representative from U.S. Instrument Company is trying to sell a new chemical compound-making process to the company. This new process will cost $200,000, have a service life of 10 years with a salvage value of $20,000, and reduce annual operating costs to $4,000. Assuming the company desires a return of 12% on all investments, should it invest in the new process?

15.15 Eight years ago a lathe was purchased for $45,000. Its operating expenses were $8,700 per year. An equipment vendor offers a new machine for $53,500, and its operating costs are $5,700 per year. An allowance of $8,500 would be made for the old machine on purchase of the new one. The old machine is expected to be scrapped at the end of 5 years. The new machine's economic service life is 5 years with a salvage value of $12,000. The new machine's O&M cost is estimated to be $4,200 for the first year, increasing at an annual rate of $500 thereafter. The firm's MARR is 12%. What option would you recommend?

15.16 The New York Taxi Cab Company has just purchased a new fleet of 2000 models. Each brand-new cab cost $12,000. From past experience, the company estimates after-tax cash returns for each cab as follows.

$$A_n = \$65,800 - 30,250(1 + 0.25)^{n-1}$$
$$S_n = \$10,000(1 - 0.35)^n$$

where, again, A_n stands for net after-tax cash flows from operation during period n, and S_n stands for the after-tax salvage value at the end of period n. The management views the replacement process as a constant and infinite chain.

(a) If the firm's MARR is 10%, and it expects no major technological and functional change in future models, what is the optimal time period (constant replacement cycle) to replace its cabs? (Ignore inflation.)

(b) What is the internal rate of return for a cab if it is retired at the end of its economic service life? What is the internal rate of return for a sequence of identical cabs if each cab in the sequence is replaced at the optimal time?

15.17 Four years ago an industrial batch oven was purchased for $23,000. It has been depreciated over a 10-year life and has a $1,000 salvage value. If sold now, the machine will bring $2,000. If sold at the end of the year, it will bring $1,500. Annual operating costs for subsequent years are $3,800. A new machine will cost $50,000 with a 12-year life and have a $3,000 salvage value. The operating cost will be $3,000 as of the end of each year with the $6,000 per year savings due to better quality control. If the firm's MARR is 10%, should the machine be purchased now?

15.18 Georgia Ceramic Company has an automatic glaze sprayer that has been used for the past 10 years. The sprayer can be used for another 10 years and will have a zero salvage value at that time. The annual operating and maintenance costs for the sprayer amount to $15,000 per year. Due to an increase in business, a new sprayer must be purchased.

- Option 1: If the old sprayer is retained, a new smaller capacity sprayer will be purchased at a cost of

$48,000, and it will have a $5,000 salvage value in 10 years. This new sprayer will have annual operating and maintenance costs of $12,000. The old sprayer has a current market value of $6,000.

- Option 2: If the old sprayer is sold, a new sprayer of larger capacity will be purchased for $84,000. This sprayer will have a $9,000 salvage value in 10 years and will have annual operating and maintenance costs of $24,000.

Which option should be selected at MARR = 12%?

Replacement Problem with a Finite Planning Horizon

15.19 The annual equivalent after-tax costs of retaining a defender machine over 4 years (physical life), or operating its challenger over 6 years (physical life), are as follows:

n	Defender	Challenger
1	−$3,200	−$5,800
2	−2,500	−4,230
3	−2,650	−3,200
4	−3,300	−3,500
5		−4,000
6		−5,500

If you need the service of either machine for only the next 10 years, what is the best replacement strategy? Assume a MARR of 12% and no technology improvement in future challengers.

15.20 The after-tax annual equivalent worth of retaining a defender over 4 years (physical life) or operating its challenger over 6 years (physical life) are as follows:

n	Defender	Challenger
1	$13,400	$12,300
2	13,500	13,000
3	13,800	13,600
4	13,200	13,400
5		13,000
6		12,500

If you need the service of either machine only for the next 8 years, what is the best replacement strategy? Assume a MARR of 12% and no technology improvement in future challengers.

15.21 An existing asset that cost $16,000 two years ago has a market value of $12,000 today, an expected salvage value of $2,000 at the end of its remaining useful life of 6 more years, and annual operating costs of $4,000. A new asset under consideration as a replacement has an initial cost of $10,000, an expected salvage value of $4,000 at the end of its economic life of 5 years, and annual operating costs of $2,000. It is assumed that this new asset could be replaced by another one identical in every respect after 3 years at a salvage value of $5,000, if desired.

By using a MARR of 11%, a 6-year study period, and PW calculations, decide whether the existing asset should be replaced by the new one.

15.22 Repeat (15.21) above based on the AE criterion.

Replacement Analysis with Tax Considerations

15.23 Problem 15.1 with the following additional information: the asset is classified as a 5-year MACRS property and has a book value of $5,760 if disposed of now. The firm's marginal tax rate is 40% and its after-tax MARR is 15%.

15.24 Problem 15.2 with the following additional information: the asset is classified as a 7-year MACRS. The firm's marginal tax rate is 40%, and its after-tax MARR is 12%.

15.25 Problem 15.3 with the following additional information:

- The current book value of the old machine is $50,000. Using the conventional straight-line methods, the old machine is being depreciated toward a zero salvage value, or by $10,000 per year.
- The new machine will be depreciated under a 7-year MACRS class.
- The company's marginal tax rate is 40%, and the firm uses a 15% of MARR after-tax.

15.26 Problem 15.4 with the following additional information:

- The old machine has been depreciated using the MACRS under the 5-year property class.
- The new machine will be depreciated under a 5-year MACRS class.
- The marginal tax rate is 35%, and the firm's after-tax MARR is 15%.

15.27 Problem 15.5 with the following additional information:

- The old stamping machine has been fully depreciated.
- For tax purpose, the entire cost of $36,500 can be depreciated according to a 5-year MACRS property class.
- The firm's marginal tax rate is 40%, and the after-tax MARR is 12%.

15.28 Problem 15.6 with the following additional information:

- The current book value of the old machine is $7,889. The anticipated book value for the following years are as follows: Year 1: $5,578, Year 2: $3,347, Year 3: $1,116, and Year 4: $0. The new machine will be depreciated under a 7-year MACRS class.

- The company's marginal tax rate is 35%, and the firm uses a 12% of MARR after-tax.

15.29 A machine has a first cost of $10,000. End-of-year book values, salvage values, and annual O&M costs are provided over its useful life as follows:

Year End	Book Value	Salvage Value	Operating Costs
1	$8,000	$5,300	$1,500
2	4,800	3,900	2,100
3	2,880	2,800	2,700
4	1,728	1,800	3,400
5	1,728	1,400	4,200
6	576	600	4,900

(a) Determine the economic life if the MARR is 15% and the marginal tax rate is 40%.

(b) Determine the economic life if the MARR is 10% and the marginal tax rate remains at 40%.

15.30 Given the following data:

$$I = \$20,000$$

$$S_n = 12,000 - 2,000n$$

$$B_n = 20,000 - 2,500n$$

$$O\&M_n = 3,000 + 1,000\,(n - 1)$$

$$t_m = 0.40,$$

where I = Asset purchase price

S_n = Market value at the end of year n

B_n = Book value at the end of year n

$O\&M_n$ = O&M cost during year n

t_m = Marginal tax rate.

(a) Determine the economic service life if $i = 10\%$.

(b) Determine the economic service life if $i = 25\%$.

(c) Assume $i = 0$: Determine the economic life mathematically (i.e., the calculus technique for finding the minimum point as described in Chapter 8).

15.31 Problem 15.8 with the following additional information:

- For tax purpose, the entire cost of $15,000 can be depreciated according to a 5-year MACRS property class.
- The firm's marginal tax rate is 40%.

15.32 Quintana Electronic Company is considering the purchase of new robot-welding equipment to perform operations currently being performed by less efficient equipment. The new machine's purchase price is $150,000, delivered and installed. A Quintana industrial engineer estimates that the new equipment will produce savings of $30,000 in labor and other direct costs annually, as compared with the present equipment. He estimates the proposed equipment's economic life at 10 years, with a zero salvage value. The present equipment is in good working order and will last, physically, for at least 10 more years. Quintana Company expects to pay income taxes of 40%, and any gains will also be taxed at 40%. Quintana uses a 10% discount rate for analysis performed on an after-tax basis. Depreciation of the new equipment for tax purposes is computed on the basis of the 7-year MACRS property class.

(a) Assuming that the present equipment has zero book value and zero salvage value, should the company buy the proposed equipment?

(b) Assuming that the present equipment is being depreciated at a straight-line rate of 10%, has a book value of $72,000 (cost, $120,000; accumulated depreciation, $48,000), and zero net salvage value today, should the company buy the proposed equipment?

(c) Assuming the present equipment has a book value of $72,000 and a salvage value today of $45,000 and that, if retained for 10 more years, its salvage value will be zero, should the company buy the proposed equipment?

(d) Assume that the new equipment will save only $15,000 a year, but that its economic life is expected to be 12 years. If other conditions are as described in (a) above, should the company buy the proposed equipment?

15.33 Quintana Company decided to purchase the equipment described in Problem 15.32 (hereafter called "Model A" equipment). Two years later, even better equipment (called "Model B") comes onto the market, which makes Model A completely obsolete, with no resale value. The Model B equipment costs $300,000 delivered and installed, but it is expected to result in annual savings of $75,000 over the cost of operating the Model A equipment. The economic life of Model B is estimated to be 10 years with a zero salvage value. (Model B also is classified as a 7-year MACRS property.)

(a) What action should the company take?

(b) If the company decides to purchase the Model B equipment, a mistake must have been made, because good equipment, bought only 2 years previously, is being scrapped. How did this mistake come about?

15.34 Problem 15.9 with the following additional information:

- The current book value of the old machine is $6,248, and the asset has been depreciated according to a 7-year MACRS property.
- The asset is also classified to a 7-year MACRS property.
- The company's marginal tax rate is 30%, and the firm uses an 8% of MARR after-tax.

15.35 Problem 15.10 with the following additional information:

- The old machine has been fully depreciated.
- The new machine will be depreciated under a 3-year MACRS class.
- The marginal tax rate is 40%, and the firm's after-tax MARR is 15%.

15.36 Problem 15.11 with the following additional information:

- The current book value of the old machine is $4,000, and the annual depreciation charge is $800, if the firm decides to keep the old machine for the additional 5 years.
- The new asset is classified to a 7-year MACRS property.
- The company's marginal tax rate is 40%, and the firm uses a 10% of MARR after-tax.

15.37 Problem 15.12 with the following additional information:

- The old machine has been fully depreciated.
- The new machine will be depreciated under a 7-year MACRS class.
- The marginal tax rate is 40%, and the firm's after-tax MARR is 15%.

15.38 Problem 15.13 with the following additional information:

- The old switching system has been fully depreciated.
- The new system falls into a 5-year MACRS property.

- The company's marginal tax rate is 40%, and the firm uses a 14% of MARR after-tax.

15.39 Five years ago a conveyor system was installed in a manufacturing plant at a cost of $35,000. It was estimated that the system, which is still in operating condition, would have a useful life of 8 years with a salvage value of $3,000. If the firm continues to operate the system, its estimated market values and operating costs for the next 3 years are as follows.

Year End	Market Value	Book Value	Operating Costs
0	$11,500	$15,000	
1	5,200	11,000	$4,500
2	3,500	7,000	5,300
3	1,200	3,000	6,100

A new system can be installed for $43,500; it would have an estimated economic life of 10 years with a salvage value of $3,500. Operating costs are expected to be $1,500 per year throughout its service life. This firm's MARR is 18%. The system belongs to the 7-year MACRS property class. The firm's marginal tax rate is 35%.

(a) Decide whether to replace the existing system now.

(b) If the decision is to replace the existing system, when should replacement occur?

15.40 Problem 15.14 with the following additional information:

- The old machine has been depreciated based on a straight-line basis.
- The new machine will be depreciated under a 7-year MACRS class.
- The marginal tax rate is 40%, and the firm's after-tax MARR is 12%.

15.41 Problem 15.15 with the following additional information:

- The old machine has been fully depreciated.
- The new machine will be depreciated under a 7-year MACRS class.
- The marginal tax rate is 35%, and the firm's after-tax MARR is 12%.

15.42 Problem 15.17 with the following additional information:

- The old machine has been depreciated according to a 7-year MACRS.
- The new machine will also be depreciated under a 7-year MACRS class.
- The marginal tax rate is 40%, and the firm's after-tax MARR is 10%.

15.43 Problem 15.18 with the following additional information:

- Option 1: The old sprayer has been fully depreciated. The new sprayer is classified as a 7-year MACRS recovery period.
- Option 2: The larger capacity sprayer is classified as a 7-year MACRS property.
- The company's marginal tax rate is 40%, and the firm uses 12% of MARR after-tax.

15.44 A 6-year-old CNC machine that originally cost $8,000 has been fully depreciated and its current market value is $1,500. If the machine is kept in service for the next 5 years, its O&M costs and salvage value are estimated as follows:

End of Year	O&M Costs		Salvage Value
	Operation and Repairs	Delays due to Breakdowns	
1	$1,300	$600	$1,200
2	1,500	800	1,000
3	1,700	1,000	500
4	1,900	1,200	0
5	2,000	1,400	0

It is suggested that the machine be replaced by a new CNC machine of improved design at a cost of $6,000. It is believed that this purchase will completely eliminate breakdowns and the resulting cost of delays and that operation and repair costs will be reduced $200 a year at each age than is the case with the old machine. Assume a 5-year life for the challenger and a $1,000 terminal salvage value. The new machine falls into a 5-year MACRS property class. The firm's MARR is 12%, and its marginal tax rate is 30%. Should the old machine be replaced now?

15.45 Problem 15.21 with the following additional information:

- The old asset has been fully depreciated according to a 5-year MACRS.
- The new asset will also be depreciated under a 5-year MACRS class.
- The marginal tax rate is 30%, and the firm's after-tax MARR is 11%.

Short Case Studies

15.46 Chevron Overseas Petroleum, Inc. entered into a 1993 joint venture agreement with the Republic of Kazakhstan, a former republic of the old Soviet Union, to develop the huge Tengiz oil field.[10] Unfortunately, the climate in the region is harsh making it difficult to keep oil flowing. The untreated oil comes out of the ground at 114°F. Even though the pipelines are insulated, as the oil gets further from the well on its way to be processed, hydrate salts begin to precipitate out of the liquid phase as the oil cools. These hydrate salts create a dangerous condition as they form plugs in the line.

The method for preventing this trap pressure condition is to inject methanol (MeOH) into the oil stream. This keeps the oil flowing and prevents hydrate salts from precipitating out of the liquid phase. The present methanol loading and storage facility is a completely manual controlled system, with no fire protection and with a rapidly deteriorating tank that causes leaks. The scope of repairs and upgrades is extensive. The storage tanks are rusting and are leaking at the riveted joints. The manual level control system causes frequent tank overfills. There is no fire protection system as water is not available at this site.

The present storage facility has been in service for 5 years. Permit requirements require upgrades to achieve minimum acceptable Kazakhstan standards. Upgrades, in the amount of $104,000, will extend the life of the current facility to about 10 years. However, upgrades will not completely stop the leaks. The expected spill and leak losses will amount to $5,000 a year. The annual operating costs are expected to $36,000.

As an alternative, a new methanol storage facility can be designed based on minimum, acceptable international oil industry practice. The new facility, which would cost $325,000, would last about 12 years before a major upgrade would be required. However, it is believed that oil transfer technology will be such that methanol will not be necessary in 10 years. The pipeline heating and insulation systems will make methanol storage and use systems obsolete. With a lower risk of leaks, spills, and evaporation loss a more closely monitored-system, the expected annual operating cost would be $12,000.

---

[10] The example was provided by Mr. Joel M. Height of the Chevron Oil Company.

(a) Assume that the storage tanks (the new as well as the upgraded ones) will have no salvage values at the end of their useful lives (after considering the removal costs), and that the tanks will be depreciated by the straight-line method according to the Kazakhstan's tax law. If Chevron's interest rate is 20% for foreign projects, which option is a better choice?

(b) How would the decision change as you consider the risk of spills (clean-up costs) and evaporation of product related to environmental impact?

15.47 National Woodwork Company, a manufacturer of window frames, is considering replacing a conventional manufacturing system with a flexible manufacturing system (FMS). The company cannot produce rapidly enough to meet demand. Some manufacturing problems identified follow:

- The present system is expected to be useful for another 5 years, but will require an estimated $105,000 per year in maintenance, which will increase $10,000 each year as parts become more scarce. The current market value of the existing system is $140,000, and the machine has been fully depreciated.

The proposed system will reduce, or entirely eliminate, the set-up times, and each window can be made as it is ordered by the customer. Customers phone their orders into head office, where details are fed into the company's main computer. These manufacturing details are then dispatched to computers on the manufacturing floor, which are, in turn, connected to a computer that controls the proposed FMS. This system eliminates the warehouse space and material

handling time that are needed when using the conventional system.

Before installing the FMS, the old equipment will be removed from the job shop floor at an estimated cost of $100,000. This cost includes the needed electrical work for the new system. The proposed FMS will cost $1,200,000. The economic life of the machine is expected to be 10 years, and the salvage value is expected to be $120,000. The change in window styles has been minimal in the past few decades and is expected to continue to remain stable in the future. The proposed equipment falls into the 7-year MACRS category. The total annual savings will be $664,243: $12,000 attributed to the reduction of defective windows, $511,043 from the elimination of 13 workers, $100,200 from the increase in productivity, and $41,000 from the near elimination of warehouse space and material handling. The O&M costs will be only $45,000, increasing by $2,000 per year. The National Woodwork's MARR is about 15%, and the expected marginal tax rate over the project years is 40%.

(a) What assumptions are required to compare the conventional system with the FMS?

(b) With the assumptions defined in (a), should the FMS be installed now?

15.48 In 2×4 and 2×6 lumber production, significant amounts of wood are present in sideboards produced after the initial cutting of logs. Instead of processing the sideboards into wood chips for the paper mill, an "edger" is used to reclaim additional lumber, thus resulting in savings for the company. An edger is capable of reclaiming lumber by any of the following three methods: (1) removing rough edges, (2) splitting large sideboards, and (3) salvaging 2 × 4 lumber from low-quality 4 × 4 boards. Union Camp Company's engineers have discovered that a significant reduction in production costs could be achieved simply by replacing the original "edger" machine with a newer laser-controlled model.

Old Edger: The old edger was placed in service 12 years ago and is fully depreciated. Any machine scrap value would offset the removal cost of the equipment. No market exists for this obsolete equipment. The old edger needs two operators. During the cutting operation, the operator makes edger settings based on his/her own judgment. The operator has no means of determining exactly what dimension of lumber could be recovered from a given sideboard and must guess at the proper setting to recover the highest grade of lumber. Furthermore, the old edger is not capable of salvaging good-quality 2 × 4s from poor-quality 4 × 4s. The defender can continue in service for another 5 years with proper maintenance.

Current market value	$0
Current book value	0
Annual maintenance cost	$2,500 in year 1, increasing at a rate of 15% each year over the previous year's cost
Annual operating costs (labor and power)	$65,000

New Laser-Controlled Edger: The new edger has numerous advantages over its defender. These advantages include laser beams that indicate where cuts should be made to obtain the maximum yield by the edger. The new edger requires a single operator, and labor savings will be reflected in lower operating and maintenance costs of $35,000 a year.

Estimated Cost	
Equipment	$35,700
Equipment installation	21,500
Building	47,200
Conveyor modification	14,500
Electrical (wiring)	16,500
Sub total	$135,400
Engineering	7,000
Construction management	20,000
Contingency	16,200
Total	$178,600

Useful life of new edger	10 years
Salvage value	
Building (tear down)	$0
Equipment	10% of the original cost
Annual O&M costs	$35,000

Depreciation Methods	
Building	39-year MACRS
Equipment and installation	7-year MACRS

Twenty-five percent of total mill volume passed through the edger. A 12% yield improvement is expected to be realized on this production, which will result in an improvement of total mill volume of $(0.25)(0.12) = 3\%$, or an annual savings of $57,895.

(a) Should the defender be replaced now if the mill's MARR and marginal tax rate are 16% and 40%, respectively?

(b) If the defender will eventually be replaced by the current challenger, when is the optimal time to replace?

15.49 Rivera Industries, a manufacturer of home heating appliances, is considering the purchase of Amada Turret Punch Press, a more advanced piece of machinery, to replace its present system that uses four old presses. Currently, the four smaller presses are used (in varying sequences, depending on the product) to produce one component of a product until a scheduled time when all machines must retool to set up for a different component. Because setup cost is high, production runs of individual components are long. This results in large inventory buildups of one component, which are necessary to prevent extended backlogging while other products are being manufactured.

- The four presses in use now were purchased 6 years ago at a price of $100,000. The manufacturing engineer expects that these machines can be used for 8 more years, but they will have no market value after that. These presses have been depreciated by the MACRS method (7-year property). The current book value is $13,387, and the present market value is estimated to be $40,000. The average setup cost, which is determined by the number of required labor hours times the labor rate for the old presses, is $80 per hour, and the number of setups per year expected by the production control department is 200. This yields a yearly setup cost of $16,000. The expected operating and maintenance cost for each year in the remaining life of this system is estimated as follows:

Year	Setup Costs	O&M Costs
1	$16,000	$15,986
2	16,000	16,785
3	16,000	17,663
4	16,000	18,630
5	16,000	19,692
6	16,000	20,861
7	16,000	22,147
8	16,000	23,562

These costs, which were estimated by the manufacturing engineer, with the aid of data provided by the vendor, represent a reduction in efficiency and an increase in needed service and repair over time.

- The price of the 2-year-old Amada turret punch press is $135,000 and would be paid for with cash from the company's capital fund. Additionally, the company would incur installation costs, which would total $1,200. An expenditure of $12,000 would be required to recondition the press to its original condition. The reconditioning would extend the Amada's economic service life to 8 years. It would have no salvage value at that time. The Amada would be depreciated under the MACRS with half-year convention as a 7-year property. The cash savings of the Amada over the present system are due to the reduced setup time. The average setup cost of the Amada is $15, and the Amada would incur 1,000 setups per year, yielding a yearly setup cost of $15,000. The savings due to the reduced setup time are experienced because of the reduction in carrying costs associated with that level of inventory where the production run and ordering quantity are reduced. The Accounting Department has estimated that at least $26,000, probably $36,000 per year, could be saved by shortening production runs. The operating and maintenance costs of the Amada as estimated by the manufacturing engineer are similar, but somewhat less, than the O&M costs for the present system.

Year	Setup Costs	O&M Costs
1	$15,000	$11,500
2	15,000	11,950
3	15,000	12,445
4	15,000	12,990
5	15,000	13,590
6	15,000	14,245
7	15,000	14,950
8	15,000	15,745

The reduction in the O&M costs is caused by the age difference of the machines and the reduced power requirements of the Amada.

- If Rivera Industries delays the replacement of the current four presses for another year, the second-hand Amada machine will no longer be available, and the company will have to buy a brand-new machine at an installed price of $200,450. The expected setup costs would be the same as those for the second-hand machine, but the annual operating and maintenance costs would be about 10% lower than the estimated O&M costs for the second-hand machine. The expected economic service life of the brand-new press would be 8 years with no salvage value. The brand-new press also falls into a 7-year MACRS.

Rivera's MARR is 12% after taxes, and the marginal income tax rate is expected to be 40% over the life of the project.

(a) Assuming that the company would need the service of either press for an indefinite period, what would you recommend?

(b) Assuming that the company would need the press for only 5 more years, what would you recommend?

15.50 Tiger Construction Company purchased its current bulldozer (Caterpillar D8H) and placed it in service 6 years ago. Since the purchase of the Caterpillar, new technology has produced changes in machines, which resulted in an increase in productivity of approximately 20%. The Caterpillar worked in a system with a fixed (required) production level to maintain overall system productivity. As the Caterpillar aged and logged more downtime, more hours had to be scheduled to maintain the required production. Tiger is considering the purchase of a new bulldozer (Komatsu K80A) to replace the Caterpillar. The following data have been collected by Tiger's civil engineer.

	Defender (Caterpillar D8H)	Challenger (Komatsu K80A)
Useful life	Not known	Not known
Purchase Price		$400,000
Salvage value, if kept for		
0 year	$75,000	$400,000
1 year	60,000	300,000
2 year	50,000	240,000
3 year	30,000	190,000
4 year	30,000	150,000
5 year	10,000	115,000
Fuel use		
(gallon/hour)	11.30	16

	Defender (Caterpillar D8H)	Challenger (Komatsu K80A)
Maintenance costs		
1	$46,800	$35,000
2	46,800	38,400
3	46,800	43,700
4	46,800	48,300
5	46,800	58,000
Operating hours (hours/year)		
1	1,800	2,500
2	1,800	2,400
3	1,700	2,300
4	1,700	2,100
5	1,600	2,000
Productivity index	1.00	1.20
Other Relevant Information		
Fuel cost ($/gallon)		$1.20
Operator's wages ($/hour)		$23.40
Market interest rate (MARR)		15%
Marginal tax rate		40%
Depreciation methods		
Defender (D8H)	Fully Depreciated	
Challenger (K80A)	MACRS (5-year)	

(a) A civil engineer notices that both machines have different working hours and hourly production capac-ities. To compare the different units of capacity, the engineer needs to devise a combined index to reflect the machine's produc-tivity as well as actual operating hours. Develop such a combined productivity index for each period.

(b) Adjust the operating and mainte-nance costs by this index.

(c) Compare the two alternatives. Should the defender be replaced now?

(d) If the following price index were Forecasted for the next 5 project years, should the defender be re-placed now?

	Forecasted Price Index			
Year	General Inflation	Fuel	Wage	Maintenance
0	100	100	100	100
1	108	110	115	108
2	116	120	125	116
3	126	130	130	124
4	136	140	135	126
5	147	150	140	128

Capital Budgeting Decisions

Coca -Cola Bottling Company stores the gasoline used by its fleet of trucks in underground storage tanks.[1] Current Environmental Protection Agency (EPA) regulations require all such tanks to be upgraded to meet 1998 standards or to be replaced by that same year. Tanks installed prior to 1976 cannot be upgraded and must be replaced by 1998. Tanks installed between 1977 and 1983 may be upgraded by 1998, and then these must be replaced within a 10-year period. Tanks installed during 1982 and thereafter may be completely upgraded to 1998 standards. These tanks would then have an expected life of 30 years. Several tanks may occupy one underground pit. All tanks in a pit are to be either upgraded or replaced simultaneously.

Prior to the upgrade or replacement, yearly inspections will be required of each tank at a cost of approximately $700 per tank. Some pit locations may have more than one pit. The pits on a given site may be upgraded or replaced individually, or as a group. Certain economies of scale are associated with upgrading or replacing all tanks on a site at the same time.

The company has over 17 pits in 11 different locations throughout the Southeast. Given the upgrade/replacement option for each pit, as well as the option of working on pits individually, or in combination with others at the same site, over 340 alternatives must be considered. The company wants to meet all EPA requirements over a 25-year time horizon.

[1] This story was written by Professor V. E. Unger of Auburn University with Mr. William Garvin at Hazelean Environmental Consultants, Inc., and Mr. Elbert Mullis of the Coca-Cola Bottling Company (reprint with thanks.)

Without budget limitations, the replacement problem would be easy: Simply select for each pit, or combination of pits, the least-cost alternative. However, to replace or upgrade all tanks over the 25-year time horizon would cost in excess of $750,000, and the company anticipates having only an annual tank-replacement budget of $200,000 or less. Because the project is to be implemented over an extended period of time, the firm's cost of capital will tend to fluctuate during the project period. In this circumstance, the choice of an appropriate interest rate (MARR) for use in the project evaluation becomes a critical issue. Given these budget and other restrictions, the company would like to determine the least-cost replacement/upgrade strategy for the 25-year period. The company is also interested in minimizing the maximum budget allocations required in any one year.

In this chapter, we present the basic framework of **capital budgeting,** which involves investment decisions related to fixed assets. Here, the term **capital budget** includes planned expenditures on fixed assets; **capital budgeting** encompasses the entire process of analyzing projects and deciding whether or not they should be included in the capital budget. In previous chapters, we focused on how to evaluate and compare investment projects—the analysis aspect of capital budgeting. In this chapter, we focus on the budgeting aspect. Proper capital budgeting decisions require choice of the method of project financing, schedule of investment opportunities, and an estimate of minimum attractive rate of return (MARR).

16.1 Methods of Financing

In previous chapters, we focused on problems relating to investment decisions. In reality, investment decisions are not always independent of the source of finance. For convenience, however, in economic analysis, investment decisions are usually separated from finance decisions—first the investment project is selected, and then the choice of financing sources is considered. After the source is chosen, appropriate modifications to the investment decision are made.

We have also assumed that the assets employed in an investment project are obtained with the firm's own capital (retained earnings) or from short-term borrowings. In practice, this arrangement is not always attractive, or even possible. If the investment calls for a significant infusion of capital, the firm may raise the needed capital by issuing stock. Alternatively, the firm may borrow the funds by issuing bonds to finance such purchases. In this section, we will first discuss how a typical firm raises new capital from external sources. Then we will discuss how external financing affects after-tax cash flows and how the decision to borrow affects the investment decision.

The two broad choices a firm has for financing an investment project are **equity financing** and **debt financing.**[2] We will look briefly at these two options for obtaining external investment funds and also examine their effects on after-tax cash flows.

16.1.1   Equity Financing

Equity financing can take two forms: (1) Use of retained earnings otherwise paid to stockholders or (2) issuance of stock. Both forms of equity financing use funds invested by the current or new owners of the company.

Until now, most of our economic analyses presumed that companies had cash on hand to make capital investments—implicitly, we were dealing with cases of financing by retained earnings. If a company had not reinvested these earnings, it might have paid them to the company's owners—the stockholders—in the form of a dividend, or it might have kept these earnings on hand for future needs.

If a company does not have sufficient cash on hand to make an investment and does not wish to borrow in order to fund it, financing can be arranged by selling common stock to raise the required funds. (Many small biotechnology and computer firms raise capital by going public and selling common stock.) To do this, the company has to decide how much money to raise, the type of securities to issue (common stock or preferred stock), and the basis for pricing the issue.

Once the company has decided to issue common stock, the firm must estimate **flotation costs**—the expenses it will incur in connection with the issue, such as investment bankers' fees, lawyers' fees, accountants' costs, and printing and engraving. Usually an investment banker will buy the issue from the company at a discount, below the price at which the stock is to be offered to the public. (The discount usually represents the *flotation costs.*) If the company is already publicly owned, the offering price will commonly be based on the existing market price of the stock. If the company is

---

[2] A hybrid financing method, known as *lease financing,* was discussed in Section 12.4.3.

going public for the first time, no established price will exist, so investment bankers have to estimate the expected market price at which the stock will sell after the stock issue. Example 16.1 illustrates how the flotation cost affects the cost of issuing common stock.

Example 16.1 Issuing Common Stock

Scientific Sports, Inc. (SSI), a golf club manufacturer, has developed a new metal club (Driver). The club is made out of titanium alloy—an extremely light and durable metal available with good vibration-damping characteristics (Figure 16.1). The company expects to acquire considerable market penetration with this new product. To produce it, the company needs a new manufacturing facility, which will cost $10 million. The company decided to raise this $10 million by selling common stock. The firm's current stock price is $30 per share. Investment bankers have informed management that the new public issue must be priced at $28 per share because of a decreasing demand, which will occur as more shares become available on the market. The flotation costs will be 6% of the issue price, so SSI will net $26.32 per share. How many shares must SSI sell to net $10 million after flotation expenses?

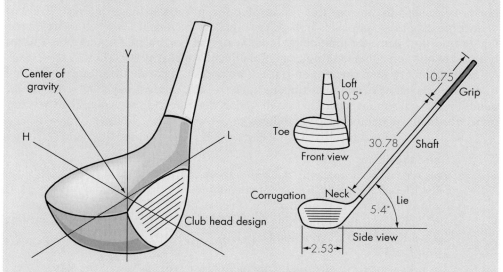

Figure 16.1   SSI's new golf club (Driver) design developed using advanced engineering materials (Example 16.1)

Solution

Let X be the number of shares to be sold. The total flotation cost will be

$$(0.06)(\$28)(X) = 1.68X.$$

To net $10 million, we establish the following relationship:

$$\text{Sales proceeds} - \text{flotation cost} = \text{Net proceeds}$$

$$28X - 1.68X = \$10,000,000$$

$$26.32X = \$10,000,000$$

$$X = 379,940 \text{ shares}.$$

Now we can figure out the flotation costs for issuing the common stock as follows:

$$1.68(379,940) = \$638,300.$$

16.1.2 Debt Financing

In addition to equity financing, the second major type of financing a company can select is **debt financing.** Debt financing includes both short-term borrowing from financial institutions and the sale of long-term bonds, where money is borrowed from investors for a fixed period. With debt financing, the interest paid on the loans or bonds is treated as an expense for income-tax purposes. Since interest is a tax-deductible expense, companies in high tax brackets may incur lower after-tax financing costs with a debt. In addition to influencing the borrowing interest rate and tax bracket, a loan-repayment method can also affect financing costs.

When the debt-financing option is used, we need to separate the interest payments from the repayment of the loan for our analysis. The interest-payment schedule depends on the repayment schedule established at the time of borrowing. The two common debt-financing methods are as follows:

1. **Bond Financing**
    This type of debt financing does not involve partial payment of principal; only interest is paid each year (or semi-annually). The principal is paid in a lump sum when the bond matures. (See Section 3.7 for bond terminologies and valuation.) Bond financing is similar to equity financing in that flotation costs are involved when issuing bonds.
2. **Term Loans**
    Term loans involve an equal repayment arrangement, where the sum of the interest payments and the principal payments is uniform; interest payments decrease while principal payments increase over the life of the loan. Term loans are usually negotiated directly between the borrowing company and a financial institution—generally a commercial bank, an insurance company, or a pension fund.

Example 16.2 illustrates how these different methods can affect the cost of issuing bond or term loans.

Example 16.2 Debt Financing

Refer to Example 16.1. Suppose SSI has instead decided to raise the $10 million by debt financing. SSI could issue a mortgage bond or secure a term loan. Conditions for each option are as follows:

- Bond financing: The flotation cost is 1.8% of the $10 million issue. The company's investment bankers have indicated that a 5-year bond issue with a face value of $1,000 can be sold at $985 per share. The bond would require annual interest payments of 12%.

- Term loan: A $10 million bank loan can be secured at an annual interest rate of 11% for 5 years; it would require five equal annual installments.

  (a) How many $1,000 par value bonds would SSI have to sell to raise the $10 million?
  (b) What are the annual payments (interest and principal) for the bond?
  (c) What are the annual payments (interest and principal) for the term loan?

Solution

(a) To net $10 million, SSI would have to sell

$$\$10,000,000/(1 - 0.018) = \$10,183,300$$

worth of bonds and pay $183,300 in flotation costs. Since the $1,000 bond will be sold at a 1.5% discount, the total number of bonds to be sold would be

$$\$10,183,300/\$985 = 10,338.38.$$

(b) For the bond financing, the annual interest is equal to

$$\$10,338,380\,(0.12) = \$1,240,606.$$

Only the interest is paid each period, and thus the principal amount owed remains unchanged.

(c) For the term loan, the annual payments are

$$\$10,000,000(A/P, 11\%, 5) = \$2,705,703.$$

The principal and interest components of each annual payment are summarized in Table 16.1.

Table 16.1
Two Common Methods of Debt Financing (Example 16.2)

	0	1	2	3	4	5
1. Bond financing: No principal repayments until end of life						
Beginning balance	$10,338,380	$10,338,380	$10,338,380	$10,338,380	$10,338,380	$10,338,380
Interest owed		1,240,606	1,240,606	1,240,606	1,240,606	1,240,606
Repayment						
Interest payment		1,240,606	1,240,606	1,240,606	1,240,606	1,240,606
Principal payment						(10,338,380)
Ending balance	$10,338,380	$10,338,380	$10,338,380	$10,338,380	$10,338,380	0
2. Term loan: Equal annual repayment [$10,000,000 (A/P, 11%, 5) = $2,705,703]						
Beginning balance	$10,000,000	$10,000,000	$ 8,394,297	$ 6,611,967	$ 4,633,580	$ 2,437,571
Interest owed		1,100,000	923,373	727,316	509,694	268,133
Repayment						
Interest payment		(1,100,000)	(923,373)	(727,316)	(509,694)	(268,133)
Principal payment		(1,605,703)	(1,782,330)	(1,978,387)	(2,196,009)	(2,437,570)
Ending balance	$10,000,000	$ 8,394,297	$ 6,611,967	$ 4,633,580	$ 2,437,571	0

16.1.3 Capital Structure

The ratio of total debt to total capital, generally called the **debt ratio,** or **capital structure,** represents the percentage of the total capital provided by borrowed funds. For example, a debt ratio of 0.4 indicates that 40% of the capital is borrowed, and the remaining funds are provided from the company's equity (retained earnings or stock offerings). This type of financing is called **mixed financing.**

Borrowing affects a firm's capital structure, and firms must determine the effects of a change in the debt ratio on their market value before making an ultimate financing decision. Even if debt financing is attractive, you should understand that companies do not simply borrow funds to finance projects. A firm usually establishes a **target capital structure,** or **target debt ratio,** after considering the effects of various financing methods. This target may change over time as business conditions vary, but a firm's management always strives to achieve this target whenever individual financing decisions are considered. If the actual debt ratio is below the target level, any new capital will probably be raised by issuing debt. On the other hand, if the debt ratio is currently above the target, expansion capital would be raised by issuing stock.

How does a typical firm set the target capital structure? This is a rather difficult question to answer, but we can list several factors that affect the capital structure policy. First, capital structure policy involves a trade-off between risk and return. As you take on more debt for business expansion, the inherent business risk[3] also increases, but investors view business expansion as a healthy indicator for a corporation with higher expected earnings. When investors perceive higher business risk, the firm's

---

[3] Unlike equity financing, where dividends are optional, debt interest and principal (face value) must be repaid on time. Also uncertainty is involved in making projections of future operating income as well as expenses. In bad times, debt can be devastating, but in good times, the tax deductibility of interest payments increases profits to owners.

stock price tends to be depressed. On the other hand, when investors perceive higher expected earnings, the firm's stock price tends to increase. The optimal capital structure is thus the one that strikes a balance between business risk and expected future earnings. The greater the firm's business risk, the lower its optimal debt ratio.

Second, a major reason for using debt is that interest is a deductible expense for business operations, which lowers the effective cost of borrowing. On the other hand, dividends paid to common stockholders are not deductible. If a company uses debt, it must pay interest on this debt, whereas if it uses equity, it pays dividends to its equity investors (shareholders). A company needs $1 in before-tax (B/T) income to pay $1 of interest, but if the company is in the 34% tax bracket, it needs $1/(1 − 0.34) = $1.52 of B/T income to pay $1 of dividend.

Third, financial flexibility, the ability to raise capital on reasonable terms from the financial market, is an important consideration. Firms need a steady supply of capital for stable operations. When money is tight in the economy, investors prefer to advance funds to companies with a healthy capital structure (lower debt ratio). These three elements (business risk, taxes, and financial flexibility) are major factors that determine the firm's optimal capital structure. Example 16.3 illustrates how a typical firm finances a large-scale engineering project by maintaining the predetermined capital structure.

Example 16.3 Project Financing Based on an Optimal Capital Structure

Reconsider SSI's $10 million venture project in Example 16.1. Suppose that SSI's optimal capital structure calls for a debt ratio of 0.5. After reviewing SSI's capital structure, the investment banker convinced management that it would be better off, in view of current market conditions, to limit the stock issue to $5 million and to raise the other $5 million as debt by issuing bonds. Because the amount of capital to be raised in each category is reduced by half, the flotation cost would also change. The flotation cost for common stock would be 8.1%, whereas the flotation cost for bonds would be 3.2%. As in Example 16.2, the 5-year, 12% bond will have a par value of $1,000 and will be sold for $985.

Assuming that the $10 million capital would be raised from the financial market, the engineering department has detailed the following financial information.

- The new venture will have a 5-year project life.
- The $10 million capital will be used to purchase land for $1 million, a building for $3 million, and equipment for $6 million. The plant site and building are already available, and production can begin during the first year. The building falls into a 39-year MACRS property class and the equipment into a 7-year MACRS class. At the end of year 5, the salvage value for each asset is as follows: The land $1.5 million, the building $2 million, and the equipment $3 million.
- For common stockholders, an annual cash dividend in the amount of $2 per share is planned over the project life. This steady cash dividend payment is deemed necessary to maintain the market value of the stock.

- The unit production cost is $50.31 (material, $22.70; labor and overhead (excluding depreciation), $10.57; and tooling, $17.04).
- The unit price is $250, and SSI expects an annual demand of 20,000 units.
- The operating and maintenance cost, including advertising expenses, would be $600,000 per year.
- An investment of $500,000 in working capital is required at the beginning of the project; the amount will be fully recovered when the project terminates.
- The firm's marginal tax rate is 40%, and this rate will remain constant throughout the project period.

(a) Determine the after-tax cash flows for this investment with external financing.

(b) Is this project justified at an interest rate of 20%?

Discussion: As the amount of financing and flotation costs change, we need to recalculate the number of shares (or bonds) to be sold in each category. For a $5 million common stock issue, the flotation cost increases to 8.1%.[4] The number of shares to be sold to net $5 million is 5,000,000/(0.919)(28) = 194,311 shares (or $5,440,708). For a $5 million bond issue, the flotation cost is 3.2%. Therefore, to net $5 million, SSI has to sell 5,000,000/(0.968)(985) = 5,243.95 units of $1,000 par value. This implies that SSI is effectively borrowing $5,243,948, upon which figure the annual bond interest will be calculated. The annual bond interest payment is $5,243,948(0.12) = $629,274.

Solution

(a) After-tax cash flows

Table 16.2 summarizes the after-tax cash flows for the new venture. The following calculations and assumptions were used in developing the cash flow table.

- Revenue: $250 \times 20,000 = $5,000,000 per year
- Costs of goods: $50.31 \times 20,000 = $1,006,200 per year
- Bond interest: $5,243,948 \times 0.12 = $629,274 per year
- Depreciation: Assuming that the building is placed in service in January, the first year's depreciation percentage is 2.4573%. Therefore, the allowed depreciation amount is $3,000,000 \times 0.024573 = $73,718. The percentages for the remaining years would be 2.5641% per year, or $76,923. Equipment is depreciated according to a 7-year MACRS.
- Gains tax:

---

[4] Flotation costs are higher for small issues than for large ones due to the existence of fixed costs—certain costs must be incurred regardless of the size of the issue, so the percentage of flotation costs increases as the size of issue gets smaller.

Property	Salvage Value	Book Value	Gains (Losses)	Gain Tax
Land	$1,500,000	$1,000,000	$500,000	$200,000
Building	2,000,000	2,621,795	(621,795)	(248,718)
Equipment	2,500,000	1,606,500	893,500	357,400
				$308,682

- Cash dividend: 194,311 shares × $2 = $388,622
- Common stock: When the project terminates and the bonds are retired, the debt ratio is no longer 0.5. If SSI wants to maintain the constant capital structure (0.5), SSI would have to repurchase the common stock in the amount of

	0	1	2	3	4	5
Income Statement						
Revenue		$ 5,000,000	$ 5,000,000	$ 5,000,000	$ 5,000,000	$ 5,000,000
Expenses						
Cost of goods		1,006,200	1,006,200	1,006,200	1,006,200	1,006,200
O&M		600,000	600,000	600,000	600,000	600,000
Bond interest		629,274	629,274	629,274	629,274	629,274
Depreciation						
Building		73,718	76,923	76,923	76,923	73,718
Equipment		857,400	1,469,400	1,049,400	749,400	267,900
Taxable income		1,833,408	1,218,203	1,638,203	1,938,203	2,422,908
Income taxes		733,363	487,281	655,281	775,281	969,163
Net income		$ 1,100,045	$ 730,922	$ 982,922	$ 1,162,922	$ 1,453,745
Cash Flow Statement						
Operating activities						
Net income		$ 1,100,045	$ 730,922	$ 982,922	$ 1,162,922	$ 1,453,745
Noncash expense		931,118	1,546,323	1,126,323	826,323	341,618
Investment activities						
Land	(1,000,000)					1,500,000
Building	(3,000,000)					2,000,000
Equipment	(6,000,000)					2,500,000
Working capital	(500,000)					500,000
Gains tax						(308,682)
Financing activities						
Common stock	5,000,000					(5,440,708)
Bond	5,000,000					(5,243,948)
Cash dividend		(388,622)	(388,622)	(388,622)	(388,622)	(388,622)
Net cash flow	$ (500,000)	$ 1,642,541	$ 1,888,623	$ 1,720,623	$ 1,600,623	$ (3,086,597)

Table 16.2
Effects of Project Financing on After-Tax Cash Flows (Example 16.3)

$5,440,708 at the prevailing market price. In developing Table 16.2, we assumed this repurchase of common stock had taken place at the end of project years. In practice, a firm may or may not repurchase the common stock. As an alternative means of maintaining the desired capital structure, the firm may use this extra debt capacity released to borrow for other projects.

- Bond: When the bonds mature at the end year 5, the total face value in the amount of $5,243,948 must be paid to the bondholders.

(b) Measure of project worth

The NPW for this project is then

$$
\begin{aligned}
PW(20\%) \quad &= \quad -\$500,000 + \$1,642,541(P/F, 20\%, 1) + \dots \\
&\quad -\$3,086,597(P/F, 20\%, 5) \\
&= \quad \$2,707,530.
\end{aligned}
$$

The investment is nonsimple, but it is a pure investment. Even though the project requires a significant amount of cash expenditure at the end of project life, it still appears to be a very profitable one.

In Example 16.3, we neither discussed the cost of capital for this project financing, nor did we explain the relationship between the cost of capital and the MARR. In the remaining sections, these issues will be discussed. As we will see later in Section 16.3, we can completely ignore the detailed cash flows related to project financing if we adjust our discount rate according to the capital structure, namely, using the weighted cost of capital.

16.2 Cost of Capital

In most of the capital budgeting examples in the earlier chapters, we assumed that the firms under consideration were financed entirely with equity funds. In those cases, the cost of capital may have represented the firm's required return on equity. However, most firms finance a substantial portion of their capital budget with long-term debt (bonds), and many also use preferred stock as a source of capital. In these cases, a firm's cost of capital must reflect the average cost of the various sources of long-term funds that the firm uses, not only the cost of equity. In this section, we will discuss the ways in which the cost of each individual type of financing (retained earnings, common stock, preferred stock, and debt) can be estimated,[5] given a firm's target capital structure.

16.2.1   Cost of Equity

Whereas debt and preferred stocks are contractual obligations that have easily determined costs, it is not easy to measure the cost of equity. In principle, the cost of equity capital involves an **opportunity cost.** In fact, the firm's after-tax cash flows belong to the stockholders. Management may either pay out these earnings in the form of divi-

---

[5] Estimating or calculating the cost of capital in any precise fashion is a very difficult task.

dends, or retain earnings and reinvest them in the business. If management decides to retain earnings, an opportunity cost is involved—stockholders could have received the earnings as dividends and invested this money in other financial assets. Therefore, the firm should earn on its retained earnings at least as much as the stockholders themselves could earn in alternative, but comparable, investments.

What rate of return can stockholders expect to earn on retained earnings? This question is difficult to answer, but the value sought is often regarded as the rate of return stockholders require on a firm's common stock. If a firm cannot invest retained earnings so as to earn at least the rate of return on equity, it should pay these funds to these stockholders and let them invest directly in other assets that do provide this return.

When investors are contemplating buying a firm's stock, they have two things in mind: (1) Cash dividends and (2) gains (share appreciation) at the time of sale. From a conceptual standpoint, investors determine market values of stocks by discounting expected future dividends at a rate that takes into account any future growth. Since investors seek growth companies, a desired growth factor for future dividends is usually included in the calculation.

To illustrate, let's take a simple numerical example. Suppose investors in the common stock of ABC Corporation expect to receive a dividend of \$5 by the end of the first year. The future annual dividends will grow at an annual rate of 10%. Investors will hold the stock for 2 more years and will expect the market price of the stock to rise to \$120 by the end of the third year. Given these hypothetical expectations, ABC expects that investors would be willing to pay \$100 for this stock in today's market. What is the required rate of return on ABC's common stock (k_r)? We may answer this question by solving the following equation for k_r:

$$\$100 = \frac{\$5}{(1 + k_r)} + \frac{\$5(1 + 0.1)}{(1 + k_r)^2} + \frac{\$5(1 + 0.1)^2 + \$120}{(1 + k_r)^3}.$$

In this case, $k_r = 11.44\%$. This implies that, if ABC finances a project by retaining its earnings or by issuing additional common stock at the going market price of \$100 per share, it must realize at least 11.44% on new investment just to provide the minimum rate of return required by the investors. Therefore, 11.44% is the specific cost of equity that should be used when calculating the weighted average cost of capital. As flotation costs are involved in issuing new stock, the cost of equity will increase. If investors view ABC's stock as risky and, therefore, are willing to buy the stock at a lower price than \$100 (but with the same expectations), the cost of equity will also increase. Now we can generalize the result as follows:

Cost of Retained Earnings

Let's assume the same hypothetical situation for ABC. Recall that ABC's retained earnings belong to holders of its common stock. If ABC's current stock is traded for a market price of P_0, with a first-year dividend[6] of D_1, but growing at the annual rate of

---

[6] When we check the stock listings in the newspaper, we do not find the expected first-year dividend, D_1. Instead, we find the dividend paid out most recently, D_0. So if we expect growth at a rate g, the dividend at the end of 1 year from now, D_1, may be estimated as

$$D_1 = D_0(1 + g).$$

g thereafter, the specific cost of retained earnings for an infinite period of holding (stocks will change hands over the years, but it does not matter who holds the stock) can be calculated as

$$P_0 = \frac{D_1}{(1 + k_r)} + \frac{D_1(1 + g)}{(1 + k_r)^2} + \frac{D_1(1 + g)^2}{(1 + k_r)^3} + \dots$$

$$= \frac{D_1}{1 + k_r} \sum_{n=0}^{\infty} \left[\frac{(1 + g)}{(1 + k_r)} \right]^n,$$

$$= \frac{D_1}{1 + k_r} \left[\frac{1}{1 - \dfrac{1 + g}{1 + k_r}} \right], \text{ where } g < k_r.$$

Solving for k_r, we obtain

$$k_r = \frac{D_1}{P_0} + g. \tag{16.1}$$

If we use k_r as the discount rate for evaluating the new project, it will have a positive NPW only if the project's IRR exceeds k_r. Therefore, any project with a positive NPW, calculated at k_r, induces a rise in the market price of the stock. Hence, by definition, k_r is the rate of return required by shareholders and should be used as the cost of the equity component when calculating the weighted average cost of capital.

Issuing New Common Stock

As flotation costs are involved in issuing new stock, we can modify the cost of retained earnings (k_r) by

$$k_r = \frac{D_1}{P_0(1 - f_c)} + g, \tag{16.2}$$

where k_e = the cost of common equity, and f_c = the flotation cost as a percentage of stock price.

Either calculation is deceptively simple because, in fact, several ways are available to determine the cost of equity. In reality, the market price fluctuates constantly, as do a firm's future earnings. Thus, future dividends may not grow at a constant rate as the model indicates. For a stable corporation with moderate growth, however, the cost of equity as calculated by evaluating either Eq. (16.1) or Eq. (16.2) serves as a good approximation.

Cost of Preferred Stock

A preferred stock is a hybrid security in the sense that it has some of the properties of bonds and other properties that are similar to common stock. Like bondholders, holders of preferred stock receive a fixed annual dividend. In fact, many firms view the payment of the preferred dividend as an obligation just like interest payments to bondholders. It is therefore relatively easy to determine the cost of preferred stock. For the purposes of calculating the weighted average cost of capital, the specific cost of a preferred stock will be defined as

$$k_r = \frac{D^*}{P^*(1 - f_c)},\qquad(16.3)$$

where D^* = the fixed annual dividend, P^* = the issuing price, and f_c as defined above.

Cost of Equity

Once we have determined the specific cost of each equity component, we can determine the weighted average cost of equity (i_e) for a new project:

$$i_e = \left(\frac{c_r}{c_e}\right)k_r + \left(\frac{c_c}{c_e}\right)k_e + \left(\frac{c_p}{c_e}\right)k_p\qquad(16.4)$$

where c_r = amount of equity financed from retained earnings, c_c = amount of equity financed from issuing new stock, c_p = amount of equity financed from issuing preferred stock, and $c_r + c_c + c_p = c_e$. Example 16.4 illustrates how we may determine the cost of equity.

Example 16.4 Determining the Cost of Equity

Alpha Corporation needs to raise $10 million for plant modernization. Alpha's target capital structure calls for a debt ratio of 0.4, indicating that $6 million has to be financed from equity.

- Alpha is planning to raise $6 million from the following equity sources:

Source	Amount	Fraction of Total Equity
Retained earnings	$1 million	0.167
New common stock	4 million	0.666
Preferred stock	1 million	0.167

- Alpha's current common stock price is $40—the market price that reflects the firm's future plant modernization. Alpha is planning to pay an annual cash dividend of $5 at the end of the first year, and the annual cash dividend will grow at an annual rate of 8% thereafter.
- Additional common stock can be sold at the same price of $40, but there will be 12.4% flotation costs.
- Alpha can issue $100 par preferred stock with a 9% dividend. (This means that Alpha will calculate the dividend based on the par value, which is $9 per share.) The stock can be sold on the market for $95, and Alpha must pay flotation costs of 6% of the market price.

Determine the cost of equity to finance the plant modernization.

Solution

We will itemize the cost of each equity component.

- Cost of retained earnings: With $D_1 = \$5$, $g = 8\%$, and $P_0 = \$40$,

$$k_r = \frac{5}{40} + 0.08 = 20.5\%.$$

- Cost of new common stock: With $D_1 = \$5$, $g = 8\%$, $f_c = 12.4\%$,

$$k_e = \frac{5}{40(1 - 0.124)} + 0.08 = 22.27\%.$$

- Cost of preferred stock: With $D^* = \$9$, $P^* = \$95$ $f_c = 0.06$,

$$k_p = \frac{9}{95(1 - 0.06)} = 10.08\%.$$

- Cost of equity: With $\dfrac{c_r}{c_e} = 0.167$, $\dfrac{c_c}{c_e} = 0.666$, and $\dfrac{c_p}{c_e} = 0.167$,

$$i_e = (0.167)(0.205) + (0.666)(0.2227) + (0.167)(0.1008)$$

$$= 19.96\%.$$

16.2.2 Cost of Debt

Now let us consider the calculation of the specific cost that is to be assigned to the debt component of the weighted average cost of capital. The calculation is relatively straightforward and simple. As we discussed in Section 16.1.2, the two types of debt financing are term loans and bonds. Because the interest payments on both are tax deductible, the effective cost of debt will be reduced.

To determine the after-tax cost of debt (i_d), we can evaluate the following expression:

$$i_d = \left(\frac{c_s}{c_d}\right)k_s(1 - t_m) + \left(\frac{c_b}{c_d}\right)k_b(1 - t_m), \tag{16.5}$$

where c_s = the amount of the term loan, k_s = the before-tax interest rate on the term loan, t_m = the firm's marginal tax rate, and k_b = the before-tax interest rate on the bond, c_b = the amount of bond financing, and $c_s + c_b = c_d$.

As for bonds, a new issue of long-term bonds incurs flotation costs. These costs reduce the proceeds to the firm, thereby raising the specific cost of the capital raised. For example, when a firm issues a $1,000 par bond, but nets only $940, the flotation cost will be 6%. Therefore, the effective after-tax cost of the bond component will be higher than the nominal interest rate specified on the bond. We will examine this problem with a numerical example.

Example 16.5 Determining the Cost of Debt

Refer to Example 16.4, and suppose that Alpha has decided to finance the remaining $4 million by securing a term loan and issuing 20-year $1,000 par bonds for the following condition.

Source	Amount	Fraction	Interest Rate	Flotation Cost
Term loan	$1 million	0.333	12% per year	
Bonds	3 million	0.667	10% per year	6%

If the bond can be sold to net $940 (after deducting the 6% flotation cost), determine the cost of debt to raise $4 million for the plant modernization. Alpha's marginal tax rate is 38%, and it is expected to remain constant in the future.

Solution

First, we need to find the effective after-tax cost of issuing the bond with a flotation cost of 6%. The before-tax specific cost is found by solving the following equivalence formula

$$\$940 = \frac{\$100}{(1 + k_b)} + \frac{\$100}{(1 + k_b)^2} + \ldots + \frac{\$100 + \$1,000}{(1 + k_b)^{20}}.$$

$$= \$100\,(P/A, k_b, 20) + \$1,000\,(P/F, k_b, 20)$$

Solving for k_b, we obtain $k_b = 10.74\%$. Note that the cost of the bond component increases from 10% to 10.74% after considering the 6% flotation cost.

The after-tax cost of debt is the interest rate on debt, multiplied by $(1 - t_m)$. In effect, the government pays part of the cost of debt because interest is tax deductible. Now we are ready to compute the after-tax cost of debt as follows:

$$i_d = (0.333)(0.12)(1 - 0.38) + (0.667)(0.1074)(1 - 0.38)$$

$$= 6.92\%.$$

16.2.3   Calculating the Cost of Capital

With the specific cost of each financing component determined, now we are ready to calculate the tax-adjusted weighted average cost of capital based on total capital. Then, we will define the marginal cost of capital that should be used in project evaluation.

Weighted Average Cost of Capital

Assuming that a firm raises capital based on the target capital structure and that the target capital structure remains unchanged in the future, we can determine a **tax-adjusted weighted-average cost of capital** (or, simply stated, the **cost of capital**). This cost of capital represents a composite index reflecting the cost of raising funds from different sources. The cost of capital is defined as

$$k = \frac{i_d c_d}{V} + \frac{i_e c_e}{V}.$$ (16.6)

where c_d =   Total debt capital (such as bonds) in dollars,

c_e =   Total equity capital in dollars,

V = $c_d + c_e$,

i_e =   Average equity interest rate per period considering all equity sources,

i_d =   After-tax average borrowing interest rate per period considering all debt sources, and

k =   Tax-adjusted weighted-average cost of capital.

Note that the cost of equity is already expressed in terms of after-tax cost, because any return to holders of either common stock or preferred stock is made after payment of income taxes.

Marginal Cost of Capital

Now we know how to calculate the cost of capital. Could a typical firm raise unlimited new capital at the same cost? The answer is no. As a practical matter, as a firm tries to attract more new dollars, the cost of raising each additional dollar will at some point rise. As this occurs, the weighted average cost of raising each additional new dollar also rises. Thus, the **marginal cost of capital** is defined as the cost of obtaining another dollar of new capital, and the marginal cost rises as more and more capital is raised during a given period. In evaluating an investment project, we are using the concept of marginal cost of capital. The formula to find the marginal cost of capital is exactly the same as Eq. (16.6). However, the costs of debt and equity in Eq. (16.6) are the interest rates on new debt and equity, not outstanding (or combined) debt or equity; in other words, we are interested in the marginal cost of capital. Our primary concern with the cost of capital is to use it in evaluating a new investment project. The rate at which the firm has borrowed in the past is less important for this purpose. Example 16.6 works through the computations for finding the cost of capital (k).

Example 16.6 Calculating the Marginal Cost of Capital

Reconsider Examples 16.4 and 16.5. The marginal income tax rate (t_m) for Alpha is expected to remain at 38% in the future. Assuming that Alpha's capital structure (debt ratio) also remains unchanged in the future, determine the cost of capital (k) of raising $10 million in addition to existing capital.

Solution

With $c_d = \$4$ million, $c_e = \$6$ million, $V = \$10$ million, $i_d = 6.92\%$, $i_e = 19.96\%$, and Eq. (16.6), we calculate

$$k = \frac{(0.0692)(4)}{10} + \frac{(0.1996)(6)}{10}$$

$$= 14.74\%.$$

This 14.74% would be the marginal cost of capital that a company with this financial structure would expect to pay to raise $10 million.

16.3 Choice of Minimum Attractive Rate of Return

Thus far, we have said little about what interest rate, or minimum attractive rate of return (MARR), is suitable for use in a particular investment situation. Choosing the MARR is a difficult problem; no single rate is always appropriate. In this section, we will discuss briefly how to select a MARR for project evaluation. Then, we will examine the relationship between capital budgeting and the cost of capital.

16.3.1   Choice of MARR when Project Financing is Known

In Chapter 12, we focused on calculating after-tax cash flows, including situations involving debt financing. When cash flow computations reflect interest, taxes, and debt repayment, what is left is called **net equity flow.** If the goal of a firm is to maximize the wealth of its stockholders, why not focus only on the after-tax cash flow to equity, instead of on the flow to all suppliers of capital? Focusing on only the equity flows will permit us to use the cost of equity as the appropriate discount rate. In fact, we have implicitly assumed that all after-tax cash flow problems, where financing flows are explicitly stated in earlier chapters represent net equity flows, so the MARR used represents the **cost of equity** (i_e). Example 16.7 illustrates project evaluation by the net equity flow method.

Example 16.7 Project Evaluation by Net Equity Flow

Suppose the Alpha Corporation, which has the capital structure described in Example 16.6, wishes to install a new set of machine tools, which is expected to increase revenues over the next 5 years. The tools require an investment of $150,000, to be financed with 60% equity and 40% debt. The equity interest rate (i_e), which combines both sources of common and preferred stocks, is 19.96%. Alpha will use a 12% short-term loan to finance the debt portion of the capital

($60,000), with the loan to be repaid in equal annual installments over 5 years. Depreciation is a MACRS over 3-year property class life, and zero salvage value is expected. Additional revenues and operating costs are expected to be

n	Revenues ($)	Operating cost
1	$68,000	$20,500
2	73,000	20,000
3	79,000	20,500
4	84,000	20,000
5	90,000	20,500

The marginal tax rate (combined federal, state, and city rate) is 38%. Evaluate this venture by using net equity flows at $i_e = 19.96\%$.

Table 16.3
After-Tax Cash Flow Analysis when Project Financing Is Known: Net Equity Cash Flow Method (Example 16.7)

End of Year	0	1	2	3	4	5
Income Statement						
Revenue		$ 68,000	$ 73,000	$ 79,000	$ 84,000	$ 90,000
Expenses						
Operating cost		20,500	20,000	20,500	20,000	20,500
Interest payment		7,200	6,067	4,797	3,376	1,783
Depreciation		50,000	66,667	22,222	11,111	0
Taxable income		(9,700)	(19,734)	31,481	49,513	67,717
Income taxes (38%)		(3,686)	(7,499)	11,963	18,815	25,732
Net income		$ (6,014)	$ (12,235)	$ 19,518	$ 30,698	$ 41,985
Cash Flow Statement						
Operating activities						
Net income		$ (6,014)	$ (12,235)	$ 19,518	$ 30,698	$ 41,985
Depreciation		50,000	66,667	22,222	11,111	0
Investment activities						
Investment	$ (150,000)					
Salvage value						0
Gains tax						0
Financing activities						
Borrowed funds	60,000					
Principal repayment		(9,445)	(10,578)	(11,847)	(13,269)	(14,861)
Net equity flow	$ (90,000)	$ 34,541	$ 43,854	$ 29,893	$ 28,540	$ 27,124

Solution

The calculations are shown in Table 16.3. The NPW and IRR calculations are as follows:

$$PW(19.96\%) = -\$90,000 + \$34,541(P/F, 19.96\%, 1)$$
$$+ \$43,854(P/F, 19.96\%, 2) + \$29,893(P/F, 19.96\%, 3)$$
$$+ \$28,540(P/F, 19.96\%, 4) + \$27,124(P/F, 19.96\%, 5)$$
$$= \$11,285$$
$$IRR = 25.91\% > 19.96\%.$$

The internal rate of return for this cash flow is 25.91%, which exceeds $i_e = 19.96\%$. Thus, the project would be profitable.

Comment: In this problem we assumed that the Alpha Corporation would be able to raise the additional equity funds at the same rate of 19.96%, so that this 19.96% can be viewed as the marginal cost of capital.

16.3.2   Choice of MARR when Project Financing Is Unknown

You might well ask why, if we use the cost of equity (i_e) exclusively, what use is the k? The answer to this question is that by using the value of k, we may evaluate investments without explicitly treating the debt flows (both interest and principal). In this case, we make a tax adjustment to the discount rate by employing the effective after-tax cost of debt. This approach recognizes that the net interest cost is effectively transferred from the tax collector to the creditor in the sense that there is a dollar-for-dollar reduction in taxes up to this amount of interest payments. Therefore, debt financing is treated implicitly. This method would be appropriate when debt financing is not identified with individual investments, but rather enables the company to engage in a set of investments. (Except where financing flows are explicitly stated, all previous examples in this book have implicitly assumed the more realistic and appropriate situation where debt financing is not identified with individual investment. Therefore, the MARRs represent the weighted cost of capital [k].) Example 16.8 provides an illustration of this concept.

Example 16.8 Project Evaluation by Marginal Cost of Capital

In Example 16.7, suppose that Alpha Corporation has not decided how the $150,000 will be financed. However, Alpha believes that the project should be financed according to its target capital structure, debt ratio of 40%. Evaluate Example 16.9 using k.

Solution

By not accounting for the cash flows related to debt financing, we calculate the after-tax cash flows as shown in Table 16.4. Notice that, when we use this procedure, interest and the resulting tax shield are ignored when deriving the net incremental after-tax cash flow. In other words, no cash flow is related to financing activity. Thus, taxable income is overstated; income taxes are also overstated. To compensate for these overstatements, the discount rate is reduced accordingly. The implicit assumption is that the tax overpayment is exactly equal to the reduction in interest implied by i_d.

The time 0 flow is simply the total investment—$150,000 in this example. Recall that Alpha's k was calculated to be 14.74%. The internal rate of return for the after-tax flow in the last line of Table 16.4 is calculated as follows:

$$PW(i) = -\$150,000 + \$48,450(P/F, i, 1)$$
$$+ \$58,193(P/F, i, 2) + \$44,714(P/F, i, 3)$$
$$+ \$43,902(P/F, i, 4) + \$43,090(P/F, i, 5)$$
$$= 0$$
$$IRR = 18.47\% > 14.74\%.$$

Since the IRR exceeds the value of k, the investment would be profitable. Here, we evaluated the after-tax flow by using the value of k, and we reached the same conclusion about the desirability of the investment.

Table 16.4
After-Tax Cash Flow Analysis when Project Financing Is Known: Cost of Capital Approach (Example 16.8)

End of Year	0	1	2	3	4	5
Income Statement						
Revenue		$ 68,000	$ 73,000	$ 79,000	$ 84,000	$ 90,000
Expenses						
Operating cost		20,500	20,000	20,500	20,000	20,500
Depreciation		50,000	66,667	22,222	11,111	0
Taxable income		$ (2,500)	$ (13,667)	$ 36,278	$ 52,889	$ 69,500
Income taxes (38%)		(950)	(5,193)	13,786	20,098	26,410
Net income		$ (1,550)	$ (8,474)	$ 22,492	$ 32,791	$ 43,090
Cash Flow Statement						
Operating activities						
Net income		$ (1,550)	$ (8,474)	$ 22,492	$ 32,791	$ 43,090
Depreciation		50,000	66,667	22,222	11,111	0
Investment activities						
Investment	$ (150,000)					
Salvage value						0
Gains tax						0
Net cash flow	$ (150,000)	$ 48,450	$ 58,193	$ 44,714	$ 43,902	$ 43,090

Comments: The net equity flow and the cost of capital methods usually lead to the same accept/reject decision for independent projects (assuming the same amortization schedule for debt repayment, such as term loans), and usually rank projects identically for mutually exclusive alternatives. Some differences may be observed as special financing arrangements may increase (or even decrease) the attractiveness of a project by manipulating tax shields and the timing of financing inflows and payments.

In summary, in cases where the exact debt-financing and repayment schedules are known, we recommend the use of the net equity flow method. The appropriate MARR would be the cost of equity, i_e. If no specific assumption is made about the exact instruments that will be used to finance a particular project (but we do assume that the given capital structure proportions will be maintained), we may determine the after-tax cash flows without incorporating any debt cash flows. Then, we use the marginal cost of capital (k) as the appropriate MARR.

16.3.3   Choice of MARR under Capital Rationing

It is important to distinguish between the cost of capital (k), as calculated in Section 16.3.1 and the MARR (i) used in project evaluation under **capital rationing. Capital rationing** refers to situations where the funds available for capital investment are not sufficient to cover potentially acceptable projects. When investment opportunities exceed the available money supply, we must decide which opportunities are preferable. Obviously, we want to ensure that all the selected projects are more profitable than the best rejected project. The best rejected project (or the worst accepted project) is the best opportunity foregone, and its value is called the **opportunity cost.** When a limit is placed on capital, the MARR is assumed to be equal to this opportunity cost, which is usually greater than the marginal cost of capital. In other words, the value of i represents the corporation's time-value trade-offs and reflects partially the available investment opportunities. Thus, there is nothing illogical about borrowing money at k and evaluating cash flows using the different rate, i. Presumably, the money will be invested to earn a rate i, or greater. In the following example, we will provide guidelines for selecting a MARR for project evaluation under capital rationing.

A company may borrow funds to invest in profitable projects, or it may return (invest) to its **investment pool** any unused funds until they are needed for other investment activities. Here, we may view the borrowing rate as a marginal cost of capital (k), as defined in (Eq. 16.6). Suppose that all available funds can be placed in investments yielding a return equal to r, the **lending rate.** We view these funds as an investment pool. The firm may withdraw funds from this investment pool for other investment purposes, but if left in the pool, the funds will earn at the rate of r (which is thus the opportunity cost). The MARR is thus related to either the borrowing interest rate or the lending interest rate. To illustrate the relationship among the borrowing interest rate, lending interest rate, and MARR, let us define the following:

$$k = \text{Borrowing rate (or the cost of capital)}$$
$$r = \text{Lending rate (or opportunity cost)}$$
$$i = \text{MARR.}$$

Generally (but not always), we might expect k to be greater than, or equal to, r. We must pay more for the use of someone else's funds than we can receive for "renting out" our own funds (unless we are running a lending institution). Then, we will find that the appropriate MARR would be found between r and k. The concept for developing a discount rate (MARR) under capital rationing will be understood best by a numerical example.

Example 16.9 Determining an Appropriate MARR as a Function of Budget

Sand Hill Corporation has identified six investment opportunities that will last 1 year. The firm draws up a list of all potentially acceptable projects. The list shows each project's required investment, projected annual net cash flows, life, and IRR. Then it ranks the projects according to their IRR, listing the highest IRR first.

Project	Cash Flow A0	A1	IRR
1	−$10,000	$12,000	20%
2	−10,000	11,500	15
3	−10,000	11,000	10
4	−10,000	10,800	8
5	−10,000	10,700	7
6	−10,000	10,400	4

Suppose $k = 10\%$, which remains constant for the budget amount up to $60,000 and $r = 6\%$. Assuming that the firm has available (a) $40,000, (b) $60,000, and (c) $0 for investments, what is the reasonable choice for the MARR in each case?

Solution

We will consider the following steps to determine the appropriate discount rate (MARR) under a capital rationing environment:

- Step 1: Develop the firm's cost of capital schedule as a function of the capital budget. For example, the cost of capital can increase as the amount of financing increases. Also determine the firm's lending rate if any unspent money is lent out or remains invested in its investment pool.

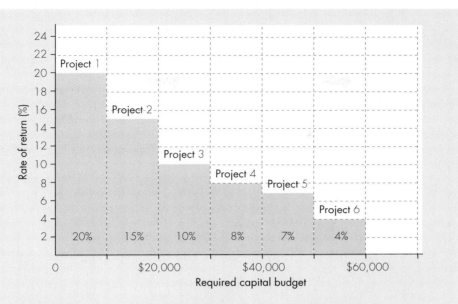

Figure 16.2   An investment opportunity schedule (Example 16.9)

- Step 2: Plot this investment opportunity schedule by showing how much money the company could invest at different rates of return as shown in Figure 16.2.

(a) If the firm has $40,000 available for investing, it should invest in projects 1, 2, 3, and 4. Clearly, it should not borrow at 10% to invest in either project 5 or 6. In these cases, the best rejected project is project 5. The worst accepted project is project 4. If you view the opportunity cost as the cost associated with accepting the worst project, the MARR could be 8%.

(b) If the firm has $60,000 available, it should invest in projects 1, 2, 3, 4, and 5. It could lend the remaining $10,000 rather than invest these funds in project 6. For this new situation, we have MARR $= r = 6\%$.

(c) If the firm has no funds available, it probably would borrow to invest in projects 1 and 2. The firm might also borrow to invest in project 3, but it would be indifferent to this alternative, unless some other consideration was involved. In this case, MARR $= k = 10\%$; therefore, we can say that $r \leq$ MARR $\leq k$ when we have a complete certainty about future investment opportunities. Figure 16.3 illustrates the concept of selecting a MARR under capital rationing.

Comments: In this example, for simplicity, we assumed that the timing of each investment is the same for all competing proposals, say period 0. If each alternative requires investments over several periods, the analysis will be significantly complicated as we have to consider both the investment amount as well as its timing in selecting the appropriate MARR. This is certainly beyond the scope of any introductory engi-

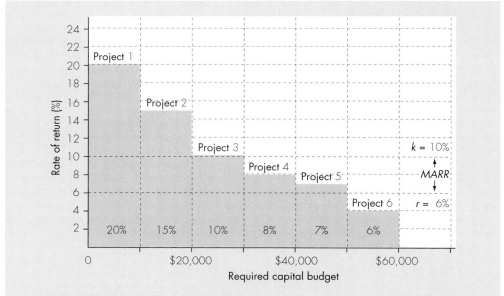

Figure 16.3   A choice of MARR under capital rationing (Example 16.9)

neering economics text, but can be found in Park C. S., Sharp-Bette G. P., *Advanced Engineering Economics*. New York: John Wiley, 1990.

Now we can generalize what we have learned. If a firm finances investments through borrowed funds, it should use MARR = k; if the firm is a lender, it should use MARR = r. A firm may be a lender in one period and a borrower in another; consequently, the appropriate rate to use may vary from period to period. In fact, whether the firm is a borrower or a lender may well depend on its investment decisions.

In practice, most firms establish a MARR for all investment projects. Note the assumption that we made in Example 16.9: **Complete certainty** about investment opportunities was assumed. Under uncertain economic environments, generally, the MARR would be much greater than k, the firm's cost of capital. For example, if $k = 10\%$, a MARR of 15% would not be considered excessive. Few firms are willing to invest in projects earning only slightly more than their cost of capital because of elements of *risk* in the project.

If the firm has a large number of current and future opportunities that will yield the desired return, we can then view the MARR as the minimum rate at which the firm is willing to invest, and we can also assume that proceeds from current investments can be reinvested to earn at the MARR. Furthermore, *if we choose the "do-nothing" alternative, all available funds are invested at the MARR.* In engineering economics, we also normally separate the risk issue from the concept of MARR. As seen in Chapter 14, we treat the effects of risk explicitly when we must. Therefore, any reference to the MARR in this book refers strictly to the risk-free interest rate.

16.4 Capital Budgeting

In this section, we will examine the process of deciding whether projects should be included in the capital budget. In particular, we will consider decision procedures that should be applied when we have to evaluate a set of multiple investment alternatives for which we have a limited capital budget.

16.4.1 Evaluation of Multiple Investment Alternatives

In Chapters 7, 8, and 9, we learned how to compare two or more mutually exclusive projects. Now, we shall extend the comparison techniques to a set of multiple decision alternatives which are not necessarily mutually exclusive. Here, we distinguish a **project** from an **investment alternative,** which is a decision option. For a single project, we have two investment alternatives: To accept or reject the project. For two independent projects, we can have four investment alternatives: (1) To accept both projects, (2) to reject both projects, (3) to accept only the first project, and (4) to accept only the second project. As we add interrelated projects, the number of investment alternatives to consider grows exponentially.

To perform a proper capital budgeting analysis, a firm must group all projects under consideration into decision alternatives. This grouping requires the firm to distinguish between projects that are independent of one another and those that are dependent on one another to formulate alternatives correctly.

Independent Projects

An **independent project** is one that may be accepted or rejected without influencing the accept-reject decision of another independent project. For example, the purchase of a milling machine, office furniture, and a forklift truck constitutes three independent projects. Only projects that are economically independent of one another can be evaluated separately. (Budget constraints may prevent us from selecting one or more of several independent projects; this external constraint does not alter the fact that the projects are independent.)

Dependent Projects

In many decision problems, several investment projects are related to one another such that the acceptance or rejection of one project influences the acceptance of others. The two such types of dependencies are as follows: Mutually exclusive projects and contingent projects. We say that two or more projects are **contingent** if the acceptance of one requires the acceptance of another. For example, the purchase of a computer printer is dependent upon the purchase of a computer, but the computer may be purchased without considering the purchase of the printer.

16.4.2 Formulation of Mutually Exclusive Alternatives

We can view the selection of investment projects as a problem of selecting a single decision alternative from a set of mutually exclusive alternatives. Note that each investment project is an investment alternative, but that a single investment alternative may entail a whole group of investment projects. The common method of handling various project relationships is to arrange the investment projects so that the selection decision involves only mutually exclusive alternatives. To obtain this set of mutually exclusive alternatives, we need to enumerate all of the feasible combinations of the projects under consideration.

Independent Projects

With a given number of independent investment projects, we can easily enumerate mutually exclusive alternatives. For example, in considering two projects, A and B, we have four decision alternatives, including a "do nothing" alternative.

Alternative	Description	X_A	X_B
1	Reject A, Reject B	0	0
2	Accept A, Reject B	1	0
3	Reject A, Accept B	0	1
4	Accept A, Accept B	1	1

In our notation, X_j is a decision variable associated with investment project j. If $X_j = 1$, project j is accepted; if $X_j = 0$, project j is rejected. Since the acceptance of one of these alternatives will exclude any other, the alternatives are considered to be mutually exclusive.

Mutually Exclusive Projects

Suppose we are considering two independent sets of projects (A and B). Within each independent set are two mutually exclusive projects (A1, A2), and (B1, B2). The selection of either A1 or A2, however, is also independent of the selection of any project from the set (B1, B2). For this set of investment projects, the mutually exclusive alternatives are

Alternative	(X_{A1}, X_{A2})	(X_{B1}, X_{B2})
1	(0, 0)	(0, 0)
2	(1, 0)	(0, 0)
3	(0, 1)	(0, 0)
4	(0, 0)	(1, 0)
5	(0, 0)	(0, 1)
6	(1, 0)	(1, 0)
7	(0, 1)	(1, 0)
8	(1, 0)	(0, 1)
9	(0, 1)	(0, 1)

Note that, with two independent sets of mutually exclusive projects, we can have nine different decision alternatives.

Contingent Projects

Suppose C is contingent on the acceptance of both A and B, and acceptance of B is contingent on acceptance of A. The possible number of decision alternatives can be formulated as follows:

Alternative	X_A	X_B	X_C
1	0	0	0
2	1	0	0
3	1	1	0
4	1	1	1

Thus, we can easily formulate a set of mutually exclusive investment alternatives with a limited number of projects that are independent, mutually exclusive, or contingent merely by arranging the projects in a logical sequence.

One difficulty with the enumeration approach is that, as the number of projects increases, the number of mutually exclusive alternatives increases exponentially. For example, for ten independent projects, the number of mutually exclusive alternatives is 2^{10}, or 1,024. For 20 independent projects, 2^{20}, or 1,048,576, alternatives exist. As the number of decision alternatives increases, we may have to resort to mathematical programming to find the solution. Fortunately, in real-world business, the number of engineering projects to consider at any one time is usually manageable, so the enumeration approach is a practical one.

16.4.3 Capital Budgeting Decisions with Limited Budgets

Recall that capital rationing refers to situations where the funds available for capital investment are not sufficient to cover all the projects. In this situation, we enumerate all investment alternatives as before, but eliminate from consideration any mutually exclusive alternatives exceeding the budget. The most efficient way to proceed in a capital rationing situation is to select the group of projects that maximizes the total NPW of future cash flows over required investment outlays. Example 16.10 illustrates the concept of an optimal capital budget under a rationing situation.

Example 16.10 Four Energy Saving Projects under Budget Constraints

The facilities department at an electronic instrument firm had under consideration four energy-efficiency projects.

Project 1 (electrical): This project requires replacing the existing standard efficiency motors in the air conditioners and exhaust blowers of a particular building with high-efficiency motors.

Project 2 (building envelope): This project involves coating the inside surface of existing fenestration in a building with low-emissivity solar film.

Project 3 (air conditioning): This project requires the installation of heat exchangers between a building's existing ventilation and relief air ducts.

Project 4 (lighting): This project requires the installation of specular reflectors and de-lamping of a building's existing ceiling grid-lighting troffers.

These projects require capital outlays in the $50,000 to $140,000 range and have useful lives of about 8 years. The facilities department's first task was to estimate the annual savings that can be realized by these energy-efficiency projects. Currently, the company pays 7.80 cents per kilowatt-hour (kWh) for electricity and $4.85 per thousand cubic feet (MCF). Assuming that the current energy prices would continue for the next 8 years, the company has estimated the cash flows and the IRR for each project as follows.

Project	Investment	Annual O&M Cost	Annual Savings (Energy)	Annual Savings (Dollars)	IRR
1	$46,800	$1,200	151,000 kWh	$11,778	15.43%
2	104,850	1,050	513,077 kWh	40,020	33.48%
3	135,480	1,350	6,700,000 CF	32,493	15.95%
4	94,230	942	385,962 kWh	30,105	34.40%

As each project could be adopted in isolation, at least as many alternatives as projects are possible. For simplicity assume that all projects are independent as opposed to being mutually exclusive, that they are equally risky, and that their risks are all equal to those of the firm's average existing assets.

(a) Determine the optimal capital budget for the energy savings projects.

(b) With $250,000 approved for energy improvement funds during the current fiscal year, the department did not have sufficient capital on hand to undertake all four projects without any additional allocation from headquarters. Enumerate the total number of decison alternatives and select the best alternative.

Discussion: The NPW calculation cannot be shown yet, as we do not know the marginal cost of capital. Therefore, our first task is to develop the **marginal cost of capital (MCC)** schedule, a graph that shows how the cost of capital changes as more and more new capital is raised during a given year. The graph in Figure 16.4 is the company's marginal cost of capital schedule. The first $100,000 would be raised at 14%, the next $100,000 at 14.5%, the next $100,000 at 15%, and for any amount over $300,000 at 15.5%. We then plot the IRR data for each project as the **investment opportunity schedule (IOS)** shown in the graph. The IOS schedule shows how much money the firm could invest at different rates of return.

Solution

(a) Optimal capital budget if projects can be accepted in part:

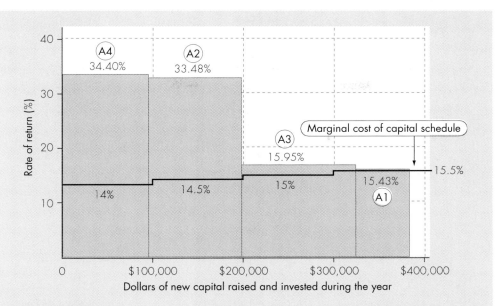

Figure 16.4 Combining the marginal cost of capital schedule and investment opportunity schedule curves to determine a firm's optimal capital budget (Example 16.10)

Consider project A4: Its IRR is 34.40%, and it can be financed with capital that costs only 14%. Consequently, it should be accepted. Projects A2, and A3 can be analyzed similarly; all are acceptable because the IRR exceeds the marginal cost of capital. Project A1, on the other hand, should be rejected because its IRR is less than the marginal cost of capital. Therefore, the firm should accept the three projects (A4, A2, and A3), which have rates of return in excess of the cost of capital that would be used to finance them if we end up with a capital budget of $334,560. This should be the amount of the *optimal capital budget.*

In Figure 16.4, even though two rate changes occur in the marginal cost of capital in funding project A3 (first change from 14.5% to 15%, second change from 15% to 15.5%), the accept/reject decision for A3 remains unchanged, as its rate of return exceeds the marginal cost of capital. What would happen if the MCC cut through project A3? For example, suppose that the marginal cost of capital for any project raised above $300,000 would cost 16% instead of 15.5%, thereby causing the MCC schedule to cut through Project A3. Should we then accept A3? If we can take A3 in part, we would take on only part of it up to 74.49%.

(b) Optimal capital budget if projects cannot be accepted in part:

If projects can be accepted in part, the budget limit does not cause any difficulty in selecting the best alternatives. For example, in Figure 16.4, we accept A4 and A2 in full and A3 in part as much as the budget allows, or 37.58% of A3.

If projects cannot be funded partially, we first need to enumerate the number of feasible investment decision alternatives within the budget limit. As shown in Table 16.5,

Table 16.5
Mutually
Exclusive
Decision Al-
ternatives
(Example
16.10)

j	Alternative	Required Budget	Combined Annual Savings
1	0	0	0
2	A1	$ (46,800)	$ 10,578
3	A2	(104,850)	38,970
4	A3	(135,480)	31,143
5	A4	(94,230)	35,691
6	A4,A1	(141,030)	46,269
7	A2,A1	(151,650)	49,548
8	A3,A1	(182,280)	41,721
9	A4,A2	(199,080)	74,661
10	A4,A3	(229,710)	66,834
11	A2,A3	(240,330)	70,113
(12)	A4,A2,A1	(245,880)	85,239
13	A4,A3,A1	(276,510)	77,412
14	A2,A3,A1	(287,130)	80,691
15	A4,A2,A3	(334,560)	105,804
16	A4,A2,A3,A1	(381,360)	116,382

Best alternative → (12)

Infeasible alternatives ← {13, 14, 15, 16}

the total number of mutually exclusive decision alternatives that can be obtained from four independent projects is 16, including the "do-nothing" alternative. However, decision alternatives 13, 14, 15, and 16 are not feasible because of a $250,000 budget limit. So, we only need to consider alternatives 1 through 12.

Now, how do we compare these alternatives as the marginal cost of capital changes for each decision alternative. Consider again Figure 16.4. If we take A1 first, it would be acceptable, because its 15.43% return would exceed the 14% cost of capital used to finance it. Why couldn't we do this? The answer is that we are seeking to maximize the excess of returns over costs, or the area above the MCC, but below the IOS. We accomplish this by accepting the most profitable projects first. This logic leads us to conclude that, as long as the budget permits, A4 should be selected first and A2 second. This will consume $199,080, which leaves us $50,920 in unspent funds. The question is what are we going to do with this left-over money. Certainly, it is not enough to take A3 in full, but we can take A1 in full. Full funding for A1 will fit the budget and the project's rate of return still exceeds the marginal cost of capital (15.43% > 15%). Unless the left-over funds earn more than 15.43% interest, alternative 12 becomes the best (Figure 16.5).

Comments: In Example 16.9, the MARR was found by applying a capital limit to the investment opportunity schedule. The firm was then allowed to borrow or lend the money as the investment situation dictates. In this example, no such borrowing is explicitly assumed.

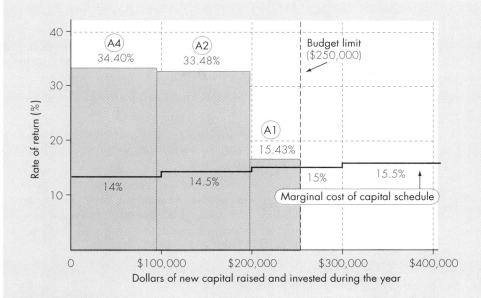

Figure 16.5   The appropriate cost of capital to be used in the capital budgeting process for decision alternative 12, with a $250,000 budget limit (Example 16.10) is 15%

16.5 Summary

- Methods of financing fall into two broad categories:

  1. **Equity financing** uses retained earnings or funds raised from an issuance of stock to finance a capital investment.

  2. **Debt financing** uses money raised through loans or by an issuance of bonds to finance a capital investment.

- Companies do not simply borrow funds to finance projects. Well-managed firms usually establish a **target capital structure** and strive to maintain the **debt ratio** when individual projects are financed.

- The selection of an appropriate MARR depends generally upon the **cost of capital**—the rate the firm must pay to various sources for the use of capital.

  1. The **cost of equity** (i_e) is used when debt-financing methods and repayment schedules are known explicitly.

  2. The **cost of capital** (k) is used when exact financing methods are unknown, but a firm keeps its capital structure on target. In this situation, a project's after-tax cash flows contain no debt cash flows such as principal and interest payment.

- The cost of the capital formula is a composite index reflecting the cost of funds raised from different sources. The formula is

$$k = \frac{i_d c_d}{V} + \frac{i_e c_e}{V}, V = c_d + c_e.$$

- The **marginal cost of capital** is defined as the cost of obtaining another dollar of new capital. The marginal cost rises as more and more capital is raised during a given period.

- Without a capital limit, the choice of MARR is dictated by the availability of financing information:

  1. In cases where the exact debt-financing and repayment schedules are known, we recommend the use of the net equity flow method. The proper MARR would be the cost of equity, i_e.

  2. If no specific assumption is made about the exact instruments that will be used to finance a particular project (but we do assume that the given capital structure proportions will be maintained), we may determine the after-tax cash flows without incorporating any debt cash flows. Then, we use the marginal cost of capital (k) as the proper MARR.

- Under conditions of capital rationing, the selection of MARR is more difficult, but generally the following possibilities exist:

Conditions	MARR
A firm borrows some capital from lending institutions at the borrowing rate, k, and some from its investment pool at the lending rate, r.	$r < \text{MARR} < k$
A firm borrows all capital from lending institutions at the borrowing rate, k.	$\text{MARR} = k$
A firm borrows all capital from its investment pool at the lending rate, r.	$\text{MARR} = r$

- The cost of capital used in the capital budgeting process is determined at the intersection of the **IOS** and **MCC** schedules. If the cost of capital at the intersection is used, then the firm will make correct accept/reject decisions, and its level of financing and investment will be optimal. This view assumes that the firm can invest and borrow at the rate where the two curves intersect.

- If a strict budget is placed in a capital budgeting problem and no projects can be taken in part, all feasible investment decision scenarios need to be enumerated. Depending upon each investment scenario, the cost of capital will also likely change. Our task is to find the best investment scenario in light of a changing cost of capital environment. As the number of projects to consider increases, we may eventually resort to a more advanced technique, such as a mathematical programming procedure.

Self-Test Questions

16s.1 In order to finance a new project, a $1,000 par value bond will be issued. The bond will be sold at discount with a market price of $970 and a coupon rate of 10%. The flotation costs for a new issue would be approximately 5%. The bond matures in 10 years and the corporate tax rate is 40%. Compute the after-tax cost of this debt financing.

(a) 10%   (b) 6%
(c) 11.36%   (d) 6.82%.

16s.2 The North Dakota Mining Company has the following capital structure.

Source	Amount
Debt	$665,000
Preferred stock	345,000
Common stock	1,200,000

The company wants to maintain the current capital structure in future project financing. If the company is allowed to use $800,000 debt, what would be the scale of the total investment?

(a) $1,144,337   (b) $2,658,647
(c) $1,333,334   (d) $2,456,785.

16s.3 The capital structure for Florida Citrus Corporation is given as follows:

Capital Structure	
Sources	Amount
Long term bonds	$3,000,000
Preferred stock	2,000,000
Common stock	5,000,000

Assuming that the firm will maintain its capital structure in the future, determine the firm's weighted average cost of capital (k) if the firm has a 7.5% cost of debt (after-tax), a 12.8% cost of preferred stock, and a 20% cost of common stock (cost of equity).

(a) 13.43%   (b) 14.81%
(c) 12.81%   (d) 11.32%.

16s.4 The Fox Corporation is considering three investments. The required investments and expected IRRs of these projects are as follows:

Investment Opportunity Schedule		
Project	Investment	IRR
A	$180,000	32%
B	250,000	18
C	120,000	15

The company intends to finance the projects by 40% debt and 60% equity. The after-tax cost of debt is 8% for the first $100,000, after which the cost will be 10%. Retained earnings (internally generated) in the amount of $150,000 are available for investment. The common stockholders' required rate of return is 20%. If new stock is issued, the cost will be 23%. What would be the marginal cost of capital to raise the first $250,000 project cost?

(a) 15.2%   (b) 17.8%
(c) 16.5%   (d) 8%.

16s.5 Using the marginal cost of capital curve defined in Problem 16s.4, decide which project(s) would be included?

(a) Project A only
(b) Projects A and B

(c) Projects A, B, and C

(d) None of the projects.

16s.6 Based on the marginal cost of the capital schedule and the investment opportunity schedule defined in Problem 16s.4, what would be the reasonable MARR (or discount rate) to use in evaluating the investments?

(a) MARR $< 17.8\%$

(b) $17.8\% < $ MARR $\leq 18\%$

(c) $18\% < $ MARR $< 32\%$

(d) MARR $= 15.2\%$.

Problems

Methods of Financing

16.1 Optical World Corporation, a manufacturer of peripheral vision storage systems, needs $10 million to market its new robotics-based vision systems. The firm is considering both financing options: common stock and bonds. If the firm decides to raise the capital through issuing common stock, the flotation costs will be 6% and the share price will be $25. If the firm decides to use debt financing, it can sell a 10-year, 12% bond with a par value of $1,000. The bond flotation costs will be 1.9%.

(a) For equity financing, determine the flotation costs and the number of shares to be sold to net $10 million.

(b) For debt financing, determine the flotation costs and the number of $1,000 par value bonds to be sold to net $10 million. What is the required annual interest payment?

16.2 Consider a project whose initial investment is $300,000, which must be financed at an interest rate of 12% per year. Assuming that the required repayment period is 6 years, determine the repayment schedule by identifying the principal as well as the interest payments for each of the following methods:

(a) Equal repayment of the principal.

(b) Equal repayment of the interest.

(c) Equal annual installments.

16.3 A chemical plant is considering the purchase of a computerized control system. The initial cost is $200,000, and the system will produce net savings of $100,000 per year. If purchased, the system will be depreciated under MACRS as a 5-year recovery property. This system will be used for 4 years at the end of which time, the firm expects to sell the system for $30,000. The firm's marginal tax rate on this investment is 35%. Any capital gains will be taxed at the same income tax rate. The firm is considering the purchase of the computer control system through either its retained earnings or through borrowing from a local bank. Two commercial banks are willing to lend the $200,000 at an interest rate of 10%, but each requires different repayment plans. Bank A requires four equal annual principal payments with interests calculated based on the unpaid balance.

Repayment Plan of Bank A		
End of Year	Principal	Interest
1	$50,000	$20,000
2	50,000	15,000
3	50,000	10,000
4	50,000	5,000

Bank B offers a payment plan which extends over 5 years with five equal annual payments.

Repayment Plan of Bank B			
End of Year	Principal	Interest	Total
1	$32,759	$20,000	$52,759
2	36,035	16,724	52,759
3	39,638	13,121	52,759
4	43,602	9,157	52,759
5	47,998	4,796	52,759

(a) Determine the cash flows if the computer control system is to be bought through its retained earnings (equity financing).

(b) Determine the cash flows if the asset is financed through either bank A or bank B.

(c) Recommend the best course of financing action. (Assume that the firm's MARR is known to be 10%.)

16.4 Edison Power Company currently owns and operates a coal-fired combustion turbine plant, which was installed 20 years ago. Because of degradation of the system, 65 forced outages occurred during the last year alone and two boiler explosions during the last 7 years. Edison is planning to scrap the current plant and install a new, improved gas turbine, which produces more energy per unit of fuel than typical coal-fired boilers.

The 50-MW gas-turbine plant, which runs on gasified coal, wood, or agricultural wastes, will cost Edison $65 million. Edison wants to raise the capital from three financing sources: 45% common stock, 10% preferred stock (which carries a 6% cash dividend when declared), and 45% borrowed funds. Edison's investment banks quote Edison the following flotation costs:

Financing Source	Flotation Costs	Selling Price	Par Value
Common stock	4.6%	$32/share	$10
Preferred stock	8.1	55/share	15
Bond	1.4	980	1,000

(a) What are the total flotation costs to raise $65 million?

(b) How many shares (both common and preferred) or bonds must be sold to raise $65 million?

(c) If Edison makes annual cash dividends of $2 per common share, and annual bond interest payments are at the rate of 12%, how much cash should Edison have available to meet both the equity and debt obligation? (Note that whenever a firm declares cash dividends to its common stockholders, the preferred stockholders are entitled to receive dividends of 6% of par value.)

Cost of Capital

16.5 Calculate the after-tax cost of debt under each of the following conditions:
(a) Interest rate, 12%; tax rate, 25%.
(b) Interest rate, 14%; tax rate, 34%.
(c) Interest rate, 15%; tax rate, 40%.

16.6 Sweeney Paper Company is planning to sell $10 million worth of long-term bonds with a 11% interest rate. The company believes that it can sell the $1,000 par value bonds at a price that will provide a yield to maturity of 13%. The flotation costs will be 1.9%. If Sweeney's marginal tax rate is 35%, what is its after-tax cost of debt?

16.7 Mobil Appliance Company's earnings, dividends, and stock price are expected to grow at an annual rate of 12%. Mobil's common stock is currently traded at $18 per share. Mobil's last cash dividend was $1.00, and its expected cash dividend for the end of this year is $1.12. Determine the cost of retained earnings (k_r).

16.8 Refer to Problem 16.7. Mobil wants to raise capital to finance a new project by issuing new common stock. With the new project, the cash dividend is expected to be $1.10 at the end of the current year, and its growth rate is 10%. The stock now sells for $18, but new common stock can be sold to net Mobil $15 per share.

(a) What is Mobil's flotation cost as a percentage?

(b) What is Mobil's cost of new common stock (k_e)?

16.9 The Callaway Company's cost of equity is 22%. Its before-tax cost of debt is 13%, and its marginal tax rate is 40%. The firm's capital structure calls for a debt-to-equity ratio of 45%. Calculate Callaway's cost of capital.

16.10 Delta Chemical Corporation is expected to have the following capital structure for the foreseeable future.

Source of Financing	Percent of Total Funds	Before-Tax Cost	After-Tax Cost
Debt	30%		
Short-term	10	14%	
Long-term	20	12	
Equity	70%		
Common stock	55		30%
Preferred stock	15		12

The flotation costs are already included in each cost component. The marginal income tax rate (t_m) for Delta is expected to remain at 40% in the future. Determine the cost of capital (k).

16.11 Charleston Textile Company is considering the acquisition of a new knitting machine at a cost of $200,000. Because of a rapid change in fashion styles, the need for this particular machine is expected to last only 5 years, after which the machine is expected to have a salvage value of $50,000. The annual operating cost is estimated at $10,000. The addition of this machine to the current production facility is expected to generate an additional revenue of $90,000 annually and will be depreciated in the 7-year MACRS property class. The income tax rate applicable for Charleston is 36%. The initial investment will be financed with 60% equity and 40% debt. The before-tax debt interest rate, which combines both short-term and long-term financing, is 12%, with the loan to be repaid in equal annual installments. The equity interest rate (i_e), which combines both sources of common and preferred stocks, is 18%.

(a) Evaluate this investment project by using net equity flows.

(b) Evaluate this investment project by using k.

16.12 The Huron Development Company is considering buying an overhead pulley system. The new system has a purchase price of $100,000, an estimated useful life and MACRS class life of 5 years, and an estimated salvage value of $30,000. It is expected to allow the company to economize on electric power usage, labor, and repair costs, as well as to reduce the number of defective products. A total annual savings of $45,000 will be realized if the new pulley system is installed. The company is in the 30% marginal tax bracket. The initial investment will be financed with 40% equity and 60% debt. The before-tax debt interest rate, which combines both short-term and long-term financing, is 15%, with the loan to be repaid in equal annual installments over the project life. The equity interest rate (i_e), which combines both sources of common and preferred stocks, is 20%.

(a) Evaluate this investment project by using net equity flows.

(b) Evaluate this investment project by using k.

16.13 Consider the following two mutually exclusive machines.

	Machine A	Machine B
Initial investment	$40,000	$60,000
Service life	6 years	6 years
Salvage value	$4,000	$8,000
Annual O&M cost	$8,000	$10,000
Annual revenues	$20,000	$28,000
MACRS property	5-year	5-year

The initial investment will be financed with 70% equity and 30% debt. The before-tax debt interest rate, which combines both short-term and long-term financing, is 10%, with the loan to be repaid in equal annual installments over the project life. The equity interest rate (i_e), which combines both sources of common and preferred stock, is 15%. The firm's marginal income tax rate is 35%.

(a) Compare the alternatives using $i_e = 15\%$. Which alternative should be selected?

(b) Compare the alternatives using k. Which alternative should be selected?

(c) Compare the results obtained in (a) and (b).

Capital Budgeting

16.14 DNA Corporation, a biotech engineering firm, has identified seven R&D projects for funding. Each project is expected to be in the R&D stage for 3 to 5 years, and the IRR figure represents the royalty income from selling the R&D results to pharmaceutical companies.

Project	Investment Type	Required IRR
1. Vaccines	$15 M	22%
2. Carbohydrate chemistry	25 M	40
3. Antisense	35 M	80
4. Chemical Synthesis	5 M	15
5. Antibodies	60 M	90
6. Peptide chemistry	23 M	30
7. Cell transplant/ gene therapy	19 M	32

DNA Corporation can only raise $100 million. DNA's borrowing rate is 18%, and its lending rate is 12%. Which R&D projects should be included in the budget?

16.15 Gene Fowler owns a house that contains 202 square feet of windows and 40 square feet of doors. Electricity usage totals 46,502 kWh: 7,960 kWh for lighting and appliances, 5,500 kWh for water heating, 30,181 kWh for space heating to 68°, and 2,861 kWh for space cooling to 78°F. The 14 energy-savings alternatives have been suggested by the local power company for Fowler's 1,620-square-foot home. Fowler can borrow money at 12% and lends money at 8%.

No.	Structural Improvement	Annual Savings	Estimated Costs	Payback Period
1	Add storm windows	$128–156	$455–556	3.5 years
2	Insulate ceilings to R-30	149–182	408–499	2.7
3	Insulate floors to R-11	158–193	327–399	2.1
4	Caulk windows and doors	25–31	100–122	4.0
5	Weather-strip windows and doors	31–38	224–274	7.2
6	Insulate ducts	184–225	1,677–2,049	9.1
7	Insulate space heating water pipes	41–61	152–228	3.7
8	Install heat retardants on E, SE, SW, W windows	37–56	304–456	8.2
9	Install heat reflecting film on E, SE, SW, W windows	21–31	204–306	9.9
10	Install heat absorbing film on E, SE, SW, W windows	5–8	204–306	39.5
11	Upgrade 6.5 EER A/C to 9.5 EER unit	21–32	772–1,158	36.6
12	Install heat pump water heating system	115–172	680–1,020	5.9
13	Install water heater jacket	26–39	32–48	1.2
14	Install clock thermostat to reduce heat from 70° to 60° for 8 hours each night	96–144	88–132	1.1

Note: EER (Energy Efficiency Ratio). R-value indicates the degree of resistance to heat. The higher the number, the greater the insulating quality.

(a) If Fowler stays in the house for the next 10 years, which alternatives would be selected with no budget constraint? Assume that his interest rate is 8%. Assume also that all installations would last 10 years. Fowler will be conservative in calculating the net present worth of each alternative (using the minimum annual savings at the maximum cost). Ignore any tax credits available to energy-saving installations.

(b) If he wants to limit his energy savings investments to $1,800, which alternatives should he include in his budget?

Short Case Studies

16.16 National Food Processing Company is considering investments in plant modernization and plant expansion. These proposed projects would be completed in 2 years, with varying requirements of money and plant engineering. Although some uncertainty exists in the data, management is willing to use the following data in selecting the best set of proposals. The resource limitations are as follows:

No.	Project	Investment Year 1	Year 2	IRR	Engineering Hours
1	Modernize production line	$300,000	0	30%	4,000
2	Build new production line	100,000	$300,000	43%	7,000
3	Provide numerical control for new production line	0	200,000	18%	2,000
4	Modernize maintenance shops	50,000	100,000	25%	6,000
5	Build raw material processing plant	50,000	300,000	35%	3,000
6	Buy present subcontractor's facilities for raw material processing	200,000	0	20%	600
7	Buy a new fleet of delivery trucks	70,000	10,000	16%	0

- First-year expenditures: $450,000
- Second-year expenditures: $420,000
- Engineering hours: 11,000 hours

The situation requires that a new or modernized production line be provided (projects 1 or 2). The numerical control (project 3) is applicable only to the new line. The company obviously does not want to both buy (project 6) and build (project 5) raw material processing facilities; it can, if desirable, rely on the present supplier as an independent firm. Neither the maintenance shop project (project 4) nor the delivery-truck purchase (project 7) is mandatory.

(a) Enumerate all possible mutually exclusive alternatives without considering the budget and engineering-hour constraints.

(b) Identify all feasible mutually exclusive alternatives.

(c) Suppose that the firm's marginal cost of capital will be 14% for raising the required capital up to $1 million. Which projects would be included in the firm's budget?

16.17 Consider the following investment projects:

n	A	B	C	D
0	-$2,000	-$3,000	-$1,000	
1	1,000	4,000	1,400	-$1,000
2	1,000		-100	1,090
3	1,000			
i*	23.38%	33.33%	32.45%	9%

Suppose that you have only $3,500 available at period 0. Neither additional budgets nor borrowing are allowed in any future budget period. However, you can lend out any remaining funds (or available funds) at 10% interest per period.

(a) If you want to maximize the future worth at period 3, which projects would you select? What is the future worth (the total amount available for lending at the end of period 3)? No partial projects are allowed.

(b) Suppose in (a) that, at period 0, you are allowed to borrow $500 at an interest rate of 13%. The loan has to be repaid at the end of year 1. Which project would you select to maximize your future worth at period 3?

(c) Considering a lending rate of 10% and a borrowing rate of 13%, what would be the reasonable MARR for project evaluation?

16.18 American Chemical Corporation (ACC) is a multinational manufacturer of industrial chemical products. ACC has made great progress in energy-cost reduction and has implemented several cogeneration projects in the United States and Puerto Rico, including the completion of a 35 megawatt (MW) unit in Chicago and a 29 MW unit at Baton Rouge. The division of ACC being considered for one of its more recent cogeneration projects is a chemical plant located in Texas. The plant has a power usage of 80 million kilowatt hours (kWh) annually. However, on the average, it uses 85% of its 10 MW capacity, which would bring the average power usage to 68 million kWh annually. Texas Electric presently charges $0.09 per kWh of electric consumption for the ACC plant, a rate that is consid-

ered high throughout the industry. Because ACC's power consumption is so large, the purchase of a cogeneration unit is considered to be desirable. Installation of the cogeneration unit would allow ACC to generate their own power and to avoid the annual $6,120,000 expense to Texas Electric. The total initial investment cost would be $10,500,000: $10,000,000 for the purchase of the power unit itself, a gas fired 10 MW Allison 571, and engineering, design, and site preparation. Five hundred thousand includes the purchase of interconnection equipment, such as poles and distribution lines that will be used to interface the cogenerator with the existing utility facilities. ACC is considering two financing options:

• ACC could finance $2,000,000 through the manufacturer at 10% for 10 years and will finance the remaining $8,500,00 through issuing common stock. The flotation cost for a common stock offering is 8.1%, and the stock will be priced at $45 per share.

• Investment bankers have indicated that 10-year 9% bonds could be sold at a price of $900 for each $1,000 bond. The flotation costs would be 1.9% to raise $10.5 million.

(a) Determine the debt-repayment schedule for the term loan from the equipment manufacturer.

(b) Determine the flotation costs and the number of common stocks to sell to raise the $8,500,000.

(c) Determine the flotation costs and the number of $1,000 par value bonds to be sold to raise $10.5 million.

16.19 (Continuation of Problem 16.18) As ACC management has decided to raise the $10.5 million by selling bonds, the company's engineers have estimated the operating costs of the cogeneration project.

The annual cash flow is comprised of many factors: maintenance, standby power, overhaul costs, and other miscellaneous expenses. Maintenance costs are projected to be approximately $500,000 per year. The unit must be overhauled every 3 years at a cost of $1.5 million. Miscellaneous expenses, such as additional personnel and insurance, are expected to total $1 million. Another annual expense is that for standby power, which is the service provided by the utility in the event of a cogeneration unit trip or scheduled maintenance outage. Unscheduled outages are expected to occur four times annually, each outage averaging 2 hours in duration at an annual expense of $6,400. Overhauling the unit takes approximately 100 hours and occurs every 3 years, requiring another triennial power cost of $100,000. Fuel (spot gas) will be consumed at a rate of $8,000 BTU per kWh, including the heat recovery cycle. At $2.00 per million BTU, the annual fuel cost will reach $1,280,000. Due to obsolescence, the expected life of the cogeneration project will be 12 years, after which Allison will pay ACC $1 million for salvage of all equipment.

A revenue will be incurred from the sale of excess electricity to the utility company at a negotiated rate. Since the chemical plant will consume on average 85% of the unit's 10 MW output, 15% of the output will be sold at $0.04 per kWh, bringing in an annual rev-

enue of $480,000. ACC's marginal tax rate (combined federal and state) is 36%, and their minimum required rate of return for any cogeneration project is 27%. The anticipated costs and revenues are summarized as follows:

Initial Investment	
Cogeneration unit, engineering, design, and site preparation (15-Year MACRS class)	$10,000,000
Interconnection equipment (5-Year MACRS class)	500,000
Salvage after 12-years' use	1,000,000
Annual expenses	
Maintenance	500,000
Misc. (Additional personnel and insurance)	1,000,000
Standby power	6,400
Fuel	1,280,000
Other operating expenses	
Overhaul every 3 years	1,500,000
Standby power during overhaul	100,000
Revenues	
Sale of excess power to Texas Electric	480,000

(a) If the cogeneration unit and other conecting equipment could be financed by issuing corporate bonds at an interest rate of 9%, compounded annually, with the flotation expenses as indicated in Problem 16.18, determine the net cash flow from the cogeneration project.

(b) If the cogeneration unit can be leased, what would be the maximum annual lease amount that ACC is willing to pay?

CHAPTER 17

Economic Analysis in the Public Sector

A committee advising Atlanta's Hartsfield International Airport[1] has this dilemma: Juggling the often-competing interests of keeping Hartsfield among the world's busiest airports while ensuring quality of life for nearby residents who'd feel the brunt of longer runways, more noise, and dislocation. The committee is looking at several options—terminals or gates and new runways—for how Hartsfield will look in year 2015, when the annual number of passengers is expected to have soared from 63 million to 121 million. Residents of Atlanta's south side for years have complained that they pay the cost for the airport's fueling the economy of the metro region. Despite the wariness of the committee, airport officials believe they can persuade its members to endorse some kind of expansion. Members will consider three options to handle the growth expected at Hartsfield over the next 15 years:

Option 1: Five runways, a new concourse adjacent to a new eastside terminal and close in parking.

Option 2: A more expensive version of the first option, putting the new concourse adjacent to the fifth runway.

Option 3: A sixth runway north of the current airport boundaries in the College-park-East Point-Hapeville corridor side terminal.

A study indicates that a north runway would displace more than 3,000 households and 978 businesses. A south-side runway could dislocate as many as 7,000 households and more than 900 businesses. Under any scenario of airport expansion, there still will be some flight delays. But some plans project longer waits than others do. The committee has 6 months to study each option before making a final recommendation. How would you conduct the economic analysis of this nature of public project?

---

[1] Office of Atlanta Regional Commission and an article in *The Atlanta Journal and Constitution,* "Committee Wary of Airport Expansion," December 13, 1997.

M any civil engineers would work on public works areas such as highway construction, airport construction and water projects. In this airport expansion scenario, each option requires a different level of investment with a different degree of benefits. One of the most important aspects of airport expansion is to quantify the cost of airport delays in dollar terms. In other words, what is the economic benefit of reducing any airport delay? From the airline's point of view, any taxiing and arrival delays means added fuel costs. For the Atlanta airport, any delays mean loss revenues in landing and departure fees. From the public's point of view, any delays mean loss of earnings, as they have to spend more time on transportation. Comparing the investment costs with potential benefits, known as benefit-cost analysis, is an important feature of the economic analysis method.

Up to this point, we have focused attention on investment decisions in the private sector; the primary objective of these investments was to increase the wealth of corporations. In the public sector, federal, state, and local governments spend hundreds of billions of dollars annually on a wide variety of public activities, such as the airport project described in this chapter opener. In addition, governments at all levels regulate the behavior of individuals and businesses by influencing the use of enormous quantities of productive resources. How can public decision-makers determine whether their decisions, which affect the use of these productive resources, are, in fact, in the best public interest?

Benefit-cost analysis is a decision-making tool used to systematically develop useful information about the desirable and undesirable effects of public projects. In a sense, we may view benefit-cost analysis in the public sector as profitability analysis in the private sector. In other words, benefit-cost analysis attempts to determine whether the social benefits of a proposed public activity outweigh the social costs. Usually, public investment decisions involve a great deal of expenditure, and their benefits are expected to occur over an extended period of time. Examples of benefit-cost analyses include studies of public transportation systems, environmental regulations on noise and pollution, public safety programs, education and training programs, public health programs, flood control, water resource development, and national defense programs.

The three types of benefit-cost analysis problems are as follows: (1) to maximize the benefits for any given set of costs (or budgets), (2) to maximize the net benefits when both benefits and costs vary, and (3) to minimize costs to achieve any given level of benefits (often called "cost-effectiveness" analysis). These three types of decision problems will be considered in this chapter.

17.1 Framework of Benefit-Cost Analysis

To evaluate public projects designed to accomplish widely differing tasks, we need to measure the benefits or costs in the same units in all projects so that we have a common perspective by which to judge the different projects. In practice this means expressing both benefits and costs in monetary units, a process that often must be performed without accurate data. In performing benefit-cost analysis, we define **"users"** as the public and **"sponsors"** as the government.

The general framework for benefit-cost analysis can be summarized as follows:

1. Identify all users' benefits expected to arise from the project.
2. Quantify, as much as possible, these benefits in dollar terms so that different benefits may be compared against one another and against the costs of attaining them.
3. Identify sponsor's costs.
4. As much as possible, quantify these costs in dollar terms to allow comparisons.
5. Determine the equivalent benefits and costs at the base period; use an interest rate appropriate for the project.
6. Accept the project if the equivalent users' benefits exceed the equivalent sponsor's costs.

We can use benefit-cost analysis to choose among such alternatives as allocating funds for construction of a mass-transit system, a dam with irrigation, highways, or an air-traffic control system. If the projects are on the same scale with respect to cost, it is merely a question of choosing the project for which the benefits exceed the costs by the greater amount. The steps outlined above are for a single (or independent) project evaluation. As is in the case for the internal rate of return criterion, when comparing mutually exclusive alternatives, an incremental benefit cost ratio must be used. Section 17.3.3 illustrates this important issue in detail.

17.2 Valuation of Benefits and Costs

In the abstract, the framework we just developed for benefit-cost analysis is no different from the one we have used throughout this text to evaluate private investment projects. The complications, as we shall discover in practice, arise in trying to identify and assign values to all the benefits and costs of a public project.

17.2.1   Users' Benefits

To begin a benefit-cost analysis, we identify all project **benefits** (favorable outcomes) and **disbenefits** (unfavorable outcomes) to the user. We should also consider the indirect consequences resulting from the project—the so-called **secondary effects**. For example, construction of a new highway will create new businesses such as gas stations, restaurants, and motels (benefits), but it will divert some traffic from the old road, and as a consequence, some businesses would be lost (disbenefits). Once the benefits and disbenefits are quantified, we define the users' benefits as follows:

$$\text{User's benefit (B)} = \text{Benefits} - \text{disbenefits.}$$

In identifying user's benefits, we should classify each one as a **primary benefit**—one directly attributable to the project—or a **secondary benefit**—one indirectly attributable to the project. As an example, at one time, the U.S. government was considering building a superconductor research laboratory in Texas. If it ever materializes, it could bring many scientists and engineers along with other supporting population to the region. Primary national benefits may include the long-term benefits that may accrue as a result of various applications of the research to U.S. businesses. Primary regional benefits may include economic benefits created by the research laboratory activities, which would generate many new supporting businesses. The secondary benefits might include the creation of new economic wealth as a consequence of a possible increase in international trade and any increase in the incomes of various regional producers attributable to a growing population.

The reason for making this distinction is that it may make our analysis more efficient. If primary benefits alone are sufficient to justify project costs, we can save time and effort by not quantifying the secondary benefits.

17.2.2   Sponsor's Costs

We determine the cost to the sponsor by identifying and classifying the expenditures required and any savings (or revenues) to be realized. The sponsor's costs should include both capital investment and annual operating costs. Any sales of products or services that take place on completion of the project will generate some revenues—for example, toll revenues on highways. These revenues reduce the sponsor's costs. Therefore, we calculate the sponsor's costs by combining these cost elements:

Sponsor's cost = Capital cost + operating and maintenance costs − revenues.

17.2.3 Social Discount Rate

As we learned in Chapter 16, the selection of an appropriate MARR for evaluating an investment project is a critical issue in the private sector. In public project analyses, we also need to select an interest rate, called the **social discount rate**, to determine equivalent benefits as well as the equivalent costs. Selection of the social discount rate in public project evaluation is as critical as selection of a MARR in the private sector.

Since present value calculations were initiated to evaluate public water resources and related land-use projects in the 1930s, a tendency to use relatively low rates of discount as compared with those existing in markets for private assets has persisted. During the 1950s and into the 1960s, the rate for water resource projects was 2.63%, which, for much of this period, was even below the yield on long-term government securities. The persistent use of a lower interest rate for water resource projects is a political issue. The point is best explained by the following newspaper article.[2]

> With a $3.5 billion price tag, the Central Arizona Project (CAP) to divert Colorado River water to central and southern Arizona is a massive undertaking by anyone's standards. But back in 1981, auditors at the General Accounting Office took a look at the interest rate that the Interior Department plans to charge on federal funds advanced for the project and discovered a far more startling statistic. By interpreting the law to require only a 3.343% interest payment rather than the market rate on the reimbursable portions of the project, the Interior Department is creating a staggering interest subsidy of $175 billion . . . over the 50-year payback period.
>
> Senator Howard Metzenbaum looks at those figures and cringes . . . "What possible reason can anyone give for providing 3.34% interest when the government itself is paying 9 and 10 and 11% for its money?"
>
> The Interior Department is justifying its low interest charge for the CAP on a legal interpretation that classifies the entire Central Arizona Project as one "unit" of the Colorado River Basin Project—meaning all features of the project, even though they will be constructed in phases over a long period of time, should enjoy the same 3.343% interest rate that was charged when the first construction began in the early 1970s.

In recent years, with the growing interest in performance budgeting and systems analysis in the 1960s, the tendency on the part of government agencies has been to examine the appropriateness of the discount rate in the public sector in relation to the efficient allocation of resources in the economic system as a whole.[3] Two views of the basis for determining the social discount rate prevail:

---

[2] *Atlanta Journal and Constitution*, Sunday, July 7, 1985. (Reprinted with permission.)
[3] Mikesell, R. F. *The Rate of Discount for Evaluating Public Projects*. American Enterprise Institute for Public Policy Research, 1977.

1. **Projects without private counterparts:** *The social discount rate should reflect only the prevailing government borrowing rate.* Projects such as dams designed purely for flood control, access roads for noncommercial uses, and reservoirs for community water supply may not have corresponding private counterparts. In those areas of government activity where benefit-cost analysis has been employed in evaluation, the rate of discount traditionally used has been the cost of government borrowing. In fact, water resource project evaluations follow this view exclusively.

2. **Projects with private counterparts:** *The social discount rate should represent the rate that could have been earned had the funds not been removed from the private sector.* If all public projects were financed by borrowing at the expense of private investment, we may focus on the opportunity cost of capital in alternative investments in the private sector to determine the social discount rate. In the case of public capital projects, similar to some in the private sector that produce a commodity or a service (such as electric power) to be sold on the market, the rate of discount employed would be the average cost of capital as discussed in Chapter 16. The reasons for using the private rate of return as the opportunity cost of capital in projects similar to those in the private sector are (1) to prevent the public sector from transferring capital from higher-yielding to lower-yielding investments, and (2) to force public project evaluators to employ market standards in justifying projects.

The Office of Management and Budget (OMB) holds the second view. Since 1972, the OMB has required that a social discount rate of 10% be used to evaluate federal public projects. Exceptions include water resource projects.

17.2.4 Quantifying Benefits and Costs

Now that we have defined the general framework for benefit-cost analyses and discussed the appropriate discount rate, we will illustrate the process of quantifying the benefits and costs associated with a public project. Our context is a motor-vehicle inspection program initiated by the state of New Jersey.[4]

Many states in the United States employ inspection systems for motor vehicles. Critics often charge that these programs lack efficacy and have a poor benefit-to-cost ratio in terms of reducing fatalities, injuries, accidents, and pollution.

Elements of Benefits and Costs

The state of New Jersey identified the primary and secondary benefits as follows:

- **Users' Benefits**

  Primary benefits: Deaths and injuries related to motor-vehicle accidents impose specific financial costs on individuals and society. Preventing such costs through the inspection program has the following primary benefits.

---

[4] Based on Loeb, P. D. and Gilad, B. "The Efficacy and Cost Effectiveness of Vehicle Inspection," *Journal of Transport Economics and Policy*, May 1984: 145–164. The original cost data, which were given in 1981 dollars, were converted to the equivalent cost data in 2000 by using the prevailing consumer price indices during the period.

1. Retention of contributions to society that might be lost due to an individual's death.

2. Retention of productivity that might be lost while an individual recuperates from an accident.

3. Savings of medical, legal, and insurance services.

4. Savings on property replacement or repair costs.

Secondary benefits: Some secondary benefits are not measurable (for example, avoidance of pain and suffering); others can be quantified. Both types of benefits should be considered. A list of secondary benefits follows:

1. Savings of income of families and friends of accidents victims who might otherwise be tending to accident victims.

2. Avoidance of air and noise pollution and savings on fuel costs.

3. Savings on enforcement and administrative costs related to the investigation of accidents.

4. Pain and suffering.

- **Users' Disbenefits**

  1. Cost of spending time to have a vehicle inspected (including travel time), as opposed to devoting that time to an alternative endeavor (opportunity cost).

  2. Cost of inspection fees.

  3. Cost of repairs that would not have been made if the inspection had not been performed.

  4. Value of time expended in repairing the vehicle (including travel time).

  5. Cost in time and direct payment for reinspection.

- **Sponsor's Costs**

  1. Capital investments in inspection facilities.

  2. Operating and maintenance costs associated with inspection facilities. These include all direct and indirect labor, personnel, and administrative costs.

- **Sponsor's Revenues or Savings**

  1. Inspection fee.

Valuation of Benefits and Costs

The aim of benefit-cost analysis is to maximize the equivalent value of all benefits less that of all costs (expressed either in present values or annual values). This objective is in line with promoting the economic welfare of citizens. In general, the benefits of public projects are difficult to measure, whereas the costs are more easily determined. For simplicity, we will only attempt to quantify the primary users' benefits and sponsor's costs on an annual basis.

(a) **Calculation of Primary Users' Benefits**

1. Benefits due to the Reduction of Deaths: The equivalent value of the average income stream lost by victims of fatal accidents[5] was estimated at $571,106 per victim in 2000 dollars. The state estimated that the inspection program would reduce the number of annual fatal accidents by 304, resulting in a potential savings of

$$(304)(\$571,106) = \$173,616,200.$$

2. Benefits due to the Reduction of Damage to Property: The average cost of damage to property per accident was estimated at $1,845. This figure includes the cost of repairs for damages to the vehicle, the cost of insurance, the cost of legal and court administration, the cost of police accident investigation, and the cost of traffic delay due to accidents. The state estimated that accidents would be reduced by 37,910 per year, and that about 63% of all accidents would result in damage to property only. Therefore, the estimated annual value of benefits due to reduction of property damage was estimated at

$$\$1,845(37,910)(0.63) = \$44,073,286.$$

The overall annual primary benefits are estimated as the sum of

Value of reduction in fatalities	$173,616,200
Value of reduction in property damage	44,073,286
Total	$217,689,486

(b) **Calculation of Primary Users' Disbenefits**

1. Opportunity Cost Associated with Time Spent Bringing Vehicles for Inspection: This cost is estimated as

$$C_1 = \text{(Number of cars inspected)}$$
$$\times \text{(average duration involved in travel)}$$
$$\times \text{(average wage rate)}.$$

With an estimated average duration of 1.02 travel-time hours per car, an average wage rate of $14.02 per hour, and 5,136,224 inspected cars per year, we obtain

$$C_1 = 5,136,224\,(1.02)\,(\$14.02)$$
$$= \$73,450,058.$$

2. Cost of Inspection Fee: This cost may be calculated as

$$C_2 = \text{(Inspection fee)} \times \text{(number of cars inspected)}.$$

---

[5] These estimates were based on the total average income that these victims could have generated had they lived. This average value on human life was calculated by considering several factors, such as age, sex, and income group.

Assuming an inspection fee of $5 is to be paid for each car, the cost of the total annual inspection cost is estimated as

$$C_2 = (\$5)(5,136,224)$$
$$= \$25,681,120.$$

3. Opportunity Cost associated with Time Spent Waiting during the Inspection Process: This cost may be calculated by the formula

$$C_3 = \text{(Average waiting time in hours)}$$
$$\times \text{(average wage rate per hour)}$$
$$\times \text{(number of cars inspected).}$$

With an average waiting time of 9 minutes (or 0.15 hours),

$$C_3 = 0.15(\$14.02)(5,136,224) = \$10,801,479.$$

4. Vehicle Usage Costs for the Inspection Process: These costs are estimated as

$$C_4 = \text{(Number of inspected cars)}$$
$$\times \text{(Vehicle operating cost per mile)}$$
$$\times \text{(average round trip miles to inspection station).}$$

Assuming that $0.35 operating cost per mile and 20 round-trip miles,

$$C_4 = 5,136,224 \, (\$0.35)(20) = \$35,953,568.$$

The overall primary annual disbenefits are estimated as

Item	Amount
C_1	$73,450,058
C_2	25,681,120
C_3	10,801,479
C_4	35,453,568
Total disbenefits,	$145,886,225
or $28.40 per vehicle	

(c) **Calculation of Primary Sponsor's Costs**
New Jersey's Division of Motor Vehicles reported an expenditure of $46,376,703 for inspection facilities (this value represents the annualized capital expenditure) and another annual operating expenditure of $10,665,600 for inspection, adding up to $57,042,303.

(d) **Calculation of Primary Sponsor's Revenues**
The sponsor's costs are offset to a large degree by annual inspection revenues; these must be subtracted to avoid double counting. Annual fee rev-

enues are the same as the direct cost of inspection incurred by the users (C_2), which was calculated as $20,339,447.

Reaching a Final Decision

Finally, a discount rate of 6% was deemed appropriate because the state of New Jersey finances most state projects by issuing a 6% long-term tax-exempt bond. The streams of costs and benefits are already discounted so as to obtain their present and annual equivalent values.

From the state's estimates, the primary benefits of inspection are valued at $217,689,486, as compared to the primary disbenefits of inspection, which total $145,886,225. Therefore, the user's net benefits are

$$\text{User's net benefits} = \$217,689,486 - \$145,886,225$$
$$= \$71,803,261.$$

Now the sponsor's net costs are

$$\text{Sponsor's net costs} = \$57,042,303 - \$20,339,447$$
$$= \$36,702,856.$$

Since all benefits and costs are expressed in annual equivalents, we can use these values directly to compute the degree of benefits that exceeds the sponsor's costs:

$$\$71,803,261 - \$36,702,856 = \$35,100,405 \text{ per year.}$$

This positive AE amount indicates that the New Jersey inspection system is economically justifiable. We can assume the AE amount would have been even greater had we also factored in secondary benefits. (For simplicity, we have not explicitly considered vehicle growth in the state of New Jersey. For a complete analysis, this growth factor must be considered to account for all related benefits and costs in equivalence calculations.)

17.2.5 Difficulties Inherent in Public Project Analysis

As we observed in the motor-vehicle inspection program in the previous section, public benefits are very difficult to quantify in a convincing manner. For example, consider the valuation of a saved human life in any category. Conceptually, the total benefit associated with saving a human life may include the avoidance of the costs of insurance administration and legal and court costs. As well, the average potential income lost, considering the factors of age and sex, because of premature death must be included. Obviously, the difficulties associated with any attempt to put precise numbers on human life are insurmountable.

Consider this example: A few years ago, a 50-year-old business executive was killed in a plane accident. The investigation indicated that the plane was not properly maintained according to the federal guidelines. The executive's family sued the airline, and the court eventually ordered the airline to pay $5,250,000 to the victim's family. The judge calculated the value of the lost human life based on the assumption that, if the executive had lived and worked in the same capacity until his retirement, his remaining lifetime earnings would have been equivalent to $5,250,000 at the time of award. This is an example of how an individual human life was assigned a dollar value, but clearly any attempt to establish an average amount that represents the general population is potentially controversial. We might even take exception to this individual case: Does the executive's salary adequately represent his worth to his family? Should we also assign a dollar value to their emotional attachment to him, and if so, how much?

Consider a situation in which a local government is planning to widen a typical municipal highway to relieve chronic traffic congestion. Knowing that the project will be financed by local and state taxes, and that many out-of-state travelers are expected to benefit, should the project be justified solely on the benefits to local residents? Which point of view should we take in measuring the benefits—the municipal level, the state level, or both? It is important that any benefit measure be done from the appropriate *point of view*.

In addition to valuation and point-of-view issues, many possibilities for tampering with the results of benefit-cost analyses may exist. Unlike in the private sector, many public projects are undertaken based on political pressure rather than on their economic benefits alone. In particular, whenever the benefit-cost ratio becomes marginal, or less than one, a potential to inflate the benefit figures to make the project look good exists.

17.3 Benefit-Cost Ratios

An alternative way of expressing the worthiness of a public project is to compare the user's benefits (B) to the sponsor's cost (C) by taking the ratio B/C. In this section, we shall define the benefit-cost (B/C) ratio, and explain the relationship between the conventional NPW criterion and the B/C ratio.

17.3.1 Definition of Benefit-Cost Ratio

For a given benefit-cost profile, let B and C be the present values of benefits and costs defined by

$$B = \sum_{n=0}^{N} b_n (1 + i)^{-n} \tag{17.1}$$

$$C = \sum_{n=0}^{N} c_n (1 + i)^{-n}, \tag{17.2}$$

where b_n = Benefit at the end of period n, $b_n \geq 0$
$\quad c_n$ = Expense at the end of period n, $c_n \geq 0$
$\quad A_n = b_n - c_n$
$\quad N$ = Project life
$\quad i$ = Sponsor's interest rate (discount rate).

The sponsor's costs (C) consist of the equivalent capital expenditure (I) and the equivalent annual operating costs (C') accrued in each successive period. (Note the sign convention we use in calculating a benefit-cost ratio. Since we are using a ratio, all benefits and cost flows are expressed in positive units. Recall that in previous equivalent worth calculation our sign convention was to explicitly assign "+" for cash inflows and "−" for cash outflows.) Let's assume that a series of initial investments is required during the first K periods, while annual operating and maintenance costs accrue in each following period. Then, the equivalent present value for each component is

$$I = \sum_{n=0}^{K} c_n (1 + i)^{-n} \tag{17.3}$$

$$C' = \sum_{n=K+1}^{N} c_n (1 + i)^{-n}, \tag{17.4}$$

and $C = I + C'$.

The B/C ratio[6] is defined as

$$BC(i) = \frac{B}{C} = \frac{B}{I + C'}, I + C' > 0. \tag{17.5}$$

If we are to accept a project, the $BC(i)$ must be greater than 1.

Note that we must express the values of B, C', and I in present worth equivalents. Alternatively, we can compute these values in terms of annual equivalents and use them in calculating the B/C ratio. The resulting B/C ratio is not affected.

Example 17.1 Benefit-Cost Ratio

A public project being considered by a local government has the following estimated benefit-cost profile (Figure 17.1).

Assume that $i = 10\%$, $N = 5$, and $K = 1$. Compute B, C, I, C', and $BC(10\%)$.

---

[6] An alternative measure, called the **net B/C ratio**, $B'C(i)$, considers only the initial capital expenditure as a cash outlay, and annual net benefits are used:

$$B'C(i) = \frac{B - C'}{I} = \frac{B'}{I}, I > 0.$$

The decision rule has not changed — the ratio must still be greater than one. It can be easily shown that a project with $BC(i) > 1$ will always have $B'C(i) > 1$, as long as both C and I are > 0, as they must be for the inequalities in the decision rules to maintain the stated senses. The magnitude of $BC(i)$ will generally be different than that for $B'C(i)$, but the magnitudes are irrelevant for making decisions. All that matters is whether the ratio exceeds the threshold value of one. However, some analysts prefer to use $B'C(i)$ because it indicates the net benefit (B') expected per dollar invested. But why do they care if the choice of ratio does not affect the decision? They may be trying to increase or decrease the magnitude of the reported ratio in order to influence audiences who do not understand the proper decision rule. People unfamiliar with benefit/cost analysis often assume that a project with a higher B/C ratio is better. This is not generally true, as is shown in 17.3.3. An incremental approach must be used to properly compare mutually exclusive alternatives.

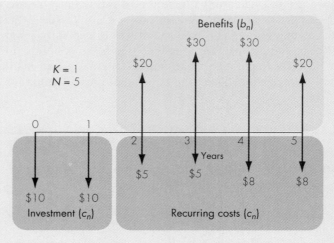

Figure 17.1 Classification of a project's cash flow elements (Example 17.1)

n	b_n	c_n	A_n
0		$10	−$10
1		10	−10
2	$20	5	15
3	30	5	25
4	30	8	22
5	20	8	12

Solution

$$B = \$20(P/F, 10\%, 2) + \$30(P/F, 10\%, 3)$$
$$+ \$30(P/F, 10\%, 4) + \$20(P/F, 10\%, 5)$$
$$= \$71.98.$$

$$C = \$10 + \$10(P/F, 10\%, 1) + \$5(P/F, 10\%, 2) + \$5(P/F, 10\%, 3)$$
$$+ \$8(P/F, 10\%, 4) + \$8(P/F, 10\%, 5)$$
$$= \$37.41.$$

$$I = \$10 + \$10(P/F, 10\%, 1)$$
$$= \$19.09.$$

$$C' = C - I$$
$$= \$18.32.$$

Using Eq. (17.5), we can compute the *B/C* ratio as

$$BC(10\%) = \frac{71.98}{\$19.09 + \$18.32}$$

$$= 1.92 > 1.$$

The *B/C* ratio exceeds 1, so the users' benefits exceed the sponsor's costs.

17.3.2   Relationship between *B/C* Ratio and NPW

The *B/C* ratio yields the same decision for a project as does the NPW criterion. Recall that the *BC*(*i*) criterion for project acceptance can be stated as

$$\frac{B}{I + C'} > 1. \tag{17.6}$$

If we multiply the term $(I + C')$ on both sides of the equation and transpose the term $(I + C')$ to the left-hand side, we have

$$B > (I + C')$$

$$B - (I + C') > 0 \tag{17.7}$$

$$PW(i) = B - C > 0, \tag{17.8}$$

which is the same decision rule[7] as that which accepts a project by the NPW criterion. This implies that we could use the benefit-cost ratio in evaluating private projects instead of using the NPW criterion, or we could use the NPW criterion in evaluating the public projects. Either approach will signal consistent project selection. Recall that, in Example 17.1, $PW(10\%) = B - C = \$34.57 > 0$; the project would be acceptable under the NPW criterion.

17.3.3   Comparing Mutually Exclusive Alternatives: Incremental Analysis

Let us now consider how we choose among mutually exclusive public projects. As we explained in Chapter 9, we must use the incremental investment approach in comparing alternatives based on any relative measure such as IRR or *B/C*.

Incremental Analysis Based on *BC*(*i*)

To apply incremental analysis, we compute the incremental differences for each term $(B, I, \text{and } C')$ and take the *B/C* ratio based on these differences. To use the *BC*(*i*) on incremental investment, we may proceed as follows:

1. If one or more alternatives have *B/C* ratios greater than 1, eliminate any alternatives with a *B/C* ratio less than 1.

---

[7] We can easily verify a similar relationship between the net *B/C* ratio and the NPW criterion.

2. Arrange the remaining alternatives in the increasing order of the denominator $(I + C')$. Thus, the alternative with the smallest denominator should be the first (j), the alternative with the second smallest (k), and so forth.
3. Compute the incremental differences for each term $(B, I, \text{and } C')$ for the paired alternatives (j, k) in the list.

$$\Delta B = B_k - B_j$$
$$\Delta I = I_k - I_j$$
$$\Delta C' = C'_k - C'_j$$

4. Compute the $BC(i)$ on incremental investment by evaluating

$$BC(i)_{k-j} = \frac{\Delta B}{\Delta I + \Delta C'}.$$

If $BC(i)_{k-j} > 1$, select the k alternative. Otherwise select the j alternative.
5. Compare the selected alternative with the next one on the list by computing the incremental benefit-cost ratio.[8] Continue the process until you reach the bottom of the list. The alternative selected during the last pairing is the best one.

We may modify the decision procedures when we encounter the following situations:

*   If $\Delta I + \Delta C' = 0$, we cannot use the benefit-cost ratio because this implies that both alternatives require the same initial investment and operating expenditure. When this happens, we simply select the alternative with the largest B value.

*   In situations where public projects with unequal service lives are to be compared but they can be repeated, we may compute all component values $(B, C', \text{and } I)$ on an annual basis and use them in incremental analysis.

Example 17.2 Incremental Benefit-Cost Ratios

Consider three investment projects, A1, A2, and A3. Each project has the same service life, and the present worth of each component value $(B, I, \text{and } C')$ is computed at 10% as follows:

	A1	Projects A2	A3
I	$5,000	$20,000	$14,000
B	12,000	35,000	21,000
C'	4,000	8,000	1,000
$PW(i)$	$3,000	$7,000	$6,000

---

[8] If we use the net B/C ratio as a basis, we need to order the alternatives in increasing order of I and compute the net B/C ratio on the incremental investment.

(a) If all three projects are independent, which projects would be selected based on $BC(i)$?

(b) If the three projects are mutually exclusive, which project would be the best alternative? Show the sequence of calculations that would be required to produce the correct results. Use the B/C ratio on incremental investment.

Solution

(a) Since $PW(i)_1$, $PW(i)_2$, and $PW(i)_3$ are positive, all projects would be acceptable if they were independent. Also, $BC(i)$ values for each project are greater than 1, so the use of the benefit-cost ratio criterion leads to the same accept-reject conclusion under the NPW criterion.

	A1	A2	A3
$BC(i)$	1.33	1.25	1.40

(b) If these projects are mutually exclusive, we must use the principle of incremental analysis. If we attempt to rank the projects according to the size of the B/C ratio, obviously we will observe a different project preference. For example, if we use the $BC(i)$ ratio on the total investment, we see that A3 appears to be the most desirable and A2 the least desirable, but selecting mutually exclusive projects on the basis of B/C ratios is incorrect. Certainly, with $PW(i)_2 > PW(i)_3 > PW(i)_1$, project A2 would be selected under the NPW criterion. By computing the incremental B/C ratios, we will select a project that is consistent with the NPW criterion.

We will first arrange the projects by increasing order of their denominator $(I + C')$ for the $BC(i)$ criterion:[9]

Ranking Base	A1	A3	A2
$I + C'$	$9,000	$15,000	$28,000

- A1 versus A3: With the do-nothing alternative, we first drop from consideration any project that has a B/C ratio smaller than 1. In our example, the B/C ratios of all three projects exceed 1, so the first incremental comparison is between A1 and A3:

$$BC(i)_{3-1} = \frac{\$21,000 - \$12,000}{(\$14,000 - \$5,000) + (\$1,000 - \$4,000)}$$

$$= 1.5 > 1.$$

---

[9] I is used as a ranking base for the $B'C(i)$ criterion. The order still remains unchanged — A1, A3, and A2.

Since the ratio is greater than 1, we prefer A3 over A1. Therefore, A3 becomes the "current best" alternative.

- A3 versus A2: Next, we must determine whether the incremental benefits to be realized from A2 would justify the additional expenditure. Therefore, we need to compare A2 and A3 as follows:

$$BC(i)_{2-3} = \frac{\$35,000 - \$21,000}{(\$20,000 - \$14,000) + (\$8,000 - \$1,000)}$$

$$= 1.08 > 1.$$

The incremental B/C ratio again exceeds 1, and therefore we prefer A2 over A3. With no further projects to consider, A2 becomes the ultimate choice.[10]

17.4 Analysis of Public Projects Based on Cost-Effectiveness

In evaluating public investment projects, we may encounter situations where competing alternatives have the same goals but the effectiveness with which those goals can be met may be, or may not be, measurable in dollars. In these situations, we compare decision alternatives directly based on their **cost-effectiveness**. Here we judge the effectiveness of an alternative in dollars or some nonmonetary measure by the extent to which that alternative, if implemented, will attain the desired objective. The preferred alternative is then either the one that produces the maximum effectiveness for a given level of cost, or the minimum cost for a fixed level of effectiveness.

17.4.1 General Procedure for Cost-Effectiveness Studies

A typical cost-effectiveness analysis procedure involves the following steps.

Step 1: Establish the goals to be achieved by the analysis.

---

[10] Using the net B/C ratio: If we had to use the net B/C ratio on this incremental investment decision, we would obtain the same conclusion. Since all $B'C(i)$ ratios exceed 1, no do-nothing alternative exists. By comparing the first pair of projects on this list, we obtain:

$$B'C(i)_{3-1} = \frac{(\$21,000 - \$12,000) - (\$1,000 - \$4,000)}{(\$14,000 - \$5,000)}$$

$$= 1.33 > 1.$$

Project A3 becomes the "current best." Next, a comparison of A2 and A3 yields

$$B'C(i)_{2-3} = \frac{(\$35,000 - \$21,000) - (\$8,000 - \$1,000)}{(\$20,000 - \$14,000)}$$

$$= 1.17 > 1.$$

Therefore, A2 becomes the best choice by the net B/C ratio criterion.

Step 2: Identify the imposed restrictions on achieving the goals, such as budget or weight.

Step 3: Identify all the feasible alternatives to achieve the goals.

Step 4: Identify the social interest rate to use in the analysis.

Step 5: Determine the equivalent life-cycle cost of each alternative, including research and development, testing, capital investment, annual operating and maintenance costs, and salvage value.

Step 6: Determine the basis for developing the cost-effectiveness index. Two approaches may be used: (1) the fixed-cost approach and (2) the fixed-effectiveness approach. If the fixed-cost approach is used, determine the amount of effectiveness obtained at a given cost. If the fixed-effectiveness approach is used, determine the cost to obtain the predetermined level of effectiveness.

Step 7: Compute the cost-effectiveness ratio for each alternative based on the selected criterion in Step 6.

Step 8: Select the alternative with the maximum cost-effective index.

When either the cost or the level of effectiveness is clearly stated in achieving the declared program objective, most cost-effectiveness studies will be relatively straightforward. If this is not the case, however, the decision-maker must come up with his or her own program objective by fixing either the cost required in the program or the level of effectiveness to be achieved. We will show how such a problem can easily evolve in many of the public projects or military applications in the following case example.

17.4.2 A Cost-Effectiveness Case Example

To illustrate the procedures involved in cost-effectiveness analysis, we shall present an example in which the most cost-effective program for developing an adverse-weather precision-guided weapon system is selected by the U.S. Air Force.[11]

Problem Statement

In a recent international conflict, precision-guided weapons demonstrated remarkable success and accuracy against a wide array of fixed and mobile targets. Such weapons rely upon laser designation of the target by an aircraft. The aircraft illuminates the target by laser; the weapon, with its on-board laser sensor, locks onto the target and flies to it. During the war, aircraft were required to fly at moderate altitude (about 10,000 feet) to escape vulnerability to anti-aircraft artillery batteries. At these altitudes, the aircraft were above the cloud/smoke levels. Unfortunately, laser beams cannot penetrate cloud cover, smoke, or fog. As a result, on those days when clouds, smoke, and fog were present, the aircraft could not to deliver the precision-guided weapons (Figure 17.2). This led to development of a weapon system that would correct these deficiencies.

---

[11] The case example is provided by Frederick A. Davis. All numbers used herein do not represent the actual values used by the U.S. Air Force.

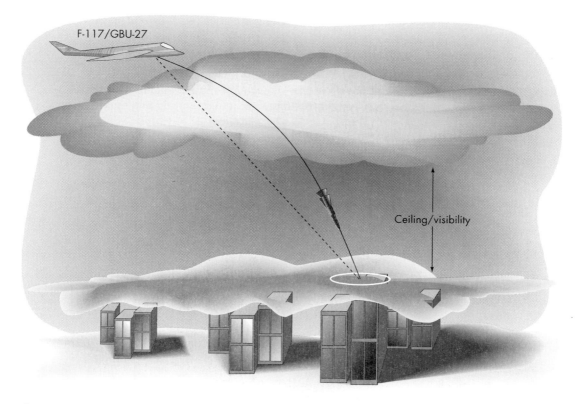

Figure 17.2   Conceptual use of an adverse weather guidance weapon by an aircraft

Defining the Goals

Selection of the best system is based on cost/kill decision criteria. **Cost/kill** is defined as the unit cost of the weapon divided by the probability of the weapon achieving its target. Mission-effectiveness studies determine this probability.

The purposes of this study are to evaluate these alternatives from a cost viewpoint and to determine the best option, based on a cost/kill decision criterion. Also, anticipating congressional scrutiny of high weapon-system costs, any option costing more than the present value of a life-cycle cost of $120K per unit cannot be considered for selection. To respond rapidly to this critical Air Force requirement, the initial operational capability (IOC) date is assumed to be 7 years. IOC (Initial Operational Capability) is defined as that point in the project life when the first block of weapons has been delivered to the field and is ready for operational use.

Description of Precision-Weapon Alternatives

Considering numerous weapon alternatives, preferred concepts were developed, and these are listed in Table 17.1. Also presented in the table are the qualitative characteristics of each alternative. These six weapon system alternatives are also considered to

Alternative A_j	Advantage	Disadvantage	Probability of Kill
A1: Inertial navigation system	Low cost, mature technology	Accuracy, target recognition	0.33
A2: Inertial navigation system: Global positioning system	Moderate cost, mature technology	Target recognition	0.70
A3: Imaging infrared (I^2R)	Accurate, target recognition	High cost, bunkered target detection	0.90
A4: Synthetic aperture radar	Accurate, target recognition	High cost	0.99
A5: Laser detection/ranging	Accurate, target recognition	High cost, technical maturity	0.99
A6: Millimeter wave (MMW)	Moderate cost, accurate	Target recognition	0.80

Table 17.1
Weapon System Alternatives

be mutually exclusive. Only one of the alternatives will be selected by the Air Force to fill the mission capability void that currently exists. Each of the alternatives is at some level of technological maturity. Some are nearly off-the-shelf, while others are just emerging from laboratory development. Because of this, some of the alternatives will require considerably more up-front development funding prior to production than others. Table 17.1 also summarizes the results (probability of kill) by the mission studies for the six guidance systems before entering into laboratory research and development.

To consider the costs associated with each option, from up-front development through complete production, the project begins with the Full Scale Development (FSD) phase, which accomplishes the up-front development prior to entering the production phase. To meet the 7-year IOC date, the FSD phase must be completed in 4 years. A 10,000-unit buy over a 5-year production life is required to meet Air Force needs. Because of the differing technological maturity of the six alternatives, widely varying FSD investments will be required.

Life-Cycle Cost for Each Alternative

The costs associated with FSD and production for each alternative will vary significantly. For systems incorporating the most current of emerging technologies, the FSD completion costs will be higher than those considered mature. Production costs will vary, depending upon the complexity of the system components. The objective of the FSD program is to rigorously test demonstrated system capability and correct any design flaws. The cost estimates used in both the FSD and the production phases were generated considering labor hours, material costs, equipment/tooling costs, subcontractor costs, travel, flight-test costs, documentation, and costs for program reviews.

Table 17.2
Life-Cycle
Costs for
Weapon De-
velopment
Alternatives

Phase	Year	Expenditures in Million Dollars					
		A1*	A2	A3	A4	A5	A6
FSD	0	$15	$19	$50	$40	$75	$28
	1	18	23	65	45	75	32
	2	19	22	65	45	75	33
	3	15	17	50	40	75	27
	4	90	140	200	200	300	150
	5	95	150	270	250	360	180
IOC	6	95	160	280	275	370	200
	7	90	150	250	275	340	200
	8	80	140	200	200	330	170
PW(10%)		$315.92	$492.22	$884.27	$829.64	$1,227.23	$613.70

*Sample calculation: Equivalent life-cycle cost for A1:

$$PW(10\%) = \$15 + \$18(P/F, 10\%, 1) + \ldots + \$80(P/F, 10\%, 8)$$
$$= \$315.92.$$

Table 17.2 summarizes the equivalent life-cycle cost for each program, estimated in constant dollars (2000), as is the standard Air Force practice in project cost estimation. The Air Force will follow the OMB guidelines in selecting the interest rate for equivalent life-cycle cost calculation, which is 10%.

Cost-Effectiveness Index

The equivalent life-cycle costs of the system need to be divided by the 10,000 units to be purchased in order to arrive at a cost/unit figure. Then, the cost/kill for each alternative is computed. The following shows the resultant cost/kill and kill/cost figures:

Type	Cost/Unit	Probability of Kill	Cost/Kill	Kill/Cost
A1	$31,592	0.33	$95,733	0.0000104
A2	49,220	0.70	70,314	0.0000142
A3	88,427	0.90	98,252	0.0000102
A4	82,964	0.99	83,802	0.0000119
A5	122,723	0.99	123,963	0.0000081
A6	61,370	0.80	76,713	0.0000130

Figure 17.3 graphically presents the cost/kill values of the six guidance systems. Obviously, the A5 system is not a feasible solution, as its unit cost exceeds the constraint of $120K. Of the feasible alternatives, the lowest cost/kill value is that of the A2 weapon system. However, it is premature to conclude that the A2 option is the best. This case example is unique in the sense that neither the costs nor the benefits are fixed. The benefit is a kill, but the probability of kill is not the same across the al-

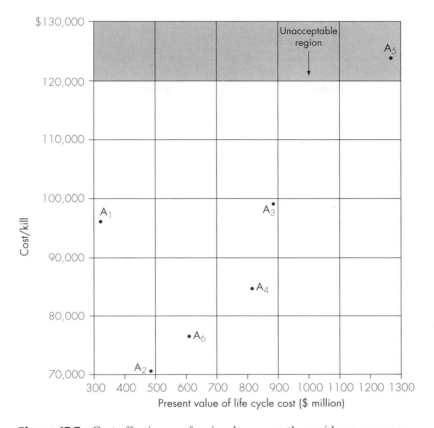

Figure 17.3 Cost effectiveness for six adverse weather guidance weapon systems

ternatives. If we rely entirely on "cost per kill" criterion, a very cheap system with a low kill rate might come out the best.

To make a valid decision, we need to either (1) fix the cost of the system, (then the objective would be to maximize the number of kills) or, equivalently, maximize the kill per cost, or (2) fix the number of kills (then the objective would be to minimize the cost of the system or, equivalently, minimize the cost per kill). The stated situation, fixing the 10,000 units to be purchased, is neither of the above. For example, if the Air Force stipulates that the minimum probability of kill be 0.90 or above, in addition to the unit cost constraint of $120K and 10,000 units, A4 becomes the best. This is the system that will meet the Air Force requirement for an adverse-weather weapon system. (Note that, if the Air Force had adopted the kill/cost criterion, the total budget commitment should have been spelled out.) Even though the cost estimates and the probability of kill were based on the best engineering judgment associated with the complexity of the system and their technological risk, project overrun costs are common, and therefore, uncertainty in the estimated costs for the six weapon alternatives can be expected.

17.5 Summary

- **Benefit-cost analysis** is commonly used to evaluate public projects; several facets unique to public project analysis are neatly addressed by benefit-cost analysis:

  1. Benefits of a nonmonetary nature can be quantified and factored into the analysis.

  2. A broad range of project users distinct from the sponsor should be considered—benefits and disbenefits to *all* these users can (and should) be taken into account.

- Difficulties involved in public project analysis include the following:

  1. Identifying all the users of the project.

  2. Identifying all the benefits and disbenefits of the project.

  3. Quantifying all the benefits and disbenefits in dollars or some other unit of measure.

  4. Selecting an appropriate interest rate at which to discount benefits and costs to a present value.

- The *B/C* ratio is defined as

$$BC(i) = \frac{B}{C} = \frac{B}{I + C'}, I + C' > 0.$$

  The decision rule is if $BC(i) \geq 1$, the project is acceptable.

- The net *B/C* ratio is defined as

$$B'C(i) = \frac{B - C'}{I} = \frac{B'}{I}, I > 0.$$

  The net *B/C* ratio expresses the net benefit expected per dollar invested. The same decision rule applies as for the *B/C* ratio.

- The **cost-effectiveness method** allows us to compare projects on the basis of cost and nonmonetary effectiveness measures. We may either maximize effectiveness for a given cost criterion or minimize cost for a given effectiveness criterion.

Self-Test Questions

17s.1 Which of the following statements is incorrect?

    (a) Both the NPW criterion and the *B/C* ratio allow you to make a consistent accept/reject decision for a given investment as long as you use the same interest rate.

    (b) As long as you use a *B/C* ratio as a base to compare mutually exclusive investment alternatives, it should be based on a *B/C* ratio on the incremental investment.

    (c) Income taxes are not considered in any public investment analyses.

    (d) Income taxes must be considered in evaluating investment projects for nonprofit organizations (such as hospitals and churches).

17s.2 A city government is considering increasing the capacity of the current wastewater treatment plant. The estimated financial data for the project is as follows:

Description	Data
Capital investment	$1,200,000
Project life	25 years
Incremental annual benefits	$250,000
Incremental annual costs	$100,000
Salvage value	$50,000
Discount rate	6%

What would be the net benefit-cost ratio for this expansion project?

    (a) 3.26   (b) 3.12

    (c) 1.61   (d) 2.23.

17s.3 Auburn Recreation and Parks Department is considering two mutually exclusive proposals for a new softball complex on a city-owned lot.

Alternative Design	Seating Capacity	Annual Benefits	Annual Costs	Required Investment
A1	3,000	$194,000	$87,500	$800,000
A2	4,000	224,000	105,000	1,000,000

The complex will be useful for 30 years and has no appreciable salvage value (regardless of seating capacity). Assuming an 8% discount rate, which of the following statements is incorrect?

    (a) Select A1 because it has the largest *B/C* ratio.

    (b) Select A1 because it has the most benefits per seating capacity.

    (c) Select A1 because it has the largest NPW.

    (d) Select A1 because the incremental benefits generated from A2 are not large enough to offset the additional investment ($200,000 over A1).

Problems

Valuation of Benefits and Costs

17.1 The state of Michigan is considering a bill that would ban the use of road salt on highways and bridges during icy conditions. Road salt is known to be toxic, costly, corrosive, and caustic. Chevron Chemical Company produces a calcium magnesium acetate de-icer (CMA) and sells it for $600 a ton as Ice-B-Gon. Road salts, on the other hand, sold for an average of $14 a ton in 1995. Michigan needs about 600,000 tons of road salt each year. (Michigan spent $9.2 million on road salt in 1995.) Chevron estimates that each ton of salt on the road costs $650 in highway corrosion, $525 in rust on vehicles, $150 in corrosion to utility lines, and $100 in damages to water supplies, for a total of $1,425. Unknown salt damage to vegetation and soil surrounding areas of highways has occurred. The state of Michigan would ban road salt (at least on expensive steel bridges or near sensitive lakes) if state studies support Chevron's cost claims.

(a) What would be the users' benefits and sponsor's costs if a complete ban on road salt were imposed in Michigan?

(b) How would you go about determining the salt damages (in dollars) to vegetation and soil?

17.2 A public school system in Ohio is considering the adoption of a 4-day school week as opposed to the current 5-day school week in high schools. The community is hesitant about the plan, but the superintendent of the school system envisions many benefits associated with the 4-day system, Wednesday being the "day off." The following pros and cons have been cited.

- Experiments with the 4-day system indicate that the "day off" in the middle of the week will cut down both on teacher and pupil absences.

- The longer hours on school days will require increased attention spans, which is not an appropriate expectation for younger children.

- The reduction in costs to the federal government should be substantial as the number of lunches served would be cut by approximately 20%.

- The state bases its expenditures on its local school systems largely on the average number of pupils attending school in the system. Since the number of absences will decrease, state expenditures on local systems should increase.

- Older students might want to work on Wednesdays. Unemployment is a problem in this region, however, and any influx of new job-seekers could aggravate an existing problem. Community centers, libraries, and other public areas may also experience increased usage on Wednesdays.

- Parents who provide transportation for their children will see a savings in fuel costs. Primarily, only those parents whose children live fewer than 2 miles from the school would be involved. Children living more than 2 miles from school are eligible for free transportation provided by the local government.

- Decreases in both public and private transportation should result in fuel conservation, decreased pollution, and less wear on the roads. Traffic congestion should ease on Wednesdays in areas where congestion caused by school traffic is a problem.

- Working parents will be forced to make child-care arrangements (and possibly payments) for one weekday per week.

- Students will benefit from wasting less time driving to and from school; Wednesdays will be available for study, thus taking the heavy demand off most nights. Bussed students will spend far less time per week waiting for buses.

- The local school board should see some ease in funding problems. The two areas most greatly impacted are the transportation system and nutritional programs.

(a) For this type of public study, what do you identify as the users' benefits and disbenefits?

(b) What items would be considered as the sponsor's costs?

(c) Discuss any other benefits or costs associated with the 4-day school week.

17.3 The Electric Department of the City of Tallahassee, Florida, operates generating and transmission facilities serving approximately 140,000 people in the city and surrounding Leon County. The city has proposed construction of a $300 million 235-MW circulating fluidized bed combustor (CFBC) at Arvah B. Hopkins Station to power a turbine generator currently receiving steam from an existing boiler fueled by gas or oil. Among the advantages associated with the use of CFBC systems are the following:

- A variety of fuels can be burned, including inexpensive low-grade fuels with high ash and high sulfur content.

- The relatively low combustion temperatures inhibit the formation of nitrogen oxides. Acid-gas emissions associated with CFBC units would be expected to be significantly lower than emissions from conventional coal-fueled units.

- The sulfur-removal method, low combustion temperatures, and high-combustion efficiency characteristic of CFBC units result in solid wastes. These are physically and chemically more amenable to land disposal than the solid wastes resulting from conventional coal-burning boilers equipped with flue-gas desulfurization equipment.

Based on the Department of Energy's (DOE) projections of growth and expected market penetration, demonstration of a 235-MW unit could lead to as much as 41,000 MW of CFBC generation being constructed by the year 2010. The proposed project would reduce the city's dependency on oil and gas fuels by converting its largest generating unit to coal-fuel capability. Consequently, substantial reductions of local acid-gas emissions could be realized in comparison to the permitted emissions associated with oil fuel. The city has requested a $50 million cost share from the DOE. Cost sharing under the Clean Coal Technology Program is considered attractive because the DOE cost share would largely offset the risk of using such a new technology. To qualify for the cost-sharing money, the city has to address the following questions for the DOE.

(a) What is the significance of the project at local and national levels?

(b) What items would constitute the users' benefits and disbenefits associated with the project?

(c) What items would constitute the sponsor's costs?

By putting yourself in the city engineer's position, respond to these questions.

Benefits and Cost Analyses

17.4 A city government is considering two types of town-dump sanitary systems. Design A requires an initial outlay of $400,000, with annual operating and maintenance costs of $50,000 for the next 15 years; design B calls for an investment of $300,000, with annual operating and maintenance costs of $80,000 per year for the next 15 years. Fee collections from the residents would be $85,000 per year. The interest rate is 8%, and no salvage value is associated with either system.

 (a) Using the benefit-cost ratio $(BC(i))$, which system should be selected?

 (b) If a new design (design C), which requires an initial outlay of $350,000 and annual operating and maintenance costs of $65,000, is proposed, would your answer in (a) change?

17.5 The U.S. government is considering building apartments for government employees working in a foreign country and living in locally owned housing. A comparison of two possible buildings indicates the following:

	Building X	Building Y
Original investment by government agencies	$8,000,000	$12,000,000
Estimated annual maintenance costs	240,000	180,000
Savings in annual rent now being paid to house employees	1,960,000	1,320,000

Assume the salvage or sale value of the apartments to be 60% of the first investment. Use 10% and a 20-year study period to compute the B/C ratio on incremental investment and make a recommendation. (Assume no do-nothing alternative.)

17.6 Three public investment alternatives are available, A1, A2, and A3. Their respective total benefits, costs, and first costs are given in present worth. These alternatives have the same service life.

		Proposals	
Present worth	A1	A2	A3
I	100	300	200
B	400	700	500
C'	100	200	150

Assuming no do-nothing alternative, which project would you select based on the benefit-cost ratio $(BC(i))$ on incremental investment?

17.7 A local city, which operates automobile parking facilities, is evaluating a proposal that it erect and operate a structure for parking in a city's downtown area. Three designs for a facility to be built on available sites have been identified. (All dollar figures are in thousands.)

	Design A	Design B	Design C
Cost of site	$240	$180	$200
Cost of building	$2,200	700	1,400
Annual fee collection	$830	750	600
Annual maintenance cost	$410	360	310
Service life	30 years	30 years	30 years

At the end of the estimated service life, whichever facility had been constructed would be torn down, and the land would be sold. It is estimated that the proceeds from the resale of the land will be equal to the cost of clearing the site. If the city's interest rate is known to be 10%, which design alternative would be selected based on the benefit-cost criterion?

17.8 The federal government is planning a hydroelectric project for a river basin. In addition to the production of electric power, this project will provide flood control, irrigation, and recreation benefits. The estimated benefits and costs expected to be derived from the three alternatives under consideration are listed below.

| | Decision Alternatives | | |
	A	B	C
Initial cost	$8,000,000	$10,000,000	$15,000,000
Annual benefits or costs			
Power sales	$1,000,000	$1,200,000	$1,800,000
Flood control savings	250,000	350,000	500,000
Irrigation benefits	350,000	450,000	600,000
Recreation benefits	100,000	200,000	350,000
O&M costs	200,000	250,000	350,000

The interest rate is 10%, and the life of each of the projects is estimated to be 50 years.

(a) Find the benefit-cost ratio for each alternative.

(b) Select the best alternative based on $BC(i)$.

17.9 Two different routes are under consideration for a new interstate highway.

	Length of Highway	First Cost	Annual Upkeep
The "long" route	22 miles	$21 million	$140,000
Transmountain shortcut	10 miles	$45 million	$165,000

For either route, the volume of traffic will be 400,000 cars per year. These cars are assumed to operate at $0.25 per mile. Assume a 40-year life for each road and an interest rate of 10%. Determine which route should be selected.

17.10 The government is considering undertaking the four projects listed below. These projects are mutually exclusive, and the estimated present worth of their costs and the present worth of their benefits are shown in millions of dollars. All projects have the same duration.

Projects	PW of Benefits	PW of Costs
A1	40	85
A2	150	110
A3	70	25
A4	120	73

Assuming no do-nothing alternative, which alternative would you select? Justify your choice by using a benefit-cost $(BC(i))$ on incremental investment.

Short Case Studies

17.11 Fast growth in the population of the city of Orlando and surrounding counties, Orange County in particular, has resulted in insurmountable traffic congestion. The county has few places to turn for extra money for road improvements except to new taxes. County officials have said that the money they receive from current taxes is insufficient to widen overcrowded roads, improve roads that don't meet modern standards, and pave dirt roads. State residents now pay 12 cents in taxes on every gallon of gas. Four cents of that goes to the federal government, 4 cents to the state, 3 cents to the county in which the tax is collected, and 1 cent to the cities. The county commissioner has suggested that the county get the money by tacking an extra penny-a-gallon tax onto gasoline, bringing the total federal and state gas tax to 13 cents a gallon. This would add about $2.6 million a year to the road-construction budget. The extra money would have a significant impact. With the additional revenue, the county could sell a $24 million bond issue. It would then have the option of spreading that amount among many smaller projects or concentrating on a major project. Assuming that voters would approve a higher gas tax, the county engineers were asked to prepare a priority list outlining which roads would be improved with the extra money. The road engineers also computed the possible public benefits associated with each road-construction project; they accounted for possible reduction in travel time, a reduction in the accident rate, land appreciation, and savings in operating costs of vehicles.

District	Project	Type of Improvement	Construction Cost	Annual O&M	Annual Benefits
	27th Street	Four-lane	$ 980,000	$ 9,800	$313,600
	Holden Avenue	Four-lane	3,500,000	35,000	850,000
I	Forest City Road	Four-lane	2,800,000	28,000	672,000
	Fairbanks Avenue	Four-lane	1,400,000	14,000	490,000
	Oak Ridge Road	Realign	2,380,000	47,600	523,600
II	University Blvd.	Four-lane	5,040,000	100,800	1,310,400
	Hiawassee Road	Four-lane	2,520,000	50,400	831,600
	Lake Avenue	Four-lane	4,900,000	98,000	1,021,000
	Apopka-Ocoee Road	Realign	1,365,000	20,475	245,700
III	Kaley Avenue	Four-lane	2,100,000	31,500	567,000
	Apoka-Vineland Road	Two-lane	1,170,000	17,550	292,000
	Washington Street	Four-lane	1,120,000	16,800	358,400
	Mercy Drive	Four-lane	2,800,000	56,000	980,000
IV	Apopka Road	Reconstruct	1,690,000	33,800	507,000
	Old Dixie Highway	Widen	975,000	15,900	273,000
	Old Apopka Road	Widen	1,462,500	29,250	424,200

Assume a 20-year planning horizon and an interest rate of 10%. Which projects would be considered for funding in (a) and (b)?

(a) Due to political pressure, each district will have the same amount of funding, say, $6 million.

(b) The funding will be based on tourist traffic volumes. Districts I and II combined will get $15 million, and Districts III and IV combined will get $9 million. It is desirable to have at least one four-lane project from each district.

17.12 The City of Portland Sanitation Department is responsible for the collection and disposal of all solid waste within the city limits. The city must collect and dispose of an average of 300 tons of garbage each day. The city is considering ways to improve the current solid-waste collection and disposal system.

- The present collection and disposal system uses Dempster Dumpmaster Frontend Loaders for collection, and incineration or landfill for disposal. Each collecting vehicle has a load capacity of 10 tons, or 24 cubic yards, and dumping is automatic. The incinerator in use was manufactured in 1942. It was designed to incinerate 150 tons per 24 hours. A natural-gas afterburner has been added in an effort to reduce air pollution; however, the incinerator still does not meet state air-pollution requirements, and it is operating under a permit from the Oregon State Air and Water Pollution Control Board. Prison-farm labor is used for the operation of the incinerator. Because the capacity of the incinerator is relatively low, some trash is not incinerated, but is taken to the city landfill. The trash landfill is located approximately 11

miles, and the incinerator approximately 5 miles, from the center of the city. The mileage and costs in man-hours for delivery to the disposal sites is excessive; a high percentage of empty vehicle miles and man-hours are required because separate methods of disposal are used, and the destination sites are remote from the collection areas. The operating cost for the present system is $905,400. This includes $624,635 to operate the prison-farm incinerator, $222,928 to operate the existing landfill, and $57,837 to maintain the current incinerator.

- The proposed system locates a number of portable incinerators, each with 100-ton-per-day capacity for the collection and disposal of refuse waste collected for three designated areas within the city. Collection vehicles will also be staged at these incineration-disposal sites with the necessary plant and support facilities for incineration operation, collection-vehicle fueling and washing, support building for stores, and shower and locker rooms for collection and site crew personnel. The pick-up and collection procedure remains essentially the same as in the existing system. The disposal-staging sites, however, are located strategically in the city based on the volume and location of wastes collected, thus eliminating long hauls and reducing the number of miles the collection vehicles must retravel from pick-up to disposal site.

Four variations of the proposed system are being considered, containing 1, 2, 3, and 4 incinerator-staging areas respectively. The type of incinerator is a modular prepackaged unit, which can be installed at several sites in the city. Such units exceed all state and federal standards on their exhaust emissions. The city of Portland needs 24 units, each with a rated capacity of 12.5 tons of garbage per 24 hours. The price per unit is $137,600, which means a capital investment of about $3,302,000. The estimated plant facilities, such as housing and foundation, were estimated to cost $200,000 per facility. This is based on a plan incorporating four incinerator plants strategically located around the city. Each plant would house eight units and be capable of handling 100 tons of garbage per day. Additional plant features, such as landscaping, were estimated to cost $60,000 per plant.

The annual operating cost of the proposed system would vary according to the type of system configuration. It takes about 1.5 to 1.7 MCF of fuel to incinerate one ton of garbage. The conservative 1.7 MCF figure was used for total cost. This means that fuel cost $4.25 per ton of garbage at a cost of $2.50 per MCF. Electric requirements at each plant will be 230 kW per day. If the plant is operating at full capacity, that means a $0.48 per ton cost for electricity. Two men can easily operate one plant, but safety factors dictate three operators at a cost of $7.14 per hour. This translates to a cost of $1.72 per ton. The maintenance cost of each plant was estimated to be $1.19 per ton. Since three plants will require fewer transportation miles, it is necessary to consider the savings accruing from this operating advantage. Three plant locations will save 6.14 miles per truck per day on the average. At an estimated $0.30 per mile cost, this would mean an annual savings of $6,750 is realized when considering minimum trips to the landfill disposer, for a total annual

savings in transportation of $15,300. A labor savings is also realized because of the shorter routes, which permit more pick-ups during the day. This results in an annual savings of $103,500. The following table summarizes all costs in thousands of dollars associated with the present and proposed systems.

| Item | Present System | Costs for Proposed Systems Site Number | | | |
		1	2	3	4
Capital costs					
Incinerators		$3,302	$3,302	$3,302	$3,302
Plant facilities		600	900	1,260	1,920
Annex buildings		91	102	112	132
Additional					
features		60	80	90	100
Total		$4,053	$4,384	$4,764	$5,454
Annual O&M					
costs	$905.4	$342	$480	$414	$408
Annual savings					
Pick-up					
transportation		$13.2	$14.7	$15.3	$17.1
Labor		87.6	99.3	103.5	119.40

A bond will be issued to provide the necessary capital investment at an interest rate of 8% with a maturity date 20 years in the future. The proposed systems are expected to last 20 years with negligible salvage values. If the current system is to be retained, the annual O&M costs would be expected to increase at an annual rate of 10%. The city will use the bond interest rate as the interest rate for any public project evaluation.

(a) Determine the operating cost of the current system in terms of dollars per ton of solid waste.

(b) Determine the economics of each solid-waste disposal alternative in terms of dollars per ton of solid waste.

17.13 Because of a rapid growth in population, a small town in Pennsylvania is considering several options to establish a waste water treatment facility that can handle a waste water flow of 2 MGD (million gallons per day). The town has five treatment options available:

Option 1—No action: This option will lead to continued deterioration of the environment. If growth continues and pollution results, fines imposed (as high as $10,000 per day) would soon exceed construction costs.

Option 2—Land treatment facility: Provide a system for land treatment of wastewater to be generated over the next 20 years. This option will require the utilization of the most land for treatment of the wastewater. In addition to finding a suitable site, pumping of the wastewater for a considerable distance out of town will be required. The land cost in the area is $3,000 per acre. The system will use spray irrigation to distribute wastewater over the site. No more than one inch of wastewater can be applied in one week per acre.

Option 3—Activated sludge-treatment facility: Provide an activated sludge-treatment facility at a site near the planning area. No pumping will be required for this alternative. Only 7 acres of land will be needed for construction of the plant at a cost of $7,000 per acre.

Option 4—Trickling filter-treatment facility: Provide a trickling filter-treatment facility at the same site selected for the activated sludge plant of option 3. The land required will be the same as used for option 3. Both facilities will provide similar levels of treatment using different units.

Option 5—Lagoon treatment system: Utilize a three-cell lagoon system for treatment. The lagoon system requires substantially more land than options 3 and 4, but less than option 2. Due to the larger land requirement, this treatment system will have to be located some distance outside of the planning area and will require pumping of the wastewater to reach the site.

The following summarizes the capital expenditures and O&M costs associated with each option:

	Land Cost for Each Option		
Option Number	Land Required (acres)	Land Cost ($)	Land Value (in 20 years)
2	800	$2,400,000	$4,334,600
3	7	49,000	88,500
4	7	49,000	88,500
5	80	400,000	722,400

The price of land is assumed to be appreciating at an annual rate of 3%.

Option Number	Capital Expenditures			
	Equipment	Structure	Pumping	Total
2	$500,000	$700,000	$100,000	$1,300,000
3	500,000	2,100,000	0	2,600,000
4	400,000	2,463,000	0	2,863,000
5	175,000	1,750,000	100,000	2,025,000

The equipment installed will require a replacement cycle of 15 years. Its replacement cost will increase at an annual rate of 5% (over the initial cost), and its salvage value at the end of the planning horizon will be 50% of the replacement cost. The structure requires replacement after 40 years and will have a salvage value of 60% of the original cost.

Option Number	Annual O & MCosts			
	Energy	Labor	Repair	Total
2	$200,000	$95,000	$30,000	$325,000
3	125,000	65,000	20,000	210,000
4	100,000	53,000	15,000	168,000
5	50,000	37,000	5,000	92,000

The cost of energy and repair will increase at an annual rate of 5% and 2%, respectively. The labor cost will increase at an annual rate of 4%.

With the following sets of assumptions, answer (a) and (b).

(a) If the interest rate (including inflation) is 10%, which option is the most cost-effective?

(b) Suppose a household discharges about 400 gallons of waste water per day through the facility selected in (a). What should be the monthly assessed bill for this household?

• Assume analysis period of 120 years.

• Replacement costs for the equipment as well as pumping facilities will increase at an annual rate of 5%.

• Replacement cost for the structure will remain constant over the planning period. However, its salvage value will be 60% of the original cost. (Because it has a 40-year replacement cycle, any increase in the future replacement cost will have very little impact on the solution.)

• The equipment's salvage value at the end of its useful life will be 50% of the original replacement cost. For example, the equipment installed for option 1 will cost $500,000. It's salvage value at the end of 15 years will be $250,000.

• All O&M cost figures are given in today's dollars. For example, the annual energy cost of $20,000 for op-

tion 2 means that the actual energy cost during the first operating year will be $200,000(1.05) = $210,000.

• Option 1 is not considered a viable alternative as its annual operating cost exceeds $36,500,00.

17.14 The Federal Highway Administration predicts that by the year 2005, Americans will be spending 8.1 billion hours per year in traffic jams. Most traffic experts believe that adding and enlarging highway systems will not alleviate the problem. As a result, current research on traffic management is focusing on three areas: (1) development of computerized dashboard navigational systems, (2) development of roadside sensors and signals that monitor and help manage the flow of traffic, and (3) development of automated steering and speed controls that might allow cars to drive themselves on certain stretches of highway.

In Los Angeles, perhaps the most traffic-congested city in the United States, a Texas Transportation Institute study found that traffic delays cost motorists $8 billion per year. But Los Angeles has already implemented a system of computerized traffic-signal controls that, by some estimates, has reduced travel time by 13.2%, fuel consumption by 12.5%, and pollution by 10%. And between Santa Monica and downtown Los Angeles, testing of an electronic traffic and navigational system—including highway sensors and cars with computerized dashboard maps—is being sponsored by federal, state, and local governments and General Motors Corporation. This test program costs $40 million; to install it throughout Los Angeles could cost $2 billion.

On a national scale, the estimates for implementing "smart" roads and vehicles is even more staggering: It would cost $18 billion to build the highways, $4 billion per year to maintain and operate them, $1 billion for research and development of driver-information aids, and $2.5 billion for vehicle-control devices. Advocates say the rewards far outweigh the costs.

(a) On a national scale, how would you identify the user's benefits and disbenefits for this type of public project?

(b) On a national scale, what would be the sponsor's cost?

(c) Suppose that the user's net benefits grow at 3% per year and the sponsor's costs grow at 4% per year. Assuming a social discount rate of 10%, what would be the B/C ratio over a 20-year study period?

17.15 Reconsider the airport expansion problem in the chapter opening. Members will consider three options to handle the growth expected at Hartsfield.

- Option 1: Five runways, a new concourse adjacent to a new eastside terminal and close in parking, Cost estimate: $1.5 billion.

- Option 2: A more expensive version of the first option, putting the new

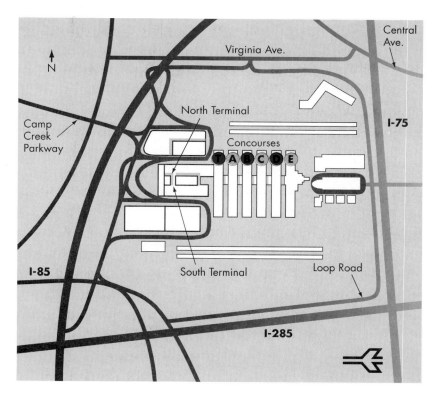

Proposed future expansion plan for Atlanta Hartsfield Airport

concourse adjacent to the fifth runway. Cost estimate: $1.6 billion.

- Option 3: A sixth runway north of the current airport boundaries in the College-Park–East Point–Hapeville corridor side terminal. Cost estimate: $5.9 billion.

A study indicates that a north runway would displace more than 3,000 households and 978 businesses. A South side runway could dislocate as many as 7,000 households and more than 900 businesses. Under any scenario of airport expansion, there still will be some flight delays. However, some plans project longer delays than others. The table summarizes some projections for the year 2015 under the no-build scenario and three scenarios that do call for some expansion and new construction.

Dealing with the Delays
(Units in minutes)

	No change	Option 1	Option 2	Option 3
Average departure delay	34.80	17.02	19.66	5.71
Average arrival delay	19.97	7.08	6.68	3.91
Average taxiing time	26.04	18.54	19.99	12.80

The committee has 6 months to study each option before making a final recommendation. What would you recommend?

Comments: This would be classified as a replacement project. Many civil engineers would work on public works areas such as highway construction, airport construction, and water projects. In this airport expansion scenario, each option requires a different level of investment with a different degree of benefits. One of the most important aspects of airport expansion is to quantify the cost of airport delays in dollar terms. In other words, what is the economic benefit of reducing any airport delay? From the airline's point of view, any taxiing and arrival delays mean added fuel costs. For Atlanta airport, any delays mean loss revenues in landing and departure fees. From the public's point of view, any delays mean loss of earnings, as they have to spend more time on transportation. Comparing the investment costs with potential benefits, known as benefit-cost analysis, is an important feature of the economic analysis method. (Source: "Committee Wary of Airport Expansion," *The Atlanta Journal/The Atlanta Constitution*, December 13, 1997)

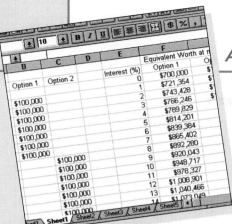

Computing IRR for Nonsimple Investments

To comprehend the nature of multiple i^*s, we need to understand the investment situation represented by any cash flow. The net investment test will indicate whether the i^* computed represents the true rate of return earned on the money invested in a project while it is actually in the project. As we shall see, the phenomenon of multiple i^*s occurs only when the net investment test fails. When multiple positive rates of return for a cash flow are found, in general none is suitable as a measure of project profitability, and we must proceed to the next analysis step: Introducing an external rate of return.

A.1 Predicting Multiple i^*s

As hinted at in Section 9.1.1, for certain series of project cash flows, we may uncover the complication of multiple i^* values that satisfy Eq. (9.1). By analyzing and classifying cash flows, we may anticipate this difficulty and adjust our analysis approach. Here we will focus on the initial problem of whether we can predict a unique i^* for a project by examining its cash flow pattern. Two useful rules allow us to focus on sign changes (1) in net cash flows and (2) in accounting net profit (accumulated net cash flows).

A.1.1 Net Cash Flow Rule of Signs

One useful method for predicting an upper limit on the number of positive i^*s of a cash flow stream is to apply the rule of signs: *The number of real i^*s that are greater than −100% for a project with N periods is never greater than the number of sign changes in the sequence of the A_n. A zero cash flow is ignored.*
    An example would be

Period	A_n	Sign Change
0	−$100	
1	−20	
2	50	1
3	0	
4	60	
5	−30	1
6	100	1

Three sign changes occur in the cash flow sequence, so three or fewer real positive i^*s exist.
    It must be emphasized that the rule of signs provides an indication only of the possibility of multiple rates of return: The rule only predicts the *maximum* number of possible i^*s. Many projects have multiple sign changes in their cash flow sequence, but still possess a unique real i^* in the $(-100\%, +\infty)$ range.

A.1.2 Accumulated Cash Flow Sign Test

The accumulated cash flow is the sum of the net cash flows up to, and including, a given time. If the rule of cash flow signs indicates multiple i^*s, we should proceed to the **accumulated cash flow sign test** to eliminate some possibility of multiple rates of return.
    If we let A_n represent the net cash flow in period n and S_n represent the accumulated cash flow up to period n, we have the following:

Period (n)	Cash Flow (A_n)	Accumulated Cash Flow (S_n)
0	A_0	$S_0 = A_0$
1	A_1	$S_1 = S_0 + A_1$
2	A_2	$S_2 = S_1 + A_2$
⋮	⋮	⋮
N	A_N	$S_N = S_{N-1} + A_N$

We then examine the sequence of accumulated cash flows ($S_0, S_1, S_2, S_3, \ldots, S_N$) to determine the number of sign changes. *If the series S_n starts negatively and changes sign only once, a unique positive i^* exists.* This cumulative cash flow sign rule is a more discriminating test for identifying the uniqueness of i^* than the previously described method.

Example A.1 Predicting the Number of i^*s

Predict the number of real positive rate(s) of return for each cash flow series:

Period	A	B	C	D
0	−$100	−$100	$ 0	−$100
1	−200	+ 50	−50	+50
2	+200	−100	+115	0
3	+200	+ 60	−66	+200
4	+200	−100		−50

Solution

Given: Four cash flow series and cumulative flow series

Find: The upper limit on number of i^*s for each series

The cash flow rule of signs indicates the following possibilities for the positive values of i^*:

Project	Number of Sign Changes in Net Cash Flows	Possible Number of Positive Values of i^*
A	1	1 or 0
B	4	4, 3, 2, 1 or 0
C	2	2, 1, or 0
D	2	2, 1, or 0

For cash flows B, C, and D, we would like to apply the more discriminating cumulative cash flow test to see if we can specify a smaller number of possible values of i^*.

Project B A_n	S_n	Project C A_n	S_n	Project D A_n	S_n
-$100	-$100	$ 0	$ 0	-$100	-$100
+50	-50	-50	-50	+50	-50
-100	-150	+115	+65	0	-50
+60	-90	-66	-1	+200	+150
-100	-190			-50	+100

Recall the test: If the series starts *negatively* and changes sign only once, a unique positive i^* exists. Only project D begins negatively and passes the test; we may predict a unique i^* value, rather than 2, 1, or 0 as predicted by the cash flow rule of signs. Project B, with no sign change in the cumulative cash flow series, has no rate of return. Project C fails the test, and we cannot eliminate the possibility of multiple i^*s. (If projects do not begin negatively, they are borrowing projects rather than investment projects.)

A.2 Net Investment Test

A project is said to be a **net investment** when the project balances computed at the project's i^* values, $PB(i^*)_n$, are either less than or equal to zero throughout the life of the investment with $A_0 < 0$. The investment is *net* in the sense that the firm does not overdraw on its return at any point and hence is *not indebted* to the project. This type of project is called a **pure investment**. [On the other hand, **pure borrowing** is defined as the situation where $PB(i^*)_n$ values are positive or zero throughout the life of the loan with $A_0 > 0$.] *Simple investments will always be pure investments.* Therefore, if a nonsimple project passes the net investment test (a pure investment), then the accept/reject decision rule will be the same as in the simple investment case given in Section 9.3.2.

If any of the project balances calculated at the project's i^* is positive, the project is not a pure investment. A positive project balance indicates that, at some time during the project life, the firm acts as a borrower [$PB(i^*)_n > 0$] rather than an investor in the project [$PB(i^*)_n < 0$]. This type of investment is called a **mixed** investment.

Example A.2 Pure versus Mixed Investments

Consider the following four investment projects with known i^* values. Determine which projects are pure investments.

n	A	B	C	D
0	−$1,000	−$1,000	−$1,000	−$1,000
1	−1,000	1,600	500	3,900
2	2,000	−300	−500	−5,030
3	1,500	−200	2,000	2,145
i*	33.64%	21.95%	29.95%	10%, 30%, 50%

Solution

Given: Four projects with cash flows and i*s as shown

Find: Which projects are pure investments

We will first compute the project balances at the projects' respective i*s. If multiple rates of return exist, we may use the largest value of i* greater than zero.[1]

Project A:

$$PB(33.64\%)_0 = -\$1,000.$$
$$PB(33.64\%)_1 = -\$1,000(1 + 0.3364) + (-\$1,000) = -\$2,336.40.$$
$$PB(33.64\%)_2 = -\$2,336.40\,(1 + 0.3364) + \$2,000 = -\$1,122.36.$$
$$PB(33.64\%)_3 = -\$1,122.36\,(1 + 0.3364) + \$1,500 = 0.$$

$(-, -, -, 0)$: Passes the net investment test (pure investment).

Project B:

$$PB(21.95\%)_0 = -\$1,000.$$
$$PB(21.95\%)_1 = -\$1,000(1 + 0.2195) + \$1,600 = \$380.50.$$
$$PB(21.95\%)_2 = +\$380.50(1 + 0.2195) - \$300 = \$164.02.$$
$$PB(21.95\%)_3 = +\$164.02(1 + 0.2195) - \$200 = 0.$$

$(-, +, +, 0)$: Fails the net investment test (mixed investment).

Project C:

$$PB(29.95\%)_0 = -\$1,000$$
$$PB(29.95\%)_1 = -\$1,000(1 + 0.2995) + \$500 = -\$799.50.$$
$$PB(29.95\%)_2 = -\$799.50(1 + 0.2995) - \$500 = -\$1,538.95.$$
$$PB(29.95\%)_3 = -\$1,538.95(1 + 0.2995) + \$2,000 = 0.$$

$(-, -, -, 0)$: Passes the net investment test (pure investment).

---

[1] In fact, it does not matter which rate we use in applying the net investment test. If one value passes the net investment test, they will all pass. If one value fails, they all fail.

Project D: (There are three rates of return. We can use any of them for the net investment test.)

$$PB(50\%)_0 = -\$1,000$$

$$PB(50\%)_1 = -\$1,000(1 + 0.50) + \$3,900 = \$2400.$$

$$PB(50\%)_2 = +\$2,400(1 + 0.50) - \$5,030 = -\$1,430.$$

$$PB(50\%)_3 = -\$1,430(1 + 0.50) + \$2,145 = 0.$$

$(-, +, -, 0)$: Fails the net investment test (mixed investment).

Comments: As shown in Figure A.1, projects A and C are the only pure investments. Project B demonstrates that the existence of a unique i^* is a necessary but not sufficient condition for a pure investment.

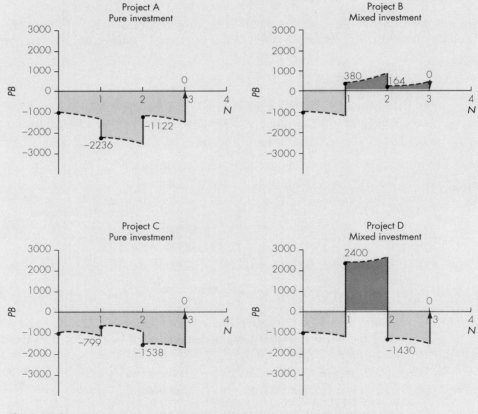

Figure A.1   Net investment test (Example A.2)

Example A.3 IRR for Nonsimple Project: Pure Investment

Consider project C in Example A.2. Assume that the project is independent and that all costs and benefits are stated explicitly. Apply the net investment test and determine the acceptability of the project using the IRR criterion.

Solution

Given: Cash flow, $i^* = 29.95\%$, MARR $= 15\%$

Find: IRR, and determine whether the project is acceptable

The net investment test was already applied in Example A.2, and it indicated that project C is a pure investment. In other words, the project balances were all less than or equal to zero, and the final balance was zero, as it must be if the interest rate used is an i^*. This proves that 29.95% is the true internal rate of return for the cash flow. At MARR $= 15\%$, the IRR $>$ MARR, thus, the project is acceptable. If we compute the NPW of this project at $i = 15\%$, we obtain

$$PW(15\%) = -\$1,000 + \$500(P/F, 15\%, 1) - \$500(P/F, 15\%, 2)$$
$$+ \$2,000(P/F, 15\%, 3)$$
$$= \$371.74.$$

Since $PW(15\%) > 0$, the project is also acceptable under the NPW criterion. The IRR and NPW criteria will always produce the same decision if the criteria are applied correctly.

Comments: In the case of a pure investment, the firm has funds committed to the project over the life of the project and at no time withdraws money from the project. The IRR is the return earned on the funds that remain internally invested in the project.

A.3 External Interest Rate for Mixed Investments

Even for a nonsimple investment, in which there is only one positive rate of return, the project may fail the net investment test, as demonstrated by project B in Example A.2. In this case, the unique i still may not be a true indicator of the project's profitability. That is, when we calculate the project balance at an i^* for mixed investments, we notice an important point—cash borrowed (released) from the project is assumed to earn the same interest rate through external investment as money that remains internally invested. In other words, in solving a cash flow for an unknown interest rate it is assumed that money released from a project can be reinvested to yield a rate of return equal to that received from the project. In fact, we have been making this assumption whether or not a cash flow produces a unique positive i^*. Note that money is borrowed only when $PB(i^*) > 0$, and the magnitude of the borrowed amount is the project balance. When $PB(i*) < 0$, no money is borrowed, even though the cash flow may be positive at that time.

In reality, it is not always possible for cash borrowed (released) from a project to be reinvested to yield a rate of return equal to that received from the project. Instead, it is likely that the rate of return available on a capital investment in the business is much different—usually higher—from the rate of return available on other external investments. Thus, it may be necessary to compute the project balances for a project's cash flow at two rates of interest—one on the internal investment and one on the external investments. As we will see later, by separating the interest rates, we can measure the **true rate of return** of any internal portion of investment project.

Because the net investment test is the only way to accurately predict project borrowing (i.e., external investment), its significance now becomes clear: In order to calculate accurately a project's true IRR, we should always test a solution by the net investment test and, when the test fails, take the further analytical step of introducing an external interest rate. Even the presence of a unique positive $i*$ is a necessary but not sufficient condition to predict net investment, so if we find a unique value we should still subject it to the net investment test.

A.4 Calculation of Return on Invested Capital for Mixed Investments

A failed net investment test indicates a combination of internal and external investment. When this combination exists, we must calculate a rate of return on the portion of capital that remains invested internally. This rate is defined as the **true IRR** for the mixed investment or commonly known as **return on invested capital (RIC)**.

How do we determine the IRR of this investment? Insofar as a project is not a net investment, one or more periods when the project has a net outflow of money (positive project balance) must later be returned to the project. This money can be put into the firm's investment pool until such time as it is needed in the project. The interest rate of this investment pool is the interest rate at which the money can in fact be invested outside the project.

Recall that the NPW method assumed that the interest rate charged to any funds withdrawn from a firm's investment pool would be equal to the MARR. In this book, we will use the MARR as an established external interest rate — (i.e., the rate earned by money invested outside of the project). We can then compute IRR, or RIC, as a function of MARR by finding the value of IRR that will make the terminal project balance equal to zero. (This implies that the firm wants to fully recover any investment made in the project and pays off any borrowed funds at the end of project life.) This way of computing rate of return is an accurate measure of the profitability of the project represented by the cash flow. The following procedure outlines the steps for determining the IRR for a mixed investment:

Step 1: Identify the MARR (or external interest rate).

Step 2: Calculate $PB(i, MARR)_n$ (or simply PB_n) according to the rule

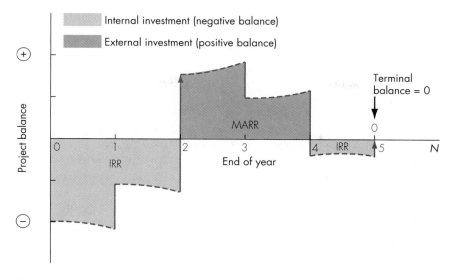

Figure A.2 Computational logic for IRR (mixed investment)

$$PB(i, MARR)_0 = A_0.$$

$$PB(i, MARR)_1 = \begin{cases} PB_0(1 + i) + A_1, \text{ if } PB_0 < 0. \\ PB_0(1 + MARR) + A_1, \text{ if } PB_0 > 0. \end{cases}$$

$$\vdots$$

$$PB(i, MARR)_n = \begin{cases} PB_{n-1}(1 + i) + A_n, \text{ if } PB_{n-1} < 0. \\ PB_{n-1}(1 + MARR) + A_n, \text{ if } PB_{n-1} > 0. \end{cases}$$

(As defined in the text, A_n stands for the net cash flow at the end of period n. Note also that the terminal project balance must be zero.)

Step 3: Determine the value of i by solving the terminal project balance equation

$$PB(i, MARR)_N = 0.$$

That interest rate is the IRR for the mixed investment.

Using the MARR as an external interest rate, we may accept a project if the IRR exceeds MARR, and should reject the project otherwise. Figure A.2 summarizes the IRR computation for a mixed investment.

Example A.4 IRR for Nonsimple Project: Mixed Investment

Reconsider the defense contractor's flight-simulator project in Example 9.6. The project was a nonsimple and mixed investment. To simplify the decision-making process, we abandoned the IRR criterion and used the NPW to make an accept/reject decision. Apply the procedures outlined to find the true IRR, or return on invested capital, or this mixed investment.

(a) Compute the IRR (RIC) for this project, assuming the MARR = 15%.

(b) Make an accept/reject decision based on the results in part (a).

Solution

Given: Cash flow shown in Example 9.6, MARR = 15%

Find: (a) IRR and (b) determine whether to accept the project

(a) As calculated in Example 9.6, the project has multiple rates of return. This is obviously not a net investment, as shown below. Because the net investment test indicates external as well as internal investment, neither 10% nor 20% represents the true internal rate of return of this government project. Since the project is a mixed investment, we need to find the IRR by applying the steps shown previously.

Net Investment Test	Using $i^* = 10\%$			Using $i^* = 20\%$		
n	0	1	2	0	1	2
Beginning balance	$0	−$1,000	$1,200	$0	−$1,000	$1,100
Return on investment	0	−100	120	0	−200	220
Payment	−1,000	2,300	−1,320	−1,000	2,300	−1,320
Ending balance	−$1,000	$1,200	0	−$1,000	$1,100	0

(Unit: $1,000)

At $n = 0$, there is a net investment to the firm so that the project balance expression becomes

$$PB(i, 15\%)_0 = -\$1,000,000.$$

The net investment of $1,000,000 that remains invested internally grows at i for the next period. With the receipt of $2,300,000 in year 1, the project balance becomes

$$PB(i, 15\%)_1 = -\$1,000,000(1 + i) + \$2,300,000$$
$$= \$1,300,000 - \$1,000,000i$$
$$= \$1,000,000(1.3 - i)$$

At this point, we do not know whether $PB(i, 15\%)_1$ is positive or negative: We want to know this in order to test for net investment and the presence of a unique i^*. It depends on the value of i, which we want to determine. Therefore, we need to consider two situations: (1) $i < 1.3$ and (2) $i > 1.3$.

- Case 1: $i < 1.3 \rightarrow PB(i, 15\%)_1 > 0$.

Since this would be a positive balance, the cash released from the project would be returned to the firm's investment pool to grow at the MARR until it is required back in the project. By the end of year 2, the cash placed in the investment pool would have grown at the rate of 15% [to $\$1,000,000(1.3 - i)(1 + 0.15)$], and must equal the investment into the project of $\$1,320,000$ required at that time. Then, the terminal balance must be

$$
\begin{aligned}
PB(i, 15\%)_2 &= \$1,000,000(1.3 - i)(1 + 0.15) - \$1,320,000 \\
&= \$175,000 - \$1,150,000i \\
&= 0.
\end{aligned}
$$

Solving for i yields

$$\text{IRR (or RIC)} = 0.1522 \text{ or } 15.22\%.$$

The computational process is shown graphically in Figure A.3.

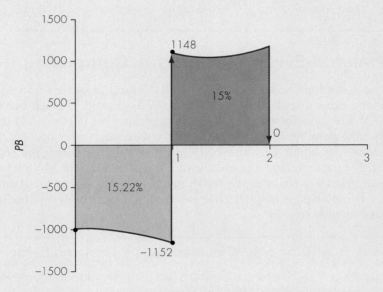

Figure A.3   Calculation of the IRR for a mixed investment (Example A.4)

- Case 2: $i > 1.3 \rightarrow PB(i, 15\%)_1 < 0$.

The firm is still in an investment mode. Therefore, the balance at the end of year 1 that remains invested will grow at the i for the next period. With the investment of $\$1,320,000$ required in year 2, and the fact that the net investment must be zero at the end of project life, the balance at the end of year 2 should be

$$PB(i, 15\%)_2 = \$1,000,000(1.3 - i)(1 + i) - \$1,320,000$$
$$= -\$20,000 + \$300,000i - \$1,000,000i^2$$
$$= 0.$$

Solving for i gives

$$IRR = 0.1 \text{ or } 0.2 < 1.3,$$

which violates the initial assumption ($i > 1.3$). Therefore, Case 1 is the correct situation.

(b) Case 1 indicates IRR > MARR, so the project would be acceptable, resulting in the same decision as obtained in Example 9.6 by applying the NPW criterion.

Comments: In this example we could have seen by inspection that Case 1 was correct. Since the project required an investment as the final cash flow, the project balance at the end of the previous period (year 1) had to be positive in order for the final balance to equal zero. Inspection does not generally work for more complex cash flows.

A.5 Trial-and-Error Method for Computing IRR

The trial-and-error approach for finding IRR(RIC) for a mixed investment is similar to the trial-and-error approach for finding i^*. We begin with a given MARR and a guess for IRR and solve for the project balance. (A value of IRR close to the MARR is a good starting point for most problems.) Since we desire the project balance to approach zero, we can adjust the value of IRR as needed after seeing the result of the initial guess. For example, for a given pair of interest rates (IRR, MARR), if the terminal project balance is positive, the IRR value is too low, so we raise it and recalculate. We can continue adjusting our IRR guesses in this way until we obtain a project balance equal or close to zero.

Example A.5 IRR for Mixed Investment by Trial and Error

Consider project D in Example A.2, which has the following cash flow. We know from an earlier calculation that this is a mixed investment.

n	A_n
0	−$1,000
1	3,900
2	−5,030
3	2,145

Compute the IRR for this project. Assume that MARR = 6%.

Solution

Given: Cash flow as stated for mixed investment, MARR = 6%
Find: IRR

For MARR = 6%, we must compute i by trial and error. Suppose we guess $i = 8\%$:

$$PB(8\%, 6\%)_0 = -\$1,000.$$

$$PB(8\%, 6\%)_1 = -\$1,000(1 + 0.08) + \$3,900 = \$2,820.$$

$$PB(8\%, 6\%)_2 = +\$2,820(1 + 0.06) - \$5,030 = -\$2,040.80.$$

$$PB(8\%, 6\%)_3 = -\$2,040.80(1 + 0.08) + \$2,145 = -\$59.06.$$

The net investment is negative at the end of the project, indicating that our trial $i = 8\%$ is in error. After several trials, we conclude that for MARR = 6%, IRR is approximately at 6.13%. To verify the results,

$$PB(6.13\%, 6\%)_0 = -\$1,000.$$

$$PB(6.13\%, 6\%)_1 = -\$1,000.00(1 + 0.0613) + \$3,900 = \$2,838.66$$

$$PB(6.13\%, 6\%)_2 = +\$2,820.66(1 + 0.0600) - \$5,030 = -\$2,021.02.$$

$$PB(6.13\%, 6\%)_3 = -\$2,021.02(1 + 0.0613) + \$2,145 = 0.$$

The positive balance at the end of year 1 indicates the need to borrow from the project during year 2. However, note that the net investment becomes zero at the end of project life, confirming that 6.13% is the IRR for the cash flow. Since IRR > MARR, the investment is acceptable. Figure A.4 is a visual representation of the occurrence of internal and external interest rates for the project.

Comments: At the end of year 1, the project releases \$2,838.66 that must be invested outside of the project at an interest rate of 6%. The money invested externally must be returned to the project at the beginning of year 2 to provide another needed disbursement of \$5,030. At the end of year 2, there is no external investment of money and, hence, no need for an external interest rate. Instead, the amount of \$2,021.02 that remains invested during year 3 has to bring in a 6.13% of return. Finally, with the receipt of \$2,145, the net investment becomes zero at the end of project life. We conclude that IRR (or RIC)= 6.13% is a correct measure of the profitability of the project.

Using the NPW criterion, the investment would be acceptable if the MARR was between zero and 10% or between 30% and 50%. The rejection region is $10\% < i < 30\%$ and $i > 50\%$. This can be verified in Figure 9.1(b). Note that the project also would be marginally accepted under the NPW analysis at MARR $= i = 6\%$:

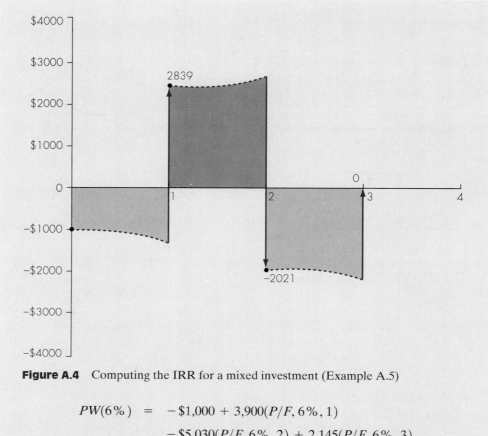

Figure A.4   Computing the IRR for a mixed investment (Example A.5)

$$PW(6\%) = -\$1,000 + 3,900(P/F, 6\%, 1)$$
$$-\$5,030(P/F, 6\%, 2) + 2,145(P/F, 6\%, 3)$$
$$= \$3.55 > 0.$$

The flowchart in Figure A.5 summarizes how you should proceed to apply the net cash flow sign test, accumulated cash flow sign test, and net investment test to calculate an IRR, and make an accept/reject decision for a single project.

Example A.6 IRR Analysis for Projects with Unequal Lives in which the Increment Is a Mixed Investment

Reconsider Example 9.13. Find the IRR on the incremental investment and select the best project. Assume that MARR = 15%, as before.

Solution

Given: Cash flows for two projects with unequal lives, MARR = 15%

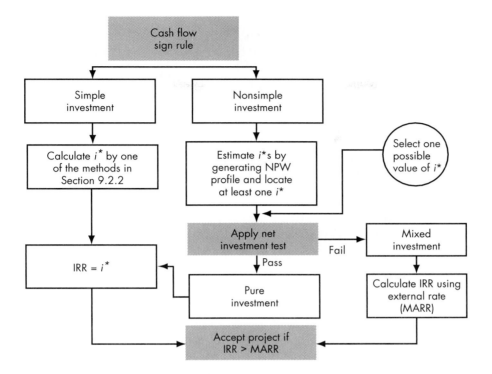

Figure A.5   Summary of IRR criteron: A flowchart that summarizes how you may proceed to apply the net cash flow sign rule and net investment test to calculate IRR for a pure, as well as a mixed investment

Find: IRR on incremental investment, and determine which is alternative is preferable

Since the analysis period is equal to the least common multiple of 12 years, we may compute the incremental cash flow over this 12-year period. As shown in Figure 9.11, we subtract cash flows of model A from those of model B to form the increment of investment. (We want the first cash flow difference to be a negative value.) We can then compute the IRR on this incremental cash flow.

Even though there are five sign changes in the cash flow, there is only one positive i^* for this problem, 63.12%. Unfortunately, however, this is not a pure investment. We need to employ an external rate to compute the IRR to make a proper accept/reject decision. Assuming that the firm's MARR is 15%, we will use a trial-and-error approach: Try $i = 20\%$.

$$PB(20\%, 15\%)_0 = -\$2,500.$$
$$PB(20\%, 15\%)_1 = -\$2,500(1.20) + \$1,000 = -\$2,000.$$
$$PB(20\%, 15\%)_2 = -\$2,000(1.20) + \$1,000 = -\$1,400.$$
$$PB(20\%, 15\%)_3 = -\$1,400(1.20) + \$11,500 = \$9,820.$$

$$PB(20\%, 15\%)_4 = \$9,820(1.15) - \$12,500 = -\$1,207.$$
$$PB(20\%, 15\%)_5 = -\$1,207(1.20) + \$1,000 = -\$448.40.$$
$$PB(20\%, 15\%)_6 = -\$448.40(1.20) + \$11,500 = \$10,961.92.$$
$$PB(20\%, 15\%)_7 = \$10,961.92(1.15) + \$1,000 = \$13,606.21.$$
$$PB(20\%, 15\%)_8 = \$13,606.21(1.15) = \$12,500 + \$3,147.14.$$
$$PB(20\%, 15\%)_9 = \$3,147.14(1.15) + \$11,500 = \$15,119.21.$$
$$PB(20\%, 15\%)_{10} = \$15,119.21(1.15) + \$1,000 = \$18,387.09$$
$$PB(20\%, 15\%)_{11} = \$18,387.09(1.15) + \$1,000 = \$22,145.16.$$
$$PB(20\%, 15\%)_{12} = \$22,145.16(1.15) + \$500 = \$25,966.93.$$

Since $PB(20\%, 15\%)_{12} > 0$, the guessed 20% is not the IRR. We may increase the value of i and repeat the calculations. After several trials, we find that the IRR is 50.68%.[2] Since $\text{IRR}_{B-A} > \text{MARR}$, model B would be selected, which is consistent with the NPW analysis. In other words, the additional investment over the years to obtain model B (–\$2,500 at $n = 0$, –\$12,500 at $n = 4$, –\$12,500 at $n = 8$) yields a satisfactory rate of return: Model B, therefore, is the preferred project. (Note that model B was picked when NPW analysis was used in Example 7.14.)

Given the complications involved in using IRR analysis to compare alternative projects, it is usually more desirable to use one of the other equivalence techniques for this purpose. As an engineering manager, you should keep in mind the intuitive appeal of the rate of return measure. Once you have selected a project on the basis of NPW or AE analysis, you may also wish to express its worth as a rate of return, for the benefit of your associates.

A.6 Summary

1. The possible presence of multiple i^*s (rates of return) can be predicted by
   - The net cash flow sign test
   - The accumulated cash flow sign test
   When multiple rates of return cannot be ruled out by the two methods, it is useful to generate an NPW profile to approximate the value of i^*.
2. All i^* values should be exposed to the **net investment test**. Passing the net investment test indicates that the i^* is an internal rate of return and is therefore a suitable measure of project profitability. Failure to pass the test indicates project borrowing, a situation that requires further analysis by use of an **external interest rate**.
3. **Return on invested capital** analysis uses one rate (the firm's MARR) on externally invested balances and solves for another rate (i^*) on internally invested balances.

---

[2] It will be tedious to solve this type of problem by a trial-and-error method on your calculator. The problem can be solved quickly by using the **EzCash** software, which can be downloaded from the book's web site.

Problems

A.1 Consider the following investment projects:

			Project Cash Flows			
n	A	B	C	D	E	F
0	−$100	−$100	−$100	−$100	−$100	−$100
1	200	470	200	300	300	300
2	−300	−720	200	−300	−250	100
3	400	360	−250	50	40	−400

(a) Apply the sign rule to predict the number of possible i^*s for each project.

(b) Plot the NPW profile as a function of i between 0 and 200% for each project.

(c) Compute the value(s) of i^* for each project.

A.2 Consider an investment project with the following cash flows:

n	Net Cash Flow
0	−$20,000
1	94,000
2	−144,000
3	72,000

(a) Find the IRR return for this investment.

(b) Plot the present worth of the cash flow as a function of *i*.

(c) Using the IRR criterion, should the project be accepted at MARR = 15%?

A.3 Consider the following sets of investment projects:

	Net Cash Flow		
n	Project 1	Project 2	Project 3
0	−$1,000	−$2,000	−$1,000
1	500	1,560	1,400
2	840	944	−100
IRR	?	?	?

Assume that MARR = 12% in the following questions:

(a) Compute the i^* for each investment. If the problem has more than one i^*, identify all of them.

(b) Compute the IRR for each project.

(c) Determine the acceptability of each investment.

A.4 Consider the following sets of investment projects:

		Project Cash Flow			
n	A	B	C	D	E
0	−$100	−$100	−$5	−$100	$200
1	100	30	10	30	100
2	24	30	30	30	−500
3		70	−40	30	−500
4		70		30	200
5				30	600

(a) Compute the i^* for A using the quadratic equation.

(b) Classify each project into either simple or nonsimple.

(c) Apply the cash flow sign rules to each project and determine the number of possible positive i^*s. Identify all projects having unique i^*.

(d) Compute the IRRs for projects B through E.

(e) Apply the net investment test to each project.

(f) With MARR = 10%, which projects will be acceptable using the IRR criterion?

A.5 Consider the following sets of investment projects:

n	Net Cash Flow Project 1	Project 2	Project 3
0	−$1,600	−$5,000	−$1,000
1	10,000	10,000	4,000
2	10,000	30,000	−4,000
3		−40,000	

Assume that MARR = 12% in the following questions:

(a) Identify the $i^*(s)$ for each investment. If the problem has more than one i^*, identify all of them.

(b) Which project(s) is (are) a mixed investment?

(c) Compute the IRR for each project.

(d) Determine the acceptability of each project.

A.6 Reconsider Problem 9.7.

(a) Compute the i^*s for projects B through E.

(b) Identify the projects that fail the net investment test.

(c) Determine the IRR for each investment.

(d) With MARR = 12%, which projects will be acceptable using the IRR criterion?

A.7 Consider the following investment projects:

n	Net Cash Flow Project A	Project B	Project C
0	−$100	−$150	−$100
1	30	50	410
2	50	50	−558
3	80	50	252
4		100	
IRR	23.24%	21.11%	20%,40%,50%

Assume the MARR = 12% for the following questions:

(a) Identify the pure investment(s).

(b) Identify the mixed investment(s).

(c) Determine the IRR for each investment.

(d) Which project would be acceptable?

A.8 The Boeing Company has received a NASA contract worth $460 million to build rocket boosters for future space missions. NASA will pay $50 million when the contract is signed, another $360 million at the end of the first year, and the $50 million balance at the end of second year. The expected cash outflows required to produce these rocket boosters are estimated to be $150 million now, $100 million during the first year, and $218 million during the second year. The firm's MARR is 12%.

n	Outflow	Inflow	Net Cash Flow
0	$150	$50	−$100
1	100	360	260
2	218	50	−168

(a) Show whether or not this project is a mixed investment.

(b) Compute the IRR for this investment.

(c) Should Boeing accept the project?

A.9 Consider the following investment projects:

| | Net Cash Flow | | |
n	Project A	Project B	Project C
0	−$100		−$100
1	−216	−150	50
2	116	100	−50
3		50	200
4		40	
i*	?	15.51%	29.95%

(a) Compute the $i*$ for project A. If there is more than one $i*$, identify all of them.

(b) Identify the mixed investment(s).

(c) Assuming that MARR = 10%, determine the acceptability of each project based on the IRR criterion.

A.10 Consider the following sets of investment projects:

| | Net Cash Flow | | | | |
n	A	B	C	D	E
0	−$1,000	−$5,000	−$2000	−$2,000	−$1,000
1	3,100	20,000	1,560	2,800	3,600
2	−2,200	−12,000	944	−200	−5,700
3		−3,000			3,600
i*	?	?	18%	32.45%	35.39%

Assume that MARR = 12% in the following questions:

(a) Compute the $i*$ for projects A and B. If the project has more than one $i*$, identify all of them.

(b) Classify each project as either a pure or mixed investment.

(c) Compute the IRR for each investment.

(d) Determine the acceptability of each project.

A.11 Consider an investment project whose cash flows are given as follows:

n	Net Cash Flow
0	−$5,000
1	10,000
2	30,000
3	−40,000

(a) Plot the present worth curve by varying i from 0% to 250%.

(b) Is this a mixed investment?

(c) Should the investment be accepted at MARR = 18%?

A.12 Consider the following two mutually exclusive investment projects.

Assume that MARR = 15%.

| | Net Cash Flow | |
n	Project A	Project B
0	−$300	−$800
1	0	1,150
2	690	40
i*	51.66%	46.31%

(a) Using the IRR criterion, which project would be selected?

(b) Sketch the $PW(i)$ function on incremental investment (B − A).

A.13 Consider the following project's cash flows:

n	Net Cash Flow
0	−$100,000
1	310,000
2	−220,000

The project's $i*$s are computed as 10% and 100%, respectively. The firm's MARR is 8%.

(a) Show why this investment project fails the net investment test.

(b) Compute the IRR, and determine the acceptability of this project.

A.14 Consider the following investment projects:

	Net Cash Flow		
n	Project 1	Project 2	Project 3
0	−$1,000	−$1,000	−$1,000
1	−1,000	1,600	1,500
2	2,000	−300	−500
3	3,000	−200	2,000

Which of the following statements is correct?

(a) All projects are nonsimple investments.

(b) Project 3 should have three real rates of return.

(c) All projects will have a unique positive real rate of return.

(d) None of the above.

Risk Simulation

In Chapter 14, we examined analytical methods of determining the NPW distributions and computing their means and variances. As we saw in Section 14.4.1, the NPW distribution offers numerous options for graphically presenting to the decision-maker probabilistic information, such as the range and likelihoods of occurrence of possible levels of NPW. Where we can adequately evaluate the risky investment problem by analytical methods, it is generally preferable to do so. However, in many investment situations, we cannot solve easily by analytical methods. In these situations, we may develop the NPW distribution through computer simulation.

B.1 Computer Simulation

Before we examine the details of risk simulation, let us consider a situation where we wish to train a new astronaut for a future space mission. Several approaches exist for training this astronaut. One (somewhat unlikely) possibility is to place the trainee in an actual space shuttle and to launch her into space. This approach certainly would be expensive; it also would be extremely risky because any human error made by the trainee would have tragic consequences. As an alternative, we can place the trainee in a flight simulator designed to mimic the behavior of the actual space shuttle in space. The advantage of this approach is that the astronaut trainee learns all the essential functions of space operation in a simulated space environment. The flight simulator generates test conditions approximating operational conditions, and any human errors made during training cause no harm to the astronaut or to the equipment being used.

The use of computer simulation is not restricted to simulating a physical phenomenon such as the flight simulator. In recent years, techniques for testing the results of some investment decisions before they are actually executed have been developed. As a result, many phases of business investment decisions have been simulated with considerable success. Now we can make an analogy between a space shuttle flight simulator and an investment simulator—a model for testing the results of some business investment decisions before they are actually executed. In fact, we can analyze BMC's transmission-housings project by building a simulation model. The general approach is to assign a subjective (or objective) probability distribution to each unknown factor and to combine these into a probability distribution for the project profitability as a whole. The essential idea is that, if we can simulate the actual state of nature for unknown investment variables on a computer, we may be able to obtain the resulting NPW distribution.

The unit demand (X) in our BMC's transmission-housing project was one of the random variables in the problem. We can know the exact value for this random variable only after the project is implemented. Is there any way to predict the actual value before we make any decision on the project?

The following logical steps are often suggested for a computer program that simulates investment scenarios:

Step 1: Identify all the variables that affect the measure of investment worth (e.g., NPW after taxes).

Step 2: Identify the relationships among all the variables. The relationships of interest here are expressed by the equations or the series of numerical computations by which we compute the NPW of an investment project. These equations make up the model we are trying to analyze.

Step 3: Classify the variables into two groups: The parameters whose values are known with certainty and the random variables for which exact values cannot be specified at the time of decision making.

Step 4: Define distributions for all the random variables.

Step 5: Perform Monte Carlo sampling and describe the resulting NPW distribution.

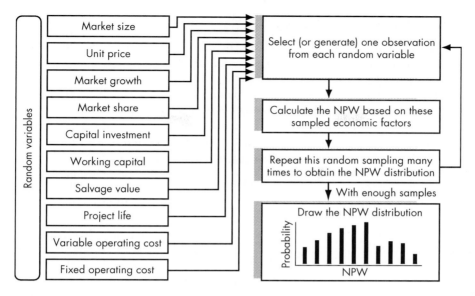

Figure B.1   Logical steps involved in simulating a risky investment

Step 6: Compute the distribution parameters and prepare graphic displays of simulation results.

Figure B.1 illustrates the logical steps involved in simulating a risky investment project. The risk simulation process we have described has two important advantages when compared with the analytical approach discussed in Chapter 14:

1.   The number of variables that can be considered is practically unlimited, and the distributions used to define the possible values for each random variable can be of any type and any shape. The distributions can be based on statistical data if they are available, or, more commonly, on subjective judgment.
2.   The method lends itself to sensitivity analyses. By defining some factors that have the most significant effect on the resulting NPW values and using different distributions (in terms of either shape or range) for each variable, we can observe the extent to which the NPW distribution is changed.

B.2 Model Building

In this section, we shall present some of the procedural details related to the first three steps (model building) outlined in Section B.1. To illustrate the typical procedure involved, we shall work with an investment setting for BMC's transmission-housings project, described in Example 14.8.

The initial step is to define the measure of investment worth and the factors that affect that measure. For our presentation, we choose the measure of investment worth as an after-tax NPW computed at a given interest rate i. In fact, we are free to choose any measure of worth, such as annual worth or future worth. In the second step, we must divide into two groups all the variables that we listed in Step 1 as affecting NPW. One group consists of all the parameters for which values are known. The second groups includes all remaining parameters for which we do not know exact values at the time of analysis. The third step is to define the relationships that tie together all the variables. These relationships may take the form of a single equation or several equations.

Example B.1 Developing a Simulation Model

Reconsider BMC's transmission-housings project in Example 14.8. Identify the input factors related to the project and develop the simulation model for the NPW distribution.

Discussion: For the BMC project, the variables that affect NPW value are investment required, unit price, demand, variable production cost, fixed production cost, tax rate, depreciation expenses, and the firm's interest rate. Some of the parameters that might be included in the known group are investment cost and interest rate (MARR). If we have already purchased the equipment or have received a price quote, then we also know the depreciation amount. Assuming that we are operating in a stable economy, we would probably know the tax rates for computing income taxes due.

The group of parameters with unknown values would usually include all the variables relating to costs and future operating expense and future demand and sales prices. These are the random variables for which we must assess the probability distributions.

For simplicity, we classify the input parameters or variables for BMC's ransmission-housings project as follows:

Assumed to Be Known Parameters	Assumed to be Unknown Parameters
MARR	Unit price
Tax rate	Demand
Depreciation amount	Salvage value
Investment amount	
Project life	
Fixed production cost	
Variable production cost	

Note that, unlike the situation in Example 14.8, here we treat the salvage value as a random variable. With these assumptions, we are now ready to build the NPW equation for the BMC project.

Solution

Recall that the basic investment parameters assumed for BMC's 5-year project in Example 14.8 were as follows.

- Investment = −$125,000,
- Marginal tax rate = 0.40,
- Annual fixed cost = $10,000,
- Variable unit production cost = $15/unit,
- MARR (i) = 15%,
- Annual depreciation amounts:

n	D_n
1	$17,863
2	30,613
3	21,863
4	15,613
5	5,575

The after-tax annual revenue is expressed in terms of functions of product demand (X) and unit price (Y):

$$R_n = XY(1 - t_m) = 0.6XY.$$

The after-tax annual expenses excluding depreciation are also expressed as a function of product demand (X):

$$
\begin{aligned}
E_n &= (\text{Fixed cost} + \text{variable cost})(1 - t_m) \\
&= (\$10,000 + 15X)(0.60) \\
&= \$6,000 + 9X.
\end{aligned}
$$

Then, the net after-tax cash revenue is

$$
\begin{aligned}
V_n &= R_n - E_n \\
&= 0.6XY - 9X - \$6,000.
\end{aligned}
$$

The present worth of the net after-tax cash inflow from revenue is

$$
\begin{aligned}
\sum_{n=1}^{5} V_n(P/F, 15\%, n) &= [0.6X(Y - 15) - \$6,000](P/A, 15\%, 5) \\
&= 0.6X(Y - 15)(3.3522) - \$20,113.
\end{aligned}
$$

We will now compute the present worth of the total depreciation credits:

$$\sum_{n=1}^{5} D_n t_m (P/F, i, n) = 0.40 \, [\$17,863 \, (P/F, 15\%, 1) + \$30,613 \, (P/F, 15\%, 2)$$
$$+ \$21,863(P/F, 15\%, 3) + \$15,613(P/F, 15\%, 4)$$
$$+ \$5,575(P/F, 15\%, 5)]$$
$$= \$25,901.$$

Since the total depreciation amount is \$91,527, the book value at the end of year 5 is \$33,473 (\$125,000 – \$91,527). Any salvage value greater than this book value is treated as a taxable gain, and this gain is taxed at t_m. In our example, the salvage value is considered to be a random variable. Thus, the amount of taxable gains (losses) also becomes a random variable. Therefore, the net salvage value after tax adjustment is

$$S - (S - \$33,473)t_m = S(1 - t_m) + 33,473 t_m$$
$$= 0.6S + \$13,389.$$

Then, the equivalent present worth of this amount is

$$(0.6S + \$13,389)(P/F, 15\%, 5) = (0.6S + \$13,389)(0.4972)$$

Now, the NPW equation can be summarized as

$$PW(15\%) = -\$125,000 + 0.6X(Y - 15)(3.3522) - \$20,113 + \$25,901$$
$$+ (0.6S + \$13,389)(0.4972)$$
$$= -\$112,555 + 2.0113X(Y - 15) + 0.2983S.$$

Note that the NPW function is now expressed in terms of three random variables X, Y, and S.

B.3 Monte Carlo Sampling

For some variables, we may base the probability distribution on objective evidence gleaned from the past if the decision-maker feels the same trend will continue to operate in the future. If not, we may use subjective probabilities as discussed in Section 14.3.1. Once we specify a distribution for a random variable, we need to determine ways to generate samples from this distribution. **Monte Carlo sampling** is a specific type of simulation method in which a random sample of outcomes is generated for a specified probability distribution. In this section, we shall discuss the Monte Carlo sampling procedure for an *independent* random variable.

B.3.1 Random Numbers

The sampling process is the key part of the analysis. It must be done such that the sequence of values sampled will be distributed in the same way as the original distribution. To accomplish this objective, we need a source of independent, identically distributed uniform random numbers between 0 and 1. We can use a table of random numbers but most digital computers have programs available to generate "equally likely (uniform)" random decimals between 0 and 1. We will use $U(0,1)$ to denote such a statistically reliable uniform random number generator, and we will use U_1, U_2, U_3, ... to represent uniform random numbers generated by this routine. (In Microsoft Excel, the RAND function can be used to generate such a random number sequence.)

B.3.2   Sampling Procedure

For any given random numbers, the question is, how are they used to sample a distribution in a simulation analysis? The first task is to convert the distribution into its corresponding cumulative frequency distribution. Then, the random number generated is set equal to its numerically equivalent percentile and is used as the entry point on the $F(x)$ axis of the cumulative frequency graph. The sampled value of the random variable is the x value corresponding to this cumulative percentile entry point.

    This method of generating random values works because choosing a random decimal between 0 and 1 is equivalent to choosing a random percentile of the distribution. Then, the random value is used to convert the random percentile to a particular value. The method is general and can be used for any cumulative probability distribution, either continuous or discrete.

Example B.2   Monte Carlo Sampling

In Example B.1, we have developed a NPW equation for BMC's transmission-housings project as a function of three random variables—demand (X), unit price (Y) and salvage value (S):

$$PW(15\%) = -\$112{,}555 + 2.0113X(Y - 15) + 0.2983S$$

- For random variable X, we will assume the same discrete distribution as defined in Table 12.5.
- For random variable Y, we will assume a triangular distribution with $L = \$48$, $H = \$53$, and $M_o = \$50$.
- For random variable S, we will assume a uniform distribution with $L = \$30{,}000$ and $H = \$50{,}000$.

With the random variables (X, Y, and S) distributed as above, and assuming that these random variables are *mutually independent* of each other, we need three uniform random numbers to sample one realization from each random variable. Determine the NPW distribution based on 200 iterations.

Discussion: As outlined previously, a simulation analysis consists of a series of repetitive computations of NPW. To perform the sequence of repeated simulation trials, we generate a sample observation for each random variable in the model and substitute these values into the NPW equation. Each trial requires that we use a different random number in the sequence to sample each distribution. Thus, if three random variables affect the NPW, we need three random numbers for each trial. After each trial,

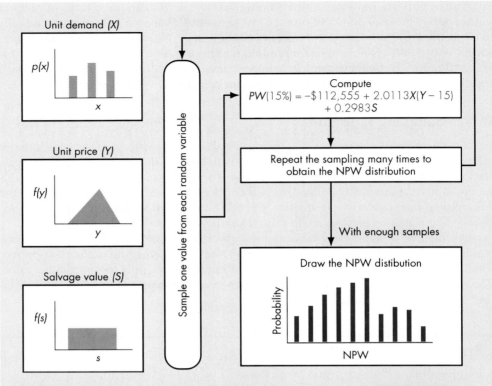

Figure B.2   A logical sequence of Monte Carlo simulation to obtain the NPW distribution for BMC's transmission-housings project (Example B.2)

the computed NPW is stored in the computer. As shown in Figure B.2, each value of NPW computed in this manner represents one state of nature. The trials are continued until a sufficient number of NPW values is available to define the NPW distribution.

Solution

Suppose the following three uniform random numbers are generated for the first iteration: $U_1 = 0.12135$ for X, $U_2 = 0.82592$ for Y, and $U_3 = 0.86886$ for S.

- Demand (X): The cumulative distribution for X is already given in Example 14.6. To generate one sample (observation) from this discrete distribution, we first find the cumulative probability function, as depicted in Figure B.3(a). On a given trial, suppose the computer gives the random number 0.12135. We then enter the vertical axis at the 12.135 percentile (the percentile numerically equivalent to the random number), read across to the cumulative curve, then read down to the x axis to find the corresponding value of the random variable X; this value is 1600. This is the value of x that we use in the NPW equation. On the next trial, we sample another value of x by obtaining another random number, entering the ordinate at the numerically equivalent percentile, and reading the corresponding value of x from the x axis.

- Price (Y): Assuming that the unit price random variable can be estimated by the three parameters, its probability distribution is shown in Fig. B.3(b). Note that Y takes a continuous value (unlike the discrete assumption in Table 14.5). The sampling procedure is again similar to the discrete situation. Using the random number $U = 0.82592$, we can approximate $y = \$51.38$ by performing a linear interpolation.

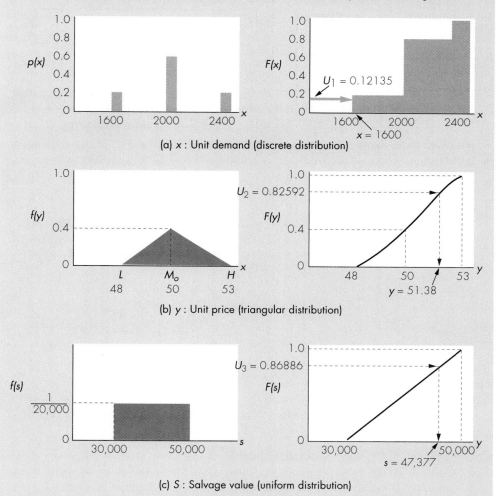

(a) x : Unit demand (discrete distribution)

(b) y : Unit price (triangular distribution)

(c) S : Salvage value (uniform distribution)

Figure B.3   Illustration of a sampling scheme for discrete and continuous random variables (Example B.2)

- Salvage (S): With the salvage value (S) distributed uniformly between \$30,000 and \$50,000, and a random number of $U = 0.86886$, the sample value is $s =$ \$47,377, or $s = \$30,000 + (50,000 - 30,000)0.86886$. The sampling scheme is shown in Fig. B.3(c).

Now we can compute the NPW equation with these sample values, yielding

$$PW\,(15\%) \;=\; -\$112{,}555 + 2.0113(1{,}600)(\$51.3841 - \$15)$$

$$+\,0.2983(\$47{,}377)$$

$$=\; \$18{,}665.$$

This result completes the first iteration of NPW_1 computation.

For the second iteration, we need to generate another set of three uniform random numbers (assume they are 0.72976, 0.79885, and 0.41879), to generate the respective sample from each distribution, and to compute $NPW_2 = \$44{,}752$. If we repeat this process for 200 iterations, we obtain the NPW values listed in Table B.1.

By ordering the observed data by increasing NPW value and tabulating the ordered NPW values, we obtain a frequency distribution shown in Table B.2. Such a tabulation results from dividing the entire range of computed NPWs into a series of sub-ranges (20 in this case), and then counting the number of computed values that fall in each of the 20 intervals. Note that the sum of all the frequencies of column 3 is the total number of trials that were made.

Column 4 simply expresses the frequencies of column 3 as a fraction of the total number of trials. At this point, all we have done is arrange the 200 numerical values of NPW into a table of relative frequencies.

$18,665	44,752	41,756	75,804	67,508
74,177	47,664	43,106	17,394	35,841
46,400	36,237	12,686	47,947	48,411
50,365	64,287	44,856	47,970	47,988
42,071	43,421	36,991	59,551	9,788
50,147	46,010	11,777	71,202	76,259
20,937	48,250	33,837	37,378	7,188
47,875	14,891	48,362	35,894	75,360
38,726	73,987	45,346	62,508	36,738
44,849	46,076	48,374	47,287	70,702
47,168	42,348	8,213	6,689	14,415

Table B.1
Observed
NPW Values
($) for
BMC's Sim-
ulation Proj-
ect (Example
B.2)

No.	U_1	U_2	U_3	x	y	s	NPW
1	0.12135	0.82592	0.86886	1600	$51.38	$47,377	$18,665
2	0.72976	0.79885	0.41879	2000	$51,26	$38,376	$44,752
⋮	⋮	⋮	⋮	⋮	⋮	⋮	⋮
200	0.57345	0.75553	0.61251	2000	$51.08	$42,250	$45,204

41,962	45,207	46,487	77,414	67,723
79,089	51,846	76,157	49,960	17,091
15,438	37,214	49,542	45,830	46,679
36,393	15,714	44,899	48,109	76,593
34,927	42,545	45,452	48,089	43,163
78,925	39,179	38,883	80,994	45,202
18,807	74,707	44,787	5,944	6,377
45,811	44,608	42,448	34,584	48,677
10,303	35,597	37,910	62,624	62,060
44,813	36,629	75,111	15,739	43,621
34,288	49,116	76,778	46,067	62,853
9,029	42,485	51,817	44,096	35,198
36,345	75,337	14,034	18,253	41,865
48,089	64,934	81,532	74,407	57,465
32,969	46,529	19,653	10,250	31,269
33,300	47,552	19,139	48,664	46,410
33,457	44,605	35,160	51,496	34,247
46,657	37,249	44,171	34,189	36,673
47,567	62,654	65,062	34,519	45,204

Table B.2
Simulated
NPW Fre-
quency Dis-
tribution for
BMC's
Transmis-
sion-
Housings
Project
(Example
B.2)

Cell No.	Cell Interval	Observed Frequency	Relative Frequency	Cumulative Frequency
1	$5,944 \leq NPW \leq \$9,723$	8	0.04	0.04
2	$9,723 < NPW \leq 13,502$	6	0.03	0.07
3	$13,502 < NPW \leq 17,281$	9	0.05	0.12
4	$17,281 < NPW \leq 21,061$	10	0.05	0.17
5	$21,061 < NPW \leq 24,840$	0	0.00	0.17
6	$24,840 < NPW \leq 28,620$	0	0.00	0.17
7	$28,620 < NPW \leq 32,399$	2	0.01	0.18
8	$32,399 < NPW \leq 36,179$	20	0.10	0.28
9	$36,179 < NPW \leq 39,958$	18	0.09	0.37
10	$39,958 < NPW \leq 43,738$	13	0.07	0.43
11	$43,378 < NPW \leq 47,517$	39	0.19	0.63
12	$47,517 < NPW \leq 51,297$	28	0.14	0.77
13	$51,297 < NPW \leq 55,076$	3	0.02	0.78
14	$55,076 < NPW \leq 58,855$	1	0.01	0.78
15	$58,855 < NPW \leq 62,635$	6	0.03	0.81
16	$62,635 < NPW \leq 66,414$	6	0.03	0.84
17	$66,414 < NPW \leq 70,194$	4	0.02	0.86
18	$70,194 < NPW \leq 73,973$	4	0.02	0.88
19	$73,973 < NPW \leq 77,752$	18	0.09	0.97
20	$77,752 < NPW \leq 81,532$	5	0.03	1.00

Cell width = $3,779; mean = $44,245; standard deviation = $18,585; minimum NPW value = $5,944; maximum NPW value = $81,532.

B.4 Simulation Output Analysis

After a sufficient number of repetitive simulation trials has been run, the analysis is essentially completed. The only remaining tasks are to tabulate the computed NPW values to determine the expected value and to make various graphic displays useful to management.

B.4.1 Interpretation of Simulation Results

Once we obtain a NPW frequency distribution (such as that shown in Table B.2), we need to make the assumption that the actual relative frequencies of column 4 in Table B.2 are representative of the probability of having a NPW in each range. That is, we assume that the relative frequencies we observed in the sampling are representative of the proportions we would have obtained had we examined all the possible combinations.

This sampling is analogous to polling the opinions of voters about a candidate for public office. We could speak to every registered voter if we had the time and resources, but a simpler procedure would be to interview a smaller group of persons selected with an unbiased sampling procedure. If 60% of this scientifically selected sample supports the candidate, it probably would be safe to assume that 60% of all registered voters support the candidate. Conceptually, we do the same thing with simulation. As long as we ensure that a sufficient number of representative trials has been made, we can rely on the simulation results.

Once we have obtained the probability distribution of the NPW, we face the crucial question: How do we use this distribution in decision-making. Recall that the probability distribution provides information regarding the probability that a random variable will attain some value x. We can use this information, in turn, to define the cumulative distribution, which expresses the probability that the random variable will attain a value smaller than or equal to some x, i.e., $F(x) = P(X \leq x)$. Thus, if the NPW distribution is known, we can also compute the probability that the NPW of a project will be negative. We use this probabilistic information in judging the profitability of the project.

With the assurance that 200 trials was a sufficient number for the BMC project, we may interpret the relative frequencies in column 4 of Table B.2 as probabilities. The NPW values range between $5,944 and $81,532, thereby indicating no loss for any situation. The NPW distribution has an expected value of $44,245 and a standard deviation of $18,585.

B.4.2   Creation of Graphic Displays

Using the output data in Table B.2, we can create the distribution in Figure B.4(a). A picture such as this can give the decision maker a feel for the ranges of possible NPWs, for the relative likelihoods of loss versus gain, for the range of NPWs that are most probable, and so on.

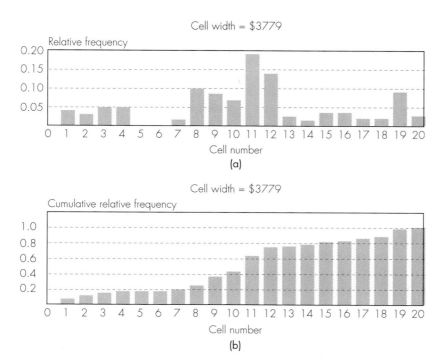

Figure B.4   Simulation result for BMC's transmission-housings project based on 200 iterations

Another useful display is the conversion of the NPW distribution to the equivalent cumulative frequency, as shown in Fig. B.4(b). Usually, a decision-maker is concerned with the likelihood of attaining at least a given level of NPW. Therefore, we construct the cumulative distribution by accumulating the areas under the distribution as the NPW decreases. The decision maker can use Fig. B.4(b) to answer many questions: For example, what is the likelihood of making at least a 15% return on investment, i.e., the likelihood that the NPW will be at least 0? In our example, this probability is virtually 100%.

B.5 Dependent Random Variables

All our simulation examples have considered independent random variables. We must recognize that some of the random variables affecting the NPW may be related to one another. If they are, we need to sample from distributions of the random variables in a manner that accounts for any dependency. For example, in the BMC project, both the demand and the unit price are not known with certainty. Both of these parameters would be on our list of variables for which we need to describe distributions, but they could be related inversely. When we describe distributions for these two parameters, we have to account for the dependency. This issue can be critical, as the results obtained from a simulation analysis can be misleading if the analysis does not account for the dependent relationships. The sampling techniques for these dependent random variables are beyond the scope of this text, but can be found in many simulation textbooks.

B6 Summary

- **Risk simulation,** in general, is the process of modeling reality to observe and weigh the likelihood of the possible outcomes of a risky undertaking.
- **Monte Carlo sampling** is a specific type of randomized sampling method in which a random sample of outcomes is generated for a specified probability distribution. Because Monte Carlo sampling and other simulation techniques often rely on generating a significant number of outcomes, they can be more conveniently performed on the computer than manually.

APPENDIX C

Electronic Spreadsheets: Summary of Built-In Financial Functions

Some of the financial functions are available in electronic spreadsheets, namely, Excel, and Lotus 1.2.3. As you will see, the keystrokes and choices are similar for both.

C.1 Nominal Versus Effective Interest Rates

A conversion from a nominal interest rate to an effective interest rate (or vice versa) is easily obtained with Excel. No comparable financial functions are provided by either Lotus or QuattroPro.

Function Description	Excel	Lotus 1.2.3
Effective interest rate	=EFFECT(**nominal_rate, npery**)	
Nominal interest rate	=NOMINAL(**effect_rate, npery**)	

Effect_rate is the effective interest rate; *Npery* is the number of compounding periods per year; *Nominal_rate* is the nominal interest rate.

C.2 Single-Sum Compounding Functions

The single-sum compounding functions deal with either the single-payment compound amount factor or the present worth factor. With these functions, the future amount when the present single-sum is given, present amount when its future amount is specified, or unknown interest (growth) rate and number of interest periods can be computed.

Function Description	Excel	Lotus 1.2.3
1. Calculating the future worth of a single payment	=FV(*i*, *N*, *0*, *P*)	
2. Calculating the present worth of a single payment	=PV(*i*, *N*, *0*, *F*)	
3. Calculating an unknown number of payment periods (*N*)	=NPER(*i*,*0*, *P*, *F*)	@CTERM(*i*, *F*, *P*)
4. Calculating an unknown interest rate	=RATE(*N*, *0*, *P*, *F*, *0*, guess)	@RATE(*F*, *P*, *N*)

i is the interest rate; *N* is the number of the interest periods; *F* is the future worth specified at the end of period *N*; *P* is the present worth at period 0; *Guess* is your estimate of interest rate.

C.3 Annuity Functions

Annuity functions provide a full range capability to calculate the future value of an annuity, the present value of an annuity, and the interest rate used in an annuity payment. If a cash payment is made at the end of period, it is an **ordinary annuity.** If a cash payment is made at the beginning of period, it is known as an **annuity due.** When using annuity functions, you can specify the choice of annuity by setting type parameters; if type = 0, or is omitted, it is an ordinary annuity. If type = 1, it is an annuity due. In annuity functions, the parameters in bold face must be specified by the user.

Function Description	**Excel**	**Lotus 1.2.3**
1. Calculating the number of payment periods (N) in an annuity	=NPER(i, A, 0, F, type)	@TERM(A, i, F)
2. Calculating the number of payment periods (N) of an annuity when its present worth is specified	=NPER(i, A, P, F, type)	@NPER(A, i, F, type, P)
3. Calculating the future worth of an annuity (equal-payment series compound amount factor)	=FV(i, N, A)	@FV(A, i, N)
4. Calculating the future worth of an annuity when its present worth is specified.	=FV(i, N, A, P, type)	@FVAL(A, i, N, type, P)
5. Calculating the periodic equal payments of an annuity (capital recovery factor)	=PMT(i, N, P)	@PMT(P, i, N)
6. Calculating the periodic equal payment of an annuity when its future worth is specified.	=PMT(i, N, P, F, type)	@PAYMT(P, i, N, type, F)
7. Calculating the present worth of an annuity (equal-payment series present worth factor)	=PV(i, N, A)	@PV(A, i, N)
8. Calculating the present worth of an annuity when its optional future worth is specified	=PV(i, N, A, F, type)	@PVAL(A, i, N, type, F)
9. Calculating the interest rate used in an annuity	=RATE(N, A, P, F, type, guess)	@IRATE(N, A, P, *type*, F, guess)

Rate (i) is the interest rate per period; **Per (n)** is the period for which you want to find the interest and must be in the range 1 to **nper; Pmt (A)** is the payment made each period and cannot change over the life of the annuity; **Nper (N)** is the total number of payment periods in an annuity; **Pv (P)** is the present value, or the lump-sum (starting) amount that a series of future payments is worth right now; **Fv (F)** is the future value, or a cash balance you want to attain after the last payment is made. If *fv* is omitted, it is assumed to be 0 (the future value of a loan, for example, is 0); **Type** is the number of 0 or 1 and indicates when payments are due. If type is omitted, it is assumed to be 0.

Set *type* equal to	If payments are due
0	At the end of the period
1	At the beginning of the period

Guess is your estimate of the interest rate.

C.4 Loan Analysis Functions

When you need to compute the monthly payments, interest and principal payments, several commands are available to facilitate a typical loan analysis.

Function Description	Excel	Lotus 1.2.3
1. Calculating the periodic loan payment size (A)	$=PMT(i, N, P,$ type$)$	@PMT(P, i, N) @PAYMT$(P, i, N,$ type$, F)$
2. Calculating the portion of loan interest payment for a given period n	$=IPMT(i, n, N, P, F,$ type$)$	@IPAYMT(P, i, N, n)
3. Calculating the cumulative interest payment between two interval periods	$=CUMIMPT(i, N, P,$ *start_period, end_period,* type$)$	@IPAYMT$(P, i, N, n,$ **start, end,** type$)$
4. Calculating portion of loan principal payment for a given period n	$=PPMT(i, n, N, P, F,$ type$)$	@PPAYMT(P, i, N, n)
5. Calculating the cumulative principal payment between two interval periods	$=CUMPRINC(i, N, P,$ *start_period, end_period,* type$)$	@PPAYMT$(P, i, N, n,$ **start, end,** type$)$

Rate (i) is the interest rate; *Nper (N)* is the total number of payment periods; *Pv (P)* is the present value; *Start_period* is the first period in the calculation. Payment periods are numbered beginning with 1; *End_period* is the last period in the calculation; *Type* is the timing of the payment.

Type	Timing
0	Payment at the end of the period
1	Payment at the beginning of the period

C.5 Bond Functions

Several financial functions are available to evaluate investments in bond. In particular, the yield calculation at bond maturity and bond pricing decision can be easily made.

Function Description	Excel	Lotus 1.2.3
1. Calculating accrued interest	=ACCRINT(*issue, first_interest, settlement, rate,* par, frequency basis)	@ACCRUED (*settlement, maturity, coupon,* par, frequency, basis)
2. Calculating bond price	=PRICE (*settlement, maturity, rate, yield,* redemption, frequency, basis)	@ACCRUED (*settlement, maturity, coupon, yield,* redemption, frequency, basis)
3. Calculating maturity yield	=YIELD(*settlement, maturity, rate, price,* redemption, frequency, basis)	@YIELD (*settlement, maturity, coupon,price,* redemption, frequency, basis)

Settlement is the secuity's settlement date, expressed as a serial date number. Use **NOW** command to convert a date to a serial number; *Maturity* is the security's maturity date, expressed as a serial date number; *First_interest* is the security's first interest date, expressed as a serial date number; *Coupon* is the security's annual coupon rate; *Issue* is a number representing the issue date; *Par* is the security's par value. If you omit par, ACCRINT uses $1000; *Price* is the security's price per $100 face value; *Redemption* is the security's redemption value per $100 face value; *Frequency* is the number of coupon payments per year. For annual payments, frequency = 1; for semi-annual payment, frequency = 2; *Yield* is the security's annual yield; *Basis* (or calendar) is the type of day count basis to use:

Basis (or Calendar)	Day count basis
0 or omitted	US(NASD) 30/360
1	Actual/actual
2	Actual/360
3	Actual/365
4	European 30/360

Firstcpn is a number representing the first coupon date.

C.6 Project Evaluation Tools

Several measures of investment worth are available to calculate the NPW, IRR, and annual equivalent of a project's cash flow series.

Function Description	Excel	Lotus 1.2.3
1. Net present value calculation	=NPV(*i, range*)	@NPV(*i, range*)
2. Rate of return calculation	=IRR(*range, guess*)	@IRR(*guess, range*)
3. Annual equivalent calculation	=PMT(*i, N, NPV(i, range)*, type),	@PMT(NPV(*i,range*), *i, N*)

i is the minimum attractive rate of return (MARR); *Range* is the cell address where the cash flow streams are stored; *Guess* is the estimated interest rate in solving IRR.

C.7 Depreciation Functions

Function Description	Excel	Lotus 1.2.3
1. Straight-line method:	=SLN(*cost, salvage, life*)	@SLN(*cost, salvage, life*)
2. Double declining balance method: Calculates 200% declining balance depreciation	=DDB(*cost, salvage, life, period,* factor)	@DDB(*cost, salvage, life, period*)
3. Variable declining balance method: Calculates the depreciation by using the variable-rate declining balance method	=VDB(*cost, salvage, life, life_start, end_period, factor,* no_switch)	@VDB(*cost, salvage, life, start_period, end_period,* depreciation_percent, switch)
4. Sum of the years' digits' method: Calculates sum of the years' digits depreciation	=SYD(*cost, salvage, life, period*)	@SYD(*cost, salvage, life,period*)

Cost (I) is the initial cost (cost basis) of the asset; *Salvage (S)* is the value at the end of the depreciable life; *Life (N)* is the number of periods over which the asset is being depreciated (known as tax life or depreciable life); *Period (n)* is the period for which you want to calculate the depreciation; *Factor (or depreciation percent)* is the rate at which the balance declines. If factor is omitted, it is assumed to be 2 (double-declining balance method). For 150% declining-balance method, enter factor = 1.5; *Start_period* is the starting period for which you want to calculate the depreciation; *End_period* is the ending period for which you want to calculate the depreciation; *No_switch* is a logical value specifying whether to switch to straight-line depreciation even when depreciation is greater than the declining balance method. *If no_switch* (or switch for Lotus) is TRUE, it does not switch to straight-line even when the depreciation is greater than the declining balance calculation; *If no_switch* is FALSE or omitted, it switches to straight-line depreciation when depreciation is greater than the declining balance calculation.

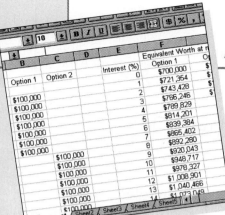

Interest Factors for Discrete Compounding

0.25%

	Single Payment		Equal Payment Series				Gradient Series		
N	Compound Amount Factor (F/P,i,N)	Present Worth Factor (P/F,i,N)	Compound Amount Factor (F/A,i,N)	Sinking Fund Factor (A/F,i,N)	Present Worth Factor (P/A,i,N)	Capital Recovery Factor (A/P,i,N)	Gradient Uniform Series (A/G,i,N)	Gradient Present Worth (P/G,i,N)	N
1	1.0025	0.9975	1.0000	1.0000	0.9975	1.0025	0.0000	0.0000	1
2	1.0050	0.9950	2.0025	0.4994	1.9925	0.5019	0.4994	0.9950	2
3	1.0075	0.9925	3.0075	0.3325	2.9851	0.3350	0.9983	2.9801	3
4	1.0100	0.9901	4.0150	0.2491	3.9751	0.2516	1.4969	5.9503	4
5	1.0126	0.9876	5.0251	0.1990	4.9627	0.2015	1.9950	9.9007	5
6	1.0151	0.9851	6.0376	0.1656	5.9478	0.1681	2.4927	14.8263	6
7	1.0176	0.9827	7.0527	0.1418	6.9305	0.1443	2.9900	20.7223	7
8	1.0202	0.9802	8.0704	0.1239	7.9107	0.1264	3.4869	27.5839	8
9	1.0227	0.9778	9.0905	0.1100	8.8885	0.1125	3.9834	35.4061	9
10	1.0253	0.9753	10.1133	0.0989	9.8639	0.1014	4.4794	44.1842	10
11	1.0278	0.9729	11.1385	0.0898	10.8368	0.0923	4.9750	53.9133	11
12	1.0304	0.9705	12.1664	0.0822	11.8073	0.0847	5.4702	64.5886	12
13	1.0330	0.9681	13.1968	0.0758	12.7753	0.0783	5.9650	76.2053	13
14	1.0356	0.9656	14.2298	0.0703	13.7410	0.0728	6.4594	88.7587	14
15	1.0382	0.9632	15.2654	0.0655	14.7042	0.0680	6.9534	102.2441	15
16	1.0408	0.9608	16.3035	0.0613	15.6650	0.0638	7.4469	116.6567	16
17	1.0434	0.9584	17.3443	0.0577	16.6235	0.0602	7.9401	131.9917	17
18	1.0460	0.9561	18.3876	0.0544	17.5795	0.0569	8.4328	148.2446	18
19	1.0486	0.9537	19.4336	0.0515	18.5332	0.0540	8.9251	165.4106	19
20	1.0512	0.9513	20.4822	0.0488	19.4845	0.0513	9.4170	183.4851	20
21	1.0538	0.9489	21.5334	0.0464	20.4334	0.0489	9.9085	202.4634	21
22	1.0565	0.9466	22.5872	0.0443	21.3800	0.0468	10.3995	222.3410	22
23	1.0591	0.9442	23.6437	0.0423	22.3241	0.0448	10.8901	243.1131	23
24	1.0618	0.9418	24.7028	0.0405	23.2660	0.0430	11.3804	264.7753	24
25	1.0644	0.9395	25.7646	0.0388	24.2055	0.0413	11.8702	287.3230	25
26	1.0671	0.9371	26.8290	0.0373	25.1426	0.0398	12.3596	310.7516	26
27	1.0697	0.9348	27.8961	0.0358	26.0774	0.0383	12.8485	335.0566	27
28	1.0724	0.9325	28.9658	0.0345	27.0099	0.0370	13.3371	360.2334	28
29	1.0751	0.9301	30.0382	0.0333	27.9400	0.0358	13.8252	386.2776	29
30	1.0778	0.9278	31.1133	0.0321	28.8679	0.0346	14.3130	413.1847	30
31	1.0805	0.9255	32.1911	0.0311	29.7934	0.0336	14.8003	440.9502	31
32	1.0832	0.9232	33.2716	0.0301	30.7166	0.0326	15.2872	469.5696	32
33	1.0859	0.9209	34.3547	0.0291	31.6375	0.0316	15.7736	499.0386	33
34	1.0886	0.9186	35.4406	0.0282	32.5561	0.0307	16.2597	529.3528	34
35	1.0913	0.9163	36.5292	0.0274	33.4724	0.0299	16.7454	560.5076	35
36	1.0941	0.9140	37.6206	0.0266	34.3865	0.0291	17.2306	592.4988	36
40	1.1050	0.9050	42.0132	0.0238	38.0199	0.0263	19.1673	728.7399	40
48	1.1273	0.8871	50.9312	0.0196	45.1787	0.0221	23.0209	1040.0552	48
50	1.1330	0.8826	53.1887	0.0188	46.9462	0.0213	23.9802	1125.7767	50
60	1.1616	0.8609	64.6467	0.0155	55.6524	0.0180	28.7514	1600.0845	60
72	1.1969	0.8355	78.7794	0.0127	65.8169	0.0152	34.4221	2265.5569	72
80	1.2211	0.8189	88.4392	0.0113	72.4260	0.0138	38.1694	2764.4568	80
84	1.2334	0.8108	93.3419	0.0107	75.6813	0.0132	40.0331	3029.7592	84
90	1.2520	0.7987	100.7885	0.0099	80.5038	0.0124	42.8162	3446.8700	90
96	1.2709	0.7869	108.3474	0.0092	85.2546	0.0117	45.5844	3886.2832	96
100	1.2836	0.7790	113.4500	0.0088	88.3825	0.0113	47.4216	4191.2417	100
108	1.3095	0.7636	123.8093	0.0081	94.5453	0.0106	51.0762	4829.0125	108
120	1.3494	0.7411	139.7414	0.0072	103.5618	0.0097	56.5084	5852.1116	120
240	1.8208	0.5492	328.3020	0.0030	180.3109	0.0055	107.5863	19398.9852	240
360	2.4568	0.4070	582.7369	0.0017	237.1894	0.0042	152.8902	36263.9299	360

0.50%

	Single Payment		Equal Payment Series				Gradient Series		
N	Compound Amount Factor (F/P,i,N)	Present Worth Factor (P/F,i,N)	Compound Amount Factor (F/A,i,N)	Sinking Fund Factor (A/F,i,N)	Present Worth Factor (P/A,i,N)	Capital Recovery Factor (A/P,i,N)	Gradient Uniform Series (A/G,i,N)	Gradient Present Worth (P/G,i,N)	N
1	1.0050	0.9950	1.0000	1.0000	0.9950	1.0050	0.0000	0.0000	1
2	1.0100	0.9901	2.0050	0.4988	1.9851	0.5038	0.4988	0.9901	2
3	1.0151	0.9851	3.0150	0.3317	2.9702	0.3367	0.9967	2.9604	3
4	1.0202	0.9802	4.0301	0.2481	3.9505	0.2531	1.4938	5.9011	4
5	1.0253	0.9754	5.0503	0.1980	4.9259	0.2030	1.9900	9.8026	5
6	1.0304	0.9705	6.0755	0.1646	5.8964	0.1696	2.4855	14.6552	6
7	1.0355	0.9657	7.1059	0.1407	6.8621	0.1457	2.9801	20.4493	7
8	1.0407	0.9609	8.1414	0.1228	7.8230	0.1278	3.4738	27.1755	8
9	1.0459	0.9561	9.1821	0.1089	8.7791	0.1139	3.9668	34.8244	9
10	1.0511	0.9513	10.2280	0.0978	9.7304	0.1028	4.4589	43.3865	10
11	1.0564	0.9466	11.2792	0.0887	10.6770	0.0937	4.9501	52.8526	11
12	1.0617	0.9419	12.3356	0.0811	11.6189	0.0861	5.4406	63.2136	12
13	1.0670	0.9372	13.3972	0.0746	12.5562	0.0796	5.9302	74.4602	13
14	1.0723	0.9326	14.4642	0.0691	13.4887	0.0741	6.4190	86.5835	14
15	1.0777	0.9279	15.5365	0.0644	14.4166	0.0694	6.9069	99.5743	15
16	1.0831	0.9233	16.6142	0.0602	15.3399	0.0652	7.3940	113.4238	16
17	1.0885	0.9187	17.6973	0.0565	16.2586	0.0615	7.8803	128.1231	17
18	1.0939	0.9141	18.7858	0.0532	17.1728	0.0582	8.3658	143.6634	18
19	1.0994	0.9096	19.8797	0.0503	18.0824	0.0553	8.8504	160.0360	19
20	1.1049	0.9051	20.9791	0.0477	18.9874	0.0527	9.3342	177.2322	20
21	1.1104	0.9006	22.0840	0.0453	19.8880	0.0503	9.8172	195.2434	21
22	1.1160	0.8961	23.1944	0.0431	20.7841	0.0481	10.2993	214.0611	22
23	1.1216	0.8916	24.3104	0.0411	21.6757	0.0461	10.7806	233.6768	23
24	1.1272	0.8872	25.4320	0.0393	22.5629	0.0443	11.2611	254.0820	24
25	1.1328	0.8828	26.5591	0.0377	23.4456	0.0427	11.7407	275.2686	25
26	1.1385	0.8784	27.6919	0.0361	24.3240	0.0411	12.2195	297.2281	26
27	1.1442	0.8740	28.8304	0.0347	25.1980	0.0397	12.6975	319.9523	27
28	1.1499	0.8697	29.9745	0.0334	26.0677	0.0384	13.1747	343.4332	28
29	1.1556	0.8653	31.1244	0.0321	26.9330	0.0371	13.6510	367.6625	29
30	1.1614	0.8610	32.2800	0.0310	27.7941	0.0360	14.1265	392.6324	30
31	1.1672	0.8567	33.4414	0.0299	28.6508	0.0349	14.6012	418.3348	31
32	1.1730	0.8525	34.6086	0.0289	29.5033	0.0339	15.0750	444.7618	32
33	1.1789	0.8482	35.7817	0.0279	30.3515	0.0329	15.5480	471.9055	33
34	1.1848	0.8440	36.9606	0.0271	31.1955	0.0321	16.0202	499.7583	34
35	1.1907	0.8398	38.1454	0.0262	32.0354	0.0312	16.4915	528.3123	35
36	1.1967	0.8356	39.3361	0.0254	32.8710	0.0304	16.9621	557.5598	36
40	1.2208	0.8191	44.1588	0.0226	36.1722	0.0276	18.8359	681.3347	40
48	1.2705	0.7871	54.0978	0.0185	42.5803	0.0235	22.5437	959.9188	48
50	1.2832	0.7793	56.6452	0.0177	44.1428	0.0227	23.4624	1035.6966	50
60	1.3489	0.7414	69.7700	0.0143	51.7256	0.0193	28.0064	1448.6458	60
72	1.4320	0.6983	86.4089	0.0116	60.3395	0.0166	33.3504	2012.3478	72
80	1.4903	0.6710	98.0677	0.0102	65.8023	0.0152	36.8474	2424.6455	80
84	1.5204	0.6577	104.0739	0.0096	68.4530	0.0146	38.5763	2640.6641	84
90	1.5666	0.6383	113.3109	0.0088	72.3313	0.0138	41.1451	2976.0769	90
96	1.6141	0.6195	122.8285	0.0081	76.0952	0.0131	43.6845	3324.1846	96
100	1.6467	0.6073	129.3337	0.0077	78.5426	0.0127	45.3613	3562.7934	100
108	1.7137	0.5835	142.7399	0.0070	83.2934	0.0120	48.6758	4054.3747	108
120	1.8194	0.5496	163.8793	0.0061	90.0735	0.0111	53.5508	4823.5051	120
240	3.3102	0.3021	462.0409	0.0022	139.5808	0.0072	96.1131	13415.5395	240
360	6.0226	0.1660	1004.5150	0.0010	166.7916	0.0060	128.3236	21403.3041	360

0.75%

	Single Payment		Equal Payment Series				Gradient Series		
	Compound Amount Factor	Present Worth Factor	Compound Amount Factor	Sinking Fund Factor	Present Worth Factor	Capital Recovery Factor	Gradient Uniform Series	Gradient Present Worth	
N	(F/P,i,N)	(P/F,i,N)	(F/A,i,N)	(A/F,i,N)	(P/A,i,N)	(A/P,i,N)	(A/G,i,N)	(P/G,i,N)	N
1	1.0075	0.9926	1.0000	1.0000	0.9926	1.0075	0.0000	0.0000	1
2	1.0151	0.9852	2.0075	0.4981	1.9777	0.5056	0.4981	0.9852	2
3	1.0227	0.9778	3.0226	0.3308	2.9556	0.3383	0.9950	2.9408	3
4	1.0303	0.9706	4.0452	0.2472	3.9261	0.2547	1.4907	5.8525	4
5	1.0381	0.9633	5.0756	0.1970	4.8894	0.2045	1.9851	9.7058	5
6	1.0459	0.9562	6.1136	0.1636	5.8456	0.1711	2.4782	14.4866	6
7	1.0537	0.9490	7.1595	0.1397	6.7946	0.1472	2.9701	20.1808	7
8	1.0616	0.9420	8.2132	0.1218	7.7366	0.1293	3.4608	26.7747	8
9	1.0696	0.9350	9.2748	0.1078	8.6716	0.1153	3.9502	34.2544	9
10	1.0776	0.9280	10.3443	0.0967	9.5996	0.1042	4.4384	42.6064	10
11	1.0857	0.9211	11.4219	0.0876	10.5207	0.0951	4.9253	51.8174	11
12	1.0938	0.9142	12.5076	0.0800	11.4349	0.0875	5.4110	61.8740	12
13	1.1020	0.9074	13.6014	0.0735	12.3423	0.0810	5.8954	72.7632	13
14	1.1103	0.9007	14.7034	0.0680	13.2430	0.0755	6.3786	84.4720	14
15	1.1186	0.8940	15.8137	0.0632	14.1370	0.0707	6.8606	96.9876	15
16	1.1270	0.8873	16.9323	0.0591	15.0243	0.0666	7.3413	110.2973	16
17	1.1354	0.8807	18.0593	0.0554	15.9050	0.0629	7.8207	124.3887	17
18	1.1440	0.8742	19.1947	0.0521	16.7792	0.0596	8.2989	139.2494	18
19	1.1525	0.8676	20.3387	0.0492	17.6468	0.0567	8.7759	154.8671	19
20	1.1612	0.8612	21.4912	0.0465	18.5080	0.0540	9.2516	171.2297	20
21	1.1699	0.8548	22.6524	0.0441	19.3628	0.0516	9.7261	188.3253	21
22	1.1787	0.8484	23.8223	0.0420	20.2112	0.0495	10.1994	206.1420	22
23	1.1875	0.8421	25.0010	0.0400	21.0533	0.0475	10.6714	224.6682	23
24	1.1964	0.8358	26.1885	0.0382	21.8891	0.0457	11.1422	243.8923	24
25	1.2054	0.8296	27.3849	0.0365	22.7188	0.0440	11.6117	263.8029	25
26	1.2144	0.8234	28.5903	0.0350	23.5422	0.0425	12.0800	284.3888	26
27	1.2235	0.8173	29.8047	0.0336	24.3595	0.0411	12.5470	305.6387	27
28	1.2327	0.8112	31.0282	0.0322	25.1707	0.0397	13.0128	327.5416	28
29	1.2420	0.8052	32.2609	0.0310	25.9759	0.0385	13.4774	350.0867	29
30	1.2513	0.7992	33.5029	0.0298	26.7751	0.0373	13.9407	373.2631	30
31	1.2607	0.7932	34.7542	0.0288	27.5683	0.0363	14.4028	397.0602	31
32	1.2701	0.7873	36.0148	0.0278	28.3557	0.0353	14.8636	421.4675	32
33	1.2796	0.7815	37.2849	0.0268	29.1371	0.0343	15.3232	446.4746	33
34	1.2892	0.7757	38.5646	0.0259	29.9128	0.0334	15.7816	472.0712	34
35	1.2989	0.7699	39.8538	0.0251	30.6827	0.0326	16.2387	498.2471	35
36	1.3086	0.7641	41.1527	0.0243	31.4468	0.0318	16.6946	524.9924	36
40	1.3483	0.7416	46.4465	0.0215	34.4469	0.0290	18.5058	637.4693	40
48	1.4314	0.6986	57.5207	0.0174	40.1848	0.0249	22.0691	886.8404	48
50	1.4530	0.6883	60.3943	0.0166	41.5664	0.0241	22.9476	953.8486	50
60	1.5657	0.6387	75.4241	0.0133	48.1734	0.0208	27.2665	1313.5189	60
72	1.7126	0.5839	95.0070	0.0105	55.4768	0.0180	32.2882	1791.2463	72
80	1.8180	0.5500	109.0725	0.0092	59.9944	0.0167	35.5391	2132.1472	80
84	1.8732	0.5338	116.4269	0.0086	62.1540	0.0161	37.1357	2308.1283	84
90	1.9591	0.5104	127.8790	0.0078	65.2746	0.0153	39.4946	2577.9961	90
96	2.0489	0.4881	139.8562	0.0072	68.2584	0.0147	41.8107	2853.9352	96
100	2.1111	0.4737	148.1445	0.0068	70.1746	0.0143	43.3311	3040.7453	100
108	2.2411	0.4462	165.4832	0.0060	73.8394	0.0135	46.3154	3419.9041	108
120	2.4514	0.4079	193.5143	0.0052	78.9417	0.0127	50.6521	3998.5621	120
240	6.0092	0.1664	667.8869	0.0015	111.1450	0.0090	85.4210	9494.1162	240
360	14.7306	0.0679	1830.7435	0.0005	124.2819	0.0080	107.1145	13312.3871	360

1.0%

	Single Payment		Equal Payment Series				Gradient Series		
N	**Compound Amount Factor (F/P,i,N)**	**Present Worth Factor (P/F,i,N)**	**Compound Amount Factor (F/A,i,N)**	**Sinking Fund Factor (A/F,i,N)**	**Present Worth Factor (P/A,i,N)**	**Capital Recovery Factor (A/P,i,N)**	**Gradient Uniform Series (A/G,i,N)**	**Gradient Present Worth (P/G,i,N)**	**N**
1	1.0100	0.9901	1.0000	1.0000	0.9901	1.0100	0.0000	0.0000	1
2	1.0201	0.9803	2.0100	0.4975	1.9704	0.5075	0.4975	0.9803	2
3	1.0303	0.9706	3.0301	0.3300	2.9410	0.3400	0.9934	2.9215	3
4	1.0406	0.9610	4.0604	0.2463	3.9020	0.2563	1.4876	5.8044	4
5	1.0510	0.9515	5.1010	0.1960	4.8534	0.2060	1.9801	9.6103	5
6	1.0615	0.9420	6.1520	0.1625	5.7955	0.1725	2.4710	14.3205	6
7	1.0721	0.9327	7.2135	0.1386	6.7282	0.1486	2.9602	19.9168	7
8	1.0829	0.9235	8.2857	0.1207	7.6517	0.1307	3.4478	26.3812	8
9	1.0937	0.9143	9.3685	0.1067	8.5660	0.1167	3.9337	33.6959	9
10	1.1046	0.9053	10.4622	0.0956	9.4713	0.1056	4.4179	41.8435	10
11	1.1157	0.8963	11.5668	0.0865	10.3676	0.0965	4.9005	50.8067	11
12	1.1268	0.8874	12.6825	0.0788	11.2551	0.0888	5.3815	60.5687	12
13	1.1381	0.8787	13.8093	0.0724	12.1337	0.0824	5.8607	71.1126	13
14	1.1495	0.8700	14.9474	0.0669	13.0037	0.0769	6.3384	82.4221	14
15	1.1610	0.8613	16.0969	0.0621	13.8651	0.0721	6.8143	94.4810	15
16	1.1726	0.8528	17.2579	0.0579	14.7179	0.0679	7.2886	107.2734	16
17	1.1843	0.8444	18.4304	0.0543	15.5623	0.0643	7.7613	120.7834	17
18	1.1961	0.8360	19.6147	0.0510	16.3983	0.0610	8.2323	134.9957	18
19	1.2081	0.8277	20.8109	0.0481	17.2260	0.0581	8.7017	149.8950	19
20	1.2202	0.8195	22.0190	0.0454	18.0456	0.0554	9.1694	165.4664	20
21	1.2324	0.8114	23.2392	0.0430	18.8570	0.0530	9.6354	181.6950	21
22	1.2447	0.8034	24.4716	0.0409	19.6604	0.0509	10.0998	198.5663	22
23	1.2572	0.7954	25.7163	0.0389	20.4558	0.0489	10.5626	216.0660	23
24	1.2697	0.7876	26.9735	0.0371	21.2434	0.0471	11.0237	234.1800	24
25	1.2824	0.7798	28.2432	0.0354	22.0232	0.0454	11.4831	252.8945	25
26	1.2953	0.7720	29.5256	0.0339	22.7952	0.0439	11.9409	272.1957	26
27	1.3082	0.7644	30.8209	0.0324	23.5596	0.0424	12.3971	292.0702	27
28	1.3213	0.7568	32.1291	0.0311	24.3164	0.0411	12.8516	312.5047	28
29	1.3345	0.7493	33.4504	0.0299	25.0658	0.0399	13.3044	333.4863	29
30	1.3478	0.7419	34.7849	0.0287	25.8077	0.0387	13.7557	355.0021	30
31	1.3613	0.7346	36.1327	0.0277	26.5423	0.0377	14.2052	377.0394	31
32	1.3749	0.7273	37.4941	0.0267	27.2696	0.0367	14.6532	399.5858	32
33	1.3887	0.7201	38.8690	0.0257	27.9897	0.0357	15.0995	422.6291	33
34	1.4026	0.7130	40.2577	0.0248	28.7027	0.0348	15.5441	446.1572	34
35	1.4166	0.7059	41.6603	0.0240	29.4086	0.0340	15.9871	470.1583	35
36	1.4308	0.6989	43.0769	0.0232	30.1075	0.0332	16.4285	494.6207	36
40	1.4889	0.6717	48.8864	0.0205	32.8347	0.0305	18.1776	596.8561	40
48	1.6122	0.6203	61.2226	0.0163	37.9740	0.0263	21.5976	820.1460	48
50	1.6446	0.6080	64.4632	0.0155	39.1961	0.0255	22.4363	879.4176	50
60	1.8167	0.5504	81.6697	0.0122	44.9550	0.0222	26.5333	1192.8061	60
72	2.0471	0.4885	104.7099	0.0096	51.1504	0.0196	31.2386	1597.8673	72
80	2.2167	0.4511	121.6715	0.0082	54.8882	0.0182	34.2492	1879.8771	80
84	2.3067	0.4335	130.6723	0.0077	56.6485	0.0177	35.7170	2023.3153	84
90	2.4486	0.4084	144.8633	0.0069	59.1609	0.0169	37.8724	2240.5675	90
96	2.5993	0.3847	159.9273	0.0063	61.5277	0.0163	39.9727	2459.4298	96
100	2.7048	0.3697	170.4814	0.0059	63.0289	0.0159	41.3426	2605.7758	100
108	2.9289	0.3414	192.8926	0.0052	65.8578	0.0152	44.0103	2898.4203	108
120	3.3004	0.3030	230.0387	0.0043	69.7005	0.0143	47.8349	3334.1148	120
240	10.8926	0.0918	989.2554	0.0010	90.8194	0.0110	75.7393	6878.6016	240
360	35.9496	0.0278	3494.9641	0.0003	97.2183	0.0103	89.6995	8720.4323	360

1.25%

	Single Payment		Equal Payment Series				Gradient Series		
N	Compound Amount Factor (F/P,i,N)	Present Worth Factor (P/F,i,N)	Compound Amount Factor (F/A,i,N)	Sinking Fund Factor (A/F,i,N)	Present Worth Factor (P/A,i,N)	Capital Recovery Factor (A/P,i,N)	Gradient Uniform Series (A/G,i,N)	Gradient Present Worth (P/G,i,N)	N
1	1.0125	0.9877	1.0000	1.0000	0.9877	1.0125	0.0000	0.0000	1
2	1.0252	0.9755	2.0125	0.4969	1.9631	0.5094	0.4969	0.9755	2
3	1.0380	0.9634	3.0377	0.3292	2.9265	0.3417	0.9917	2.9023	3
4	1.0509	0.9515	4.0756	0.2454	3.8781	0.2579	1.4845	5.7569	4
5	1.0641	0.9398	5.1266	0.1951	4.8178	0.2076	1.9752	9.5160	5
6	1.0774	0.9282	6.1907	0.1615	5.7460	0.1740	2.4638	14.1569	6
7	1.0909	0.9167	7.2680	0.1376	6.6627	0.1501	2.9503	19.6571	7
8	1.1045	0.9054	8.3589	0.1196	7.5681	0.1321	3.4348	25.9949	8
9	1.1183	0.8942	9.4634	0.1057	8.4623	0.1182	3.9172	33.1487	9
10	1.1323	0.8832	10.5817	0.0945	9.3455	0.1070	4.3975	41.0973	10
11	1.1464	0.8723	11.7139	0.0854	10.2178	0.0979	4.8758	49.8201	11
12	1.1608	0.8615	12.8604	0.0778	11.0793	0.0903	5.3520	59.2967	12
13	1.1753	0.8509	14.0211	0.0713	11.9302	0.0838	5.8262	69.5072	13
14	1.1900	0.8404	15.1964	0.0658	12.7706	0.0783	6.2982	80.4320	14
15	1.2048	0.8300	16.3863	0.0610	13.6005	0.0735	6.7682	92.0519	15
16	1.2199	0.8197	17.5912	0.0568	14.4203	0.0693	7.2362	104.3481	16
17	1.2351	0.8096	18.8111	0.0532	15.2299	0.0657	7.7021	117.3021	17
18	1.2506	0.7996	20.0462	0.0499	16.0295	0.0624	8.1659	130.8958	18
19	1.2662	0.7898	21.2968	0.0470	16.8193	0.0595	8.6277	145.1115	19
20	1.2820	0.7800	22.5630	0.0443	17.5993	0.0568	9.0874	159.9316	20
21	1.2981	0.7704	23.8450	0.0419	18.3697	0.0544	9.5450	175.3392	21
22	1.3143	0.7609	25.1431	0.0398	19.1306	0.0523	10.0006	191.3174	22
23	1.3307	0.7515	26.4574	0.0378	19.8820	0.0503	10.4542	207.8499	23
24	1.3474	0.7422	27.7881	0.0360	20.6242	0.0485	10.9056	224.9204	24
25	1.3642	0.7330	29.1354	0.0343	21.3573	0.0468	11.3551	242.5132	25
26	1.3812	0.7240	30.4996	0.0328	22.0813	0.0453	11.8024	260.6128	26
27	1.3985	0.7150	31.8809	0.0314	22.7963	0.0439	12.2478	279.2040	27
28	1.4160	0.7062	33.2794	0.0300	23.5025	0.0425	12.6911	298.2719	28
29	1.4337	0.6975	34.6954	0.0288	24.2000	0.0413	13.1323	317.8019	29
30	1.4516	0.6889	36.1291	0.0277	24.8889	0.0402	13.5715	337.7797	30
31	1.4698	0.6804	37.5807	0.0266	25.5693	0.0391	14.0086	358.1912	31
32	1.4881	0.6720	39.0504	0.0256	26.2413	0.0381	14.4438	379.0227	32
33	1.5067	0.6637	40.5386	0.0247	26.9050	0.0372	14.8768	400.2607	33
34	1.5256	0.6555	42.0453	0.0238	27.5605	0.0363	15.3079	421.8920	34
35	1.5446	0.6474	43.5709	0.0230	28.2079	0.0355	15.7369	443.9037	35
36	1.5639	0.6394	45.1155	0.0222	28.8473	0.0347	16.1639	466.2830	36
40	1.6436	0.6084	51.4896	0.0194	31.3269	0.0319	17.8515	559.2320	40
48	1.8154	0.5509	65.2284	0.0153	35.9315	0.0278	21.1299	759.2296	48
50	1.8610	0.5373	68.8818	0.0145	37.0129	0.0270	21.9295	811.6738	50
60	2.1072	0.4746	88.5745	0.0113	42.0346	0.0238	25.8083	1084.8429	60
72	2.4459	0.4088	115.6736	0.0086	47.2925	0.0211	30.2047	1428.4561	72
80	2.7015	0.3702	136.1188	0.0073	50.3867	0.0198	32.9822	1661.8651	80
84	2.8391	0.3522	147.1290	0.0068	51.8222	0.0193	34.3258	1778.8384	84
90	3.0588	0.3269	164.7050	0.0061	53.8461	0.0186	36.2855	1953.8303	90
96	3.2955	0.3034	183.6411	0.0054	55.7246	0.0179	38.1793	2127.5244	96
100	3.4634	0.2887	197.0723	0.0051	56.9013	0.0176	39.4058	2242.2411	100
108	3.8253	0.2614	226.0226	0.0044	59.0865	0.0169	41.7737	2468.2636	108
120	4.4402	0.2252	275.2171	0.0036	61.9828	0.0161	45.1184	2796.5694	120
240	19.7155	0.0507	1497.2395	0.0007	75.9423	0.0132	67.1764	5101.5288	240
360	87.5410	0.0114	6923.2796	0.0001	79.0861	0.0126	75.8401	5997.9027	360

1.5%

	Single Payment		Equal Payment Series				Gradient Series		
	Compound Amount Factor	Present Worth Factor	Compound Amount Factor	Sinking Fund Factor	Present Worth Factor	Capital Recovery Factor	Gradient Uniform Series	Gradient Present Worth	
N	(F/P,i,N)	(P/F,i,N)	(F/A,i,N)	(A/F,i,N)	(P/A,i,N)	(A/P,i,N)	(A/G,i,N)	(P/G,i,N)	N
1	1.0150	0.9852	1.0000	1.0000	0.9852	1.0150	0.0000	0.0000	1
2	1.0302	0.9707	2.0150	0.4963	1.9559	0.5113	0.4963	0.9707	2
3	1.0457	0.9563	3.0452	0.3284	2.9122	0.3434	0.9901	2.8833	3
4	1.0614	0.9422	4.0909	0.2444	3.8544	0.2594	1.4814	5.7098	4
5	1.0773	0.9283	5.1523	0.1941	4.7826	0.2091	1.9702	9.4229	5
6	1.0934	0.9145	6.2296	0.1605	5.6972	0.1755	2.4566	13.9956	6
7	1.1098	0.9010	7.3230	0.1366	6.5982	0.1516	2.9405	19.4018	7
8	1.1265	0.8877	8.4328	0.1186	7.4859	0.1336	3.4219	25.6157	8
9	1.1434	0.8746	9.5593	0.1046	8.3605	0.1196	3.9008	32.6125	9
10	1.1605	0.8617	10.7027	0.0934	9.2222	0.1084	4.3772	40.3675	110
11	1.1779	0.8489	11.8633	0.0843	10.0711	0.0993	4.8512	48.8568	11
12	1.1956	0.8364	13.0412	0.0767	10.9075	0.0917	5.3227	58.0571	12
13	1.2136	0.8240	14.2368	0.0702	11.7315	0.0852	5.7917	67.9454	13
14	1.2318	0.8118	15.4504	0.0647	12.5434	0.0797	6.2582	78.4994	14
15	1.2502	0.7999	16.6821	0.0599	13.3432	0.0749	6.7223	89.6974	15
16	1.2690	0.7880	17.9324	0.0558	14.1313	0.0708	7.1839	101.5178	16
17	1.2880	0.7764	19.2014	0.0521	14.9076	0.0671	7.6431	113.9400	17
18	1.3073	0.7649	20.4894	0.0488	15.6726	0.0638	8.0997	126.9435	18
19	1.3270	0.7536	21.7967	0.0459	16.4262	0.0609	8.5539	140.5084	19
20	1.3469	0.7425	23.1237	0.0432	17.1686	0.0582	9.0057	154.6154	20
21	1.3671	0.7315	24.4705	0.0409	17.9001	0.0559	9.4550	169.2453	21
22	1.3876	0.7207	25.8376	0.0387	18.6208	0.0537	9.9018	184.3798	22
23	1.4084	0.7100	27.2251	0.0367	19.3309	0.0517	10.3462	200.0006	23
24	1.4295	0.6995	28.6335	0.0349	20.0304	0.0499	10.7881	216.0901	24
25	1.4509	0.6892	30.0630	0.0333	20.7196	0.0483	11.2276	232.6310	25
26	1.4727	0.6790	31.5140	0.0317	21.3986	0.0467	11.6646	249.6065	26
27	1.4948	0.6690	32.9867	0.0303	22.0676	0.0453	12.0992	267.0002	27
28	1.5172	0.6591	34.4815	0.0290	22.7267	0.0440	12.5313	284.7958	28
29	1.5400	0.6494	35.9987	0.0278	23.3761	0.0428	12.9610	302.9779	29
30	1.5631	0.6398	37.5387	0.0266	24.0158	0.0416	13.3883	321.5310	30
31	1.5865	0.6303	39.1018	0.0256	24.6461	0.0406	13.8131	340.4402	31
32	1.6103	0.6210	40.6883	0.0246	25.2671	0.0396	14.2355	359.6910	32
33	1.6345	0.6118	42.2986	0.0236	25.8790	0.0386	14.6555	379.2691	33
34	1.6590	0.6028	43.9331	0.0228	26.4817	0.0378	15.0731	399.1607	34
35	1.6839	0.5939	45.5921	0.0219	27.0756	0.0369	15.4882	419.3521	35
36	1.7091	0.5851	47.2760	0.0212	27.6607	0.0362	15.9009	439.8303	36
40	1.8140	0.5513	54.2679	0.0184	29.9158	0.0334	17.5277	524.3568	40
48	2.0435	0.4894	69.5652	0.0144	34.0426	0.0294	20.6667	703.5462	48
50	2.1052	0.4750	73.6828	0.0136	34.9997	0.0286	21.4277	749.9636	50
60	2.4432	0.4093	96.2147	0.0104	39.3803	0.0254	25.0930	988.1674	60
72	2.9212	0.3423	128.0772	0.0078	43.8447	0.0228	29.1893	1279.7938	72
80	3.2907	0.3039	152.7109	0.0065	46.4073	0.0215	31.7423	1473.0741	80
84	3.4926	0.2863	166.1726	0.0060	47.5786	0.0210	32.9668	1568.5140	84
90	3.8189	0.2619	187.9299	0.0053	49.2099	0.0203	34.7399	1709.5439	90
96	4.1758	0.2395	211.7202	0.0047	50.7017	0.0197	36.4381	1847.4725	96
100	4.4320	0.2256	228.8030	0.0044	51.6247	0.0194	37.5295	1937.4506	100
108	4.9927	0.2003	266.1778	0.0038	53.3137	0.0188	39.6171	2112.1348	108
120	5.9693	0.1675	331.2882	0.0030	55.4985	0.0180	42.5185	2359.7114	120
240	35.6328	0.0281	2308.8544	0.0004	64.7957	0.0154	59.7368	3870.6912	240
360	212.7038	0.0047	14113.5854	0.0001	66.3532	0.0151	64.9662	4310.7165	360

1.75%

	Single Payment		Equal Payment Series				Gradient Series		
N	Compound Amount Factor (F/P,i,N)	Present Worth Factor (P/F,i,N)	Compound Amount Factor (F/A,i,N)	Sinking Fund Factor (A/F,i,N)	Present Worth Factor (P/A,i,N)	Capital Recovery Factor (A/P,i,N)	Gradient Uniform Series (A/G,i,N)	Gradient Present Worth (P/G,i,N)	N
1	1.0175	0.9828	1.0000	1.0000	0.9828	1.0175	0.0000	0.0000	1
2	1.0353	0.9659	2.0175	0.4957	1.9487	0.5132	0.4957	0.9659	2
3	1.0534	0.9493	3.0528	0.3276	2.8980	0.3451	0.9884	2.8645	3
4	1.0719	0.9330	4.1062	0.2435	3.8309	0.2610	1.4783	5.6633	4
5	1.0906	0.9169	5.1781	0.1931	4.7479	0.2106	1.9653	9.3310	5
6	1.1097	0.9011	6.2687	0.1595	5.6490	0.1770	2.4494	13.8367	6
7	1.1291	0.8856	7.3784	0.1355	6.5346	0.1530	2.9306	19.1506	7
8	1.1489	0.8704	8.5075	0.1175	7.4051	0.1350	3.4089	25.2435	8
9	1.1690	0.8554	9.6564	0.1036	8.2605	0.1211	3.8844	32.0870	9
10	1.1894	0.8407	10.8254	0.0924	9.1012	0.1099	4.3569	39.6535	10
11	1.2103	0.8263	12.0148	0.0832	9.9275	0.1007	4.8266	47.9162	11
12	1.2314	0.8121	13.2251	0.0756	10.7395	0.0931	5.2934	56.8489	12
13	1.2530	0.7981	14.4565	0.0692	11.5376	0.0867	5.7573	66.4260	13
14	1.2749	0.7844	15.7095	0.0637	12.3220	0.0812	6.2184	76.6227	14
15	1.2972	0.7709	16.9844	0.0589	13.0929	0.0764	6.6765	87.4149	15
16	1.3199	0.7576	18.2817	0.0547	13.8505	0.0722	7.1318	98.7792	16
17	1.3430	0.7446	19.6016	0.0510	14.5951	0.0685	7.5842	110.6926	17
18	1.3665	0.7318	20.9446	0.0477	15.3269	0.0652	8.0338	123.1328	18
19	1.3904	0.7192	22.3112	0.0448	16.0461	0.0623	8.4805	136.0783	19
20	1.4148	0.7068	23.7016	0.0422	16.7529	0.0597	8.9243	149.5080	20
21	1.4395	0.6947	25.1164	0.0398	17.4475	0.0573	9.3653	163.4013	21
22	1.4647	0.6827	26.5559	0.0377	18.1303	0.0552	9.8034	177.7385	22
23	1.4904	0.6710	28.0207	0.0357	18.8012	0.0532	10.2387	192.5000	23
24	1.5164	0.6594	29.5110	0.0339	19.4607	0.0514	10.6711	207.6671	24
25	1.5430	0.6481	31.0275	0.0322	20.1088	0.0497	11.1007	223.2214	25
26	1.5700	0.6369	32.5704	0.0307	20.7457	0.0482	11.5274	239.1451	26
27	1.5975	0.6260	34.1404	0.0293	21.3717	0.0468	11.9513	255.4210	27
28	1.6254	0.6152	35.7379	0.0280	21.9870	0.0455	12.3724	272.0321	28
29	1.6539	0.6046	37.3633	0.0268	22.5916	0.0443	12.7907	288.9623	29
30	1.6828	0.5942	39.0172	0.0256	23.1858	0.0431	13.2061	306.1954	30
31	1.7122	0.5840	40.7000	0.0246	23.7699	0.0421	13.6188	323.7163	31
32	1.7422	0.5740	42.4122	0.0236	24.3439	0.0411	14.0286	341.5097	32
33	1.7727	0.5641	44.1544	0.0226	24.9080	0.0401	14.4356	359.5613	33
34	1.8037	0.5544	45.9271	0.0218	25.4624	0.0393	14.8398	377.8567	34
35	1.8353	0.5449	47.7308	0.0210	26.0073	0.0385	15.2412	396.3824	35
36	1.8674	0.5355	49.5661	0.0202	26.5428	0.0377	15.6399	415.1250	36
40	2.0016	0.4996	57.2341	0.0175	28.5942	0.0350	17.2066	492.0109	40
48	2.2996	0.4349	74.2628	0.0135	32.2938	0.0310	20.2084	652.6054	48
50	2.3808	0.4200	78.9022	0.0127	33.1412	0.0302	20.9317	693.7010	50
60	2.8318	0.3531	104.6752	0.0096	36.9640	0.0271	24.3885	901.4954	60
72	3.4872	0.2868	142.1263	0.0070	40.7564	0.0245	28.1948	1149.1181	72
80	4.0064	0.2496	171.7938	0.0058	42.8799	0.0233	30.5329	1309.2482	80
84	4.2943	0.2329	188.2450	0.0053	43.8361	0.0228	31.6442	1387.1584	84
90	4.7654	0.2098	215.1646	0.0046	45.1516	0.0221	33.2409	1500.8798	90
96	5.2882	0.1891	245.0374	0.0041	46.3370	0.0216	34.7556	1610.4716	96
100	5.6682	0.1764	266.7518	0.0037	47.0615	0.0212	35.7211	1681.0886	100
108	6.5120	0.1536	314.9738	0.0032	48.3679	0.0207	37.5494	1816.1852	108
120	8.0192	0.1247	401.0962	0.0025	50.0171	0.0200	40.0469	2003.0269	120
240	64.3073	0.0156	3617.5602	0.0003	56.2543	0.0178	53.3518	3001.2678	240
360	515.6921	0.0019	29410.9747	0.0000	57.0320	0.0175	56.4434	3219.0833	360

2.0%

	Single Payment		Equal Payment Series				Gradient Series		
	Compound Amount Factor	Present Worth Factor	Compound Amount Factor	Sinking Fund Factor	Present Worth Factor	Capital Recovery Factor	Gradient Uniform Series	Gradient Present Worth	
N	(F/P,i,N)	(P/F,i,N)	(F/A,i,N)	(A/F,i,N)	(P/A,i,N)	(A/P,i,N)	(A/G,i,N)	(P/G,i,N)	N
1	1.0200	0.9804	1.0000	1.0000	0.9804	1.0200	0.0000	0.0000	1
2	1.0404	0.9612	2.0200	0.4950	1.9416	0.5150	0.4950	0.9612	2
3	1.0612	0.9423	3.0604	0.3268	2.8839	0.3468	0.9868	2.8458	3
4	1.0824	0.9238	4.1216	0.2426	3.8077	0.2626	1.4752	5.6173	4
5	1.1041	0.9057	5.2040	0.1922	4.7135	0.2122	1.9604	9.2403	5
6	1.1262	0.8880	6.3081	0.1585	5.6014	0.1785	2.4423	13.6801	6
7	1.1487	0.8706	7.4343	0.1345	6.4720	0.1545	2.9208	18.9035	7
8	1.1717	0.8535	8.5830	0.1165	7.3255	0.1365	3.3961	24.8779	8
9	1.1951	0.8368	9.7546	0.1025	8.1622	0.1225	3.8681	31.5720	9
10	1.2190	0.8203	10.9497	0.0913	8.9826	0.1113	4.3367	38.9551	110
11	1.2434	0.8043	12.1687	0.0822	9.7868	0.1022	4.8021	46.9977	11
12	1.2682	0.7885	13.4121	0.0746	10.5753	0.0946	5.2642	55.6712	12
13	1.2936	0.7730	14.6803	0.0681	11.3484	0.0881	5.7231	64.9475	13
14	1.3195	0.7579	15.9739	0.0626	12.1062	0.0826	6.1786	74.7999	14
15	1.3459	0.7430	17.2934	0.0578	12.8493	0.0778	6.6309	85.2021	15
16	1.3728	0.7284	18.6393	0.0537	13.5777	0.0737	7.0799	96.1288	16
17	1.4002	0.7142	20.0121	0.0500	14.2919	0.0700	7.5256	107.5554	17
18	1.4282	0.7002	21.4123	0.0467	14.9920	0.0667	7.9681	119.4581	18
19	1.4568	0.6864	22.8406	0.0438	15.6785	0.0638	8.4073	131.8139	19
20	1.4859	0.6730	24.2974	0.0412	16.3514	0.0612	8.8433	144.6003	20
21	1.5157	0.6598	25.7833	0.0388	17.0112	0.0588	9.2760	157.7959	21
22	1.5460	0.6468	27.2990	0.0366	17.6580	0.0566	9.7055	171.3795	22
23	1.5769	0.6342	28.8450	0.0347	18.2922	0.0547	10.1317	185.3309	23
24	1.6084	0.6217	30.4219	0.0329	18.9139	0.0529	10.5547	199.6305	24
25	1.6406	0.6095	32.0303	0.0312	19.5235	0.0512	10.9745	214.2592	25
26	1.6734	0.5976	33.6709	0.0297	20.1210	0.0497	11.3910	229.1987	26
27	1.7069	0.5859	35.3443	0.0283	20.7069	0.0483	11.8043	244.4311	27
28	1.7410	0.5744	37.0512	0.0270	21.2813	0.0470	12.2145	259.9392	28
29	1.7758	0.5631	38.7922	0.0258	21.8444	0.0458	12.6214	275.7064	29
30	1.8114	0.5521	40.5681	0.0246	22.3965	0.0446	13.0251	291.7164	30
31	1.8476	0.5412	42.3794	0.0236	22.9377	0.0436	13.4257	307.9538	31
32	1.8845	0.5306	44.2270	0.0226	23.4683	0.0426	13.8230	324.4035	32
33	1.9222	0.5202	46.1116	0.0217	23.9886	0.0417	14.2172	341.0508	33
34	1.9607	0.5100	48.0338	0.0208	24.4986	0.0408	14.6083	357.8817	34
35	1.9999	0.5000	49.9945	0.0200	24.9986	0.0400	14.9961	374.8826	35
36	2.0399	0.4902	51.9944	0.0192	25.4888	0.0392	15.3809	392.0405	36
40	2.2080	0.4529	60.4020	0.0166	27.3555	0.0366	16.8885	461.9931	40
48	2.5871	0.3865	79.3535	0.0126	30.6731	0.0326	19.7556	605.9657	48
50	2.6916	0.3715	84.5794	0.0118	31.4236	0.0318	20.4420	642.3606	50
60	3.2810	0.3048	114.0515	0.0088	34.7609	0.0288	23.6961	823.6975	60
72	4.1611	0.2403	158.0570	0.0063	37.9841	0.0263	27.2234	1034.0557	72
80	4.8754	0.2051	193.7720	0.0052	39.7445	0.0252	29.3572	1166.7868	80
84	5.2773	0.1895	213.8666	0.0047	40.5255	0.0247	30.3616	1230.4191	84
90	5.9431	0.1683	247.1567	0.0040	41.5869	0.0240	31.7929	1322.1701	90
96	6.6929	0.1494	284.6467	0.0035	42.5294	0.0235	33.1370	1409.2973	96
100	7.2446	0.1380	312.2323	0.0032	43.0984	0.0232	33.9863	1464.7527	100
108	8.4883	0.1178	374.4129	0.0027	44.1095	0.0227	35.5774	1569.3025	108
120	10.7652	0.0929	488.2582	0.0020	45.3554	0.0220	37.7114	1710.4160	120
240	115.8887	0.0086	5744.4368	0.0002	49.5686	0.0202	47.9110	2374.8800	240
360	1247.5611	0.0008	62328.0564	0.0000	49.9599	0.0200	49.7112	2483.5679	360

3.0%

	Single Payment		Equal Payment Series				Gradient Series		
N	Compound Amount Factor (F/P,i,N)	Present Worth Factor (P/F,i,N)	Compound Amount Factor (F/A,i,N)	Sinking Fund Factor (A/F,i,N)	Present Worth Factor (P/A,i,N)	Capital Recovery Factor (A/P,i,N)	Gradient Uniform Series (A/G,i,N)	Gradient Present Worth (P/G,i,N)	N
1	1.0300	0.9709	1.0000	1.0000	0.9709	1.0300	0.0000	0.0000	1
2	1.0609	0.9426	2.0300	0.4926	1.9135	0.5226	0.4926	0.9426	2
3	1.0927	0.9151	3.0909	0.3235	2.8286	0.3535	0.9803	2.7729	3
4	1.1255	0.8885	4.1836	0.2390	3.7171	0.2690	1.4631	5.4383	4
5	1.1593	0.8626	5.3091	0.1884	4.5797	0.2184	1.9409	8.8888	5
6	1.1941	0.8375	6.4684	0.1546	5.4172	0.1846	2.4138	13.0762	6
7	1.2299	0.8131	7.6625	0.1305	6.2303	0.1605	2.8819	17.9547	7
8	1.2668	0.7894	8.8923	0.1125	7.0197	0.1425	3.3450	23.4806	8
9	1.3048	0.7664	10.1591	0.0984	7.7861	0.1284	3.8032	29.6119	9
10	1.3439	0.7441	11.4639	0.0872	8.5302	0.1172	4.2565	36.3088	110
11	1.3842	0.7224	12.8078	0.0781	9.2526	0.1081	4.7049	43.5330	11
12	1.4258	0.7014	14.1920	0.0705	9.9540	0.1005	5.1485	51.2482	12
13	1.4685	0.6810	15.6178	0.0640	10.6350	0.0940	5.5872	59.4196	13
14	1.5126	0.6611	17.0863	0.0585	11.2961	0.0885	6.0210	68.0141	14
15	1.5580	0.6419	18.5989	0.0538	11.9379	0.0838	6.4500	77.0002	15
16	1.6047	0.6232	20.1569	0.0496	12.5611	0.0796	6.8742	86.3477	16
17	1.6528	0.6050	21.7616	0.0460	13.1661	0.0760	7.2936	96.0280	17
18	1.7024	0.5874	23.4144	0.0427	13.7535	0.0727	7.7081	106.0137	18
19	1.7535	0.5703	25.1169	0.0398	14.3238	0.0698	8.1179	116.2788	19
20	1.8061	0.5537	26.8704	0.0372	14.8775	0.0672	8.5229	126.7987	20
21	1.8603	0.5375	28.6765	0.0349	15.4150	0.0649	8.9231	137.5496	21
22	1.9161	0.5219	30.5368	0.0327	15.9396	0.0627	9.3186	148.5094	22
23	1.9736	0.5067	32.4529	0.0308	16.4436	0.0608	9.7093	159.6566	23
24	2.0328	0.4919	34.4265	0.0290	16.9355	0.0590	10.0954	170.9711	24
25	2.0938	0.4776	36.4593	0.0274	17.4131	0.0574	10.4768	182.4336	25
26	2.1566	0.4637	38.5530	0.0259	17.8768	0.0559	10.8535	194.0260	26
27	2.2213	0.4502	40.7096	0.0246	18.3270	0.0546	11.2255	205.7309	27
28	2.2879	0.4371	42.9309	0.0233	18.7641	0.0533	11.5930	217.5320	28
29	2.3566	0.4243	45.2189	0.0221	19.1885	0.0521	11.9558	229.4137	29
30	2.4273	0.4120	47.5754	0.0210	19.6004	0.0510	12.3141	241.3613	30
31	2.5001	0.4000	50.0027	0.0200	20.0004	0.0500	12.6678	253.3609	31
32	2.5751	0.3883	52.5028	0.0190	20.3888	0.0490	13.0169	265.3993	32
33	2.6523	0.3770	55.0778	0.0182	20.7658	0.0482	13.3616	277.4642	33
34	2.7319	0.3660	57.7302	0.0173	21.1318	0.0473	13.7018	289.5437	34
35	2.8139	0.3554	60.4621	0.0165	21.4872	0.0465	14.0375	301.6267	35
40	3.2620	0.3066	75.4013	0.0133	23.1148	0.0433	15.6502	361.7499	40
45	3.7816	0.2644	92.7199	0.0108	24.5187	0.0408	17.1556	420.6325	45
50	4.3839	0.2281	112.7969	0.0089	25.7298	0.0389	18.5575	477.4803	50
55	5.0821	0.1968	136.0716	0.0073	26.7744	0.0373	19.8600	531.7411	55
60	5.8916	0.1697	163.0534	0.0061	27.6756	0.0361	21.0674	583.0526	60
65	6.8300	0.1464	194.3328	0.0051	28.4529	0.0351	22.1841	631.2010	65
70	7.9178	0.1263	230.5941	0.0043	29.1234	0.0343	23.2145	676.0869	70
75	9.1789	0.1089	272.6309	0.0037	29.7018	0.0337	24.1634	717.6978	75
80	10.6409	0.0940	321.3630	0.0031	30.2008	0.0331	25.0353	756.0865	80
85	12.3357	0.0811	377.8570	0.0026	30.6312	0.0326	25.8349	791.3529	85
90	14.3005	0.0699	443.3489	0.0023	31.0024	0.0323	26.5667	823.6302	90
95	16.5782	0.0603	519.2720	0.0019	31.3227	0.0319	27.2351	853.0742	95
100	19.2186	0.0520	607.2877	0.0016	31.5989	0.0316	27.8444	879.8540	100

4.0%

	Single Payment		Equal Payment Series				Gradient Series		
	Compound Amount Factor	Present Worth Factor	Compound Amount Factor	Sinking Fund Factor	Present Worth Factor	Capital Recovery Factor	Gradient Uniform Series	Gradient Present Worth	
N	(F/P,i,N)	(P/F,i,N)	(F/A,i,N)	(A/F,i,N)	(P/A,i,N)	(A/P,i,N)	(A/G,i,N)	(P/G,i,N)	N
1	1.0400	0.9615	1.0000	1.0000	0.9615	1.0400	0.0000	0.0000	1
2	1.0816	0.9246	2.0400	0.4902	1.8861	0.5302	0.4902	0.9246	2
3	1.1249	0.8890	3.1216	0.3203	2.7751	0.3603	0.9739	2.7025	3
4	1.1699	0.8548	4.2465	0.2355	3.6299	0.2755	1.4510	5.2670	4
5	1.2167	0.8219	5.4163	0.1846	4.4518	0.2246	1.9216	8.5547	5
6	1.2653	0.7903	6.6330	0.1508	5.2421	0.1908	2.3857	12.5062	6
7	1.3159	0.7599	7.8983	0.1266	6.0021	0.1666	2.8433	17.0657	7
8	1.3686	0.7307	9.2142	0.1085	6.7327	0.1485	3.2944	22.1806	8
9	1.4233	0.7026	10.5828	0.0945	7.4353	0.1345	3.7391	27.8013	9
10	1.4802	0.6756	12.0061	0.0833	8.1109	0.1233	4.1773	33.8814	110
11	1.5395	0.6496	13.4864	0.0741	8.7605	0.1141	4.6090	40.3772	11
12	1.6010	0.6246	15.0258	0.0666	9.3851	0.1066	5.0343	47.2477	12
13	1.6651	0.6006	16.6268	0.0601	9.9856	0.1001	5.4533	54.4546	13
14	1.7317	0.5775	18.2919	0.0547	10.5631	0.0947	5.8659	61.9618	14
15	1.8009	0.5553	20.0236	0.0499	11.1184	0.0899	6.2721	69.7355	15
16	1.8730	0.5339	21.8245	0.0458	11.6523	0.0858	6.6720	77.7441	16
17	1.9479	0.5134	23.6975	0.0422	12.1657	0.0822	7.0656	85.9581	17
18	2.0258	0.4936	25.6454	0.0390	12.6593	0.0790	7.4530	94.3498	18
19	2.1068	0.4746	27.6712	0.0361	13.1339	0.0761	7.8342	102.8933	19
20	2.1911	0.4564	29.7781	0.0336	13.5903	0.0736	8.2091	111.5647	20
21	2.2788	0.4388	31.9692	0.0313	14.0292	0.0713	8.5779	120.3414	21
22	2.3699	0.4220	34.2480	0.0292	14.4511	0.0692	8.9407	129.2024	22
23	2.4647	0.4057	36.6179	0.0273	14.8568	0.0673	9.2973	138.1284	23
24	2.5633	0.3901	39.0826	0.0256	15.2470	0.0656	9.6479	147.1012	24
25	2.6658	0.3751	41.6459	0.0240	15.6221	0.0640	9.9925	156.1040	25
26	2.7725	0.3607	44.3117	0.0226	15.9828	0.0626	10.3312	165.1212	26
27	2.8834	0.3468	47.0842	0.0212	16.3296	0.0612	10.6640	174.1385	27
28	2.9987	0.3335	49.9676	0.0200	16.6631	0.0600	10.9909	183.1424	28
29	3.1187	0.3207	52.9663	0.0189	16.9837	0.0589	11.3120	192.1206	29
30	3.2434	0.3083	56.0849	0.0178	17.2920	0.0578	11.6274	201.0618	30
31	3.3731	0.2965	59.3283	0.0169	17.5885	0.0569	11.9371	209.9556	31
32	3.5081	0.2851	62.7015	0.0159	17.8736	0.0559	12.2411	218.7924	32
33	3.6484	0.2741	66.2095	0.0151	18.1476	0.0551	12.5396	227.5634	33
34	3.7943	0.2636	69.8579	0.0143	18.4112	0.0543	12.8324	236.2607	34
35	3.9461	0.2534	73.6522	0.0136	18.6646	0.0536	13.1198	244.8768	35
40	4.8010	0.2083	95.0255	0.0105	19.7928	0.0505	14.4765	286.5303	40
45	5.8412	0.1712	121.0294	0.0083	20.7200	0.0483	15.7047	325.4028	45
50	7.1067	0.1407	152.6671	0.0066	21.4822	0.0466	16.8122	361.1638	50
55	8.6464	0.1157	191.1592	0.0052	22.1086	0.0452	17.8070	393.6890	55
60	10.5196	0.0951	237.9907	0.0042	22.6235	0.0442	18.6972	422.9966	60
65	12.7987	0.0781	294.9684	0.0034	23.0467	0.0434	19.4909	449.2014	65
70	15.5716	0.0642	364.2905	0.0027	23.3945	0.0427	20.1961	472.4789	70
75	18.9453	0.0528	448.6314	0.0022	23.6804	0.0422	20.8206	493.0408	75
80	23.0498	0.0434	551.2450	0.0018	23.9154	0.0418	21.3718	511.1161	80
85	28.0436	0.0357	676.0901	0.0015	24.1085	0.0415	21.8569	526.9384	85
90	34.1193	0.0293	827.9833	0.0012	24.2673	0.0412	22.2826	540.7369	90
95	41.5114	0.0241	1012.7846	0.0010	24.3978	0.0410	22.6550	552.7307	95
100	50.5049	0.0198	1237.6237	0.0008	24.5050	0.0408	22.9800	563.1249	100

5.0%

	Single Payment		Equal Payment Series				Gradient Series		
	Compound Amount Factor	Present Worth Factor	Compound Amount Factor	Sinking Fund Factor	Present Worth Factor	Capital Recovery Factor	Gradient Uniform Series	Gradient Present Worth	
N	(F/P,i,N)	(P/F,i,N)	(F/A,i,N)	(A/F,i,N)	(P/A,i,N)	(A/P,i,N)	(A/G,i,N)	(P/G,i,N)	N
1	1.0500	0.9524	1.0000	1.0000	0.9524	1.0500	0.0000	0.0000	1
2	1.1025	0.9070	2.0500	0.4878	1.8594	0.5378	0.4878	0.9070	2
3	1.1576	0.8638	3.1525	0.3172	2.7232	0.3672	0.9675	2.6347	3
4	1.2155	0.8227	4.3101	0.2320	3.5460	0.2820	1.4391	5.1028	4
5	1.2763	0.7835	5.5256	0.1810	4.3295	0.2310	1.9025	8.2369	5
6	1.3401	0.7462	6.8019	0.1470	5.0757	0.1970	2.3579	11.9680	6
7	1.4071	0.7107	8.1420	0.1228	5.7864	0.1728	2.8052	16.2321	7
8	1.4775	0.6768	9.5491	0.1047	6.4632	0.1547	3.2445	20.9700	8
9	1.5513	0.6446	11.0266	0.0907	7.1078	0.1407	3.6758	26.1268	9
10	1.6289	0.6139	12.5779	0.0795	7.7217	0.1295	4.0991	31.6520	10
11	1.7103	0.5847	14.2068	0.0704	8.3064	0.1204	4.5144	37.4988	11
12	1.7959	0.5568	15.9171	0.0628	8.8633	0.1128	4.9219	43.6241	12
13	1.8856	0.5303	17.7130	0.0565	9.3936	0.1065	5.3215	49.9879	13
14	1.9799	0.5051	19.5986	0.0510	9.8986	0.1010	5.7133	56.5538	14
15	2.0789	0.4810	21.5786	0.0463	10.3797	0.0963	6.0973	63.2880	15
16	2.1829	0.4581	23.6575	0.0423	10.8378	0.0923	6.4736	70.1597	16
17	2.2920	0.4363	25.8404	0.0387	11.2741	0.0887	6.8423	77.1405	17
18	2.4066	0.4155	28.1324	0.0355	11.6896	0.0855	7.2034	84.2043	18
19	2.5270	0.3957	30.5390	0.0327	12.0853	0.0827	7.5569	91.3275	19
20	2.6533	0.3769	33.0660	0.0302	12.4622	0.0802	7.9030	98.4884	20
21	2.7860	0.3589	35.7193	0.0280	12.8212	0.0780	8.2416	105.6673	21
22	2.9253	0.3418	38.5052	0.0260	13.1630	0.0760	8.5730	112.8461	22
23	3.0715	0.3256	41.4305	0.0241	13.4886	0.0741	8.8971	120.0087	23
24	3.2251	0.3101	44.5020	0.0225	13.7986	0.0725	9.2140	127.1402	24
25	3.3864	0.2953	47.7271	0.0210	14.0939	0.0710	9.5238	134.2275	25
26	3.5557	0.2812	51.1135	0.0196	14.3752	0.0696	9.8266	141.2585	26
27	3.7335	0.2678	54.6691	0.0183	14.6430	0.0683	10.1224	148.2226	27
28	3.9201	0.2551	58.4026	0.0171	14.8981	0.0671	10.4114	155.1101	28
29	4.1161	0.2429	62.3227	0.0160	15.1411	0.0660	10.6936	161.9126	29
30	4.3219	0.2314	66.4388	0.0151	15.3725	0.0651	10.9691	168.6226	30
31	4.5380	0.2204	70.7608	0.0141	15.5928	0.0641	11.2381	175.2333	31
32	4.7649	0.2099	75.2988	0.0133	15.8027	0.0633	11.5005	181.7392	32
33	5.0032	0.1999	80.0638	0.0125	16.0025	0.0625	11.7566	188.1351	33
34	5.2533	0.1904	85.0670	0.0118	16.1929	0.0618	12.0063	194.4168	34
35	5.5160	0.1813	90.3203	0.0111	16.3742	0.0611	12.2498	200.5807	35
40	7.0400	0.1420	120.7998	0.0083	17.1591	0.0583	13.3775	229.5452	40
45	8.9850	0.1113	159.7002	0.0063	17.7741	0.0563	14.3644	255.3145	45
50	11.4674	0.0872	209.3480	0.0048	18.2559	0.0548	15.2233	277.9148	50
55	14.6356	0.0683	272.7126	0.0037	18.6335	0.0537	15.9664	297.5104	55
60	18.6792	0.0535	353.5837	0.0028	18.9293	0.0528	16.6062	314.3432	60
65	23.8399	0.0419	456.7980	0.0022	19.1611	0.0522	17.1541	328.6910	65
70	30.4264	0.0329	588.5285	0.0017	19.3427	0.0517	17.6212	340.8409	70
75	38.8327	0.0258	756.6537	0.0013	19.4850	0.0513	18.0176	351.0721	75
80	49.5614	0.0202	971.2288	0.0010	19.5965	0.0510	18.3526	359.6460	80
85	63.2544	0.0158	1245.0871	0.0008	19.6838	0.0508	18.6346	366.8007	85
90	80.7304	0.0124	1594.6073	0.0006	19.7523	0.0506	18.8712	372.7488	90
95	103.0347	0.0097	2040.6935	0.0005	19.8059	0.0505	19.0689	377.6774	95
100	131.5013	0.0076	2610.0252	0.0004	19.8479	0.0504	19.2337	381.7492	100

6.0%

	Single Payment		Equal Payment Series				Gradient Series		
	Compound Amount Factor	Present Worth Factor	Compound Amount Factor	Sinking Fund Factor	Present Worth Factor	Capital Recovery Factor	Gradient Uniform Series	Gradient Present Worth	
N	(F/P,i,N)	(P/F,i,N)	(F/A,i,N)	(A/F,i,N)	(P/A,i,N)	(A/P,i,N)	(A/G,i,N)	(P/G,i,N)	N
1	1.0600	0.9434	1.0000	1.0000	0.9434	1.0600	0.0000	0.0000	1
2	1.1236	0.8900	2.0600	0.4854	1.8334	0.5454	0.4854	0.8900	2
3	1.1910	0.8396	3.1836	0.3141	2.6730	0.3741	0.9612	2.5692	3
4	1.2625	0.7921	4.3746	0.2286	3.4651	0.2886	1.4272	4.9455	4
5	1.3382	0.7473	5.6371	0.1774	4.2124	0.2374	1.8836	7.9345	5
6	1.4185	0.7050	6.9753	0.1434	4.9173	0.2034	2.3304	11.4594	6
7	1.5036	0.6651	8.3938	0.1191	5.5824	0.1791	2.7676	15.4497	7
8	1.5938	0.6274	9.8975	0.1010	6.2098	0.1610	3.1952	19.8416	8
9	1.6895	0.5919	11.4913	0.0870	6.8017	0.1470	3.6133	24.5768	9
10	1.7908	0.5584	13.1808	0.0759	7.3601	0.1359	4.0220	29.6023	10
11	1.8983	0.5268	14.9716	0.0668	7.8869	0.1268	4.4213	34.8702	11
12	2.0122	0.4970	16.8699	0.0593	8.3838	0.1193	4.8113	40.3369	12
13	2.1329	0.4688	18.8821	0.0530	8.8527	0.1130	5.1920	45.9629	13
14	2.2609	0.4423	21.0151	0.0476	9.2950	0.1076	5.5635	51.7128	14
15	2.3966	0.4173	23.2760	0.0430	9.7122	0.1030	5.9260	57.5546	15
16	2.5404	0.3936	25.6725	0.0390	10.1059	0.0990	6.2794	63.4592	16
17	2.6928	0.3714	28.2129	0.0354	10.4773	0.0954	6.6240	69.4011	17
18	2.8543	0.3503	30.9057	0.0324	10.8276	0.0924	6.9597	75.3569	18
19	3.0256	0.3305	33.7600	0.0296	11.1581	0.0896	7.2867	81.3062	19
20	3.2071	0.3118	36.7856	0.0272	11.4699	0.0872	7.6051	87.2304	20
21	3.3996	0.2942	39.9927	0.0250	11.7641	0.0850	7.9151	93.1136	21
22	3.6035	0.2775	43.3923	0.0230	12.0416	0.0830	8.2166	98.9412	22
23	3.8197	0.2618	46.9958	0.0213	12.3034	0.0813	8.5099	104.7007	23
24	4.0489	0.2470	50.8156	0.0197	12.5504	0.0797	8.7951	110.3812	24
25	4.2919	0.2330	54.8645	0.0182	12.7834	0.0782	9.0722	115.9732	25
26	4.5494	0.2198	59.1564	0.0169	13.0032	0.0769	9.3414	121.4684	26
27	4.8223	0.2074	63.7058	0.0157	13.2105	0.0757	9.6029	126.8600	27
28	5.1117	0.1956	68.5281	0.0146	13.4062	0.0746	9.8568	132.1420	28
29	5.4184	0.1846	73.6398	0.0136	13.5907	0.0736	10.1032	137.3096	29
30	5.7435	0.1741	79.0582	0.0126	13.7648	0.0726	10.3422	142.3588	30
31	6.0881	0.1643	84.8017	0.0118	13.9291	0.0718	10.5740	147.2864	31
32	6.4534	0.1550	90.8898	0.0110	14.0840	0.0710	10.7988	152.0901	32
33	6.8406	0.1462	97.3432	0.0103	14.2302	0.0703	11.0166	156.7681	33
34	7.2510	0.1379	104.1838	0.0096	14.3681	0.0696	11.2276	161.3192	34
35	7.6861	0.1301	111.4348	0.0090	14.4982	0.0690	11.4319	165.7427	35
40	10.2857	0.0972	154.7620	0.0065	15.0463	0.0665	12.3590	185.9568	40
45	13.7646	0.0727	212.7435	0.0047	15.4558	0.0647	13.1413	203.1096	45
50	18.4202	0.0543	290.3359	0.0034	15.7619	0.0634	13.7964	217.4574	50
55	24.6503	0.0406	394.1720	0.0025	15.9905	0.0625	14.3411	229.3222	55
60	32.9877	0.0303	533.1282	0.0019	16.1614	0.0619	14.7909	239.0428	60
65	44.1450	0.0227	719.0829	0.0014	16.2891	0.0614	15.1601	246.9450	65
70	59.0759	0.0169	967.9322	0.0010	16.3845	0.0610	15.4613	253.3271	70
75	79.0569	0.0126	1300.9487	0.0008	16.4558	0.0608	15.7058	258.4527	75
80	105.7960	0.0095	1746.5999	0.0006	16.5091	0.0606	15.9033	262.5493	80
85	141.5789	0.0071	2342.9817	0.0004	16.5489	0.0604	16.0620	265.8096	85
90	189.4645	0.0053	3141.0752	0.0003	16.5787	0.0603	16.1891	268.3946	90
95	253.5463	0.0039	4209.1042	0.0002	16.6009	0.0602	16.2905	270.4375	95
100	339.3021	0.0029	5638.3681	0.0002	16.6175	0.0602	16.3711	272.0471	100

7.0%

	Single Payment		Equal Payment Series				Gradient Series		
	Compound Amount Factor	Present Worth Factor	Compound Amount Factor	Sinking Fund Factor	Present Worth Factor	Capital Recovery Factor	Gradient Uniform Series	Gradient Present Worth	
N	(F/P,i,N)	(P/F,i,N)	(F/A,i,N)	(A/F,i,N)	(P/A,i,N)	(A/P,i,N)	(A/G,i,N)	(P/G,i,N)	N
1	1.0700	0.9346	1.0000	1.0000	0.9346	1.0700	0.0000	0.0000	1
2	1.1449	0.8734	2.0700	0.4831	1.8080	0.5531	0.4831	0.8734	2
3	1.2250	0.8163	3.2149	0.3111	2.6243	0.3811	0.9549	2.5060	3
4	1.3108	0.7629	4.4399	0.2252	3.3872	0.2952	1.4155	4.7947	4
5	1.4026	0.7130	5.7507	0.1739	4.1002	0.2439	1.8650	7.6467	5
6	1.5007	0.6663	7.1533	0.1398	4.7665	0.2098	2.3032	10.9784	6
7	1.6058	0.6227	8.6540	0.1156	5.3893	0.1856	2.7304	14.7149	7
8	1.7182	0.5820	10.2598	0.0975	5.9713	0.1675	3.1465	18.7889	8
9	1.8385	0.5439	11.9780	0.0835	6.5152	0.1535	3.5517	23.1404	9
10	1.9672	0.5083	13.8164	0.0724	7.0236	0.1424	3.9461	27.7156	10
11	2.1049	0.4751	15.7836	0.0634	7.4987	0.1334	4.3296	32.4665	11
12	2.2522	0.4440	17.8885	0.0559	7.9427	0.1259	4.7025	37.3506	12
13	2.4098	0.4150	20.1406	0.0497	8.3577	0.1197	5.0648	42.3302	13
14	2.5785	0.3878	22.5505	0.0443	8.7455	0.1143	5.4167	47.3718	14
15	2.7590	0.3624	25.1290	0.0398	9.1079	0.1098	5.7583	52.4461	15
16	2.9522	0.3387	27.8881	0.0359	9.4466	0.1059	6.0897	57.5271	16
17	3.1588	0.3166	30.8402	0.0324	9.7632	0.1024	6.4110	62.5923	17
18	3.3799	0.2959	33.9990	0.0294	10.0591	0.0994	6.7225	67.6219	18
19	3.6165	0.2765	37.3790	0.0268	10.3356	0.0968	7.0242	72.5991	19
20	3.8697	0.2584	40.9955	0.0244	10.5940	0.0944	7.3163	77.5091	20
21	4.1406	0.2415	44.8652	0.0223	10.8355	0.0923	7.5990	82.3393	21
22	4.4304	0.2257	49.0057	0.0204	11.0612	0.0904	7.8725	87.0793	22
23	4.7405	0.2109	53.4361	0.0187	11.2722	0.0887	8.1369	91.7201	23
24	5.0724	0.1971	58.1767	0.0172	11.4693	0.0872	8.3923	96.2545	24
25	5.4274	0.1842	63.2490	0.0158	11.6536	0.0858	8.6391	100.6765	25
26	5.8074	0.1722	68.6765	0.0146	11.8258	0.0846	8.8773	104.9814	26
27	6.2139	0.1609	74.4838	0.0134	11.9867	0.0834	9.1072	109.1656	27
28	6.6488	0.1504	80.6977	0.0124	12.1371	0.9824	9.3289	113.2264	28
29	7.1143	0.1406	87.3465	0.0114	12.2777	0.0814	9.5427	117.1622	29
30	7.6123	0.1314	94.4608	0.0106	12.4090	0.0806	9.7487	120.9718	30
31	8.1451	0.1228	102.0730	0.0098	12.5318	0.0798	9.9471	124.6550	31
32	8.7153	0.1147	110.2182	0.0091	12.6466	0.0791	10.1381	128.2120	32
33	9.3253	0.1072	118.9334	0.0084	12.7538	0.0784	10.3219	131.6435	33
34	9.9781	0.1002	128.2588	0.0078	12.8540	0.0778	10.4987	134.9507	34
35	10.6766	0.0937	138.2369	0.0072	12.9477	0.0772	10.6687	138.1353	35
40	14.9745	0.0668	199.6351	0.0050	13.3317	0.0750	11.4233	152.2928	40
45	21.0025	0.0476	285.7493	0.0035	13.6055	0.0735	12.0360	163.7559	45
50	29.4570	0.0339	406.5289	0.0025	13.8007	0.0725	12.5287	172.9051	50
55	41.3150	0.0242	575.9286	0.0017	13.9399	0.0717	12.9215	180.1243	55
60	57.9464	0.0173	813.5204	0.0012	14.0392	0.0712	13.2321	185.7677	60
65	81.2729	0.0123	1146.7552	0.0009	14.1099	0.0709	13.4760	190.1452	65
70	113.9894	0.0088	1614.1342	0.0006	14.1604	0.0706	13.6662	193.5185	70
75	159.8760	0.0063	2269.6574	0.0004	14.1964	0.0704	13.8136	196.1035	75
80	224.2344	0.0045	3189.0627	0.0003	14.2220	0.0703	13.9273	198.0748	80
85	314.5003	0.0032	4478.5761	0.0002	14.2403	0.0702	14.0146	199.5717	85
90	441.1030	0.0023	6287.1854	0.0002	14.2533	0.0702	14.0812	200.7042	90
95	618.6697	0.0016	8823.8535	0.0001	14.2626	0.0701	14.1319	201.5581	95
100	867.7163	0.0012	12381.6618	0.0001	14.2693	0.0701	14.1703	202.2001	100

8.0%

	Single Payment		Equal Payment Series				Gradient Series		
	Compound Amount Factor	Present Worth Factor	Compound Amount Factor	Sinking Fund Factor	Present Worth Factor	Capital Recovery Factor	Gradient Uniform Series	Gradient Present Worth	
N	(F/P,i,N)	(P/F,i,N)	(F/A,i,N)	(A/F,i,N)	(P/A,i,N)	(A/P,i,N)	(A/G,i,N)	(P/G,i,N)	N
1	1.0800	0.9259	1.0000	1.0000	0.9259	1.0800	0.0000	0.0000	1
2	1.1664	0.8573	2.0800	0.4808	1.7833	0.5608	0.4808	0.8573	2
3	1.2597	0.7938	3.2464	0.3080	2.5771	0.3880	0.9487	2.4450	3
4	1.3605	0.7350	4.5061	0.2219	3.3121	0.3019	1.4040	4.6501	4
5	1.4693	0.6806	5.8666	0.1705	3.9927	0.2505	1.8465	7.3724	5
6	1.5869	0.6302	7.3359	0.1363	4.6229	0.2163	2.2763	10.5233	6
7	1.7138	0.5835	8.9228	0.1121	5.2064	0.1921	2.6937	14.0242	7
8	1.8509	0.5403	10.6366	0.0940	5.7466	0.1740	3.0985	17.8061	8
9	1.9990	0.5002	12.4876	0.0801	6.2469	0.1601	3.4910	21.8081	9
10	2.1589	0.4632	14.4866	0.0690	6.7101	0.1490	3.8713	25.9768	10
11	2.3316	0.4289	16.6455	0.0601	7.1390	0.1401	4.2395	30.2657	11
12	2.5182	0.3971	18.9771	0.0527	7.5361	0.1327	4.5957	34.6339	12
13	2.7196	0.3677	21.4953	0.0465	7.9038	0.1265	4.9402	39.0463	13
14	2.9372	0.3405	24.2149	0.0413	8.2442	0.1213	5.2731	43.4723	14
15	3.1722	0.3152	27.1521	0.0368	8.5595	0.1168	5.5945	47.8857	15
16	3.4259	0.2919	30.3243	0.0330	8.8514	0.1130	5.9046	52.2640	16
17	3.7000	0.2703	33.7502	0.0296	9.1216	0.1096	6.2037	56.5883	17
18	3.9960	0.2502	37.4502	0.0267	9.3719	0.1067	6.4920	60.8426	18
19	4.3157	0.2317	41.4463	0.0241	9.6036	0.1041	6.7697	65.0134	19
20	4.6610	0.2145	45.7620	0.0219	9.8181	0.1019	7.0369	69.0898	20
21	5.0338	0.1987	50.4229	0.0198	10.0168	0.0998	7.2940	73.0629	21
22	5.4365	0.1839	55.4568	0.0180	10.2007	0.0980	7.5412	76.9257	22
23	5.8715	0.1703	60.8933	0.0164	10.3711	0.0964	7.7786	80.6726	23
24	6.3412	0.1577	66.7648	0.0150	10.5288	0.0950	8.0066	84.2997	24
25	6.8485	0.1460	73.1059	0.0137	10.6748	0.0937	8.2254	87.8041	25
26	7.3964	0.1352	79.9544	0.0125	10.8100	0.0925	8.4352	91.1842	26
27	7.9881	0.1252	87.3508	0.0114	10.9352	0.0914	8.6363	94.4390	27
78	8.6271	0.1159	95.3388	0.0105	11.0511	0.0905	8.8289	97.5687	28
29	9.3173	0.1073	103.9659	0.0096	11.1584	0.0896	9.0133	100.5738	29
30	10.0627	0.0994	113.2832	0.0088	11.2578	0.0888	9.1897	103.4558	30
31	10.8677	0.0920	123.3459	0.0081	11.3498	0.0881	9.3584	106.2163	31
32	11.7371	0.0852	134.2135	0.0075	11.4350	0.0875	9.5197	108.8575	32
33	12.6760	0.0789	145.9506	0.0069	11.5139	0.0869	9.6737	111.3819	33
34	13.6901	0.0730	158.6267	0.0063	11.5869	0.0863	9.8208	113.7924	34
35	14.7853	0.0676	172.3168	0.0058	11.6546	0.0858	9.9611	116.0920	35
40	21.7245	0.0460	259.0565	0.0039	11.9246	0.0839	10.5699	126.0422	40
45	31.9204	0.0313	386.5056	0.0026	12.1084	0.0826	11.0447	133.7331	45
50	46.9016	0.0213	573.7702	0.0017	12.2335	0.0817	11.4107	139.5928	50
55	68.9139	0.0145	848.9232	0.0012	12.3186	0.0812	11.6902	144.0065	55
60	101.2571	0.0099	1253.2133	0.0008	12.3766	0.0808	11.9015	147.3000	60
65	148.7798	0.0067	1847.2481	0.0005	12.4160	0.0805	12.0602	149.7387	65
70	218.6064	0.0046	2720.0801	0.0004	12.4428	0.0804	12.1783	151.5326	70
75	321.2045	0.0031	4002.5566	0.0002	12.4611	0.0802	12.2658	152.8448	75
80	471.9548	0.0021	5886.9354	0.0002	12.4735	0.0802	12.3301	153.8001	80
85	693.4565	0.0014	8655.7061	0.0001	12.4820	0.0801	12.3772	154.4925	85
90	1018.9151	0.0010	12723.9386	0.0001	12.4877	0.0801	12.4116	154.9925	90
95	1497.1205	0.0007	18701.5069	0.0001	12.4917	0.0801	12.4365	155.3524	95
100	2199.7613	0.0005	27484.5157	0.0000	12.4943	0.0800	12.4545	155.6107	100

9.0%

	Single Payment		Equal Payment Series				Gradient Series		
N	Compound Amount Factor (F/P,i,N)	Present Worth Factor (P/F,i,N)	Compound Amount Factor (F/A,i,N)	Sinking Fund Factor (A/F,i,N)	Present Worth Factor (P/A,i,N)	Capital Recovery Factor (A/P,i,N)	Gradient Uniform Series (A/G,i,N)	Gradient Present Worth (P/G,i,N)	N
1	1.0900	0.9174	1.0000	1.0000	0.9174	1.0900	0.0000	0.0000	1
2	1.1881	0.8417	2.0900	0.4785	1.7591	0.5685	0.4785	0.8417	2
3	1.2950	0.7722	3.2781	0.3051	2.5313	0.3951	0.9426	2.3860	3
4	1.4116	0.7084	4.5731	0.2187	3.2397	0.3087	1.3925	4.5113	4
5	1.5386	0.6499	5.9847	0.1671	3.8897	0.2571	1.8282	7.1110	5
6	1.6771	0.5963	7.5233	0.1329	4.4859	0.2229	2.2498	10.0924	6
7	1.8280	0.5470	9.2004	0.1087	5.0330	0.1987	2.6574	13.3746	7
8	1.9926	0.5019	11.0285	0.0907	5.5348	0.1807	3.0512	16.8877	8
9	2.1719	0.4604	13.0210	0.0768	5.9952	0.1668	3.4312	20.5711	9
10	2.3674	0.4224	15.1929	0.0658	6.4177	0.1558	3.7978	24.3728	10
11	2.5804	0.3875	17.5603	0.0569	6.8052	0.1469	4.1510	28.2481	11
12	2.8127	0.3555	20.1407	0.0497	7.1607	0.1397	4.4910	32.1590	12
13	3.0658	0.3262	22.9534	0.0436	7.4869	0.1336	4.8182	36.0731	13
14	3.3417	0.2992	26.0192	0.0384	7.7862	0.1284	5.1326	39.9633	14
15	3.6425	0.2745	29.3609	0.0341	8.0607	0.1241	5.4346	43.8069	15
16	3.9703	0.2519	33.0034	0.0303	8.3126	0.1203	5.7245	47.5849	16
17	4.3276	0.2311	36.9737	0.0270	8.5436	0.1170	6.0024	51.2821	17
18	4.7171	0.2120	41.3013	0.0242	8.7556	0.1142	6.2687	54.8860	18
19	5.1417	0.1945	46.0185	0.0217	8.9501	0.1117	6.5236	58.3868	19
20	5.6044	0.1784	51.1601	0.0195	9.1285	0.1095	6.7674	61.7770	20
21	6.1088	0.1637	56.7645	0.0176	9.2922	0.1076	7.0006	65.0509	21
22	6.6586	0.1502	62.8733	0.0159	9.4424	0.1059	7.2232	68.2048	22
23	7.2579	0.1378	69.5319	0.0144	9.5802	0.1044	7.4357	71.2359	23
24	7.9111	0.1264	76.7898	0.0130	9.7066	0.1030	7.6384	74.1433	24
25	8.6231	0.1160	84.7009	0.0118	9.8226	0.1018	7.8316	76.9265	25
26	9.3992	0.1064	93.3240	0.0107	9.9290	0.1007	8.0156	79.5863	26
27	10.2451	0.0976	102.7231	0.0097	10.0266	0.0997	8.1906	82.1241	27
28	11.1671	0.0895	112.9682	0.0089	10.1161	0.0989	8.3571	84.5419	28
29	12.1722	0.0822	124.1354	0.0081	10.1983	0.0981	8.5154	86.8422	29
30	13.2677	0.0754	136.3075	0.0073	10.2737	0.0973	8.6657	89.0280	30
31	14.4618	0.0691	149.5752	0.0067	10.3428	0.0967	8.8083	91.1024	31
32	15.7633	0.0634	164.0370	0.0061	10.4062	0.0961	8.9436	93.0690	32
33	17.1820	0.0582	179.8003	0.0056	10.4644	0.0956	9.0718	94.9314	33
34	18.7284	0.0534	196.9823	0.0051	10.5178	0.0951	9.1933	96.6935	34
35	20.4140	0.0490	215.7108	0.0046	10.5668	0.0946	9.3083	98.3590	35
40	31.4094	0.0318	337.8824	0.0030	10.7574	0.0930	9.7957	105.3762	40
45	48.3273	0.0207	525.8587	0.0019	10.8812	0.0919	10.1603	110.5561	45
50	74.3575	0.0134	815.0836	0.0012	10.9617	0.0912	10.4295	114.3251	50
55	114.4083	0.0087	1260.0918	0.0008	11.0140	0.0908	10.6261	117.0362	55
60	176.0313	0.0057	1944.7921	0.0005	11.0480	0.0905	10.7683	118.9683	60
65	270.8460	0.0037	2998.2885	0.0003	11.0701	0.0903	10.8702	120.3344	65
70	416.7301	0.0024	4619.2232	0.0002	11.0844	0.0902	10.9427	121.2942	70
75	641.1909	0.0016	7113.2321	0.0001	11.0938	0.0901	10.9940	121.9646	75
80	986.5517	0.0010	10950.5741	0.0001	11.0998	0.0901	11.0299	122.4306	80
85	1517.9320	0.0007	16854.8003	0.0001	11.1038	0.0901	11.0551	122.7533	85
90	2335.5266	0.0004	25939.1842	0.0000	11.1064	0.0900	11.0726	122.9758	90
95	3593.4971	0.0003	39916.6350	0.0000	11.1080	0.0900	11.0847	123.1287	95
100	5529.0408	0.0002	61422.6755	0.0000	11.1091	0.0900	11.0930	123.2335	100

10.0%

	Single Payment		Equal Payment Series				Gradient Series		
N	Compound Amount Factor (F/P,i,N)	Present Worth Factor (P/F,i,N)	Compound Amount Factor (F/A,i,N)	Sinking Fund Factor (A/F,i,N)	Present Worth Factor (P/A,i,N)	Capital Recovery Factor (A/P,i,N)	Gradient Uniform Series (A/G,i,N)	Gradient Present Worth (P/G,i,N)	**N**
1	1.1000	0.9091	1.0000	1.0000	0.9091	1.1000	0.0000	0.0000	1
2	1.2100	0.8264	2.1000	0.4762	1.7355	0.5762	0.4762	0.8264	2
3	1.3310	0.7513	3.3100	0.3021	2.4869	0.4021	0.9366	2.3291	3
4	1.4641	0.6830	4.6410	0.2155	3.1699	0.3155	1.3812	4.3781	4
5	1.6105	0.6209	6.1051	0.1638	3.7908	0.2638	1.8101	6.8618	5
6	1.7716	0.5645	7.7156	0.1296	4.3553	0.2296	2.2236	9.6842	6
7	1.9487	0.5132	9.4872	0.1054	4.8684	0.2054	2.6216	12.7631	7
8	2.1436	0.4665	11.4359	0.0874	5.3349	0.1874	3.0045	16.0287	8
9	2.3579	0.4241	13.5795	0.0736	5.7590	0.1736	3.3724	19.4215	9
10	2.5937	0.3855	15.9374	0.0627	6.1446	0.1627	3.7255	22.8913	10
11	2.8531	0.3505	18.5312	0.0540	6.4951	0.1540	4.0641	26.3963	11
12	3.1384	0.3186	21.3843	0.0468	6.8137	0.1468	4.3884	29.9012	12
13	3.4523	0.2897	24.5227	0.0408	7.1034	0.1408	4.6988	33.3772	13
14	3.7975	0.2633	27.9750	0.0357	7.3667	0.1357	4.9955	36.8005	14
15	4.1772	0.2394	31.7725	0.0315	7.6061	0.1315	5.2789	40.1520	15
16	4.5950	0.2176	35.9497	0.0278	7.8237	0.1278	5.5493	43.4164	16
17	5.0545	0.1978	40.5447	0.0247	8.0216	0.1247	5.8071	46.5819	17
18	5.5599	0.1799	45.5992	0.0219	8.2014	0.1219	6.0526	49.6395	18
19	6.1159	0.1635	51.1591	0.0195	8.3649	0.1195	6.2861	52.5827	19
20	6.7275	0.1486	57.2750	0.0175	8.5136	0.1175	6.5081	55.4069	20
21	7.4002	0.1351	64.0025	0.0156	8.6487	0.1156	6.7189	58.1095	21
22	8.1403	0.1228	71.4027	0.0140	8.7715	0.1140	6.9189	60.6893	22
23	8.9543	0.1117	79.5430	0.0126	8.8832	0.1126	7.1085	63.1462	23
24	9.8497	0.1015	88.4973	0.0113	8.9847	0.1113	7.2881	65.4813	24
25	10.8347	0.0923	98.3471	0.0102	9.0770	0.1102	7.4580	67.6964	25
26	11.9182	0.0839	109.1818	0.0092	9.1609	0.1092	7.6186	69.7940	26
27	13.1100	0.0763	121.0999	0.0083	9.2372	0.1083	7.7704	71.7773	27
28	14.4210	0.0693	134.2099	0.0075	9.3066	0.1075	7.9137	73.6495	28
29	15.8631	0.0630	148.6309	0.0067	9.3696	0.1067	8.0489	75.4146	29
30	17.4494	0.0573	164.4940	0.0061	9.4269	0.1061	8.1762	77.0766	30
31	19.1943	0.0521	181.9434	0.0055	9.4790	0.1055	8.2962	78.6395	31
32	21.1138	0.0474	201.1378	0.0050	9.5264	0.1050	8.4091	80.1078	32
33	23.2252	0.0431	222.2515	0.0045	9.5694	0.1045	8.5152	81.4856	33
34	25.5477	0.0391	245.4767	0.0041	9.6086	0.1041	8.6149	82.7773	34
35	28.1024	0.0356	271.0244	0.0037	9.6442	0.1037	8.7086	83.9872	35
40	45.2593	0.0221	442.5926	0.0023	9.7791	0.1023	9.0962	88.9525	40
45	72.8905	0.0137	718.9048	0.0014	9.8628	0.1014	9.3740	92.4544	45
50	117.3909	0.0085	1163.9085	0.0009	9.9148	0.1009	9.5704	94.8889	50
55	189.0591	0.0053	1880.5914	0.0005	9.9471	0.1005	9.7075	96.5619	55
60	304.4816	0.0033	3034.8164	0.0003	9.9672	0.1003	9.8023	97.7010	60
65	490.3707	0.0020	4893.7073	0.0002	9.9796	0.1002	9.8672	98.4705	65
70	789.7470	0.0013	7887.4696	0.0001	9.9873	0.1001	9.9113	98.9870	70
75	1271.8954	0.0008	12708.9537	0.0001	9.9921	0.1001	9.9410	99.3317	75
80	2048.4002	0.0005	20474.0021	0.0000	9.9951	0.1000	9.9609	99.5606	80
85	3298.9690	0.0003	32979.6903	0.0000	9.9970	0.1000	9.9742	99.7120	85
90	5313.0226	0.0002	53120.2261	0.0000	9.9981	0.1000	9.9831	99.8118	90
95	8556.6760	0.0001	85556.7605	0.0000	9.9988	0.1000	9.9889	99.8773	95
100	13780.6123	0.0001	137796.1234	0.0000	9.9993	0.1000	9.9927	99.9202	100

11.0%

	Single Payment		Equal Payment Series				Gradient Series		
	Compound Amount Factor	Present Worth Factor	Compound Amount Factor	Sinking Fund Factor	Present Worth Factor	Capital Recovery Factor	Gradient Uniform Series	Gradient Present Worth	
N	(F/P,i,N)	(P/F,i,N)	(F/A,i,N)	(A/F,i,N)	(P/A,i,N)	(A/P,i,N)	(A/G,i,N)	(P/G,i,N)	N
1	1.1100	0.9009	1.0000	1.0000	0.9009	1.1100	0.0000	0.0000	1
2	1.2321	0.8116	2.1100	0.4739	1.7125	0.5839	0.4739	0.8116	2
3	1.3676	0.7312	3.3421	0.2992	2.4437	0.4092	0.9306	2.2740	3
4	1.5181	0.6587	4.7097	0.2123	3.1024	0.3223	1.3700	4.2502	4
5	1.6851	0.5935	6.2278	0.1606	3.6959	0.2706	1.7923	6.6240	5
6	1.8704	0.5346	7.9129	0.1264	4.2305	0.2364	2.1976	9.2972	6
7	2.0762	0.4817	9.7833	0.1022	4.7122	0.2122	2.5863	12.1872	7
8	2.3045	0.4339	11.8594	0.0843	5.1461	0.1943	2.9585	15.2246	8
9	2.5580	0.3909	14.1640	0.0706	5.5370	0.1806	3.3144	18.3520	9
10	2.8394	0.3522	16.7220	0.0598	5.8892	0.1698	3.6544	21.5217	10
11	3.1518	0.3173	19.5614	0.0511	6.2065	0.1611	3.9788	24.6945	11
12	3.4985	0.2858	22.7132	0.0440	6.4924	0.1540	4.2879	27.8388	12
13	3.8833	0.2575	26.2116	0.0382	6.7499	0.1482	4.5822	30.9290	13
14	4.3104	0.2320	30.0949	0.0332	6.9819	0.1432	4.8619	33.9449	14
15	4.7846	0.2090	34.4054	0.0291	7.1909	0.1391	5.1275	36.8709	15
16	5.3109	0.1883	39.1899	0.0255	7.3792	0.1355	5.3794	39.6953	16
17	5.8951	0.1696	44.5008	0.0225	7.5488	0.1325	5.6180	42.4095	17
18	6.5436	0.1528	50.3959	0.0198	7.7016	0.1298	5.8439	45.0074	18
19	7.2633	0.1377	56.9395	0.0176	7.8393	0.1276	6.0574	47.4856	19
20	8.0623	0.1240	64.2028	0.0156	7.9633	0.1256	6.2590	49.8423	20
21	8.9492	0.1117	72.2651	0.0138	8.0751	0.1238	6.4491	52.0771	21
22	9.9336	0.1007	81.2143	0.0123	8.1757	0.1223	6.6283	54.1912	22
23	11.0263	0.0907	91.1479	0.0110	8.2664	0.1210	6.7969	56.1864	23
24	12.2392	0.0817	102.1742	0.0098	8.3481	0.1198	6.9555	58.0656	24
25	13.5855	0.0736	114.4133	0.0087	8.4217	0.1187	7.1045	59.8322	25
26	15.0799	0.0663	127.9988	0.0078	8.4881	0.1178	7.2443	61.4900	26
27	16.7386	0.0597	143.0786	0.0070	8.5478	0.1170	7.3754	63.0433	27
28	18.5799	0.0538	159.8173	0.0063	8.6016	0.1163	7.4982	64.4965	28
29	20.6237	0.0485	178.3972	0.0056	8.6501	0.1156	7.6131	65.8542	29
30	22.8923	0.0437	199.0209	0.0050	8.6938	0.1150	7.7206	67.1210	30
31	25.4104	0.0394	221.9132	0.0045	8.7331	0.1145	7.8210	68.3016	31
32	28.2056	0.0355	247.3236	0.0040	8.7686	0.1140	7.9147	69.4007	32
33	31.3082	0.0319	275.5292	0.0036	8.8005	0.1136	8.0021	70.4228	33
34	34.7521	0.0288	306.8374	0.0033	8.8293	0.1133	8.0836	71.3724	34
35	38.5749	0.0259	341.5896	0.0029	8.8552	0.1129	8.1594	72.2538	35
40	65.0009	0.0154	581.8261	0.0017	8.9511	0.1117	8.4659	75.7789	40
45	109.5302	0.0091	986.6386	0.0010	9.0079	0.1110	8.6763	78.1551	45
50	184.5648	0.0054	1668.7712	0.0006	9.0417	0.1106	8.8185	79.7341	50
55	311.0025	0.0032	2818.2042	0.0004	9.0617	0.1104	8.9135	80.7712	55
60	524.0572	0.0019	4755.0658	0.0002	9.0736	0.1102	8.9762	81.4461	60

12.0%

	Single Payment		Equal Payment Series				Gradient Series		
N	Compound Amount Factor (F/P,i,N)	Present Worth Factor (P/F,i,N)	Compound Amount Factor (F/A,i,N)	Sinking Fund Factor (A/F,i,N)	Present Worth Factor (P/A,i,N)	Capital Recovery Factor (A/P,i,N)	Gradient Uniform Series (A/G,i,N)	Gradient Present Worth (P/G,i,N)	N
1	1.1200	0.8929	1.0000	1.0000	0.8929	1.1200	0.0000	0.0000	1
2	1.2544	0.7972	2.1200	0.4717	1.6901	0.5917	0.4717	0.7972	2
3	1.4049	0.7118	3.3744	0.2963	2.4018	0.4163	0.9246	2.2208	3
4	1.5735	0.6355	4.7793	0.2092	3.0373	0.3292	1.3589	4.1273	4
5	1.7623	0.5674	6.3528	0.1574	3.6048	0.2774	1.7746	6.3970	5
6	1.9738	0.5066	8.1152	0.1232	4.1114	0.2432	2.1720	8.9302	6
7	2.2107	0.4523	10.0890	0.0991	4.5638	0.2191	2.5515	11.6443	7
8	2.4760	0.4039	12.2997	0.0813	4.9676	0.2013	2.9131	14.4714	8
9	2.7731	0.3606	14.7757	0.0677	5.3282	0.1877	3.2574	17.3563	9
10	3.1058	0.3220	17.5487	0.0570	5.6502	0.1770	3.5847	20.2541	10
11	3.4785	0.2875	20.6546	0.0484	5.9377	0.1684	3.8953	23.1288	11
12	3.8960	0.2567	24.1331	0.0414	6.1944	0.1614	4.1897	25.9523	12
13	4.3635	0.2292	28.0291	0.0357	6.4235	0.1557	4.4683	28.7024	13
14	4.8871	0.2046	32.3926	0.0309	6.6282	0.1509	4.7317	31.3624	14
15	5.4736	0.1827	37.2797	0.0268	6.8109	0.1468	4.9803	33.9202	15
16	6.1304	0.1631	42.7533	0.0234	6.9740	0.1434	5.2147	36.3670	16
17	6.8660	0.1456	48.8837	0.0205	7.1196	0.1405	5.4353	38.6973	17
18	7.6900	0.1300	55.7497	0.0179	7.2497	0.1379	5.6427	40.9080	81
19	8.6128	0.1161	63.4397	0.0158	7.3658	0.1358	5.8375	42.9979	19
20	9.6463	0.1037	72.0524	0.0139	7.4694	0.1339	6.0202	44.9676	20
21	10.8038	0.0926	81.6987	0.0122	7.5620	0.1322	6.1913	46.8188	21
22	12.1003	0.0826	92.5026	0.0108	7.6446	0.1308	6.3514	48.5543	22
23	13.5523	0.0738	104.6029	0.0096	7.7184	0.1296	6.5010	50.1776	23
24	15.1786	0.0659	118.1552	0.0085	7.7843	0.1285	6.6406	51.6929	24
25	17.0001	0.0588	133.3339	0.0075	7.8431	0.1275	6.7708	53.1046	25
26	19.0401	0.0525	150.3339	0.0067	7.8957	0.1267	6.8921	54.4177	26
27	21.3249	0.0469	169.3740	0.0059	7.9426	0.1259	7.0049	55.6369	27
28	23.8839	0.0419	190.6989	0.0052	7.9844	0.1252	7.1098	56.7674	28
29	26.7499	0.0374	214.5828	0.0047	8.0218	0.1247	7.2071	57.8141	29
30	29.9599	0.0334	241.3327	0.0041	8.0552	0.1241	7.2974	58.7821	30
31	33.5551	0.0298	271.2926	0.0037	8.0850	0.1237	7.3811	59.6761	31
32	37.5817	0.0266	304.8477	0.0033	8.1116	0.1233	7.4586	60.5010	32
33	42.0915	0.0238	342.4294	0.0029	8.1354	0.1229	7.5302	61.2612	33
34	47.1425	0.0212	384.5210	0.0026	8.1566	0.1226	7.5965	61.9612	34
35	52.7996	0.0189	431.6635	0.0023	8.1755	0.1223	7.6577	62.6052	35
40	93.0510	0.0107	767.0914	0.0013	8.2438	0.1213	7.8988	65.1159	40
45	163.9876	0.0061	1358.2300	0.0007	8.2825	0.1207	8.0572	66.7342	45
50	289.0022	0.0035	2400.0182	0.0004	8.3045	0.1204	8.1597	67.7624	50
55	509.3206	0.0020	4236.0050	0.0002	8.3170	0.1202	8.2251	68.4082	55
60	897.5969	0.0011	7471.6411	0.0001	8.3240	0.1201	8.2664	68.8100	60

13.0%

	Single Payment		Equal Payment Series				Gradient Series		
	Compound Amount Factor	Present Worth Factor	Compound Amount Factor	Sinking Fund Factor	Present Worth Factor	Capital Recovery Factor	Gradient Uniform Series	Gradient Present Worth	
N	(F/P,i,N)	(P/F,i,N)	(F/A,i,N)	(A/F,i,N)	(P/A,i,N)	(A/P,i,N)	(A/G,i,N)	(P/G,i,N)	N
1	1.1300	0.8850	1.0000	1.0000	0.8850	1.1300	0.0000	0.0000	1
2	1.2769	0.7831	2.1300	0.4695	1.6681	0.5995	0.4695	0.7831	2
3	1.4429	0.6931	3.4069	0.2935	2.3612	0.4235	0.9187	2.1692	3
4	1.6305	0.6133	4.8498	0.2062	2.9745	0.3362	1.3479	4.0092	4
5	1.8424	0.5428	6.4803	0.1543	3.5172	0.2843	1.7571	6.1802	5
6	2.0820	0.4803	8.3227	0.1202	3.9975	0.2502	2.1468	8.5818	6
7	2.3526	0.4251	10.4047	0.0961	4.4226	0.2261	2.5171	11.1322	7
8	2.6584	0.3762	12.7573	0.0784	4.7988	0.2084	2.8685	13.7653	8
9	3.0040	0.3329	15.4157	0.0649	5.1317	0.1949	3.2014	16.4284	9
10	3.3946	0.2946	18.4197	0.0543	5.4262	0.1843	3.5162	19.0797	10
11	3.8359	0.2607	21.8143	0.0458	5.6869	0.1758	3.8134	21.6867	11
12	4.3345	0.2307	25.6502	0.0390	5.9176	0.1690	4.0936	24.2244	12
13	4.8980	0.2042	29.9847	0.0334	6.1218	0.1634	4.3573	26.6744	13
14	5.5348	0.1807	34.8827	0.0287	6.3025	0.1587	4.6050	29.0232	14
15	6.2543	0.1599	40.4175	0.0247	6.4624	0.1547	4.8375	31.2617	15
16	7.0673	0.1415	46.6717	0.0214	6.6039	0.1514	5.0552	33.3841	16
17	7.9861	0.1252	53.7391	0.0186	6.7291	0.1486	5.2589	35.3876	17
18	9.0243	0.1108	61.7251	0.0162	6.8399	0.1462	5.4491	37.2714	18
19	10.1974	0.0981	70.7494	0.0141	6.9380	0.1441	5.6265	39.0366	19
20	11.5231	0.0868	80.9468	0.0124	7.0248	0.1424	5.7917	40.6854	20
21	13.0211	0.0768	92.4699	0.0108	7.1016	0.1408	5.9454	42.2214	21
22	14.7138	0.0680	105.4910	0.0095	7.1695	0.1395	6.0881	43.6486	22
23	16.6266	0.0601	120.2048	0.0083	7.2297	0.1383	6.2205	44.9718	23
24	18.7881	0.0532	136.8315	0.0073	7.2829	0.1373	6.3431	46.1960	24
25	21.2305	0.0471	155.6196	0.0064	7.3300	0.1364	6.4566	47.3264	25
26	23.9905	0.0417	176.8501	0.0057	7.3717	0.1357	6.5614	48.3685	26
27	27.1093	0.0369	200.8406	0.0050	7.4086	0.1350	6.6582	49.3276	27
28	30.6335	0.0326	227.9499	0.0044	7.4412	0.1344	6.7474	50.2090	28
29	34.6158	0.0289	258.5834	0.0039	7.4701	0.1339	6.8296	51.0179	29
30	39.1159	0.0256	293.1992	0.0034	7.4957	0.1334	6.9052	51.7592	30
31	44.2010	0.0226	332.3151	0.0030	7.5183	0.1330	6.9747	52.4380	31
32	49.9471	0.0200	376.5161	0.0027	7.5383	0.1327	7.0385	53.0586	32
33	56.4402	0.0177	426.4632	0.0023	7.5560	0.1323	7.0971	53.6256	33
34	63.7774	0.0157	482.9034	0.0021	7.5717	0.1321	7.1507	54.1430	34
35	72.0685	0.0139	546.6808	0.0018	7.5856	0.1318	7.1998	54.6148	35
40	132.7816	0.0075	1013.7042	0.0010	7.6344	0.1310	7.3888	56.4087	40
45	244.6414	0.0041	1874.1646	0.0005	7.6609	0.1305	7.5076	57.5148	45
50	450.7359	0.0022	3459.5071	0.0003	7.6752	0.1303	7.5811	58.1870	50
55	830.4517	0.0012	6380.3979	0.0002	7.6830	0.1302	7.6260	58.5909	55
60	1530.0535	0.0007	11761.9498	0.0001	7.6873	0.1301	7.6531	58.8313	60

14.0%

	Single Payment		Equal Payment Series				Gradient Series		
	Compound Amount Factor	Present Worth Factor	Compound Amount Factor	Sinking Fund Factor	Present Worth Factor	Capital Recovery Factor	Gradient Uniform Series	Gradient Present Worth	
N	(F/P,i,N)	(P/F,i,N)	(F/A,i,N)	(A/F,i,N)	(P/A,i,N)	(A/P,i,N)	(A/G,i,N)	(P/G,i,N)	N
1	1.1400	0.8772	1.0000	1.0000	0.8772	1.1400	0.0000	0.0000	1
2	1.2996	0.7695	2.1400	0.4673	1.6467	0.6073	0.4673	0.7695	2
3	1.4815	0.6750	3.4396	0.2907	2.3216	0.4307	0.9129	2.1194	3
4	1.6890	0.5921	4.9211	0.2032	2.9137	0.3432	1.3370	3.8957	4
5	1.9254	0.5194	6.6101	0.1513	3.4331	0.2913	1.7399	5.9731	5
6	2.1950	0.4556	8.5355	0.1172	3.8887	0.2572	2.1218	8.2511	6
7	2.5023	0.3996	10.7305	0.0932	4.2883	0.2332	2.4832	10.6489	7
8	2.8526	0.3506	13.2328	0.0756	4.6389	0.2156	2.8246	13.1028	8
9	3.2519	0.3075	16.0853	0.0622	4.9464	0.2022	3.1463	15.5629	9
10	3.7072	0.2697	19.3373	0.0517	5.2161	0.1917	3.4490	17.9906	10
11	4.2262	0.2366	23.0445	0.0434	5.4527	0.1834	3.7333	20.3567	11
12	4.8179	0.2076	27.2707	0.0367	5.6603	0.1767	3.9998	22.6399	12
13	5.4924	0.1821	32.0887	0.0312	5.8424	0.1712	4.2491	24.8247	13
14	6.2613	0.1597	37.5811	0.0266	6.0021	0.1666	4.4819	26.9009	14
15	7.1379	0.1401	43.8424	0.0228	6.1422	0.1628	4.6990	28.8623	15
16	8.1372	0.1229	50.9804	0.0196	6.2651	0.1596	4.9011	30.7057	16
17	9.2765	0.1078	59.1176	0.0169	6.3729	0.1569	5.0888	32.4305	17
18	10.5752	0.0946	68.3941	0.0146	6.4674	0.1546	5.2630	34.0380	18
19	12.0557	0.0829	78.9692	0.0127	6.5504	0.1527	5.4243	35.5311	19
20	13.7435	0.0728	91.0249	0.0110	6.6231	0.1510	5.5734	36.9135	20
21	15.6676	0.0638	104.7684	0.0095	6.6870	0.1495	5.7111	38.1901	21
22	17.8610	0.0560	120.4360	0.0083	6.7429	0.1483	5.8381	39.3658	22
23	20.3616	0.0491	138.2970	0.0072	6.7921	0.1472	5.9549	40.4463	23
24	23.2122	0.0431	158.6586	0.0063	6.8351	0.1463	6.0624	41.4371	24
25	26.4619	0.0378	181.8708	0.0055	6.8729	0.1455	6.1610	42.3441	25
26	30.1666	0.0331	208.3327	0.0048	6.9061	0.1448	6.2514	43.1728	26
27	34.3899	0.0291	238.4993	0.0042	6.9352	0.1442	6.3342	43.9289	27
28	39.2045	0.0255	272.8892	0.0037	6.9607	0.1437	6.4100	44.6176	28
29	44.6931	0.0224	312.0937	0.0032	6.9830	0.1432	6.4791	45.2441	29
30	50.9502	0.0196	356.7868	0.0028	7.0027	0.1428	6.5423	45.8132	30
31	58.0832	0.0172	407.7370	0.0025	7.0199	0.1425	6.5998	46.3297	31
32	66.2148	0.0151	465.8202	0.0021	7.0350	0.1421	6.6522	46.7979	32
33	75.4849	0.0132	532.0350	0.0019	7.0482	0.1419	6.6998	47.2218	33
34	86.0528	0.0116	607.5199	0.0016	7.0599	0.1416	6.7431	47.6053	34
35	98.1002	0.0102	693.5727	0.0014	7.0700	0.1414	6.7824	47.9519	35
40	188.8835	0.0053	1342.0251	0.0007	7.1050	0.1407	6.9300	49.2376	40
45	363.6791	0.0027	2590.5648	0.0004	7.1232	0.1404	7.0188	49.9963	45
50	700.2330	0.0014	4994.5213	0.0002	7.1327	0.1402	7.0714	50.4375	50

15.0%

	Single Payment		Equal Payment Series				Gradient Series		
	Compound Amount Factor	Present Worth Factor	Compound Amount Factor	Sinking Fund Factor	Present Worth Factor	Capital Recovery Factor	Gradient Uniform Series	Gradient Present Worth	
N	(F/P,i,N)	(P/F,i,N)	(F/A,i,N)	(A/F,i,N)	(P/A,i,N)	(A/P,i,N)	(A/G,i,N)	(P/G,i,N)	N
1	1.1500	0.8696	1.0000	1.0000	0.8696	1.1500	0.0000	0.0000	1
2	1.3225	0.7561	2.1500	0.4651	1.6257	0.6151	0.4651	0.7561	2
3	1.5209	0.6575	3.4725	0.2880	2.2832	0.4380	0.9071	2.0712	3
4	1.7490	0.5718	4.9934	0.2003	2.8550	0.3503	1.3263	3.7864	4
5	2.0114	0.4972	6.7424	0.1483	3.3522	0.2983	1.7228	5.7751	5
6	2.3131	0.4323	8.7537	0.1142	3.7845	0.2642	2.0972	7.9368	6
7	2.6600	0.3759	11.0668	0.0904	4.1604	0.2404	2.4498	10.1924	7
8	3.0590	0.3269	13.7268	0.0729	4.4873	0.2229	2.7813	12.4807	8
9	3.5179	0.2843	16.7858	0.0596	4.7716	0.2096	3.0922	14.7548	9
10	4.0456	0.2472	20.3037	0.0493	5.0188	0.1993	3.3832	16.9795	10
11	4.6524	0.2149	24.3493	0.0411	5.2337	0.1911	3.6549	19.1289	11
12	5.3503	0.1869	29.0017	0.0345	5.4206	0.1845	3.9082	21.1849	12
13	6.1528	0.1625	34.3519	0.0291	5.5831	0.1791	4.1438	23.1352	13
14	7.0757	0.1413	40.5047	0.0247	5.7245	0.1747	4.3624	24.9725	14
15	8.1371	0.1229	47.5804	0.0210	5.8474	0.1710	4.5650	26.6930	15
16	9.3576	0.1069	55.7175	0.0179	5.9542	0.1679	4.7522	28.2960	16
17	10.7613	0.0929	65.0751	0.0154	6.0472	0.1654	4.9251	29.7828	17
18	12.3755	0.0808	75.8364	0.0132	6.1280	0.1632	5.0843	31.1565	18
19	14.2318	0.0703	88.2118	0.0113	6.1982	0.1613	5.2307	32.4213	19
20	16.3665	0.0611	102.4436	0.0098	6.2593	0.1598	5.3651	33.5822	20
21	18.8215	0.0531	118.8101	0.0084	6.3125	0.1584	5.4883	34.6448	21
22	21.6447	0.0462	137.6316	0.0073	6.3587	0.1573	5.6010	35.6150	22
23	24.8915	0.0402	159.2764	0.0063	6.3988	0.1563	5.7040	36.4988	23
24	28.6252	0.0349	184.1678	0.0054	6.4338	0.1554	5.7979	37.3023	24
25	32.9190	0.0304	212.7930	0.0047	6.4641	0.1547	5.8834	38.0314	25
26	37.8568	0.0264	245.7120	0.0041	6.4906	0.1541	5.9612	38.6918	26
27	43.5353	0.0230	283.5688	0.0035	6.5135	0.1535	6.0319	39.2890	27
28	50.0656	0.0200	327.1041	0.0031	6.5335	0.1531	6.0960	39.8283	28
29	57.5755	0.0174	377.1697	0.0027	6.5509	0.1527	6.1541	40.3146	29
30	66.2118	0.0151	434.7451	0.0023	6.5660	0.1523	6.2066	40.7526	30
31	76.1435	0.0131	500.9569	0.0020	6.5791	0.1520	6.2541	41.1466	31
32	87.5651	0.0114	577.1005	0.0017	6.5905	0.1517	6.2970	41.5006	32
33	100.6998	0.0099	664.6655	0.0015	6.6005	0.1515	6.3357	41.8184	33
34	115.8048	0.0086	765.3654	0.0013	6.6091	0.1513	6.3705	42.1033	34
35	133.1755	0.0075	881.1702	0.0011	6.6166	0.1511	6.4019	42.3586	35
40	267.8635	0.0037	1779.0903	0.0006	6.6418	0.1506	6.5168	43.2830	40
45	538.7693	0.0019	3585.1285	0.0003	6.6543	0.1503	6.5830	43.8051	45
50	1083.6574	0.0009	7217.7163	0.0001	6.6605	0.1501	6.6205	44.0958	50

16.0%

	Single Payment		Equal Payment Series				Gradient Series		
N	Compound Amount Factor (F/P,i,N)	Present Worth Factor (P/F,i,N)	Compound Amount Factor (F/A,i,N)	Sinking Fund Factor (A/F,i,N)	Present Worth Factor (P/A,i,N)	Capital Recovery Factor (A/P,i,N)	Gradient Uniform Series (A/G,i,N)	Gradient Present Worth (P/G,i,N)	N
1	1.1600	0.8621	1.0000	1.0000	0.8621	1.1600	0.0000	0.0000	1
2	1.3456	0.7432	2.1600	0.4630	1.6052	0.6230	0.4630	0.7432	2
3	1.5609	0.6407	3.5056	0.2853	2.2459	0.4453	0.9014	2.0245	3
4	1.8106	0.5523	5.0665	0.1974	2.7982	0.3574	1.3156	3.6814	4
5	2.1003	0.4761	6.8771	0.1454	3.2743	0.3054	1.7060	5.5858	5
6	2.4364	0.4104	8.9775	0.1114	3.6847	0.2714	2.0729	7.6380	6
7	2.8262	0.3538	11.4139	0.0876	4.0386	0.2476	2.4169	9.7610	7
8	3.2784	0.3050	14.2401	0.0702	4.3436	0.2302	2.7388	11.8962	8
9	3.8030	0.2630	17.5185	0.0571	4.6065	0.2171	3.0391	13.9998	9
10	4.4114	0.2267	21.3215	0.0469	4.8332	0.2069	3.3187	16.0399	10
11	5.1173	0.1954	25.7329	0.0389	5.0286	0.1989	3.5783	17.9941	11
12	5.9360	0.1685	30.8502	0.0324	5.1971	0.1924	3.8189	19.8472	12
13	6.8858	0.1452	36.7862	0.0272	5.3423	0.1872	4.0413	21.5899	13
14	7.9875	0.1252	43.6720	0.0229	5.4675	0.1829	4.2464	23.2175	14
15	9.2655	0.1079	51.6595	0.0194	5.5755	0.1794	4.4352	24.7284	15
16	10.7480	0.0930	60.9650	0.0164	5.6685	0.1764	4.6086	26.1241	16
17	12.4677	0.0802	71.6730	0.0140	5.7487	0.1740	4.7676	27.4074	17
18	14.4625	0.0691	84.1407	0.0119	5.8178	0.1719	4.9130	28.5828	18
19	16.7765	0.0596	98.6032	0.0101	5.8775	0.1701	5.0457	29.6557	19
20	19.4608	0.0514	115.3797	0.0087	5.9288	0.1687	5.1666	30.6321	20
21	22.5745	0.0443	134.8405	0.0074	5.9731	0.1674	5.2766	31.5180	21
22	26.1864	0.0382	157.4150	0.0064	6.0113	0.1664	5.3765	32.3200	22
23	30.3762	0.0329	183.6014	0.0054	6.0442	0.1654	5.4671	33.0442	23
24	35.2364	0.0284	213.9776	0.0047	6.0726	0.1647	5.5490	33.6970	24
25	40.8742	0.0245	249.2140	0.0040	6.0971	0.1640	5.6230	34.2841	25
26	47.4141	0.0211	290.0883	0.0034	6.1182	0.1634	5.6898	34.8114	26
27	55.0004	0.0182	337.5024	0.0030	6.1364	0.1630	5.7500	35.2841	27
28	63.8004	0.0157	392.5028	0.0025	6.1520	0.1625	5.8041	35.7073	28
29	74.0085	0.0135	456.3032	0.0022	6.1656	0.1622	5.8528	36.0856	29
30	85.8499	0.0116	530.3117	0.0019	6.1772	0.1619	5.8964	36.4234	30
31	99.5859	0.0100	616.1616	0.0016	6.1872	0.1616	5.9356	36.7247	31
32	115.5196	0.0087	715.7475	0.0014	6.1959	0.1614	5.9706	36.9930	32
33	134.0027	0.0075	831.2671	0.0012	6.2034	0.1612	6.0019	27.2318	33
34	155.4432	0.0064	965.2698	0.0010	6.2098	0.1610	6.0299	37.4441	34
35	180.3141	0.0055	1120.7130	0.0009	6.2153	0.1609	6.0548	37.6327	35
40	378.7212	0.0026	2360.7572	0.0004	6.2335	0.1604	6.1441	38.2992	40
45	795.4438	0.0013	4965.2739	0.0002	6.2421	0.1602	6.1934	38.6598	45
50	1670.7038	0.0006	10435.6488	0.0001	6.2463	0.1601	6.2201	38.8521	50

18.0%

	Single Payment		Equal Payment Series				Gradient Series		
N	Compound Amount Factor (F/P,i,N)	Present Worth Factor (P/F,i,N)	Compound Amount Factor (F/A,i,N)	Sinking Fund Factor (A/F,i,N)	Present Worth Factor (P/A,i,N)	Capital Recovery Factor (A/P,i,N)	Gradient Uniform Series (A/G,i,N)	Gradient Present Worth (P/G,i,N)	N
1	1.1800	0.8475	1.0000	1.0000	0.8475	1.1800	0.0000	0.0000	1
2	1.3924	0.7182	2.1800	0.4587	1.5656	0.6387	0.4587	0.7182	2
3	1.6430	0.6086	3.5724	0.2799	2.1743	0.4599	0.8902	1.9354	3
4	1.9388	0.5158	5.2154	0.1917	2.6901	0.3717	1.2947	3.4828	4
5	2.2878	0.4371	7.1542	0.1398	3.1272	0.3198	1.6728	5.2312	5
6	2.6996	0.3704	9.4420	0.1059	3.4976	0.2859	2.0252	7.0834	6
7	3.1855	0.3139	12.1415	0.0824	3.8115	0.2624	2.3526	8.9670	7
8	3.7589	0.2660	15.3270	0.0652	4.0776	0.2452	2.6558	10.8292	8
9	4.4355	0.2255	19.0859	0.0524	4.3030	0.2324	2.9358	12.6329	9
10	5.2338	0.1911	23.5213	0.0425	4.4941	0.2225	3.1936	14.3525	10
11	6.1759	0.1619	28.7551	0.0348	4.6560	0.2148	3.4303	15.9716	11
12	7.2876	0.1372	34.9311	0.0286	4.7932	0.2086	3.6470	17.4811	12
13	8.5994	0.1163	42.2187	0.0237	4.9095	0.2037	3.8449	18.8765	13
14	10.1472	0.0985	50.8180	0.0197	5.0081	0.1997	4.0250	20.1576	14
15	11.9737	0.0835	60.9653	0.0164	5.0916	0.1964	4.1887	21.3269	15
16	14.1290	0.0708	72.9390	0.0137	5.1624	0.1937	4.3369	22.3885	16
17	16.6722	0.0600	87.0680	0.0115	5.2223	0.1915	4.4708	23.3482	17
18	19.6733	0.0508	103.7403	0.0096	5.2732	0.1896	4.5916	24.2123	18
19	23.2144	0.0431	123.4135	0.0081	5.3162	0.1881	4.7003	24.9877	19
20	27.3930	0.0365	146.6280	0.0068	5.3527	0.1868	4.7978	25.6813	20
21	32.3238	0.0309	174.0210	0.0057	5.3837	0.1857	4.8851	26.3000	21
22	38.1421	0.0262	206.3448	0.0048	5.4099	0.1848	4.9632	26.8506	22
23	45.0076	0.0222	244.4868	0.0041	5.4321	0.1841	5.0329	27.3394	23
24	53.1090	0.0188	289.4945	0.0035	5.4509	0.1835	5.0950	27.7725	24
25	62.6686	0.0160	342.6035	0.0029	5.4669	0.1829	5.1502	28.1555	25
26	73.9490	0.0135	405.2721	0.0025	5.4804	0.1825	5.1991	28.4935	26
27	87.2598	0.0115	479.2211	0.0021	5.4919	0.1821	5.2425	28.7915	27
28	102.9666	0.0097	566.4809	0.0018	5.5016	0.1818	5.2810	29.0537	28
29	121.5005	0.0082	669.4475	0.0015	5.5098	0.1815	5.3149	29.2842	29
30	143.3706	0.0070	790.9480	0.0013	5.5168	0.1813	5.3448	29.4864	30
31	169.1774	0.0059	934.3186	0.0011	5.5227	0.1811	5.3712	29.6638	31
32	199.6293	0.0050	1103.4960	0.0009	5.5277	0.1809	5.3945	29.8191	32
33	235.5625	0.0042	1303.1253	0.0008	5.5320	0.1808	5.4149	29.9549	33
34	277.9638	0.0036	1538.6878	0.0006	5.5356	0.1806	5.4328	30.0736	34
35	327.9973	0.0030	1816.6516	0.0006	5.5386	0.1806	5.4485	30.1773	35
40	750.3783	0.0013	4163.2130	0.0002	5.5482	0.1802	5.5022	30.5269	40
45	1716.6839	0.0006	9531.5771	0.0001	5.5523	0.1801	5.5293	30.7006	45
50	3927.3569	0.0003	21813.0937	0.0000	5.5541	0.1800	5.5428	30.7856	50

20.0%

	Single Payment		Equal Payment Series				Gradient Series		
N	Compound Amount Factor (F/P,i,N)	Present Worth Factor (P/F,i,N)	Compound Amount Factor (F/A,i,N)	Sinking Fund Factor (A/F,i,N)	Present Worth Factor (P/A,i,N)	Capital Recovery Factor (A/P,i,N)	Gradient Uniform Series (A/G,i,N)	Gradient Present Worth (P/G,i,N)	N
1	1.2000	0.8333	1.0000	1.0000	0.8333	1.2000	0.0000	0.0000	1
2	1.4400	0.6944	2.2000	0.4545	1.5278	0.6545	0.4545	0.6944	2
3	1.7280	0.5787	3.6400	0.2747	2.1065	0.4747	0.8791	1.8519	3
4	2.0736	0.4823	5.3680	0.1863	2.5887	0.3863	1.2742	3.2986	4
5	2.4883	0.4019	7.4416	0.1344	2.9906	0.3344	1.6405	4.9061	5
6	2.9860	0.3349	9.9299	0.1007	3.3255	0.3007	1.9788	6.5806	6
7	3.5832	0.2791	12.9159	0.0774	3.6046	0.2774	2.2902	8.2551	7
8	4.2998	0.2326	16.4991	0.0606	3.8372	0.2606	2.5756	9.8831	8
9	5.1598	0.1938	20.7989	0.0481	4.0310	0.2481	2.8364	11.4335	9
10	6.1917	0.1615	25.9587	0.0385	4.1925	0.2385	3.0739	12.8871	10
11	7.4301	0.1346	32.1504	0.0311	4.3271	0.2311	3.2893	14.2330	11
12	8.9161	0.1122	39.5805	0.0253	4.4392	0.2253	3.4841	15.4667	12
13	10.6993	0.0935	48.4966	0.0206	4.5327	0.2206	3.6597	16.5883	13
14	12.8392	0.0779	59.1959	0.0169	4.6106	0.2169	3.8175	17.6008	14
15	15.4070	0.0649	72.0351	0.0139	4.6755	0.2139	3.9588	18.5095	15
16	18.4884	0.0541	87.4421	0.0114	4.7296	0.2114	4.0851	19.3208	16
17	22.1861	0.0451	105.9306	0.0094	4.7746	0.2094	4.1976	20.0419	17
18	26.6233	0.0376	128.1167	0.0078	4.8122	0.2078	4.2975	20.6805	18
19	31.9480	0.0313	154.7400	0.0065	4.8435	0.2065	4.3861	21.2439	19
20	38.3376	0.0261	186.6880	0.0054	4.8696	0.2054	4.4643	21.7395	20
21	46.0051	0.0217	225.0256	0.0044	4.8913	0.2044	4.5334	22.1742	21
22	55.2061	0.0181	271.0307	0.0037	4.9094	0.2037	4.5941	22.5546	22
23	66.2474	0.0151	326.2369	0.0031	4.9245	0.2031	4.6475	22.8867	23
24	79.4968	0.0126	392.4842	0.0025	4.9371	0.2025	4.6943	23.1760	24
25	95.3962	0.0105	471.9811	0.0021	4.9476	0.2021	4.7352	23.4276	25
26	114.4755	0.0087	567.3773	0.0018	4.9563	0.2018	4.7709	23.6460	26
27	137.3706	0.0073	681.8528	0.0015	4.9636	0.2015	4.8020	23.8353	27
28	164.8447	0.0061	819.2233	0.0012	4.9697	0.2012	4.8291	23.9991	28
29	197.8136	0.0051	984.0680	0.0010	4.9747	0.2010	4.8527	24.1406	29
30	237.3763	0.0042	1181.8816	0.0008	4.9789	0.2008	4.8731	24.2628	30
31	284.8516	0.0035	1419.2579	0.0007	4.9824	0.2007	4.8908	24.3681	31
32	341.8219	0.0029	1704.1095	0.0006	4.9854	0.2006	4.9061	24.4588	32
33	410.1863	0.0024	2045.9314	0.0005	4.9878	0.2005	4.9194	24.5368	33
34	492.2235	0.0020	2456.1176	0.0004	4.9898	0.2004	4.9308	24.6038	34
35	590.6682	0.0017	2948.3411	0.0003	4.9915	0.2003	4.9406	24.6614	35
40	1469.7716	0.0007	7343.8578	0.0001	4.9966	0.2001	4.9728	24.8469	40
45	3657.2620	0.0003	18281.3099	0.0001	4.9986	0.2001	4.9877	24.9316	45

25.0%

	Single Payment		Equal Payment Series				Gradient Series		
	Compound Amount Factor	Present Worth Factor	Compound Amount Factor	Sinking Fund Factor	Present Worth Factor	Capital Recovery Factor	Gradient Uniform Series	Gradient Present Worth	
N	(F/P,i,N)	(P/F,i,N)	(F/A,i,N)	(A/F,i,N)	(P/A,i,N)	(A/P,i,N)	(A/G,i,N)	(P/G,i,N)	N
1	1.2500	0.8000	1.0000	1.0000	0.8000	1.2500	0.0000	0.0000	1
2	1.5625	0.6400	2.2500	0.4444	1.4400	0.6944	0.4444	0.6400	2
3	1.9531	0.5120	3.8125	0.2623	1.9520	0.5123	0.8525	1.6640	3
4	2.4414	0.4096	5.7656	0.1734	2.3616	0.4234	1.2249	2.8928	4
5	3.0518	0.3277	8.2070	0.1218	2.6893	0.3718	1.5631	4.2035	5
6	3.8147	0.2621	11.2588	0.0888	2.9514	0.3388	1.8683	5.5142	6
7	4.7684	0.2097	15.0735	0.0663	3.1611	0.3163	2.1424	6.7725	7
8	5.9605	0.1678	19.8419	0.0504	3.3289	0.3004	2.3872	7.9469	8
9	7.4506	0.1342	25.8023	0.0388	3.4631	0.2888	2.6048	9.0207	9
10	9.3132	0.1074	33.2529	0.0301	3.5705	0.2801	2.7971	9.9870	10
11	11.6415	0.0859	42.5661	0.0235	3.6564	0.2735	2.9663	10.8460	11
12	14.5519	0.0687	54.2077	0.0184	3.7251	0.2684	3.1145	11.6020	12
13	18.1899	0.0550	68.7596	0.0145	3.7801	0.2645	3.2437	12.2617	13
14	22.7374	0.0440	86.9495	0.0115	3.8241	0.2615	3.3559	12.8334	14
15	28.4217	0.0352	109.6868	0.0091	3.8593	0.2591	3.4530	13.3260	15
16	35.5271	0.0281	138.1085	0.0072	3.8874	0.2572	3.5366	13.7482	16
17	44.4089	0.0225	173.6357	0.0058	3.9099	0.2558	3.6084	14.1085	17
18	55.5112	0.0180	218.0446	0.0046	3.9279	0.2546	3.6698	14.4147	18
19	69.3889	0.0144	273.5558	0.0037	3.9424	0.2537	3.7222	14.6741	19
20	86.7362	0.0115	342.9447	0.0029	3.9539	0.2529	3.7667	14.8932	20
21	108.4202	0.0092	429.6809	0.0023	3.9631	0.2523	3.8045	15.0777	21
22	135.5253	0.0074	538.1011	0.0019	3.9705	0.2519	3.8365	15.2326	22
23	169.4066	0.0059	673.6264	0.0015	3.9764	0.2515	3.8634	15.3625	23
24	211.7582	0.0047	843.0329	0.0012	3.9811	0.2512	3.8861	15.4711	24
25	264.6978	0.0038	1054.7912	0.0009	3.9849	0.2509	3.9052	15.5618	25
26	330.8722	0.0030	1319.4890	0.0008	3.9879	0.2508	3.9212	15.6373	26
27	413.5903	0.0024	1650.3612	0.0006	3.9903	0.2506	3.9346	15.7002	27
28	516.9879	0.0019	2063.9515	0.0005	3.9923	0.2505	3.9457	15.7524	28
29	646.2349	0.0015	2580.9394	0.0004	3.9938	0.2504	3.9551	15.7957	29
30	807.7936	0.0012	3227.1743	0.0003	3.9950	0.2503	3.9628	15.8316	30
31	1009.7420	0.0010	4034.9678	0.0002	3.9960	0.2502	3.9693	15.8614	31
32	1262.1774	0.0008	5044.7098	0.0002	3.9968	0.2502	3.9746	15.8859	32
33	1577.7218	0.0006	6306.8872	0.0002	3.9975	0.2502	3.9791	15.9062	33
34	1972.1523	0.0005	7884.6091	0.0001	3.9980	0.2501	3.9828	15.9229	34
35	2465.1903	0.0004	9856.7613	0.0001	3.9984	0.2501	3.9858	15.9367	35
40	7523.1638	0.0001	30088.6554	0.0000	3.9995	0.2500	3.9947	15.9766	40

30.0%

	Single Payment		Equal Payment Series				Gradient Series		
	Compound Amount Factor	Present Worth Factor	Compound Amount Factor	Sinking Fund Factor	Present Worth Factor	Capital Recovery Factor	Gradient Uniform Series	Gradient Present Worth	
N	(F/P,i,N)	(P/F,i,N)	(F/A,i,N)	(A/F,i,N)	(P/A,i,N)	(A/P,i,N)	(A/G,i,N)	(P/G,i,N)	N
1	1.3000	0.7692	1.0000	1.0000	0.7692	1.3000	0.0000	0.0000	1
2	1.6900	0.5917	2.3000	0.4348	1.3609	0.7348	0.4348	0.5917	2
3	2.1970	0.4552	3.9900	0.2506	1.8161	0.5506	0.8271	1.5020	3
4	2.8561	0.3501	6.1870	0.1616	2.1662	0.4616	1.1783	2.5524	4
5	3.7129	0.2693	9.0431	0.1106	2.4356	0.4106	1.4903	3.6297	5
6	4.8268	0.2072	12.7560	0.0784	2.6427	0.3784	1.7654	4.6656	6
7	6.2749	0.1594	17.5828	0.0569	2.8021	0.3569	2.0063	5.6218	7
8	8.1573	0.1226	23.8577	0.0419	2.9247	0.3419	2.2156	6.4800	8
9	10.6045	0.0943	32.0150	0.0312	3.0190	0.3312	2.3963	7.2343	9
10	13.7858	0.0725	42.6195	0.0235	3.0915	0.3235	2.5512	7.8872	10
11	17.9216	0.0558	56.4053	0.0177	3.1473	0.3177	2.6833	8.4452	11
12	23.2981	0.0429	74.3270	0.0135	3.1903	0.3135	2.7952	8.9173	12
13	30.2875	0.0330	97.6250	0.0102	3.2233	0.3102	2.8895	9.3135	13
14	39.3738	0.0254	127.9125	0.0078	3.2487	0.3078	2.9685	9.6437	14
15	51.1859	0.0195	167.2863	0.0060	3.2682	0.3060	3.0344	9.9172	15
16	66.5417	0.0150	218.4722	0.0046	3.2832	0.3046	3.0892	10.1426	16
17	86.5042	0.0116	285.0139	0.0035	3.2948	0.3035	3.1345	10.3276	17
18	112.4554	0.0089	371.5180	0.0027	3.3037	0.3027	3.1718	10.4788	18
19	146.1920	0.0068	483.9734	0.0021	3.3105	0.3021	3.2025	10.6019	19
20	190.0496	0.0053	630.1655	0.0016	3.3158	0.3016	3.2275	10.7019	20
21	247.0645	0.0040	820.2151	0.0012	3.3198	0.3012	3.2480	10.7828	21
22	321.1839	0.0031	1067.2796	0.0009	3.3230	0.3009	3.2646	10.8482	22
23	417.5391	0.0024	1388.4635	0.0007	3.3254	0.3007	3.2781	10.9009	23
24	542.8008	0.0018	1806.0026	0.0006	3.3272	0.3006	3.2890	10.9433	24
25	705.6410	0.0014	2348.8033	0.0004	3.3286	0.3004	3.2979	10.9773	25
26	917.3333	0.0011	3054.4443	0.0003	3.3297	0.3003	3.3050	11.0045	26
27	1192.5333	0.0008	3971.7776	0.0003	3.3305	0.3003	3.3107	11.0263	27
28	1550.2933	0.0006	5164.3109	0.0002	3.3312	0.3002	3.3153	11.0437	28
29	2015.3813	0.0005	6714.6042	0.0001	3.3317	0.3001	3.3189	11.0576	29
30	2619.9956	0.0004	8729.9855	0.0001	3.3321	0.3001	3.3219	11.0687	30
31	3405.9943	0.0003	11349.9811	0.0001	3.3324	0.3001	3.3242	11.0775	31
32	4427.7926	0.0002	14755.9755	0.0001	3.3326	0.3001	3.3261	11.0845	32
33	5756.1304	0.0002	19183.7681	0.0001	3.3328	0.3001	3.3276	11.0901	33
34	7482.9696	0.0001	24939.8985	0.0000	3.3329	0.3000	3.3288	11.0945	34
35	9727.8604	0.0001	32422.8681	0.0000	3.3330	0.3000	3.3297	11.0980	35

35.0%

	Single Payment		Equal Payment Series				Gradient Series		
N	Compound Amount Factor (F/P,i,N)	Present Worth Factor (P/F,i,N)	Compound Amount Factor (F/A,i,N)	Sinking Fund Factor (A/F,i,N)	Present Worth Factor (P/A,i,N)	Capital Recovery Factor (A/P,i,N)	Gradient Uniform Series (A/G,i,N)	Gradient Present Worth (P/G,i,N)	N
1	1.3500	0.7407	1.0000	1.0000	0.7407	1.3500	0.0000	0.0000	1
2	1.8225	0.5487	2.3500	0.4255	1.2894	0.7755	0.4255	0.5487	2
3	2.4604	0.4064	4.1725	0.2397	1.6959	0.5897	0.8029	1.3616	3
4	3.3215	0.3011	6.6329	0.1508	1.9969	0.5008	1.1341	2.2648	4
5	4.4840	0.2230	9.9544	0.1005	2.2200	0.4505	1.4220	3.1568	5
6	6.0534	0.1652	14.4384	0.0693	2.3852	0.4193	1.6698	3.9828	6
7	8.1722	0.1224	20.4919	0.0488	2.5075	0.3988	1.8811	4.7170	7
8	11.0324	0.0906	28.6640	0.0349	2.5982	0.3849	2.0597	5.3515	8
9	14.8937	0.0671	39.6964	0.0252	2.6653	0.3752	2.2094	5.8886	9
10	20.1066	0.0497	54.5902	0.0183	2.7150	0.3683	2.3338	6.3363	10
11	27.1439	0.0368	74.6967	0.0134	2.7519	0.3634	2.4364	6.7047	11
12	36.6442	0.0273	101.8406	0.0098	2.7792	0.3598	2.5205	7.0049	12
13	49.4697	0.0202	138.4848	0.0072	2.7994	0.3572	2.5889	7.2474	13
14	66.7841	0.0150	187.9544	0.0053	2.8144	0.3553	2.6443	7.4421	14
15	90.1585	0.0111	254.7385	0.0039	2.8255	0.3539	2.6889	7.5974	15
16	121.7139	0.0082	344.8970	0.0029	2.8337	0.3529	2.7246	7.7206	16
17	164.3138	0.0061	466.6109	0.0021	2.8398	0.3521	2.7530	7.8180	17
18	221.8236	0.0045	630.9247	0.0016	2.8443	0.3516	2.7756	7.8946	18
19	299.4619	0.0033	852.7483	0.0012	2.8476	0.3512	2.7935	2.9547	19
20	404.2736	0.0025	1152.2103	0.0009	2.8501	0.3509	2.8075	8.0017	20
21	545.7693	0.0018	1556.4838	0.0006	2.8519	0.3506	2.8186	8.0384	21
22	736.7886	0.0014	2102.2532	0.0005	2.8533	0.3505	2.8272	8.0669	22
23	994.6646	0.0010	2839.0418	0.0004	2.8543	0.3504	2.8340	8.0890	23
24	1342.7973	0.0007	3833.7064	0.0003	2.8550	0.3503	2.8393	8.1061	24
25	1812.7763	0.0006	5176.5037	0.0002	2.8556	0.3502	2.8433	8.1194	25
26	2447.2480	0.0004	6989.2800	0.0001	2.8560	0.3501	2.8465	8.1296	26
27	3303.7848	0.0003	9436.5280	0.0001	2.8563	0.3501	2.8490	8.1374	27
28	4460.1095	0.0002	12740.3128	0.0001	2.8565	0.3501	2.8509	8.1435	28
29	6021.1478	0.0002	17200.4222	0.0001	2.8567	0.3501	2.8523	8.1481	29
30	8128.5495	0.0001	23221.5700	0.0000	2.8568	0.3500	2.8535	8.1517	30

40.0%

	Single Payment		Equal Payment Series				Gradient Series		
N	Compound Amount Factor (F/P,i,N)	Present Worth Factor (P/F,i,N)	Compound Amount Factor (F/A,i,N)	Sinking Fund Factor (A/F,i,N)	Present Worth Factor (P/A,i,N)	Capital Recovery Factor (A/P,i,N)	Gradient Uniform Series (A/G,i,N)	Gradient Present Worth (P/G,i,N)	N
1	1.4000	0.7143	1.0000	1.0000	0.7143	1.4000	0.0000	0.0000	1
2	1.9600	0.5102	2.4000	0.4167	1.2245	0.8167	0.4167	0.5102	2
3	2.7440	0.3644	4.3600	0.2294	1.5889	0.6294	0.7798	1.2391	3
4	3.8416	0.2603	7.1040	0.1408	1.8492	0.5408	1.0923	2.0200	4
5	5.3782	0.1859	10.9456	0.0914	2.0352	0.4914	1.3580	2.7637	5
6	7.5295	0.1328	16.3238	0.0613	2.1680	0.4613	1.5811	3.4278	6
7	10.5414	0.0949	23.8534	0.0419	2.2628	0.4419	1.7664	3.9970	7
8	14.7579	0.0678	34.3947	0.0291	2.3306	0.4291	1.9185	4.4713	8
9	20.6610	0.0484	49.1526	0.0203	2.3790	0.4203	2.0422	4.8585	9
10	28.9255	0.0346	69.8137	0.0143	2.4136	0.4143	2.1419	5.1696	10
11	40.4957	0.0247	98.7391	0.0101	2.4383	0.4101	2.2215	5.4166	11
12	56.6939	0.0176	139.2348	0.0072	2.4559	0.4072	2.2845	5.6106	12
13	79.3715	0.0126	195.9287	0.0051	2.4685	0.4051	2.3341	5.7618	13
14	111.1201	0.0090	275.3002	0.0036	2.4775	0.4036	2.3729	5.8788	14
15	155.5681	0.0064	386.4202	0.0026	2.4839	0.4026	2.4030	5.9688	15
16	217.7953	0.0046	541.9883	0.0018	2.4885	0.4018	2.4262	6.0376	16
17	304.9135	0.0033	759.7837	0.0013	2.4918	0.4013	2.4441	6.0901	17
18	426.8789	0.0023	1064.6971	0.0009	2.4941	0.4009	2.4577	6.1299	18
19	597.6304	0.0017	1491.5760	0.0007	2.4958	0.4007	2.4682	6.1601	19
20	836.6826	0.0012	2089.2064	0.0005	2.4970	0.4005	2.4761	6.1828	20
21	1171.3556	0.0009	2925.8889	0.0003	2.4979	0.4003	2.4821	6.1998	21
22	1639.8978	0.0006	4097.2445	0.0002	2.4985	0.4002	2.4866	6.2127	22
23	2295.8569	0.0004	5737.1423	0.0002	2.4989	0.4002	2.4900	6.2222	23
24	3214.1997	0.0003	8032.9993	0.0001	2.4992	0.4001	2.4925	6.2294	24
25	4499.8796	0.0002	11247.1990	0.0001	2.4994	0.4001	2.4944	6.2347	25
26	6299.8314	0.0002	15747.0785	0.0001	2.4996	0.4001	2.4959	6.2387	26
27	8819.7640	0.0001	22046.9099	0.0000	2.4997	0.4000	2.4969	6.2416	27
28	12347.6696	0.0001	30866.6739	0.0000	2.4998	0.4000	2.4977	6.2438	28
29	17286.7374	0.0001	43214.3435	0.0000	2.4999	0.4000	2.4983	6.2454	29
30	24201.4324	0.0000	60501.0809	0.0000	2.4999	0.4000	2.4988	6.2466	30

50.0%

	Single Payment		Equal Payment Series				Gradient Series		
N	Compound Amount Factor (F/P,i,N)	Present Worth Factor (P/F,i,N)	Compound Amount Factor (F/A,i,N)	Sinking Fund Factor (A/F,i,N)	Present Worth Factor (P/A,i,N)	Capital Recovery Factor (A/P,i,N)	Gradient Uniform Series (A/G,i,N)	Gradient Present Worth (P/G,i,N)	N
1	1.5000	0.6667	1.0000	1.0000	0.6667	1.5000	0.0000	0.0000	1
2	2.2500	0.4444	2.5000	0.4000	1.1111	0.9000	0.4000	0.4444	2
3	3.3750	0.2963	4.7500	0.2105	1.4074	0.7105	0.7368	1.0370	3
4	5.0625	0.1975	8.1250	0.1231	1.6049	0.6231	1.0154	1.6296	4
5	7.5938	0.1317	13.1875	0.0758	1.7366	0.5758	1.2417	2.1564	5
6	11.3906	0.0878	20.7813	0.0481	1.8244	0.5481	1.4226	2.5953	6
7	17.0859	0.0585	32.1719	0.0311	1.8829	0.5311	1.5648	2.9465	7
8	25.6289	0.0390	49.2578	0.0203	1.9220	0.5203	1.6752	3.2196	8
9	38.4434	0.0260	74.8867	0.0134	1.9480	0.5134	1.7596	3.4277	9
10	57.6650	0.0173	113.3301	0.0088	1.9653	0.5088	1.8235	3.5838	10
11	86.4976	0.0116	170.9951	0.0058	1.9769	0.5058	1.8713	3.6994	11
12	129.7463	0.0077	257.4927	0.0039	1.9846	0.5039	1.9068	3.7842	12
13	194.6195	0.0051	387.2390	0.0026	1.9897	0.5026	1.9329	3.8459	13
14	291.9293	0.0034	581.8585	0.0017	1.9931	0.5017	1.9519	3.8904	14
15	437.8939	0.0023	873.7878	0.0011	1.9954	0.5011	1.9657	3.9224	15
16	656.8408	0.0015	1311.6817	0.0008	1.9970	0.5008	1.9756	3.9452	16
17	985.2613	0.0010	1968.5225	0.0005	1.9980	0.5005	1.9827	3.9614	17
18	1477.8919	0.0007	2953.7838	0.0003	1.9986	0.5003	1.9878	3.9729	18
19	2216.8378	0.0005	4431.6756	0.0002	1.9991	0.5002	1.9914	3.9811	19
20	3325.2567	0.0003	6648.5135	0.0002	1.9994	0.5002	1.9940	3.9868	20

Answers to Self-Test Questions

Chapter 2

2s.1 1. (F); 2. (F); 3. (F); 4. (T)

2s.2 Income statement; The balance sheet; Cash flow statement; Operating activities, Investing activities, and Financing activities; Treasury account

2s.3 (7), (8), (1), (11), (3), (9)

2s.4 Accounts receivable = $150, current liabilities = $134, current assets = $430, total assets = $710, long-term debt = $76, profit margin = 30%

Chapter 3

3s.1 1. (T); 2. (T); 3. (F); 4. (T); 5. (F)

3s.2 1. (product, variable), 2. (prime), 3. (product, period), 4. (fixed), 5. (product, cost of goods sold), 6. (inventory), 7. (period, variable), 8. (sunk), 9. (fixed, indirect), 10. (period)

3s.3 (a) increase, decrease, (b) increase, decrease, (c) decrease, increase, (d) increase, decrease, (e) increase, decrease

Chapter 4

4s.1 (b), $800(F/P, 10\%, 10) + $1,500(F/A, 10\%, 9)(1.10) = $24,481

4s.2 (a)

4s.3 (c), $P = -$1,500(P/F, 10\%, 7) + $1,000(P/A, 10\%, 10) = $5,375

4s.4 (d), $C(P/G, 12\%, 6) = $800(F/A, 12\%, 4) + ($1,000 - $200(P/G, 12\%, 4))(F/P, 12\%, 4), C = $458.90

4s.5 (a), $A(F/A, 10\%, 12)(1.1)= $25,000(P/A, 10\%, 4) + $2,000 (P/G, 10\%,4), A = $3741

4s.6 (c), Simple interest = $1,000(0.10)(3) = $300, compound interest = $1,000(1 + 0.095)^3 - $,1000 = $312.90

4s.7 (b), $F = $2,000(F/P, 12\%, 7) = $4,422

4s.8 (c), $C(F/A, 6\%, 10) = $5,000(P/A, 6\%, 15) + $1,000(P/G, 6\%, 15), C = $8,054

4s.9 (b), $P = $11,000(P/F, 10\%, 1) + $12,100(P/F, 10\%, 2) + $14,641(P/F, 10\%, 4) = $30,000

4s.10 (d)

4s.11 (c), $F = [$450(F/P, 10\%, 7) + $200(P/A, 10\%, 7)](F/P, 10\%, 1) = $3,052

4s.12 (d), $F_3 = $100(F/A, 10\%, 3)(F/P, 10\%, 1) + $50 (P/A, 10\%, 3) = $488

4s.13 (c), $A + A(P/A, 10\%, 7) = [$200 (P/G, 10\%, 4) + $1,000(P/A, 10\%, 4)](P/F, 10\%, 3)$; Solve for A. $A = $518

4s.14 (c), $I = $2,000(0.10)(5) = $1,000

4s.15 (b), $100(1 + i) + $120 = $50(1 + i) + $180, i = 20\%$

4s.16 (d), $A(P/A, 15\%, 6) - A(P/F, 15\%, 3) = $10,000, A = $3,198

4s.17 (a), $100(F/A, 12\%, 5) + $108(P/A1, 8\%, 12\%, 5) = $1,083

4s.18 (c), $100(1.1) + 100 + 150(P/A, 10\%, 4) = C(1.1) + C(P/A, 10\%, 4), C = 160$
4s.19 (d), $400(P/A, 10\%, 5) - 100(P/F, 10\%, 4) = X(P/A, 10\%, 5) - (400 + X)(P/F, 10\%, 4), X = 554$
4s.20 (c), $F = 200(F/A, 21\%, 5) = 1,517$

Chapter 5

5s.1 (c)
5s.2 (d), $F = 2500(F/A, 2.2669\%, 40) = 160,058$
5s.3 (c), $A = 10,000(A/P, 0.75\%, 60) = 207.58, P = 207.58(P/A, 0.75\%, 36) = 6,528$
5s.4 (b), Option 1: $F = 1,000(F/A, 1.5\%, 40)(F/P, 1.5\%, 60) = 132,587$; Option 2: $F = 6,000(F/A, 6.136\%, 15) = 141,110$
5s.5 (c)
5s.6 (a), $(1 + i)^2 = 1.12, i = 5.83\%$ per semi-annual, $1,000(P/A, 5.83\%, 6) = 4944$
5s.7 (d), $2 = (1 + r/4)^{20}, r = (3.526\%)(4) = 14.11\%$
5s.8 (c), $F = 1,000(F/A, 2.26692\%, 12) = 13,615$
5s.9 (c), $i_a = (1 + 0.018)^{12} = 23.87\%$
5s.10 (b), $P = 1,000(P/A, 2.020\%, 20) = 16,319$
5s.11 (c), $A = 25,000(A/P, 0.75\%, 36) = 795$
5s.12 (c), $70,000 = 1000(P/A, 1\%, N), N = 121$ months
5s.13 (a), $20,000 = 922.90(P/A, i, 24), i = 0.8333\%$ per month, $P = 922.90(P/A, 0.8333\%, 12) = 10,498$
5s.14 (c), $20,000 = 5548.19(P/A, i, 5), i = 12\%$
5s.15 (a), $2 = 1(F/P, 9.3083\%, N), N = 8$ years
5s.16 (c), $0.0887 = (1 + r/365)^{365} - 1, r = 8.5\%$
5s.17 (c), $100(P/F, 8\%, 1) + 300(P/F, 8\%, 2) + 500(P/F, 8\%, 3) + X(P/F, 8\%, 4) = 1,000, X = 345$
5s.18 (b)
5s.19 (a), Bank A: $i_a = (1 + 0.085/4)^4 - 1 = 8.885\%$, Bank B: $i_a = e^{0.084} - 1 = 8.763\%$

5s.20 (a), $F(P/F, 8\%, 4) = 500(1.08) + 500 + 1,000(P/A, 8\%, 2), 0.7750F = 2,823.30, F = 3,841$
5s.21 (d), $F = 20,000(F/P, 1\%, 48) = 32,244$
5s.22 (d), $A = 10,000(A/P, 1\%, 36) = 332, B_{12} = 332(P/A, 1\%, 24) = 7,046.48$, New installment $(A') = 7,046.48(A/P, 3.03\%, 8) = 1,005, i_p = (1 + 0.01)^3 - 1 = 3.03\%$
5s.23 (b), $F = 20,000(F/P, 3\%, 16) = 32,094$
5s.24 (c), $A(F/A, 4\%, 30) = 30,000(P/A, 8.16\%, 15), A = 4,534$
5s.25 (b)
5s.26 (d), $10,000 = 200(P/A, 0.5\%, N), N = 58$ months
5s.27 (a), $P = 5,000(P/F, 0.75\%, 30)(P/F, 1\%, 30) = 2,965$

Chapter 6

6s.1 (d)
6s.2 (b)
6s.3 (e)
6s.4 (d)
6s.5 (c)

Chapter 7

7s.1 (a), $n = P/0.125P = 8$ years
7s.2 (c)
7s.3 (a)
7s.4 (c), $PW(9\%) = 866.51$
7s.5 (a), $PW(9\%) = 1,386$
7s.6 (b), $PW(8\%) = -20,000 + (25,000 - 5,000)(P/A, 8\%, 10) + 225,000 (P/F, 8\%,10) = 218,420$ (at maximum)
7s.7 (d), $PW(15\%) = 1,000,000(P/A_1, -25\%, 15\%, 4) = 2,047,734$
7s.8 (b), $CE(10\%) = 400/0.1 + (100/0.1)(P/F, 10\%, 10) = 4,386$
7s.9 (a), $CE(10\%) = [100 + 100(A/F, 10\%, 2)]/0.1 = 1476$
7s.10 (a)
7s.11 (d), Project balance at $n = 6 : -2,492$, project balance at $n = 7: 2,134$

7s.12 (c), $CE(6\%) = -\$2,000,000 - \$1,000,000(A/F, 6\%, 30)/0.06 - \$100,000/0.06 = -\$3,877,482$

7s.13 (b), $PW(12\%)_A = \$140.87$, $PW(12\%)_B = \$197.68$

7s.14 (b), Let M = annual operating hours; $AE(6\%)_1 = [150\ HP\ (0.7457\ kW/HP)$ (M hours/year) ($\$0.05/kWh$)]/0.83 + $4,500$ $(A/P, 6\%, 10) + 0.15(\$4,500) = 6.7383M + 1286$; $AE(6\%)_2 = [150\ HP\ (0.7457\ kW/HP)$ (M hours/year) ($\$0.05/kWh$)]/0.80 + $\$3,600(A/P, 6\%, 10) + 0.15(\$3,600) = 6.9909M + 1029$ Let $AE(6\%)_1 = AE(6\%)_2$, solve for M. $M = 1,018$ hours

7s.15 (a), $\$100,000 = -P + \$10,000(P/A, 10\%, 9) + \$5,000(P/G, 10\%, 9) + (\$52,000/0.1)(P/F, 10\%, 9) - [\$40,000(A/F, 10\%, 10)]/0.1$, $P = \$250,149$

7s.16 (b), $-\$1000 + \$900(P/F, 10\%, 1) + (\$800 + S)(P/F, 10\%, 2) = -\$2,000 + \$2,500(P/F, 10\%, 1) + \$1,000(P/F, 10\%, 2)$; Solve for S; S = $750

7s.17 (a), $PW(12\%)_A = -\$3,113$, $PW(12\%)_B = -\$2,768$

7s.18 (d), Base period = year 0; Equivalent investment amount = $\$10(F/A, 20\%, 5) + \$30 = \$104.42$; Equivalent future benefit = $\$100(P/A, 20\%, 10) = \$419$; Starting asking price = $\$104.42 + \$419 = \$523.42$

7s.19 (c), Compute the project balance at year 3: $PB(10\%)_3 = \$600(1.1) + A_3 = \$900$, $A_3 = \$240$

7s.20 (d), $CE(10\%) = \$60,000 + \$8,000/0.1 + \$20,000/0.331 = \$200,423$ (Note that a 10% annual compounding for 3 years is equivalent to $i_p = (1 + 0.1)^3 - 1 = 33.1\%$ once for every 3 years.)

7s.21 (b), $PW(12\%)_{E1} = -\$400 - \$300(P/A, 12\%, 3) + \$150(P/F, 12\%, 3) = -\$1,014$, $PW(12\%)_{E2} = -\$600 - \$250(P/A, 12\%, 3) + \$320(P/F, 12\%, 3) = -\$972$; Select E2.

7s.22 (a), $-\$1,000(1 + i) + \$200 = -\$900$, $i = 10\%$

7s.23 (b), Compute the effective annual interest rates; 12% – annual, 12.095% – semi-annual, 12.006% – quarterly, 11.849% – monthly

Chapter 8

8s.1 (a)

8s.2 (b), $AE(10\%)_A = \$5,000(A/P, 10\%, 10) = \$813$, $AE(10\%)_B = \$3,000(A/P, 10\%, 5) = \$791$

8s.3 (b), $AE(10\%) = \$400 + (\$100/0.1)(P/F, 10\%, 10)(0.1) = \$438.55$

8s.4 (c), $CR(20\%) = (\$100,000 - S)(A/P, 20\%, 8) + S(0.20) = \$25,455$, $S = \$9,982$

8s.5 (d), $CR(15\%) = (\$18,000 - \$3,000)(A/P, 15\%, 10) + \$3,000(0.15) = \$3,440$

8s.6 (d), $AE(10\%) = \$500 + (\$500/0.1)(P/F, 10\%, 10)(0.1) = \$693$

8s.7 (a), $AE(15\%) = (\$56,000 - \$5,000)(A/P, 15\%, 5) + \$5,000(0.15) + \$6,000 = \$21,964$ machine cost per hour = $21,964/2500$ hours = 8.79/hour

8s.8 (c), $AE(5\%) = \$800,000(0.05) + \$120,000 + \$13,000 - \$32,000 + \$50,000(A/F, 5\%, 5) = \$150,049$, ticket cost per person = $150,049/40,000 = 3.75$ per person

8s.9 (c), $AE(14\%) = [-\$100K + 30K(P/A1, 3\%, 14\%, 5)](A/P, 14\%, 5) = \$2,481$, savings per hour = $2,481/3000$ hours = 0.827/hour

8s.10 (d), $(\$100,000 - \$12,000)(A/P, 12\%, 8) + \$12,000(0.12) + \$10,000 = \$29,155$, cost per book binding = $29,155/1000 = 29.15$ per book

8s.11 (c), $\$29,155/X = \$25$, $X = 1166$

8s.12 (b),

	3200 kWh/month	6700 kWh/month
First 1500 kWh @ $0.025	$37.50	$37.50
next 1250 kWh @ $0.015	18.75	18.75
next 3000 kWh @ $0.009	4.05	27.00
All over 5750 kWh @ $0.008		7.60
Total	$60.30	$90.85

Difference with an additional 3,500 kWh = $90.85 - \$60.30 = \$30.55$

8s.13 (b), Equivalent worth for the first cycle at $n = 2$. $V_2 = \$100(1.1) + X$; $CE(21\%) = (V_2)/0.21 = \$1,426$, $X = \$200$

8s.14 (d), $CR(8\%) = (\$50,000 - \$5,000)(A/P, 8\%, 12) + 0.08(\$5,000) = \$6,371$, Annual Eq. O&M cost = \$8,000, $AE(8\%) = \$6,371 + \$8,000 = \$14,371$

8s.15 (c), $AE(12\%)_X = (\$4,500 - \$250)(A/P, 12\%, 10) + 0.12(\$250) + \$300 + \{150(0.746)/0.83\}(2,000)(0.05) = \$14,564$, unit cost = \$14,564/2,000 = \$7.28 per hour; $AE(12\%)_Y = (\$3,600 - \$100)(A/P, 12\%, 10) + 0.12(\$100) + \$500 + \{150(0.746)/0.80\}(2,000)(0.05) = \$15,119$, unit cost = \$15,119/2,000 = \$7.56 per hour

8s.16 (b)

8s.17 (d), $AE(10\%) = \$500 + \{(\$500/0.1)(P/F, 10\%, 10)\}(0.10) = \$500 + \$193 = \$693$

8s.18 (b), $AE(25\%) = (\$500 - \$120)(A/P, 25\%, 10) + \$120(0.25) + \$175 = \$311.43$, unit cost = \$311.43/75 = \$4.152

Chapter 9

9s.1 (c), $-\$2,000 + \$1,000(P/F, 10\%, 1) + X(P/F, 10\%, 2) + \$1,200(P/F, 10\%, 3) = 0$, $X = \$230$

9s.2 (b), $(\$150,000 - \$15,000)(A/P, 18\%, 10) + \$15,000(0.18) + \$50,000 = \$82,739$

9s.3 (c), $-\$15,459 + \$3,000/i = 0$, $i = \$3,000/\$15,459 = 19.41\%$

9s.4 (c), Rate of return for Fidelity Mutual Funds: $-\$52451\ \$80,810(P/F, i, 15) = 0$, $i = 20\%$; worth of the Wal–Mart investment at the end of 10 years = $\$1,650(F/P, 35\%, 10) = \$33,176$; worth of the Fidelity mutual funds after 15 years = $\$33,176(F/P, 20\%, 15) = \$511,140$

9s.5 (a)

9s.6 (d)

9s.7 (b), $\$14,762 = \$1,000/i + \$1,000(A/F, i, 2)/i$

9s.8 (c)

9s.9 (c)

9s.10 (c), $PW(10\%) = -\$1,500 + X(P/F,10\%,1) + \$650(P/F, 10\%, 2) + X(P/F, 10\%, 3) = 0$, $1.6604X = \$962.84$, $X = \$580$

9s.11 (c), From $\text{IRR}_{B-A} = 30\%$, A $\Leftrightarrow$ B; from $\text{BRR}_{B-C} = 45\%$, B $\Rightarrow$ C; from $\text{BRR}_{A-C} = 40\%$, A $\Rightarrow$ C; C is the preferred choice.

9s.12 (c)

9s.13 (d), $\$100(1 + i)^3 = \$337.50$, $i = 50\%$

9s.14 (c)

9s.15 (b), From the relation, $CE(i) = A/i$, or $A = i\,CE(i)$, $A = \$16,205\,(0.15) = \$2,430.75$; Find the annual equivalent worth for the first cycle, $A' = \$2,000 + (X - \$2,000)(F/A, 15\%, 2)(A/F, 15\%, 4) = 0.43065X - \$861.29 + \$2,000$; Let $A = A'$ and solve for X, or $X = \$3,004$

Chapter 10

10s.1 (d)

10s.2 (b), ($a = 200\%(1/4) = 0.5$; $D_1 = \$45,000(0.5) = \$22,500$; $D_2 = (\$45,000 - D_1)(0.5) = \$11,250$

10s.3 (c), SOYD = 10; $D_2 = (\$45,000 - \$5,000)(3/10) = \$12,000$

10s.4 (a), Total amount depreciated = \$21,360 or 71.20%, with the 5-year MACRS

10s.5 (c)

10s.6 (d), Cost basis = \$170,000 + \$30,000 = \$200,000; Total amount depreciated = $\$200,000(0.1429 + 0.2449 + 0.1749 + 0.1249/2) = \$125,030$; Book value = \$200,000 − \$125,030 = \$74,970

Chapter 11

11s.1 (b)

11s.2 (c), Total amount depreciated with the half-year convertion = $(88.48\%)(\$50,000) = \$44,240$, Book value = \$50,000 − \$44,240 = \$5760

11s.3 (c), Taxable income = \$200,000 − \$84,000 − \$4,000 = \$112,000; net income = $(1 - 0.30)\,\$112,000 = \$78,400$

11s.4 (a), Net cash generated = \$78,400 + \$4,000 = \$82,400

11s.5 (a)

11s.6 (b), Taxable income = (\$120,000 − \$40,000 − \$15,000) = \$65,000; Income taxes = $0.15(\$50,000) + 0.25(\$15,000) = \$11,250$; Average tax rate = \$11,250/\$65,000 = 17.31%

11s.7 (c), Net income = $0.6(\$110,000 - X)$; Net cash flow = $0.6(\$110,000 - X) + \$20,000(0.6) = \$30,000$; $X = \$80,000$ (Note that the asset has been fully depreciated, so that $D_5 = 0$ and $BV_5 = 0$)

Chapter 12

12s.1 (a), Net cash flow in year 10 = $(1 - 0.40)$ $(\$150,000 - \$50,000) + \$15,000$ $(1 - 0.40) = \$69,000$

12s.2 (d), Net cash flow in year 10 = $(1 - 0.40)$ $(\$150,000 - \$50,000 - \$1,480) + \$15,000(1 - 0.40) - \$14,795 = \$53,317$

12s.3 (c), 100% equity financing, $PW(12\%)$ = $-\$100,000 + \$500,000(P/A, 12\%, 5) + \$600,000(P/F, 12\%, 6) = \$2,006,366$; 100% debt financing, $PW(12\%) = 0 + (\$500,000 - \$7200)(P/A, 12\%, 6) = \$1,975,436$; difference = $\$30,929$

12s.4 (a), $(1 - 0.40)(\$130,000 - \$20,000)(P/A, 12\%, 4) + 0.40[0.3333X$ $(P/F,12\%, 1) + 0.4445X(P/F, 12\%, 2) + 0.1481X(P/F, 12\%, 3) + 0.0741X(P/F, 12\%, 4)] = X$; $X = \$295,582$

12s.5 (b), Cash flow series for model A: $\{-\$25,000, -1,360, -160, -1,440, -2,208, -2,208, -2,784, -3,360, -3,360, -3,360, -3,360\}$; Cash flow series for model B: $\{-\$35,000, 700, 2,380, 588, -487, -487, -1,294, -2,100, -2,100, -2,100, 300\}$; Rate of return on incremental investment (model B – model A) = 14.12%

12s.6 (c), $\$300 = \{\$20,000 + \$500 - (S + \$500)\}(A/P, 0.75\%, 36) + 0.0075(S + \$500)$, Solve for S, or $S = \$13,674$

12s.7 (b), $X = (\$120,000)(A/P, 14\%, 4) + \$20,000 = \$61,185$

12s.8 (a), Amount of loan = $\$150,000(0.60)$ = $\$90,000$; Annual installment = $\$90,000(A/P, 10\%, 5) = \$23,742$; The interest payment during the first year = $\$90,000(0.10)$ = $\$9,000$; Principal payment = $\$23,742 - \$9,000 = \$14,740$

12s.9 (d), $X = [0.6(\$200,000 - \$80,000) + 0.4(0.2X)](P/A, 15\%, 5)$, where X = the

amount of capital investment; $0.73182\ X = \$241,358.40$, $X = \$329,821$

12s.10 (c), $A_5 = 0.6(\$300,000 - \$180,000) + 0.4(\$11,520) + \$50,000 - (\$50,000 - \$23,040)(0.40) + \$40,000 = \$155,824$

12s.11 (a), $A_2 = 0.6(\$300,000 - \$180,000 - \$10,000) + 0.40(0.32)(\$200,000) = \$91,600$

Chapter 13

13s.1 (c), $P = \$6,000(P/A_1, 5\%, 6.73\%, 5) = \$27,207$, $i\ \square = [(1 + i)/(1 + f)] - 1 = 6.73\%$

13s.2 (b), Constant series: $P = \$100(P/A, 1.9048\%, 3) = \$288.92$; Actual dollar series: $P = \$105(P/A, 7\%, 3) = \$275.55$

13s.3 (c), $153.6 = 136.2(F/P, f, 4)$, $f = 3.0513\%$

13s.4 (b), Rule of 72: $72/9 = 8$ years; Exact solution: $0.5 = 1\ (P/F, 9\%, n)$, $n = 8.03$ years

13s.5 (b)

13s.6 (a), Cash flow series: $\{-\$1,000, \$95, \$95 + \$1,080\}$; yield in actual dollars = 13.26%; yield in constant (real) dollars = $[(1 + 0.1326)/(1 + 0.04)] - 1 = 8.9\%$

13s.7 (c)

13s.8 (a), $i = i'' + f + i'f$, $i' = (i - f)/(1 + f) = 0.05769$, $PW(5.769\%) = -\$10,000 + \$6,000(P/A, 5.769\%, 5) = \$15,434$

13s.9 (d), $V_{10} = \$1,000(F/A, 2\%, 40) = \$60,402$, required college expenses for freshman year in actual dollars = $\$20,000(F/P, 5\%, 10) = \$32,577$, sophomore expenses = $\$20,000(F/P, 5\%, 11) = \$34,207$; Funds balance after meeting first year expense = $\$60,402 - \$32,577 = \$27,825$, Funds balance at the beginning sophomore year, $\$27,825(F/P, 2\%, 4) = \$30,118$; required funds = $\$34,207 - \$30,118 = \$4,088$

13s.10 (a)

Chapter 14

14s.1 (c), Base : NPW = $\$10,450(40 - 10) - 7890 = \$305,610$; 10% increase in X: NPW = $\$10,450(44 - 10) - 7890 = \$347, 410$; % change = $\$41,800/\$305,610 = 13.68\%$

14s.2 (b), $1000 = (1 - 0.30)\$100(P/A, 15\%, 7)$ $(P/F, 15\%, 3) + [S - (S - \$1,000)(0.20)](P/F, 15\%, 15);$ $F = \$7975$

14s.3 (d), Book value = $12,000 - \$10,618 = \$1382$; gains tax = $(\$3,500 - \$1,382)(0.40) = \$847$; net proceeds from sale = $\$3,500 - \$847 = \$2,653$; capital cost = $(\$12,000 - \$2,653)(A/P, 15\%, 5) + \$2,653(0.15) = \$3,186$; equivalent annual depreciation tax credit = $912 (You obtain this figure by computing the depreciation tax credit each year, $0.4D_n$, finding the total Present Worth of these credits, and annualizing the Present Worth amount over 5 years.); break-even equation: $\$3,186 + 0.15(1 - 0.40)X + 960(1 - 0.4) - 912 = (1 - 0.4)0.25X$; solving for X yields $X = 47,506$ miles

14s.4 (c), $E[FW] = 0.1(-\$12,000) + 0.2(\$4,000) + (0.4)(\$12,000) + 0.2(\$20,000) + 0.1(\$30,000) = \$11,400$

14s.5 (a), $Var[FW] = (0.1) (-\$12,000 - \$11,400)^2 + \ldots + (0.1)(\$30,000 - \$11,400)^2 = 115,240,000; s([FW]) = \$10,735$

14s.6 (a), You lose 25 cents for each play, so the total expected loss will be $25.

14s.7 (d), Let $Y = 2X - 11$, $E[Y] = 2E[X] - 11 = 2(30) - 11 = 49$, $V[Y] = 4V[X] = 4(33) = 132$, $NPW = 40.28VY - 77,860$, $V[NPW] = (40.28)^2 Var[VY] = (E[V])^2V[Y] + (E[Y])^2Var[V] + Var[V]Var[Y] = 1.118549 \times 10^{13}$

14s.8 (c), $e_1 = (1000, 20) \Rightarrow NPW = \$1,090,260$; $e_2 = (1000, 40) \Rightarrow NPW = \$2,701,460$; $e_3 = (6000, 20) \Rightarrow NPW = \$6,930,860$; $e_4 = (6000, 40) \Rightarrow NPW = \$16,598,060$; only two events satisfy the condition, namely e_3 and e_4 and their joint probabilities are $P(e_3) = (0.6)(0.7) = 0.42$ and $P(e_4) = (0.6)(0.3) = 0.18$. So $P(e_3 + e_4) = 0.42 + 0.18 = 0.60$

Chapter 15

15s.1 (b), $\$2,500(0.34) = \$850$
15s.2 (c), $\$2,500 - \$850 = \$1,650$
15s.3 (d), $\$4,000 - \$2,500 = \$1,500$

15s.4 (a)
15s.5 (b), Cash flow with the defender: $\{0, -\$6,600, -\$6,600, -\$6,600\}$; cash flow with the challenger: $\{-\$12,350, -\$844, -\$1,028, +\$1352\}$; incremental cash flow with the replacement: $\{-\$12,350, \$5,756, \$5,572, \$7,952\}$, $AE(15\%) = \$918$

15s.6 (c), Cash flow series (retaining for 2 years): $\{-\$15,000, -\$600, \$9,000\}$, $AE(15\%) = -\$5,362$

15s.7 (c), Replace now: $PW(12\%) = \$3,000/0.12 = \$25,000$; Replace 1 year later: $PW(12\%) = (\$2,500) \$25,000)(P/F,12\%,1) = \$24,553$; Replace 2 years later: $PW(12\%) = \$2500 (P/A, 12\%, 2) + \$25,000(P/F, 12\%, 2) = \$24,155$; Replace 3 years later: $PW(12\%) = \$2,800 (P/A, 12\%, 3) + \$25,000(P/F, 12\%, 3) = \$24,520$

15s.8 (a), sunk cost related to old equipment = $\$12,000 - \$2,000 = \$10,000$

15s.9 (b), The asset has been fully depreciated, so BV = 0. Net proceeds from sale = $\$2,000 - (\$2,000 - 0)(0.35) = \$1,300$

15s.10 (a), BV = $\$15,000 - (\$5,000 - \$6,668 - \$1,110) = \$2,222$; Taxable gains = $\$5,000 - \$2,222 = \$2,778$; Net proceeds from sale = $\$5,000 - \$2,778(0.35) = \$4,028$

15s.11 (b), $PW(15\%)_{\text{Option 1}} = \$1,300 + 0.65(\$10,000)(P/A, 15\%, 3) = \$16,141$; $PW(15\%)_{\text{Option 2}} = \$15,000 + 0.65(\$3,000)(P/A,15\%,3) - 0.35[\$5,000(P/F, 15\%, 1) + \$6,668(P/F, 15\%, 2) + \$1,110(P/F, 15\%, 3)] - \$4,028(P/F, 15\%, 3) = \$13,262$; $\Delta = \$16,141 - \$13,262 = \$2,879$ in favor for challenger.

15s.12 (b), $n = 0$: $PW(10\%) = \$3,000/0.10 = \$30,000$; $n = 1$: $PW(10\%) = \$3,500(P/F, 10\%, 1) + \$30,000(P/F, 10\%, 1) = \$30,455$; $n = 2$: $PW(10\%) = \$3,500(P/F, 10\%, 1) + (\$2,500 + \$30,000)(P/F, 10\%, 2) = \$30,041$; $n = 3$: $PW(10\%) = \$3,500(P/F, 10\%, 1) + \$2,500(P/F, 10\%, 2) + (\$1,500 + \$30,000)(P/F, 10\%, 3) = \$28,914$; When $n^* = 2$, it has the maximum NPW, as we are dealing with revenue figures.

Chapter 16

16s.1 (d), $920 = $100(P/A, k_d, 10) + $1000 (P/F, k_d, 10); k_d = 11.39\%$; after-tax $k_d = (1 - 0.4)(11.39\%) = 6.83\%$

16s.2 (b), debt ratio = 30.09%; $800,000/0.3009 = $2,658,647$

16s.3 (b), $k = (0.3)(0.075) + (0.2)(0.128) + (0.5)(0.20) = 14.81\%$ (Note: the debt interest is already given in after-tax basis.)

16s.4 (a), Amount of debt financing: $250,000(0.40) = $100,000$; amount of equity financing: $250,000(0.60) = $150,000; k = (0.40)(0.08)+(0.6)(0.20) = 15.20\%$ up to $250,000

16s.5 (b), $k = (0.40)(0.10) + (0.60)(0.23) = 17.8\%$ for any financing over $250,000

16s.6 (b)

Chapter 17

17s.1 (d)

17s.2 (c), $I = $1,200,000 - $50,000(P/F, 6\%, 25) = $1,188,350, C' = $100,000(P/A, 6\%, 25) = $1,278,336, B = $250,000(P/A, 6\%, 25) = $3,195,834; B/C$ ratio = $3,195,834/($1,188,350 + $1,278,336) = 1.29$; net B/C ratio = ($3,195,834 - $1,278,336)/$1,188,350 = 1.61$

17s.3 (a), $PW(8\%)_{A1} = $398,954, PW(8\%)_{A2} = $339,676$; net B/C ratio: A1 = 1.5, A2 = 1.34

Index

Methods of Calculating Interests on Your Credit Card

Method	Description	Interest You owe
Adjusted Balance	The bank subtracts the amount of your payment from the beginning balance and charges you interest on the remainder. This method costs you the least.	Your beginning balance is $3,000. With the $1,000 payment, your new balance will be $2,000. You pay 1.5% on this new balance, which will be $30.
Average Daily Balance	The bank charges you interest on the average of the amount you owe each day during the period. So the larger the payment you make, the lower the interest you pay.	Your beginning balance is $3,000. With your $1,000 payment at the 15th day, your balance will be reduced to $2,000. Therefore, your average balance will be ($3,000 + $2,000)/2 = $2,500. The bank will charge 1.5% on this amount: (1.5%)($2,500) = $37.50
Previous Balance	The bank does not subtract any payments you make from your previous balance. You pay interest on the total amount you owe at the beginning of the period. This method costs you the most.	Regardless of your payment size, the bank will charge 1.5% on your beginning balance $3,000: (1.5%)($3,000) = $45